NOV   2000

DATE DUE

TECHNICAL MATHEMATICS

FOURTH EDITION

TECHNICAL MATHEMATICS

PAUL A. CALTER

Professor Emeritus
Vermont Technical College

MICHAEL A. CALTER, Ph.D.

Associate Professor
University of Rochester

John Wiley & Sons, Inc.
New York • Chichester • Weinheim • Brisbane • Singapore • Toronto

This book was set in Times Roman by The Clarinda Company and printed and bound by R. R. Donnelley, Willard. The cover was printed by Phoenix Color. Cover art © by Bob Eddy.

Library of Congress Cataloging in Publication Data:
Calter, Paul.
    Technical mathematics. — 4th ed. / Paul A. Calter, Michael A.
  Calter.
        p.   cm.
    Includes indexes.
    ISBN 0-471-36903-9 (casebound)
    1. Engineering mathematics.   I. Calter, Michael A.   II. Title.
TA330.C343   2000
510—dc21   98-43765
                                                           CIP

Printed in the United States of America

10 9 8 7 6 5 4 3 2 1

To Margaret Jolind and Kimberley Ann

Preface

The primary aim of this fourth edition of *Technical Mathematics* was to carry out suggestions for improvements made by many reviewers and colleagues, as well as those that occurred to the authors while using the preceding edition. In this edition we have tried to follow as closely as possible the guidelines in the NCTM *Standards* and *Guidelines,* the AMATYC *Crossroads in Mathematics,* and the AMATYC *Position Statement on Undergraduate Textbooks.*

Features of the Book

Each chapter begins with a listing of **Chapter Objectives,** which state specifically what the student should be able to do upon completion of the chapter. Following that, we have tried to present the material as clearly as possible, preferring an intuitive approach rather than an over-rigorous one. Realizing that a mathematics book is not easy reading, we have given information in small segments, included many illustrations, and designed each page with care.

The numerous **Examples** form the backbone of the textbook. They are fully worked out and are chosen to help the student to do the exercises. Examples have markers above and below to separate them clearly from the text discussion.

The examples, and often the text itself, include discussions of many **Technical Applications,** such as the use of mathematics for analyzing motion, or for electric circuits. They are included for classes that wish to cover those topics, as well as for motivation, to show that mathematics has real uses. Because space does not permit a full discussion of each application, judgment must be used in assigning applications problems. It is not intended that every student be able to solve every application. We assume that students have sufficient background before attempting the more difficult problems in their own technical area, and that they will get technical help not offered in this text. The **Index to Applications** will help in finding specific applications.

We've tried to avoid contrived "school" problems with neat solutions, and include many **Problems with Approximate Solutions.** These include expressions and equations with approximate constants, but those that do not yield to many of the exact methods we teach and must be tackled with an approximate method.

We've all seen wild answers on homework and exams, such as "the cost of each pencil is $300." To avoid that, we have tried to show students how to **Estimate an Answer** in order to check their work. We give suggestions for estimation in the chapter on word problems. Thereafter many applications examples begin with an estimation step, or end with a check, either by graphing, by computer, by calculator, by an alternate solution, by making a physical model, or simply to **Examine the Answer for Reasonableness.**

Formulas used in the text are boxed and numbered and are also listed in the Appendix as the **Summary of Facts and Formulas.** This listing can function as a "handbook" for a calculus course and for other courses as well, and provides a common thread between chapters. We hope it will also help students to see interconnections that might otherwise be overlooked. The formulas are grouped logically in the Summary of Facts and Formulas and are numbered sequentially there.

Therefore, the formulas do not necessarily appear in numerical order in the text. In addition to mathematical formulas, we include some from technology, motion, electric circuits, and so on. These are grouped together at the end of the Formula Summary and have formula numbers starting with the letter A.

Also in the text are **Marginal Notes,** which are used for encouragement, to give historical notes, for reminders of things already covered, for peeks ahead at things to come, and so forth. **Common Error Boxes** emphasize some of the pitfalls and traps that "get" students year after year.

To give students the essential practice they need to learn mathematics, we include thousands of **Exercises.** Exercises are given after each section, graded by difficulty and grouped by type, to allow practice in a particular area. These are indicated by title as well as by number. The **Chapter Review Problems** are scrambled as to type and difficulty. Answers to all odd-numbered problems are given in the **Answer Key** in the Appendix and in the **Annotated Instructor's Edition.** Complete solutions to every problem are contained in the **Instructor's Solutions Manual.** Complete solutions to every other odd problem are given in the **Student's Solutions Manual.**

For those instructors who like to assign occasional **Writing Questions** but need ideas about what to ask, we have provided at least one such question per chapter in the Chapter Review. For those interested in collaborative learning, where students work in groups, we have placed at least one **Team Project** at the end of every chapter. These, like the writing questions, can serve as models for others that teacher and students can make up themselves. There is an introductory project in Chapter 1 which gives general guidelines that can be followed in later projects. An **Index to Writing Questions** as well as an **Index to Team Projects** are given in the Appendix.

Use of a **Graphics Calculator** has been fully integrated throughout. Calculator instruction and examples are given in the text, where appropriate, and calculator problems are given in the exercises. Many problems can be practically solved *only* by graphics calculator, and the graphics calculator is sometimes used to verify a solution found by another method. However, we have still retained most of the noncalculator methods, such as manual graphing by plotting of point pairs, for those who want to present these methods. An **Index to Graphics Calculator Instructions and Exercises** is given in the Appendix.

We give projects for the **Computer** where appropriate. These include writing programs in **BASIC** or using **Spreadsheets** or **Computer Algebra Systems,** or CAS. This material has sometimes been placed in the text, but more often appears in the Exercises. We include many **Numerical Methods for the Computer,** but these topics are included in such a way that they can be skipped without harm. Those calculations best done by computer are given in the text itself. Elsewhere, computer problems are suggested at the end of exercises as enrichment activities. Examples of numerical methods included are the finding of roots of equations by the midpoint method, by the false position method, and by simple iteration; solving systems of equations by the Gauss-Seidel method and by matrix inversion, and numerical evaluation of series. A complete list of these numerical methods can be found in the General Index, and an **Index to Computer Instructions and Exercises** is given in the Appendix.

New Features to the Fourth Edition

- Presentation of concepts has been refined at many places to enhance clarity.
- Many Team Projects have been added to further enhance collaborative learning.
- Many exercises have been changed and updated to capture interest of students and maintain relevancy.
- New examples have been included throughout to increase student understanding.

- A clearer, cleaner design has been implemented, particularly within the figures and marginal notes.
- Additional Writing Questions have been added to the Chapter Review Problems.
- Additional Computer Algebra System problems in Maple, Mathematica, and Derive have been added.

Teaching Resources

There are supplements to aid both the instructor and the student. An **Annotated Instructor's Edition** (AIE) of this text contains answers to every exercise and problem. The answers are printed in red right in the exercise or problem. The AIE also has red marginal notes to the instructor which give teaching tips, applications, and practice problems.

An **Instructor's Solutions Manual** contains worked-out solutions to every problem in the text and a list of all computer programs. The **Student's Solutions Manual** gives the solution to every other odd problem. They are usually worked in more detail than in the Instructor's Solutions Manual. Both were prepared by Susan Porter and Laurel Technical Services.

A **Computerized Test Item File** is a bank of test questions with answers. Questions may be mixed, sorted, changed, or deleted. It consists of a test file disk and a test generator disk, ready to run.

Also available are **PowerPoint transparencies** that simplify the job of presenting complicated figures in class.

Acknowledgments

We are extremely grateful to reviewers of this edition and of the earlier editions of the book, reviewers of the supplements and the writing questions, and participants in group discussions about the book. They are:

A. David Allen, Ricks College

Byron Angell, Vermont Technical College

David Bashaw, New Hampshire Technical Institute

Jim Beam, Savannah Area Vo-Tech

Elizabeth Bliss, Trident Technical College

Franklin Blou, Essex County College

Donna V. Boccio, Queensboro Community College

Jacquelyn Briley, Guilford Technical Community College

Frank Caldwell, York Technical College

James H. Carney, Lorain County Community College

Cheryl Cleaves, State Technical Institute at Memphis

Ray Collings, Tri-County Technical College

Miriam Conlon, Vermont Technical College

Kati Dana, Norwich University

Linda Davis, Vermont Technical College

John Eisley, Mott Community College

Walt Granter, Vermont Technical College

Crystal Gromer, Vermont Technical College

Richard Hanson, Burnsville, MN

Tommy Hinson, Forsythe Community College

Margie Hobbs, University of Mississippi

Martin Horowitz, Queensborough Community College

Glenn Jacobs, Greenville Technical College

Wendell Johnson, Akron, OH

Joseph Jordan, John Tyler Community College

Frank L. Juszli

Rob Kimball, Wake Technical Community College

John Knox, Vermont Technical College

Bruce Koopika, Northeast Wisconsin Technical College

Ellen Kowalczyk, Madison Area Technical College

Fran Leach, Delaware Technical College

Jon Luke, Indiana University-Purdue University

Michelle Maclenar, Terra Community College

Paul Maini, Suffolk County Community College

Edgar M. Meyer, St. Cloud State, MN

David Nelson, Western Wisconsin Technical College

Don Nevin, Vermont Technical College

Mary Beth Orrange, Erie Community College

Harold Oxsen, Walnut Creek, CA

Ursula Rodin, Nashville State Technical Institute

Donald Reichman, Mercer County Community College

Bob Rosenfeld, Nassau Community College and University of Vermont

Nancy J. Sattler, Terra Technical College

Frank Scalzo, Queensborough Community College

Blin Scatterday, University of Akron Community and Technical College

Edward W. Seabloom, Lane Community College

Robert Seaver, Lorain Community College

Saeed Shaikh, Miami-Dade Community College

Thomas Stark, Cincinnati Technical College

Dale H. Thielker, Ranken Technical College

William N. Thomas, Jr., Thomas & Associates Group

Joel Turner, Blackhawk Technical Institute

Roy A. Wilson, Cerritos College

Douglas Wolansky, North Alberta Institute of Technology

Henry Zatkis, New Jersey Institute of Technology

The tedious job of solving many of the exercise problems was done by our students, Brett Benner, Keith Crowe, Jim Davis, Nancy Davis, Kelly Dennehy, Mel Emerson, Ellie Germain, Robert Morel, Jeffery Sloan, and Ray Wells. The solutions to all problems new to this edition were checked by Susan Porter, who also proofread the galleys and prepared the answer key and the Solutions Manuals. George Seki and Terri Bittner of Laurel Technical Services carefully solved and rechecked every example and exercise in the book. We also want to express our gratitude to John Morin for his many constructive suggestions on the first edition. Thank you all.

Michael A. Calter
Rochester, New York

Paul A. Calter
Randolph, Vermont

About the Authors

Paul Calter is Professor Emeritus of Mathematics at Vermont Technical College and Visiting Scholar at Dartmouth College. A graduate of The Cooper Union, New York, he received his M.S. from Columbia University and a MFA from Norwich University. Professor Calter has taught technical mathematics for over twenty-five years. In 1987, he was the recipient of the Vermont State College Faculty Fellow Award.

He is member of the American Mathematical Association of Two Year Colleges, the Mathematical Association of America, the National Council of Teachers of Mathematics, the International Society for the Arts, Sciences, and Technology, the College Art Association, and the Author's Guild.

Calter is involved in the *Mathematics Across the Curriculum* movement, and has developed and taught a course called *Geometry in Art and Architecture* at Dartmouth College under an NSF grant.

Professor Calter is the author of several other mathematics textbooks, among which are the *Schaum's Outline of Technical Mathematics, Problem Solving with Computers, Practical Math Handbook for the Building Trades, Practical Math for Electricity and Electronics, Mathematics for Computer Technology, Introductory Algebra and Trigonometry,* and *Technical Calculus.*

Michael Calter is Associate Professor at the University of Rochester. He received his B.S. from the University of Vermont. After receiving his Ph.D. from Harvard University, he completed a post-doctorial fellowship at the University of California at Irvine and was Assistant Professor at Virginia Polytechnic Institute and State University.

Michael has been working on his father's mathematics texts since 1983, when he completed a set of programs to accompany *Technical Mathematics with Calculus.* Since that time, he has become progressively more involved with his father's writing endeavors, culminating with becoming co-author on the second edition of *Technical Calculus* and the fourth edition of *Technical Mathematics with Calculus.* Michael also enjoys the applications of mathematical techniques to chemical and physical problems as part of his academic research.

Michael is a member of the American Mathematical Association of Two Year Colleges, the American Association for the Advancement of Science, and the American Chemical Society.

Michael and Paul enjoy hiking and camping trips together. These have included an expedition up Mt. Washington in January, a hike across Vermont, a walk across England on Hadrian's Wall, and many sketching trips into the mountains.

About the Cover

Paul Calter is also a sculptor whose work often has mathematical themes. The cover shows details of his *Armillary VII* on the campus of Vermont Technical College.

Contents

Indexes

Numerical Computation

••• **OBJECTIVES** ••

When you have completed this chapter, you should be able to:

- Perform basic arithmetic operations on signed numbers.
- Perform basic arithmetic operations on approximate numbers.
- Take powers, roots, and reciprocals of signed and approximate numbers.
- Perform combined arithmetic operations to obtain a numerical result.
- Convert numbers between decimal and scientific notation.
- Perform basic arithmetic operations on numbers in scientific notation.
- Convert units of measurement from one denomination to another.
- Substitute given values into equations and formulas.
- Solve common percentage problems.

•••

In this first chapter we cover ordinary arithmetic, but in a different way. The basic operations are explained here in terms of the *calculator*. Although the calculator makes arithmetic easier than before, it also introduces a complication: that of knowing how many of the digits shown in a calculator display should be kept. We will see that it is usually incorrect to keep them all. Then, after performing the algebraic operations on pairs of numbers, we will show how to link these operations to evaluate more complex expressions.

Also in this chapter, we will learn some rules that will help us when we get to algebra.

We also learn *scientific notation*—a convenient way to handle very large or very small numbers without having to write down lots of zeros. It is also the way in which calculators and computers display such numbers.

We then consider the *units* in which quantities are measured: how to convert from one unit to another and how to substitute into equations and formulas and have the units work out right.

Next we learn how to *substitute values* into equations and formulas and to *evaluate the result,* an operation of great practical importance in technical work.

Finally, we cover *percentage,* one of the most often used mathematical ideas in technology as well as in everyday life.

When you finish this chapter, you should be able to use most of the keys on your calculator to do some fairly difficult computations. The keys for trigonometry and logarithms are discussed in the chapters on those topics.

1–1   The Real Numbers

Before we start our calculator practice, we must learn a few definitions. In mathematics, as in many fields, you will have trouble understanding the material if you do not clearly understand the meanings of the words that are used.

Integers

The *integers*

$$\ldots, -4, -3, -2, -1, 0, 1, 2, 3, 4, \ldots$$

are the whole numbers, including zero and negative values.

The three dots indicate that the sequence of numbers continues indefinitely.

Rational and Irrational Numbers

The *rational numbers* include the integers and all other numbers that can be expressed as the quotient of two integers, for example,

$$\frac{1}{2}, \quad -\frac{3}{5}, \quad \frac{57}{23}, \quad -\frac{98}{99}, \quad \text{and} \quad 7$$

Numbers that cannot be expressed as the quotient of two integers are called *irrational.* Some irrational numbers are

$$\sqrt{2}, \quad \sqrt[3]{5}, \quad \sqrt{7}, \quad \pi, \quad \text{and} \quad e$$

where π is approximately equal to 3.1416 and e is approximately equal to 2.7182.

Real Numbers

The rational and irrational numbers together make up the *real numbers.*

Numbers such as $\sqrt{-4}$ do not belong to the real number system. They are called *imaginary numbers* and are discussed in Chapter 21. Except when otherwise noted, all of the numbers we will work with are real numbers.

Decimal Numbers

Most of our computations are with numbers written in the familiar *decimal* system. The names of the places relative to the *decimal point* are shown in Fig. 1–1. We say that the decimal system uses a *base of 10* because it takes 10 units in any place to equal 1 unit in the next-higher place. For example, 10 units in the hundreds position equals 1 unit in the thousands position.

Systems having bases other than 10 are used in computer science: *binary* (base 2), *octal* (base 8), and *hexadecimal* (base 16) (see Chapter 23).

Positional Number Systems

A *positional* number system is one in which the *position* of a digit determines its value. Our decimal system is positional.

◆◆◆ **Example 1:** In the number 351.3, the digit 3 on the right has the value $\frac{3}{10}$, but the digit 3 on the left has the value 300. Thus

$$351.3 = 300 + 50 + 1 + 0.3$$
$$= 3(10^2) + 5(10^1) + 1(10^0) + 3(10^{-1})$$

The position immediately to the left of the decimal point is called position 0. The position numbers then increase to the left and decrease to the right of the 0 position. ◆◆◆

Place Value

Each position in a number has a *place value* equal to the base of the number system raised to the position number. The place values in the decimal number system, as well as the place names, are shown in Fig. 1–1.

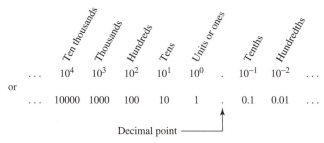

or

FIGURE 1–1 Values of the positions in a decimal number.

The numbers 10^2, 10^3, etc., are called *powers of 10*. Don't worry if they are unfamiliar to you. We will explain them in Sec. 1–7.

Signed Numbers

A *positive* number is a number that is *greater* than zero, and a *negative* number is *less* than zero. On the *number line* it is customary to show the positive numbers to the right of zero and the negative numbers to the left of zero (Fig. 1–2). Negative numbers may be integers, fractions, rational numbers, or irrational numbers. To distinguish negative numbers from positive numbers, we always place a negative sign (−) in front of a negative number.

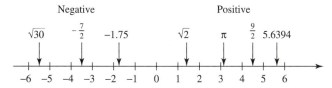

FIGURE 1–2 The number line.

◆◆◆ **Example 2:** Some negative numbers are

$$-5, \quad -6.293, \quad -\tfrac{2}{3}, \quad -2\tfrac{7}{8}, \quad -\sqrt{5}$$ ◆◆◆

We usually omit writing the positive sign (+) in front of a positive number. Thus a number without a sign is always assumed to be positive. In this chapter we will often write in a + sign for emphasis.

◆◆◆ **Example 3:** Some positive numbers are

$$5, \quad +5, \quad \tfrac{2}{3}, \quad +\tfrac{2}{3}, \quad \sqrt{5}$$ ◆◆◆

The Opposite of a Number

The *opposite* of a number n is the number which, when added to n, gives a sum of zero.

◆◆◆ **Example 4:** The opposite of 2 is -2, because $2 + (-2) = 0$. The opposite of -6 is $+6$.   ◆◆◆

Geometrically, the opposite of a number n lies on the opposite side of the zero point of the number line from n, and at an equal distance (see Fig. 1–2).

The opposite of a number is also called the *additive inverse* of that number.

Symbols of Equality and Inequality

Several symbols are used to show the relative positions of two quantities on the number line.

$a = b$ means that a *equals* b and that a and b occupy the same position on the number line.

$a \neq b$ means that a and b are *not equal* and have different locations on the number line.

$a > b$ means that a is *greater than* b and a lies to the right of b on the number line.

$a < b$ means that a is *less than* b and a lies to the left of b on the number line.

$a \cong b$ means that a is *approximately equal to* b and that a and b are *near* each other on the number line.

Absolute Value

The *absolute value* of a number n is its *magnitude* regardless of its algebraic sign. It is written $|n|$. It is the distance from n to zero on the number line, without regard to direction.

Many calculators have a key for evaluating absolute values. Try to use it to evaluate any of these expressions.

◆◆◆ **Example 5:**

(a) $|5| = 5$
(b) $|-9| = 9$
(c) $|3 - 7| = |-4| = 4$
(d) $-|-4| = -4$
(e) $-|7 - 21| - |13 - 19| = -|-14| - |-6| = -14 - 6 = -20$ ◆◆◆

Approximate Numbers

Most of the numbers we deal with in technology are *approximate*.

◆◆◆ **Example 6:**

(a) All numbers that represent *measured* quantities are approximate. A certain shaft, for example, is approximately 1.75 inches in diameter.
(b) Many *fractions* can be expressed only approximately in decimal form. Thus $\frac{2}{3}$ is approximately equal to 0.6667.
(c) *Irrational numbers* can be written only approximately in decimal form. The number $\sqrt{3}$ is approximately equal to 1.732.   ◆◆◆

Exact Numbers

Exact numbers are those that *have no uncertainty.*

◆◆◆ Example 7:

(a) There are exactly 24 hours in a day, no more, no less.
(b) An automobile has exactly four wheels.
(c) Exact numbers are usually integers, but not always. For example, there are *exactly* 2.54 cm in an inch, by definition.
(d) On the other hand, not all integers are exact. For example, a certain town has a population of *approximately* 3500 people.   ◆◆◆

Significant Digits

In a decimal number, zeros are sometimes used just to locate the decimal point. When zeros are used in that way, we say that they are *not significant*. The remaining digits in the number, including zeros, are called *significant digits*.

◆◆◆ Example 8:

(a) The numbers 497.3, 39.05, 8003, and 2.008 each have *four* significant digits.
(b) The numbers 1570, 24,900, 0.0583, and 0.000583 each have *three* significant digits. The zeros in these numbers serve only to locate the decimal point.
(c) The numbers 18.50, 1.490, and 2.000 each have *four* significant digits. The zeros here are not needed to locate the decimal point. They are placed there to show that those digits are in fact zeros, and not something else.   ◆◆◆

An overscore is sometimes placed over the last trailing zero that is significant. Thus the numbers 3250 and 735,000 each have four significant digits.

Accuracy and Precision

The *accuracy* of a decimal number is given by the number of *significant digits* in the number; the *precision* of a decimal number is given by the position of the rightmost significant digit.

◆◆◆ Example 9:

(a) The number 1.255 is accurate to four significant digits, and precise to three decimal places. We also say that it is precise to the nearest thousandth.
(b) The number 23,800 is accurate to three significant digits, and precise to the nearest hundred.   ◆◆◆

Rounding

We will see, in the next few sections, that the numbers we get from a computation often contain *worthless digits* that must be *thrown away*. Whenever we do this, we must *round* our answer.

Round down (do not change the last retained digit) when the first discarded digit is 4 or less. *Round up* (increase the last retained digit by 1) when the first discarded digit is 6 or more, or a 5 followed by a nonzero digit in any of the decimal places to the right.

◆◆◆ Example 10:

Number	Rounded to Three Decimal Places
4.3654	4.365
4.3656	4.366
4.365501	4.366
1.764999	1.765
1.927499	1.927

◆◆◆

This is just a convention. We could just as well round to the nearest odd number.

When the discarded portion is 5 *exactly,* it usually does not matter whether you round up or down. The exception is when you are adding or subtracting a long column of figures, as in statistical computations. If, when discarding a 5, you always rounded up, you could bias the result in that direction. To avoid that, you want to round up about as many times as you round down, and a simple way to do that is to always *round to the nearest even number.*

◆◆◆ **Example 11:**

Number	Rounded to Two Decimal Places
4.365	4.36
4.355	4.36
7.76500	7.76
7.75500	7.76

◆◆◆

Exercise 1 ◆ The Real Numbers

Equality and Inequality Signs

Insert the proper sign of equality or inequality ($=$, $\cong$, $>$, $<$) between each pair of numbers.

1. 7 and 10   **2.** 9 and -2 **3.** -3 and 4
4. -3 and -5 **5.** $\frac{3}{4}$ and 0.75   **6.** $\frac{2}{3}$ and 0.667

Absolute Value

Evaluate each expression.

7. $-|9 - 23| - |-7 + 3|$ **8.** $|12 - 5 + 8| - |-6| + |15|$
9. $-|3 - 9| - |5 - 11| + |21 + 4|$

Significant Digits

Determine the number of significant digits in each approximate number.

10. 78.3   **11.** 9274   **12.** 4.008
13. 9400   **14.** 20,000   **15.** 5000.0
16. 0.9972   **17.** 1.0000

Round each number to two decimal places.

18. 38.468   **19.** 1.996   **20.** 96.835001
21. 55.8650   **22.** 398.372   **23.** 2.9573
24. 2985.339   **25.** 278.382

Round each number to one decimal place.

26. 13.98   **27.** 745.62   **28.** 5.6501
29. 0.482   **30.** 398.36   **31.** 34.927
32. 9839.2857   **33.** 0.847

Round each number to the nearest hundred.

34. 28,583   **35.** 7550
36. 3,845,240   **37.** 274,837

Round each number to three significant digits.

38. 9.284　　　　**39.** 2857　　　　**40.** 0.04825
41. 483,982　　　**42.** 0.08375　　　**43.** 29.555
44. 29.45001　　　**45.** 8372

Calculator

46. Use the absolute value key on your calculator, probably marked ABS, to evaluate the expressions in problems 7 through 9.

1–2　Addition and Subtraction

Adding and Subtracting Integers

To add two numbers by calculator, simply enter the first number; press +; enter the second number; and press the *equals* key =, the *execute* key EXE, or the *enter* key ENTER, depending upon your particular calculator. The number we get is called the *sum*.

◆◆◆ **Example 12:** Evaluate 2845 + 1172 by calculator.

Solution: There are so many types of calculators in use that it would be confusing for us to show keystrokes for any particular one. You must consult your operator's manual. However, the keystrokes for the basic operations are simple and easily learned. For this example you should get the result

$$2845 + 1172 = 4017$$　　　◆◆◆

If we had wanted to *subtract* 1172 from 2845, we would have simply pressed the − key instead of the + key.

Horizontal and Vertical Addition

Numbers to be added may be arranged vertically or horizontally. Of course, when you are using the calculator, it does not make any difference how the numbers are arranged, as they must be keyed in one at a time anyway.

◆◆◆ **Example 13:** The addition of 335, 103, and 224 may be represented horizontally as

$$335 + 103 + 224 = 662$$

or vertically as

$$\begin{array}{r} 335 \\ 103 \\ \underline{224} \\ 662 \end{array}$$　　　◆◆◆

Adding Signed Numbers

Let us say that we have a shoebox (Fig. 1–3) into which we toss all of our uncashed checks and unpaid bills until we have time to deal with them. Let us further assume that the total checks minus the total bills in the shoebox is $500.

We can think of the amount of a check as a positive number because it increases our wealth, and the amount of a bill as a negative number because it decreases our wealth. We thus represent a check for $100 as (+100), and a bill for $200 as (−200).

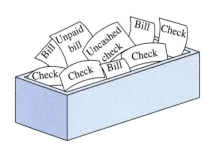

FIGURE 1–3　The shoebox.

Now, let's *add a check* for $100 to the box. If we had $500 at first, we must now have $600.

$$500 + (+100) = 600$$

or

$$500 + 100 = 600$$

Here we have added a positive number, and our total has increased by that amount. That is easy to understand. But what does it mean to *add a negative number?*

To find out, let us now *add a bill* for $100 to the box. If we had $500 at first, we must now have $100 less, or $400. Representing the bill by (-100), we have

$$500 + (-100) = 400$$

It seems clear that to add a negative number is no different than subtracting the absolute value of that number.

$$500 + (-100) = 500 - 100$$

This gives us our rule of signs for addition.

If b is a positive quantity, then the operation of adding a negative quantity $(-b)$ to the quantity a is equivalent to the operation of subtracting the positive quantity b from a.

All boxed and numbered formulas are tabulated in numerical order in Appendix A. There they are arranged in logical order, by type, and are numbered consecutively. Since the formulas often appear in the text in a different order than in Appendix A, they may not be in numerical order here in the text.

Rule of Signs for Addition	$a + (-b) = a - b$	6

◆◆◆ Example 14:

(a) $7 + (-2) = 7 - 2 = 5$
(b) $-8 + (-3) = -8 - 3 = -11$
(c) $9.92 + (-15.36) = 9.92 - 15.36 = -5.44$ ◆◆◆

Subtracting Signed Numbers

Let us return to our shoebox problem. But now instead of adding checks or bills to the box, we will *subtract* (remove) checks or bills from the box.

First we remove (subtract) a check for $100 from the box. If we had $500 at first, we must now have $400.

$$500 - (+100) = 400$$

or

$$500 - 100 = 400$$

Here we have subtracted a positive number, and our total has decreased by the amount subtracted, as expected.

Now let us see what it means to *subtract a negative number.* We will remove (subtract) a bill for $100 from the box. If we had $500 at first, we must now have 100 more, or $600, since we have removed a bill. Representing the bill by (-100), we have

$$500 - (-100) = 600$$

It seems clear that to subtract a negative number is the same as to add the absolute value of that number.

$$500 - (-100) = 500 + 100$$

Thus if b is a positive quantity, the operation of subtracting a negative quantity $(-b)$ from a quantity a is equivalent to the operation of adding the positive quantity b to a.

Rule of Signs for Subtraction	$a - (-b) = a + b$	7

◆◆◆ Example 15:

(a) $15 - (-3) = 15 + 3 = 18$
(b) $-5 - (-9) = -5 + 9 = 4$
(c) $-25.62 - (-5.15) = -25.62 + 5.15 = -20.47$ ◆◆◆

Subtracting Negative Numbers by Calculator

Note that the $(-)$ sign is used for two different things:

1. For the operation of subtraction.
2. To enter a negative quantity.

This difference is clear on the calculator, which has separate keys for these two functions. The $\boxed{-}$ key is used only for subtraction; the $\boxed{(-)}$ key is used to enter a negative quantity.

Common Error	These two keys look almost alike, so be careful not to confuse them. Note that the key used to enter a negative quantity has parentheses around the negative sign.

To enter a negative number, simply press the $\boxed{(-)}$ key and then key in the number.

On some calculators a negative number is entered by first keying in the number and then changing its sign by using a *change-sign key,* marked $\boxed{+/-}$ or $\boxed{CHS}$. Check your manual to see how this key is marked on your calculator.

Try the following examples on your calculator and see if you get the correct answers.

◆◆◆ Example 16:

(a) $15 - (-3) = 15 + 3 = 18$
(b) $-5 - (-9) = -5 + 9 = 4$
(c) $-25 - (-5) = -25 + 5 = -20$ ◆◆◆

Commutative and Associative Laws

The *commutative law* simply says that you can add numbers in *any order.*

Commutative Law for Addition	$a + b = b + a$	1

These laws are surely familiar to you, even if you do not recognize their names. We will run into them again when studying algebra.

✦✦✦ Example 17:

$$2 + 3 = 3 + 2$$
$$= 5 \qquad \text{✦✦✦}$$

The *associative law* says that you can group numbers to be added in several ways.

Associative Law for Addition	$a + (b + c) = (a + b) + c$ $\qquad\qquad = (a + c) + b$	**3**

✦✦✦ Example 18:

$$2 + 3 + 4 = 2 + (3 + 4) \quad 2 + 7 = 9$$
$$= (2 + 3) + 4 = 5 + 4 = 9$$
$$= (2 + 4) + 3 = 6 + 3 = 9 \qquad \text{✦✦✦}$$

Addition and Subtraction of Approximate Numbers

Addition and subtraction of integers are simple enough. But now let us tackle the problem mentioned earlier: How many digits do we keep in our answer when adding or subtracting *approximate numbers?*

Rule	When adding or subtracting approximate numbers, keep as many decimal places in your answer as contained in the number having the fewest decimal places.

✦✦✦ Example 19:

$$32.4 \text{ cm} + 5.825 \text{ cm} = 38.2 \text{ cm} \quad (\textit{not } 38.225 \text{ cm})$$

We do not use the symbol $\cong$ when dealing with approximate numbers. We would not write, for example, 32.4 cm + 5.825 cm $\cong$ 38.2 cm.

Here we can see the reason for our rule for rounding. In one of the original numbers (32.4 cm), we do not know what digit is to the right of the 4, in the hundredths place. We cannot assume that it is zero, for if we *knew* that it was zero, it would have been written in (as 32.**40**). This uncertainty in the hundredths place in an original number causes uncertainty in the hundredths place in the answer, so we discard that digit and any to the right of it. ✦✦✦

Students *hate* to throw away those last digits. Remember that by keeping worthless digits, you are telling whoever reads that number that it is more precise than it really is.

✦✦✦ Example 20:

$$
\begin{array}{r}
25.8 \\
18.3\,125 \\
\underline{5.4\,07} \\
49.5\,|195
\end{array}
$$

discard ✦✦✦

✦✦✦ Example 21: A certain stadium contains about 3500 people. It starts to rain, and 372 people leave. How many are left in the stadium?

Solution: Subtracting, we obtain

$$3500 - 372 = 3128$$

which we round to 3100, because 3500 here is known to only two significant digits. ◆◆◆

It is safer to round the answer *after* adding, rather than to round the original numbers before adding. If you do round before adding, it is prudent to round each original number to *one more* decimal place than you expect in the rounded answer.

Combining Exact and Approximate Numbers

When you are combining an exact number and an approximate one, the accuracy of the result will be limited by the approximate number. Thus round the result to the number of decimal places found in the approximate number, even though the exact number may *appear* to have fewer decimal places.

◆◆◆ **Example 22:** Express 2 hr and 35.8 min in minutes.

Solution: We must add an exact number (120) and an approximate number (35.8).

$$
\begin{array}{r}
120 \ \text{ min} \\
+ \ \ 35.8 \ \text{min} \\
\hline
155.8 \ \text{min}
\end{array}
$$

Since 120 is exact, we do *not* round our answer to the nearest 10 minutes, but we retain as many decimal places as in the approximate number. Our answer is thus 155.8 min. ◆◆◆

Common Error	Be sure to recognize which numbers in a computation are exact; otherwise, you may perform drastic rounding by mistake.

Exercise 2 ◆ Addition and Subtraction

Combine as indicated.

1. −955	**2.** 8275	**3.** −748
+212	−2163	−212
−347	− 874	−156

Add each column of figures.

4. $99.84	**5.** 96256	**6.** 98304
24.96	6016	6144
6.24	376	384
1.56	141	24576
12.48	188	3072
0.98	1504	144
3.12	752	49152

Combine as indicated.

7. $926 + 863$

8. $274 + (-412)$

9. $-576 + (-553)$

10. $-207 + (-819)$

11. $-575 - 275$

12. $-771 - (-976)$

13. $1123 - (-704)$

14. $818 - (-207) + 318$

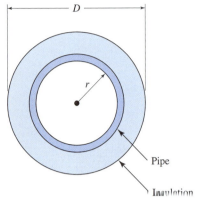

FIGURE 1–4   An insulated pipe.

Equation numbers with the prefix "A" are applications. They are listed toward the end of Appendix A.

Combine each set of approximate numbers as indicated. Round your answer.

15. 4857 + 73.8

16. 39.75 + 27.4

17. 296.44 + 296.997

18. 385.28 − 692.8

19. 0.000583 + 0.0008372 − 0.00173

20. Mt. Blanc is 15,572 ft high, and Pike's Peak is about 14,000 ft high. What is the difference in their heights?

21. California contains 158,933 square miles (mi^2), and Texas 237,321 mi^2. How much larger is Texas than California?

22. A man willed $125,000 to his wife and two children. To one child he gave $44,675, to the other $26,380, and to his wife the remainder. What was his wife's share?

23. A circular pipe has an inside radius r of 10.6 cm and a wall thickness of 2.125 cm. It is surrounded by insulation having a thickness of 4.8 cm (Fig. 1–4). What is the outside diameter D of the insulation?

24. A batch of concrete is made by mixing 267 kg of aggregate, 125 kg of sand, 75.5 kg of cement, and 25.25 kg of water. Find the total weight of the mixture.

25. Three resistors, having values of 27.3 ohms (Ω), 4.0155 Ω, and 9.75 Ω, are wired in series. What is the total resistance? (See Eq. A63, which says that the total series resistance is the sum of the individual resistances.)

1–3   Multiplication

Factors and Product

The numbers we multiply to get a *product* are called *factors*. For example,

$$3 \times 5 = 15$$

factors ———— ↑ ↑   ↑ ———— product

Multiplication by Calculator

Simply use the $\boxed{\times}$ key on your calculator, as instructed in your calculator manual.

◆◆◆ **Example 23:** Use your calculator to multiply 183 by 27.

Solution: You should get

$$183 \text{ by } 27 = 4941$$

◆◆◆

Commutative, Associative, and Distributive Laws

The *commutative law* states that the *order* of multiplication is not important.

Commutative Law for Multiplication	$ab = ba$	2

◆◆◆ **Example 24:** It is no surprise that

$$2 \times 3 = 3 \times 2$$

◆◆◆

The *associative law* allows us to group in any way the numbers to be multiplied.

Associative Law for Multiplication	$a(bc) = (ab)c = (ac)b = abc$	4

◆◆◆ **Example 25:**

$$2 \times 3 \times 4 = 2\,(3 \times 4) = 2(12) = 24$$
$$= (2 \times 3)4 = (6)4 = 24$$
$$= (2 \times 4)3 = (8)3 = 24 \qquad ◆◆◆$$

The *distributive law* shows how a factor may be *distributed* among several terms.

Distributive Law for Multiplication	$a(b + c) = ab + ac$	5

◆◆◆ **Example 26:**

$$2(3 + 4) = 2(7) = 14$$

But the distributive law enables us to do the same computation in a different way.

$$2(3 + 4) = 2(3) + 2(4) = 6 + 8$$
$$= 14 \quad \text{(as before)} \qquad ◆◆◆$$

Multiplying Signed Numbers

To get our rules of signs for multiplication, we use the idea of *multiplication as repeated addition.* For example, to multiply 3 by 4 means to add four 3's (or three 4's).

$$3 \times 4 = 3 + 3 + 3 + 3$$

or

$$3 \times 4 = 4 + 4 + 4$$

Let us return to our shoebox example. Recall that it contains uncashed checks and unpaid bills. Let's first *add 5 checks* ($+5$), each worth \$100 ($+100$), to the box. The value of the contents of the box then increases by \$500. Multiplying, we have

(number of checks) $\times$ (value of one check) $=$ change in value of contents
$$(+5)(+100) = +500$$

Thus a positive number times a positive number gave a positive product. This is nothing new.

Now let's *add 5 bills* to the box, thus decreasing its value by \$500. To show this multiplication, we use ($+5$) for the number of bills, (-100) for the value of each bill, and (-500) for the change in value of the box contents.

$$(+5)(-100) = -500$$

Here, a positive number times a negative number gives a negative product.

Next we *remove 5 checks* from the box, thus decreasing its value by $500.

$$(-5)(+100) = -500$$

Here again, the product of a positive number and a negative number is negative. Thus it doesn't matter whether the negative number is the first or the second. This, of course, is what you would expect from the commutative property, which applies to negative numbers as well as to positive numbers.

Finally, we *remove 5 bills* from the box, causing its value to increase by $500.

$$(-5)(-100) = +500$$

Here, the product of two negative numbers is positive.

We summarize these findings to get our *rules of signs for multiplication.*

If a and b are two positive numbers, then the product of two quantities of like sign is positive.

Rule of Signs for Multiplication	$(+a)(+b) = (-a)(-b) = +ab$	8

Also, the product of two quantities of unlike sign is negative.

Rule of Signs for Multiplication	$(+a)(-b) = (-a)(+b) = -ab$	9

◆◆◆ **Example 27:**

(a) $2(-3) = -6$
(b) $(-2)3 = -6$
(c) $(-2)(-3) = 6$

Multiplying a String of Numbers

When we multiply two negative numbers, we get a positive product. So when we are multiplying a *string* of numbers, if an even number of them are negative, the answer will be positive, and if an odd number of them are negative, the answer will be negative.

◆◆◆ **Example 28:**

(a) $2(-3)(-1)(-2) = -12$
(b) $2(-3)(-1)(2) = 12$ ◆◆◆

Multiplying Negative Numbers by Calculator

As we mentioned earlier, you enter negative numbers into the calculator by using a key marked $\boxed{(-)}$, $\boxed{+/-}$, or $\boxed{\text{CHS}}$. On some calculators you first enter the number and then change its sign using the proper key; on other calculators you enter the minus sign first and then the number, as in the following example.

◆◆◆ **Example 29:** Use your calculator to multiply -96 by -83.

Solution: You should get

$$(-96)(-83) = 7968$$ ◆◆◆

A simpler way to do the last problem would be to multiply +96 and +83 and determine the sign by inspection.

Common Error	Do not try to use the $\boxed{-}$ key to enter negative numbers into your calculator. The $\boxed{-}$ key is only for subtraction.

Multiplication of Approximate Numbers

Rule	When multiplying two or more approximate numbers, round the result to as many digits as in the factor having the fewest significant digits.

◆◆◆ **Example 30:**

$$12.1 \quad \times \quad 15.6 \quad = \quad 189$$

three digits · · · three digits · · · three digits ◆◆◆

When the factors have different numbers of significant digits, keep the same number of digits in your answer as is contained in the factor that has the *fewest* significant digits.

◆◆◆ **Example 31:**

$$123.56 \quad \times \quad 2.21 \quad = \quad 273$$

five digits · · · three digits · · · keep three digits ◆◆◆

Common Error	Do not confuse *significant digits* with *decimal places.* The number 274.56 has *five* significant digits and *two* decimal places. Decimal places determine how we round after adding or subtracting. Significant figures determine how we round after multiplying and, as we will soon see, after dividing, raising to a power, or taking roots.

Multiplication with Exact Numbers

When using *exact numbers* in a computation, treat them as if they had *more* significant figures than any of the approximate numbers in that computation.

◆◆◆ **Example 32:** If a certain car tire weighs 32.2 lb, how much will four such tires weigh?

Solution: Multiplying, we obtain

$$32.2(4) = 128.8 \text{ lb}$$

Since the 4 is an exact number, we retain as many significant figures as contained in 32.2, and round our answer to 129 lb.   ◆◆◆

Exercise 3 ◆ Multiplication

Multiply each approximate number and retain the proper number of digits in your answer.

1. 3967×1.84

2. 4.900×59.3

3. 93.9×0.0055908

4. $4.97 \times 9.27 \times 5.78$

5. $69.0 \times (-258)$

6. $-385 \times (-2.2978)$

7. $2.86 \times (4.88 \times 2.97) \times 0.553$

8. $(5.93 \times 7.28) \times (8.26 \times 1.38)$

Word Problems

9. What is the cost of 52.5 tons of cement at $63.25 a ton?

10. If 108 tons of rail is needed for 1 mi of track, how many tons will be required for 476 mi, and what will be its cost at $925 a ton?

11. Three barges carry 26 tons of gravel each, and a fourth carries 35 tons. What is the value, to the nearest dollar, of the whole shipment, at $12.75 per ton?

12. Two cars start from the same place and travel in opposite directions, one at the rate of 45 km/h, the other at 55 km/h. How far apart will they be at the end of 6.0 h?

13. What will be the cost of installing a telephone line 274 km long, at $5723 per kilometer?

14. The current to a projection lamp is measured at 4.7 A when the line voltage is 115.45 V. Using Eq. A65 (power = voltage × current), find the power dissipated in the lamp.

15. A gear in a certain machine rotates at the speed of 1808 rev/min. How many revolutions will it make in 9.500 min?

16. How much will 1000 washers weigh if each weighs 2.375 g?

17. One inch equals exactly 2.54 cm. Convert 385.84 in. to centimeters.

18. If there are 360 degrees per revolution, how many degrees are there in 4.863 revolutions?

1–4   Division

Definitions

The *dividend,* when divided by the *divisor,* gives us the *quotient.*

$$\text{dividend} \div \text{divisor} = \text{quotient}$$

or

$$\frac{\text{dividend}}{\text{divisor}} = \text{quotient}$$

> A quantity a/b is also a *fraction* and can also be referred to as the *ratio* of a to b. Fractions and ratios are discussed in Chapter 8.

Division by Calculator

Here you would use the $\boxed{\div}$ key on your calculator, as instructed in your calculator manual.

◆◆◆ **Example 33:** Use your calculator to divide 861 by 123.

Solution: You should get

$$861 \div 123 = 7$$ ◆◆◆

When we multiplied two integers, we always got an integer for an answer. This is not always the case when dividing.

◆◆◆ **Example 34:** When we divide 2 by 3, we get 0.666666666. . ... We must choose how many digits we wish to retain, and we must round our answer. Rounding to, say, three significant digits, we obtain

$$2 \div 3 \cong 0.667$$

Here it is appropriate to use the $\cong$ symbol. ◆◆◆

Division of Approximate Numbers

The rule for rounding with division is almost the same as with multiplication.

Rule	After dividing one approximate number by another, round the quotient to as many digits as there are in the original number having the fewest significant digits.

◆◆◆ **Example 35:**

Solution: By calculator,

$$846.2 \div 4.75 = 178.1473684$$

Since 4.75 has three significant digits, we round our quotient to 178. ◆◆◆

◆◆◆ **Example 36:** Divide 846.2 into three equal parts.

Solution: We divide by the integer 3, and since we consider integers to be exact, we retain in our answer the same number of significant digits as in 846.2.

$$846.2 \div 3 = 282.1$$ ◆◆◆

Dividing Signed Numbers

We will use the rules of signs for multiplication to get the rules of signs for division.

We know that the product of a negative number and a positive number is negative. Thus

$$(-2)(+3) = -6$$

If we divide both sides of this equation by $(+3)$, we get

$$-2 = \frac{-6}{+3}$$

From this we see that *a negative number divided by a positive number gives a negative quotient.*

Again starting with

$$(-2)(+3) = -6$$

we divide both sides by (-2) and get

$$+3 = \frac{-6}{-2}$$

Here we see that *a negative number divided by a negative number gives a positive product.*

We also know that the product of two negative numbers is positive. Thus

$$(-2)(-3) = +6$$

Dividing both sides by (-3), we get

$$-2 = \frac{+6}{-3}$$

Thus *a positive number divided by a negative number gives a negative quotient.* We combine these findings with the fact that the quotient of two positive numbers is positive and get our *rules of signs for division.*

The quotient is positive when dividend and divisor have the same sign.

Rule of Signs for Division	$\dfrac{+a}{+b} = \dfrac{-a}{-b} = \dfrac{a}{b}$	10

The quotient is negative when dividend and divisor have opposite signs.

Rule of Signs for Division	$\dfrac{+a}{-b} = \dfrac{-a}{+b} = -\dfrac{a}{b}$	11

◆◆◆ **Example 37:**

(a) $8 \div (-4) = -2$
(b) $-8 \div 4 = -2$
(c) $-8 \div (-4) = 2$ ◆◆◆

◆◆◆ **Example 38:** Divide 85.4 by -2.5386 on the calculator.

Solution: You should get

$$85.4 \div (-2.5386) = -33.6405893012$$

which we round to three digits, getting -33.6.   ◆◆◆

As with multiplication, the sign could also have been found by inspection.

Zero

Zero divided by any quantity (except zero) is zero. But division *by* zero is not defined. It is an illegal operation in mathematics.

◆◆◆ **Example 39:** Using your calculator, divide 5 by zero.

Solution: You should get an error message on your calculator.   ◆◆◆

Reciprocals

The *reciprocal* of any number n is $1/n$. Thus the product of a quantity and its reciprocal is equal to 1.

◆◆◆ **Example 40:**

(a) The reciprocal of 10 is 1/10.
(b) The reciprocal of 1/2 is 2.
(c) The reciprocal of $-\frac{3}{4}$ is $-\frac{4}{3}$.   ◆◆◆

Some keys, such as the reciprocal key, might be a *second function* on your calculator, and some other operations might be available only from a menu.

Reciprocals by Calculator

Simply enter the number and press the $\boxed{1/x}$ or the $\boxed{x^{-1}}$ key. Keep as many digits in your answer as there are significant digits in the original number.

◆◆◆ Example 41:

(a) The reciprocal of 6.38 is 0.157.
(b) The reciprocal of −2.754 is −0.3631.

◆◆◆

Exercise 4 ◆ Division

Divide, and then round your answer to the proper number of digits.

1. 947 ÷ 5.82
2. 0.492 ÷ 0.00478
3. −99.4 ÷ 286.5
4. −4.8 ÷ −2.557
5. 5836 ÷ 8264
6. 5.284 ÷ 3.827
7. 94,840 ÷ 1.33876
8. 3.449 ÷ (−6.837)
9. 2,497,000 ÷ 150,000
10. 2.97 ÷ 4.828

Word Problems Involving Division

11. A stretch of roadway 1858.54 m long is to be divided into 5 equal sections. Find the length of each section.

12. At the rate of 24.5 km in 8.25 h, how many kilometers would a person walk in 12.75 h? (Distance = rate × time.)

13. If 5 masons can build a wall in 8 days, how many masons are needed to build it in 4 days?

14. If 867 shares of railroad stock are valued at $84,099, what is the value of each share?

Reciprocals

Find the reciprocal of each number, retaining the proper number of digits in your answer.

15. 693
16. 0.00630
17. −396,000
18. 39.74
19. −0.00573
20. 938.4
21. 4.992
22. −6.93
23. 11.1
24. −375
25. 1.007
26. 3.98

Word Problems Involving Reciprocals

27. Using Eq. A64, find the equivalent resistance of a 475-Ω resistor and a 928-Ω resistor, connected in parallel.

28. When an object is placed 126 cm in front of a certain thin lens having a focal length f, the image will be formed 245 cm from the lens. The distances are related by

$$\frac{1}{f} = \frac{1}{126} + \frac{1}{245}$$

Find f.

29. The sine of an angle θ (written sin θ) is equal to the reciprocal of the cosecant of θ (csc θ). Find sin θ if csc θ = 3.58.

30. If two straight lines are perpendicular, the slope of one line is the negative reciprocal of the slope of the other. If the slope of a line is −2.55, find the slope of a perpendicular to that line.

1–5   Powers and Roots

Definitions

In the expression

$$2^4$$

the number 2 is called the *base,* and the number 4 is called the *exponent.* The expression is read "two to the fourth power." Its value is

$$2^4 = 2 \cdot 2 \cdot 2 \cdot 2 = 16$$

Powers by Calculator

To square a number on the calculator, simply enter the number and press the $\boxed{x^2}$ key. To raise a number to other powers, use the $\boxed{y^x}$, $\boxed{x^y}$, or $\boxed{\wedge}$ key, as in the following example. Round your result to the number of significant digits contained *in the base,* not the exponent.

◆◆◆ **Example 42:** Use your calculator to verify that

$$(3.85)^3 = 57.1 \quad \text{(rounded to three digits)} \qquad \text{◆◆◆}$$

Negative Base

A negative base raised to an *even* power gives a *positive* number. A negative base raised to an *odd* power gives a *negative* number.

◆◆◆ **Example 43:**

(a) $(-2)^2 = (-2)(-2) = 4$
(b) $(-2)^3 = (-2)(-2)(-2) = -8$
(c) $(-1)^{24} = 1$
(d) $(-1)^{25} = -1$ ◆◆◆

If you try to do these problems on your calculator, you will probably get an error indication. Some calculators will not work with a negative base, even though this is a valid operation.

Then how do you do it? Simply enter the base as *positive,* find the power, and determine the sign by inspection.

◆◆◆ **Example 44:** Find $(-1.45)^5$.

Solution: From the calculator,

$$(+1.45)^5 = 6.41$$

Since we know that a negative number raised to an odd power is negative, we write

$$(-1.45)^5 = -6.41 \qquad \text{◆◆◆}$$

Negative Exponent

A number can be raised to a negative exponent on the calculator with the $\boxed{y^x}$, $\boxed{x^y}$, or $\boxed{\wedge}$ key.

◆◆◆ **Example 45:** Use your calculator to verify that

$$(3.85)^{-3} = 0.0175 \quad \text{(rounded to three digits)}$$

Notice that raising a positive number to a negative power does *not* result in a negative number.

◆◆◆

Fractional Exponents

We'll see later that fractional exponents are another way of writing radicals. For now, we'll just evaluate fractional exponents on the calculator. The keystrokes are the same as before, but we must be sure that the fractional exponent is enclosed in parentheses, as shown in the following example.

◆◆◆ **Example 46:** Use your calculator to verify that

$$8^{2/3} = 4$$

◆◆◆

Roots

If $a^n = b$, then

$$\sqrt[n]{b} = a$$

which is read "the *n*th root of *b* equals *a*." The symbol $\sqrt{}$ is a *radical sign*, *b* is the *radicand*, and *n* is the *index* of the radical.

◆◆◆ **Example 47:**

(a) $\sqrt{4} = 2$ because $2^2 = 4$
(b) $\sqrt[3]{8} = 2$ because $2^3 = 8$
(c) $\sqrt[4]{81} = 3$ because $3^4 = 81$

◆◆◆

Principal Root

The *principal root* of a positive number is defined as the *positive* root. Thus $\sqrt{4} = +2$, not ± 2.

    The principal root is *negative* when we take an *odd* root of a *negative* number.

◆◆◆ **Example 48:**

$$\sqrt[3]{-8} = -2$$

because $(-2)(-2)(-2) = -8$.

◆◆◆

> When we speak about the *root* of a number, we mean, unless otherwise stated, the *principal root*. The principal root of a number has the same sign as the number itself.

Roots by Calculator

To find square roots, we simply use the $\boxed{\sqrt{}}$ key. To find other roots, use the $\boxed{\wedge}$ key. Retain as many significant digits in the answer as there are significant digits in the original radicand.

    Some calculators require that the radicand be entered first, and some require the index to be entered first. Again, consult your calculator manual.

◆◆◆ **Example 49:** Use your calculator to verify that

$$\sqrt[5]{28.4} = 1.952826537$$

which we round to 1.95.

◆◆◆

Odd Roots of Negative Numbers by Calculator

An *even* root of a negative number is *imaginary* (such as $\sqrt{-4}$). We will study these in Chapter 21. But an *odd* root of a negative number is *not* imaginary. It is a real, negative, number. As with powers, some calculators will not accept a negative radicand. Fortunately, we can outsmart our calculators and take odd roots of negative numbers anyway.

◆◆◆ **Example 50:** Find $\sqrt[5]{-875}$.

Solution: We know that an odd root of a negative number is real and negative. So we take the fifth root of +875, by calculator,

$$\sqrt[5]{+875} = 3.88 \quad \text{(rounded)}$$

and we only have to place a minus sign before the number.

$$\sqrt[5]{-875} = -3.88$$

◆◆◆

Exercise 5 ◆ Powers and Roots

Powers

Powers of 10 will be needed for scientific notation later. Arrange these powers of 10 in order, and try to invent a rule that will enable you to write the value of a power of 10 without doing the computation.

Evaluate each power without using a calculator. Do not round your answers.

1. 2^3	**2.** 5^3	**3.** $(-2)^3$	**4.** 9^2
5. 1^3	**6.** $(-1)^2$	**7.** $(-1)^{40}$	**8.** 3^2
9. 1^8	**10.** $(-1)^3$	**11.** $(-1)^{41}$	**12.** $(-3)^2$

Evaluate each power of 10.

13. 10^2	**14.** 10^1	**15.** 10^0	**16.** 10^{-4}
17. 10^3	**18.** 10^5	**19.** 10^{-2}	**20.** 10^{-1}
21. 10^4	**22.** 10^{-3}	**23.** 10^{-5}	

Evaluate each expression, retaining the correct number of digits in your answer.

24. $(8.55)^3$	**25.** $(1.007)^5$
26. $(9.55)^3$	**27.** $(-4.82)^3$
28. $(-77.2)^2$	**29.** $(8.28)^{-2}$
30. $(0.0772)^{0.426}$	**31.** $(5.28)^{-2.15}$
32. $(35.2)^{1/2}$	**33.** $(462)^{2/3}$
34. $(88.2)^{-2}$	**35.** $(-37.3)^{-3}$

Word Problems Involving Powers

36. The distance traveled by a falling body, starting from rest, is equal to $16t^2$, where t is the elapsed time. In 5.448 s, the distance fallen is $16(5.448)^2$ ft. Evaluate this quantity. (Treat 16 here as an approximate number.)

37. The power dissipated in a resistance R through which is flowing a current I is equal to I^2R. Therefore the power in a 365-Ω resistor carrying a current of 0.5855 A is $(0.5855)^2(365)$ W. Evaluate this power.

38. The volume of a cube of side 35.8 cm (Fig. 1–5) is $(35.8)^3$. Evaluate this volume.

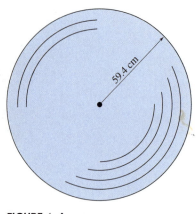

FIGURE 1–5

39. The volume of a 59.4-cm-radius sphere (Fig. 1–6) is $\frac{4}{3}\pi(59.4)^3$ cm^3. Find this volume.

40. An investment of $2000 at a compound interest rate of $6\frac{1}{4}\%$, left for $7\frac{1}{2}$ years, will be worth $2000(1.0625)^{7.5}$ dollars. Find this amount, to the nearest cent.

Roots

Find each principal root without using your calculator.

41. $\sqrt{25}$	**42.** $\sqrt[3]{27}$	**43.** $\sqrt{49}$
44. $\sqrt[3]{-27}$	**45.** $\sqrt[3]{-8}$	**46.** $\sqrt[5]{-32}$

FIGURE 1–6

Evaluate each radical by calculator, retaining the proper number of digits in your answer.

47. $\sqrt{49.2}$ **48.** $\sqrt{1.863}$ **49.** $\sqrt[3]{88.3}$

50. $\sqrt{772}$ **51.** $\sqrt{3875}$ **52.** $\sqrt[3]{7295}$

53. $\sqrt[3]{-386}$ **54.** $\sqrt[5]{-18.4}$ **55.** $\sqrt[3]{-2.774}$

Applications of Roots

56. The period T (time for one swing) of a simple pendulum (Fig. 1–7) 2.55 ft long is

$$T = 2\pi \sqrt{\frac{2.55}{32.0}} \text{ seconds}$$

Evaluate T.

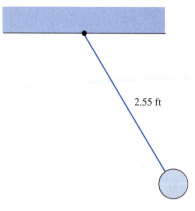

2.55 ft

FIGURE 1–7

57. The magnitude Z of the impedance in a circuit having a resistance of 3540 Ω and a reactance of 2750 Ω is

$$Z = \sqrt{(3540)^2 + (2750)^2} \text{ ohms}$$

Find Z.

58. The geometric mean B between 3.75 and 9.83 is

$$B = \sqrt{(3.75)(9.83)}$$

Evaluate B.

1–6 Combined Operations

Order of Operations

If the expression to be evaluated does not contain parentheses, perform the operations in the following order:

1. *Powers and roots, in any order.*
2. *Multiplications and divisions, from left to right.*
3. *Additions and subtractions, from left to right.*

Our first group of calculations will be with integers only, and later we will do some problems that require rounding. We first show a problem containing both addition and multiplication.

♦♦♦ **Example 51:** Evaluate $7 + 3 \times 4$.

Solution: The multiplication is done before the addition.

$$7 + 3 \times 4 = 7 + 12 = 19 \qquad \text{♦♦♦}$$

Next we do a calculation having both a power and multiplication.

♦♦♦ **Example 52:** Evaluate 5×3^2.

Solution: We raise to the power before multiplying:

$$5 \times 3^2 = 5 \times 9 = 45 \qquad \text{♦♦♦}$$

Be sure to repeat each of these computations on your calculator, consulting your manual where the operations are not clear.

Parentheses

When an expression contains parentheses, evaluate first the expression within the parentheses and then the entire expression.

◆◆◆ Example 53: Evaluate $(7 + 3) \times 4$.

Solution:

$$(7 + 3) \times 4 = 10 \times 4 = 40$$ ◆◆◆

If the sum or difference of more than one number is to be raised to a power, those numbers must be enclosed in parentheses.

◆◆◆ Example 54: Evaluate $(5 + 2)^2$.

Solution: We combine the numbers inside the parentheses before squaring.

$$(5 + 2)^2 = 7^2 = 49$$ ◆◆◆

◆◆◆ Example 55: Evaluate $(2 + 6)(7 + 9)$.

Solution: Evaluate the two quantities in parentheses before multiplying.

$$(2 + 6)(7 + 9) = 8 \times 16 = 128$$ ◆◆◆

◆◆◆ Example 56: Evaluate $\dfrac{8 + 4}{9 - 3}$.

Solution: Here the fraction line acts like parentheses, grouping the 8 and 4, as well as the 9 and 3. Written on a single line, this problem would be

$$(8 + 4) \div (9 - 3)$$

or

$$12 \div 6 = 2$$ ◆◆◆

Combined Operations with Approximate Numbers

Combined operations with approximate numbers are done the same way. However, we must round our answer properly using the rules given earlier in this chapter.

◆◆◆ Example 57: Evaluate the expression

$$\left(\frac{118.8 + 4.23}{\sqrt{136}} \right)^3$$

Solution: By calculator, we get a result of 1174.153047. But how many digits should we keep? In the numerator, we added a number with one decimal place to another with two decimal places, so we are allowed to keep just one. Thus the numerator, after addition, is good to one decimal place, or, in this case, four significant digits. The denominator, however, has just three significant digits, so we round our answer to three significant digits, getting 1170. ◆◆◆

33.74 ft

x

21.8 ft

FIGURE 1–8   A rectangular courtyard.

◆◆◆ Example 58: A rectangular courtyard (Fig. 1–8) having sides of 21.8 ft and 33.74 ft has a diagonal measurement x given by the expression

$$x = \sqrt{(21.8)^2 + (33.74)^2}$$

Evaluate the expression to find x.

Solution: By calculator we get a result of 40.16998. We round our answer to three significant digits, because 21.8 has only three, and we get $x = 40.2$ ft. ◆◆◆

Exercise 6 ◆ Combined Operations

Combined Operations with Exact Numbers

Perform each computation by calculator.

1. $(37)(28) + (36)(64)$

2. $(22)(53) - (586)(4) + (47)(59)$

3. $(63 + 36)(37 - 97)$

4. $(89 - 74 + 95)(87 - 49)$

5. $\dfrac{219}{73} + \dfrac{194}{97}$

6. $\dfrac{228}{38} - \dfrac{78}{26} + \dfrac{364}{91}$

7. $\dfrac{647 + 688}{337 + 108}$

8. $\dfrac{809 - 463 + 1858}{958 - 364 + 508}$

9. $(5 + 6)^2$

10. $(422 + 113 - 533)^4$

11. $(423 - 420)^3$

12. $\left(\dfrac{853 - 229}{874 - 562}\right)^2$

13. $\left(\dfrac{141}{47}\right)^3$

14. $\sqrt{434 + 466}$

15. $\sqrt{(8)(72)}$

16. $\sqrt[3]{657 + 553 - 1085}$

17. $\sqrt[4]{(27)(768)}$

18. $\sqrt{\dfrac{2404}{601}}$

19. $\sqrt[4]{\dfrac{1136}{71}}$

20. $\sqrt{961} + \sqrt{121}$

21. $\sqrt[4]{625} + \sqrt{961} - \sqrt[3]{216}$

22. $\sqrt[4]{256} \times \sqrt{49}$

Combined Operations with Approximate Numbers

Perform the computations, keeping the proper number of digits in your answer.

23. $(7.37)(3.28) + (8.36)(2.64)$

24. $(522)(9.53) - (586)(4.70) + (847)(7.59)$

25. $(63.5 + 83.6)(8.37 - 1.72)$

26. $(8.93 - 3.74 + 9.05)(68.70 - 64.90)$

27. $\dfrac{583}{473} + \dfrac{946}{907}$

28. $\dfrac{6.73}{8.38} - \dfrac{5.97}{8.06} + \dfrac{8.63}{1.91}$

29. $\dfrac{6.47 + 8.604}{3.37 + 90.8}$

30. $\dfrac{809 - 463 + 744}{758 - 964 + 508}$

31. $(5.37 + 2.36)^2$

32. $(4.25 + 4.36 - 5.24)^4$

33. $(6.423 + 1.05)^2$

34. $\left(\dfrac{45.3 - 8.34}{8.74 - 5.62}\right)^{2.5}$

35. $\left(\dfrac{8.90}{4.75}\right)^2$

36. $\sqrt{4.34 + 4.66}$

37. $\sqrt[3]{657 + 553 - 842}$

38. $\sqrt{(28.1)(5.94)}$

39. $\sqrt[5]{(9.06)(4.86)(7.93)}$

40. $\sqrt{\dfrac{653}{601}}$

41. $\sqrt[4]{\dfrac{4.50}{7.81}}$

42. $\sqrt{9.74} + \sqrt{12.5}$

43. $\sqrt[4]{528} + \sqrt{94.2} - \sqrt[3]{284}$

44. $\sqrt[4]{653} \times \sqrt{55.3}$

1–7   Scientific Notation

Definitions

Let us multiply two large numbers on the calculator: 500,000 and 300,000. We get a display like $\boxed{1.5\ E + 11}$ or $\boxed{1.5\quad 11}$, depending on the calculator. What has happened?

Our answer (150,000,000,000) is too large to fit the calculator display, so the machine has automatically switched to *scientific notation.* Our calculator display actually contains *two* numbers: a decimal number (1.5) and an integer (11). Our answer is equal to the decimal number multiplied by 10 raised to the value of the integer.

Calculator display: $\boxed{1.5\quad 11}$
Scientific notation: 1.5×10^{11}

decimal part ———————┘   └——— power of 10

Ten raised to a power (such as 10^{11}) is called a *power of 10.*

A number is said to be in *scientific notation* when it is written as a number whose absolute value is between 1 and 10, multiplied by a power of 10.

◆◆◆ **Example 59:** The following numbers are written in scientific notation:

(a) 2.74×10^3 (b) 8.84×10^9
(c) 5.4×10^{-6} (d) -1.2×10^{-5} ◆◆◆

Engineering Notation

Engineering notation is similar to scientific notation. The difference is that

- the exponent is a multiple of three; and
- there can be one, two, or three digits to the left of the decimal point, rather than just one digit.

Having an exponent that is a multiple of 3 makes it easier to use the *metric prefixes* described in Sect. 1–8.

◆◆◆ **Example 60:** Some examples of numbers written in engineering notation are as follows:

(a) 66.3×10^3 (b) 8.14×10^9
(c) 725×10^{-6} (d) 28.72×10^{-12} ◆◆◆

Evaluating Powers of 10

We did some work with powers in Sec. 1–5. We saw, for example, that 2^3 meant

$$2^3 = 2 \cdot 2 \cdot 2 = 8$$

Here, the power 3 tells how many 2's are to be multiplied to give the product. For powers of 10, the power tells how many 10's are to be multiplied to give the product.

◆◆◆ **Example 61:**

(a) $10^2 = 10 \times 10 = 100$
(b) $10^3 = 10 \times 10 \times 10 = 1000$ ◆◆◆

Negative powers, as before, are calculated by means of Eq. 35, $x^{-a} = 1/x^a$.

♦♦♦ **Example 62:**

(a) $10^{-2} = \dfrac{1}{10^2} = \dfrac{1}{100} = 0.01$

(b) $10^{-5} = \dfrac{1}{10^5} = \dfrac{1}{100,000} = 0.00001$ ♦♦♦

Some powers of 10 are summarized in the following table:

Positive Powers	Negative Powers		
$1,000,000 = 10^6$	0.1	$= 1/10$	$= 10^{-1}$
$100,000 = 10^5$	0.01	$= 1/10^2$	$= 10^{-2}$
$10,000 = 10^4$	0.001	$= 1/10^3$	$= 10^{-3}$
$1,000 = 10^3$	0.0001	$= 1/10^4$	$= 10^{-4}$
$100 = 10^2$	0.00001	$= 1/10^5$	$= 10^{-5}$
$10 = 10^1$	0.000001	$= 1/10^6$	$= 10^{-6}$
$1 = 10^0$			

Converting Numbers to Scientific Notation

First rewrite the given number with a single digit to the left of the decimal point, discarding any nonsignificant zeros. Then multiply this number by the power of 10 that will make it equal to the original number.

♦♦♦ **Example 63:**

$$346 = 3.46 \times 100$$
$$= 3.46 \times 10^2$$ ♦♦♦

♦♦♦ **Example 64:**

$$2700 = 2.7 \times 1000$$
$$= 2.7 \times 10^3$$

Note that we have discarded the two nonsignificant zeros. ♦♦♦

When we are converting a number whose absolute value is less than 1, our power of 10 will be negative, as in Example 65.

♦♦♦ **Example 65:**

$$0.00000950 = 9.50 \times 0.000001$$
$$= 9.50 \times 10^{-6}$$ ♦♦♦

The sign of the exponent has nothing to do with the sign of the original number. You can convert a negative number to scientific notation just as you would a positive number, and put the minus sign on afterward.

♦♦♦ **Example 66:** Convert $-34,720$ to scientific notation.

Solution: Converting $+34,720$ to scientific notation, we obtain

$$34,720 = 3.472 \times 10,000$$
$$= 3.472 \times 10^4$$

Then multiplying by -1 gives

$$-34,720 = -3.472 \times 10^4$$ ♦♦♦

Converting Numbers from Scientific Notation

To convert *from* scientific notation, simply reverse the process.

◆◆◆ Example 67:

$$4.82 \times 10^5 = 4.82 \times 100{,}000$$
$$= 482{,}000$$

◆◆◆

◆◆◆ Example 68:

$$8.25 \times 10^{-3} = 8.25 \times 0.001$$
$$= 0.00825$$

◆◆◆

Converting Numbers to Engineering Notation

Converting to engineering notation is simple if the digits of the decimal number are grouped by commas into sets of three, in the usual way.

◆◆◆ Example 69:

(a) $21{,}840 = 21.84 \times 10^{-3}$
(b) $548{,}000 = 548 \times 10^3$
(c) $72{,}560{,}000 = 72.56 \times 10^6$

◆◆◆

For numbers less than 1, it helps to first separate the digits following the decimal point into groups of three.

◆◆◆ Example 70:

(a) $0.87217 = 0.872\ 17 = 872.17 \times 10^{-3}$
(b) $0.000736492 = 0.000\ 736\ 492 = 736.492 \times 10^{-6}$
(c) $0.0000000472 = 0.000\ 000\ 047\ 2 = 47.2 \times 10^{-9}$

◆◆◆

Addition and Subtraction

If two or more numbers to be added or subtracted have the *same power of 10,* simply combine the numbers and keep the same power of 10.

This is another application of the distributive law, $ab + ac = a(b + c)$. Thus

$2 \times 10^5 + 3 \times 10^5$

$$= (2 + 3)10^5$$
$$= 5 \times 10^5$$

◆◆◆ Example 71:

(a) $(2 \times 10^5) + (3 \times 10^5) = 5 \times 10^5$
(b) $(8 \times 10^3) - (5 \times 10^3) + (3 \times 10^3) = 6 \times 10^3$

◆◆◆

If the powers of 10 are different, *they must be made equal* before the numbers can be combined. A shift of the decimal point of one place to the *left* will *increase* the exponent by 1. Conversely, a shift of the decimal point one place to the *right* will *decrease* the exponent by 1.

◆◆◆ Example 72:

(a) $(1.5 \times 10^4) + (3 \times 10^3) = (1.5 \times 10^4) + (0.3 \times 10^4)$
$$= 1.8 \times 10^4$$
(b) $(1.25 \times 10^5) - (2 \times 10^4) + (4 \times 10^3)$
$$= (1.25 \times 10^5) - (0.2 \times 10^5) + (0.04 \times 10^5)$$
$$= 1.09 \times 10^5$$

◆◆◆

Multiplication

We multiply powers of 10 by *adding their exponents.*

◆◆◆ Example 73:

(a) $10^3 \cdot 10^4 = 10^{3+4} = 10^7$

(b) $10^{-2} \cdot 10^5 = 10^{-2+5} = 10^3$ ◆◆◆

> We are really using Eq. 29, $x^a \cdot x^b = x^{a+b}$. This equation is one of the *laws of exponents* that we will study later.

To multiply two numbers in scientific notation, multiply the decimal parts and the powers of 10 *separately.*

◆◆◆ Example 74:

$$(2 \times 10^5)(3 \times 10^2) = (2 \times 3)(10^5 \times 10^2)$$
$$= 6 \times 10^{5+2} = 6 \times 10^7$$ ◆◆◆

Division

We divide powers of 10 by subtracting the exponent of the denominator from the exponent of the numerator.

◆◆◆ Example 75:

(a) $\dfrac{10^5}{10^3} = 10^{5-3} = 10^2$

(b) $\dfrac{10^{-4}}{10^{-2}} = 10^{-4-(-2)} = 10^{-2}$ ◆◆◆

As with multiplication, we divide the decimal parts and the powers of 10 separately.

◆◆◆ Example 76:

(a) $\dfrac{8 \times 10^5}{4 \times 10^2} = \dfrac{8}{4} \times \dfrac{10^5}{10^2} = 2 \times 10^{5-2} = 2 \times 10^3$

(b) $\dfrac{12 \times 10^3}{4 \times 10^5} = 3 \times 10^{3-5} = 3 \times 10^{-2}$ ◆◆◆

Scientific Notation on Computer or Calculator

A computer or calculator will switch to scientific notation by itself when the printout or display gets too long. For example, a computer might print the number

$$2.75E\text{-}6$$

which we interpret as

$$2.75 \times 10^{-6} \quad \text{or} \quad 0.00000275$$

On a calculator, we can enter numbers in scientific notation with the *enter exponent* key, usually marked $\boxed{\text{EE}}$, $\boxed{\text{EXP}}$, or $\boxed{\text{EEX}}$.

Common Error

Students often enter powers of 10 (such as 10^4) incorrectly into their calculators, forgetting that 10^4 is really 1×10^4.
Thus to enter 10^4, we press

$$1 \ \boxed{\text{EE}} \ 4$$

and not

$$10 \ \boxed{\text{EE}} \ 4$$

Alternately, we can switch the calculator into scientific notation or engineering notation mode (you may have to select this from a menu; consult your manual). In either case, we then use the same keystrokes as for calculations with decimal numbers, and let the calculator worry about the powers of 10.

Exercise 7 Scientific Notation

Powers of 10

Write each number as a power of 10.

1. 100	**2.** 1,000,000	**3.** 0.0001
4. 0.001	**5.** 100,000,000	

Write each power of 10 as a decimal number.

6. 10^5	**7.** 10^{-2}	**8.** 10^{-5}
9. 10^{-1}	**10.** 10^4	

Scientific Notation

Write each number in scientific notation and engineering notation.

11. 186,000	**12.** 0.0035
13. 25,742	**14.** $80\overline{0}0$
15. 98.3×10^3	**16.** 0.0775×10^{-2}

Convert each number from scientific notation to decimal notation.

17. 2.85×10^3	**18.** 1.75×10^{-5}	**19.** 9×10^4
20. 9.00×10^4	**21.** 3.667×10^{-3}	

Multiplication and Division

Multiply the following powers of 10.

22. $10^5 \cdot 10^2$	**23.** $10^4 \cdot 10^{-3}$	**24.** $10^{-5} \cdot 10^{-4}$
25. $10^{-2} \cdot 10^5$	**26.** $10^{-1} \cdot 10^{-4}$	

Divide the following powers of 10.

27. $10^8 \div 10^5$	**28.** $10^4 \div 10^6$	**29.** $10^5 \div 10^{-2}$
30. $10^{-3} \div 10^5$	**31.** $10^{-2} \div 10^{-4}$	

Multiply without using a calculator.

32. $(3.0 \times 10^3)(5.0 \times 10^2)$	**33.** $(5 \times 10^4)(8 \times 10^{-3})$
34. $(2 \times 10^{-2})(4 \times 10^{-5})$	**35.** $(7.0 \times 10^4)(3\overline{0},000)$

Divide without using a calculator.

36. $(8 \times 10^4) \div (2 \times 10^2)$	**37.** $(6 \times 10^4) \div 0.03$
38. $(3 \times 10^3) \div (6 \times 10^5)$	**39.** $(8 \times 10^{-4}) \div 400,000$
40. $(9 \times 10^4) \div (3 \times 10^{-2})$	**41.** $49,000 \div (7.0 \times 10^{-2})$

Addition and Subtraction

Combine without using a calculator.

42. $(3.0 \times 10^4) + (2.1 \times 10^5)$
43. $(75.0 \times 10^2) + 32\overline{0}0$
44. $(1.557 \times 10^2) + (9.000 \times 10^{-1})$
45. $0.037 - (6.0 \times 10^{-3})$
46. $(7.2 \times 10^4) + (1.1 \times 10^4)$

Scientific Notation on the Calculator

Perform the following computations in scientific notation. Combine the powers of 10 by hand or with your calculator.

47. $(1.58 \times 10^2)(9.82 \times 10^3)$
48. $(9.83 \times 10^5) \div (2.77 \times 10^3)$
49. $(3.87 \times 10^{-2})(5.44 \times 10^5)$
50. $(2.74 \times 10^3) \div (9.13 \times 10^5)$
51. $(5.6 \times 10^2)(3.1 \times 10^{-1})$
52. $(7.72 \times 10^8) \div (3.75 \times 10^{-9})$

Applications

53. Three resistors, having resistances of 4.98×10^5 Ω, 2.47×10^4 Ω, and 9.27×10^6 Ω, are wired in series (Fig. 1–9). Find the total resistance, using Eq. A63.

54. Find the equivalent resistance if the three resistors of problem 53 are wired in parallel (Fig. 1–10). Use Eq. A64.

55. Find the power dissipated in a resistor if a current of 3.75×10^{-3} A produces a voltage drop of 7.24×10^{-4} V across the resistor. Use Eq. A65.

56. The voltage across an 8.35×10^5-Ω resistor is 2.95×10^{-3} V. Find the power dissipated in the resistor, using Eq. A66.

57. Three capacitors, 8.26×10^{-6} farad (F), 1.38×10^{-7} F, and 5.93×10^{-5} F, are wired in parallel. Find the equivalent capacitance by Eq. A73.

58. A wire 4.75×10^3 cm long when loaded is seen to stretch 9.55×10^{-2} cm. Find the strain in the wire, using Eq. A53.

59. How long will it take a rocket traveling at a rate of 3.2×10^6 m/h to travel the 3.8×10^8 m from the earth to the moon? Use Eq. A17.

60. The oil shale reserves of the United States are estimated at 2.0×10^9 tons. How long would this supply last at a rate of consumption of 9.0×10^7 tons/yr?

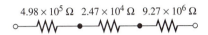

FIGURE 1–9   Resistors in series.

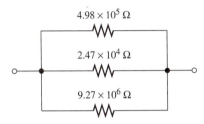

FIGURE 1–10   Resistors in parallel.

1–8 Units of Measurement

A *unit* is a standard of measurement, such as the meter, inch, hour, or pound. The two main systems of units in use are the *British* system (feet, pounds, gallons, etc.) and the SI or *metric* system (meters, kilograms, liters, etc.). SI stands for Le Système International d'Unites, or the *International System of Units*. In addition, some special units, such as a *square* of roofing material, and some obsolete units, such as *rods* and *chains,* must occasionally be dealt with.

Numbers having units of measure are sometimes called *denominate* numbers. Our main task in this section is to learn how to convert such a number from one unit of measurement to another, say, from feet to meters.

The word *denominate* comes from the root *nominare,* to name.

Systems of Units

Each physical quantity can be expressed in any one of a bewildering variety of units. Length, for example, can be measured in meters, inches, nautical miles, angstroms, feet, and so on. In this section we will convert from one British unit to another. In the next section we will convert from one metric unit to another, and also convert between British and metric units.

Abbreviations and Symbols

Most units have an abbreviated form, so that we do not have to write the full word. Thus the abbreviation for millimeters is mm, and the symbol for ohms is Ω. In this section we will usually give the full word *and* the abbreviation. The abbreviations for all of the units in this text are given in a table of conversion factors in Appendix B.

Conversion Factors

We can convert from any unit of length to any other unit of length by multiplying by a suitable number, called a *conversion factor.*

◆◆◆ **Example 77:** Convert 1.530 miles (mi) to feet (ft).

Solution: From Appendix B we find the relation between miles and feet.

$$5280 \text{ ft} = 1 \text{ mi}$$

Dividing both sides by 1 mile, we get the conversion factor.

$$\frac{5280 \text{ ft}}{1 \text{ mi}} = 1$$

We know that we can multiply any quantity by 1 without changing the value of that quantity. Thus if we multiply our original quantity (1.530 mi) by the conversion factor (5280 ft/mi), we do not change the value of the original quantity. We will, however, change the units. Multiplying yields

$$1.530 \text{ mi} = 1.530 \text{ mi} \times \frac{5280 \text{ ft}}{1 \text{ mi}} = 8078 \text{ ft}$$

Note that we have rounded our answer to four significant digits, because all numbers used in the calculation have at least four significant digits (the 5280 is exact).   ◆◆◆

Suppose that in the first step of Example 77, we had divided both sides by 5280 ft instead of by 1 mi. We could have gotten another conversion factor:

$$\frac{1 \text{ mi}}{5280 \text{ ft}} = 1$$

Thus each relation between two units of measurement gives us *two* conversion factors. Each of these is sometimes called a *unity ratio*—that is, a fraction that is equal to unity, or 1. Since each of these conversion factors is equal to 1, we may multiply any quantity by a conversion factor without changing the value of that quantity. We will, however, cause the units to change. But which of the two conversion factors should we use? It is simple. *Multiply by the conversion factor that will cancel the units you wish to eliminate.*

Significant Digits

You should try to use a conversion factor that is exact, or one that contains as many significant digits as (or, preferably, one more than) your original number. Then you should round your answer to as many significant digits as in the original number.

◆◆◆ **Example 78:** Convert 934 acres to square miles (mi^2).

Solution: From Appendix B we find the equation

$$1 \text{ mi}^2 = 640 \text{ acres}$$

where 640 is an exact number. We must write our conversion factor so that the unwanted unit (acres) is in the denominator, so the acres will cancel. Our conversion factor is thus

$$\frac{1 \text{ mi}^2}{640 \text{ acres}} = 1$$

Multiplying, we obtain

$$934 \text{ acres} = 934 \text{ acres} \times \frac{1 \text{ mi}^2}{640 \text{ acres}}$$

$$= 1.46 \text{ mi}^2$$

The conversion factor used here is exact, so we have rounded our answer to three significant digits.　◆◆◆

Using More Than One Conversion Factor

Sometimes you may not be able to find a *single* conversion factor linking the units you want to convert. You may have to use *more than one*.

◆◆◆ **Example 79:** Convert 7375 yards (yd) to nautical miles (nau mi).

Solution: In Appendix B we find no conversion factor between nautical miles and yards, but we see that

$$1 \text{ nau mi} = 6076 \text{ ft} \quad \text{and} \quad 3 \text{ ft} = 1 \text{ yd}$$

So

$$7375 \text{ yd} = 7375 \text{ yd} \times \frac{3 \text{ ft}}{1 \text{ yd}} \times \frac{1 \text{ nau mi}}{6076 \text{ ft}} = 3.641 \text{ nautical mi} \qquad ◆◆◆$$

Metric Units

The *metric system* is a system of weights and measures that was developed in France in 1793 and that has since been adopted by most countries of the world. It is widely used in scientific work in the United States.

The basic unit of length in the metric system is the *meter* (m). The unit of area is the *are,* or 100 square meters (m^2). The unit of volume is the *liter* (L), the volume of a cube one-tenth of a meter on a side. The unit of weight is the *gram* (g), the theoretical weight of a cube of distilled water measuring $\frac{1}{100}$ of a meter on a side.

Metric Prefixes

Converting between metric units is made easy because larger and smaller metric units are related to the basic units by *factors of 10*. These larger or smaller units are indicated by placing a *prefix* before the basic unit. A prefix is a group of letters placed at the beginning of a word to modify the meaning of that word. For example, the prefix *kilo* means 1000, or 10^3. Thus a *kilo*gram is 1000 grams. Other metric prefixes are given in Table 1–1.

◆◆◆ **Example 80:**

(a) A *kilo*meter (km) is a thousand meters, because *kilo* means one thousand.

$$1 \text{ km} = 1000 \text{ m}$$

TABLE 1–1　Metric prefixes.

Amount	Multiples and Submultiples	Prefix	Symbol	Pronunciation	Meaning
1 000 000 000 000	10^{12}	tera	T	ter′à	One trillion times
1 000 000 000	10^{9}	giga	G	ji′ga	One billion times
1 000 000	10^{6}	mega	M*	meg′à	One million times
1 000	10^{3}	kilo	k*	kil′o	One thousand times
100	10^{2}	hecto	h	hek′to	One hundred times
10	10	deka	da	dek′a	Ten times
0.1	10^{-1}	deci	d	des′i	One tenth of
0.01	10^{-2}	centi	c*	sen′ti	One hundredth of
0.001	10^{-3}	milli	m*	mil′i	One thousandth of
0.000 001	10^{-6}	micro	μ*	mi′kro	One millionth of
0.000 000 001	10^{-9}	nano	n	nan′o	One billionth of
0.000 000 000 001	10^{-12}	pico	p	pē′co	One trillionth of
0.000 000 000 000 001	10^{-15}	femto	f	fem′to	One quadrillionth of
0.000 000 000 000 000 001	10^{-18}	atto	a	at′to	One quintillionth of

*Most commonly used.

(b) A *centi*meter (cm) is one-hundredth of a meter, because *centi* means one-hundredth.

$$1 \text{ cm} = 1/100 \text{ m}$$

(c) A *milli*meter (mm) is one-thousandth of a meter, because *milli* means one-thousandth.

$$1 \text{ mm} = 1/1000 \text{ m}$$　　　　◆◆◆

Converting from One Metric Unit to Another

Converting from one metric unit to another is usually a matter of multiplying or dividing by a power of 10. Most of the time, the names of the units will tell how they are related, so we do not even have to look them up.

◆◆◆ **Example 81:**　Convert 72,925 meters (m) to kilometers (km).

Solution:　A *kilo*meter is a thousand meters.

$$\frac{1 \text{ km}}{1000 \text{ m}} = 1$$

So, as before,

$$72,925 \text{ m} = 72,925 \text{ m} \times \frac{1 \text{ km}}{1000 \text{ m}} = 72.925 \text{ km}$$　　　　◆◆◆

For more unusual metric units, simply look up the conversion factor in a table and convert as we did for British units.

◆◆◆ **Example 82:**　Convert 2.75 newtons (N) to dynes.

Solution:　These two metric units of force do not have any basic units in their names, nor any prefixes. Thus we cannot tell just from their names how they are related to each other. However, from Appendix B we find that

$$1 \text{ newton} = 10^5 \text{ dynes}$$

Converting in the usual way, we obtain

$$2.75 \text{ newton} = 2.75 \text{ newton} \times \frac{10^5 \text{ dynes}}{1 \text{ newton}} = 2.75 \times 10^5 \text{ dynes}$$

or 275,000 dynes.   ◆◆◆

Converting between British and Metric Units

We convert between British and metric units in the same way that we converted within each system.

◆◆◆ **Example 83:** Convert 2.84 U.S. gallons (gal) to liters (L).

Solution: From Appendix B we find

$$1 \text{ gal (U.S.)} = 3.785 \text{ L}$$

Converting gives

$$2.84 \text{ gal} = 2.84 \text{ gal} \times \frac{3.785 \text{ L}}{1 \text{ gal}} = 10.7 \text{ L}$$

rounded to three significant digits.   ◆◆◆

Converting Areas and Volumes

Length may be given in, say, centimeters (cm), but an *area* may be given in square centimeters (cm^2). Similarly, a *volume* may be in cubic centimeters (cm^3). So to get a conversion factor for area or volume, if not found in Appendix B, simply square or cube the conversion factor for length.
    If we take the equation

$$2.54 \text{ cm} = 1 \text{ in.}$$

and square both sides, we get

$$(2.54 \text{ cm})^2 = (1 \text{ in.})^2$$

or

$$6.4516 \text{ cm}^2 = 1 \text{ in.}^2$$

This gives us a conversion between square centimeters and square inches.

◆◆◆ **Example 84:** Convert 864 yd^2 to acres.

Solution: Appendix B has no conversion for square yards. However,

$$1 \text{ yd} = 3 \text{ ft}$$

Squaring yields

$$1 \text{ yd}^2 = (3 \text{ ft})^2 = 9 \text{ ft}^2$$

Also from the table,

$$1 \text{ acre} = 43,560 \text{ ft}^2$$

So

$$864 \text{ yd}^2 = 864 \text{ yd}^2 \times \frac{9 \text{ ft}^2}{1 \text{ yd}^2} \times \frac{1 \text{ acre}}{43,560 \text{ ft}^2} = 0.179 \text{ acre}$$ ◆◆◆

Converting Rates to Other Units

A *rate* is the amount of one quantity expressed *per unit of some other quantity*. Some rates, with typical units, are:

rate of travel (mi/h) or (km/h)   flow rate (gal/min) or (m³/s)

application rate (lb/acre)   unit price (dollars/lb)

Each rate contains *two* units of measure; miles per hour, for example, has *miles* in the numerator and *hours* in the denominator. It may be necessary to convert *either* or *both* of those units to other units. Sometimes a single conversion factor can be found (such as 1 m/h = 1.466 ft/s), but more often you will have to convert each unit with a *separate* conversion factor.

◆◆◆ **Example 85:** A certain chemical is to be added to a pool at the rate of 3.74 oz per gallon of water. Convert this to pounds of chemical per cubic foot of water.

Solution: We write the original quantity as a fraction and multiply by the appropriate factors, themselves written as fractions.

$$3.74 \text{ oz/gal} = \frac{3.74 \ \cancel{oz}}{\cancel{gal}} \times \frac{1 \text{ lb}}{16 \ \cancel{oz}} \times \frac{7.481 \ \cancel{gal}}{\text{ft}^3} = 1.75 \text{ lb/ft}^3$$ ◆◆◆

Exercise 8 ◆ Units of Measurement

Convert the following British units.

1. 152 inches to feet
2. 0.153 mile to yards
3. 762.0 feet to inches
4. 627 feet to yards
5. 29 tons to pounds
6. 88.90 pounds to ounces
7. 89,600 lb to tons
8. 8552 ounces to pounds

Convert the following metric units. Write your answer in scientific notation if the numerical value is greater than 1000 or less than 0.1.

9. 364,000 meters to kilometers
10. 0.000473 volt to millivolts
11. 735,900 grams to kilograms
12. 7.68×10^{-5} kilowatt to watts
13. 6.2×10^9 ohms to megohms
14. 825×10^4 newtons to kilonewtons
15. 9348 picofarads to microfarads
16. 84,398 nanoseconds to milliseconds

Convert between the given British and metric units.

17. 364.0 meters to feet
18. 6.83 inches to millimeters
19. 7.35 pounds to newtons
20. 2.55 horsepower to kilowatts
21. 4.66 gallons to liters
22. 825×10^4 dynes to pounds
23. 3.94 yards to meters
24. 834 cubic centimeters to gallons

Convert the following areas and volumes.

25. 2840 square yards to acres
26. 1636 square meters to ares

Be sure to write the original quantity as a built-up fraction, $\frac{a}{b}$, rather than on a single line, a/b. This will greatly reduce your chances of making an error.

Take gallons in this exercise to mean U.S. gallons.

27. 24.8 square feet to square meters
28. 3.72 square meters to square feet
29. 0.982 square kilometer to acres
30. 5.93 acres to square meters
31. 7.360 cubic feet to cubic inches
32. 4.83 cubic meters to cubic yards
33. 73.8 cubic yards to cubic meters
34. 8.220 gallons to cubic feet
35. 267 cubic millimeters to cubic inches
36. 112 liters to gallons

Convert units on the following time rates.

37. 4.86 feet per second to miles per hour
38. 777 gallons per minute to cubic meters per hour
39. 66.2 miles per hour to kilometers per hour
40. 52.0 knots to miles per minute
41. 953 births per year to births per week

Convert units on the following unit prices.

42. $1.25 per gram to dollars per kilogram
43. $800 per acre to cents per square meter
44. $3.54 per pound to cents per ounce
45. $4720 per ton to cents per pound

Applications

46. Convert all of the dimensions for the part in Fig. 1–11 to inches.

47. The jet fuel tank in Fig. 1–12 has a volume of 15.7 cubic feet. How many gallons of jet fuel will it hold?
48. A certain circuit board weighs 0.176 pound. Find its weight in ounces.
49. A certain laptop computer weighs 6.35 kilograms. What is its weight in pounds?
50. A generator has an output of 5.34×10^6 millivolts. What is the output in kilovolts?
51. Convert all of the dimensions in Fig. 1–13 to centimeters.

52. The surface area of a certain lake, shown in Fig. 1–14, is 7361 square yards. Convert this to square meters.

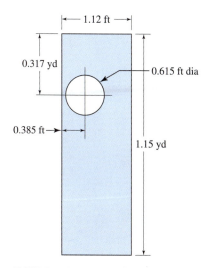

FIGURE 1–11

FIGURE 1–12

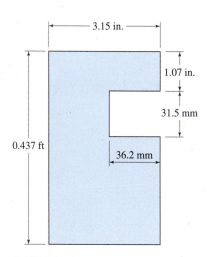

FIGURE 1–13

FIGURE 1–14   **FIGURE 1–15**

53. A solar collector, shown in Fig. 1–15, has an area of 8834 square inches. Convert this to square meters.

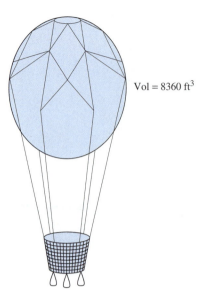

FIGURE 1–16   FIGURE 1–17

54. The volume of a balloon, shown in Fig. 1–16, is 8360 cubic feet. Convert this to cubic inches.

55. The volume of a certain gasoline tank, shown in Fig. 1–17, is 9274 cubic centimeters. Convert this to gallons.

56. An airplane is cruising at a speed of 785 miles per hour. Convert this speed to kilometers per hour.

1–9   Substituting into Equations and Formulas

Substituting into Equations

We get an *equation* when two expressions are set equal to each other.

◆◆◆ **Example 86:** $x = 5a - 2b + 3c$ is an equation that enables us to find x if we know a, b, and c.   ◆◆◆

We will study equations in detail later, but for now we will simply substitute into equations and use our calculators to compute the result. To *substitute into an equation* means to replace the letter quantities in an equation by their given numerical values and to perform the computation indicated.

◆◆◆ **Example 87:** Substitute the values $a = 5$, $b = 3$, and $c = 6$ into the equation

$$x = \frac{3a + b}{c}$$

Solution: Substituting, we obtain

$$x = \frac{3(5) + 3}{6} = \frac{18}{6} = 3$$ ◆◆◆

When substituting approximate numbers, be sure to round your answer to the proper number of digits. Treat any integers in the equation as exact numbers.

Substituting by Calculator

Problems such as in Example 87 can easily be done on a graphics calculator. In the following example, the key $\boxed{\rightarrow}$ or $\boxed{\text{STO}}$ sends a number to a memory location. Thus

$$5 \boxed{\rightarrow} A$$

means to store the number 5 in memory location A.

◆◆◆ **Example 88:** Repeating Example 87 by calculator yields the following:

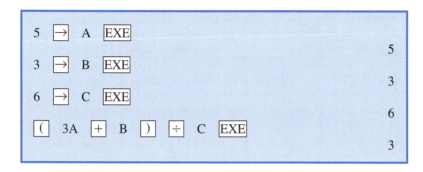

It is now possible to go back and replace the given values 5, 3, and 6 with a new set of numbers, if needed, to obtain a new answer.   ◆◆◆

Substituting into Formulas

A *formula* is an equation expressing some general mathematical or physical fact, such as the formula for the area of a circle of radius *r*.

Area of a Circle	$A = \pi r^2$	114

We substitute into formulas just as we substituted into equations, except that we now carry *units* along with the numerical values. You will often need conversion factors to make the units cancel properly, so that the answer will be in the desired units.

◆◆◆ **Example 89:** A tensile load of 4500 lb is applied to a bar that is 5.2 yd long and has a cross-sectional area of 11.6 cm^2 (Fig. 1–18). The elongation is 0.38 mm. Using Eq. A54, find the modulus of elasticity E in pounds per square inch.

Solution: Substituting the values *with units* into Eq. A54, we obtain

$$E = \frac{PL}{ae} = \frac{4500 \text{ lb} \times 5.2 \text{ yd}}{11.6 \text{ cm}^2 \times 0.38 \text{ mm}}$$

Notice that we have a length (5.2 yd) in the numerator and a length (0.38 mm) in the denominator. To make these units cancel, we use the conversion factors

$$25.4 \text{ mm} = 1 \text{ in.}$$

and

$$36 \text{ in.} = 1 \text{ yd}$$

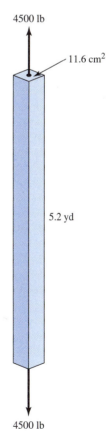

4500 lb

11.6 cm^2

5.2 yd

4500 lb

FIGURE 1–18

If the units to be used in a certain formula are specified, convert all quantities to those specified units before substituting into the formula.

Also, our answer is to have square inches in the denominator, not square centimeters. So we use another conversion factor.

$$6.452 \text{ cm}^2 = 1 \text{ in.}^2$$

$$E = \frac{4500 \text{ lb} \times 5.2 \text{ yd}}{11.6 \text{ cm}^2 \times 0.38 \text{ mm}} \times \frac{25.4 \text{ mm}}{\text{in.}} \times \frac{36 \text{ in.}}{\text{yd}} \times \frac{6.452 \text{ cm}^2}{\text{in.}^2}$$

$$= 31{,}000{,}000 \text{ lb/in.}^2 \quad \text{(rounded to two digits)} \qquad \blacklozenge\blacklozenge\blacklozenge$$

Common Error	Students often neglect to include *units* when substituting into a formula, with the result that the units often do not cancel properly.

Exercise 9 ◆ Substituting into Equations and Formulas

Substitute the given integers into each equation. Do not round your answer.

1. $y = 5x + 2$ $(x = 3)$
2. $y = 2m^2 - 3m + 5$ $(m = -2)$
3. $y = 2a - 3x^2$ $(x = 3, a = -5)$
4. $y = 3x^3 - 2x^2 + 4x - 7$ $(x = 2)$
5. $y = 2b + 3w^2 - 5z^3$ $(b = 3, w = -4, z = 2)$

6. $y = \dfrac{r^2}{x} - \dfrac{x^3}{r} + \dfrac{w}{x^2}$ $(x = 5, w = 3, r = -4)$

Substitute the given approximate numbers into each equation. Treat the constants in the equations as exact numbers. Round your answer to the proper number of digits.

7. $y = 7x - 5$ $(x = 2.73)$
8. $y = 2w^2 - 3x^2$ $(x = -11.5, w = 9.83)$
9. $y = 8 - x + 3x^2$ $(x = -8.49)$
10. $y = \sqrt[3]{8x + 7w}$ $(x = 1.255, w = 2.304)$
11. $y = \sqrt{x^3 - 3x}$ $(x = 4.25)$
12. $y = (w - 2x)^{1.6}$ $(x = 1.8, w = 7.2)$

13. Use Eq. A9 to find to the nearest dollar the amount to which $3000 will accumulate in 5 years at a simple interest rate of 6.5%.

14. Using Eq. A18, find the displacement, after 1.30 s, of a body thrown downward with a speed of 12.0 ft/s. Use $g = 32.2$ ft/s².

15. Using Eq. A50, convert 128°F to degrees Celsius.

16. A bar 15.2 m long having a cross-sectional area of 12.7 cm² is subject to a tensile load of 22,500 N (Fig. 1–19). The elongation is 2.75 mm. Use Eq. A54 to find the modulus of elasticity in newtons per square centimeter.

17. Use Eq. A10 to find to the nearest dollar the amount y obtained when $9570 is allowed to accumulate for 5 years at a compound interest rate of $6\frac{3}{4}\%$.

18. The resistance of a copper coil is 775 Ω at 20.0°C. The temperature coefficient of resistance is 0.00393 at 20.0°C. Use Eq. A70 to find the resistance at 80.0°C.

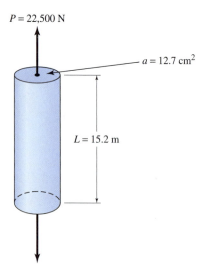

$P = 22{,}500$ N

$a = 12.7$ cm²

$L = 15.2$ m

FIGURE 1–19   A bar in tension.

1–10 Percentage

Definition of Percent

The word *percent* means *by the hundred,* or *per hundred.* A percent thus gives the number of parts in every hundred.

◆◆◆ **Example 90:** If we say that a certain concrete mix is 12% cement by weight, we mean that 12 lb out of every 100 lb of mix will be cement. ◆◆◆

Rates

The word *rate* is often used to indicate a percent, or percentage rate, as in "rejection rate," "rate of inflation," or "growth rate."

◆◆◆ **Example 91:** A failure rate of 2% means that, on average, 2 parts out of every 100 would be expected to fail. ◆◆◆

Percent as a Fraction

Percent is another way of expressing a *fraction* having 100 as the denominator.

◆◆◆ **Example 92:** If we say that a builder has finished 75% of a house, we mean that he has finished $\frac{75}{100}$ (or $\frac{3}{4}$) of the house. ◆◆◆

Converting Decimals to Percent

Before working some percentage problems, let us first get some practice in converting decimals and fractions to percents, and vice versa. To convert decimals to percent, simply move the decimal point two places to the right and affix the percent symbol (%).

◆◆◆ **Example 93:**

(a) $0.75 = 75\%$
(b) $3.65 = 365\%$
(c) $0.003 = 0.3\%$
(d) $1.05 = 105\%$ ◆◆◆

Converting Fractions or Mixed Numbers to Percent

First write the fraction or mixed number as a decimal, and then proceed as above.

◆◆◆ **Example 94:**

(a) $\frac{1}{4} = 0.25 = 25\%$

(b) $\frac{5}{2} = 2.5 = 250\%$

(c) $1\frac{1}{4} = 1.25 = 125\%$ ◆◆◆

Converting Percent to Decimals

To convert percent to decimals, move the decimal point two places to the left and remove the percent sign.

◆◆◆ Example 95:

(a) $13\% = 0.13$

(b) $4.5\% = 0.045$

(c) $155\% = 1.55$

(d) $27\frac{3}{4}\% = 0.2775$

(e) $200\% = 2$

◆◆◆

Converting Percent to a Fraction

Write a fraction with 100 in the denominator and the percent in the numerator. Remove the percent sign and reduce the fraction to lowest terms.

◆◆◆ Example 96:

(a) $75\% = \dfrac{75}{100} = \dfrac{3}{4}$

(b) $87.5\% = \dfrac{87.5}{100} = \dfrac{875}{1000} = \dfrac{7}{8}$

(c) $125\% = \dfrac{125}{100} = \dfrac{5}{4} = 1\frac{1}{4}$

◆◆◆

Amount, Base, and Rate

Percentage problems always involve three quantities:

1. The *percent rate, P.*
2. The *base, B*: the quantity we are taking the percent of.
3. The *amount, A,* that we get when we take the percent of the base, also called the *percentage.*

In a percentage problem, you will know two of these three quantities (amount, base, or rate), and you will be required to find the third. This is easily done, for the rate, base, and amount are related by the following equation:

Percentage	amount = rate × base $A = PB$ where P is expressed as a decimal	**12**

Finding the Amount When the Base and the Rate Are Known

We substitute the given base and rate into Eq. 12 and solve for the amount.

◆◆◆ Example 97: What is 35.0 percent of 80.0?

Solution: In this problem the rate is 35.0%, so

$$P = 0.350$$

But is 80.0 the amount or the base?

Tip	In a particular problem, if you have trouble telling which number is the base and which is the amount, look for the key phrase *percent of.* The quantity following this phrase is *always the base.*

Thus we look for the key phrase "percent of."

base

What is 35.0 percent of $\boxed{80.0}$?

Since 80.0 immediately follows *percent of,*

$$B = 80.0$$

From Eq. 12,

$$A = PB = (0.350)80.0 = 28.0$$ ◆◆◆

Common Error	Do not forget to convert the percent rate to a *decimal* when using Eq. 12.

◆◆◆ **Example 98:** Find 3.74% of 5710.

Solution: We substitute into Eq. 12 with

$$P = 0.0374 \quad \text{and} \quad B = 5710$$

So

$$A = PB = (0.0374)(5710) = 214$$

after rounding to three significant digits. ◆◆◆

Finding the Base When a Percent of It Is Known

We see from Eq. 12 that the base equals the amount divided by the rate (expressed as a decimal), or $B = A/P$.

◆◆◆ **Example 99:** 12% of what number is 78?

Solution: First find the key phrase.

12 percent of [what number] is 78?
base

It is clear that we are looking for the base. So

$$A = 78 \quad \text{and} \quad P = 0.12$$

By Eq. 12,

$$B = \frac{A}{P} = \frac{78}{0.12} = 650$$ ◆◆◆

◆◆◆ **Example 100:** 140 is 25% of what number?

Solution: From Eq. 12,

$$B = \frac{A}{P} = \frac{140}{0.25} = 560$$ ◆◆◆

Finding the Percent That One Number Is of Another Number

From Eq. 12, the rate equals the amount divided by the base, or $P = A/B$.

◆◆◆ **Example 101:** 42.0 is what percent of 405?

Solution: By Eq. 12, with $A = 42.0$ and $B = 405$,

$$P = \frac{A}{B} = \frac{42.0}{405} = 0.104 = 10.4\%$$ ◆◆◆

◆◆◆ **Example 102:** What percent of 1.45 is 0.357?

Solution: From Eq. 12,

$$P = \frac{A}{B} = \frac{0.357}{1.45} = 0.246 = 24.6\%$$ ◆◆◆

Percent Change

Percentages are often used to compare two quantities. You often hear statements such as the following:

The price of steel rose 3% over last year's price.

The weights of two cars differed by 20%.

Production dropped 5% from last year.

When the two numbers being compared involve a *change* from one to the other, the *original value* is usually taken as the base.

Percent Change	percent change $= \dfrac{\text{new value} - \text{original value}}{\text{original value}} \times 100$	**13**

◆◆◆ **Example 103:** A certain price rose from $1.55 to $1.75. Find the percentage change in price.

Solution: We use the original value, $1.55, as the base. From Eq. 13,

$$\text{percent change} = \frac{1.75 - 1.55}{1.55} \times 100 = 12.9\% \text{ increase}$$ ◆◆◆

A common type of problem is to *find the new value* when the original value is changed by a given percent. We see from Eq. 13 that

$$\text{new value} = \text{original value} + (\text{original value}) \times (\text{percent change})$$

Be sure to show the *direction* of change with a plus or a minus sign, or with words such as *increase* or *decrease*.

◆◆◆ **Example 104:** Find the cost of a $156.00 suit after the price increases by $2\frac{1}{2}\%$.

Solution: The original value is 156.00, and the percent change, expressed as a decimal, is 0.025. So

$$\text{new value} = 156.00 + 156.00(0.025) = \$159.90$$ ◆◆◆

Percent Efficiency

The power output of any machine or device is always *less* than the power input, because of inevitable power losses within the device. The *efficiency* of the device is a measure of those losses.

Percent Efficiency	percent efficiency $= \dfrac{\text{output}}{\text{input}} \times 100$	**16**

◆◆◆ **Example 105:** A certain electric motor consumes 865 W and has an output of 1.12 hp (Fig. 1–20). Find the efficiency of the motor. (1 hp = 746 W.)

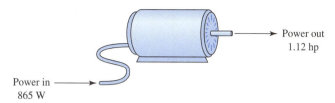

Power out
1.12 hp

Power in
865 W

FIGURE 1–20

Solution: Since output and input must be in the same units, we must convert either to horsepower or to watts. Converting the output to watts, we obtain

$$\text{output} = 1.12 \text{ hp} \left(\frac{746 \text{ W}}{\text{hp}} \right) = 836 \text{ W}$$

By Eq. 16,

$$\text{percent efficiency} = \frac{836}{865} \times 100 = 96.6\% \qquad \text{◆◆◆}$$

Percent Error

The accuracy of measurements is often specified by the *percent error.* The percent error is the difference between the measured value and the known or "true" value, expressed as a percent of the known value.

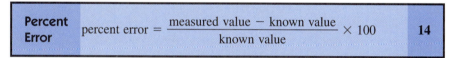

Percent Error	$\text{percent error} = \dfrac{\text{measured value} - \text{known value}}{\text{known value}} \times 100$	**14**

◆◆◆ **Example 106:** A laboratory weight that is certified to be 500.0 g is placed on a scale (Fig. 1–21). The scale reading is 507.0 g. What is the percent error in the reading?

Solution: From Eq. 14,

$$\text{percent error} = \frac{507.0 - 500.0}{500.0} \times 100 = 1.4\% \text{ high} \qquad \text{◆◆◆}$$

Percent Concentration

The following equation applies to a mixture of two or more ingredients:

Percent Concentration	$\text{percent concentration of ingredient } A = \dfrac{\text{amount of } A}{\text{amount of mixture}} \times 100$	**15**

◆◆◆ **Example 107:** A certain fuel mixture contains 18.9 liters of alcohol and 84.7 liters of gasoline. Find the percentage of gasoline in the mixture.

Solution: The total amount of mixture is

$$18.9 + 84.7 = 103.6 \text{ liters}$$

So by Eq. 15,

$$\text{percent gasoline} = \frac{84.7}{103.6} \times 100 = 81.8\% \qquad \text{◆◆◆}$$

Common Error	The denominator in Eq. 15 must be the *total amount* of mixture, or the sum of *all* of the ingredients. Do not use just one of the ingredients.

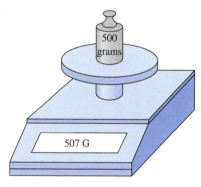

500 grams

507 G

FIGURE 1–21

As with percent change, be sure to specify the *direction* of the error.

Exercise 10 ◆ Percentage

Conversions

Convert each decimal to a percent.

1. 3.72	**2.** 0.877	**3.** 0.0055	**4.** 0.563

Convert each fraction to a percent. Round to three significant digits.

5. $\dfrac{2}{5}$ **6.** $\dfrac{3}{4}$ **7.** $\dfrac{7}{10}$ **8.** $\dfrac{3}{7}$

Convert each percent to a decimal.

9. 23%   **10.** 2.97%   **11.** $287\frac{1}{3}\%$ **12.** $6\frac{1}{4}\%$

Convert each percent to a fraction.

13. 37.5%   **14.** $12\frac{1}{2}\%$ **15.** 150%   **16.** 3%

Finding the Amount

Find:

17. 41.1% of 255 tons.   **18.** 15.3% of 326 mi.

19. 33.3% of 662 kg.   **20.** 12.5% of 72.0 gal.

21. 35.0% of 343 liters.   **22.** 50.8% of $245.

23. A resistance, now 7250 Ω, is to be increased by 15.0%. How much resistance should be added?

24. It is estimated that $\frac{1}{2}\%$ of the earth's surface receives more energy than the total projected needs for the year 2000. Assuming the earth's surface area to be 1.97×10^8 mi², find the required area in acres.

25. As an incentive to install solar equipment, a tax credit of 42% of the first $1100 and 25% of the next $6400 spent on solar equipment is proposed. How much credit, to the nearest dollar, would a homeowner get when installing $5500 worth of solar equipment?

26. How much metal will be obtained from 375 tons of ore if the metal is 10.5% of the ore?

Finding the Base

Find the number of which:

27. 86.5 is 16.7%.   **28.** 45.8 is 1.46%.

29. 1.22 is 1.86%.   **30.** 55.7 is 25.2%.

31. 66.6 is 66.6%.   **32.** 58.2 is 75.4%.

33. A Department of Energy report on an experimental electric car gives the range of the car as 161 km and states that this is "49.5% better than on earlier electric vehicles." What was the range of earlier electric vehicles?

34. A man withdrew 25% of his bank deposits and spent 33% of the money withdrawn in the purchase of a radio worth $25. How much money did he have in the bank?

35. Solar panels provide 65% of the heat for a certain building. If $225 per year is now spent for heating oil, what would have been spent if the solar panels were not used?

36. If the United States imports 9.14 billion barrels of oil per day, and if this is 48.2% of its needs, how much oil is needed per day?

Finding the Rate

What percent of:

37. 26.8 is 12.3?   **38.** 36.3 is 12.7?   **39.** 44.8 is 8.27?

40. 844 is 428?   **41.** 455 h is 152 h?   **42.** 483 tons is 287 tons?

43. A 50,500-liter-capacity tank contains 5840 liters of water (Fig. 1–22). Express the amount of water in the tank as a percentage of the total capacity.
44. In a journey of 1560 km, a person traveled 195 km by car and the rest of the distance by rail. What percent of the distance was traveled by rail?
45. A power supply has a dc output of 51 V with a ripple of 0.75 V peak to peak. Express the ripple as a percentage of the dc output voltage.
46. The construction of a factory cost $136,000 for materials and $157,000 for labor. What percentage of the total was the labor cost?

Percent Change

Find the percent change when a quantity changes:

47. from 29.3 to 57.6.
48. from 107 to 23.75.
49. from 227 to 298.
50. from 0.774 to 0.638.
51. The temperature in a building rose from 19.0°C to 21.0°C during the day. Find the percent change in temperature.
52. A casting initially weighing 115 lb has 22.0% of its material machined off. What is its final weight?
53. A certain common stock rose from a value of $35\frac{1}{2}$ per share to $37\frac{5}{8}$ per share. Find the percent change in value.
54. A house that costs $635 per year to heat has insulation installed in the attic, causing the fuel bill to drop to $518 per year. Find the percent change in fuel cost.

Percent Efficiency

55. A certain device (Fig. 1–23) consumes 18.5 hp and delivers 12.4 hp. Find its efficiency.
56. An electric motor consumes 1250 W. Find the horsepower it can deliver if it is 85.0% efficient. (1 hp = 746 W.)
57. A water pump requires an input of 0.50 hp and delivers $\overline{1}0,000$ lb of water per hour to a house 72 ft above the pump. Find its efficiency. (1 hp = 550 ft·lb/s.)
58. A certain speed reducer delivers 1.7 hp with a power input of 2.2 hp. Find the percent efficiency of the speed reducer.

Percent Error

59. A certain quantity is measured at 125.0 units but is known to be actually 128.0 units. Find the percent error in the measurement.
60. A shaft is known to have a diameter of 35.000 mm. You measure it and get a reading of 34.725 mm. What is the percent error of your reading?
61. A certain capacitor has a working voltage of 125.0 V dc − 10%, +150%. Between what two voltages would the actual working voltage lie?
62. A resistor is labeled as $550\overline{0}$ Ω with a tolerance of ±5%. Between what two values is the actual resistance expected to lie?

Percent Concentration

63. A solution is made by mixing 75.0 liters of alcohol with 125 liters of water. Find the percent concentration of alcohol.
64. 8.0 cubic feet of cement is contained in a concrete mixture that is 12% cement by volume. What is the volume of the total mixture?
65. How many liters of alcohol are contained in 455 liters of a gasohol mixture that is 5.5% alcohol by volume?
66. How many liters of gasoline are there in 155 gal of a methanol–gasoline blend that is 10.0% methanol by volume?

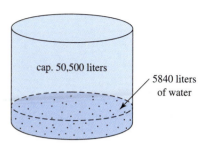

cap. 50,500 liters

5840 liters of water

FIGURE 1–22

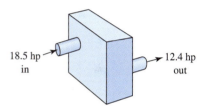

18.5 hp in

12.4 hp out

FIGURE 1–23

◆◆◆ CHAPTER 1 REVIEW PROBLEMS ◆◆◆◆◆◆◆◆◆◆◆◆◆◆◆◆◆◆◆◆◆◆◆◆◆◆◆◆◆◆◆◆◆◆◆◆◆◆

1. Combine: $1.435 - 7.21 + 93.24 - 4.1116$
2. Give the number of significant digits in:
   (a) 9.886   (b) 1.002   (c) 0.3500   (d) $15,0\overline{0}0$
3. Multiply: $21.8(3.775 \times 1.07)$
4. Divide: $88.25 \div 9.15$
5. Find the reciprocal of 2.89.
6. Evaluate: $-|-4 + 2| - |-9 - 7| + 5$
7. Evaluate: $(9.73)^2$
8. Evaluate: $(7.75)^{-2}$
9. Evaluate: $\sqrt{29.8}$
10. Evaluate: $(123)(2.75) - (81.2)(3.24)$
11. Evaluate: $(91.2 - 88.6)^2$
12. Evaluate: $\left(\dfrac{77.2 - 51.4}{21.6 - 11.3}\right)^2$
13. Evaluate: $y = 3x^2 - 2x$ when $x = -2.88$
14. Evaluate: $y = 2ab - 3bc + 4ac$ when $a = 5$, $b = 2$, and $c = -6$
15. Evaluate: $y = 2x - 3w + 5z$ when $x = 7.72$, $w = 3.14$, and $z = 2.27$
16. Round to two decimal places.
    (a) 7.977   (b) 4.655   (c) 11.845   (d) 1.004
17. Round to three significant digits.
    (a) 179.2   (b) 1.076   (c) 4.8550   (d) 45,725
18. A news report states that a new hydroelectric generating station in Holyoke, Massachusetts, will produce 47 million kWh/yr and that this power, for 20 years of operation, is equivalent to 2.0 million barrels of oil. Using these figures, how many kilowatthours is each barrel of oil equivalent to?
19. A certain generator has a power input of 2.50 hp and delivers 1310 W. Find its percent efficiency.
20. Using Eq. A52, find the stress in pounds per square inch for a force of 1.17×10^3 N distributed over an area of 3.14×10^3 mm^2.
21. Combine: $(8.34 \times 10^5) + (2.85 \times 10^6) - (5.29 \times 10^4)$
22. A train running at 25 mi/h increases its speed $12\frac{1}{2}\%$. How fast does it then go?
23. The average solar radiation in the continental United States is about 7.4×10^5 joules per square meter per hour (J/m^2·h). How many kilowatts would be collected by 15 acres of solar panels?
24. An item rose in price from $29.35 to $31.59. Find the percent increase.
25. Find the percent concentration of alcohol if 2.0 liters of alcohol is added to 15 gal of gasoline.
26. A bar, known to be 2.0000 inches in diameter, is measured at 2.0064 in. Find the percent error in the measurement.
27. The Department of Energy estimates that there are 700 billion barrels of oil in the oil shale deposits of Colorado, Wyoming, and Utah. Express this amount in scientific notation.
28. Multiply: $(7.23 \times 10^5) \times (1.84 \times 10^{-3})$
29. Divide: -39.2 by -0.003826
30. Convert 6930 Btu/h to foot-pounds per minute.
31. Divide: 8.24×10^{-3} by 1.98×10^7
32. What percent of 40.8 is 11.3?
33. Evaluate: $\sqrt[5]{82.8}$
34. Multiply: $(4.92 \times 10^6) \times (9.13 \times 10^{-3})$

35. Insert the proper sign of equality or inequality between $-\frac{2}{3}$ and -0.660.

36. Convert 0.000426 mA to microamperes.

37. Find 49.2% of 4827.

38. Combine: $-385 - (227 - 499) - (-102) + (-284)$

39. Find the reciprocal of -0.582.

40. Find the percent change in a voltage that increased from $11\overline{0}$ V to 118 V.

41. A homeowner added insulation, and her yearly fuel consumption dropped from 628 gal to 405 gal. Her present oil consumption is what percent of the former?

42. Write in decimal notation: 5.28×10^4

43. Convert 49.3 pounds to newtons.

44. Evaluate: $(45.2)^{-0.45}$

45. Using Eq. A18, find the distance in feet traveled by a falling object in 5.25 s, thrown downward with an initial velocity of 284 m/min.

46. Write in scientific notation: 0.000374

47. 8460 is what percent of 38,400?

48. The U.S. energy consumption of 37 million barrels oil equivalent per day is expected to climb to 48 million in 6 years. Find the percent increase in consumption.

49. The population of a certain town is 8118, which is $12\frac{1}{2}\%$ more than it was 3 years ago. What was the population then?

50. The temperature of a room rose from 68.0°F to 73.0°F. Find the percent increase.

51. Combine: $4.928 + 2.847 - 2.836$

52. Give the number of significant digits in 2003.0.

53. Multiply: 2.84(38.4)

54. Divide: $48.3 \div 2.841$

55. Find the reciprocal of 4.82.

56. Evaluate: $|-2| - |3 - 5|$

57. Evaluate: $(3.84)^2$

58. Evaluate: $(7.62)^{-2}$

59. Evaluate: $\sqrt{38.4}$

60. Evaluate: $(49.3 - 82.4)(2.84)$

61. Evaluate: $x^2 - 3x + 2$ when $x = 3$

62. Round 45.836 to one decimal place.

63. Round 83.43 to three significant digits.

64. Multiply: $(7.23 \times 10^5) \times (1.84 \times 10^{-3})$

65. Convert 36.82 in. to centimeters.

66. What percent of 847 is 364?

67. Evaluate: $\sqrt[3]{746}$

68. Find 35.8% of 847.

69. 746 is what percent of 992?

70. Write 0.00274 in scientific notation.

71. Write 73.7×10^{-3} in decimal notation.

72. Evaluate: $(47.3)^{-0.26}$

73. Combine: $6.128 + 8.3470 - 7.23612$

74. Give the number of significant digits in 6013.00.

75. Multiply: 7.184(16.8)

76. Divide: $78.7 \div 8.251$

77. Find the reciprocal of 0.825.

78. Evaluate: $|-5| - |2 - 7| + |-6|$

79. Evaluate: $(0.55)^2$

80. Write 23,800 in scientific notation and in engineering notation.

Writing

81. Suppose you have submitted a report that contains calculations in which you have rounded the answers according to the rules given in this chapter. Jones, your company rival, has sharply attacked your work, calling it "inaccurate" because you did not keep enough digits, and your boss seems to agree.

Write a memo to your boss defending your rounding practices. Point out why it is misleading to retain too many digits. Do not write more than one page. You may use numerical examples to prove your point.

Team Projects

You will find a *team project* suggested at the end of most chapters. These projects are generally longer and more complicated than the usual exercise, and they are meant to be done by several people working as a team over a few days.

If your class decides to do one or more of these, you should divide into teams of about three to six students. Then each team should:

Study the problem.

Estimate the answer(s).

List possible tools and methods for solution; then choose one.

Solve the problem.

Check the answer.

Write a brief report.

Students now have more computational tools at their disposal than anyone has had before. Depending on the problem, you might choose from the following:

Manual computation

Trial and error

Scientific calculator

Graphics calculator

Computer, using BASIC, a spreadsheet such as *Lotus,* mathematical computation software such as *MathCad,* or a computer algebra program such as *Derive, Maple,* or *Mathematica.*

You might choose *more than one* tool, perhaps doing the computation by calculator and checking the results by computer. For some projects you may want to borrow measuring equipment such as a tape measure or a transit, or you may want to consult reference books. Sometimes it may be useful to make a model. Be sure to consider *any* aids that may help you.

Your first team project is problem 82, the simple one of finding a missing dimension on a template.

82. Starting with the template in Fig. 1–24, you are to make a new template by increasing dimension A by 25.0% and decreasing dimension B by 30.0%. Find the dimension x (in millimeters) for the new template.

83. Make a drawing of a cylindrical steel bar, 1 inch in diameter and 3 in. long. Label the diameter as 1.00 in. Take your drawing to a machine shop and ask for a cost estimate for each of six bars, having lengths of:

3 in.	3.000 in.
3.0 in.	3.0000 in.
3.00 in.	3.00000 in.

Before you go, have each member of your team make cost estimates.

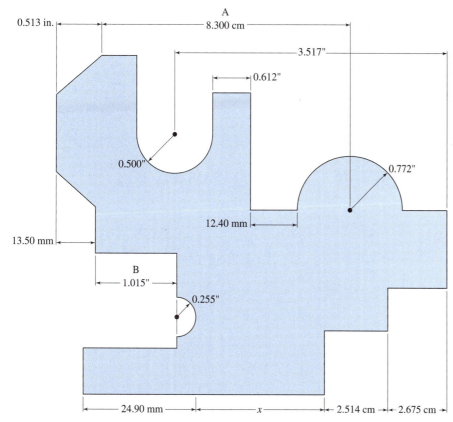

FIGURE 1–24   A template, not drawn to scale.

Internet

84. Feel free to contact the authors with any comments, corrections, or questions about this textbook. You can reach us at:

PCalter@vtc.vsc.edu
MCalter@vt.edu

85. For more information about the authors, please visit:

http://www.sover.net/~pcalter/
http://www.chem.vt.edu/chem-dept/calter/calter.html

2

Introduction to Algebra

•••• **OBJECTIVES** ••

When you have completed this chapter, you should be able to:

- Determine the number of terms in an expression, identify the coefficient of each term, and state the degree of the expression.
- Simplify expressions by removing symbols of grouping and by combining like terms.
- Use the laws of exponents to simplify and combine expressions containing powers.
- Add, subtract, multiply, and divide algebraic expressions.

••

Algebra is a generalization of arithmetic. For example, the statement

$$2 \cdot 2 \cdot 2 \cdot 2 = 2^4$$

can be generalized to

$$a \cdot a \cdot a \cdot a = a^4$$

where *a* can be *any* number, not just 2. We can go further and say that

$$\underbrace{a \cdot a \cdot a \cdot a \ldots a}_{n \text{ factors}} = a^n$$

where *a* can be any number, as before, and *n* can be *any* positive integer, not just 4.

We will learn many new words in this chapter. Every field has its special terms, and algebra is no exception. But since algebra is generalized arithmetic, some of what was said in Chapter 1 (such as rules of signs) will be repeated here.

We will redo the basic operations of addition, subtraction, and so on, but now with symbols rather than numbers. Learn this material well, for it is the foundation on which later chapters rest.

2–1 Algebraic Expressions

Mathematical Expressions

A *mathematical expression* is a grouping of mathematical symbols, such as signs of operation, numbers, and letters.

◆◆◆ **Example 1:** The following are mathematical expressions:

(a) $x^2 - 2x + 3$
(b) $4 \sin 3x$
(c) $5 \log x + e^{2x}$ ◆◆◆

Algebraic Expressions

An *algebraic expression* is one containing only algebraic symbols and operations (addition, subtraction, multiplication, division, roots, and powers), such as in Example 1(a). All other expressions are called *transcendental*, such as Examples 1(b) and (c).

Equations

None of the expressions in Example 1 contains an equal sign ($=$). When two expressions are set equal to each other, we get an *equation*.

◆◆◆ **Example 2:** The following are equations:

(a) $2x^2 + 3x - 5 = 0$
(b) $6x - 4 = x + 1$
(c) $y = 3x - 5$
(d) $3 \sin x = 2 \cos x$ ◆◆◆

We mentioned equations briefly in Chapter 1 when we substituted into equations. We treat them in detail in Chapter 3.

Constants and Variables

A *constant* is a quantity that does not change in value in a particular problem. It is usually a number, such as 8, 4.67, or π.

A *variable* is a quantity that may change during a particular problem. A variable is usually represented by a letter from the end of the alphabet (x, y, z, etc.).

◆◆◆ **Example 3:** The constants in the expression

$$3x^2 + 4x + 5$$

are 3, 4, and 5, and the variable is x. ◆◆◆

A constant can also be represented by a letter. Such a letter is usually chosen from the beginning of the alphabet (a, b, c, etc.). The letter k is often used as a constant. An expression in which the constants are represented by letters is called a *literal* expression.

◆◆◆ **Example 4:** The constants in the literal expression

$$ax^2 + bx + c$$

are a, b, and c, and the variable is x. ◆◆◆

◆◆◆ **Example 5:** In the expression

$$ax + by + cz$$

the letters a, b, and c would usually represent constants, and x, y, and z would represent variables.   ◆◆◆

Symbols of Grouping

We will show how to *remove* symbols of grouping in Sec. 2–2. We will also see in Sec. 2–4 that symbols of grouping are used to indicate *multiplication*.

Mathematical expressions often contain parentheses ( ), brackets [ ], and braces { }. These are used to group parts of the expression together, and they affect the meaning of the expression.

◆◆◆ **Example 6:** The expression

$$\{[(3 + x^2) - 2] - (2x + 3)\}(x - 3)$$

shows the use of symbols of grouping.   ◆◆◆

◆◆◆ **Example 7:** The value of the expression

$$2 + 3(4 + 5)$$

is *different* from the value of

$$(2 + 3)4 + 5$$

The first expression has a value of 29, and the second expression has a value of 25. As you can see, the placement of symbols of grouping *is* important.   ◆◆◆

Terms

The plus and the minus signs divide an expression into *terms*.

◆◆◆ **Example 8:** The expression

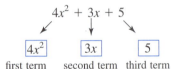

has *three* terms.   ◆◆◆

An exception is when the plus or minus sign is *within* a symbol of grouping.

◆◆◆ **Example 9:** The expression

$$\underbrace{[(x + 2x^2) + 3]}_{\text{first term}} - \underbrace{(2x + 2)}_{\text{second term}}$$

has *two* terms.   ◆◆◆

Factors

Any divisor of a term is called a *factor* of that term.

The number 1 and the entire expression $3axy$ are also factors. They usually are not stated. See Chapter 8 for a more thorough discussion of factors.

◆◆◆ **Example 10:** Some factors of $3axy$ are 3, a, x, and y.   ◆◆◆

◆◆◆ **Example 11:** The expression

$$2x + 3yz$$

has two *terms*, $2x$ and $3yz$. The first term has the *factors* 2 and x, and the second term has the factors 3, y, and z. ◆◆◆

Coefficient

The *coefficient* of a term is the constant part of the term.

◆◆◆ **Example 12:** The coefficient in the term $3axy^2$ is $3a$, where a is a constant. We say that $3a$ is the coefficient of xy^2. ◆◆◆

When we use the word *coefficient* in this book, it will always refer to the constant part of the term (such as $3a$ in Example 12), also called the *numerical* coefficient. If there is no numerical coefficient written before a term, it is understood to be 1.

◆◆◆ **Example 13:** In the expression $\dfrac{w}{2} + x - y - 3z$:

The coefficient of w is $\frac{1}{2}$.
The coefficient of x *is* 1.
The coefficient of y is -1.
The coefficient of z is -3. ◆◆◆

Do not forget to include the minus sign with the coefficient.

Degree

The *degree* of a term refers to the integer power to which the variable is raised.

◆◆◆ **Example 14:**

(a) $2x$ is a first-degree term.
(b) $3x^2$ is a second-degree term.
(c) $5y^9$ is a ninth-degree term. ◆◆◆

The term *degree* is used only when the exponents are *positive integers*. You would not say that $x^{1/2}$ is a "half-degree" term.

If there is more than one variable, we add their powers to obtain the degree of the term.

◆◆◆ **Example 15:**

(a) $2x^2y^3$ is of fifth degree.
(b) $3xyz^2$ is of fourth degree. ◆◆◆

The degree of an *expression* is the same as that of the term having the highest degree.

◆◆◆ **Example 16:** $3x^2 - 2x + 4$ is a second-degree expression. ◆◆◆

A *multinomial* is an algebraic expression having *more than one term*.

◆◆◆ **Example 17:** Some multinomials are

(a) $3x + 5$
(b) $2x^3 - 3x^2 + 7$
(c) $\dfrac{1}{x} + \sqrt{2x}$ ◆◆◆

A *polynomial* is a monomial or multinomial in which the powers to which the unknown is raised are all *nonnegative integers*. The first two expressions in Example 17 are polynomials, but the third is not.

A *binomial* is a polynomial with *two* terms, and a *trinomial* is a polynomial having *three* terms. In Example 17, the first expression is a binomial, and the second is a trinomial.

Exercise 1 ◆ Algebraic Expressions

How many terms are there in each expression?

1. $x^2 - 3x$ **2.** $7y - (y^2 + 5)$

3. $(x + 2)(x - 1)$ **4.** $3(x - 5) + 2(x + 1)$

Write the coefficients of each term. Assume that the letters x, y, and z are variables and that all other letters are constants.

5. $5x^3$ **6.** $2ay^2$ **7.** $\dfrac{bx^3}{4}$

8. $\frac{1}{4}(bx)$ **9.** $\dfrac{3y^2}{2a}$ **10.** $\dfrac{2c}{a}(4x^2)$

2–2   Addition and Subtraction of Algebraic Expressions

Like Terms

Like terms are those that differ only in their coefficients.

◆◆◆ Example 18:

(a) $2wx$ and $-3wx$ are like terms.
(b) $2wx$ and $-3wx^2$ are *not* like terms.   ◆◆◆

Algebraic expressions are added and subtracted by *combining like terms*. Like terms are added by adding their coefficients.

This process is also referred to as *collecting terms*.

◆◆◆ Example 19:

(a) $7x + 5x = 12x$
(b) $8w + 2w - 4w - w = 5w$
(c) $9y - 3y = 6y$
(d) $2x - 3y - 5x + 2y = -3x - y$
(e) $4.21x + 1.23x - 3.11x = 2.33x$ ◆◆◆

In Example 19 we combined the terms on a single line. This method is sometimes called *horizontal* addition and subtraction. For more complicated problems, you may prefer *vertical* addition and subtraction, as shown in the following example.

◆◆◆ Example 20: From the sum of $3x^3 + 2x - 5$ and $x^3 - 3x^2 + 7$, subtract $2x^3 + 3x^2 - 4x - 7$.

Solution: Combine the first two expressions.

$$
\begin{array}{r}
3x^3 \qquad\quad + 2x - 5 \\
+ \quad x^3 - 3x^2 \qquad\quad + 7 \\
\hline
4x^3 - 3x^2 + 2x + 2
\end{array}
$$

Then subtract the third expression from their sum.

$$
\begin{array}{r}
4x^3 - 3x^2 + 2x + 2 \\
- \quad (2x^3 + 3x^2 - 4x - 7) \\
\hline
2x^3 - 6x^2 + 6x + 9
\end{array}
$$

◆◆◆

Commutative Law of Addition

We mentioned the commutative law (Eq. 1) when adding numbers in Chapter 1. It simply means that the order of addition does not affect the sum $(2 + 3 = 3 + 2)$. The same law applies, of course, when we have letters instead of numbers.

◆◆◆ **Example 21:**

(a) $a + b = b + a$
(b) $2x + 3x = 3x + 2x$
$\qquad\quad = 5x$

◆◆◆

This law enables us to *rearrange the terms* of an expression for our own convenience.

◆◆◆ **Example 22:** Simplify the expression

$$4y - 2x + 3z - 7z + 4x - 6y$$

Solution: Using the commutative law, we rearrange the expression so as to get like terms together.

$$-2x + 4x + 4y - 6y + 3z - 7z$$

Collecting terms, we obtain

$$2x - 2y - 4z$$

With practice, you will soon be able to omit the first step and collect terms by inspecting the original expression.

◆◆◆

Removal of Parentheses

To combine quantities within parentheses with other quantities not within those parentheses, we must first remove the parentheses. If the parentheses (or other symbol of grouping) are preceded only by a + sign (or no sign), the parentheses may be removed.

◆◆◆ **Example 23:**

(a) $(a + b - c) = a + b - c$
(b) $(x + y) + (w - z) = x + y + w - z$

◆◆◆

When the parentheses are preceded only by a $(-)$ sign, change the sign of every term within the parentheses upon removing the parentheses.

◆◆◆ **Example 24:**

(a) $-(a + b - c) = -a - b + c$
(b) $-(x + y) - (x - y) = -x - y - x + y = -2x$
(c) $(2w - 3x - 2y) - (w - 4x + 5y) - (3w - 2x - y)$
$\quad = 2w - 3x - 2y - w + 4x - 5y - 3w + 2x + y$
$\quad = 2w - w - 3w - 3x + 4x + 2x - 2y - 5y + y$
$\quad = -2w + 3x - 6y$

◆◆◆

When there are groups within groups, start simplifying the *innermost* groups and work outward.

◆◆◆ **Example 25:**

$$3x - [2y + 5z - (y + 2)] + 3 - 7z = 3x - [2y + 5z - y - 2] + 3 - 7z$$
$$= 3x - [y + 5z - 2] + 3 - 7z$$
$$= 3x - y - 5z + 2 + 3 - 7z$$
$$= 3x - y - 5z + 5 - 7z$$
$$= 3x - y - 12z + 5 \qquad ◆◆◆$$

We'll do more complicated problems of this type later.

Good Working Habits

Start now to develop careful working habits. Why make things even harder for yourself by scribbling? Form the symbols with care; work in sequence, from top of page to bottom; do not crowd; use a sharp pencil, not a pen; and erase mistakes instead of crossing them out. What chance do you have if you cannot read your own work?

Common Error	Do not switch between capitals and lowercase letters without reason. Although b and B are the same alphabetic letter, in a math problem they could stand for entirely different quantities.

Exercise 2 ◆ Addition and Subtraction of Algebraic Expressions

Combine as indicated, and simplify.

1. $7x + 5x$
2. $2x + 5x - 4x + x$
3. $6ab - 7ab - 9ab$
4. $9.4x - 3.7x + 1.4x$
5. Add $7a - 3b + m$ and $3b - 7a - c + m$.
6. What is the sum of $6ab + 12bc - 8cd$, $3cd - 7cd - 9bc$, and $12cd - 2ab - 5bc$?
7. What is the sum of $3a + b - 10$, $c - d - a$, and $-4c + 2a - 3b - 7$?
8. Add $7m + 3n - 11p$, $3a - 9n - 11m$, $8n - 4m + 5p$, and $6n - m + 3p$.
9. Add $8ax + 2(x + a) + 3b$, $9ax + 6(x + a) - 9b$, and $11x + 6b - 7ax - 8(x + a)$.
10. Add $a - 9 - 8a^2 + 16a^3$, $5 + 15a^3 - 12a - 2a^2$, and $6a^2 - 10a^3 + 11a - 13$.
11. Subtract $41x^3 - 2x^2 + 13$ from $15x^3 + x - 18$.
12. Subtract $3b - 6d - 10c + 7a$ from $4d + 12a - 13c - 9b$.
13. Add $x - y - z$ and $y - x + z$.
14. What is the sum of $a + 2b - 3c - 10$, $3b - 4a + 5c + 10$, and $5b - c$?
15. Add $72ax^4 - 8ay^3$, $-38ax^4 - 3ay^4 + 7ay^3$, $8 + 12ay^4$, $-6ay^3 + 12$, and $-34ax^4 + 5ay^3 - 9ay^4$.
16. Add $2a(x - y^2) - 3mz^2$, $4a(x - y^2) - 5mz^2$, and $5a(x - y^2) + 7mz^2$.
17. What is the sum of $9b^2 - 3ac + d$, $4b^2 + 7d - 4ac$, $3d - 4b^2 + 6ac$, $5b^2 - 2ac - 12d$, and $4b^2 - d$?
18. From the sum of $2a + 3b - 4c$ and $3b + 4c - 5d$, subtract the sum of $5c - 6d - 7a$ and $-7d + 8a + 9b$.
19. What is the sum of $7ab - m^2 + q$, $-4ab - 5m^3 - 3q$, $12ab + 14m^2 - z$, and $-6m^2 - 2q$?

20. Add $14(x + y) - 17(y + z)$, $4(y + z) - 19(z + x)$, and $-7(z + x) - 3(x + y)$.

21. Add $3.52(a + b)$, $4.15(a + b)$, and $-1.84(a + b)$.

Symbols of Grouping

Remove symbols of grouping and collect terms.

22. $(3x + 5) + (2x - 3)$
23. $3a^2 - (3a - x + b)$
24. $40xy - (30xy - 2b^2 + 3c - 4d)$
25. $(6.4 - 1.8x) - (7.5 + 2.6x)$
26. $7m^2 + 2bc - (3m^2 - bc - x)$
27. $a^2 - a - (4a - y - 3a^2 - 1)$
28. $(2x^4 + 3x^3 - 4) + (3x^4 - 2x^3 - 8)$
29. $a + b - m - (m - a - b)$
30. $(3xy - 2y + 3z) + (-3y + 8z) - (-3xy - z)$
31. $3m - z - y - (2z - y - 3m)$
32. $(2x - 3y) - (5x - 3y - 2z)$
33. $(w + 2z) - (-3w - 5x) + (x + z) - (z - w)$
34. $2a + \{-6b - [3c + (-4b - 6c + a)]\}$
35. $4a - \{a - [-7a - [8a - (5a + 3)] - (-6a - 2a - 9)]\}$
36. $9m - \{3n + [4m - (n - 6m)] - [m + 7n]\}$

Applications

37. The surface area of the box in Fig. 2–1 is

$$2[2w^2 + 3w^2 + 6w^2]$$

Simplify this expression.

38. If a person invests \$5000, x dollars at 12% interest and the rest at 8% interest, the total earnings will be

$$0.12x + 0.08(5000 - x)$$

Simplify this expression.

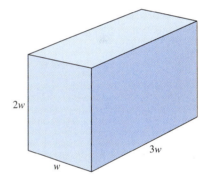

$2w$

$3w$

w

FIGURE 2–1

Computer

Programs are available that let us do *symbolic* algebra on the computer, in addition to numerical computation. These are called *computer algebra systems* (CAS). Some programs currently available are *Derive*, *Mathematica*, and *Maple*. Each of these systems uses the command *Simplify* to add or subtract algebraic expressions.

39. Using whatever symbolic algebra program is available to you, repeat any of the problems in this exercise set. Compare the results with what you got by hand.

The CAS problems in this chapter are clearly trivial. Our reason for including them is to provide a slow and easy introduction to CAS. More interesting problems will come in later chapters.

2–3 Integral Exponents

Definitions

In this section we deal only with expressions that have integers (positive or negative, and zero) as exponents.

◆◆◆ **Example 26:** We study such expressions as

$$x^3 \quad a^{-2} \quad y^0 \quad x^2y^3w^{-1} \quad (m + n)^5 \qquad \text{◆◆◆}$$

A positive *exponent* shows how many times the *base* is to be multiplied by itself.

◆◆◆ **Example 27:** In the expression 3^4, 3 is the base and 4 is the exponent.

$$\text{base} \longrightarrow 3^4 \longleftarrow \text{exponent}$$

Its meaning is

$$3^4 = 3(3)(3)(3) = 81$$ ◆◆◆

In general, the following equation applies:

Positive Integral Exponent	a^n means $\underbrace{a \cdot a \cdot a \cdot a \cdots \cdots a}_{n \text{ factors}}$	28

Common Errors	An exponent applies *only* to the symbol directly before it. Thus $$2x^2 \neq 2^2 x^2$$ However, $$(2x)^2 = (2x)(2x) = 4x^2$$
	Also, $$-2^2 \neq 4$$ However, $$(-2)^2 = (-2)(-2) = 4 \quad \text{and} \quad -2^2 = -(2)(2) = -4$$

Multiplying Powers

Let us multiply the quantity x^2 by x^3. From Eq. 28, we know that

$$x^2 = x \cdot x$$

and that

$$x^3 = x \cdot x \cdot x$$

Multiplying, we obtain

$$x^2 \cdot x^3 = (x \cdot x)(x \cdot x \cdot x)$$
$$= x \cdot x \cdot x \cdot x \cdot x$$
$$= x^5$$

Notice that the exponent in the result is the *sum of the two original exponents.*

$$x^2 \cdot x^3 = x^{2+3} = x^5$$

This will always be so. We summarize this rule as our first *law of exponents.*

Products	$x^a \cdot x^b = x^{a+b}$	29

When multiplying powers of the same base, we keep the same base and **add the exponents.**

◆◆◆ Example 28:

(a) $x^4(x^3) = x^{4+3} = x^7$
(b) $x^5(x^a) = x^{5+a}$
(c) $y^2(y^n)(y^3) = y^{2+n+3} = y^{5+n}$ ◆◆◆

Every quantity has an exponent of 1, even though it is usually not written.

◆◆◆ Example 29:

(a) $x(x^3) = x^{1+3} = x^4$
(b) $a^2(a^4)(a) = a^{2+4+1} = a^7$
(c) $x^a(x)(x^b) = x^{a+1+b}$ ◆◆◆

Quotients

Let us divide x^5 by x^3. By Eq. 29,

$$\frac{x^5}{x^3} = \frac{x \cdot x \cdot x \cdot x \cdot x}{x \cdot x \cdot x} = \frac{x \cdot x \cdot x}{x \cdot x \cdot x} \cdot x \cdot x$$

$$= x \cdot x = x^2$$

The "invisible 1" appears again. We saw it before as the unwritten coefficient of every term, and now as the unwritten exponent. It is also in the denominator.

$$x = \frac{1x^1}{1}$$

Do not forget about those invisible 1's. *We use them all the time.*

We could have obtained the same result by *subtracting exponents:*

$$\frac{x^5}{x^3} = x^{5-3} = x^2$$

This always works, and we state it as another law of exponents.

Quotients	$\dfrac{x^a}{x^b} = x^{a-b} \quad (x \neq 0)$	30

When dividing powers of the same base, we keep the same base and **subtract the exponents.**

◆◆◆ Example 30:

(a) $\dfrac{x^5}{x^4} = x^{5-4} = x^1 = x$

(b) $\dfrac{y^4}{y^6} = y^{4-6} = y^{-2}$

(c) $\dfrac{a^4 b^3}{ab^2} = a^{4-1}b^{3-2} = a^3 b$

(d) $\dfrac{y^{n-m}}{y^{n+m}} = y^{n-m-(n+m)} = y^{-2m}$ ◆◆◆

Common Error	Remember when using Rules 29 or 30 that the quantities to be multiplied or divided **must have the same base.**

Power Raised to a Power

Let us take a quantity raised to a power, say, x^2, and raise the entire expression to *another* power, say, 3.

$$(x^2)^3$$

By Eq. 29,

$$(x^2)^3 = (x^2)(x^2)(x^2)$$
$$= (x \cdot x)(x \cdot x)(x \cdot x)$$
$$= x \cdot x \cdot x \cdot x \cdot x \cdot x = x^6$$

a result that could have been obtained by *multiplying the exponents*.

$$(x^2)^3 = x^{2(3)} = x^6$$

In general, the following formula applies for a power raised to a power:

Powers	$(x^a)^b = x^{ab} = (x^b)^a$	31

When raising a power to a power, we keep the same base and **multiply the exponents**.

◆◆◆ Example 31:

(a) $(2^2)^3 = 2^{2(3)} = 2^6 = 64$
(b) $(w^5)^2 = w^{5(2)} = w^{10}$
(c) $(a^{-3})^2 = a^{(-3)(2)} = a^{-6}$
(d) $(10^4)^3 = 10^{4(3)} = 10^{12}$
(e) $(b^{x+2})^3 = b^{3x+6}$ ◆◆◆

Product Raised to a Power

We now raise a product, such as xy, to some power, say, 3.

$$(xy)^3$$

By Eq. 28,

$$(xy)^3 = (xy)(xy)(xy)$$
$$= x \cdot y \cdot x \cdot y \cdot x \cdot y$$
$$= x \cdot x \cdot x \cdot y \cdot y \cdot y$$
$$= x^3 y^3$$

In general, the following equation applies to a product raised to a power:

Product Raised to a Power	$(xy)^n = x^n \cdot y^n$	32

When a product is raised to a power, each factor may be **separately** raised to the power.

◆◆◆ Example 32:

(a) $(xyz)^5 = x^5 y^5 z^5$
(b) $(2x)^3 = 2^3 x^3 = 8x^3$
(c) $(3.75 \times 10^3)^2 = (3.75)^2 \times (10^3)^2$
$$= 14.1 \times 10^6$$
(d) $(3x^2 y^n)^3 = 3^3 (x^2)^3 (y^n)^3$
$$= 27 x^6 y^{3n}$$
(e) $(-x^2 y)^3 = (-1)^3 (x^2)^3 y^3 = -x^6 y^3$ ◆◆◆

Common Error	There is *no* similar rule for the *sum* of two quantities raised to a power. $$(x + y)^n \neq x^n + y^n$$

A good way to test a "rule" that you are not sure of is to *try it with numbers*. In this case, does $(2 + 3)^2$ equal $2^2 + 3^2$? Evaluating each expression, we obtain

$$(5)^2 \overset{?}{=} 4 + 9$$
$$25 \neq 13$$

Quotient Raised to a Power

Using the same steps as in the preceding section, see if you can show that

$$\left(\frac{x}{y}\right)^3 \overset{?}{=} \frac{x^3}{y^3}$$

In general, the following equation applies:

Quotient Raised to a Power	$\left(\dfrac{x}{y}\right)^n = \dfrac{x^n}{y^n} \quad (y \neq 0)$	33

When a quotient is raised to a power, the numerator and the denominator may be **separately** raised to the power.

◆◆◆ Example 33:

(a) $\left(\dfrac{2}{3}\right)^2 = \dfrac{2^2}{3^2} = \dfrac{4}{9}$

(b) $\left(\dfrac{x}{5}\right)^2 = \dfrac{x^2}{5^2} = \dfrac{x^2}{25}$

(c) $\left(\dfrac{3a}{2b}\right)^3 = \dfrac{3^3 a^3}{2^3 b^3} = \dfrac{27a^3}{8b^3}$

(d) $\left(\dfrac{2x^2}{5y^3}\right)^3 = \dfrac{2^3(x^2)^3}{5^3(y^3)^3} = \dfrac{8x^6}{125y^9}$

(e) $\left(-\dfrac{a}{b^2}\right)^3 = (-1)^3 \dfrac{a^3}{(b^2)^3} = -\dfrac{a^3}{b^6}$

◆◆◆

Zero Exponent

If we divide x^n by itself we get, by Eq. 30,

$$\frac{x^n}{x^n} = x^{n-n} = x^0$$

But any expression divided by itself equals 1, so we have the following law:

Zero Exponent	$x^0 = 1 \quad (x \neq 0)$	34

Any quantity (except 0) raised to the zero power equals 1.

◆◆◆ Example 34:

(a) $(xyz)^0 = 1$

(b) $3862^0 = 1$

(c) $(x^2 - 2x + 3)^0 = 1$

(d) $5x^0 = 5(1) = 5$

◆◆◆

Negative Exponent

We now divide x^0 by x^a. By Eq. 30,

$$\frac{x^0}{x^a} = x^{0-a} = x^{-a}$$

Since $x^0 = 1$, we get the following equation:

Negative Exponent	$x^{-a} = \dfrac{1}{x^a} \quad (x \neq 0)$	35

When taking the reciprocal of a base raised to a power, **change the sign of the exponent**.

We can, of course, also give Eq. 35 in the form

$$x^a = \frac{1}{x^{-a}} \quad (x \neq 0)$$

For practice, let's first rewrite an expression containing negative exponents as one without a negative exponent.

◆◆◆ **Example 35:** Rewrite each expression without negative exponents, using fractions where necessary.

(a) $5^{-1} = \dfrac{1}{5}$

(b) $x^{-4} = \dfrac{1}{x^4}$

(c) $\dfrac{1}{xy^{-2}} = \dfrac{y^2}{x}$

(d) $\dfrac{w^{-3}}{z^{-2}} = \dfrac{z^2}{w^3}$

(e) $\left(\dfrac{3x}{2y^2}\right)^{-b} = \left(\dfrac{2y^2}{3x}\right)^{b}$

◆◆◆

We can also use negative exponents to rewrite an expression *without fractions*.

◆◆◆ **Example 36:** Rewrite each expression without fractions. Use negative exponents where necessary.

(a) $\dfrac{1}{y} = y^{-1}$

(b) $\dfrac{1}{y^4} = y^{-4}$

(c) $\dfrac{a}{b} = ab^{-1}$

(d) $\dfrac{a^2}{b^3} = a^2 b^{-3}$

(e) $\dfrac{w}{xy^2} = wx^{-1}y^{-2}$

(f) $\dfrac{x^2 y^{-3}}{w^4 z^{-2}} = x^2 y^{-3} w^{-4} z^2$

Summary of the Laws of Exponents

Positive Integral Exponent	a^n means $\underbrace{a \cdot a \cdot a \cdot \cdots \cdot a}_{n \text{ factors}}$	28
Products	$x^a \cdot x^b = x^{a+b}$	29

Quotients	$\dfrac{x^a}{x^b} = x^{a-b} \quad (x \neq 0)$	30
Powers	$(x^a)^b = x^{ab} = (x^b)^a$	31
Product Raised to a Power	$(xy)^n = x^n \cdot y^n$	32
Quotient Raised to a Power	$\left(\dfrac{x}{y}\right)^n = \dfrac{x^n}{y^n} \quad (y \neq 0)$	33
Zero Exponent	$x^0 = 1 \quad (x \neq 0)$	34
Negative Exponent	$x^{-a} = \dfrac{1}{x^a} \quad (x \neq 0)$	35

Exercise 3 ◆ Integral Exponents

Multiply.

1. $10^2 \cdot 10^3$
2. $10^a \cdot 10^b$
3. $x \cdot x^2$
4. $x^3 \cdot x^2$
5. $y^3 \cdot y^4$
6. $w^2 \cdot w^a$
7. $x^2 \cdot x \cdot x^4$
8. $x^2 \cdot x^b \cdot x^3$
9. $w \cdot w^2 \cdot w^3$
10. $y^3 \cdot y^2 \cdot y^4$
11. $x^{n+1} \cdot x^2$
12. $y^{a+2} \cdot y^{2a-1}$

Divide. Write your answers without negative exponents.

13. $\dfrac{y^5}{y^2}$
14. $\dfrac{5^5}{5^3}$
15. $\dfrac{x^{n+2}}{x^{n+1}}$

16. $\dfrac{10^5}{10}$
17. $\dfrac{10^{x+5}}{10^{x+3}}$
18. $\dfrac{10^2}{10^{-3}}$

19. $\dfrac{x^{-2}}{x^{-3}}$
20. $\dfrac{a^{-5}}{a}$

Simplify.

21. $(x^3)^4$
22. $(9^2)^3$
23. $(a^x)^y$
24. $(x^{-2})^{-2}$
25. $(x^{a+1})^2$

Raise to the power indicated.

26. $(xy)^2$
27. $(2x)^3$
28. $(3x^2y^3)^2$

29. $(3abc)^3$
30. $\left(\dfrac{3}{5}\right)^2$
31. $\left(-\dfrac{1}{3}\right)^3$

32. $\left(\dfrac{x}{y}\right)^5$
33. $\left(\dfrac{2a}{3b^2}\right)^3$
34. $\left(\dfrac{3x^2y}{4wz^3}\right)^2$

Write each expression with positive exponents only.

35. a^{-2}
36. $(-x)^{-3}$
37. $\left(\dfrac{3}{y}\right)^{-3}$

38. $a^{-2}bc^{-3}$ **39.** $\left(\dfrac{2a}{3b^3}\right)^{-2}$ **40.** xy^{-4}

41. $2x^{-2} + 3y^{-3}$ **42.** $\left(\dfrac{x}{y}\right)^{-1}$

Express without fractions, using negative exponents where needed.

43. $\dfrac{1}{x}$ **44.** $\dfrac{3}{y^2}$ **45.** $\dfrac{x^2}{y^2}$

46. $\dfrac{x^2y^{-3}}{z^{-2}}$ **47.** $\dfrac{a^{-3}}{b^2}$ **48.** $\dfrac{x^{-2}y^{-3}}{w^{-1}z^{-4}}$

Evaluate.

49. $(a + b + c)^0$ **50.** $8x^0y^2$ **51.** $\dfrac{a^0}{9}$

52. $\dfrac{y}{x^0}$ **53.** $\dfrac{x^{2n} \cdot x^3}{x^{3+2n}}$ **54.** $5\left(\dfrac{x}{y}\right)^0$

Applications

55. We can find the volume of the box in Fig. 2–1 by multiplying length by width by height, getting

$$(w)(2w)(3w)$$

Simplify this expression.

56. A freely falling body, starting from rest, falls a distance of $16.1t^2$ feet in t seconds. In twice that time it will fall

$$16.1(2t)^2 \text{ ft}$$

Simplify this expression.

57. The power in a resistor of resistance R in which flows a current i is i^2R. If the current is reduced to one-third its former value, the power will be

$$\left(\frac{i}{3}\right)^2 R$$

Simplify this expression.

58. The resistance R of two resistors R_1 and R_2 wired in parallel (Fig. 2–2) is found from

$$\frac{1}{R} = \frac{1}{R_1} + \frac{1}{R_2}$$

Write this equation without fractions.

FIGURE 2–2

Computer

59. Use a computer algebra system (CAS) to repeat any of the problems in this exercise set. Compare your results with what you got by hand.

2–4   Multiplication of Algebraic Expressions

Symbols and Definitions

Multiplication is indicated in several ways: by the usual $\times$ symbol; by a dot; or by parentheses, brackets, or braces. Thus the product of b and d could be written

$$b \cdot d \qquad b \times d \qquad b(d) \qquad (b)d \qquad (b)(d)$$

Avoid using the $\times$ symbol because it could get confused with the letter x.

The most common convention is to use no symbol at all. The product of b and d would usually be written bd. A calculator or computer usually uses the asterisk or star (*) to indicate multiplication.

We get a *product* when we multiply two or more *factors*.

$$(\text{factor})(\text{factor})(\text{factor}) = \text{product}$$

Rules of Signs

The product of two factors having the *same* sign is *positive*; the product of two factors having *different* signs is *negative*. If a and b are positive quantities, the following rules apply:

Rules of Signs		
	$(+a)(+b) = (-a)(-b) = +ab$	8
	$(+a)(-b) = (-a)(+b) = -(+a)(+b) = -ab$	9

When we are multiplying more than two factors, the product of every *pair* of negative terms will be positive. Thus if there are an even number of negative factors, the final result will be positive; if there are an odd number of negative factors, the final result will be negative.

◆◆◆ Example 37:

(a) $x(-y)(-z) = xyz$
(b) $(-a)(-b)(-c) = -abc$
(c) $(-p)(-q)(-r)(-s) = pqrs$ ◆◆◆

Commutative Law

As we saw when multiplying numbers, the *order* of multiplication does not make any difference. In symbols, the law is stated as follows:

Commutative Law of Multiplication		
	$ab = ba$	2

◆◆◆ Example 38: Multiply $3y$ by $-2x$.

Solution:

$$(3y)(-2x) = 3(y)(-2)(x)$$

By Eq. 2,

$$= 3(-2)(y)(x)$$
$$= -6yx$$

or

$$= -6xy$$

since it is common practice to write the letters in alphabetical order. ◆◆◆

Multiplying Monomials

A *monomial* is an algebraic expression having *one term*.

♦♦♦ **Example 39:** Some monomials are

(a) $2x^3$ (b) $\dfrac{3wxy}{4}$ (c) $(2b)^3$ ♦♦♦

To multiply monomials, we use the commutative law (Eq. 2), the rules of signs (Eqs. 8 and 9), and the law of exponents for products.

Products	$x^a \cdot x^b = x^{a+b}$	29

♦♦♦ **Example 40:**

$$(4a^3b)(-3ab^2) = (4)(-3)(a^3)(a)(b)(b^2)$$
$$= -12a^{3+1}b^{1+2}$$
$$= -12a^4b^3$$ ♦♦♦

♦♦♦ **Example 41:** Multiply $5x^2$, $-3xy$, and $-xy^3z$.

Solution: By Eq. 2,

$$5x^2(-3xy)(-xy^3z) = 5(-3)(-1)(x^2)(x)(x)(y)(y^3)(z)$$

By Eq. 29,

$$= 15x^{2+1+1}y^{1+3}z$$
$$= 15x^4y^4z$$ ♦♦♦

> With some practice you should be able to omit some steps in many of these solutions. But do not rush it. Use as many steps as you need to get it right.

♦♦♦ **Example 42:** Multiply $3x^2$ by $2ax^n$.

Solution: By Eq. 2,

$$3x^2(2ax^n) = 3(2a)(x^2)(x^n)$$

By Eq. 29,

$$= 6ax^{2+n}$$ ♦♦♦

Multiplying a Multinomial by a Monomial

To multiply a monomial and a multinomial, we use the *distributive law*.

Distributive Law of Multiplication	$a(b + c) = ab + ac$	5

> Later we will use this law to remove common factors.

The product of a monomial and a polynomial is equal to the sum of the products of the monomial and each term of the polynomial.

♦♦♦ **Example 43:**

$$2x(x - 3x^2) = 2x(x) + 2x(-3x^2)$$
$$= 2x^2 - 6x^3$$ ♦♦♦

The distributive law, although written for a monomial times a binomial, can be extended for a multinomial having any number of terms. Simply *multiply every term in the multinomial by the monomial*.

◆◆◆ **Example 44:**

(a) $-3x^3(3x^2 - 2x + 4) = -3x^3(3x^2) + (-3x^3)(-2x) + (-3x^3)(4)$
$$= -9x^5 + 6x^4 - 12x^3$$
(b) $4xy(x^2 - 3xy + 2y^2) = 4x^3y - 12x^2y^2 + 8xy^3$
(c) $3.75x(1.83x^2 + 2.27x - 1.49) = 6.86x^3 + 8.51x^2 - 5.59x$ ◆◆◆

Removing Symbols of Grouping

Symbols of grouping, in addition to indicating that the enclosed expression is to be taken as a whole, also indicate multiplication.

◆◆◆ **Example 45:**

(a) $x(y)$ is the product of x and y.
(b) $x(y + 2)$ is the product of x and $(y + 2)$.
(c) $(x - 1)(y + 2)$ is the product of $(x - 1)$ and $(y + 2)$.
(d) $-(y + 2)$ is the product of -1 and $(y + 2)$. ◆◆◆

The parentheses may be removed after the multiplication has been performed.

◆◆◆ **Example 46:**

(a) $x(y + 2) = xy + 2x$
(b) $5 + 3(x - 2) = 5 + 3x - 6 = 3x - 1$ ◆◆◆

Common Errors	Do not forget to multiply *every* term within the grouping by the preceding factor.
	$$-(2x + 5) \neq -2x + 5$$
	Multiply the terms within the parentheses *only* by the factor directly preceding it.
	$$x + 2(a + b) \neq (x + 2)(a + b)$$

If there are groupings within groupings, start simplifying with the *innermost* groupings.

◆◆◆ **Example 47:**

(a) $x - 3[2 - 4(y + 1)] = x - 3[2 - 4y - 4]$
$$= x - 3[-2 - 4y]$$
$$= x + 6 + 12y$$
(b) $3\{2[4(w - 4) - (x + 3)] - 2\} - 6x$
$$= 3\{2[4w - 16 - x - 3] - 2\} - 6x$$
$$= 3\{2[4w - x - 19] - 2\} - 6x$$
$$= 3\{8w - 2x - 38 - 2\} - 6x$$
$$= 3\{8w - 2x - 40\} - 6x$$
$$= 24w - 6x - 120 - 6x$$
$$= 24w - 12x - 120$$ ◆◆◆

Multiplying a Multinomial by a Multinomial

Using the distributive law, we multiply one of the multinomials by each term in the other multinomial. We then use the distributive law again to remove the remaining parentheses, and simplify.

◆◆◆ **Example 48:**

$$(x + 4)(x - 3) = x(x - 3) + 4(x - 3)$$
$$= x^2 - 3x + 4x - 12$$
$$= x^2 + x - 12$$

◆◆◆

Some prefer a vertical arrangement for problems like the last one.

◆◆◆ **Example 49:**

$$
\begin{array}{r}
x + \ 4 \\
x - \ 3 \\
\hline
- 3x - 12 \\
x^2 + 4x \\
\hline
x^2 + \ x - 12
\end{array}
$$

as before.

◆◆◆

◆◆◆ **Example 50:**

(a) $(x - 3)(x^2 - 2x + 1)$
$$= x(x^2) + x(-2x) + x(1) - 3(x^2) - 3(-2x) - 3(1)$$
$$= x^3 - 2x^2 + x - 3x^2 + 6x - 3$$
$$= x^3 - 5x^2 + 7x - 3$$

(b) $(2x + 3y - 4z)(x - 2y - 3z)$
$$= 2x^2 - 4xy - 6xz + 3xy - 6y^2 - 9yz - 4xz + 8yz + 12z^2$$
$$= 2x^2 - 6y^2 + 12z^2 - xy - 10xz - yz$$

(c) $(1.24x + 2.03)(3.16x^2 - 2.61x + 5.36)$
$$= 3.92x^3 - 3.24x^2 + 6.65x + 6.41x^2 - 5.30x + 10.9$$
$$= 3.92x^3 + 3.17x^2 + 1.35x + 10.9$$

◆◆◆

When multiplying *more than two* expressions, first multiply any pair, and then multiply that product by the third expression, and so on.

◆◆◆ **Example 51:** Multiply $(x - 2)(x + 1)(2x - 3)$.

Solution: Multiplying the last two binomials yields

$$(x - 2)(x + 1)(2x - 3) = (x - 2)(2x^2 - x - 3)$$

Then multiplying that product by $(x - 2)$ gives

$$(x - 2)(2x^2 - x - 3) = 2x^3 - x^2 - 3x - 4x^2 + 2x + 6$$
$$= 2x^3 - 5x^2 - x + 6$$

◆◆◆

To raise an expression to a power, simply multiply the expression by itself the proper number of times.

◆◆◆ **Example 52:**

$$(x - 1)^3 = (x - 1)(x - 1)(x - 1)$$
$$= (x - 1)(x^2 - 2x + 1)$$
$$= x^3 - 2x^2 + x - x^2 + 2x - 1$$
$$= x^3 - 3x^2 + 3x - 1$$

◆◆◆

Product of the Sum and Difference of Two Terms

Let us multiply the binomials $(a - b)$ and $(a + b)$. We get

$$(a - b)(a + b) = a^2 + ab - ab - b^2$$

The two middle terms cancel, leaving us with the following rule:

Difference of Two Squares	$(a - b)(a + b) = a^2 - b^2$	41

When we multiply two binomials, we get expressions that occur frequently enough in our later work for us to take special note of here. These are called *special products*. We will need them especially when we study factoring.

Thus the product is the *difference of two squares*: the squares of the original numbers.

◆◆◆ **Example 53:**

$$(x + 2)(x - 2) = x^2 - 2^2 = x^2 - 4$$ ◆◆◆

Product of Two Binomials Having the Same First Term

When we multiply $(x + a)$ and $(x + b)$, we get

$$(x + a)(x + b) = x^2 + ax + bx + ab$$

Combining like terms, we obtain the following:

Trinomial, Leading Coefficient = 1	$(x + a)(x + b) = x^2 + (a + b)x + ab$	45

◆◆◆ **Example 54:**

$$(w + 3)(w - 4) = w^2 + (3 - 4)w + 3(-4)$$
$$= w^2 - w - 12$$ ◆◆◆

General Quadratic Trinomial

When we multiply two binomials of the form $ax + b$ and $cx + d$, we get

$$(ax + b)(cx + d) = acx^2 + adx + bcx + bd$$

Combining like terms, we obtain the following formula:

Some people like the *FOIL* rule. Multiply the *First* terms, then *Outer*, *Inner*, and *Last* terms.

General Quadratic Trinomial	$(ax + b)(cx + d) = acx^2 + (ad + bc)x + bd$	46

◆◆◆ **Example 55:**

$$(3x - 2)(2x + 5) = 3(2)x^2 + [(3)(5) + (-2)(2)]x + (-2)(5)$$
$$= 6x^2 + 11x - 10$$ ◆◆◆

Powers of Multinomials

As with a monomial raised to a power that is a positive integer, we raise a *multinomial* to a power by multiplying that multinomial by itself the proper number of times.

◆◆◆ **Example 56:**

$$(x - 3)^2 = (x - 3)(x - 3)$$
$$= x^2 - 3x - 3x + 9$$
$$= x^2 - 6x + 9$$

◆◆◆

◆◆◆ **Example 57:**

$$(x - 3)^3 = (x - 3)(x - 3)(x - 3)$$
$$= (x - 3)(x^2 - 6x + 9) \quad \text{(from Example 56)}$$
$$= x^3 - 6x^2 + 9x - 3x^2 + 18x - 27$$

Combining like terms gives us

$$(x - 3)^3 = x^3 - 9x^2 + 27x - 27$$

◆◆◆

Perfect Square Trinomial

When we *square a binomial* such as $(a + b)$, we get

$$(a + b)^2 = (a + b)(a + b) = a^2 + ab + ab + b^2$$

Collecting terms, we obtain the following formula:

Perfect Square Trinomial	$(a + b)^2 = a^2 + 2ab + b^2$	47

This trinomial is called a *perfect square trinomial* because it is the square of a binomial. Its first and last terms are the squares of the two terms of the binomial; its middle term is twice the product of the terms of the binomial. Similarly, when we square $(a - b)$, we get the following:

Perfect Square Trinomial	$(a - b)^2 = a^2 - 2ab + b^2$	47

◆◆◆ **Example 58:**

$$(3x - 2)^2 = (3x)^2 - 2(3x)(2) + 2^2$$
$$= 9x^2 - 12x + 4$$

◆◆◆

Exercise 4 ◆ Multiplication of Algebraic Expressions

Multiply the following monomials, and simplify.

1. $x^3(x^2)$
2. $x^2(3axy)$
3. $(-5ab)(-2a^2b^3)$
4. $(3.52xy)(-4.14xyz)$
5. $(6ab)(2a^2b)(3a^3b^3)$
6. $(-2a)(-3abc)(ac^2)$
7. $2p(-4p^2q)(pq^3)$
8. $(-3xy)(-2w^2x)(-xy^3)(-wy)$

Multiply the polynomial by the monomial, and simplify.

9. $2a(a - 5)$
10. $(x^2 + 2x)x^3$
11. $xy(x - y + xy)$
12. $3a^2b^3(ab - 2ab^2 + b)$
13. $-4pq(3p^2 + 2pq - 3q^2)$
14. $2.8x(1.5x^3 - 3.2x^2 + 5.7x - 4.6)$

15. $3ab(2a^2b^2 - 4a^2b + 3ab^2 + ab)$
16. $-3xy(2xy^3 - 3x^2y + xy^2 - 4xy)$

Multiply the following binomials, and simplify.

17. $(a + b)(a + c)$
18. $(2x + 2)(5x + 3)$
19. $(x + 3y)(x^2 - y)$
20. $(a - 5)(a + 5)$
21. $(m + n)(9m - 9n)$
22. $(m^2 + 2c)(m^2 - 5c)$
23. $(7cd^2 + 4y^3z)(7cd^2 - 4y^3z)$
24. $(3a + 4c)(2a - 5c)$
25. $(x^2 + y^2)(x^2 - y^2)$
26. $(3.50a^2 + 1.20b)(3.80a^2 - 2.40b)$

Multiply the following binomials and trinomials, and simplify.

27. $(a - c - 1)(a + 1)$
28. $(x - y)(2x + 3y - 5)$
29. $(m^2 - 3m - 7)(m - 2)$
30. $(-3x^2 + 3x - 9)(x + 3)$
31. $(x^6 + x^4 + x^2)(x^2 - 1)$
32. $(m - 1)(2m - m^2 + 1)$
33. $(1 - z)(z^2 - z + 2)$
34. $(3a^3 - 2ab - b^2)(2a - 4b)$
35. $(a^2 - ay + y^2)(a + y)$
36. $(y^2 - 7.5y + 2.8)(y + 1.6)$
37. $(a^2 + ay + y^2)(a - y)$
38. $(a^2 + ay - y^2)(a - y)$

Multiply the following binomials and polynomials, and simplify.

39. $[x^2 - (m - n)x - mn](x - p)$
40. $(x^2 + ax - bx - ab)(x + b)$
41. $(m^4 + m^3 + m^2 + m + 1)(m - 1)$

Multiply the following monomials and binomials, and simplify.

42. $(2a + 3x)(2a + 3x)9$
43. $(3a - b)(3a - b)x$
44. $c(m - n)(m + n)$
45. $(x + 1)(x + 1)(x - 2)$
46. $(m - 2.5)(m - 1.3)(m + 3.2)$
47. $(x + a)(x + b)(x - c)$
48. $(x - 4)(x - 5)(x + 4)(x + 5)$
49. $(1 + c)(1 + c)(1 - c)(1 + c^2)$

Multiply the following trinomials, and simplify.

50. $(a + b - c)(a - b + c)$
51. $(2a^3 + 5a)(2a^3 - 5ac^2 + 2a)$
52. $[3(a + b)^2 - (a + b) + 2][4(a + b)^2 - (a + b) - 5]$

53. $(a + b + c)(a - b + c)(a + b - c)$

Multiply the following polynomials, and simplify.

54. $(x^3 - 6x^2 + 12x - 8)(x^2 + 4x + 4)$

55. $(n - 5n^2 + 2 + n^3)(5n + n^2 - 10)$

56. $(mx + my - nx - ny)(mx - my + nx - ny)$

Powers of Algebraic Expressions

Square each binomial.

57. $(a + c)$ **58.** $(p + q)$

59. $(m - n)$ **60.** $(x - y)$

61. $(A + B)$ **62.** $(A - C)$

63. $(3.80a - 2.20x)$ **64.** $(m + c)$

65. $(2c - 3d)$ **66.** $(x^2 - x)$

67. $(a - 1)$ **68.** $(a^2 x - ax^3)$

69. $(y^2 - 20)$ **70.** $(x^n \quad y^2)$

Square each trinomial.

71. $(a + b + c)$

72. $(a - b - c)$

73. $(1 + x - y)$

74. $(x^2 + 2x - 3)$

75. $(x^2 + xy + y^2)$

76. $(x^3 - 2x - 1)$

Cube each binomial.

77. $(a + b)$

78. $(2x + 1)$

79. $(p - 3q)$

80. $(x^2 - 1)$

81. $(a - b)$

82. $(3m - 2n)$

Symbols of Grouping

Remove symbols of grouping, and simplify.

83. $-3[-(a + b) - c]$

84. $-[x + (y - 2) - 4]$

85. $(p - q) - [p - (p - q) - q]$

86. $-(x - y) - [y - (y - x) - x]$

87. $-5[-2(3y - z) - z] + y$

88. $[(m + 2n) - (2m - n)][(2m + n) - (m - 2n)]$

89. $(x - 1) - \{[x - (x - 3)] - x\}$

90. $(a - b)(a^3 + b^3)[a(a + b) + b^2]$

91. $y - \{4x - [y - (2y - 9) - x] + 2\}$

92. $[2x^2 + (3x - 1)(4x + 5)][5x^2 - (4x + 3)(x - 2)]$

Applications

93. A rectangle has its length L increased by 4.0 units and its width W decreased by 3.0 units. Write an expression for the area of the rectangle, and multiply out.

94. A car traveling at a rate R for a time T will go a distance equal to RT. If the rate is decreased by 3.60 mi/h and the time is increased by 4.10 h, write an expression for the new distance traveled, and multiply out.

95. A square of side x has each side increased by 3.50 units. Write an expression for the area of the new square, and multiply out.

96. When a current i flows through a resistance R, the power in the resistance is i^2R. If the current is increased by 3.20 amperes, write an expression for the new power, and multiply out.

97. If the radius of a sphere of radius r is decreased by 3.60 units, write an expression for the new volume of the sphere, and multiply out.

Computer

98. Use a computer algebra system (CAS) to repeat any of the problems in this exercise set. Compare your results with what you got by hand.

2–5 Division of Algebraic Expressions

Symbols for Division

Division may be indicated by any of the following symbols:

$$x \div y \qquad \frac{x}{y} \qquad x/y$$

The names of the parts are

$$\text{quotient} = \frac{\text{dividend}}{\text{divisor}} = \frac{\text{numerator}}{\text{denominator}}$$

The quantity x/y is called a *fraction*. It is also called the *ratio* of x to y.

Reciprocals

As we saw in Sec. 1–4, the *reciprocal* of a number is 1 divided by that number. The reciprocal of n is $1/n$. We can use the idea of a reciprocal to show how division is related to multiplication. We may write the quotient of $x \div y$ as

$$\frac{x}{y} = \frac{x}{1} \cdot \frac{1}{y} = x \cdot \frac{1}{y}$$

We see that to *divide by a number* is the same thing as to *multiply by its reciprocal*. This fact will be especially useful for dividing by a fraction.

Division by Zero

Division by zero is not a permissible operation.

◆◆◆ **Example 59:** In the fraction

$$\frac{x + 5}{x - 2}$$

x cannot equal 2, or the illegal operation of division by zero will result.   ◆◆◆

> If division by zero were allowed, we could, for example, divide 2 by zero and get a quotient x:
>
> $$\frac{2}{0} = x$$
>
> or
>
> $$2 = 0 \cdot x$$
>
> but there is no number x which, when multiplied by zero, gives 2, so we cannot allow this operation.

Rules of Signs

The quotient of two terms of *like* sign is *positive*.

$$\frac{+a}{+b} = \frac{-a}{-b} = \frac{a}{b}$$

The quotient of two terms of *unlike* sign is *negative*.

$$\frac{+a}{-b} = \frac{-a}{+b} = -\frac{a}{b}$$

The fraction itself carries a third sign, which, when negative, reverses the sign of the quotient. These three ideas are summarized in the following rules:

Rules of Signs for Division	$$\frac{+a}{+b} = \frac{-a}{-b} = -\frac{-a}{+b} = -\frac{+a}{-b} = \frac{a}{b}$$	10
	$$\frac{+a}{-b} = \frac{-a}{+b} = -\frac{-a}{-b} = -\frac{a}{b}$$	11

These rules show that any *pair* of negative signs may be removed without changing the value of the fraction.

◆◆◆ **Example 60:** Simplify $-\dfrac{ax^2}{-y}$.

Solution: Removing the pair of minus signs, we obtain

$$-\frac{ax^2}{-y} = \frac{ax^2}{y}$$

◆◆◆

Common Error	Removal of pairs of negative signs pertains only to negative signs that are factors of the numerator, the denominator, or the fraction as a whole. Do not try to remove pairs of signs that apply only to single terms. Thus $$\frac{x-2}{x-3} \neq \frac{x+2}{x+3}$$

Dividing a Monomial by a Monomial

Any quantity (except zero) divided by itself equals *one*. So if the same factor appears in both the divisor and the dividend, it may be eliminated.

◆◆◆ **Example 61:** Divide $6ax$ by $3a$.

Solution:

$$\frac{6ax}{3a} = \frac{6}{3} \cdot \frac{a}{a} \cdot x = 2x$$

◆◆◆

To divide quantities having exponents, we use the law of exponents for division.

Quotients	$$\frac{x^a}{x^b} = x^{a-b} \quad (x \neq 0)$$	30

◆◆◆ **Example 62:** Divide y^5 by y^3.

Solution: By Eq. 30,

$$\frac{y^5}{y^3} = y^{5-3} = y^2$$

◆◆◆

When there are numerical coefficients, divide them separately.

◆◆◆ **Example 63:**

(a) $\dfrac{15x^6}{3x^4} = \dfrac{15}{3} \cdot \dfrac{x^6}{x^4} = 5x^{6-4} = 5x^2$

(b) $\dfrac{8.35y^5}{3.72y} = \dfrac{8.35}{3.72} \cdot \dfrac{y^5}{y} = 2.24y^{5-1} = 2.24y^4$ ◆◆◆

If there is more than one unknown, treat each separately.

◆◆◆ **Example 64:** Divide $18x^5y^2z^4$ by $3x^2yz^3$.

Solution:

$$\frac{18x^5y^2z^4}{3x^2yz^3} = \frac{18}{3} \cdot \frac{x^5}{x^2} \cdot \frac{y^2}{y} \cdot \frac{z^4}{z^3}$$
$$= 6x^3yz$$ ◆◆◆

Sometimes negative exponents will be obtained.

◆◆◆ **Example 65:**

$$\frac{6x^2}{x^5} = 6x^{2-5} = 6x^{-3}$$

The answer may be left in this form, or, as is usually done, we may use Eq. 35 to eliminate the negative exponent. Thus

$$6x^{-3} = \frac{6}{x^3}$$ ◆◆◆

◆◆◆ **Example 66:** Simplify the fraction

$$\frac{3x^2yz^5}{9xy^4z^2}$$

> The process of dividing a monomial by a monomial is also referred to as *simplifying a fraction*, or *reducing a fraction to lowest terms*. We discuss this in Chapter 9.

Solution: The procedure is no different than if we had been asked to divide $3x^2yz^5$ by $9xy^4z^2$.

$$\frac{3x^2yz^5}{9xy^4z^2} = \frac{3}{9}x^{2-1}y^{1-4}z^{5-2}$$
$$= \frac{1}{3}xy^{-3}z^3$$

or

$$= \frac{xz^3}{3y^3}$$ ◆◆◆

Do not be dismayed if the expressions to be divided have negative exponents. Apply Eq. 30, as before.

◆◆◆ **Example 67:** Divide $21x^2y^{-3}z^{-1}$ by $7x^{-4}y^2z^{-3}$.

Solution: Proceeding as before, we obtain

$$\frac{21x^2y^{-3}z^{-1}}{7x^{-4}y^2z^{-3}} = \frac{21}{7}x^{2-(-4)}y^{-3-2}z^{-1-(-3)}$$
$$= 3x^6y^{-5}z^2$$
$$= \frac{3x^6z^2}{y^5}$$ ◆◆◆

Dividing a Multinomial by a Monomial

Divide each term of the multinomial by the monomial.

◆◆◆ Example 68: Divide $9x^2 + 3x - 2$ by $3x$.

Solution:

$$\frac{9x^2 + 3x - 2}{3x} = \frac{9x^2}{3x} + \frac{3x}{3x} - \frac{2}{3x}$$

This is really a consequence of the distributive law (Eq. 5).

$$\frac{9x^2 + 3x - 2}{3x} = \frac{1}{3x}(9x^2 + 3x - 2) = \left(\frac{1}{3x}\right)(9x^2) + \left(\frac{1}{3x}\right)(3x) + \left(\frac{1}{3x}\right)(-2)$$

Each of these terms is now simplified as in the preceding section

$$\frac{9x^2}{3x} + \frac{3x}{3x} - \frac{2}{3x} = 3x + 1 - \frac{2}{3x}$$

◆◆◆

Common Error	Do not forget to divide *every* term of the multinomial by the monomial. $$\frac{x^2 + 2x + 3}{x} \neq x + 2 + 3$$

◆◆◆ Example 69:

(a)
$$\frac{3ab + 2a^2b - 5ab^2 + 3a^2b^2}{15ab} = \frac{3ab}{15ab} + \frac{2a^2b}{15ab} - \frac{5ab^2}{15ab} + \frac{3a^2b^2}{15ab}$$

$$= \frac{1}{5} + \frac{2a}{15} - \frac{b}{3} + \frac{ab}{5}$$

(b)
$$\frac{6x^3y - 8xy^2 - 2x^4y^4}{-2xy} = -3x^2 + 4y + x^3y^3$$

(c)
$$\frac{8.26x^3 - 6.24x^2 + 7.37x}{2.82x} = 2.93x^2 - 2.21x + 2.61$$

◆◆◆

Common Error	There is no similar rule for dividing a monomial by a multinomial. Thus $$\frac{a}{b + c} \neq \frac{a}{b} + \frac{a}{c}$$

Dividing a Polynomial by a Polynomial

Write the divisor and the dividend in the order of descending powers of the variable. Supply any missing terms, using a coefficient of zero. Set up as a long-division problem, as in the following example.

This method is used only for *polynomials:* expressions in which the powers are all *positive integers.*

◆◆◆ Example 70: Divide $x^2 + 4x^4 - 2$ by $1 + x$.

Solution:

1. Write the dividend in descending order of the powers.

$$4x^4 + x^2 - 2$$

2. Supply the missing terms with coefficients of zero.

$$4x^4 + 0x^3 + x^2 + 0x - 2$$

3. Write the divisor in descending order of the powers, and set up in long-division format.

$$(x + 1)\overline{)4x^4 + 0x^3 + x^2 + 0x - 2}$$

The method works just as well if we write the terms in order of *ascending* powers of the variable.

4. Divide the first term in the dividend $(4x^4)$ by the first term in the divisor (x). Write the result $(4x^3)$ above the dividend, in line with the term having the same power. It is the first term of the quotient.

$$\begin{array}{r} 4x^3 \\ (x + 1)\overline{)4x^4 + 0x^3 + x^2 + 0x - 2} \end{array}$$

5. Multiply the divisor by the first term of the quotient. Write the result below the dividend. Subtract it from the dividend.

$$\begin{array}{r} 4x^3 \\ (x + 1)\overline{)4x^4 + 0x^3 + x^2 + 0x - 2} \\ \underline{4x^4 + 4x^3 } \\ - 4x^3 + x^2 + 0x - 2 \end{array}$$

6. Repeat Steps 4 and 5, each time using the new dividend obtained, until the degree of the remainder is less than the degree of the divisor.

$$\begin{array}{r} 4x^3 - 4x^2 + 5x - 5 \\ (x + 1)\overline{)4x^4 + 0x^3 + x^2 + 0x - 2} \\ \underline{4x^4 + 4x^3 } \\ -4x^3 + x^2 + 0x - 2 \\ \underline{-4x^3 - 4x^2 } \\ 5x^2 + 0x - 2 \\ \underline{5x^2 + 5x } \\ - 5x - 2 \\ \underline{- 5x - 5} \\ 3 \end{array}$$

The result is written

$$\frac{4x^4 + x^2 - 2}{x + 1} = 4x^3 - 4x^2 + 5x - 5 + \frac{3}{x + 1}$$

or

$$4x^3 - 4x^2 + 5x - 5 \quad (R3)$$

to indicate a remainder of 3. ◆◆◆

Common Error	Errors are often made during the subtraction step (Step 5 in Example 70). $$\begin{array}{r} 4x^3 \\ (x + 1)\overline{)4x^4 + 0x^3 + x^2 + 0x - 2} \\ \underline{4x^4 + 4x^3 } \\ 4x^3 + x^2 + 0x - 2 \end{array}$$ No! Should be $-4x^3$.

◆◆◆ **Example 71:** Divide $x^4 - 5x^3 + 9x^2 - 7x + 2$ by $x^2 - 3x + 2$.

Solution: Using long division, we get

$$
\begin{array}{r}
x^2 - 2x + 1 \\
x^2 - 3x + 2 \,\overline{\big)\, x^4 - 5x^3 + 9x^2 - 7x + 2} \\
\underline{x^4 - 3x^3 + 2x^2} \\
-2x^3 + 7x^2 - 7x + 2 \\
\underline{-2x^3 + 6x^2 - 4x} \\
x^2 - 3x + 2 \\
\underline{x^2 - 3x + 2}
\end{array}
$$

◆◆◆

Exercise 5 ◆ Division of Algebraic Expressions

Divide the following monomials.

1. z^5 by z^3

2. $45y^3$ by $15y^2$

3. $-25x^2y^2z^3 \div 5xyz^2$

4. $20a^5b^5c \div 10abc$

5. $30cd^2f \div 15cd^2$

6. $36ax^2y \div 18ay$

7. $-18x^2yz \div 9xy$

8. $\dfrac{56y^3w^2}{7yw}$

9. $15axy^3$ by $-3ay$

10. $-18a^3x$ by $-3ay$

11. $6acdxy^2$ by $2adxy^2$

12. $12a^2x^2$ by $-3a^2x$

13. $15ay^2$ by $-3ay$

14. $\dfrac{21a^5b}{7a^4b}$

15. $-21vwz^2 \div 7vz^2$

16. $-33r^2s \div 11rs^2$

17. $35m^2nx \div 5m^2x$

18. $20x^3y^3z^3 \div 10x^3yz^3$

19. $16ab$ by $4a$

20. $21acd$ by $7c$

21. ab^2c by ac

22. $(a + b)^4$ by $(a + b)$

23. x^m by x^n

24. $-14a^2x^3y^4 \div 7axy^2$

25. $32r^2s^2q \div 8r^2sq$

26. $-18v^2x^2y \div v^2xy$

27. $24a^{2n}bc^n \div (-a^{2n}bc^n)$

28. $36a^{2n}xy^{3n} \div (-4a^nxy^{2n})$

29. $25xyz^2 \div (-5x^2y^2z)$

30. $-28y^2z^{2m} \div 4y^2z^3x$

31. $-30n^2x^2 \div 6m^2x^3$

32. $28x^2y^2z^2 \div 7xyz$

33. $6abc$ by $2c$

34. ax^3 by ax^2

35. $3mx^6$ by mx

36. $210c^3b$ by $7cd$

37. $42xy$ by xy^2

38. $-21ac$ by $-7a$

39. $-12xy$ by $3y$

40. $72abc$ by $-8c$

Divide the polynomial by the monomial.

41. $2a^3 - a^2$ by a

42. $42a^5 - 6a^2$ by $6a$

43. $21x^4 - 3x^2$ by $3x^2$

44. $35m^4 - 7p^2$ by 7

45. $27x^6 - 45x^4$ by $9x^2$

46. $24x^6 - 8x^3$ by $-8x^3$

47. $34x^3 - 51x^2$ by $17x$

48. $5x^5 - 10x^3$ by $-5x^3$

49. $-3a^2 - 6ac$ by $-3a$

50. $-5x^3 + x^2y$ by $-x^2$

51. $ax^2y - 2xy^2$ by xy

52. $3xy^2 - 3x^2y$ by xy

53. $4x^3y^2 + 2x^2y^3$ by $2x^2y^2$

54. $3a^2b^2 - 6ab^3$ by $3ab$

55. $abc^2 - a^2b^2c$ by $-abc$

56. $9x^2y^2z + 3xyz^2$ by $3xyz$

57. $x^2y^2 - x^3y - xy^3$ by xy

58. $a^3 - a^2b - ab^2$ by $-a$

59. $a^2b - ab + ab^2$ by $-ab$

60. $xy - x^2y^2 + x^3y^3$ by $-xy$

61. $a^2 - 3ab + ac^2$ by a

62. $x^2y - xy^2 + x^2y^3$ by xy

63. $x^2 - 2xy + y^2$ by x

64. $z^2 - 3xz + 3z^3$ by z

65. $m^2n + 2mn - 3m^2$ by mn
66. $c^2d - 3cd^2 + 4d^3$ by cd

Divide the polynomial by the binomial.

67. $x^2 + 15x + 56$ by $x + 7$
68. $x^2 - 15x + 56$ by $x - 7$
69. $2a^2 + 11a + 5$ by $2a + 1$
70. $6a^2 - 7a - 3$ by $2a - 3$
71. $27a^3 - 8b^3$ by $3a - 2b$
72. $4a^2 + 23a + 15$ by $4a + 3$
73. $x^2 - 4x + 3$ by $x + 2$
74. $2 + 4x - x^2$ by $4 - x$
75. $4 + 2x - 5x^2$ by $3 - x$
76. $2x^2 - 5x + 4$ by $x + 1$

Divide each polynomial by the given trinomial.

77. $a^2 - 2ab + b^2 - c^2$ by $a - b - c$
78. $a^2 + 2ab + b^2 - c^2$ by $a + b + c$
79. $x^2 - y^2 + 2yz - z^2$ by $x - y + z$
80. $c^4 + 2c^2 - c + 2$ by $c^2 - c + 1$
81. $x^2 - 4y^2 - 4yz - z^2$ by $x + 2y + z$
82. $x^5 + 37x^2 - 70x + 50$ by $x^2 - 2x + 10$

Computer

83. Use a computer algebra system (CAS) to repeat any of the problems in this exercise set. Compare your results with what you got by hand.

••• CHAPTER 2 REVIEW PROBLEMS ••••••••••••••••••••••••••••••••••

1. Multiply: $(b^4 + b^2x^3 + x^4)(b^2 - x^2)$
2. Square: $(x + y - 2)$
3. Evaluate: $(7.28 \times 10^4)^3$
4. Square: $(xy + 5)$
5. Multiply: $(3x - m)(x^2 + m^2)(3x - m)$
6. Divide: $-x^6 - 2x^5 - x^4$ by $-x^4$
7. Cube: $(2x + 1)$
8. Simplify: $7x - \{-6x - [-5x - (-4x - 3x) - 2]\}$
9. Multiply: $3ax^2$ by $2ax^3$
10. Simplify: $\left(\dfrac{3a^2}{2b^3}\right)^3$
11. Divide: $a^2x - abx - acx$ by ax
12. Divide: $3x^5y^3 - 3x^4y^3 - 3x^2y^4$ by $3x^3y^2$
13. Square: $(4a - 3b)$
14. Multiply: $(x^3 - xy + y^2)(x + y)$
15. Multiply: $(xy - 2)(xy - 4)$
16. Divide: $x^{m+1} + x^{m+2} + x^{m+3} + x^{m+4}$ by x^4
17. Square: $(2a - 3b)$
18. Multiply: $(a^2 - 3a + 8)(a + 3)$
19. Divide: $a^3b^2 - a^2b^5 - a^4b^2$ by a^2b

20. Multiply: $(2x - 5)(x + 2)$

21. Multiply: $(2m - c)(2m + c)(4m^2 + c^2)$

22. Simplify: $\left(\dfrac{8x^5y^{-2}}{4x^3y^{-3}}\right)^3$

23. Divide: $2a^6$ by a^4

24. Multiply: $(a^2 + a^2y + ay^2 + y^3)(a - y)$

25. Simplify: $y - 3[y - 2(4 - y)]$

26. Divide: $-a^7$ by a^5

27. Multiply: $(2x^2 + xy - 2y^2)(3x + 3y)$

28. Divide: $x^4 - \frac{1}{2}x^3 - \frac{1}{3}x^2 - 2x - 1$ by $2x$

29. Simplify: $-2[w - 3(2w - 1)] + 3w$

30. Multiply: $(a^2 + b)(a + b^2)$

31. Multiply: $(a^4 - 2a^3c + 4a^2c^2 - 8ac^3 + 16c^4)(a + 2c)$

32. Divide: $16x^3$ by $4x$

33. Divide: $(a - c)^m$ by $(a - c)^2$

34. Divide: $7 - 8c^2 + 5c^3 + 8c$ by $5c - 3$

35. Divide: $-x^2y - xy^2$ by $-xy$

36. Combine: $(-2x^2 - x + 6) - (7x^2 - 2x + 4) + (x^2 - 3)$

37. Cube: $(b - 3)$

38. Simplify: $(2x^2y^3z^{-1})^3$

39. Divide: $2x^{-2}y^3$ by $4x^{-4}y^6$

40. Simplify: $-\{-[-(a - 3) - a]\} + 2a$

41. Multiply: $(x - 2)(x + 4)$

42. Square: $(x^2 + 2)$

43. Divide: $a^2b^2 - 2ab - 3ab^3$ by ab

44. Divide: $3a^3c^3 + 3a^2c - 3ac^2$ by $3ac$

45. Divide: $6a^3x^2 - 15a^4x^2 + 30a^3x^3$ by $-3a^3x^2$

46. Divide: $20x^2y^4 - 14xy^3 + 8x^2y^2$ by $2x^2y^2$

47. Evaluate: $(2.83 \times 10^3)^2$

48. Simplify: $(x - 3) - [x - (2x + 3) + 4]$

49. Multiply: $2xy^3$ by $5x^2y$

50. Divide: $27xy^5z^2$ by $3x^2yz^2$

51. Multiply: $(x - 1)(x^2 + 4x)$

52. Square: $(z^2 - 3)$

53. Evaluate: $(1.33 \times 10^4)^2$

54. Simplify: $(y + 1) - [y(y + 1) + (3y - 1) + 5]$

55. Multiply: $2ab^2$ by $3a^2b$

56. Divide: $64ab^4c^3$ by $8a^2bc^3$

57. Divide: $x^8 + x^4 + 1$ by $x^4 - x$

58. Divide: $1 - a^3b^3$ by $1 - ab$

59. To make 750 pounds of a new alloy, we take x pounds of alloy A, which contains 85% copper, and for the remainder use alloy B, which contains 72% copper. The pounds of copper in the final batch will be

$$0.85x + 0.72(750 - x)$$

Simplify this expression.

60. The area of a circle of radius r is πr^2. If the radius is tripled, the area will be

$$\pi(3r)^2$$

Simplify this expression.

61. A freely falling body, starting from rest, falls a distance of $16.1t^2$ feet in t seconds. In half that time it will fall

$$16.1\left(\frac{t}{2}\right)^2 \text{ ft}$$

Simplify this expression.

62. The power in a resistor of resistance R which has a voltage V across it is V^2/R. If the voltage is doubled, the power will be

$$\frac{(2V)^2}{R}$$

Simplify this expression.

Writing

63. Suppose your friend is sick and missed the introduction to algebra and is still out of class. Write a note to your friend explaining in your own words what you think algebra is and how it is related to the arithmetic that you both just finished studying.

64. In simplifying a fraction, we are careful not to change its value. If that's the case, and the new fraction is "the same" as the old, then why bother? Write a few paragraphs explaining why you think it's valuable to simplify fractions (and other expressions as well) or whether there is some advantage to just leaving them alone.

65. Your friend refuses to learn the FOIL rule. "*I want to learn math, not memorize a bunch of tricks*!" he declares. What do you think? Write a paragraph or so giving your opinion on the value or harm in learning devices such as the FOIL rule.

66. In the following chapter we will solve simple equations, something you have probably done before. Without peeking ahead, write down in your own words whatever you remember about solving equations. You may give a list of steps or a description in paragraph form.

Team Projects

67. Write an hour exam that will test for the most important skills covered in this chapter. Be sure that it is not too long and hard, or not too short and easy. Then work the test and produce an answer key, to be handed in with the test.

68. Fig. 2–3 shows a geometric representation of $(a + b)^2$, the square of the binomial $(a + b)$. Use the figure to evaluate $(a + b)^2$. Then make a similar sketch to evaluate each of the following:

   (a) $(2a + b)^2$
   (b) $(a + 2)(a + 3)$
   (c) $(a + b)(c + d)$
   (d) $(a + b + 1)^2$

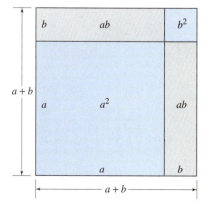

FIGURE 2–3 A geometric representation of $(a + b)^2$.

3

Simple Equations and Word Problems

•••• **OBJECTIVES** ••

When you have completed this chapter, you should be able to:

- Solve simple linear equations.
- Check an apparent solution to a linear equation.
- Solve simple fractional equations.
- Write an algebraic expression to describe a given verbal statement.
- Set up and solve word problems that lead to linear equations.
- Solve number problems, financial problems, mixture problems, and statics problems.

•••

When two mathematical expressions are set equal to each other, we get an *equation*. Much of our work in technical mathematics is devoted to *solving equations*. We start in this chapter with the simplest types, and in later chapters we cover more difficult ones (quadratic equations, exponential equations, trigonometric equations, etc.).

Why is it so important to solve equations? The main reason is that equations are used to describe or *model* the way certain things happen in the world. Solving an equation often tells us something important that we would not otherwise know. For example, the note produced by a guitar string is not a whim of nature but can be predicted (solved for) when you know the length, the mass, and the tension in the string, and you know the *equation* relating the pitch, length, mass, and tension.

Thousands of equations exist that link together the various quantities in the physical world, in chemistry, in finance, in factories, and so on—and their number is still increasing. To be able to solve and to manipulate such equations is essential for anyone who has to deal with these quantities on the job.

3–1 Solving First-Degree Equations

Equations

An equation has two *sides*, or members, and an *equal sign*.

$$3x^2 - 4x = 2x + 5$$

left side   equal sign   right side

Conditional Equations

A *conditional equation* is one whose sides are equal only for certain values of the unknown.

♦♦♦ **Example 1:** The equation

$$x - 5 = 0$$

is a conditional equation because the sides are equal only when $x = 5$. When we say "equation," we will mean "conditional equation." Other equations that are true for any value of the unknown, such as $x(x + 2) = x^2 + 2x$, are called *identities*.

♦♦♦

> The symbol $\equiv$ is used for identities. We would write $x(x + 2) \equiv x^2 + 2x$.

The Solution

The *solution* to an equation is that value (or those values) of the unknown that makes the sides equal.

♦♦♦ **Example 2:** The solution to the equation

$$x - 5 = 0$$

is

$$x = 5$$

The value $x = 5$ is also called the *root* of the equation. We also say that it *satisfies* the equation. ♦♦♦

First-Degree Equations

A *first-degree* equation (also called a *linear* equation) is one in which the terms containing the unknown are all of first degree.

♦♦♦ **Example 3:** The equations $2x + 3 = 9 - 4x$, $3x/2 = 6x + 3$, and $3x + 5y - 6z = 0$ are all of first degree. The first and second equations have one variable, but the third equation has more than one variable. ♦♦♦

Solving an Equation

To solve an equation, we *perform the same mathematical operation on each side of the equation*. The object is to get the unknown standing alone on one side of the equation.

♦♦♦ **Example 4:** Solve $3x = 8 + 2x$.

Solution: Subtracting $2x$ from both sides, we obtain

$$3x - 2x = 8 + 2x - 2x$$

Combining like terms yields

$$x = 8$$

◆◆◆

Checking

To check an apparent solution, substitute it back into the original equation.

Get into the habit of checking your work, for errors creep in everywhere.

◆◆◆ **Example 5:** Checking the apparent solution to Example 4, we substitute 8 for x in the original equation.

$$3x = 8 + 2x$$
$$3(8) \stackrel{?}{=} 8 + 2(8)$$
$$24 = 8 + 16 \quad \text{(checks)}$$

◆◆◆

Common Error	Check your solution only in the *original* equation. Later versions may already contain errors.

Checking by Calculator

To check the solution by calculator, enter the left side (L) of the original equation, the right side (R), and the apparent solution. Thus for Example 4, we would enter

$$\text{L} = 3x$$
$$\text{R} = 8 + 2x$$
$$x = 8$$

Then recall the value for L and for R. The calculator should return the same value (24 in this example) for L and for R.

◆◆◆ **Example 6:** Solve the equation $3x - 5 = x + 1$.

Solution: We first subtract x from both sides and then add 5 to both sides.

$$
\begin{array}{rcl}
3x - 5 & = & x + 1 \\
\underline{-x + 5} & & \underline{-x + 5} \\
2x & = & 6
\end{array}
$$

Dividing both sides by 2, we obtain

$$x = 3$$

Check: Substituting into the original equation yields

$$3(3) - 5 \stackrel{?}{=} 3 + 1$$
$$9 - 5 = 4 \quad \text{(checks)}$$

◆◆◆

Symbols of Grouping

When the equation contains symbols of grouping, remove them early in the solution.

◆◆◆ **Example 7:** Solve the equation $3(3x + 1) - 6 = 5(x - 2) + 15$.

Solution: First, remove the parentheses:

$$9x + 3 - 6 = 5x - 10 + 15$$

Combine like terms:
$$9x - 3 = 5x + 5$$

Add $-5x + 3$ to both sides:
$$
\begin{array}{rcl}
\underline{-5x + 3} & & \underline{-5x + 3} \\
4x & = & 8
\end{array}
$$

Divide by 4:
$$x = 2$$

Check:
$$3(6 + 1) - 6 \ \overset{?}{=} \ 5(2 - 2) + 15$$
$$21 - 6 \ \overset{?}{=} \ 0 + 15$$
$$15 \ = \ 15 \quad \text{(checks)} \qquad \blacklozenge\blacklozenge\blacklozenge$$

The mathematical operations that you perform must be done *to both sides* of the equation in order to preserve the equality.

The mathematical operations that you perform on both sides of an equation must be done to each side as a whole—*not term by term.*

Fractional Equations

A *fractional equation* is one that contains one or more fractions. When an equation contains a single fraction, the fraction can be eliminated by *multiplying both sides by the denominator of the fraction.*

◆◆◆ **Example 8:** Solve:

$$\frac{x}{3} - 2 = 5$$

Solution: Multiplying both sides by 3 gives us

$$3\left(\frac{x}{3} - 2\right) \ = \ 3(5)$$

$$3\left(\frac{x}{3}\right) - 3(2) \ = \ 3(5)$$

Add 6 to both sides:
$$\begin{array}{rcr} x - 6 &=& 15 \\ +6 && +\ 6 \\ \hline x &=& 21 \end{array}$$

Check:

$$\frac{21}{3} - 2 \ \overset{?}{=} \ 5$$

$$7 - 2 \ = \ 5 \quad \text{(checks)} \qquad \blacklozenge\blacklozenge\blacklozenge$$

When there are two or more fractions, multiplying both sides by the product of the denominators (called a *common* denominator) will clear the fractions.

In Chapter 9 we treat fractional equations in detail. There we will see how to find the *least* common denominator.

◆◆◆ **Example 9:** Solve:

$$\frac{x}{2} + 3 = \frac{x}{3}$$

Solution: Multiplying by 2(3), or 6, we have

$$6\left(\frac{x}{2} + 3\right) = 6\left(\frac{x}{3}\right)$$

$$6\left(\frac{x}{2}\right) + 6(3) = 6\left(\frac{x}{3}\right)$$

$$3x + 18 = 2x$$

We now subtract $2x$ from both sides, and also subtract 18 from both sides.

$$3x - 2x = -18$$
$$x = -18$$

Check:

$$\frac{-18}{2} + 3 \stackrel{?}{=} \frac{-18}{3}$$

$$-9 + 3 \stackrel{?}{=} -6$$
$$-6 = -6 \quad \text{(checks)} \qquad \blacklozenge\blacklozenge\blacklozenge$$

Strategy

While solving an equation, you should keep in mind the objective of *getting* x *by itself on one side of the equal sign*, *with no* x *on the other side*. Use any valid operation toward this end.

Equations can be of so many different forms that we cannot give a step-by-step procedure for their solution, but the following tips should help:

1. Eliminate fractions by multiplying both sides by the common denominator of both sides.
2. Remove any parentheses by performing the indicated multiplication.
3. All terms containing x should be moved to one side of the equation, and all other terms moved to the other side.
4. Like terms on the same side of the equation should be combined at any stage of the solution.
5. Any coefficient of x may be removed by dividing both sides by that coefficient.
6. Check the answer in the original equation.

The following example shows the use of these steps.

◆◆◆ **Example 10:** Solve:

$$\frac{3x - 5}{2} = \frac{2(x - 1)}{3} + 4$$

Solution:

1. We eliminate the fractions by multiplying by 6.

$$6\left(\frac{3x - 5}{2}\right) = 6\left[\frac{2(x - 1)}{3}\right] + 6(4)$$

$$3(3x - 5) = 4(x - 1) + 6(4)$$

2. Then we remove parentheses and combine like terms.

$$9x - 15 = 4x - 4 + 24$$
$$9x - 15 = 4x + 20$$

3. We get all x terms on one side by adding 15 and subtracting $4x$ from both sides.

$$9x - 4x = 20 + 15$$

4. Combining like terms gives us

$$5x = 35$$

5. Dividing by the coefficient of x yields

$$x = 7$$

Check:

$$\frac{3(7) - 5}{2} \stackrel{?}{=} \frac{2(7 - 1)}{3} + 4$$

$$\frac{21 - 5}{2} \stackrel{?}{=} \frac{12}{3} + 4$$

$$\frac{16}{2} \stackrel{?}{=} 4 + 4$$

$$8 = 8 \quad \text{(checks)} \qquad \qquad \blacklozenge\blacklozenge\blacklozenge$$

♦♦♦ **Example 11:** Solve:

$$3(x + 2)(2 - x) = 3(x - 2)(x - 3) + 2x(4 - 3x)$$

Solution: We remove parentheses in two steps. First,

$$3(2x - x^2 + 4 - 2x) = 3(x^2 - 5x + 6) + 8x - 6x^2$$

Then we remove the remaining parentheses.

$$-3x^2 + 12 = 3x^2 - 15x + 18 + 8x - 6x^2$$

Next we combine like terms.

$$-3x^2 + 12 = -3x^2 - 7x + 18$$

Adding $3x^2$ and subtracting 18 from both sides gives

$$12 - 18 = -7x$$
$$-6 = -7x$$

Dividing both sides by -7 and switching sides, we get

$$x = \frac{6}{7} \qquad \qquad \blacklozenge\blacklozenge\blacklozenge$$

Other Variables

Up to now we have used only the familiar x as our variable. Of course, any letter can represent the unknown quantity. When we say to "solve" an equation, we mean to isolate the letter quantity, if only one is present. If there is more than one letter, we must be told which to solve for.

♦♦♦ **Example 12:** Solve the equation

$$4(p - 3) = 3(p + 4) - 5$$

Solution: Removing parentheses and combining like terms, we have

$$4p - 12 = 3p + 12 - 5$$
$$4p - 12 = 3p + 7$$

Adding 12 to and subtracting $3p$ from both sides and combining like terms gives

$$4p - 3p = 7 + 12$$
$$p = 19 \qquad \qquad \blacklozenge\blacklozenge\blacklozenge$$

Equations with Approximate Numbers

All of our equations so far have had coefficients and constants that were integers, and we assumed them to be exact numbers. But in technology we must often solve equations containing approximate numbers. The method of solution is no different than with exact numbers. Just be sure to round following the rules given in Chapter 1.

◆◆◆ **Example 13:** Solve for x:

$$5.72(x - 1.15) = 3.22x - 7.73$$

Solution: Removing parentheses gives

$$5.72x - 6.58 = 3.22x - 7.73$$

Next, collect terms.

$$5.72x - 3.22x = -7.73 + 6.58$$
$$2.50x = -1.15$$

Dividing both sides by 2.50, we get

$$x = \frac{-1.15}{2.50} = -0.460$$

◆◆◆

Literal Equations

A *literal equation* is one in which some or all of the constants are represented by letters.

◆◆◆ **Example 14:** The following is a literal equation:

$$3x + b = 5$$

◆◆◆

To solve a literal equation for a given quantity means to isolate that quantity on one side of the equal sign. The other side of the equation will, of course, contain the other letter quantities. We do this by following the same procedures as for numerical equations.

◆◆◆ **Example 15:** Solve the equation $3x + b = 5$ for x.

Solution: Subtracting b from both sides gives

$$3x = 5 - b$$

Dividing by 3, we obtain

$$x = \frac{5 - b}{3}$$

◆◆◆

In this chapter we will solve only the simplest kinds of literal equations. Our main treatment will be in Sec. 9–7.

Exercise 1 ◆ Solving First-Degree Equations

Solve and check each equation.

Treat any integers in these equations as exact numbers. Leave your answers in fractional, rather than decimal, form. When the equation has decimal coefficients, keep as many significant digits in your answer as contained in those coefficients.

1. $x - 5 = 28$ **2.** $4x + 2 = 18$
3. $8 - 3x = 7x$ **4.** $9x + 7 = 25$
5. $9x - 2x = 3x + 2$ **6.** $x + 8 = 12 - 3x$
7. $x + 4 = 5x + 2$ **8.** $3y - 2 = 2y$
9. $3x = x + 8$ **10.** $3x = 2x + 5$
11. $3x + 4 = x + 10$ **12.** $y - 4 = 7 + 23y$
13. $2.80 - 1.30y = 4.60$ **14.** $3.75x + 7.24 = 5.82x$
15. $4x + 6 = x + 9$ **16.** $7x - 19 = 5x + 7$

17. $p - 7 = -3 + 9p$
18. $-3x + 7 = -15 + 2x$
19. $4x - 11 = 2x - 7$
20. $3x - 8 = 12x - 7$
21. $8x + 7 = 4x + 27$
22. $3x + 10 = x + 20$

23. $\frac{1}{2}x = 14$
24. $\frac{x}{5} = 3x - 25$

25. $5x + 22 - 2x = 31$
26. $4x + 20 - 6 = 34$

27. $\frac{n}{3} + 5 = 2n$
28. $5x - \frac{2x}{5} = 3$

29. $2x + 5(x - 4) = 6 + 3(2x + 3)$
30. $6x - 5(x - 2) = 12 - 3(x - 4)$
31. $4(3x + 2) - 4x = 5(2 - 3x) - 7$
32. $6 - 3(2x + 4) - 2x = 7x + 4(5 - 2x) - 8$
33. $3(6 - x) + 2(x - 3) = 5 + 2(3x + 1) - x$
34. $3x - 2 + x(3 - x) = (x - 3)(x + 2) - x(2x + 1)$
35. $(2x - 5)(x + 3) - 3x = 8 - x + (3 - x)(5 - 2x)$
36. $x + (1 + x)(3x + 4) = (2x + 3)(2x - 1) - x(x - 2)$
37. $3 - 6(x - 1) = 9 - 2(1 + 3x) + 2x$
38. $7x + 6(2 - x) + 3 = 4 + 3(6 + x)$
39. $3 - (4 + 3x)(2x + 1) = 6x - (3x - 2)(x + 1) - (6 - 3x)(1 - x)$
40. $6.10 - (x + 2.30)(3.20x - 4.30) - 5.20x$
     $= 3.30x - x(3.10 + 3.20x) - 4.70$
41. $7r - 2r(2r - 3) - 2 = 2r^2 - (r - 2)(3 + 6r) - 8$
42. $2w - 3(w - 6)(2w - 2) + 6 = 3w - (2w - 1)(3w + 2) + 6$
43. $8.27x - 2.22(2.57 + x)(3.35x - 6.64)$
     $= 3.24 - 3.35(2.22x + 5.64)(x - 1.24) - 2.64(2.11x - 3.85)$
44. $1.30x - (x - 2.30)(3.10x - 5.40) - 2.50$
     $= 4.60 - (3.10x - 1.20)(x - 5.30) - 4.30x$

Simple Literal Equations

Solve for x.

45. $ax + 4 = 7$
46. $6 + bx = b - 3$

47. $c(x - 1) = 5$
48. $ax + 4 = b - 3$

49. $c - bx = a - 3b$

Computer

50. If you have a computer algebra system (CAS), such as *Derive*, *Maple*, or *Mathematica*, it can be used to solve quickly any of the equations in this exercise set. Each of these systems uses the *Solve* command to solve an equation. Try some problems, and compare the answers with those you get by hand computation.

3–2   Solving Word Problems

Problems in Verbal Form

We have already done some simple word problems in earlier chapters, and we will have more of them with each new topic, but their main treatment will be

Math skills are important, but so are reading skills. Most students find word problems difficult, but if you have an *unusual* amount of trouble with them, you should seek out a reading teacher and get help.

here. After looking at these word problems you may ask, "When will I ever have to solve problems like that on the job?"

Maybe never. But you will most likely have to deal with technical material in written form: instruction manuals, specifications, contracts, building codes, handbooks, and so on. This chapter will aid you in coping with precise technical documents, where a mistaken meaning of a word could cost dollars, your job, or even lives.

Some tips on how to approach any word problem are given in the following paragraphs.

Picture the Problem

Try to *visualize* the situation described in the problem. Form a picture in your mind. *Draw a diagram* showing as much of the given information as possible, including the unknown.

Understand the Words

Look up the meanings of unfamiliar words in a dictionary, handbook, or textbook.

◆◆◆ **Example 16:** A certain problem contains the following statement:

A simply supported beam carries a concentrated load midway between its center of gravity and one end.

To understand this statement, you must know what a "simply supported beam" is, and what "concentrated load" and "center of gravity" mean. ◆◆◆

Locate the words and expressions standing for mathematical operations, such as

. . . *the difference of* . . .

. . . *is equal to* . . .

. . . *the quotient of* . . .

◆◆◆ **Example 17:**

(a) "The difference of a and b" means $a - b$ (or $b - a$)
(b) "Four less than x" means $x - 4$
(c) "Three greater than c" means $c + 3$
(d) "The sum of x and 4 is equal to three times x," in symbols, is

$$x + 4 = 3x$$

(e) "Three consecutive integers, starting with x," are

$$x, \qquad x + 1, \qquad x + 2$$ ◆◆◆

Many problems will contain a *rate*, either as the unknown or as one of the given quantities. The word *per* is your tipoff. Look for "miles *per* hour," "dollars *per* pound," "pounds *per* gallon," "dollars *per* mile," and the like. The word *per* also suggests *division*. Miles per gallon, for example, means miles traveled *divided* by gallons consumed.

Identify the Unknown

Be sure that you know exactly *what is to be found* in a particular problem. Find the sentence that ends with a question mark, or the sentence that starts with "Find . . ." or "Calculate . . ." or "What is . . ." or "How much . . ." or similar phrases.

Label the unknown with a statement such as "Let x = rate of the train, mi/h." Be sure to include *units of measure* when defining the unknown.

◆◆◆ **Example 18:** Fifty-five liters of a mixture that is 24% alcohol by volume is mixed with 74 liters of a mixture that is 76% alcohol. Find the amount of alcohol in the final mixture.

Solution: You may be tempted to label the unknown in the following ways:

Let x = alcohol	This is too vague. Are we speaking of volume of alcohol, or weight, or some other property?
Let x = volume of alcohol	The units are missing.
Let x = liters of alcohol	Do we mean liters of alcohol in one of the *initial* mixtures or in the *final* mixture?
Let x = number of liters of alcohol in final mixture	Good.

We will do complete solutions of this type of problem in Sec. 3–4.   ◆◆◆

This step of labeling the unknown annoys many students; they see it as a nit-picking thing that teachers make them do just for show, when they would rather start solving the problem. But realize that this step *is* the start of the problem. It is the crucial first step and *should never be omitted.*

Define Other Unknowns

If there is more than one unknown, you will often be able to define the additional unknowns *in terms of the original unknown*, rather than introducing another symbol.

◆◆◆ **Example 19:** A word problem contains the following statement: "Find two numbers whose sum is 80 and, ..."

Solution: Here we have two unknowns. We can label them with separate symbols.

Let x = first number
Let y = second number

But another way is to label the second unknown in terms of the first, thus avoiding the use of a second variable. Since the sum of the two numbers x and y is 80,

$$x + y = 80$$

from which

$$y = 80 - x$$

This enables us to label both unknowns in terms of the single variable x.

Let x = first number
Let $80 - x$ = second number   ◆◆◆

Sometimes a careful reading of the word problem is needed to *identify* the unknowns before you can define them in terms of one of the variables.

◆◆◆ **Example 20:** A word problem contains the following statement: "A person invests part of his $10,000 savings in a bank, at 6%, and the remainder in a certificate of deposit, at 8%. . . ." We are then given more information and are asked to find the amount invested at each rate.

Solution: Since the investment is in two parts, both of which are unknown, we can say

Let x = amount invested at 6%

and

$10,000 - x$ = amount invested at 8%

We'll finish this problem in Sec. 3–3.

But isn't it equally correct to say

$$\text{Let } x = \text{amount invested at 8\%}$$

and

$$10{,}000 - x = \text{amount invested at 6\%?}$$

Yes. You can define the variables either way, as long as you are consistent and *do not change the definition partway through the problem.*   ◆◆◆

Estimate the Answer

It is a good idea to *estimate* or *guess* the answer before solving the problem, so that you will have some idea whether the answer you finally get is reasonable.

The kinds of problems you may encounter are so diverse that it is not easy to give general rules for estimating. However, we usually make *simplifying assumptions*, or try to *bracket the answer.*

To make a simplifying assumption means to solve a problem simpler than the given one and then use its answer as an estimate for the given problem.

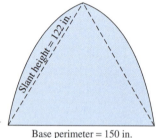

Base perimeter = 150 in.

FIGURE 3–1   Aircraft nose cone.

◆◆◆ **Example 21:** To estimate the surface area of the aircraft nose cone in Fig. 3–1, we could assume that its shape is a right circular cone (shown by dashed lines) rather than the complex shape given. This would enable us to estimate the area using the following simple formula (Eq. 131):

$$\text{area} = \tfrac{1}{2}(\text{perimeter of base}) \times (\text{slant height})$$

For the nose cone,

$$\text{area} \approx \tfrac{1}{2}(150)(122) = 9150 \text{ in.}^2$$

We should be suspicious if our "exact" calculation gives an answer much different from this.   ◆◆◆

To "bracket" an answer means to find two numbers between which your answer must lie. Don't worry that your two values are far apart. This method will catch more errors than you may suppose.

◆◆◆ **Example 22:** It is clear that the length of the bridge cable of Fig. 3–2 must be greater than the straight line distances $AB + BC$, but less than the straight line distances $AD + DE + EC$. So the answer must lie between 894 ft and 1200 ft.

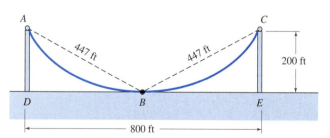

FIGURE 3–2   Suspension bridge.   ◆◆◆

Write and Solve the Equation

The unknown quantity must now be related to the given quantities by means of an *equation.* Where does one get an equation? Either the equation is given verbally, right in the problem statement, or else the relationship between the quantities in the problem is one you are expected to know (or to be able to find). These could be mathematical relationships, or they could be from any branch of technology. For example:

For good problem-solving techniques, read *How to Solve It,* by G. Polya.

> *How is the area of a circle related to its radius?*
> *How is the current in a circuit related to the circuit resistance?*
> *How is the deflection of a beam related to its depth?*
> *How is the distance traveled by a car related to its speed?*

and so on. The relationships you will need for the problems in this book can be found in Appendix A, "Summary of Facts and Formulas." Of course, each branch of technology has many more formulas than can be shown there. Problems that need some of these formulas are treated later in this chapter. Sometimes, as in the number problems we will solve later in this section, the equation is given verbally in the statement of the problem.

Check Your Answer

There are several things we can do to check an answer.

1. Does the answer look *reasonable?* If the problem is a complicated technical one, you may not yet have the experience to know whether an answer is reasonable, but in many cases you can spot a strange answer.

◆◆◆ **Example 23:** Does your result show an automobile going 600 miles per hour? Did a person's age calculate to be 150 years? Did the depth of the bridge girder come out to be 2 millimeters?　　◆◆◆

Don't just write down the answer and run. Look at it hard to see if it makes sense.

2. Compare your answer with your initial estimate or guess. Do they agree within reason? If you have found two bracketing numbers within which the answer *must* fall, then you must recheck any answers outside that range.

◆◆◆ **Example 24:** If a calculation showed the length of the bridge cable in Fig. 3–2 to be 1250 ft, you could say with certainty that there is an error, because we saw that the length cannot exceed 1200 ft.　　◆◆◆

3. Check your answer in the original problem statement. Be sure that the answer(s) meets all of the conditions in the problem statement.

◆◆◆ **Example 25:** A problem asked for three consecutive odd integers whose sum is 39. Four students gave answers of (a) 12, 13, 14; (b) 13, 15, 17; (c) 9, 11, 19; and (d) 11, 13, 15. Determine which, if any, are correct, without actually solving the problem.

Solution:

(a) 12, 13, 14 are consecutive integers whose sum is 39. However, they are not all odd.
(b) 13, 15, 17 are consecutive odd integers, but their sum is not 39.
(c) 9, 11, 19 are odd integers whose sum is 39, but they are not consecutive.
(d) 11, 13, 15 are consecutive odd integers whose sum is 39, and thus (a) is the correct answer.　　◆◆◆

Number Problems

Number problems, although of little practical value, will give you practice at picking out the relationship between the quantities in a problem.

◆◆◆ **Example 26:** If twice a certain number is increased by 9, the result is equal to 2 less than triple the number. Find the number.

Solution: Let x = the number. From the problem statement,

$$2x + 9 = 3x - 2$$

Solving for x, we obtain

$$9 + 2 = 3x - 2x$$
$$x = 11$$

Check: We check our answer in the original problem statement. If the number is 11, then twice the number is 22. Thus twice the number increased by 9 is 31. Further, triple the number is 33. Two less than triple the number is 31. So our answer checks.

◆◆◆

Common Error	Checking an answer by substituting in your *equation* is not good enough. The equation itself may be wrong. Check your answer with the problem statement.

Often the solution of a problem involves finding *more than one quantity.*

◆◆◆ **Example 27:** Find three consecutive odd integers whose sum is 39.

Solution: We first define our unknowns. Since any odd integer is 2 greater than the preceding odd integer, we can say

$$\text{Let } x = \text{first odd integer}$$

Then

$$x + 2 = \text{second consecutive odd integer}$$

and

$$x + 4 = \text{third consecutive odd integer}$$

Since the problem states that the sum of the three integers is 39, we write

$$x + (x + 2) + (x + 4) = 39$$

Solving for x gives

$$3x + 6 = 39$$
$$3x = 33$$
$$x = 11$$

But don't stop here—we are not finished. We're looking for *three* numbers. So

$$x + 2 = 13 \quad \text{and} \quad x + 4 = 15$$

Our three consecutive odd integers are then 11, 13, and 15.

Check:

$$11 + 13 + 15 = 39 \quad \text{(as required)}$$

◆◆◆

Sometimes there may seem to be too many unknowns to solve a problem. If so, carefully choose one variable to call x, so that the other unknowns can be expressed in terms of x. Then use the remaining information to write your equation.

◆◆◆ **Example 28:** In a group of 102 employees, there are three times as many on the day shift as on the night shift, and two more on the swing shift than on the night shift. How many are on each shift?

Estimate: Let's assume that the night and swing shifts each have the *same* number N of employees. Then the day shift would have $3N$ employees, making $5N$ in all. So N would be $102 \div 5$, or 20, rounded. So we might expect about 20 on the night shift, a few more than 20 on the swing shift, and about 60 on the day shift.

Solution: Since the numbers of employees on the day and swing shifts are given in terms of the number on the night shift, we write

$$\text{Let } x = \text{number of employees on the night shift}$$

Then

$$3x = \text{number on the day shift}$$

and

$$x + 2 = \text{number on the swing shift}$$

Since we are told that there are 102 employees in all,

$$x + 3x + (x + 2) = 102$$

from which

$$
\begin{aligned}
5x &= 100 \\
x &= 20 \quad \text{on the night shift} \\
3x &= 60 \quad \text{on the day shift} \\
x + 2 &= \underline{22} \quad \text{on the swing shift} \\
& 102 \quad \text{total}
\end{aligned}
$$

These numbers agree well with our estimate.   ◆◆◆

The usual steps followed in solving a word problem are summarized here.

---

1. **Study the problem.** Look up unfamiliar words. Make a sketch. Try to visualize the situation in your mind.
2. **Estimate the answer.** Make simplifying assumptions, or try to bracket the answer.
3. **Identify the unknown(s).** Give it a symbol, such as "Let $x = \ldots$." If there is more than one unknown, try to label the others in terms of the first. Include units.
4. **Write and solve an equation.** Look for a relationship between the unknown and the known quantities that will lead to an equation. Write and solve that equation for the unknown. Include units in your answer. If there is a second unknown, be sure to find it also.
5. **Check your answer.** See if the answer looks reasonable. See if it agrees with your estimate. Be sure to do a numerical check in the verbal statement itself.

Exercise 2 ◆ Solving Word Problems

Verbal Statements

Rewrite each expression as one or more algebraic expressions.

1. Eight more than 5 times a certain number.
2. Four consecutive integers.
3. Two numbers whose sum is 83.
4. Two numbers whose difference is 17.
5. The amounts of antifreeze and of water in 6 gal of an antifreeze–water solution.

6. A fraction whose denominator is 7 more than 5 times its numerator.

7. The angles in a triangle, if one angle is half another.

8. The number of gallons of antifreeze in a radiator containing x gallons of a mixture that is 9% antifreeze.

9. The distance traveled in x hours by a car going 82 km/h.

10. If x equals the length of a rectangle, write an expression for the width of the rectangle if (a) the perimeter is 58 and (b) the area is 200.

Number Problems

11. Nine more than 4 times a certain number is 29. Find the number.

12. Find a number such that the sum of 15 and that number is 4 times that number.

13. Find three consecutive integers whose sum is 174.

14. The sum of 56 and some number equals 9 times that number. Find it.

15. Twelve less than 3 times a certain number is 30. Find the number.

16. When 9 is added to 7 times a certain number, the result is equal to subtracting 3 from 10 times that number. Find the number.

17. The denominator of a fraction is 2 greater than 3 times its numerator, and the reduced value of the fraction is $\frac{3}{10}$. Find the numerator.

18. Find three consecutive odd integers whose sum is 63.

19. Seven less than 4 times a certain number is 9. Find the number.

20. Find the number whose double exceeds 70 by as much as the number itself is less than 80.

21. If $2x - 3$ stands for 29, for what number does $4 + x$ stand?

22. Separate 197 into two parts such that the smaller part divides into the greater 5 times, with a remainder of 23.

23. After a certain number is diminished by 8, the amount remaining is multiplied by 8; the result is the same as if the number were diminished by 6 and the amount remaining multiplied by 6. What is the number?

24. Thirty-one times a certain number is as much above 40 as 9 times the number is below 40. Find the number.

25. Three times the amount by which a certain number exceeds 6 is equal to the number plus 144. Find the number.

26. In a group of 90 persons, composed of men, women, and children, there are 3 times as many children as men, and twice as many women as men. How many are there of each?

Some of these problems are really stupid! Take problem 26 for example. Wouldn't it be easier just to count the people, and anyway, who cares.

If you would like to tell the author just what you think of these word problems, see problem 71 at the end of this chapter.

3–3   Financial Problems

We turn now from number problems to ones that have more application to technology: financial, mixture, and statics problems. Uniform motion problems, which often require the solution of a fractional equation, are left for the chapter on fractions.

◆◆◆ **Example 29:** A consultant had to pay income taxes of $4867 plus 28% of the amount by which her taxable income exceeded $32,450. Her tax bill was $7285. What was her taxable income? Work to the nearest dollar.

Solution: Let x = taxable income (dollars). The amount by which her income exceeded $32,450 is then

$$x - 32,450$$

Her tax is 28% of that amount, plus $4867, so

$$\text{tax} = 4867 + 0.28(x - 32,450) = 7285$$

Solving for x, we get

$$0.28(x - 32{,}450) = 7285 - 4867 = 2418$$

$$x - 32{,}450 = \frac{2418}{0.28} = 8636$$

$$x = 8636 + 32{,}450 = \$41{,}086$$

Check: Her income exceeds $32,450 by ($41,086 − $32,450) or $8636. This amount is taxed at 28%, or 0.28($8636) = $2418. To this we add the $4867 tax, getting

$$\$2418 + \$4867 = \$7285$$

as required.   ♦♦♦

We usually work financial problems to the nearest penny or nearest dollar, regardless of the number of significant digits in the problem.

♦♦♦ **Example 30:** A person invests part of his $10,000 savings in a bank, at 6%, and part in a certificate of deposit, at 8%, both simple interest. He gets a total of $750 per year in interest from the two investments. How much is invested at each rate?

Solution: Does this look familiar? It is the problem we started in Example 20. There we defined our variables as

$$x = \text{amount invested at } 6\%$$

and

$$10{,}000 - x = \text{amount invested at } 8\%$$

If x dollars are invested at 6%, the interest on that investment is $0.06x$ dollars. Similarly, $0.08(10{,}000 - x)$ dollars are earned on the other investment. Since the total earnings are $750, we write

$$0.06x + 0.08(10{,}000 - x) = 750$$
$$0.06x + 800 - 0.08x = 750$$
$$-0.02x = -50$$
$$x = \$2500 \quad \text{at } 6\%$$

and

$$10{,}000 - x = \$7500 \quad \text{at } 8\%$$

Check: Does this look reasonable? Suppose that half of his savings ($5000) were invested at each rate. Then the 6% deposit would earn $300 and the 8% deposit would earn $400, for a total of $700. But he got more than that ($750) in interest, so we would expect more than $5000 to be invested at 8% and less than $5000 at 6%, as we have found.   ♦♦♦

Exercise 3 ♦ Financial Problems

1. The labor costs for a certain project were $3345 per day for 17 technicians and helpers. If the technicians earned $210/day and the helpers $185/day, how many technicians were employed on the project?

2. A company has $86,500 invested in bonds and earns $6751 in interest annually. Part of the money is invested at 7.4%, and the remainder at 8.1%, both simple interest. How much is invested at each rate?

3. How much, to the nearest dollar, must a company earn in order to have $95,000 left after paying 27% in taxes?

4. Three equal batches of fiberglass insulation were bought for $408: the first for $17 per ton, the second for $16 per ton, and the third for $18 per ton. How many tons of each were bought?

Financial problems make heavy use of percentage, so you may want to review Sec. 1–10.

5. A water company changed its rates from $1.95 per 1000 gal to $1.16 per 1000 gal plus a service charge of $45 per month. How much water (to the nearest 1000 gal) can you purchase before your new monthly bill will equal the bill under the former rate structure?

6. What salary should a person receive in order to take home $40,000 after deducting 23% for taxes? Work to the nearest dollar.

7. A person was $450.00 in debt, owing a certain sum to a brother, twice that amount to an uncle, and twice as much to a bank as to the uncle. How much was owed to each?

8. A student sold used skis and boots for $210, getting 4 times as much for the boots as for the skis. What was the price of each?

9. A person used part of a $100,000 inheritance to build a house and invested the remainder for 1 year, 1/3 of it at 6% and 2/3 at 5%, both simple interest. The income from both investments was $320. Find the cost of the house.

10. A used sports utility vehicle (SUV) and a snowplow attachment are worth $7200, the SUV being worth 7 times as much as the plow. Find the value of each.

11. A person spends 1/4 of her annual income for board, 1/12 for clothes, and 1/2 for other expenses, and saves $10,000. What is her income?

3–4 Mixture Problems

Basic Relationships

The total amount of mixture is, obviously, equal to the sum of the amounts of the ingredients.

$$\boxed{\text{total amount of mixture} = \text{amount of } A + \text{amount of } B + \cdots \qquad \textbf{A1}}$$

$$\boxed{\begin{array}{l}\text{final amount of each ingredient} \\ \qquad = \text{initial amount} + \text{amount added} - \text{amount removed}\end{array} \qquad \textbf{A2}}$$

◆◆◆ **Example 31:** From 100.0 lb of solder, half lead and half zinc, 20.0 lb is removed. Then 30.0 lb of lead is added. How much lead is contained in the final mixture?

Solution:

$$\text{initial weight of lead} = 0.5(100.0) = 50.0 \text{ lb}$$
$$\text{amount of lead removed} = 0.5(20.0) = 10.0 \text{ lb}$$
$$\text{amount of lead added} = 30.0 \text{ lb}$$

By Eq. A2,

$$\text{final amount of lead} = 50.0 - 10.0 + 30.0 = 70.0 \text{ lb} \qquad \text{◆◆◆}$$

Percent Concentration

The percent concentration of each ingredient is given by the following equation:

$$\boxed{\text{percent concentration of ingredient } A = \frac{\text{amount of } A}{\text{amount of mixture}} \times 100 \qquad \textbf{15}}$$

◆◆◆ **Example 32:** The total weight of the solder in Example 31 is

$$100.0 - 20.0 + 30.0 = 110.0 \text{ lb}$$

so the percent concentration of lead (of which there is 70.0 lb) is

$$\text{percent lead} = \frac{70.0}{110.0} \times 100 = 63.6\%$$

where 110.0 is the total weight (lb) of the final mixture. ◆◆◆

Two Mixtures

When *two mixtures* are combined to make a *third* mixture, the amount of any ingredient A in the final mixture is given by the following equation:

final amount of A = amount of A in first mixture + amount of A in second mixture	**A3**

◆◆◆ **Example 33:** One hundred liters of gasohol containing 12% alcohol is mixed with 200 liters of gasohol containing 8% alcohol. The volume of alcohol in the final mixture is

$$\text{final amount of alcohol} = 0.12(100) + 0.08(200)$$
$$= 12 + 16 = 28 \text{ liters}$$ ◆◆◆

A Typical Mixture Problem

We now use these ideas about mixtures to solve a typical mixture problem.

◆◆◆ **Example 34:** How much steel containing 5.25% nickel must be combined with another steel containing 2.84% nickel to make 3.25 tons of steel containing 4.15% nickel?

Estimate: The final steel needs 4.15% of 3.25 tons of nickel, about 0.135 ton. If we assume that *equal amounts* of each steel were used, the amount of nickel would be 5.25% of 1.625 tons or 0.085 ton from the first alloy, and 2.84% of 1.625 tons or 0.046 ton from the second alloy. This gives a total of 0.131 ton of nickel in the final steel. This is not enough (we need 0.135 ton). Thus *more than half* of the final alloy must come from the higher-nickel alloy, or between 1.625 tons and 3.25 tons.

Solution: Let x = tons of 5.25% steel needed. Fig. 3–3 shows the three alloys and the amount of nickel in each, with the nickel drawn as if it were separated from the rest of the steel. The weight of the 2.84% steel is

$$3.25 - x$$

The weight of nickel that it contains is

$$0.0284(3.25 - x)$$

The weight of nickel in x tons of 5.25% steel is

$$0.0525x$$

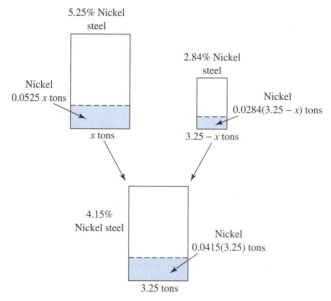

FIGURE 3–3

The sum of these must give the weight of nickel in the final mixture.

$$0.0525x + 0.0284(3.25 - x) = 0.0415(3.25)$$

Clearing parentheses, we have

$$0.0525x + 0.0923 - 0.0284x = 0.1349$$
$$0.0241x = 0.0426$$
$$x = 1.77 \text{ tons of } 5.25\% \text{ steel}$$
$$3.25 - x = 1.48 \text{ tons of } 2.84\% \text{ steel}$$

Check: First we see that more than half of the final alloy comes from the higher-nickel steel, as predicted in our estimate. Now let us see if the final mixture has the proper percentage of nickel.

$$\text{final tons of nickel} = 0.0525(1.77) + 0.0284(1.48)$$
$$= 0.135 \text{ ton}$$

$$\text{percent nickel} = \frac{0.135}{3.25} \times 100 = 0.0415$$

or 4.15%, as required.

Alternate Solution: This problem can also be set up in tabular form, as follows:

	Percent Nickel	×	Amount (tons)	= Amount of Nickel (tons)
Steel with 5.25% nickel	5.25%		x	$0.0525x$
Steel with 2.84% nickel	2.84%		$3.25 - x$	$0.0284(3.25 - x)$
Final steel	4.15%		3.25	$0.0415(3.25)$

We then equate the sum of the amounts of nickel in the original steels with the amount in the final steel, getting

$$0.0525x + 0.0284(3.25 - x) = 0.0415(3.25)$$

as before.

◆◆◆

<table>
<tr><td rowspan="2">Common Error</td><td>If you wind up with an equation that looks like this:</td></tr>
<tr><td>( ) lb nickel + ( ) lb iron = ( ) lb nickel

you know that something is wrong. When you are using Eq. A3, all of the terms must be *for the same ingredient*.</td></tr>
</table>

Exercise 4 ◆ Mixture Problems

1. Two different mixtures of gasohol are available, one with 5% alcohol and the other containing 12% alcohol. How many gallons of the 12% mixture must be added to 250 gal of the 5% mixture to produce a mixture containing 9% alcohol?

2. How many metric tons of chromium must be added to 2.50 metric tons of stainless steel to raise the percent of chromium from 11% to 18%?

3. How many kilograms of nickel silver containing 18% zinc and how many kilograms of nickel silver containing 31% zinc must be melted together to produce 700 kg of a new nickel silver containing 22% zinc?

4. A certain bronze alloy containing 4% tin is to be added to 350 lb of bronze containing 18% tin to produce a new bronze containing 15% tin. How many pounds of the 4% bronze are required?

5. How many kilograms of brass containing 63% copper must be melted with 1100 kg of brass containing 72% copper to produce a new brass containing 67% copper?

6. A certain chain saw requires a fuel mixture of 5.5% oil and the remainder gasoline. How many liters of 2.5% mixture and how many of 9% mixture must be combined to produce 40.0 liters of 5.5% mixture?

7. A certain automobile cooling system contains 11.0 liters of coolant that is 15% antifreeze. How many liters of mixture must be removed so that, when it is replaced with pure antifreeze, a mixture of 25% antifreeze will result?

8. A vat contains 4000 liters of wine with an alcohol content of 10%. How much of this wine must be removed so that, when it is replaced with wine with a 17% alcohol content, the alcohol content in the final mixture will be 12%?

9. A certain paint mixture weighing 300 lb contains 20% solids suspended in water. How many pounds of water must be allowed to evaporate to raise the concentration of solids to 25%?

10. Fifteen liters of fuel containing 3.2% oil is available for a certain two-cycle engine. This fuel is to be used for another engine requiring a 5.5% oil mixture. How many liters of oil must be added?

11. A concrete mixture is to be made which contains 35% sand by weight, and 640 lb of mixture containing 29% sand is already on hand. How many pounds of sand must be added to this mixture to arrive at the required 35%?

12. How many liters of a solution containing 18% sulfuric acid and how many liters of another solution containing 25% sulfuric acid must be mixed together to make 550 liters of solution containing 23% sulfuric acid? (All percentages are by volume.)

Treat the percents given in this exercise as exact numbers, and work to three significant digits.

3–5   Statics Problems

Moments

The *moment of a force* about some point a (M_a) is the product of the force F and the perpendicular distance d from the force to the point.

Moment of a Force about Point a	$M_a = Fd$	A12

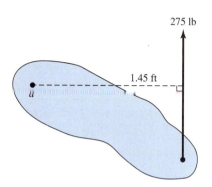

FIGURE 3–4   Moment of a force.

◆◆◆ **Example 35:** The moment of the force in Fig. 3–4 about point a is

$$M_a = 275 \text{ lb}(1.45 \text{ ft}) = 399 \text{ ft·lb}$$ ◆◆◆

Equations of Equilibrium

The following equations apply to a body that is in *equilibrium* (is not moving, or moves with a constant velocity):

This is Newton's first law of motion.

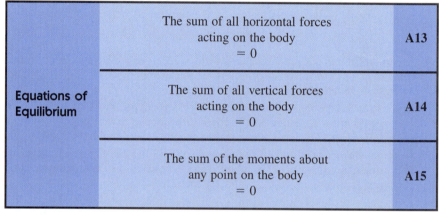

Equations of Equilibrium	The sum of all horizontal forces acting on the body = 0	A13
	The sum of all vertical forces acting on the body = 0	A14
	The sum of the moments about any point on the body = 0	A15

◆◆◆ **Example 36:** A horizontal uniform beam of negligible weight is 6.35 m long and is supported by columns at either end. A concentrated load of 525 N is applied to the beam. At what distance from one end must this load be located so that the vertical reaction at that same end is 315 N, and what will be the reaction at the other end?

Estimate: If the 525-N load were at the midspan, the two reactions would have equal values of ½(525) or 262.5 N. Since the left reaction (315 N) is greater than that, we deduce that the load is to the left of the midspan and that the reaction at the right will be less than 262.5 N.

Solution: We draw a diagram (Fig. 3–5) and label the required distance as x. By Eq. A14,

$$R + 315 = 525$$
$$R = 210 \text{ N}$$

Taking moments about p, we set the moments that tend to turn the bar in a clockwise (CW) direction equal to the moments that tend to turn the bar in the counterclockwise (CCW) direction. By Eq. A15,

$$525x = 210(6.35)$$
$$x = \frac{210(6.35)}{525} = 2.54 \text{ m}$$ ◆◆◆

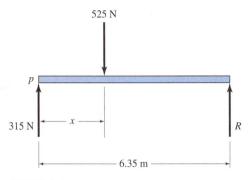

FIGURE 3–5

◆◆◆ **Example 37:** Find the reactions R_1 and R_2 in Fig. 3–6.

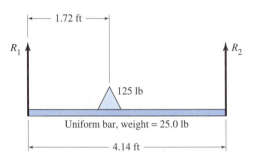

FIGURE 3–6

Estimate: Let's assume that the 125-lb weight is centered on the bar. Then each re-action would equal half the total weight (150 lb ÷ 2 = 75 lb). But since the weight is to the left of center, we expect R_1 to be a bit larger than 75 lb and R_2 to be a bit smaller than 75 lb.

Solution: In a statics problem, we may consider all of the weight of an object to be concentrated at a single point (called the *center of gravity*) on that object. For a uniform bar, the center of gravity is just where you would expect it to be, at the mid-point. Replacing the weights by forces gives the simplified diagram, Fig. 3–7.

The moment of the 125-lb force about p is, by Eq. A12,

$$125(1.72) \text{ ft·lb clockwise}$$

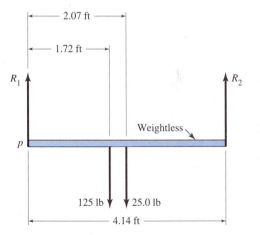

FIGURE 3–7

Similarly, the other moments about p are

$$25.0(2.07) \text{ ft·lb clockwise}$$

and

$$4.14 \, R_2 \text{ ft·lb counterclockwise}$$

But Eq. A15 says that the sum of the clockwise moments must equal the sum of the counterclockwise moments, so

$$4.14 R_2 = 125(1.72) + 25.0(2.07)$$
$$= 266.8$$
$$R_2 = 64.4 \text{ lb}$$

Also, Eq. A14 says that the sum of the upward forces (R_1 and R_2) must equal the sum of the downward forces (125 lb and 25.0 lb), so

$$R_1 + R_2 = 125 + 25.0$$
$$R_1 = 125 + 25.0 - 64.4$$
$$= 85.6 \text{ lb} \qquad \qquad \text{◆◆◆}$$

Exercise 5 ◆ Statics Problems

1. A horizontal beam of negligible weight is 18.0 ft long and is supported by columns at either end. A vertical load of 14,500 lb is applied to the beam at a distance x from the left end. Find x so that the reaction at the left column is 10,500 lb.

2. For the beam of problem 1, find the reaction at the right column.

3. A certain beam of negligible weight has an additional support 13.1 ft from the left end, as shown in Fig. 3–8. The beam is 17.3 ft long and has a concentrated load of 2350 lb at the free end. Find the vertical reactions R_1 and R_2.

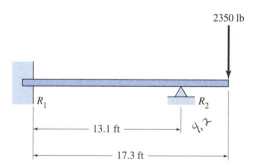

FIGURE 3–8

With technical problems involving approximate numbers, be sure to retain the proper number of significant digits in your answer.

4. A horizontal bar of negligible weight has a 55.1-lb weight hanging from the left end and a 72.0-lb weight hanging from the right end. The bar is seen to balance 97.5 in. from the left end. Find the length of the bar.

5. A horizontal bar of negligible weight hangs from two cables, one at each end. When a 624-N force is applied vertically downward from a point 185 cm from the left end of the bar, the right cable is seen to have a tension of 341 N. Find the length of the bar.

6. A uniform horizontal beam is 9.74 ft long and weighs 386 lb. It is supported by columns at either end. A vertical load of 3814 lb is applied to the beam at a distance x from the left end. Find x so that the reaction at the right column is 2000 lb.

7. A uniform horizontal beam is 19.80 ft long and weighs $136\overline{0}$ lb. It is supported at either end. A vertical load of 13,510 lb is applied to the beam 8.450 ft from the left end. Find the reaction at each end of the beam.

8. A bar of uniform cross section is 82.3 in. long and weighs 10.5 lb. A weight of 27.2 lb is suspended from one end. The bar and weight combination is to be suspended from a cable attached at the balance point. How far from the weight should the cable be attached, and what is the tension in the cable?

••• CHAPTER 3 REVIEW PROBLEMS ••••••••••••••••••••••••••••••••••••

Solve each equation.

1. $2x - (3 + 4x - 3x + 5) = 4$

2. $5(2 - x) + 7x - 21 = x + 3$

3. $3(x - 2) + 2(x - 3) + (x - 4) = 3x - 1$

4. $x + 1 + x + 2 + x + 4 = 2x + 12$

5. $(2x - 5) - (x - 4) + (x - 3) = x - 4$

6. $4 - 5w - (1 - 8w) = 63 - w$

7. $3z - (z + 10) - (z - 3) = 14 - z$

8. $(2x - 9) - (x - 3) = 0$

9. $3x + 4(3x - 5) = 12 - x$

10. $6(x - 5) = 15 + 5(7 - 2x)$

11. $x^2 - 2x - 3 = x^2 - 3x + 1$

12. $(x^2 - 9) - (x^2 - 16) + x = 10$

13. $x^2 + 8x - (x^2 - x - 2) = 5(x + 3) + 3$

14. $x^2 + x - 2 + x^2 + 2x - 3 = 2x^2 - 7x - 1$

15. $10x - (x - 5) = 2x + 47$

16. $7x - 5 - (6 - 8x) + 2 = 3x - 7 + 106$

17. $3p + 2 = \dfrac{p}{5}$

18. $\dfrac{4}{n} = 3$

19. $\dfrac{2x}{3} - 5 = \dfrac{3x}{2}$

20. $5.90x - 2.80 = 2.40x + 3.40$

21. $4.50(x - 1.20) = 2.80(x + 3.70)$

22. $\dfrac{x - 4.80}{1.50} = 6.20x$

23. $6x + 3 - (3x + 2) = (2x - 1) + 9$

24. $3(x + 10) + 4(x + 20) + 5x - 170 = 15$

25. $20 - x + 4(x - 1) - (x - 2) = 30$

26. $5x + 3 - (2x - 2) + (1 - x) = 6(9 - x)$

27. Subdivide a meter of tape into two parts so that one part will be 6 cm longer than the other part.

28. A certain mine yields low-grade oil shale containing 18.0 gal of oil per ton of rock, and another mine has shale yielding 30.0 gal/ton. How many tons of each must be sent each day to a processing plant that processes 25,000 tons of rock per day, so that the overall yield will be 23.0 gal/ton?

29. A certain automatic soldering machine requires a solder containing half tin and half lead. How much pure tin must be added to 55 kg of a solder containing 61% lead and 39% tin to raise the tin content to 50%?

30. A person owed to A a certain sum, to B four times as much, to C eight times as much, and to D six times as much. A total of \$570 would pay all of the debts. What was the debt to A?

31. A technician spends $\frac{2}{3}$ of his salary for board and $\frac{2}{3}$ of the remainder for clothing, and saves \$2500 per year. What is his salary?

32. Find four consecutive odd numbers such that the product of the first and third will be 64 less than the product of the second and fourth.

33. The front and rear wheels of a tractor are 10 ft and 12 ft, respectively, in circumference. How many feet will the tractor have traveled when the front wheel has made 250 revolutions more than the rear wheel?

Solve each equation.

34. $3x - (x - 4) + (x + 1) = x - 7$
35. $4(x - 3) = 21 + 7(2x - 1)$
36. $8.20(x - 2.20) = 1.30(x + 3.30)$
37. $5(x - 1) - 3(x + 3) = 12$
38. $(x - 2)(x + 3) = x(x + 7) - 8$
39. $5x + (2x - 3) + (x + 9) = x + 1$
40. $1.20(x - 5.10) = 7.30(x - 1.30)$
41. $2(x - 7) = 11 + 2(4x - 5)$
42. $4.40(x - 1.90) = 8.30(x + 1.10)$
43. $8x - (x - 5) + (3x + 2) = x - 14$
44. $7(x + 1) = 13 + 4(3x - 2)$

Solve for x.

45. $bx + 8 = 2$

46. $c + ax = b - 7$

47. $2(x - 3) = 4 + a$

48. $3x + a = b - 5$

49. $8 - ax = c - 5a$

50. Find the reactions R_1 and R_2 in Fig. 3–9.

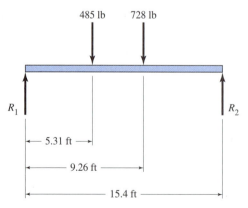

FIGURE 3–9

51. A carpenter estimates that a certain deck needs \$4285 worth of lumber, if there is no waste. How much should he buy if he estimates the waste on the amount bought is 7%?

52. How many liters of olive oil costing $4.86 per liter must be mixed with 136 liters of corn oil costing $2.75 per liter to make a blend costing $3.00 per liter?

53. The labor costs for a certain brick wall were $1118 per day for 10 masons and helpers. If a mason earns $125/day and a helper $92/day, how many masons were on the job?

54. According to a tax table, for your filing status, if your taxable income is over $38,000 but not over $91,850, your tax is $5700 plus 28% of the amount over $38,100. If your tax bill was $8126, what was your taxable income? Work to the nearest dollar.

55. A casting weighs 875 kg and is made of a brass that contains 87.5% copper. How many kilograms of copper are in the casting?

56. How much must a person earn to have $45,824 after paying 28% in taxes?

57. 235 gallons of fuel for a two-cycle engine contains 3.15% oil. To this is added 186 gallons of fuel containing 5.05% oil. How many gallons of oil are in the final mixture?

58. How much steel containing 1.15% chromium must be combined with another steel containing 1.50% chromium to make 8.00 tons of steel containing 1.25% chromium?

59. Two gasohol mixtures are available, one with 6.35% alcohol and the other with 11.28% alcohol. How many gallons of the 6.35% mixture must be added to 25.0 gallons of the other mixture to make a final mixture containing 8.00% alcohol?

60. How many tons of tin must be added to 2.75 tons of bronze to raise the percentage of tin from 10.5% to 16.0%?

61. A student sold a computer and a printer for a total of $995, getting $1\frac{1}{2}$ times as much for the printer as for the computer. What was the price of each?

62. How many pounds of nickel silver containing 12.5% zinc must be melted with 248 kg of nickel silver containing 16.4% zinc to make a new alloy containing 15.0% zinc?

63. A company has $223,821 invested in two separate accounts and earns $14,817 in simple interest annually from these investments. Part of the money is invested at 5.94%, and the remainder at 8.56%. How much is invested at each rate?

64. From 12.5 liters of coolant containing 13.4% antifreeze, 5.50 liters is removed. If 5.50 liters of antifreeze is then added, how much antifreeze will be contained in the final mixture?

65. A person had $125,815 invested, part in a mutual fund that earned 10.25% per year and the remainder in a bank account that earned 4.25% per year, both simple interest. The amount earned from both investments combined was $11,782. How much was in each investment?

66. The labor costs for a certain project were $3971 per day for 15 technicians and helpers. If a technician earns $325/day and a helper $212/day, how many technicians were on the job?

67. 15.6 gallons of cleaner containing 17.8% acid is added to 25.5 gallons of cleaner containing 10.3% acid. How many gallons of acid are in the final mixture?

68. How much brass containing 75.5% copper must be combined with another brass containing 86.3% copper to make 5250 kg of brass containing 80.0% copper?

Writing

69. Make up a word problem. It can be very similar to one given in this chapter or, better, very different. Try to solve it yourself. Then swap with a classmate and solve each other's problem. Note down where the problem may be unclear, unrealistic, or ambiguous. Finally, each of you should rewrite your problem if needed.

70. You probably have some idea what field you plan to enter later: business, engineering, and so on. Look at a book from that field, and find at least one formula commonly used. Write it out; describe what it is used for; and explain the quantities it contains, including their units.

71. Write down the reasons why you think the word problems in this book are stupid and hardly worth studying, and E-mail them to the authors. We promise to answer you. If you don't want to mail your observations, just file them in your portfolio.

    If, on the other hand, you think the problems are stupid but still worth studying, state your reasons. Those too can be sent to the author or filed with your notes.

Team Project

72. Fig. 3–10 shows three beams, each carrying a box of a given weight. The weight of each beam is given, which can be considered to be at the midpoint of each beam. The right end of each of the two upper beams rests on the box on the beam below it. A force of 3150 lb is needed to support the right end of the lowest beam.

    Find the reactions R_1, R_2, and R_3, and the distance x.

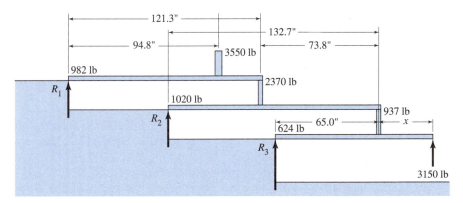

FIGURE 3–10

Functions

OBJECTIVES

When you have completed this chapter, you should be able to:

- Distinguish between relations and functions.
- Identify types of relations and functions.
- Find the domain and the range of a function.
- Use functional notation to manipulate, combine, and evaluate functions.

The equations we have been solving in the last few chapters have contained only *one variable*. For example,

$$x(x - 3) = x^2 + 7$$

contains the single variable x.

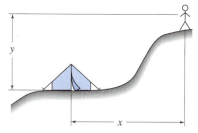

FIGURE 4–1

But many situations involve *two* (or more) variables that are somehow related to each other. Look, for example, at Fig. 4–1, and suppose that you leave your campsite and walk a path up the hill. As you walk, both the horizontal distance x from your camp, and the vertical distance y above your camp, will change. You cannot change x without changing y, and vice versa (unless you jump into the air or dig a hole). The variables x and y are *related*.

In this chapter we study the relation between two variables and introduce the concept of a *function*. The idea of a function provides us with a different way of speaking about mathematical relationships. We could say, for example, that the formula for the area of a circle *as a function of* its radius is $A = \pi r^2$.

It may become apparent as you study this chapter that these same problems could be solved without ever introducing the idea of a function. Does the function concept, then, merely give us new jargon for the same old ideas? Not really.

A new way of *speaking* about something can lead to a new way of *thinking* about that thing, and so it is with functions. It also will lead to the powerful and convenient *functional notation*, which will be especially useful when you study calculus and computer programming. In addition, this introduction to functions will prepare us for the later study of *functional variation*.

4–1   Functions and Relations

If we have an equation that enables us to find exactly one y for any given x, then y *is said to be a function of x.*

◆◆◆ **Example 1:** The equation

$$y = 3x - 5$$

is a function. For any x, say, $x = 2$, there corresponds exactly one y (in this case, $y = 1$).   ◆◆◆

Thus a function may be in the form of an equation. But *not every equation is a function.*

◆◆◆ **Example 2:** The equation $y = \pm\sqrt{x}$ is not a function. Each value of x yields *two* values of y. This equation is called a *relation.*   ◆◆◆

We see from Examples 1 and 2 that a relation need not have *one* value of y for each value of x, as is required for a function. Thus only some equations that are relations can also be called functions.

◆◆◆ **Example 3:** The equation

$$y^2 = 2x + 4$$

is a relation but is not a function, because for some values of x there are two values of y. When $x = 0$, for example, y has the values 2 and -2.   ◆◆◆

The fact that a relation is not a function does not imply second-class status. Relations are no less useful when they are not functions. In fact, for practical work we often do not care whether or not our relation is a function.

The word *function* is commonly used to describe certain types of equations.

Some older textbooks do not distinguish between relations and functions, but instead distinguish between *single-valued* functions and *two-valued* or *multiple-valued* functions (what we now call *relations*).

◆◆◆ **Example 4:**

$y = 5x + 3$	is a linear function
$y = 3x^3$	is a power function
$y = 7x^2 + 2x - 5$	is a quadratic function
$y = 3 \sin 2x$	is a trigonometric function
$y = 5^{2x}$	is an exponential function
$y = \log(x + 2)$	is a logarithmic function

◆◆◆

A More General Definition

When we think about a function or use one, it will most often be in the form of an equation. However, we do not want to limit our definition of a function to equations only, but we would like to include other kinds of associations between numbers, such as tables of data, graphs, and verbal statements.

A more general definition of a function that includes all of these is usually given in terms of sets. A *set* is a collection of particular things, such as the set of all automobiles. The objects or members of a set are called *elements*. The only sets we will consider here are those that contain as elements all or some of the real numbers.

We do not intend to study sets here, or to introduce set notation, but we will use the idea of a set to define a function, as follows:

> Let x be an element in set A, and y an element in set B. We say that y is a function of x if there is a rule that associates exactly one y in set B with each x in set A.

Here, set A is called the *domain* and set B is called the *range*.

We can picture a function as in Fig. 4–2. Each set is represented by a shaded area, and each element by a point within the shaded area. The function f associates exactly one y in the range B with each x in the domain A.

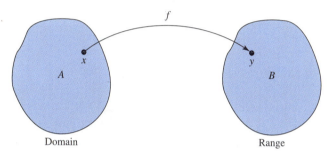

Domain   Range

FIGURE 4–2

Some find it helpful to think of a function as a machine (Fig. 4–3). If the machine receives an input of x (within the domain of x), it uses its built-in rules to produce an output of a single corresponding value of y.

FIGURE 4–3   A "function machine."

◆◆◆ Example 5: The equation from Example 1,

$$y = 3x - 5$$

can be thought of as a rule that associates exactly one y with any given x. The domain of x is the set of all real numbers, and the range of y is the set of all real numbers. ◆◆◆

Other Forms of a Function

We have seen that certain equations are functions. But a function may also be (1) a set of ordered pairs, (2) a verbal statement, or (3) a graph. In technical work we must often deal with *pairs* of numbers rather than with single values, and these pairs often make up a function.

◆◆◆ Example 6: In the experiment shown in Fig. 4–4, we change the load on the spring, and for each load we measure and record the distance that the spring has stretched from its unloaded position. If a load of 6 kg causes a stretch of 3 cm, then 6 kg and 3 cm are called *corresponding values*. The value 3 cm has no meaning *by itself*, but is meaningful only when it is *paired* with the load that produced it (6 kg). If we always write the pair of numbers in the *same order* (the load first and the distance second, in this example), it is then called an *ordered pair* of numbers. It is written (6, 3). ◆◆◆

A set or table of ordered pairs is a function if for each x value there is only one y value given. The domain of x is the set of all x values in the table, and the range is the set of all y values.

FIGURE 4–4

◆◆◆ **Example 7:** Suppose that the spring experiment of Example 6 gave the following results:

Load (kg)	0	1	2	3	4	5	6	7	8
Stretch (cm)	0	0.5	0.9	1.4	2.1	2.4	3.0	3.6	4.0

This table of ordered pairs is a function. The domain is the set of numbers (0, 1, 2, 3, 4, 5, 6, 7, 8), and the range is the set of numbers (0, 0.5, 0.9, 1.4, 2.1, 2.4, 3.0, 3.6, 4.0). The table associates exactly one number in the range with each number in the domain.   ◆◆◆

A set of ordered pairs is not always given in table form.

◆◆◆ **Example 8:** The function of Example 7 can also be written

(0, 0), (1, 0.5), (2, 0.9), (3, 1.4), (4, 2.1), (5, 2.4), (6, 3.0), (7, 3.6), (8, 4.0)

◆◆◆

Functions may also be given in verbal form.

◆◆◆ **Example 9:** "The shipping charges are 55¢/lb for the first 50 lb and 45¢/lb thereafter." This verbal statement is a function relating the shipping costs to the weight of the item.   ◆◆◆

We sometimes want to switch from verbal form to an equation, or vice versa.

◆◆◆ **Example 10:** Write y as a function of x if y equals twice the cube of x diminished by half of x.

Solution: We replace the verbal statement by the equation

$$y = 2x^3 - \frac{x}{2}$$

◆◆◆

◆◆◆ **Example 11:** The equation $y = 5x^2 + 9$ can be stated verbally as, "y equals the sum of 9 and 5 times the square of x."   ◆◆◆

There are, of course, many different ways to express this same relationship.

◆◆◆ **Example 12:** Express the volume V of a cone having a base area of 75 units *as a function of* its altitude H.

Solution: This is another way of saying, "Write an equation for *the volume of a cone in terms of its base area and altitude*."

The formula for the volume of a cone is given in Appendix A, "Summary of Facts and Formulas."

$$V = \tfrac{1}{3}(\text{base area})(\text{altitude}) = \tfrac{1}{3}(75)H$$

So

$$V = 25H$$

is the required expression.   ◆◆◆

We see, then, that a function or relation can be expressed in several different ways (Fig. 4–5): as an equation, as a table or set of ordered pairs, or as a verbal statement. A function or relation can also be expressed as a *graph*, as we'll see in Sec. 5–2.

Different forms of a relation or function	
Equation: $y = x^2 - 3$	Table of values: $\begin{array}{c\|cccc} x & 0 & 1 & 2 & 3 \\ \hline y & -3 & -2 & 1 & 6 \end{array}$
Verbal statement: "y is equal to the square of x diminished by 3."	
Set of ordered pairs: $(1, -2), \quad (3, 6),$ $(0, -3), \quad (2, 1)$	Graph:

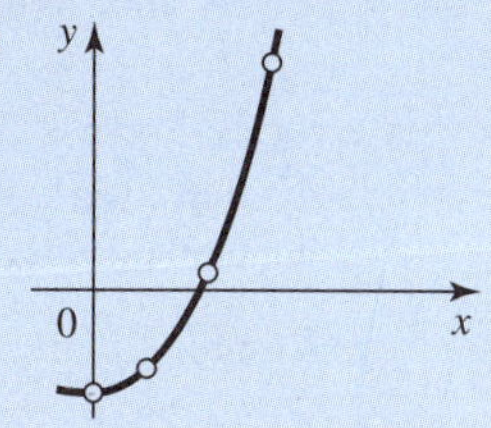

FIGURE 4–5

Finding Domain and Range

To be strictly correct, the domain should be stated whenever an equation is written. However, this is often not done, so we follow this convention:

The domain of the function is assumed to be the largest set of real x values that yield real values of y.

◆◆◆ **Example 13:** The function

$$y = x^2$$

gives a real y value for every real x value. Thus the domain is all of the real numbers, which we can write

Domain: $-\infty < x < \infty$

But notice that there is no x that will make y negative. Thus the range of y includes all of the positive numbers and zero. This can be written

Range: $y \geq 0$ ◆◆◆

Our next example is one in which certain values of x result in a *negative number under a radical sign.*

◆◆◆ **Example 14:** Find the domain and the range of the function

$$y = \sqrt{x - 2}$$

Solution: Our method is to see what values of x and y "do not work" (give a nonreal result or an illegal operation); then the domain and the range will be those values that "do work."

Any value of x less than 2 will make the quantity under the radical sign negative, resulting in an imaginary y. Thus the domain of x is all of the positive numbers equal to or greater than 2.

Domain: $x \geq 2$

An x equal to 2 gives a y of zero. Any x larger than 2 gives a real y greater than zero. Thus the range of y is

Range: $y \geq 0$ ◆◆◆

Our next example shows some values of x that result in *division by zero*.

◆◆◆ **Example 15:** Find the domain and the range of the function

$$y = \frac{1}{x}$$

Solution: Here any value of x but zero will give a real y. Thus the domain is

Domain: $x \neq 0$

Notice here that it is more convenient to state which values of x *do not* work, rather than those that do.

Now what happens to y as x varies over its domain? We see that large x values give small y values, and conversely that small x values give large y values. Also, negative x's give negatives y's. However, there is no x that will make y zero, so

Range: $y \neq 0$ ◆◆◆

Our final example shows both division by zero and negative numbers under a radical sign, for certain x values.

◆◆◆ **Example 16:** Find the domain and the range of the function

$$y = \frac{9}{\sqrt{4 - x}}$$

Solution: Since the denominator cannot be zero, x cannot be 4. Also, any x greater than 4 will result in a negative quantity under the radical sign. The domain of x is then

$$x < 4$$

Restricted to these values, the quantity $4 - x$ is positive and ranges from very small (when x is nearly 4) to very large (when x is large and negative). Thus the denominator is positive, since we allow only the principal (positive) root and we can vary from near zero to infinity. The range of y, then, includes all of the values greater than zero.

Range: $y > 0$ ◆◆◆

Exercise 1 ◆ Functions and Relations

Which of the following relations are also functions?

1. $y = 3x^2 - 5$ **2.** $y = \sqrt{2x}$
3. $y = \pm\sqrt{2x}$ **4.** $y^2 = 3x - 5$
5. $2x^2 = 3y^2 - 4$ **6.** $x^2 - 2y^2 - 3 = 0$
7. Is the set of ordered pairs (1, 3), (2, 5), (3, 8), (4, 12) a function? Explain.
8. Is the set of ordered pairs (0, 0), (1, 2), (1, −2), (2, 3), (2, −3) a function? Explain.

Functions in Verbal Form

For each of the following, write y as a function of x, where the value of y is equal to the given expression.

9. The cube of x
10. The square root of x, diminished by 5
11. x increased by twice the square of x
12. The reciprocal of the cube of x
13. Two-thirds of the amount by which x exceeds 4

Replace each function by a verbal statement of the type given in problems 9 through 13.

14. $y = 3x^2$

15. $y = 5 - x$

16. $y = \dfrac{1}{x} + x$

17. $y = 2\sqrt[3]{x}$

18. $y = 5(4 - x)$

Write the equation called for in each following statement. Refer to Appendix A, "Summary of Facts and Formulas," if necessary.

19. Express the area A of a triangle as a function of its base b and altitude h.

20. Express the hypotenuse c of a right triangle as a function of its legs, a and b.

21. Express the volume V of a sphere as a function of the radius r.

22. Express the power P dissipated in a resistor as a function of its resistance R and the current I through the resistor.

Write the equations called for in each statement.

23. A car is traveling at a speed of 55 mi/h. Write the distance d traveled by the car as a function of time t.

24. To ship its merchandise, a mail-order company charges 65¢/lb plus \$2.25 for handling and insurance. Express the total shipping charge s as a function of the item weight w.

25. A projectile is shot upward with an initial velocity of 125 m/s. Express the height H of the projectile as a function of time t. (See Eq. A18.)

Domain and Range

State the domain and the range of each function.

26. (0, 2), (1, 4), (2, 8), (3, 16), (4, 32)

27. (−10, 20), (5, 7), (−7, 10), (10, 20), (0, 3)

28.

x	2	4	6	8	10
y	0	−2	−5	−9	−15

Find the domain and the range for each function.

29. $y = \sqrt{x - 7}$

30. $y = \dfrac{3}{\sqrt{x - 2}}$

31. $y = x - \dfrac{1}{x}$

32. $y = \sqrt{x^2 - 25}$

33. $y = \dfrac{8 - x}{9 - x}$

34. $y = \dfrac{11}{(x + 4)(x + 2)}$

35. $y = \dfrac{4}{\sqrt{1 - x}}$

36. $y = \dfrac{x + 1}{x - 1}$

37. $y = \sqrt{x - 1}$

38. $y = \dfrac{5}{(x + 2)(x - 4)}$

39. Find the range of the function $y = 3x^2 - 5$ whose domain is $0 \le x \le 5$.

Calculator

40. Many graphics calculators enable you to recall an equation that was entered in the preceding calculation. This enables you to use the equation again with new values of the variables. Some calculators have an $\boxed{\text{ENTRY}}$ key for bringing back the last entry, and some use an arrow key for that purpose. Check your manual.

Your calculator or computer can be your own "function machine."

Use your calculator to compute a set of ordered pairs for the function

$$y = 2x^2 - 3x + 5$$

for integer values of x from 0 to 5.

41. If you have a programmable calculator, program it to execute the computation of problem 40 *automatically*.

42. Many graphics calculators can display a *table of values* for an entered function, as well as a graph. The value of x appears in the first column, and the corresponding value of the function appears in the second column. You can also choose the starting value for x and the increment for x.

Consult your calculator manual to see if you have this feature, and use it to create a table of values for any function in this exercise set.

Computer:

43. The following is a program to compute and print in BASIC a table of ordered pairs for the function in problem 40:

```
10    FOR X = 0 to 5
20    LET Y = 2*X↑2 − 3*X + 5
30    PRINT X, Y
40    NEXT X
```

We don't intend to teach computer programming in this text, but we include an occasional problem for those who know simple programming.

Try this program on your computer, if you have BASIC, or modify it for the language you have available. Then change it for different values of x, different step sizes (say, FOR X = −10 TO 10 STEP .5), and for a different function.

4–2   Functional Notation

Implicit and Explicit Forms

When one variable in an equation is isolated on one side of the equal sign, the equation is said to be in *explicit form*.

◆◆◆ **Example 17:** The following equations are all in explicit form:

$$y = 2x^3 + 5$$
$$z = ay + b$$
$$x = 3z^2 + 2z - z$$

◆◆◆

When a variable is *not* isolated, the equation is in *implicit form*.

◆◆◆ **Example 18:** The following equations are all in implicit form:

$$y = x^2 + 4y$$
$$x^2 + y^2 = 25$$
$$w + x = y + z$$

◆◆◆

Dependent and Independent Variables

In the equation

$$y = x + 5$$

y is called the *dependent variable* because its value depends on the value of x, and x is called the *independent* variable. Of course, the same equation can be written $x = y - 5$, so that x becomes the dependent variable and y the independent variable.

The terms *dependent* and *independent* are used only for an equation in explicit form.

◆◆◆ **Example 19:** In the implicit equation $y - x = 5$, neither x nor y is called dependent or independent. ◆◆◆

Functional Notation

Just as we use the symbol x to represent a *number*, without saying which number we are specifying, we use the notation

$$f(x)$$

to represent a *function* without having to specify which particular function we are talking about.

We can also use functional notation to designate a *particular* function, such as

$$f(x) = x^3 - 2x^2 - 3x + 4$$

A functional relation between two variables x and y, in explicit form, such as

$$y = 5x^2 - 6$$

could be written

$$y = f(x)$$

or

$$f(x) = 5x^2 - 6$$

The expression

$$y = f(x)$$

is read "y is a function of x."

Common Error	The expression $y = f(x)$ **does not** mean "y equals f times x."

The independent variable x is sometimes referred to as the *argument* of the function.

◆◆◆ **Example 20:** We may know that the horsepower P of an engine depends (somehow) on the engine displacement d. We can express this fact by

$$P = f(d)$$

We are saying that P is a function of d, even though we do not know (or perhaps even care, for now) what the relationship is. ◆◆◆

The letter f is usually used to represent a function, but *other letters* can, of course, be used (g and h being common). Subscripts are also used to distinguish one function from another.

◆◆◆ **Example 21:** You may see functions written as

$$y = f_1(x) \qquad y = g(x)$$
$$y = f_2(x) \qquad y = h(x)$$
$$y = f_3(x)$$

◆◆◆

The letter y itself is often used to represent a function.

◆◆◆ **Example 22:** The function

$$y = x^2 - 3x$$

will often be written

$$y(x) = x^2 - 3x$$

to emphasize the fact that y is a function of x.   ◆◆◆

Implicit functions can also be represented in functional notation.

◆◆◆ **Example 23:** The equation $x - 3xy + 2y = 0$ can be represented in functional notation by $f(x, y) = 0$.   ◆◆◆

Functions relating *more than two variables* can be represented in functional notation, as in the following example.

◆◆◆ **Example 24:**

$y = 2x + 3z$	can be written	$y = f(x, z)$
$z = x^2 - 2y + w^2$	can be written	$z = f(w, x, y)$
$x^2 + y^2 + z^2 = 0$	can be written	$f(x, y, z) = 0$

◆◆◆

Manipulating Functions

Functional notation provides us with a convenient way of indicating what is to be done with a function or functions. This includes solving an equation for a different variable; changing from implicit to explicit form, or vice versa; combining two or more functions to make another function; or substituting numerical or literal values into an equation.

◆◆◆ **Example 25:** Write the equation $y = 2x - 3$ in the form $x = f(y)$.

Solution: We are being asked to write the given equation with x, instead of y, as the dependent variable. Solving for x, we obtain

$$2x = y + 3$$

$$x = \frac{y + 3}{2}$$

◆◆◆

Remember that not every function $y = f(x)$ can be solved for x to get a function

$$x = f(y)$$

Thus the function $y = x^2$ when solved for x gives

$$x = \pm\sqrt{y}$$

which is a relation but not a function.

◆◆◆ **Example 26:** Write the equation $x = \dfrac{3}{2y - 7}$ in the form $y = f(x)$.

Solution: Here we are asked to rewrite the equation with y as the dependent variable, so we solve for y. Multiplying both sides by $2y - 7$ and dividing by x gives

$$2y - 7 = \frac{3}{x}$$

or

$$2y = \frac{3}{x} + 7$$

Dividing by 2 gives

$$y = \frac{3}{2x} + \frac{7}{2}$$

◆◆◆

◆◆◆ **Example 27:** Write the equation $y = 3x^2 - 2x$ in the form $f(x, y) = 0$.

Solution: We are asked here to go from explicit to implicit form. Rearranging gives us

$$3x^2 - 2x - y = 0 \qquad \text{◆◆◆}$$

Substituting into Functions

If we have a function, say, $f(x)$, then the notation $f(a)$ means to replace x by a in the same function.

◆◆◆ **Example 28:** Given $f(x) = x^3 - 5x$, find $f(2)$.

Solution: The notation $f(2)$ means that 2 is to be *substituted for* x in the given function. Wherever an x appears, we replace it with 2.

$$\begin{array}{ccc} f(x) = & x^3 - & 5x \\ \updownarrow & \updownarrow & \updownarrow \\ f(2) = & (2)^3 - & 5(2) \\ & = 8 - & 10 = -2 \end{array} \qquad \text{◆◆◆}$$

◆◆◆ **Example 29:** Given $y(x) = 3x^2 - 2x$, find $y(5)$.

Solution: The given notation means to substitute 5 for x, so

$$\begin{aligned} y(5) &= 3(5)^2 - 2(5) \\ &= 75 - 10 = 65 \end{aligned} \qquad \text{◆◆◆}$$

Often, we must substitute *several values* into a function and combine them as indicated.

◆◆◆ **Example 30:** If $f(x) = x^2 - 3x + 4$, find

$$\frac{f(5) - 3f(2)}{2f(3)}$$

Solution:

$$\begin{aligned} f(2) &= 2^2 - 3(2) + 4 = 2 \\ f(3) &= 3^2 - 3(3) + 4 = 4 \\ f(5) &= 5^2 - 3(5) + 4 = 14 \end{aligned}$$

so

$$\begin{aligned} \frac{f(5) - 3f(2)}{2f(3)} &= \frac{14 - 3(2)}{2(4)} \\ &= \frac{8}{8} = 1 \end{aligned} \qquad \text{◆◆◆}$$

The substitution might involve *literal* quantities instead of numerical values.

◆◆◆ **Example 31:** Given $f(x) = 3x^2 - 2x + 3$, find $f(5a)$.

Solution: We substitute $5a$ for x.

$$\begin{aligned} f(5a) &= 3(5a)^2 - 2(5a) + 3 \\ &= 3(25a^2) - 10a + 3 \\ &= 75a^2 - 10a + 3 \end{aligned} \qquad \text{◆◆◆}$$

◆◆◆ **Example 32:** If $f(x) = 5x - 2$, find $f(2w)$.

Solution:

$$f(2w) = 5(2w) - 2 = 10w - 2 \qquad \text{◆◆◆}$$

◆◆◆ **Example 33:** If $f(x) = x^2 - 2x$, find $f(w^2)$.

Solution:

$$f(w^2) = (w^2)^2 - 2(w^2) = w^4 - 2w^2 \qquad \text{◆◆◆}$$

It can be confusing when the expression to be substituted contains the same variable as in the original function.

◆◆◆ **Example 34:** If $f(x) = 5x - 2$, then

$$f(x) = 5 \quad x \quad - 2$$

$$f(x + a) = 5\,(x + a) - 2$$
$$= 5x + 5a - 2 \qquad \text{◆◆◆}$$

◆◆◆ **Example 35:** If $f(x) = x^2 - 2x$, then

$$f(x - 1) = (x - 1)^2 - 2(x - 1)$$
$$= x^2 - 2x + 1 - 2x + 2$$
$$= x^2 - 4x + 3 \qquad \text{◆◆◆}$$

There may be *more than one function* in a single problem.

◆◆◆ **Example 36:** Given three different functions,

$$f(x) = 3x \qquad g(x) = x^2 \qquad h(x) = \sqrt{x}$$

evaluate

$$\frac{2g(3) + 4h(9)}{f(5)}$$

Solution: First substitute into each function.

$$f(5) = 3(5) = 15 \qquad g(3) = 3^2 = 9 \qquad h(9) = \sqrt{9} = 3$$

Then combine these as indicated.

$$\frac{2g(3) + 4h(9)}{f(5)} = \frac{2(9) + 4(3)}{15} = \frac{18 + 12}{15} = \frac{30}{15} = 2 \qquad \text{◆◆◆}$$

Sometimes we must substitute into a function containing *more than one variable.*

◆◆◆ **Example 37:** Given $f(x, y, z) = 2y - 3z + x$, find $f(3, 1, 2)$.

Solution: We substitute the given numerical values for the variables. Be sure that the numerical values are taken in the *same order* as the variable names in the functional notation.

$$f(x, y, z)$$
$$\updownarrow \ \updownarrow \ \updownarrow$$
$$f(3, 1, 2)$$

Substituting, we obtain

$$f(3, 1, 2) = 2(1) - 3(2) + 3 = 2 - 6 + 3$$
$$= -1$$

♦♦♦

Exercise 2 ♦ Functional Notation

Implicit and Explicit Forms

Which equations are in explicit form and which in implicit form?

1. $y = 5x - 8$
2. $x = 2xy + y^2$
3. $3x^2 + 2y^2 = 0$
4. $y = wx + wz + xz$

Dependent and Independent Variables

Label the variables in each equation as dependent or independent.

5. $y = 3x^2 + 2x$
6. $x = 3y - 8$
7. $w = 3x + 2y$
8. $xy = x + y$
9. $x^2 + y^2 = z$

Manipulating Functions

10. If $y = 5x + 3$, write $x = f(y)$

11. If $x = \dfrac{2}{y - 3}$, write $y = f(x)$.

12. If $y = \dfrac{1}{x} - \dfrac{1}{5}$, write $x = f(y)$.

13. If $x^2 + y = x - 2y + 3x^2$, write $y = f(x)$.

14. If $2xw - 3w^2 = 6 + 3x$, write $x = f(w)$.

15. If $5p - q = q - p^2$, write $q = f(p)$.

16. The power P dissipated in a resistor is given by $P = I^2R$. Write $R = f(P, I)$.

17. Young's modulus E is given by

$$E = \frac{PL}{ae}$$

Write $e = f(P, L, a, E)$.

Substituting into Functions

Substitute the given numerical value(s) into each function.

18. If $f(x) = 2x^2 + 4$, find $f(3)$.
19. If $f(x) = 5x + 1$, find $f(1)$.
20. If $f(x) = 15x + 9$, find $f(3)$.
21. If $f(x) = 5 - 13x$, find $f(2)$.
22. If $g(x) = 9 - 3x^2$, find $g(-2)$.
23. If $h(x) = x^3 - 2x + 1$, find $h(2.55)$.
24. If $f(x) = 7 + 2x$, find $f(3)$.
25. If $f(x) = x^2 - 9$, find $f(-2)$.
26. If $f(x) = 2x + 7$, find $f(-1)$.

27. If $f_1(z) = z^2 - 6.43$, find $f_1(3.02)$.

28. If $g_3(w) = 6.83 + 2.74w$, find $g_3(1.74)$.

Substitute and combine as indicated.

29. Given $f(x) = 2x - x^2$, find $f(5) + 3f(2)$.

30. Given $f(x) = x^2 - 5x + 2$, find $f(2) + f(3) - f(1)$.

31. Given $f(x) = x^2 - 4$, find $\dfrac{f(2) + f(3)}{4}$.

32. Given (a) $f(x) = 2x$ and (b) $f(x) = x^2$, evaluate

$$\frac{f(x + d) - f(x)}{d}$$

for each given function.

We deal with expressions of this form in calculus, when studying the derivative.

Substitute the literal values into each function.

33. If $f(x) = 2x^2 + 4$, find $f(a)$.

34. If $f(x) = 2x - \dfrac{1}{x} + 4$, find $f(2a)$.

35. If $f(x) = 5x + 1$, find $f(a + b)$.

36. If $f(x) = 5 - 13x$, find $f(-2c)$.

Substitute in each function and combine.

37. If $f(x) = x^2$ and $g(x) = \dfrac{1}{x}$, find $f(3) + g(2)$.

38. If $f(x) = \dfrac{2}{x}$ and $g(x) = 3x$, find $f(4) - g\left(\dfrac{1}{3}\right)$.

39. If $f(x) = 8x^2 - 2x + 1$ and $g(x) = x - 5$, find $2f(3) - 4g(3)$.

40. If $h(z) = 2x$ and $g(w) = w^2$, find the following:

(a) $\dfrac{h(4) + g(1)}{5}$ (b) $h[g(3)]$ (c) $g[h(2)]$

Functions of More Than One Variable

Substitute the values into each function.

41. If $f(x, y) = 3x + 2y^2 - 4$, find $f(2, 3)$.

42. If $f(x, y, z) = 3z - 2x + y^2$, find $f(3, 1, 5)$.

43. If $g(a, b) = 2b - 3a^2$, find $g(4, -2)$.

44. If $f(x, y) = y - 3x$, find $3f(2, 1) + 2f(3, 2)$.

45. If $g(a, b, c) = 2b - a^2 + c$, find $\dfrac{3g(1, 1, 1) + 2g(1, 2, 3)}{g(2, 1, 3)}$.

46. If $h(p, q, r) = 6r - 2p + q$, find $h(3, 2, 1) + \dfrac{2h(1, 1, 2) - 3h(3, 4, 5)}{4h(2, 2, 2)}$.

47. If $g(k, l, m) = \dfrac{k + m}{l}$, find $g(2.38, 1.25, 5.38) + 3g(4.27, 8.26, 2.48)$.

Applications

48. The distance traveled by a freely falling body is a function of the elapsed time t:

$$f(t) = V_0 t + \tfrac{1}{2}gt^2 \text{ ft}$$

where V_0 is the initial velocity and g is the acceleration due to gravity (32.2 ft/s^2). If V_0 is 55.0 ft/s, find $f(10.0)$, $f(15.0)$, and $f(20.0)$.

49. The resistance R of a conductor is a function of temperature:

$$f(t) = R_0(1 + \alpha t)$$

where R_0 is the resistance at 0°C and α is the temperature coefficient of resistance (0.00427 for copper). If the resistance of a copper coil is 9800 Ω at 0°C, find $f(20.0)$, $f(25.0)$, and $f(30.0)$.

50. The maximum deflection in inches of a certain cantilever beam, with a concentrated load applied r feet from the fixed end, is a function of r.

$$f(r) = 0.000030r^2(8\overline{0} - r) \text{ in.}$$

Find the deflections $f(1\overline{0})$ and $f(15)$.

51. The length of a certain steel bar at temperature t is given by

$$L = f(t) = 96.0(1 + 0.0635t) \text{ in.}$$

Find $f(115)$, $f(155)$, and $f(215)$.

52. The power P in a resistance R in which flows a current I is given by

$$P = f(I) = I^2 R \quad \text{W}$$

If $R = 25.6 \ \Omega$, find $f(1.59)$, $f(2.37)$, and $f(3.17)$.

4–3 Composite Functions and Inverse Functions

Composite Functions

Just as we can substitute a constant or a variable into a given function, so we can substitute a *function* into a function.

◆◆◆ **Example 38:** If $g(x) = x + 1$, find

(a) $g(2)$ (b) $g(z^2)$
(c) $g[f(x)]$

Solution:

(a) $g(2) = 2 + 1 = 3$ (b) $g(z^2) = z^2 + 1$
(c) $g[f(x)] = f(x) + 1$ ◆◆◆

The function $g[f(x)]$ (which we read "g of f of x"), being made up of the two functions $g(x)$ and $f(x)$, is called a *composite function*. If we think of a function as a machine, it is as if we are using the output $f(x)$ of the function machine f as the input of a second function machine g.

$$x \rightarrow \boxed{f} \rightarrow f(x) \rightarrow \boxed{g} \rightarrow g[f(x)]$$

We thus obtain $g[f(x)]$ by replacing x in $g(x)$ by the function $f(x)$.

In calculus, we will take derivatives of composite functions by means of the *chain rule*.

◆◆◆ **Example 39:** Given the functions $g(x) = x + 1$ and $f(x) = x^3$, write the composite function $g[f(x)]$.

Solution: In the function $g(x)$, we replace x by $f(x)$.

$$\begin{align} g(x) &= x + 1 \\ g[f(x)] &= f(x) + 1 \\ &= x^3 + 1 \end{align}$$

since $f(x) = x^3$.   ◆◆◆

As we have said, the notation $g[f(x)]$ means to substitute $f(x)$ into the function $g(x)$. On the other hand, the notation $f[g(x)]$ means to substitute $g(x)$ into $f(x)$.

$$x \to \boxed{g} \to g(x) \to \boxed{f} \to f[g(x)]$$

In general, $f[g(x)]$ *will not be the same as* $g[f(x)]$.

◆◆◆ **Example 40:** Given $g(x) = x^2$ and $f(x) = x + 1$, find the following:

(a) $f[g(x)]$ (b) $g[f(x)]$
(c) $f[g(2)]$ (d) $g[f(2)]$

Solution:

(a) $f[g(x)] = g(x) + 1 = x^2 + 1$
(b) $g[f(x)] = [f(x)]^2 = (x + 1)^2$
(c) $f[g(2)] = 2^2 + 1 = 5$
(d) $g[f(2)] = (2 + 1)^2 = 9$

Notice that here $f[g(x)]$ is not equal to $g[f(x)]$.   ◆◆◆

Inverse of a Function

Consider a function f that, given a value of x, returns some value of y.

$$x \to \boxed{f} \to y$$

If that y is now put into a function g that *reverses the operations* in f so that its output is the original x, then g is called the *inverse* of f.

$$x \to \boxed{f} \to y \to \boxed{g} \to x$$

The inverse of a function $f(x)$ is often designated by $f^{-1}(x)$.

$$x \to \boxed{f} \to y \to \boxed{f^{-1}} \to x$$

Common Error	Do not confuse f^{-1} with $1/f$.

◆◆◆ **Example 41:** Two such inverse operations are "cube" and "cube root."

$$x \to \boxed{\text{cube } x} \to x^3 \to \boxed{\text{take cube root}} \to x \qquad ◆◆◆$$

Thus if a function $f(x)$ has an inverse $f^{-1}(x)$ that reverses the operations in $f(x)$, then the *composite* of $f(x)$ and $f^{-1}(x)$ should have no overall effect. If the input is x, then the output must also be x. In symbols, if $f(x)$ and $f^{-1}(x)$ are inverse functions, then

$$f^{-1}[f(x)] = x$$

and

$$f[f^{-1}(x)] = x$$

Conversely, if $g[f(x)] = x$ and $f[g(x)] = x$, then $f(x)$ and $g(x)$ are inverse functions.

◆◆◆ **Example 42:** Using the example of the cube and cube root, if

$$f(x) = x^3 \quad \text{and} \quad g(x) = \sqrt[3]{x}$$

then

$$g[f(x)] = \sqrt[3]{f(x)} = \sqrt[3]{x^3} = x$$

and

$$f[g(x)] = [g(x)]^3 = (\sqrt[3]{x})^3 = x$$

This shows that here $f(x)$ and $g(x)$ are indeed inverse functions. ◆◆◆

To find the inverse of a function $y = f(x)$:

1. Solve the given equation for x.
2. Interchange x and y.

◆◆◆ **Example 43:** We use the cube and cube root example one more time. Find the inverse $g(x)$ of the function

$$y = f(x) = x^3$$

Solution: We solve for x and get

$$x = \sqrt[3]{y}$$

It is then customary to interchange variables so that the dependent variable is (as usual) y. This gives

$$y = f^{-1}(x) = \sqrt[3]{x}$$

Thus $f^{-1}(x) = \sqrt[3]{x}$ is the inverse of $f(x) = x^3$, as verified earlier. ◆◆◆

◆◆◆ **Example 44:** Find the inverse $f^{-1}(x)$ of the function

$$y = f(x) = 2x + 5$$

Solution: Solving for x gives

$$x = \frac{y - 5}{2}$$

Interchanging x and y, we obtain

$$y = f^{-1}(x) = \frac{x - 5}{2}$$ ◆◆◆

Sometimes the inverse of a function will *not* be a function itself, but it may be a relation.

◆◆◆ **Example 45:** Find the inverse of the function $y = x^2$.

Solution: Take the square root of both sides.

$$x = \pm\sqrt{y}$$

Interchanging x and y, we get

$$y = \pm\sqrt{x}$$

Thus a single value of x (say, 4) gives *two* values of y ($+2$ and -2), so our inverse does not meet the definition of a function. It is, however, a relation. ◆◆◆

Sometimes the inverse of a function gets a special name. The inverse of the sine function, for example, is called the *arcsin*, and the inverse of an exponential function is a *logarithmic* function.

We cover the inverse trigonometric functions in Secs. 7–3, 15–1, and 16–1, and graph them in Sec. 18–6. The exponential function and its inverse, the logarithmic function, are covered in Chapter 20.

Exercise 3 ◆ Composite Functions and Inverse Functions

Composite Functions

1. Given the functions $g(x) = 2x + 3$ and $f(x) = x^2$, write the composite function $g[f(x)]$.

2. Given the functions $g(x) = x^2 - 1$ and $f(x) = 3 + x$, write the composite function $f[g(x)]$.
3. Given the functions $g(x) = 1 - 3x$ and $f(x) = 2x$, write the composite function $g[f(x)]$.
4. Given the functions $g(x) = x - 4$ and $f(x) = x^2$, write the composite function $f[g(x)]$.

Given $g(x) = x^3$ and $f(x) = 4 - 3x$, find:

5. $f[g(x)]$ 6. $g[f(x)]$
7. $f[g(3)]$ 8. $g[f(3)]$

Inverse Functions

Find the inverse of:

9. $y = 8 - 3x$
10. $y = 5(2x - 3) + 4x$

11. $y = 7x + 2(3 - x)$

12. $y = (1 + 2x) + 2(3x - 1)$

13. $y = 3x - (4x + 3)$

14. $y = 2(4x - 3) - 3x$

15. $y = 4x + 2(5 - x)$
16. $y = 3(x - 2) - 4(x + 3)$

Computer

Program languages usually have a command for defining a function. BASIC, for example, uses the DEF FN statement, as in

$$50 \qquad \text{DEF FND(X)} = 3*X + 5$$

Here the *name* of the function is *D,* and the *independent variable or argument* is *x.* Functions of more than one variable can also be defined, as in

$$100 \qquad \text{DEF FNB(X,Y,Z)} = 3*X + Y/2 - Z$$

17. Using the DEFINE instructions on your machine, compute and print

$$y = \frac{4f_1(x)}{f_2(x) + 3}$$

for integer values of *x* from 1 to 10, where

$$f_1(x) = 2x - 5 \quad \text{and} \quad f_2(x) = 3x^2 + 2x - 3$$

◆◆◆ CHAPTER 4 REVIEW PROBLEMS ◆◆◆◆◆◆◆◆◆◆◆◆◆◆◆◆◆◆◆◆◆◆◆◆◆◆◆◆◆

1. Which of the following relations are also functions?
   (a) $y = 5x^3 - 2x^2$ (b) $x^2 + y^2 = 25$
   (c) $y = \pm\sqrt{2x}$
2. Write *y* as a function of *x* if *y* is equal to half the cube of *x*, diminished by twice *x.*
3. Write an equation to express the surface area *S* of a sphere as a function of its radius *r.*

4. Find the largest possible domain and range for each function.

   (a) $y = \dfrac{5}{\sqrt{3-x}}$ (b) $y = \dfrac{1+x}{1-x}$

5. Label each function as implicit or explicit. If it is explicit, name the dependent and independent variables.

   (a) $w = 3y - 7$
   (b) $x - 2y = 8$

6. Given $y = 3x - 5$, write $x = f(y)$.

7. Given $x^2 + y^2 + 2w = 3$, write $w = f(x, y)$.

8. Write the inverse of the function $y = 9x - 5$.

9. If $y = 3x^2 + 2z$ and $z = 2x^2$, write $y = f(x)$.

10. If $f(x) = 5x^2 - 7x + 2$, find $f(3)$.

11. If $f(x) = 9 - 3x$, find $2f(3) + 3f(1) - 4f(2)$.

12. If $f(x) = 7x + 5$ and $g(x) = x^2$, find $3f(2) - 5g(3)$.

13. If $f(x, y, z) = x^2 + 3xy + z^3$, find $f(3, 2, 1)$.

14. If $f(x) = 8x + 3$ and $g(u) = u^2 - 4$, write $f[g(u)]$.

15. If $f(x) = 5x$, $g(x) = 1/x$, and $h(x) = x^3$, find

$$\frac{5h(1) - 2g(3)}{3f(2)}$$

16. Is the relation $x^2 + 3xy + y^2 = 1$ a function? Why?

17. Replace the function $y = 5x^3 - 7$ by a verbal statement.

18. What are the domain and the range of the function $(10, -8)$, $(20, -5)$, $(30, 0)$, $(40, 3)$, and $(50, 7)$?

19. If $y = 6 - 3x^2$, write $x = f(y)$.

20. If $x = 6w - 5y$ and $y = 3z + 2w$, write $x = f(w, z)$.

21. Write the inverse of the function $y = 3x + 4(2 - x)$.

22. If $xw - 4w^2 = 5 + 4x$, write $x = f(w)$.

23. If $7p - 2q = 3q - p^2$, write $q = f(p)$.

24. Given the functions $g(x) = x + x^2$ and $f(x) = 5x$, write the composite function $g[f(x)]$.

25. If $h(p, q, r) = 5r + p + 7q$, find $h(1, 2, 3) + \dfrac{2h(1, 1, 1) - 3h(2, 2, 2)}{h(3, 3, 3)}$

26. If $g_3(w) = 1.74 + 9.25w$, find $g_3(1.44)$.

27. Write the inverse of the function $y = 5(7 - 4x) - x$.

28. If $g(k, l, m) = \dfrac{k + 3m}{2l}$, find $g(4.62, 1.39, 7.26)$.

29. Write the inverse of the function $y = 9x + (x + 5)$.

30. The length of a certain steel bar at temperature t is given by

$$L = f(t) = 112(1 + 0.0655t) \text{ in.}$$

   Find $f(112)$, $f(176)$, and $f(195)$.

31. If $f_1(z) = 3.82z^2 - 2.46$, find $f_1(5.27)$.

32. The power P in a resistance R in which flows a current I is given by

$$P = f(I) = I^2 R \quad \text{W}$$

   If $R = 325\ \Omega$, find $f(11.2)$, $f(15.3)$, and $f(21.8)$.

33. Given the functions $g(x) = 3x - 5$ and $f(x) = 7x^2$, write the composite function $f[g(x)]$.

34. Given $g(x) = 5x^2$ and $f(x) = 7 - 2x$ find:

    (a) $f[g(x)]$ **(b)** $g[f(x)]$

    (c) $f[g(5)]$ **(d)** $g[f(5)]$

Writing

35. State, in your own words, what is meant by "function" and "relation," and describe how a function differs from a relation. You may use examples to help describe these terms.

Graphs

When you have completed this chapter, you should be able to:

- Graph points and empirical data in rectangular coordinates.
- Make a complete graph of a relation or a function.
- Graph parametric functions in rectangular coordinates.
- Solve equations graphically.

In this chapter we study *graphing*, a topic that has seen renewed interest with the availability of the graphics calculator. We start by explaining the *rectangular coordinate system* and the plotting of single points, and then we continue with the graphing of functions and relations. The idea of a *complete graph* is stressed.

The *equation of a straight line* is also introduced in this chapter. We conclude by showing how approximately to solve any equation by a graphical method. This is extremely important because it enables us to do solutions that would not be possible by any other method.

5–1   Rectangular Coordinates

The Rectangular Coordinate System

Rectangular coordinates are also called *Cartesian* coordinates, after the French mathematician René Descartes (1596–1650). Another type of coordinate system we will use is called the *polar* coordinate system.

In Chapter 1 we plotted numbers on the number line. Suppose, now, that we take a second number line and place it at right angles to the first one, so that each intersects the other at the zero mark, as in Fig. 5–1. We call this a *rectangular coordinate system*.

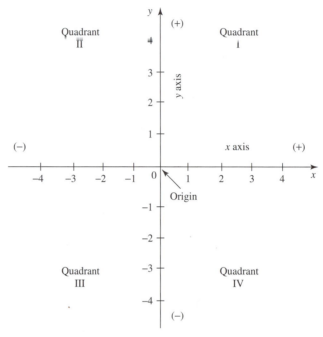

FIGURE 5–1   The rectangular coordinate system.

The horizontal number line is called the *x axis*, and the vertical line is called the *y axis*. They intersect at the *origin*. These two axes divide the plane into four *quadrants*, numbered counterclockwise, as in Fig. 5–1.

Graphing Ordered Pairs

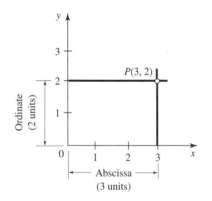

FIGURE 5–2   Is it clear from this figure why we call these *rectangular coordinates?*

Figure 5–2 shows a point P in the first quadrant. Its horizontal distance from the origin, called the *x coordinate* or *abscissa* of the point, is 3 units. Its vertical distance from the origin, called the *y coordinate* or *ordinate* of the point, is 2 units. The numbers in the *ordered pair* (3, 2) are called the *rectangular coordinates* (or simply *coordinates*) of the point. They are always written in the same order, with the x coordinate first. The letter identifying the point is sometimes written before the coordinates, as in $P(3, 2)$.

To plot any ordered pair (h, k), simply place a point at a distance h units from the y axis and k units from the x axis. Remember that negative values of x are located to the left of the origin and that negative y values are below the origin.

◆◆◆ **Example 1:**   The points

$$P(4, 1) \quad Q(-2, 3) \quad R(-1, -2) \quad S(2, -3) \quad T(1.3, 2.7)$$

are shown plotted in Fig. 5–3.

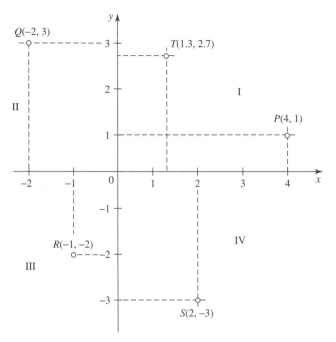

FIGURE 5–3

Notice that the abscissa is negative in the second and third quadrants and that the ordinate is negative in the third and fourth quadrants. Thus the signs of the coordinates of a point tell us the quadrant in which the point lies. ◆◆◆

◆◆◆ **Example 2:** The point $(-3, -5)$ lies in the third quadrant, for that is the only quadrant in which the abscissa and the ordinate are both negative. ◆◆◆

Exercise 1 ◆ Rectangular Coordinates

Rectangular Coordinates

If h and k are positive quantities, in which quadrants would the following points lie?

1. $(h, -k)$ **2.** (h, k)

3. $(-h, k)$ **4.** $(-h, -k)$

5. Which quadrant contains points having a positive abscissa and a negative ordinate?

6. In which quadrants is the ordinate negative?

7. In which quadrants is the abscissa positive?

8. The ordinate of any point on a certain straight line is -5. Give the coordinates of the point of intersection of that line and the y axis.

9. Find the abscissa of any point on a vertical straight line that passes through the point $(7, 5)$.

Graphing Ordered Pairs

10. Write the coordinates of points, A, B, C, and D in Fig. 5–4.

11. Write the coordinates of points E, F, G, and H in Fig. 5–4.

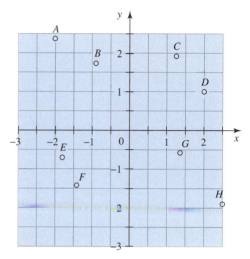

FIGURE 5–4

12. Write the coordinates of the points A, B, C, and D in Fig. 5–5.

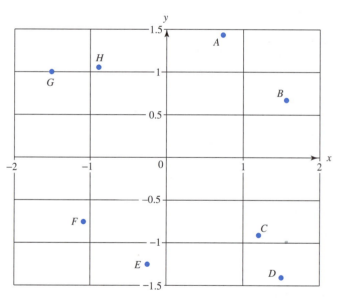

FIGURE 5–5

13. Write the coordinates of the points E, F, G, and H in Fig. 5–5.

14. Graph each point.
   (a) (3, 5)   (b) (4, −2)   (c) (−2.4, −3.8)
   (d) (−3.75, 1.42)   (e) (−4, 3)   (f) (−1, −3)

Graph each set of points, connect them, and identify the geometric figure formed.

15. (0.7, 2.1), (2.3, 2.1), (2.3, 0.5), and (0.7, 0.5)
16. $(2, -\frac{1}{2})$, $(3, -1\frac{1}{2})$, $(1\frac{1}{2}, -3)$, and $(\frac{1}{2}, -2)$
17. $(-1\frac{1}{2}, 3)$, $(-2\frac{1}{2}, \frac{1}{2})$, and $(-\frac{1}{2}, \frac{1}{2})$
18. $(-3, -1)$, $(-1, -\frac{1}{2})$, $(-2, -3)$, and $(-4, -3\frac{1}{2})$

19. Three corners of a rectangle have the coordinates $(-4, 9)$, $(8, 3)$, and $(-8, 1)$. Graphically find the coordinate of the fourth corner.

5–2   The Graph of a Function

If the function is given as a set of ordered pairs, simply plot each ordered pair. We usually connect the points with a smooth curve unless we have reason to believe that there are sharp corners, breaks, or gaps in the graph.

If the function is in the form of an *equation*, we obtain a table of ordered pairs by first selecting values of x over the required domain and then computing corresponding values of y. Since we are usually free to select any x values we like, we pick "easy" integer values. We then plot the set of ordered pairs.

Our first graph will be of the straight line.

◆◆◆ **Example 3:** Graph the function $y = f(x) = 2x - 1$ for values of x from -2 to 2.

Solution: Substituting into the equation, we obtain

$$f(-2) = 2(-2) - 1 = -5$$
$$f(-1) = 2(-1) - 1 = -3$$
$$f(0) \quad = 2(0) \quad - 1 = -1$$
$$f(1) \quad = 2(1) \quad - 1 = \quad 1$$
$$f(2) \quad = 2(2) \quad - 1 = \quad 3$$

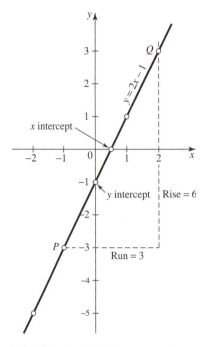

FIGURE 5–6   A first-degree equation will always plot as a straight line— hence the name *linear* equation.

Thus our points are $(-2, -5)$, $(-1, -3)$, $(0, -1)$, $(1, 1)$, and $(2, 3)$. Note that each of these pairs of numbers satisfies the given equation.

These points plot as a straight line (Fig. 5–6). Had we known in advance that the graph would be a straight line, we could have saved time by plotting just two points, with perhaps a third as a check.   ◆◆◆

In Sec. 5–3 we will discuss the straight line in more detail and show another way to graph it.

Our next graph will be of a curve called the *parabola*. We graph it, and any other function, in the same way that we graphed the straight line: Make a table of point pairs, plot the points, and connect them.

◆◆◆ **Example 4:** Graph the function $y = f(x) = x^2 - 4x - 3$ for values of x from -1 to 5.

Solution: Substituting into the equation, we obtain

$$f(-1) = (-1)^2 - 4(-1) - 3 = 1 + 4 - 3 = 2$$
$$f(0) \quad = 0^2 - 0 - 3 = -3$$
$$f(1) \quad = 1^2 - 4(1) - 3 = 1 - 4 - 3 = -6$$
$$f(2) \quad = 2^2 - 4(2) - 3 = 4 - 8 - 3 = -7$$
$$f(3) \quad = 3^2 - 4(3) - 3 = 9 - 12 - 3 = -6$$
$$f(4) \quad = 4^2 - 4(4) - 3 = 16 - 16 - 3 = -3$$
$$f(5) \quad = 5^2 - 4(5) - 3 = 25 - 20 - 3 = 2$$

If we had plotted many more points than these—say, billions of them—they would be crowded so close together that they would seem to form a continuous line. The curve can be thought of as a *collection* of all points that satisfy the equation. Such a curve (or the set of points) is called a *locus* of the equation.

The points obtained are plotted in Fig. 5–7.

The straight line and the parabola are covered in much greater detail in analytic geometry in Chapter 22.

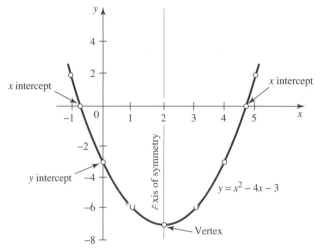

FIGURE 5–7   This curve is called a *parabola*, which we will meet again later. Note the *vertex* and the *axis of symmetry*.

◆◆◆

Common Error	Be especially careful when substituting negative values into an equation. It is easy to make an error.

Exercise 2 ◆ The Graph of a Function

Graphing Sets of Ordered Pairs

Graph each set of ordered pairs. Connect them with a curve that seems to you to fit the data best.

1. $(-3, -2)$, $(9, 6)$, $(3, 2)$, $(-6, -4)$
2. $(-7, 3)$, $(0, 3)$, $(4, 10)$, $(-6, 1)$, $(2, 6)$, $(-4, 0)$
3. $(-10, 9)$, $(-8, 7)$, $(-6, 5)$, $(-4, 3)$, $(-2, 4)$, $(0, 5)$, $(2, 6)$, $(4, 7)$
4. $(0, 4)$, $(3, 3.2)$, $(5, 2)$, $(6, 0)$, $(5, -2)$, $(3, -3.2)$, $(0, -4)$

Graph of a Straight Line

Graph each straight line by plotting two points on the line plus a third as a check.

5. $y = 3x + 1$ 6. $y = 2x - 2$
7. $y = 3 - 2x$ 8. $y = -x + 2$

Graph of Any Function

For each function, make a table of ordered pairs, taking integer values of x from -3 to 3. Plot the point pairs and connect them with a smooth curve.

9. $y = x^2$ 10. $y = 4 - 2x^2$

11. $y = \dfrac{x^2}{x + 3}$ 12. $y = x^2 - 7x + 10$

13. $y = x^2 - 1$ 14. $y = 5x - x^2$
15. $y = x^3$ 16. $y = x^3 - 2$

Computer

17. Write a program to generate a table of (x, y) point pairs for any function that you enter. You should also be able to enter the domain of x and the interval between your points. Use your program to obtain plotting points for any of the equations in this exercise, and plot the points by hand.

18. Many graphing programs are available for the computer. Most mathematical software, such as *MathCAD* and *Derive*, have built-in graphing capabilities. Some spreadsheet programs, such as *Lotus 1-2-3*, are able to create graphs of data in the spreadsheet. If you have any such program available, learn how to use it. Practice with any of the graphs in this exercise.

5–3   Graphing the Straight Line

We graphed a straight line in Sec. 5–2 by finding the coordinates of two points on the line, plotting those points, and connecting them. Here, we will show how to plot a straight line by using the steepness of the line, called the *slope*, and the point where it crosses the *y* axis, called the *y intercept*.

This is still introductory material; our main study of the straight line will come in Chapter 22.

Slope

The horizontal distance between any two points on a straight line is called the *run*, and the vertical distance between the same points is called the *rise*. For points P and Q in Fig. 5–6, the rise is 6 in a run of 3. The rise divided by the run is called the *slope* of the line and is given the symbol m.

$$\text{slope} = m = \frac{\text{rise}}{\text{run}}$$

◆◆◆ **Example 5:** The straight line in Fig. 5–6 has a rise of 6 in a run of 3, so

$$\text{slope } m = \frac{\text{rise}}{\text{run}} = \frac{6}{3} = 2$$

◆◆◆

◆◆◆ **Example 6:** Find the slope of a straight line that drops half a unit as we move 1 unit in the positive x direction.

Solution: The rise is $-1/2$ in a run of 1, so the slope is

$$\text{slope } m = \frac{-1/2}{1} = -1/2$$

◆◆◆

◆◆◆ **Example 7:** Find the slope of the line in Fig. 5–8.

Solution: The rise is $30 - 10 = 20$ and the run is $2 - 0 = 2$, so

$$m = \frac{20}{2} = 10$$

(Do not be thrown off by the different scales on each axis).

◆◆◆

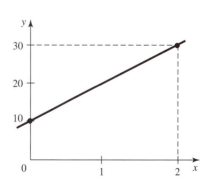

FIGURE 5–8

If (x_1, y_1) and (x_2, y_2) are two points on the line, the slope is given by the following formula:

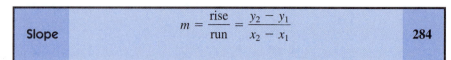

| Slope | $m = \dfrac{\text{rise}}{\text{run}} = \dfrac{y_2 - y_1}{x_2 - x_1}$ | 284 |

The slope is equal to the rise divided by the run.

◆◆◆ Example 8: Find the slope of the line connecting the points $(-3, 5)$ and $(4, -6)$ in Fig. 5–9.

Solution. We will see that it does not matter which is called point 1 and which is point 2. Let us choose

$$x_1 = -3, \qquad y_1 = 5, \qquad x_2 = 4, \qquad y_2 = -6$$

Then by Eq. 284,

$$\text{slope } m = \frac{-6 - 5}{4 - (-3)} = \frac{-11}{7} = -\frac{11}{7}$$

If we had chosen $(x_1 = 4, y_1 = -6)$ and $(x_2 = -3, y_2 = 5)$, the computation for slope,

$$\text{slope } m = \frac{5 - (-6)}{-3 - 4} = \frac{11}{-7} = -\frac{11}{7}$$

would have given the same result.   **◆◆◆**

| Common Error | Be careful not to mix up the subscripts. $m \neq \dfrac{y_2 - y_1}{x_1 - x_2}$ |

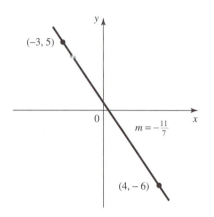

(−3, 5)

$m = -\frac{11}{7}$

(4, −6)

FIGURE 5–9

Horizontal and Vertical Lines

For any two points on a horizontal line, the values of y_1 and y_2 in Eq. 284 are equal, making the slope equal to zero. For a vertical line, the values of x_1 and x_2 are equal, giving division by zero. Hence the slope is undefined for a vertical line. The slopes of various lines are shown in Fig. 5–10.

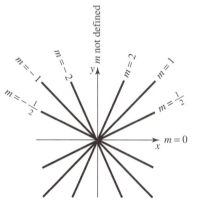

FIGURE 5–10

Some graphics calculators can automatically find the y intercept of a plotted function. Check your calculator manual.

Intercepts

The *intercepts* of a curve are the places where the curve crosses the x and y axes. The term *intercept* is used for any curve, not just the straight line.

◆◆◆ Example 9: The x intercept for the straight line in Fig. 5–6 is $(1/2, 0)$, and the y intercept is $(0, -1)$.   **◆◆◆**

Equation of a Straight Line

Suppose that $P(x, y)$ is any point on a straight line. We seek an equation that links x and y in a functional relationship so that, for any x, a value of y can be found. We can get such an equation by applying the definition of slope to our point P and some *known* point on the line. Let us use the y intercept $(0, b)$ as the coordinates for the known point. For P, a general point on the line, we use coordinates (x, y).

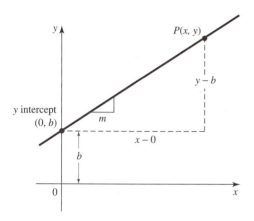

FIGURE 5–11

For a line that intersects the y axis b units from the origin (Fig. 5–11), the rise is $y - b$ and the run is $x - 0$, so by Eq. 284,

$$m = \frac{y - b}{x - 0}$$

Simplifying, we have $mx = y - b$, also written in the following form:

Straight Line, Slope-Intercept Form	$y = mx + b$	289

This is called the slope-intercept form of the equation of a straight line because the slope m and the y intercept b are easily identified once the equation is in this form. For example, in the equation $y = 2x + 1$,

$$y = 2\underset{\text{slope}}{x} + \underset{\text{y intercept}}{1}$$

Here m, the slope, is 2 and b, the y intercept, is 1.

◆◆◆ **Example 10:** Write the equation of the straight line, in slope-intercept form, that has a slope of 2 and a y intercept of 3. Make a graph.

Solution: Substituting $m = 2$ and $b = 3$ into Eq. 289 gives

$$y = 2x + 3$$

To graph the line, we first locate the y intercept, 3 units up from the origin. The slope of the line is 2, so we get another point on the line by moving 1 unit to the right and 2 units up from the y intercept. This brings us to (1, 5). Connecting this point to the y intercept gives us our line as shown in Fig. 5–12. ◆◆◆

◆◆◆ **Example 11:** Find the slope and the y intercept of the line $y = 3x - 4$. Make a graph.

Solution: By inspection, we see that the slope is the coefficient of x and that the y intercept is the constant term. So

$$m = 3 \quad \text{and} \quad b = -4$$

We plot the y intercept 4 units below the origin, and through it we draw a line with a rise of 3 units in a run of 1 unit, as shown in Fig. 5–13. ◆◆◆

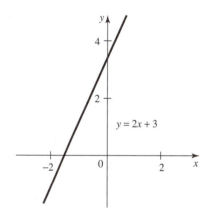

FIGURE 5–12

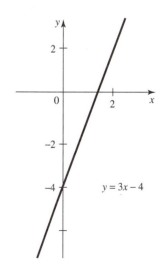

FIGURE 5–13

Exercise 3 ◆ Graphing the Straight Line

Find the slope of each straight line.

1. Rise = 4; run = 2
2. Rise = 6; run = 4
3. Rise = −4; run = 4
4. Rise = −9; run = −3
5. Connecting (2, 4) and (5, 7)
6. Connecting (5, 2) and (3, 6)
7. Connecting (−2, 5) and (5, −6)
8. Connecting (3, −3) and (−6, 2)

Write the equation of each straight line in slope-intercept form, and make a graph.

9. Slope = 4; y intercept = −3
10. Slope = −1; y intercept = 2
11. Slope = 3; y intercept = −1
12. Slope = −2; y intercept = 3
13. Slope = 2.3; y intercept = −1.5
14. Slope = −1.5; y intercept = 3.7

Find the slope and the y intercept for each equation, and make a graph.

15. $y = 3x - 5$ **16.** $y = 7x + 2$
17. $y = -\frac{1}{2}x - \frac{1}{4}$ **18.** $y = -3x + 2$

5–4   The Graphics Calculator

The *graphics calculator* will help us to quickly make graphs. To do this we must:

1. Set the domain and the range.
2. Enter the function.

The names for the keys given in this section are generic. The exact names for the keys on your calculator may differ slightly. However, the functions should be exactly the same.

You should now look at your calculator manual to find out which keys are used. To set the domain and the range, look for a key marked $\boxed{\text{Range}}$. To enter the function, you will need a key probably marked $\boxed{Y =}$ or $\boxed{\text{Graph}}$.

◆◆◆ **Example 12:** Graph the parabola of Example 4 using a graphics calculator. Use the same domain and range as in that example.

Solution: We first set the domain and the range. Pressing the $\boxed{\text{Range}}$ key will give you a screen that will look something like this:

> Range
> Xmin = −10
> Xmax = 10
> Xscl = 1
> Ymin = −10
> Ymax = 10
> Yscl = 1

Here, Xmin and Xmax define the domain, and Ymin and Ymax give the range. This screen shows both domain and range set to *default values* of −10 to 10 (your calculator may show different values). The Xscl value shows the spacing of the tick marks on the *x* axis. With an Xscl setting of 1, there will be a tick mark at each unit. The Yscl value does the same thing for the *y* axis.

We will now change these values. We know from Example 4 that *y* will vary between -7 to 2, when *x* has the given domain, so we set Ymin to -8 and Ymax to 3. If we were graphing this curve for the first time, we would not have known these values and would have to find them by trial and error. We'll do this in a later example.

The calculator *cursor* should be at the Xmin value (-10). Key in -1 and press [ENTER] [EXE]. The cursor should now be at the Xmax value. Key in 5 and press [ENTER]. Continue through the list. If the value shown is not the one you want, simply replace it. If it is already correct, press [ENTER] to proceed to the next line.

Set Xscl and Yscl each to 1. When you finish, your screen should look something like this:

Range
Xmin = -1
Xmax = 5
Xscl = 1
Ymin = -8
Ymax = 3
Yscl = 1

Next we enter the function. Press the [Y =] key or the [Graph] key, and the screen will display something like

Y =

Enter the function $x^2 - 4x - 3$ to the right of the equal sign. Your screen should look something like this:

Y = X^2 − 4X − 3

To see the graph, press the [Graph] key. You should get the same graph as in Example 4.   ◆◆◆

Complete Graph

In the preceding examples we were told which *x* values to use for the graph. But if we want to graph a function to see how it behaves, we would not bother with those regions where the curve is simply rising or falling without change, but we would include those *x* values that show *all of the features of interest*. Those features of interest include

Intercepts on the *x* or *y* axes.

Peaks on the curve, called *maximum points*, such as *A* in Fig. 5–14.

Valleys on the curve, called *minimum points*, such as *B* in Fig. 5–14.

The *domain* of the function, and those regions where the curve does not exist, such as the shaded regions in Fig. 5–15.

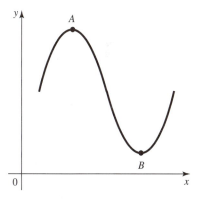

FIGURE 5–14

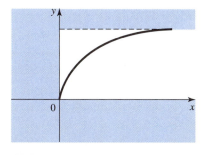

FIGURE 5–15

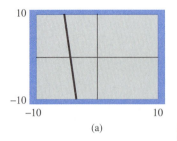

(a)

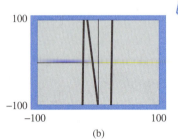

(b)

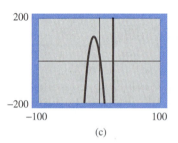

(c)

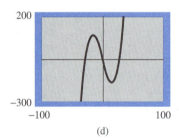

(d)

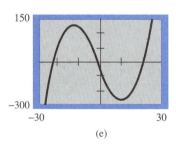

(e)

FIGURE 5–16

Most graphing calculators have a ⬛zoom⬛ feature which allows you to quickly make both ranges larger or smaller. Check your manual to see how it is used.

A graph showing all of the features of interest is sometimes called a *complete graph*. With some experience you will often be able to tell just by looking at a function if its graph will have certain features, such as how many maximum and minimum points it will have. But until you have such experience, you can find the features of interest simply by trial and error. The graphics calculator is ideal for such exploration, as shown in the following example.

◆◆◆ **Example 13:** Make a complete graph of the function

$$y = 0.054x^3 - 23x - 80$$

Solution: Let us set our ranges for both x and y from -10 to 10. We key in the given function and get the graph in Fig. 5–16(a). We notice an x intercept at about $x = -3$.

Suspecting that there is more to the graph than what shows in this viewing window, we reset both ranges from -100 to $+100$. We get the graph of Fig. 5–16(b) and see two more x intercepts. However, the curve seems to be "clipped" at the top and bottom, so let us increase the y range to -200 to $+200$, keeping the x range unchanged. We get Fig. 5–16(c) and observe a maximum point in the second quadrant and a y intercept below the origin. The curve still seems clipped at the bottom, so let us change Ymin to -300. This reveals a minimum point, Fig. 5–16(d). We are fairly sure now (but not positive) that all of the features of interest are within the viewing window. Let us finally adjust the ranges so that the curve nicely fills the viewing window. We set the x range from -30 to 30 and the y range from -300 to 150.

We have been ignoring the Xscl and Yscl settings until we found suitable ranges. Let us now set the x scale to 10 and the y scale to 50, getting the graph of Fig. 5–16(e).

We see three x intercepts, one y intercept, a maximum point, and a minimum point. We'll show how to get the approximate coordinates of those points in the following section.   ◆◆◆

Using the ⬛TRACE⬛ Key

When the ⬛TRACE⬛ key is pressed, a small cursor will appear on the curve in the viewing window. Further, the x and y coordinates of the cursor will be displayed on the screen. (Some calculators show only the x coordinate. Use the ⬛X<>Y⬛ key to show the y coordinate.) Pressing the left and right arrow keys will cause the cursor to move along the curve, with the new coordinate(s) displayed as you move. This provides an easy way of finding the approximate coordinates of any point of interest on the graph.

◆◆◆ **Example 14:** Using the ⬛TRACE⬛ key, find the approximate coordinates of the intercepts and the maximum and minimum points on the curve of Example 13.

Solution: Going from left to right along the curve, we read

$$x \text{ intercept at } x = -18.6$$
$$\text{Maximum at } (-11.7, 102.6)$$
$$x \text{ intercept at } x = -3.5$$
$$y \text{ intercept at } y = -72.7$$
$$\text{Minimum at } (11.7, -262.6)$$
$$x \text{ intercept at } x = 22.4$$

◆◆◆

Note that the coordinates in Example 14 are not very accurate, because the scale for this graph is very large. But we need not be satisfied with this accuracy.

All we have to do is to *zoom in* to any particular point of interest to get more accuracy. In fact, *we can zoom in repeatedly* to get whatever degree of accuracy we want.

Exercise 4 ◆ The Graphics Calculator

Graph these functions from Exercise 2, using a graphics utility. Compare your graphs with those you got manually.

1. $y = x^2$

2. $y = 4 - 2x^2$

3. $y = \dfrac{x^2}{x + 3}$

4. $y = x^2 - 7x + 10$

5. $y = x^2 - 1$

6. $y = 5x - x^2$

7. $y = x^3$

8. $y = x^3 - 2$

Make a complete graph of each function, choosing a viewing window that includes all features of interest. Use the $\boxed{\text{TRACE}}$ key to find the approximate coordinates of any x intercepts, maximum points, or minimum points.

9. $y = 2x^2 - 14x + 22$

10. $y = 7x^2 + 9x - 14$

11. $y = 4x^3 - 4x^2 + 11x - 24$

12. $y = 5x^4 + 13x^2 - 31$

5–5 Graphing Empirical Data, Formulas, and Parametric Equations

Graphing Empirical Data

Data obtained from an experiment or by observation are called *empirical* data. Empirical data are graphed just as any other set of point pairs. Be sure to label the graph completely, including units on the axes.

◆◆◆ **Example 15:** Graph the following data for the temperature rise in a certain oven:

Time (h)	0	1	2	3	4	5	6	7	8	9	10
Temperature (°F)	102	463	748	1010	1210	1370	1510	1590	1710	1770	1830

Solution: We plot each point, connect the points, and label the graph, as shown in Fig. 5–17.

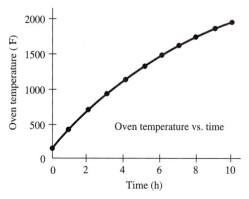

FIGURE 5–17 Note the use of different scales on each axis. ◆◆◆

Graphing Formulas

A formula is graphed in the same way that we graph a function: We substitute suitable values of the independent variable into the formula and compute the corresponding values of the dependent variable. We then plot the ordered pairs and connect the points obtained with a smooth curve. But with formulas we must be careful to handle the units properly and to label the graph more fully.

♦♦♦ **Example 16:** The formula for the power P dissipated in a resistor carrying a current of i amperes (A) is

$$P = i^2 R \quad \text{watts (W)}$$

where R is the resistance in ohms. Graph P versus i for a resistance of 10,000 Ω. Take i from 0 to 10 A.

Solution: Let us choose values of i of 0, 1, 2, 3, . . . , 10 and make a table of ordered pairs by substituting these into the formula.

i (A)	0	1	2	3	$\cdots$
P (W)	0	10,000	40,000	90,000	$\cdots$

At this point we notice that the figures for wattage are so high that it will be more convenient to work in kilowatts (kW), where 1 kW = 1000 W.

i (A)	0	1	2	3	4	5	6	7	8	9	10
P (kW)	0	10	40	90	160	250	360	490	640	810	1000

These points are plotted in Fig. 5–18. Note the labeling of the graph and of the axes.   ♦♦♦

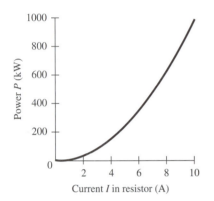

FIGURE 5–18   Power dissipated in a 10,000-Ω resistor calculated using $P = i^2R$ with $R = 10,000\ \Omega$.

Graphing Parametric Equations

For the equations we have graphed so far, y has been expressed as a function of x.

$$y = f(x)$$

But x and y can also be related to each other by means of a third variable, say, t, if both x and y are given as functions of t.

$$x = g(t)$$

and

$$y = h(t)$$

Such equations are called *parametric equations*. The third variable t is called the *parameter*.

To graph parametric equations, we assign values to the parameter t and compute x and y for each t. We then plot the table of (x, y) pairs.

♦♦♦ **Example 17:** Graph the parametric equations

$$x = 2t \quad \text{and} \quad y = t^2 - 2$$

for $t = -3$ to 3.

Solution: We make a table with rows for t, x, and y. We take values of t from -3 to 3, and for each we compute x and y.

t	-3	-2	-1	0	1	2	3
x	-6	-4	-2	0	2	4	6
y	7	2	-1	-2	-1	2	7

In a later chapter we'll plot parametric equations in *polar* coordinates. We'll also use polar coordinates to describe the *trajectory* of a projectile.

We now plot the (x, y) pairs, $(-6, 7)$, $(-4, 2)$, . . . , $(6, 7)$ and connect them with a smooth curve (Fig. 5–19). The curve obtained is a parabola, as in Example 4, but obtained here with parametric equations. ◆◆◆

We can quickly plot parametric equations with a graphics calculator.

◆◆◆ **Example 18:** Plot the parametric equations from Example 17 with a graphics calculator.

Solution: We first put the calculator into parametric equation mode. This option may have to be selected from a menu; check your calculator manual.
  Next we enter the equations.

X1T = 2T
Y1T = T² − 2

Finally, we set the range:

Tmin = −3
Tmax = 3
Tstep = .1
Xmin = −6
Xmax = 6
Xscl = 1
Ymin = −2
Ymax = 7
Yscl = 1

The graph obtained will be the same as in Fig. 5–19. ◆◆◆

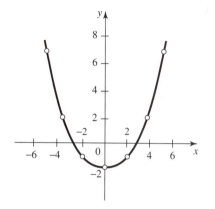

FIGURE 5–19

Exercise 5 ◆ Graphing Empirical Data, Formulas, and Parametric Equations

Graphing Empirical Data

Graph the following experimental data. Label the graph completely. Take the first quantity in each table as the abscissa, and the second quantity as the ordinate. Connect the points with a smooth curve.

1. The current I (mA) through a tungsten lamp and the voltage V (V) that is applied to the lamp, related as shown in the following table:

V	10	20	30	40	50	60	70	80	90	100	110	120
I	158	243	306	367	420	470	517	559	598	639	676	710

2. The melting point T (°C) of a certain alloy and the percent of lead P in the alloy, related as shown in the following table:

P	40	50	60	70	80	90
T	186	205	226	250	276	304

3. A steel wire in tension, with the stress σ (lb/in.^{2}) and the strain ϵ (in./in.), related as follows:

ε	0	0.00019	0.00057	0.00094	0.00134	0.00173	0.00216	0.00256
σ	5000	10,000	20,000	30,000	40,000	50,000	60,000	70,000

Graphing Formulas

4. A milling machine having a purchase price P of \$15,600 has an annual depreciation A of \$1600. Graph the book value y at the end of each year, for $t = 10$ years, using the equation $y = P - At$.

5. The force f required to pull a block along a rough surface is given by $f = \mu N$, where N is the normal force and μ is the coefficient of friction. Plot f for values of N from 0 to 1000 N, taking μ as 0.45.

6. A 2580-Ω resistor is wired in parallel with a resistor R_1. Graph the equivalent resistance R for values of R_1 from 0 to 5000 Ω, in steps of 500 Ω. Use Eq. A64.

7. Use Eq. A67 to graph the power P dissipated in a 2500-Ω resistor for values of the current I from 0 to 10 A.

8. A resistance of 5280 Ω is placed in series with a device that has a reactance of X ohms (Fig. 5–20). Using Eq. A99, plot the absolute value of the impedance Z for values of X from 0 to 10,000 Ω.

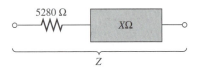

FIGURE 5–20

Graphing Parametric Equations

Plot the following parametric equations for values of t from -3 to 3.

9. $x = t,\ y = t$ **10.** $x = 3t,\ y = t^2$

11. $x = -t,\ y = 2t^2$ **12.** $x = -2t,\ y = t^2 + 1$

5–6   Graphical Solution of Equations

Solving Equations Graphically

We can get two different expressions when we put the given equation into the form $f(x) = 0$. Thus if $x^2 = 2x + 2$, we can write either

$$f(x) = x^2 - 2x - 2$$

or

$$f(x) = -x^2 + 2x + 2$$

It doesn't matter which we use because both cross the x axis at the same points.

We can use our knowledge of graphing functions to solve equations of the form $f(x) = 0$.

We mentioned earlier that a point at which a graph of a function $y = f(x)$ crosses or touches the x axis is called an x intercept. Such an x intercept is also called a *zero* of that function.

In Fig. 5–21 there are two zeros, since there are two x values for which $y = 0$, and hence $f(x) = 0$. Those x values for which $f(x) = 0$ are called *roots* or *solutions* to the equation $f(x) = 0$.

Thus if we were to graph the function $y = f(x)$, any value of x at which y is equal to zero would be a solution to $f(x) = 0$. So to solve an equation graphically, we simply put it into the form $f(x) = 0$ and then graph the function $y = f(x)$. Each x intercept is then an approximate solution to the equation.

◆◆◆ **Example 19:** Graphically find the approximate root(s) of the equation

$$4.1x^3 - 5.9x^2 - 3.8x + 7.5 = 0 \qquad (1)$$

Solution: Let us represent the left side of the given equation by $f(x)$.

$$f(x) = 4.1x^3 - 5.9x^2 - 3.8x + 7.5$$

Any value of x for which $f(x) = 0$ will clearly be a solution to Equation (1), so we simply plot $f(x)$ and look for the x intercepts. Not knowing the shape of the curve, we compute $f(x)$ at various values of x until we are satisfied that we have located each region in which an x intercept is located. We then make a table of point pairs for this region.

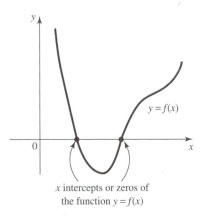

x intercepts or zeros of the function $y = f(x)$

FIGURE 5–21

x	-2	-1	0	1	2	3
$f(x)$	-41.3	1.3	7.5	1.9	9.1	53.7

We graph the function (Fig. 5–22) and read the approximate value of the x intercept.

$$x \cong -1.1$$

This, then, is an approximate solution to the given equation. Have we found *every* root? If we computed y for values of x outside the region that we have graphed, we would see that the curve moves even farther from the x axis, so we are reasonably sure that we have found the only root. With other functions it may not be so easy to tell if the curve will reverse direction somewhere and cross the x axis again. ◆◆◆

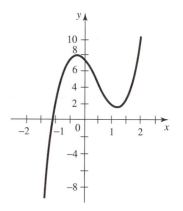

FIGURE 5–22 The shape of this curve is typical of a third-degree (cubic) function.

Solving Equations with the Graphics Calculator

In Sec. 5–2, we showed how to use the TRACE features of a graphics calculator to find intercepts. Since the graphical solution of an equation requires finding an x intercept, the same method will work here.

◆◆◆ **Example 20:** Repeat Example 19 using TRACE on a graphics calculator.

Solution: We key in the function and display the graph in the standard viewing window, with the x range set at -10 to 10. We see that the curve crosses the x axis at about $x = -1$. Let us reduce the x range in steps, each time using TRACE to locate the x intercept as closely as we can. Also, as we reduce the x range, we also reduce the y range (values not shown) so that the curve is steep enough to clearly see where it appears to cross the x axis. The cursor jumps in steps and may not land right on the intercept, so we get as close as we can. The following table shows the coordinates of the point closest to the intercept, for the various x ranges tried:

x range	x	y
-2 to 0	-1.052632	0.18056568
-1.1 to -1	-1.061053	-0.00814060
-1.07 to -1.05	-1.060526	0.00373216
-1.061 to -1.060	-1.060695	-0.00006597

Some graphics calculators can automatically find a root when the cursor is moved to a point on the curve near that root. Again, read your calculator manual for the proper instructions to use.

Now, if we were exactly *at* the intercept, the value of y would be *zero*. Thus the smaller the value of y, the closer we are to the intercept. We can make the value of y as small as we please, within the limits of the calculator, by zooming in even further.

To get a measure of the accuracy of the root, subtract the x value at any point from the x value one step to the right. This will give you the *step size* in the x direction.

Thus for the final x range in our example, suppose that the step size is found to be 0.000011. That shows that the fifth decimal place in our intercept has some uncertainty but that any digits to the right of that are unreliable. Thus we give our root as

$$x = -1.0607$$

◆◆◆

Computer Solution of Equations by the Midpoint Method

There are several methods for finding the roots of equations by computer. In Chapter 14 we show the method of *simple iteration*, and when you study calculus, you will learn the popular *Newton's method*. Here we cover the simple *midpoint* method.

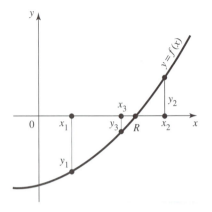

FIGURE 5–23 The midpoint method for finding roots.

This method is also called the *half-interval method* or the *bisection method*.

A technique such as this one, in which we repeat a computation many times, each time getting an approximate result that is (with luck) closer to the true value, is called *iteration*. Iteration methods are used often on the computer, and we give several in this text.

Figure 5–23 shows a graph of some function $y = f(x)$. It is a smooth, continuous curve that crosses the x axis at R, which is a root of the equation $f(x) = 0$. We determine two values x_1 and x_2 that lie to the left and right of R, a fact that we verify by testing whether y_1 and y_2 have opposite signs. We then locate x_3 midway between x_1 and x_2, and then we determine whether it is to the left or to the right of R by comparing the sign of y_3 with that of y_1. Thus x_3 becomes a new endpoint (either the left or the right as determined by the test), and we repeat the computation, halving the interval again and again until we get as close to R as we wish.

Following is an algorithm for the midpoint method:

1. Make two initial guesses, x_1 and x_2, which must lie on either side of the root R.
2. Go to Step 3 if y_1 and y_2 have opposite signs. Otherwise, complete Step 1.
3. Compute the abscissa x_3 of the point midway between the two original points, with

$$x_3 = \frac{x_1 + x_2}{2}$$

4. Compute y_1 and y_3 by substituting x_1 and x_3 into the equation of the curve.
5. If the absolute value of y_3 is "small," then print the value of x_3 and end.
6. Determine whether x_3 is to the right or to the left of the root by comparing the signs of y_1 and y_3.
7. If y_1 and y_3 have the same signs, replace x_1 with x_3. Otherwise, replace x_2 with x_3.
8. Return to Step 3 and repeat the computation.

Using this algorithm as a guide, try to program the midpoint method on your computer. Save your program carefully because you can use it to check your manual solutions to the various equations in this text.

Exercise 6 ◆ Graphical Solution of Equations

The following equations all have at least one root between $x = -10$ and $x = 10$. Put each equation into the form $y = f(x)$, and plot just enough of each to find the approximate value of the root(s). (Your graph may appear *inverted* when compared to others, depending on which side of the equals sign the zero term has been placed. The roots, however, will be the same.)

1. $2.4x^3 - 7.2x^2 - 3.3 = 0$
2. $9.4x = 4.8x^3 - 7.2$
3. $25x^2 - 19 = 48x + x^3$
4. $1.2x + 3.4x^3 = 2.8$
5. $6.4x^4 - 3.8x = 5.5$
6. $621x^4 - 284x^3 - 25 = 0$

In each of problems 7 through 10, two equations are given. Graph each equation as if the other were not there, but as if they were on the same coordinate axes. Each equation will graph as a straight line. Give the approximate coordinates of the point of intersection of the two lines.

Can you guess the significance of the point of intersection? Peek ahead at Sec. 10–1.

7. $y = 2x + 1$
   $y = -x + 2$
8. $x - y = 2$
   $x + y = 6$
9. $y = -3x + 5$
   $y = x - 4$
10. $2x + 3y = 9$
    $3x - y = -4$

Computer

11. Write a program for solving equations by the midpoint method. Use it to solve any of the equations in problems 1 through 6.

••• **CHAPTER 5 REVIEW PROBLEMS** •••••••••••••••••••••••••••••••••••••

1. Graph the following points and connect them with a smooth curve:

x	0	1	2	3	4	5	6
y	2	$2\frac{1}{4}$	3	4	6	9	13

2. Plot the function $y = x^3 - 2x$ for $x = -3$ to $+3$. Label any roots or intercepts.

3. Graph the function $y = 3x - 2x^2$ from $x = -3$ to $x = 3$. Label any roots or intercepts.

4. Graphically find the approximate value of the roots of the equation

$$(x + 3)^2 = x - 2x^2 + 4$$

5. Graphically find the point of intersection of the following equations:

$$2.5x - 3.7y = 5.2$$
$$6.3x + 4.2y = 2.6$$

6. Make a graph of the pressure p (lb/in.^{2}) in an engine cylinder, where v is the volume (in.^{3}) above the piston.

p	39.6	44.7	53.8	73.5	85.8	113.2	135.8	178.2
v	10.61	9.73	8.55	7.00	6.23	5.18	4.59	3.87

For problems 7 and 8, plot the parametric equations for $t = -2$ to 2.

7. $x = 3t$
   $y = t^3$
8. $x = 2t^2$
   $y = 4t$

For problems 9 through 12, make a complete graph of each function, choosing a domain and a range that include all of the features of interest.

9. $y = 5x^2 + 24x - 12$
10. $y = 9x^2 + 22x - 17$
11. $y = 6x^3 + 3x^2 - 14x - 21$
12. $y = 3x^4 + 21x^2 + 21$

For problems 13 and 14, write the equation of each straight line, and make a graph.

13. Slope $= -2$; y intercept $= 4$
14. Slope $= 4$; y intercept $= -6$

For problems 15 and 16, find the slope and the y intercept of each straight line, and make a graph.

15. $y = 12x - 4$
16. $y = -2x + 11$

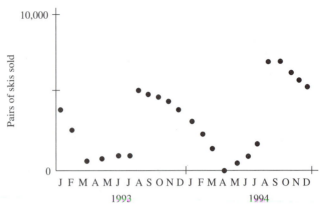

FIGURE 5–24   Ski sales for 1998–99

Writing

17. The sales of skis by a ski manufacturer are shown for a two-year period in Fig. 5–24. Study the graph and write a verbal summary of the sales pattern for the period shown.

Team Project

18. Graph the function

$$y = 1.24x^3 + 0.482x^2 + 2.85x + 0.394$$

Find, to the nearest hundredth, any roots or maximum and minimum points.

6

Geometry

OBJECTIVES

When you have completed this chapter, you should be able to:

- Find the angles formed by intersecting straight lines.
- Solve practical problems that require finding the sides and angles of right triangles.
- Solve practical problems in which the area of a triangle or a quadrilateral must be found.
- Solve problems involving the circumference, diameter, area, or tangent to a circle.
- Compute surface areas and volumes of spheres, cylinders, cones, and other solid figures.

We present in this chapter some of the more useful facts from geometry, together with examples of their use in solving practical problems. Geometrical *theorems*—provable statements—are treated as facts and are presented almost "cookbook" style, with no proofs given and with the emphasis on the applications of these facts. Our main concern will be the computation of lengths, areas, and volumes of various geometric figures.

Much of this material on angles, triangles, and radian measure will be good preparation for the later chapters on trigonometry.

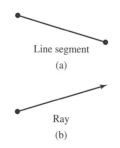

Line segment

(a)

Ray

(b)

FIGURE 6–1

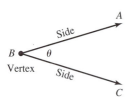

FIGURE 6–2   An angle.

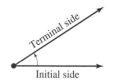

FIGURE 6–3   An angle formed by rotation.

6–1   Straight Lines and Angles

Straight Line

A *line segment* is that portion of a straight line lying between two endpoints [Fig. 6–1(a)]. A *ray*, or *half-line*, is the portion of a line lying to one side of a point (endpoint) on the line [Fig. 6–1(b)].

Angles

An *angle* is formed when two rays intersect at their endpoints (Fig. 6–2). The point of intersection is called the *vertex* of the angle, and the two rays are called the *sides* of the angle.

The angle shown in Fig. 6–2 can be *designated* in any of the following ways:

angle ABC angle CBA angle B angle θ

The symbol $\angle$ means *angle*, so $\angle B$ means angle *B*.

An angle can also be thought of as having been *generated* by a ray turning from some *initial position* to a *terminal position* (Fig. 6–3).

One *revolution* is the amount a ray would turn to return to its original position.

The *units of angular measure* in common use are the *degree* and the *radian* (Sec. 6–4). The *measure* of an angle is the number of units of measure it contains. Two angles are *equal* if they have the same measure.

The degree (°) is a unit of angular measure equal to 1/360 of a revolution. Thus there are 360° in one complete revolution.

Figure 6–4 shows a *right* angle ($\frac{1}{4}$ revolution, or 90°), usually marked with a small square at the vertex; an *acute* angle (less than $\frac{1}{4}$ revolution); an *obtuse* angle (greater than $\frac{1}{4}$ revolution but less than $\frac{1}{2}$ revolution); and a *straight* angle ($\frac{1}{2}$ revolution, or 180°).

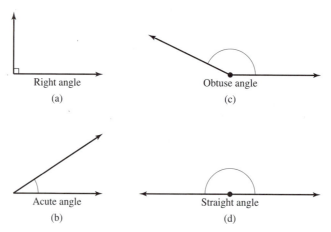

Right angle

(a)

Obtuse angle

(c)

Acute angle

(b)

Straight angle

(d)

FIGURE 6–4   Types of angles.

Two lines at right angles to each other are said to be *perpendicular*. Two angles are called *complementary* if their sum is a right angle, and *supplementary* if their sum is a straight angle.

Two more words we use in reference to angles are *intercept* and *subtend*. In Fig. 6–5 we say that the angle θ *intercepts* the section *PQ* of the curve. Conversely, we say that angle θ is *subtended by PQ*.

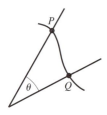

FIGURE 6–5

Angles between Intersecting Lines

Angles *A* and *B* of Fig. 6–6 are called *opposite* or *vertical* angles. It should be apparent that *A* and *B* are equal.

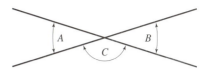

FIGURE 6–6

Opposite angles of two intersecting straight lines are equal.	**104**

Two angles are called *adjacent* when they have a common side and a common vertex, such as angles *A* and *C* of Fig. 6–6. When two lines intersect, the adjacent angles are *supplementary*.

◆◆◆ **Example 1:** Find the measure (in degrees) of angles *A* and *B* in the structure shown in Fig. 6–7.

Solution: We see from Fig. 6–7 that *RQ* and *PS* are straight intersecting lines (we will often assume relationships from the diagram rather than proving them). Since angle *A* is opposite the 34° angle, angle *A* = 34°. Angle *B* and the 34° angle are supplementary, so

$$B = 180° - 34° = 146°$$

◆◆◆

For brevity, we say "angle *A*" instead of the cumbersome but more correct "the measure of angle *A*." We will learn how to convert angles from one system of measurement to another (such as degrees to radians) in Chapter 7.

Families of Lines

A *family* of lines is a group of lines that are related to each other in some way. Figure 6–8(a) shows a family of lines, none of which intersect. They are called *parallel*. Figure 6–8(b) shows a family of lines that all pass through the same point but have different slopes.

In a later chapter we will talk about *families of curves*.

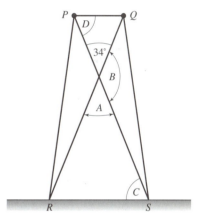

FIGURE 6–7

Transversals

A *transversal* is a line that intersects a family or system of lines. In Fig. 6–9, the lines labeled *T* are transversals. In Fig. 6–9(a), two parallel lines, L_1 and L_2, are cut

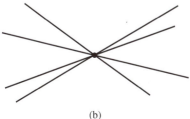

(a)

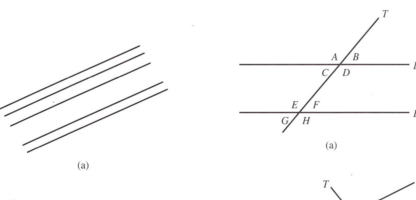

(a)

(b)

FIGURE 6–8 Families of lines.

FIGURE 6–9 Transversals.

by transversal *T*. Angles *A*, *B*, *G*, and *H* are called *exterior* angles, and *C*, *D*, *E*, and *F* are called *interior* angles. Angles *A* and *E* are called *corresponding* angles. Other corresponding angles in Fig. 6–9(a) are *C* and *G*, *B* and *F*, and *D* and *H*. Angles *C* and *F* are called *alternate interior angles*, as are *D* and *E*. We have a theorem:

> If two parallel straight lines are cut by a transversal, corresponding angles are equal, and alternate interior angles are equal.   **105**

Thus in Fig. 6–9(a), $\angle A = \angle E$, $\angle C = \angle G$, $\angle B = \angle F$, and $\angle D = \angle H$. Also,

$$\angle D = \angle E \quad \text{and} \quad \angle C = \angle F$$

◆◆◆ **Example 2:** The top girder *PQ* in the structure of Fig. 6–7 is parallel to the ground, and angle *C* is 73°. Find angle *D*.

Solution: We have two parallel lines, *PQ* and *RS*, cut by transversal *PS*. By statement 105,

$$\angle D = \angle C = 73°$$ ◆◆◆

Another useful theorem applies when a number of parallel lines are cut by *two* transversals, such as in Fig. 6–10. The portions of the transversals lying between the same parallels are called *corresponding segments*. (In Fig. 6–10, *a* and *b* are corresponding segments, *c* and *d* are corresponding segments, and *e* and *f* are corresponding segments.)

> When a number of parallel lines are cut by two transversals, the ratios of corresponding segments on the transversals are equal.   **106**

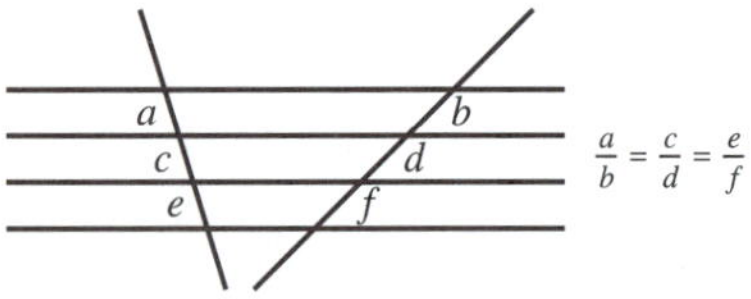

In Fig. 6–10,

$$\frac{a}{b} = \frac{c}{d} = \frac{e}{f}$$

◆◆◆ **Example 3:** A portion of a street map is shown in Fig. 6–11. Find the distances *PQ* and *QR*.

$$\frac{a}{b} = \frac{c}{d} = \frac{e}{f}$$

FIGURE 6–10

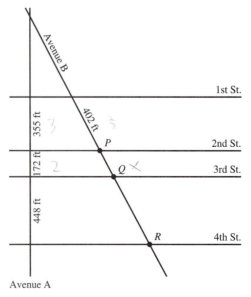

Avenue A

FIGURE 6–11

Solution: From statement 106,

$$\frac{PQ}{172} = \frac{402}{355}$$

$$PQ = \frac{402}{355}(172) = 195 \text{ ft}$$

Similarly,

$$\frac{QR}{448} = \frac{402}{355}$$

$$QR = \frac{402}{355}(448) = 507 \text{ ft}$$

◆◆◆

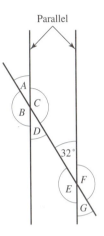

FIGURE 6–13

Exercise 1 ◆ Straight Lines and Angles

1. Find angle θ in Figs. 6–12 (a), (b), and (c).

 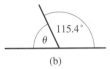

(a)   (b)   (c)

FIGURE 6–12

2. Find angles A, B, C, D, E, F, and G in Fig. 6–13.
3. Find distance x in Fig. 6–14.
4. Find angles A and B in Fig. 6–15.
5. On a certain day, the angle of elevation of the sun is 46.3° (Fig. 6–16). Find the angles A, B, C, and D that a ray of sunlight makes with a horizontal sheet of glass.
6. Three parallel steel cables hang from a girder to the deck of a bridge (Fig. 6–17). Find distance x.

The *angle of elevation* of some object above the observer is the angle between the line of sight to that object and the horizontal, as in Fig. 7–12.

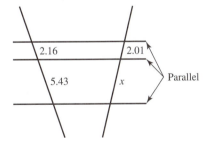

FIGURE 6–14

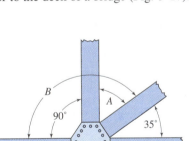

FIGURE 6–15

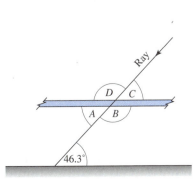

FIGURE 6–16

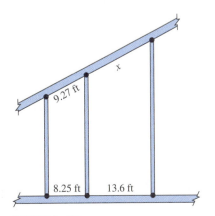

FIGURE 6–17

6–2 Triangles

Polygons

A *polygon* is a plane figure formed by three or more line segments, called the *sides* of the polygon, joined at their endpoints, as in Fig. 6–18. The points where the sides meet are called *vertices*. We say that the sides of a polygon are equal if their measures (lengths) are equal. If all of the sides and angles of a polygon are equal, it is called a *regular* polygon, as in Fig. 6–19. The *perimeter* of a polygon is simply the sum of its sides.

Modern definitions of plane figures exclude the *interior* as part of the figure. Thus in Fig. 6–19, point *A* is *not* on the square, while point *B* is on the square. The interior is referred to as a *region*.

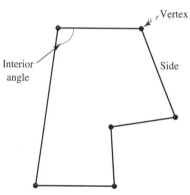

FIGURE 6–18   A polygon.

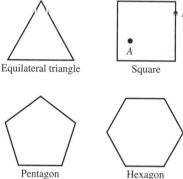

FIGURE 6–19   Some regular polygons.

Sum of Interior Angles

A polygon of *n* sides has *n* interior angles, such as those shown in Fig. 6–20. Their sum is equal to the following:

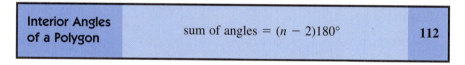

Interior Angles of a Polygon	sum of angles = $(n - 2)180°$	112

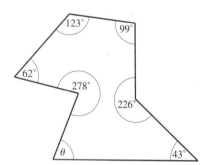

FIGURE 6–20

◆◆◆ **Example 4:** Find angle θ in Fig. 6–20.

Solution: The polygon shown in Fig. 6–20 has seven sides, so $n = 7$. By Eq. 112,

$$\text{sum of angles} = (7 - 2)(180°) = 900°$$

Adding the six given angles gives us

$$278° + 62° + 123° + 99° + 226° + 43° = 831°$$

So

$$\theta = 900° - 831° = 69°$$

◆◆◆

Triangles

A *triangle* is a polygon having three *sides*. The angles between the sides are the *interior angles* of the triangle, usually referred to as simply the *angles* of the triangle.

As shown in Fig. 6–21, a *scalene* triangle has no equal sides; an *isosceles* triangle has two equal sides; and an *equilateral* triangle has three equal sides.

An *acute* triangle has three acute angles; an *obtuse* triangle has one obtuse angle and two acute angles; and a *right* triangle has one right angle and two acute angles.

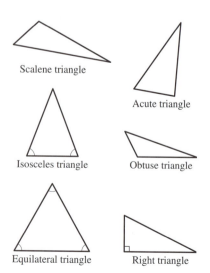

FIGURE 6–21   Types of triangles.

Altitude and Base

The *altitude* of a triangle is the perpendicular distance from a vertex to the opposite side called the *base*, or an extension of that side (Fig. 6–22).

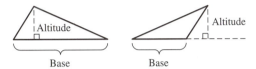

FIGURE 6–22

Area of a Triangle

We use the following familiar formula:

| **Area of a Triangle** | Area equals one-half the product of the base and the altitude to that base. $$A = \frac{bh}{2}$$ | 137 |

The measuring of areas, volumes, and lengths is sometimes referred to as *mensuration*.

◆◆◆ Example 5: Find the area of the triangle in Fig. 6–22 if the base is 52.0 and the altitude is 48.0.

Solution: By Eq. 137,

$$\text{area} = \frac{52.0(48.0)}{2} = 1250 \quad \text{(rounded)} \qquad \text{◆◆◆}$$

If the altitude is not known but we have instead the lengths of the three sides, we may use *Hero's formula*. If a, b, and c are the lengths of the sides, the following formula applies:

This formula is named for Hero (or Heron) of Alexandria, a Greek mathematician and physicist of the 1st century A.D.

| **Hero's Formula** | $$\text{area of triangle} = \sqrt{s(s - a)(s - b)(s - c)}$$ where s is half the perimeter, or $$s = \frac{a + b + c}{2}$$ | 138 |

◆◆◆ Example 6: Find the area of a triangle having sides of lengths 3.25, 2.16, and 5.09.

Solution: We first find s, which is half the perimeter.

$$s = \frac{3.25 + 2.16 + 5.09}{2} = 5.25$$

Thus the area is

$$\text{area} = \sqrt{5.25(5.25 - 3.25)(5.25 - 2.16)(5.25 - 5.09)} = 2.28 \qquad \text{◆◆◆}$$

Sum of the Angles

An extremely useful relationship exists between the interior angles of any triangle.

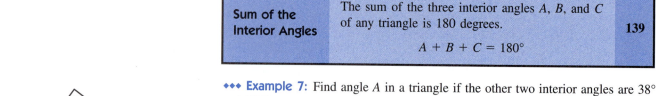

	The sum of the three interior angles A, B, and C of any triangle is 180 degrees.	
Sum of the Interior Angles	$A + B + C = 180°$	**139**

◆◆◆ Example 7: Find angle A in a triangle if the other two interior angles are 38° and 121°.

Solution: By Eq. 139,

$$A = 180 - 121 - 38 = 21°$$ ◆◆◆

Exterior Angles

An *exterior* angle is the angle between the side of a triangle and an extension of the adjacent side, such as angle θ in Fig. 6–23. The following theorem applies to exterior angles:

	An exterior angle equals the sum of the two opposite interior angles.	
Exterior Angle of a Triangle	$\theta = A + B$	**142**

Congruent Triangles

Two triangles (or any other polygons, for that matter) are said to be *congruent* if the angles and sides of one are equal to the angles and sides of the other, as in Fig. 6–24.

Similar Triangles

Two triangles are said to be *similar* if they have the *same shape*, even if one triangle is larger than the other. This means that the angles of one of the triangles must equal the angles of the other triangle, as in Fig. 6–25. Sides that lie between the same pair of equal angles are called *corresponding sides*, such as sides a and d. Sides b and e, as well as sides c and f, are also corresponding sides. Thus we have the following two theorems:

	If two angles of a triangle equal two angles of another triangle, their third angles must be equal, and the triangles are similar.	
Similar Triangles		**143**
	If two triangles are similar, their corresponding sides are in proportion.	**144**

◆◆◆ Example 8: Two beams, AB and CD, in the framework of Fig. 6–26 are parallel. Find distance AE.

Solution: By statement 104, we know that angle AEB equals angle DEC. Also, by statement 105, angle ABE equals angle ECD. Thus triangle ABE is similar to triangle CDE. Since AE and ED are corresponding sides,

$$\frac{AE}{5.87} = \frac{5.14}{7.25}$$

$$AE = 5.87\left(\frac{5.14}{7.25}\right) = 4.16 \text{ m}$$ ◆◆◆

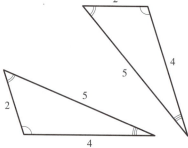

FIGURE 6–23   An exterior angle.

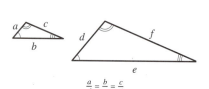

FIGURE 6–24   Congruent triangles.

$$\frac{a}{d} = \frac{b}{e} = \frac{c}{f}$$

FIGURE 6–25   Similar triangles.

We will see in a later chapter that these relationships also hold for similar figures *other than* triangles, and for similar solids as well.

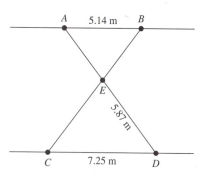

FIGURE 6–26

Right Triangles

In a right triangle (Fig. 6–27), the side opposite the right angle is called the *hypotenuse*, and the other two sides are called *legs*. The legs and the hypotenuse are related by the well-known *Pythagorean theorem*.

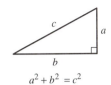

| Pythagorean Theorem | The square of the hypotenuse of a right triangle is equal to the sum of the squares of the two legs.

$$a^2 + b^2 = c^2$$ | 145 |

FIGURE 6–27 The Pythagorean theorem, named for the Greek mathematician Pythagoras (ca. 580–500 B.C.).

◆◆◆ **Example 9:** A right triangle has legs of length 6 units and 11 units. Find the length of the hypotenuse.

Solution: Letting c = the length of the hypotenuse, we have

$$c^2 = 6^2 + 11^2 = 36 + 121 = 157$$
$$c = \sqrt{157} = 12.5 \quad \text{(rounded)}$$ ◆◆◆

◆◆◆ **Example 10:** A right triangle has a hypotenuse 5 units in length and a leg of 3 units. Find the length of the other leg.

Solution:

$$a^2 = c^2 - b^2$$
$$= 5^2 - 3^2 = 16$$
$$a = \sqrt{16} = 4$$ ◆◆◆

The 3–4–5 right triangle is a good one to remember.

| Common Error | Remember that the Pythagorean theorem applies *only to right triangles*. Later we will use trigonometry to find the sides and angles of *oblique* triangles (triangles with no right angles). |

Some Special Triangles

In a *30–60–90 right triangle* [Fig. 6–28(a)], the side opposite the 30° angle is half the length of the hypotenuse.

A *45° right triangle* [Fig. 6–28(b)] is also *isosceles*, and the hypotenuse is $\sqrt{2}$ times the length of either side.

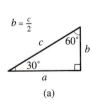
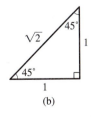
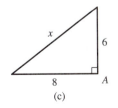

(a)   (b)   (c)

FIGURE 6–28 Some special right triangles.

A *3–4–5 triangle* is a right triangle in which the sides are in the ratio of 3 to 4 to 5.

◆◆◆ **Example 11:** In Fig. 6–28(c), side x is 10 cm. ◆◆◆

Exercise 2 ◆ Triangles

For practice, do these problems using only geometry, even if you know some trigonometry.

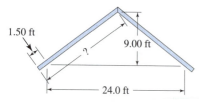

1.50 ft   9.00 ft   ?   24.0 ft

FIGURE 6–29

1. What is the cost, to the nearest dollar, of a triangular piece of land whose base is 828 ft and altitude 412 ft, at $1125 an acre?

2. The house shown in Fig. 6–29 is 24.0 ft wide. The ridge is 9.00 ft higher than the side walls, and the rafters project 1.50 ft beyond the sides of the house. How long are the rafters?

3. What is the length of a handrail to a flight of stairs (Fig. 6–30), each step 30.5 cm wide and 22.3 cm high?

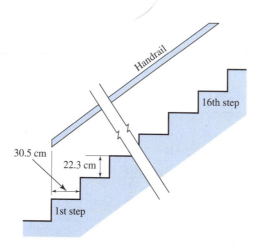

Handrail   16th step   30.5 cm   22.3 cm   1st step

FIGURE 6–30

To help you solve each problem, draw a diagram and label it completely. Look for rectangles, special triangles, or right triangles contained in the diagram. Be sure to look up any word that may be unfamiliar, such as *diagonal*, *radius*, or *diameter*.

4. At $3.50 a square yard, find to the nearest dollar the cost of paving a triangular court, its base being 105 ft and its altitude 21.0 yd.

5. A vertical pole 45.0 ft high is supported by three guy wires attached to the top and reaching the ground at distances of 60.0 ft, 108.0 ft, and 200.0 ft from the foot of the pole. What are the lengths of the wires?

6. A ladder 39.0 ft long reaches to the top of a building when its foot stands 15.0 ft from the building. How high is the building?

7. Two streets, one 16.2 m and the other 31.5 m wide, cross at right angles. What is the diagonal distance between the opposite corners?

8. A room is 20.0 ft long, 16.0 ft wide, and 12.0 ft high. What is the diagonal distance from one of the lower corners to the opposite upper corner?

9. What is the side of a square whose diagonal is 50.0 m?

10. A rectangular park 125 m long and 233 m wide has a straight walk running through it from opposite corners. What is the length of the walk?

11. A ladder 32.0 ft long stands flat against the side of a building. How many feet must it be drawn out at the bottom so that the top may be lowered 4.00 ft?

12. The slant height of a cone (Fig. 6–31) is 21.8 in., and the diameter of the base is 18.4 in. How high is the cone?

13. A house (Fig. 6–32) that is 50.0 ft long and 40.0 ft wide has a pyramidal roof whose height is 15.0 ft. Find the length PQ of a hip rafter that reaches from a corner of the building to the vertex of the roof.

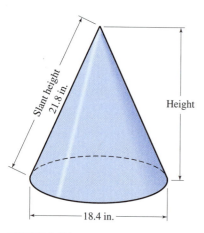

Slant height 21.8 in.   Height   18.4 in.

FIGURE 6–31

14. An antenna, 125 m high, is supported by cables of length L that reach the ground at a distance D from the base of the antenna. If D is $\frac{3}{4}$ the length of a cable, find L and D.

15. A hex head bolt (Fig. 6–33) measures 0.500 in. across the flats. Find the distance x across the corners.

16. A 1.00-m-long piece of steel hexagonal stock measuring 30.0 mm across the flats has a 10.0-mm-diameter hole running lengthwise from end to end. Find the mass of the bar. (Density = 7.85 g/cm^3.)

We have not studied the hexagon as such, but you can solve problems 15 and 16 by dividing the hexagon into right triangles or a rectangle and right triangles. This also works for other regular polygons.

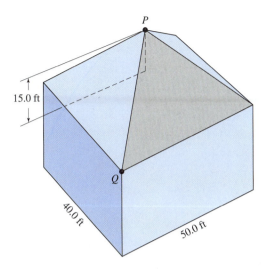

FIGURE 6–32

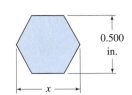

FIGURE 6–33

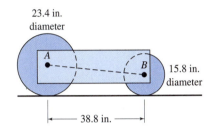

17. Find the distance AB between the centers of the two rollers in Fig. 6–34.

FIGURE 6–34

18. A 0.500-m-long hexagonal concrete shaft measuring 20.0 cm across the corners has a triangular hole running lengthwise from end to end (Fig. 6–35). The triangular hole is in the shape of an equilateral triangle with sides 10.0 cm long. What is the total surface area of the shaft?

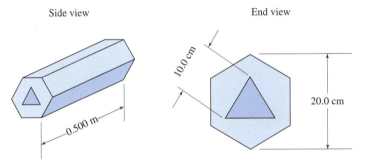

FIGURE 6–35

19. A highway (Fig. 6–36) cuts a corner from a parcel of land. Find the number of acres in the triangular lot ABC. (1 acre = 43,560 ft².)

20. A surveyor starts at A in Fig. 6–37 and lays out lines AB, BC, and CA. Find the three interior angles of the triangle.

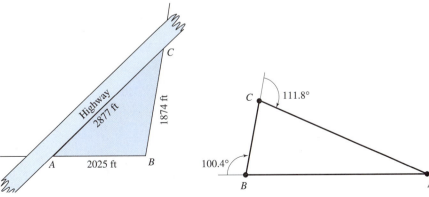

FIGURE 6–36 Note: Angle B is not a right angle.

FIGURE 6–37

21. A beam *AB* is supported by two crossed beams (Fig. 6–38). Find distance *x*.

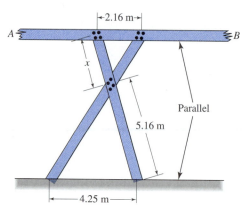

FIGURE 6–38

22. Find dimension *x* in Fig. 6–39.
23. Find dimension *x* on the beveled end of the shaft in Fig. 6–40.

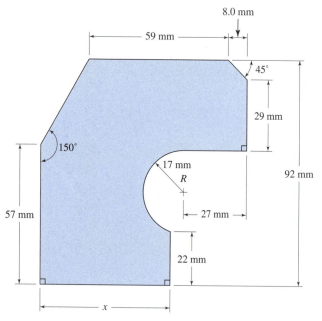

FIGURE 6–39

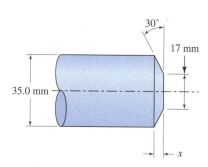

FIGURE 6–40

 Computer

24. Write a program that will accept as input any two sides of a right triangle and that will then display or print the length of the third side.
25. Write a program that will accept as input the three sides of any triangle and that will compute and display the area by Hero's formula.

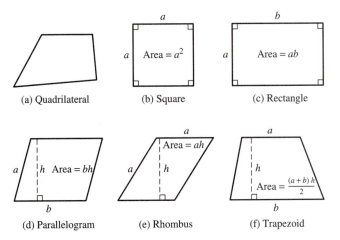

FIGURE 6–41   Quadrilaterals.

6–3   Quadrilaterals

A *quadrilateral* is a polygon having four sides. They are the familiar figures shown in Fig. 6–41. The formula for the area of each region is given right on the figure. Lengths of line segments are measured in units of length (such as inches, meters, and centimeters), and areas of regions are measured in *square* units (such as square inches, square meters, and square centimeters).

For the *parallelogram*, opposite sides are parallel and equal. Opposite angles are equal, and each diagonal cuts the other diagonal into two equal parts (they *bisect* each other).

The *rhombus* is also a parallelogram, so the previous facts apply to it as well. In addition, its diagonals bisect each other at *right angles* and bisect the angles of the rhombus.

The *trapezoid* has two parallel sides, which are called the *bases*, and the altitude is the distance between the bases.

◆◆◆ **Example 12:** A solar collector array consists of six rectangular panels, each 45.3 in. × 92.5 in., each with a blocked rectangular area (needed for connections) measuring 4.70 in. × 8.80 in. Find the total collecting area in square feet.

Estimate: If the panels were 4 ft by 8 ft, or 32 ft^2 each, six of them would be 6×32 or 192 ft^2. Since we have estimated each dimension higher than it really is, and have also ignored the blocked areas, we would expect this answer to be higher than the actual area.

Solution:

$$\text{area of each panel} = (45.3)(92.5) = 4190 \text{ in.}^2$$
$$\text{blocked area} = (4.70)(8.80) = 41.0 \text{ in.}^2$$

Subtracting yields

$$\text{collecting area per panel} = 4190 - 41.0 = 4149 \text{ in.}^2$$

There are six panels, so

$$\text{total collecting area} = 6(4149) = 24{,}890 \text{ in.}^2$$
$$= 173 \text{ ft}^2$$

since there are 144 square inches in a square foot. This is lower than our estimate, as expected.   ◆◆◆

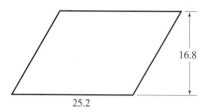

FIGURE 6–42

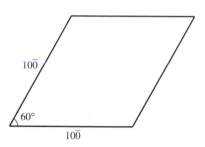

FIGURE 6–43

Work to the nearest dollar in this exercise.

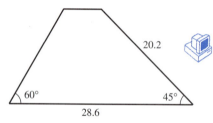

FIGURE 6–44

Exercise 3 ◆ Quadrilaterals

1. Find the area of the parallelogram in Fig. 6–42.
2. Find the area of the rhombus in Fig. 6–43.
3. Find the area of the trapezoid in Fig. 6–44.
4. Find the cost of shingling a flat roof, 64 ft 9 in. long and 45 ft wide, at 95¢/ft².

5. What will be the cost of flagging a sidewalk 312 ft long and 6.5 ft wide, at $13.50 per square yard?
6. How many 9-in.-square tiles will cover a floor 48 ft by 12 ft?
7. What will it cost to carpet a floor, 6.25 m by 7.18 m, at $7.75 per square meter?
8. How many rolls of paper, each 8.00 yd long and 18.0 in. wide, will paper the sides of a room 16.0 ft by 14.0 ft and 10.0 ft high, deducting 124 ft² for doors and windows?
9. What is the cost of plastering the walls and ceiling of a room 40 ft long, 36 ft wide, and 22 ft high, at $8.50 per square yard, allowing 1375 ft² for doors, windows, and baseboard?
10. What will it cost to cement the floor of a cellar 25.3 ft long and 18.4 ft wide, at $0.35 per square foot?
11. Find the cost of lining a topless rectangular tank 68 in. long, 54 in. wide, and 48 in. deep, with zinc, weighing 5.2 lb per square foot, at $1.55 per pound installed.

Computer

12. A certain building supplies store has stocks of window glass having widths of 10 in., 12 in., 14 in., . . . , 28 in. Each of these widths is available in lengths of 20 in., 24 in., 28 in., . . . , 40 in. The price of glass is $2.18 per square foot. Write a program to compute and display the area in square feet and the price of each size of glass available.

6–4   The Circle

Circumference and Area

For a circle of diameter d and radius r (Fig. 6–45), the following formulas apply:

Circle of Radius r and Diameter d $\pi \cong 3.1416$	circumference $= 2\pi r = \pi d$	113
	area $= \pi r^2 = \dfrac{\pi d^2}{4}$	114

◆◆◆ **Example 13:** The circumference and the area of a 25.70-cm-radius circle are

$$C = 2\pi(25.70) = 161.5 \text{ cm}$$

and

$$A = \pi(25.70)^2 = 2075 \text{ cm}^2 \qquad \text{◆◆◆}$$

Radian Measure

We cover radian measure, and compute areas of sectors and segments, in Chapter 16.

A *central angle* (Fig. 6–45) is one whose vertex is at the center of the circle. An *arc* is a portion of the circle. If an arc is laid off along a circle with a length equal

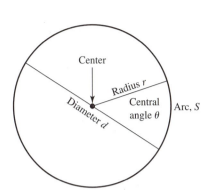

FIGURE 6–45 A circle.

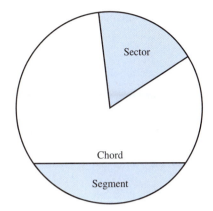

FIGURE 6–46

to the radius of the circle, the central angle subtended by this arc is defined as one *radian*.

Other Parts of a Circle

A *sector* (Fig. 6–46) is the region bounded by two radii and one of the arcs that they intercept. A *chord* is a line segment connecting two points on the circle. A *segment* is the region bounded by a chord and one of the arcs that it intercepts. When a circle has two intersecting chords, the following theorem applies:

Intersecting Chords	If two chords in a circle intersect, the product of the parts of one chord is equal to the product of the parts of the other chord. $$ab = cd$$	121

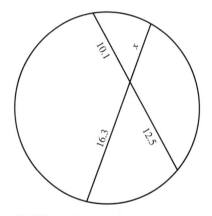

FIGURE 6–47

◆◆◆ Example 14: Find the length x in Fig. 6–47.

Solution: By statement 121,

$$16.3x = 10.1(12.5)$$
$$x = \frac{10.1(12.5)}{16.3} = 7.75$$

◆◆◆

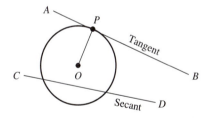

FIGURE 6–48 Tangent and secant.

Tangents and Secants

A *tangent* (Fig. 6–48) is a straight line that touches a circle in just one point. A *secant* cuts the circle in *two* points. A chord is that portion of a secant lying within the circle.

We have two very useful theorems about tangents. First:

Tangents to a Circle	A tangent is perpendicular to the radius drawn through the point of contact.	119

Thus in Fig. 6–48, angle *OPB* is a right angle.

◆◆◆ Example 15: Find the distance *OP* in Fig. 6–49.

Solution: By statement 119, we know that angle *PQO* is a right angle. Also,

$$PQ = 340 - 115 = 225 \text{ cm}$$

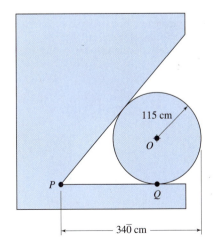

FIGURE 6–49

So, by the Pythagorean theorem,

$$OP = \sqrt{(115)^2 + (225)^2} = 253 \text{ cm}$$

◆◆◆

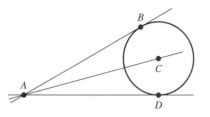

FIGURE 6–50

The second theorem follows:

Tangents to a Circle	Two tangents drawn to a circle from a point outside the circle are equal, and they make equal angles with a line from the point to the center of the circle.	120

Thus, in Fig. 6–50,

$$AB = AD \quad \text{and} \quad \angle BAC = \angle CAD$$

Semicircle

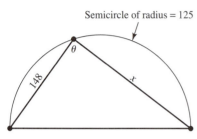

FIGURE 6–51

	Any angle inscribed in a semicircle is a right angle.	118

◆◆◆ **Example 16:** Find the distance x in Fig. 6–51.

Solution: By statement 118, we know that $\theta = 90°$. Then, by Eq. 145,

$$x = \sqrt{(250)^2 - (148)^2} = 201$$

◆◆◆

Exercise 4 ◆ The Circle

1. In a park is a circular fountain whose basin is 22.5 m in circumference. What is the diameter of the basin?
2. How much larger is the side of a square circumscribing a circle 155 cm in diameter than a square inscribed in the same circle?
3. The area of the bottom of a circular pan is 196 in.². What is its diameter?
4. Find the diameter of a circular solar pond that has an area of 125 m².
5. What is the circumference of a circular lake 33.0 m in diameter?
6. The radius of a circle is 5.00 m. Find the diameter of another circle containing 4 times the area of the first.
7. The distance around a circular park is 2.50 mi. How many acres does it contain?
8. A woodcutter uses a tape measure and finds the circumference of a tree to be 95.0 in. Assuming the tree to be circular, what length of chain-saw bar is needed to fell the tree?
9. Find the distance x in Fig. 6–52.
10. In Fig. 6–53, distance OP is 8.65 units. Find the distance PQ.
11. Figure 6–54 shows a semicircle with a diameter of 105. Find distance PQ.

By cutting from both sides, one can fell a tree whose diameter is twice the length of the chain-saw bar.

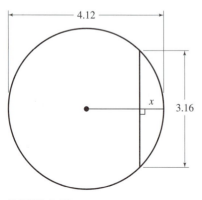

FIGURE 6–52

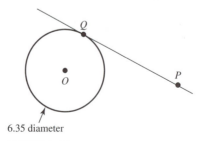

FIGURE 6–53

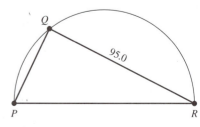

FIGURE 6–54

12. What must be the diameter d of a cylindrical piston so that a pressure of 125 lb/in.^{2} on its circular end will result in a total force of 3600 lb?

13. Find the length of belt needed to connect the three 15.0-cm-diameter pulleys in Fig. 6–55. *Hint:* The total *curved* portion of the belt is equal to the circumference of one pulley.

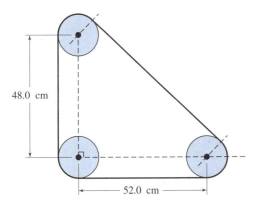

48.0 cm

52.0 cm

FIGURE 6–55

14. Seven cables of equal diameter are contained within a circular conduit, as in Fig. 6–56. If the inside diameter of the conduit is 25.6 cm, find the cross-sectional area *not* occupied by the cables.

15. A 60° screw thread is measured by placing three wires on the thread and measuring the distance T, as in Fig. 6–57. Find the distance D if the wire diameters are 0.1000 in. and the distance T is 1.500 in., assuming that the root of the thread is a sharp V shape.

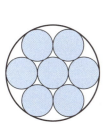

FIGURE 6–56

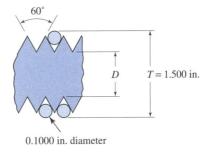

60°

D $T = 1.500$ in.

0.1000 in. diameter

FIGURE 6–57   Measuring a screw thread "over wires."

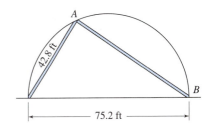

A

42.8 ft

B

75.2 ft

FIGURE 6–58

16. Figure 6–58 shows two of the supports for a hemispherical dome. Find the length of girder AB.

17. A certain car tire is 78.5 cm in diameter. How far will the car move forward with one revolution of the wheel?

18. Archimedes claimed that the area of a circle is equal to the area of a triangle that has an altitude equal to the radius of the circle and a base equal to the circumference of the circle. Use the formulas of this chapter to show that this is true.

Computer

19. A lens in a certain optical instrument has a diameter of 75.0 mm. This diameter is reduced by a series of circular "stops" (Fig. 6–59), with each stop reducing the area of the opening by 15% of its previous area. Write a program or use a spreadsheet that will compute the required diameter of each stop and that will display both the stop diameter and the area in mm^2. Continue until the stop diameter falls below 20 mm.

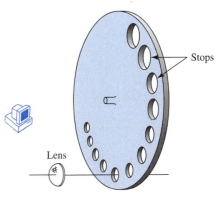

Stops

Lens

FIGURE 6–59

6–5   Volumes and Areas of Solids

Volume

The volume of a solid is a measure of the space it occupies or encloses. It is measured in cubic units (cm^3, ft^3, m^3, etc.) or, usually for liquids, in liters or gallons. One of our main tasks in this chapter is to compute the volumes of various solids.

Area

We speak about three different kinds of areas. *Surface area* will be the *total area* of the surface of a solid, including the ends, or bases. The *lateral area*, which will be defined later for each solid, does *not* include the areas of the bases. The *cross-sectional area* is the area of the plane figure obtained when we slice the solid in a specified way.

The formulas for the areas and volumes of some common solids are given in Fig. 6–60.

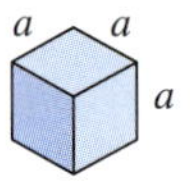

Cube		Volume = a^3	122
		Surface area = $6a^2$	123
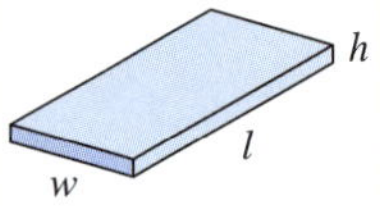 Rectangular parallelepiped		Volume = lwh	124
		Surface area = $2(lw + hw + lh)$	125
Any cylinder or prism		Volume = (area of base)(altitude)	126
Right cylinder or prism		Lateral area = (perimeter of base)(altitude) (not incl. bases)	127
Sphere		Volume = $\frac{4}{3}\pi r^2$	128
		Surface area = $4\pi r^3$	129
Any cone or pyramid		Volume = $\frac{h}{3}$(area of base)	130
Right circular cone or regular pyramid		Lateral area = $\frac{s}{2}$ (perimeter of base)	131
Any cone or pyramid		Volume = $\frac{h}{3}(A_1 + A_2 + \sqrt{A_1 A_2})$	132
Right circular cone or regular pyramid		Lateral area = $\frac{s}{2}$(sum of base perimeters) = $\frac{s}{2}(P_1 + P_2)$	133

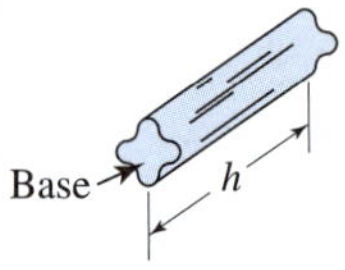

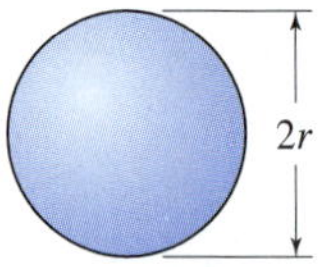

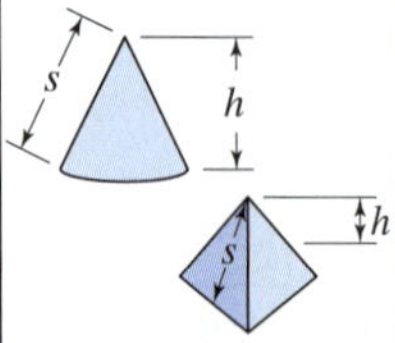

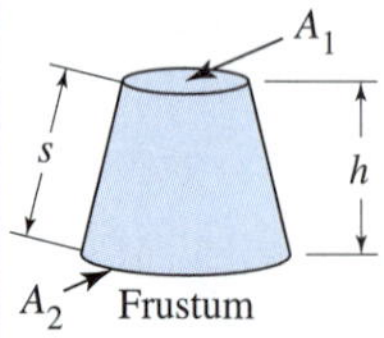

FIGURE 6–60   Some solids.

◆◆◆ Example 17: Find the volume of a cone having a base area of 125 cm² and an altitude of 11.2 cm.

Solution: By Eq. 130,

$$\text{volume} = \frac{125(11.2)}{3} = 467 \text{ cm}^3$$

◆◆◆

> Note that the formulas for the cone and frustum of a cone are the same as for the pyramid and frustum of a pyramid.

Dimension

A geometric figure that has length but no area or volume (such as a line or a curve) is said to be of *one dimension*. A geometric figure having area but not volume (such as a circle) is said to have *two* dimensions, or to be *two dimensional*. A figure having volume (such as a sphere) is said to have *three* dimensions.

Fractals

Can you imagine a curve so complex that it is misleading to call it one dimensional? The coastline of England is often given as an example. Another example is the *Koch Snowflake*, shown in Fig. 6–61(d)

> The snowflake is named for the Swedish mathematician Helge von Koch who constructed this curve in 1904.

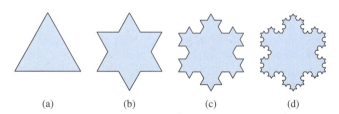

(a)   (b)   (c)   (d)

FIGURE 6–61

We start with an equilateral triangle, as in Figure 6–61(a). On each of the three sides we place another equilateral triangle whose side is $\frac{1}{3}$ that of the original triangle [Fig. 6–61(b)]. Our figure now has 12 sides and a perimeter $\frac{4}{3}$ times the original perimeter. Next, on each of the 12 new sides we place an equilateral triangle whose side is $\frac{1}{3}$ the length of the preceding side [Fig. 6–61(c)]. We continue indefinitely and get a boundary that has a dimension not of 1, but of approximately 1.2619.

The snowflake curve is called a *fractal*.

> The term *fractal* was coined by Benoit B. Mandelbrot to describe objects that have dimensions other than one, two, or three.

Exercise 5 ◆ Volumes and Areas of Solids

Prism and Rectangular Parallelepiped

> For some of these problems, we need Eq. A43: The weight (or mass) equals the volume of the solid times the density of the material.

1. Find the volume of the triangular prism in Fig. 6–62.
2. A 1.00-in. cube of steel is placed in a surface grinding machine, and the vertical feed is set so that 0.0050 in. of metal is removed from the top of the cube at each cut. How many cuts are needed to reduce the weight of the cube by 0.600 oz?
3. A rectangular tank is being filled with liquid, with each cubic meter of liquid increasing the depth by 2.0 cm. The length of the tank is 12.0 m.
   (a) What is the width of the tank?
   (b) How many cubic meters will be required to fill the tank to a depth of 3.0 m?

4. How many loads of gravel will be needed to cover 2.0 mi of roadbed, 35 ft wide, to a depth of 3.0 in. if one truckload contains 8.0 yd³ of gravel?

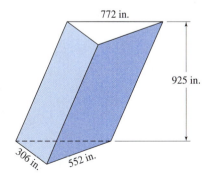

772 in.

925 in.

306 in.   552 in.

FIGURE 6–62

Cylinder

5. Find the volume of the cylinder in Fig. 6–63.
6. Find the volume and the lateral area of a right circular cylinder having a base radius of 128 and a height of 285.
7. A 7.00-ft-long piece of iron pipe has an outside diameter of 3.50 in. and weighs 126 lb. Find the wall thickness.
8. A certain bushing is in the shape of a hollow cylinder 18.0 mm in diameter and 25.0 mm long, with an axial hole 12.0 mm in diameter. If the density of the material from which they are made is 2.70 g/cm³, find the mass of 1000 bushings.
9. A steel gear is to be lightened by drilling holes through the gear. The gear is 3.50 in. thick. Find the diameter d of the holes if each is to remove 12 oz.
10. A certain gasoline engine has four cylinders, each with a bore of 82.0 mm and a piston stroke of 95.0 mm. Find the engine displacement in liters.

The density of iron varies between 440 and 490 lb/ft³, depending upon processing, and that of steel varies between 480 and 490 lb/ft³. We will use a density of 485 lb/ft³ for both iron and steel.

The engine *displacement* is the total volume swept out by all of the pistons.

Cone and Pyramid

11. The circumference of the base of a right circular cone is 40.0 in., and the slant height is 38.0 in. What is the area of the lateral surface?
12. Find the volume of a circular cone whose altitude is 24.0 cm and whose base diameter is 30.0 cm.
13. Find the weight of the mast of a ship, its height being $5\overline{0}$ ft, the circumference at one end 5.0 ft and at the other 3.0 ft, if the density of the wood is 58.5 lb/ft³.
14. The slant height of a right pyramid is 11.0 in., and the base is a 4.00-in. square. Find the area of the entire surface.
15. How many cubic feet are in a piece of timber 30.0 ft long, one end being a 15.0-in. square and the other a 12.0-in. square?
16. Find the total surface area and volume of a tapered steel roller 12.0 ft long and having end diameters of 12.0 in. and 15.0 in.

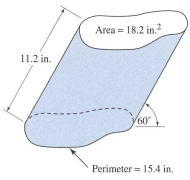

Area = 18.2 in.²

11.2 in.

60°

Perimeter = 15.4 in.

FIGURE 6–63

Sphere

17. Find the volume and the surface area of a sphere having a radius of 744.

18. Find the volume and the radius of a sphere having a surface area of 462.

19. Find the surface area and the radius of a sphere that has a volume of 5.88.

20. How many great circles of a sphere would have the same area as that of the surface of the sphere?

21. Find the weight in pounds of 100 steel balls each 2.50 inches in diameter.

22. A spherical radome encloses a volume of $90\overline{0}0$ m³. Assume that the sphere is complete.
    (a) Find the radome radius, r.
    (b) If the radome is constructed of a material weighing 2.00 kg/m², find its weight.

A *great circle* of a sphere is one that has the same diameter as the sphere.

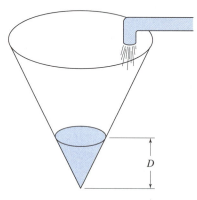

D

FIGURE 6–64

Computer

23. A conical tank (Fig. 6–64) that has a base diameter equal to its height is being filled with water at the rate of 2.25 m³/h. Write a program or use a spreadsheet that will compute and display the volume of water in the tank (m³) and the depth D (m), for every 5 min, up to 1 h.

⬩⬩⬩ CHAPTER 6 REVIEW PROBLEMS ⬩⬩⬩⬩⬩⬩⬩⬩⬩⬩⬩⬩⬩⬩⬩⬩⬩⬩⬩⬩⬩⬩⬩⬩⬩⬩⬩⬩

1. A rocket ascends at an angle of 60.0° with the horizontal. After 1.00 min, it is directly over a point that is a horizontal distance of 12.0 mi from the launch point. Find the speed of the rocket.

2. A rectangular beam 16 in. thick is cut from a log 20 inches in diameter. Find the greatest depth beam that can be obtained.

3. A cylindrical tank 4.00 m in diameter is placed with its axis vertical and is partially filled with water. A spherical diving bell is completely immersed in the tank, causing the water level to rise 1.00 m. Find the diameter of the diving bell.

4. Two vertical piers are 2̄40 ft apart and support a circular bridge arch. The highest point of the arch is 30.0 ft higher than the piers. Find the radius of the arch.

5. Two antenna masts are 10 m and 15 m high and are 12 m apart. How long a wire is needed to connect the tops of the two masts?

6. Find the area and the side of a rhombus whose diagonals are 1̄00 and 140.

7. Find the area of a triangle that has sides of length 573, 638, and 972.

8. Two concentric circles have radii of 5.00 and 12.0. Find the length of a chord of the larger circle, which is tangent to the smaller circle.

9. A belt that does not cross goes around two pulleys, each with a radius of 4.00 in. and whose centers are 9.00 in. apart. Find the length of the belt.

10. A regular triangular pyramid has an altitude of 12.0 m, and the base is 4.00 m on a side. Find the area of a section made by a plane parallel to the base and 4.00 m from the vertex.

11. A fence parallel to one side of a triangular field cuts a second side into segments of 15.0 m and 21.0 m long. The length of the third side is 42.0 m. Find the length of the shorter segment of the third side.

12. When Figure 6–65 is rotated about axis *AB*, we generate a cone inscribed in a hemisphere which is itself inscribed in a cylinder. Show that the volumes of these three solids are in the ratio 1 : 2 : 3. (Archimedes was so pleased with this discovery, it is said, that he ordered this figure to be engraved on his tomb.)

13. Four interior angles of a certain irregular pentagon are 38°, 96°, 112°, and 133°. Find the fifth interior angle.

14. Find the area of a trapezoid whose bases have lengths of 837 m and 583 m and are separated by a distance of 746 m.

15. A plastic drinking cup has a base diameter of 49.0 mm, is 63.0 mm wide at the top, and is 86.0 mm tall. Find the volume of the cup.

16. A spherical balloon is 17.4 ft in diameter and is made of a material that weighs 2.85 lb per 1̄00 ft^2 of surface area. Find the volume and the weight of the balloon.

17. Find the area of a triangle whose base is 38.4 in. and whose altitude is 53.8 in.

18. Find the volume of a sphere of radius 33.8 cm.

19. Find the area of a parallelogram of base 39.2 m and height 29.3 m.

20. In *The Musgrave Ritual*, Sherlock Holmes calculates the length of the shadow of an elm tree that is no longer standing. He does know that the elm was 64 ft high and that the shadow was cast at the instant that the sun was grazing the top of a certain oak tree.

    Holmes held a 6-ft-long fishing rod vertical and measured the length of its shadow at the proper instant. It was 9 ft long. He then said, "Of course the calculation now was a simple one. If a rod of six feet threw a shadow of nine, a tree of sixty-four feet would throw one of _____." How long was the shadow of the elm?

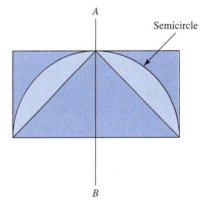

A

Semicircle

B

FIGURE 6–65

21. Find the volume of a cylinder with base radius 22.3 cm and height 56.2 cm.

22. Find the surface area of a sphere having a diameter of 39.2 in.

23. Find the volume of a right circular cone with base diameter 2.84 ft and height 5.22 ft.

24. Find the volume of a box measuring 35.8 in. × 37.4 in. × 73.4 in.

25. Find the volume of a triangular prism having a length of 4.65 cm if the area of one end is 24.6 cm².

26. To find a circle diameter when the center is inaccessible, you can place a scale as in Fig. 6–66 and measure the chord c and the perpendicular distance h. Find the radius of the curve if the chord length is 8.25 cm and h is 1.16 cm.

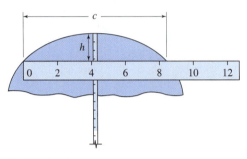

FIGURE 6–66

Writing

27. Write a section in an instruction manual for machinists on how to find the root diameter of a screw thread by measuring "over wires" (Fig. 6–57). Your entry should have two parts: First, a "how-to" section giving step-by-step instructions, and then a "theory" section explaining why this method works. Keep the entry to one page or less.

Team Projects

28. Actually use the method that Sherlock Holmes used in *The Musgrave Ritual*, problem 20, to find the height of a flagpole or building on your campus. Find some way of checking your result.

29. Estimate which weighs more, a million dollars in dollar bills or a million dollars in gold.

30. Writing about trees, Leonardo da Vinci said that "every division of the branches, when joined together, adds up to the thickness of the branch at their meeting point" (from the *Codex Urbino*). In other words, the sum of the diameters at *B* in Fig. 6–67 should be approximately equal to the diameter at *A*, and should also be equal to the sum of the diameters at *C*. Try to verify this by actual measurement on some tree branches, assuming that the branch cross sections are circular. Also compare the sum of the *areas* at *A*, *B*, and *C*.

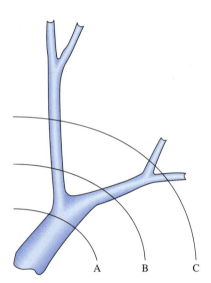

A   B   C

FIGURE 6–67

7

Right Triangles and Vectors

◆◆◆ **OBJECTIVES** ◆◆◆◆◆◆◆◆◆◆◆◆◆◆◆◆◆◆◆◆◆◆◆◆◆◆◆◆◆◆◆◆◆◆◆◆◆◆◆

When you have completed this chapter, you should be able to:

- Convert angles between decimal degrees and radians, and degrees, minutes, and seconds.
- Find the trigonometric functions of an angle.
- Find the acute angle that has a given trigonometric function.
- Find the missing sides and angles of a right triangle.
- Solve practical problems involving the right triangle.
- Resolve a vector into components and, conversely, combine components into a resultant vector.
- Solve practical problems using vectors.

◆◆◆

With this chapter we begin our study of *trigonometry*, the branch of mathematics that enables us to solve triangles. The *trigonometric functions* are introduced here and are used to solve right triangles. Other kinds of triangles (*oblique* triangles) are discussed in Chapter 15, and other applications of trigonometry are given in Chapters 16, 17, and 18.

We build on what we learned about triangles in Chapter 5, mainly the Pythagorean theorem and the fact that the sum of the angles of a triangle is 180°. Here we make use of *coordinate axes*, described in Sec. 5–1.

Also introduced in this chapter are *vectors*, the study of which will be continued in Chapter 15.

7–1   Angles and Their Measures

Angles Formed by Rotation

We can think of an angle as the figure generated by a ray rotating from some initial position to some terminal position (Fig. 7–1). We usually consider angles formed by rotation in the *counterclockwise* direction as *positive*, and angles formed by *clockwise* rotation as *negative*. A rotation of one *revolution* brings the ray back to its initial position.

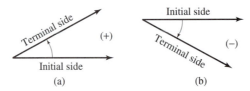

(a)   (b)

FIGURE 7–1   Angles formed by rotation.

Degrees, Minutes, and Seconds

The degree (°) is a unit of angular measure equal to 1/360 of a revolution; thus 360° = one revolution. A fractional part of a degree may be expressed as a common fraction (such as $36\frac{1}{2}°$), as a decimal (28.74°), or as *minutes* and *seconds*.

A *minute* (′) is equal to 1/60 of a degree; a *second* (″) is equal to 1/60 of a minute, or 1/3600 of a degree.

Note that minutes or seconds less than 10 are written with an initial zero. Thus 6′ is written 06′.

◆◆◆ **Example 1:** Some examples of angles written in degrees, minutes, and seconds are

$$85°18′42″ \qquad 62°12′ \qquad 69°55′25.4″ \qquad 75°06′03″$$ ◆◆◆

We will sometimes abbreviate "degrees, minutes, and seconds" as DMS.

Radian Measure

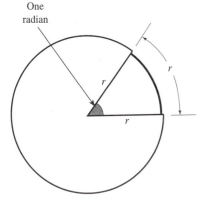

One radian

FIGURE 7–2

A *central angle* in a circle is one whose vertex is at the center of the circle. An *arc* is a portion of the circle. If an arc is laid off along a circle with a length equal to the radius of the circle, the central angle subtended by this arc is defined as one *radian* (Fig. 7–2). A radian (rad) is a unit of angular measure commonly used in trigonometry. We will study radian measure in detail in Chapter 16.

An arc having a length of twice the radius will subtend a central angle of 2 radians. Similarly, an arc with a length of 2π times the radius (the circumference) will subtend a central angle of 2π radians. Thus our conversion factor is

$$2\pi \text{ radians} = 1 \text{ revolution} = 360°$$

Conversion of Angles

We convert angles in the same way as we convert any other units. The following is a summary of the needed conversions:

You may prefer to use π rad = 180°. Some calculators have special keys for angle conversions. Check your manual.

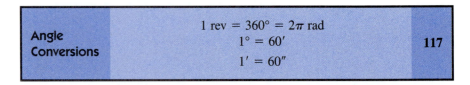

Angle Conversions	1 rev = 360° = 2π rad 1° = 60′ 1′ = 60″	**117**

◆◆◆ **Example 2:** Convert 47.6° to radians and revolutions.

Solution: By Eq. 117,

$$47.6° \left(\frac{2\pi \text{ rad}}{360°} \right) = 0.831 \text{ rad}$$

Note that we keep the *same number of significant digits* as we had in the original angle. We convert to revolutions as follows:

$$47.6° \left(\frac{1 \text{ rev}}{360°} \right) = 0.132 \text{ rev}$$

◆◆◆

◆◆◆ **Example 3:** Convert 1.8473 rad to degrees and revolutions.

Solution: By Eq. 117,

$$1.8473 \text{ rad} \left(\frac{360°}{2\pi \text{ rad}} \right) = 105.84°$$

and

$$1.8473 \text{ rad} \left(\frac{1 \text{ rev}}{2\pi \text{ rad}} \right) = 0.29401 \text{ rev}$$

◆◆◆

◆◆◆ **Example 4:** Convert 1.0000 radian to degrees.

Solution:

$$1.0000 \text{ rad} \left(\frac{360°}{2\pi \text{ rad}} \right) = 57.296°$$

This is a good conversion factor to memorize.

◆◆◆

Conversions involving degrees, minutes, and seconds require several steps. Further, to know how many digits to retain, we note the following:

$$1 \text{ min} = 1/60 \text{ degree} \approx 0.02°$$

and

$$1 \text{ sec} = 1/3600 \text{ degree} \approx 0.0003°$$

Thus an angle known to the nearest minute is about as accurate as a decimal angle known to two decimal places. Also, an angle known to the nearest second is about as accurate as a decimal angle known to the fourth decimal place. Therefore we would treat an angle such as 28°17′37″ as if it had four decimal places and six significant digits.

> Note that 1 second of an arc is an extremely small angle. Angle measurements accurate to the nearest second are rare and require special equipment and techniques.

◆◆◆ **Example 5:** Convert 28°17′37″ to decimal degrees and radians.

Solution: We separately convert the minutes and the seconds to degrees, and add them. Since the given angle is known to the nearest second, we will work to four decimal places.

$$37 \text{ sec} \left(\frac{1 \text{ deg}}{3600 \text{ sec}} \right) = 0.0103°$$

$$17 \text{ min} \left(\frac{1 \text{ deg}}{60 \text{ min}} \right) = 0.2833°$$

$$28° = \underline{28.0000°}$$

Add: 28.2936°

Now, by Eq. 117,

$$28.2936° \left(\frac{2\pi \text{ rad}}{360 \text{ deg}} \right) = 0.493817 \text{ rad}$$

◆◆◆

◆◆◆ **Example 6:** Convert 1.837520 rad to degrees, minutes, and seconds.

Solution: We first convert to decimal degrees.

$$1.837520 \text{ rad} \left(\frac{360°}{2\pi \text{ rad}} \right) = 105.2821°$$

Then we convert the decimal part (0.2821°) to minutes,

$$0.2821° \left(\frac{60'}{1°} \right) = 16.93'$$

and convert the decimal part of 16.93′ to seconds,

$$0.93' \left(\frac{60''}{1 \text{ min}} \right) = 56''$$

So

$$1.837520 \text{ rad} = 105°16'56''$$

◆◆◆

Another unit of angular measure is the *grad*. There are 400 grads in 1 revolution.

Common Error	When you are reading an angle quickly, it is easy to mistake decimal degrees for degrees and minutes. Don't mistake 28°50′ for 28.50°

Exercise 1 ◆ Angles and Their Measures

Angle Conversions

Convert to radians.

1. 27.8°	**2.** 38.7°	**3.** 35°15′
4. 270°27′25″	**5.** 0.55 rev	**6.** 0.276 rev

Convert to revolutions.

7. 4.772 rad	**8.** 2.38 rad	**9.** 68.8°
10. 3.72°	**11.** 77°18′	**12.** 135°27′42″

Convert to degrees (decimal).

13. 2.83 rad	**14.** 4.275 rad	**15.** 0.475 rev
16. 0.236 rev	**17.** 29°27′	**18.** 275°18′35″

Convert to degrees, minutes, and seconds.

19. 4.2754 rad	**20.** 1.773 rad	**21.** 0.44975 rev
22. 0.78426 rev	**23.** 185.972°	**24.** 128.259°

Computer

25. Write a program that will accept an angle in decimal degrees, and compute and print the angle in degrees, minutes, and seconds.

7–2   The Trigonometric Functions

Definitions

We start with coordinate axes upon which we draw an angle, as shown in Fig. 7–3. We always draw such angles the same way, with the vertex at the origin (O) and with one side along the x axis. Such an angle is said to be in *standard position*. We can think of the angle as being generated by a line rotating counterclockwise, whose initial position is on the x axis. We now select any point P on the terminal side of the angle θ, and we let the coordinates of P be x and y. We form a *right triangle OPQ* by dropping a perpendicular from P to the x axis. The side OP is the *hypotenuse* of the right triangle; we label its length r. Note that by the Pythagorean theorem, $r^2 = x^2 + y^2$. Also, side OQ is *adjacent* to angle θ and has a length x, and side PQ is *opposite* to angle θ and has a length y.

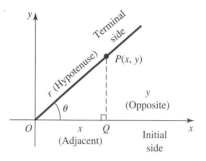

FIGURE 7–3   An angle in standard position.

If we had chosen to place point P farther out on the terminal side of the angle, the lengths x, y, and r would all be greater than they are now. But the *ratio* of any two sides will be the same regardless of where we place P, because all of the triangles thus formed are similar to each other. Such a ratio will change only if we change the angle θ. A ratio between any two sides of a right triangle is called a *trigonometric ratio* or a *trigonometric function.*

The six trigonometric functions are defined as follows:

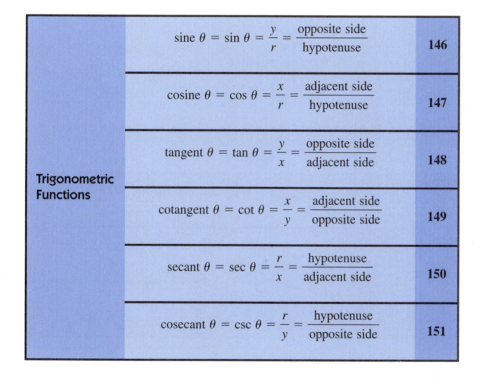

Trigonometric Functions

$$\text{sine } \theta = \sin \theta = \frac{y}{r} = \frac{\text{opposite side}}{\text{hypotenuse}} \qquad \textbf{146}$$

$$\text{cosine } \theta = \cos \theta = \frac{x}{r} = \frac{\text{adjacent side}}{\text{hypotenuse}} \qquad \textbf{147}$$

$$\text{tangent } \theta = \tan \theta = \frac{y}{x} = \frac{\text{opposite side}}{\text{adjacent side}} \qquad \textbf{148}$$

$$\text{cotangent } \theta = \cot \theta = \frac{x}{y} = \frac{\text{adjacent side}}{\text{opposite side}} \qquad \textbf{149}$$

$$\text{secant } \theta = \sec \theta = \frac{r}{x} = \frac{\text{hypotenuse}}{\text{adjacent side}} \qquad \textbf{150}$$

$$\text{cosecant } \theta = \csc \theta = \frac{r}{y} = \frac{\text{hypotenuse}}{\text{opposite side}} \qquad \textbf{151}$$

We discuss only *acute* angles here and discuss larger angles in Chapter 15.

The cotangent is sometimes abbreviated ctn.

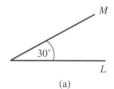

(a)

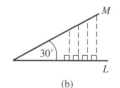

(b)

Note that each of the trigonometric ratios is *related to* the angle θ. Take the ratio y/r, for example. The values of y/r depend only on the measure of θ, not on the lengths of y or r. To convince yourself of this, draw an angle of, say, 30°, as in Fig. 7–4(a). Then form a right triangle by drawing a perpendicular from any point on side L or from any point on side M [Fig. 7–4(b) and (c)]. No matter how you draw the triangle, the side opposite the 30° angle *will always be half the hypotenuse* (sin 30° = ½). You would get a similar result for any other angle and for any other trigonometric ratio. *The value of the trigonometric ratio depends only on the measure of the angle.* We have a *relation,* such as the relations we studied in Chapter 4.

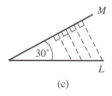

(c)

FIGURE 7–4

Furthermore, for any angle θ, there exists *only one* value for each trigonometric ratio. Thus *each ratio is a function of* θ, which is why we call them the *trigonometric functions* of θ. In functional notation, we could write

$$\sin \theta = f(\theta)$$
$$\cos \theta = g(\theta)$$
$$\tan \theta = h(\theta)$$

and so forth. The angle θ is called the *argument* of the function.

Note that while there is one and only one value of each trigonometric function for a given angle θ, we will see that for any given value of a trigonometric function (say, $\sin \theta = \frac{1}{2}$), there are *indefinitely many* values of θ. We treat this idea more fully in Sec. 15–1.

The trigonometric functions are sometimes called *circular* functions.

Reciprocal Relationships

From the trigonometric functions we see that

$$\sin \theta = \frac{y}{r}$$

and that

$$\csc \theta = \frac{r}{y}$$

Obviously, $\sin \theta$ and $\csc \theta$ are reciprocals.

$$\sin \theta = \frac{1}{\csc \theta}$$

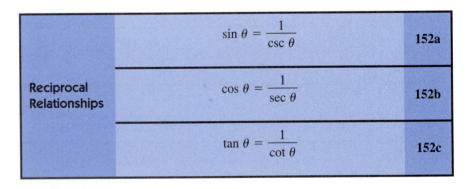

Inspection of the trigonometric functions shows two more sets of reciprocals (see also Fig. 7–5).

Reciprocal Relationships		
	$\sin \theta = \dfrac{1}{\csc \theta}$	152a
	$\cos \theta = \dfrac{1}{\sec \theta}$	152b
	$\tan \theta = \dfrac{1}{\cot \theta}$	152c

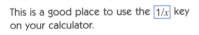

sin θ **cos θ** **tan θ** **cot θ** **sec θ** **csc θ**

FIGURE 7–5 This diagram shows which functions are reciprocals of each other.

This is a good place to use the $\boxed{1/x}$ key on your calculator.

◆◆◆ **Example 7:** Find $\tan \theta$ if $\cot \theta = 0.638$.

Solution: From the reciprocal relationships,

$$\tan \theta = \frac{1}{\cot \theta} = \frac{1}{0.638} = 1.57$$

◆◆◆

Finding the Trigonometric Functions of an Angle

The six trigonometric functions of an angle can be written if we know the angle or if we know two of the sides of a right triangle containing the angle.

••• **Example 8:** A right triangle has sides of 4.37 and 2.83 (Fig. 7–6). Write the six trigonometric functions of angle θ.

Solution: We first find the hypotenuse by the Pythagorean theorem.

$$c = \sqrt{(2.83)^2 + (4.37)^2} = \sqrt{27.1}$$
$$= 5.21$$

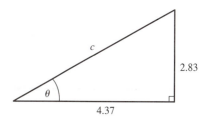

FIGURE 7–6

We also note that the opposite side is 2.83 and the adjacent side is 4.37. Then, by Eqs. 146 through 151,

$$\sin \theta = \frac{\text{opposite side}}{\text{hypotenuse}} = \frac{2.83}{5.21} = 0.543$$

$$\cos \theta = \frac{\text{adjacent side}}{\text{hypotenuse}} = \frac{4.37}{5.21} = 0.839$$

$$\tan \theta = \frac{\text{opposite side}}{\text{adjacent side}} = \frac{2.83}{4.37} = 0.648$$

$$\cot \theta = \frac{\text{adjacent side}}{\text{opposite side}} = \frac{4.37}{2.83} = 1.54$$

$$\sec \theta = \frac{\text{hypotenuse}}{\text{adjacent side}} = \frac{5.21}{4.37} = 1.19$$

$$\csc \theta = \frac{\text{hypotenuse}}{\text{opposite side}} = \frac{5.21}{2.83} = 1.84$$

•••

Common Error	Do not omit the angle when writing a trigonometric function. To write "sin = 0.543," for example, has no meaning.

••• **Example 9:** A point P on the terminal side of an angle B (in standard position) has the coordinates (8.84, 3.15). Write the sine, cosine, and tangent of angle B.

Solution: We are given $x = 8.84$ and $y = 3.15$. We find r from the Pythagorean theorem.

$$r = \sqrt{(8.84)^2 + (3.15)^2} = 9.38$$

Then, by Eqs. 146 through 151,

$$\sin B = \frac{y}{r} = \frac{3.15}{9.38} = 0.336$$

$$\cos B = \frac{x}{r} = \frac{8.84}{9.38} = 0.942$$

$$\tan B = \frac{y}{x} = \frac{3.15}{8.84} = 0.356$$

•••

Trigonometric Functions by Calculator

A trigonometric function is found by calculator by using a trig function key: sin, cos, or tan. On some calculators you first enter the angle and then press the trig function key; on other calculators you must press the trig function key first. You must find out which way your calculator works. Check your calculator manual to find out which way your own calculator works, and see if you can get the same results as in the following examples.

◆◆◆ **Example 10:** Find (a) sin 38.6° and (b) tan 1.174 rad by calculator to four decimal places.

Solution: By calculator, in degree mode, we get

(a) sin 38.6° = 0.6239

Then switching to radian mode, we have

(b) tan 1.174 rad = 2.3865   ◆◆◆

To find the cotangent, secant, or cosecant, first find the reciprocal function (tangent, cosine, or sine), and then press the $\boxed{1/x}$ or $\boxed{x^{-1}}$ key to get the reciprocal.

◆◆◆ **Example 11:** Find (a) sec 72.8° and (b) csc 0.748 rad by calculator to three decimal places.

Solution: By calculator, in degree mode, we get

(a) sec 72.8° = 3.382

Then switching to radian mode, we have

(b) csc 0.748 rad = 1.470   ◆◆◆

Common Error	It is easy to forget to set your calculator in the proper $\boxed{\text{Degree/Radian}}$ mode. Be sure to check it each time.

Evaluating Trigonometric Expressions

We can combine trigonometric functions with numbers or variables to get more complex expressions, such as

$$2 \sin 35° - 5 \cos 12°$$

In this section we will evaluate some very simple expressions of this kind. Later, we will cover more complex ones.

Simply perform the indicated operations. Just remember that angles are assumed to be in *radians* unless marked otherwise.

◆◆◆ **Example 12:** Evaluate the following to three significant digits:

$$7.82 - 3.15 \cos 67.8°$$

Solution: By calculator, we find that cos 67.8° = 0.3778, so our expression becomes

$$7.82 - 3.15(0.3778) = 6.63$$

As usual when doing such calculations, we keep all of the figures in the calculator whenever possible. If you must write down an intermediate result, as we did here just to show the steps, retain 1 or 2 digits more than required in the final answer.

◆◆◆

◆◆◆ **Example 13:** Evaluate the following to four significant digits:

$$1.836 \sin 2 + 2.624 \tan 3$$

Solution: Here no angular units are indicated, so we assume them to be radians. Switching to radian mode, we get

$$1.836 \sin 2 + 2.624 \tan 3 = 1.836(0.90930) + 2.624(-0.14255)$$
$$= 1.295$$ ◆◆◆

Exercise 2 ◆ The Trigonometric Functions

Using a protractor, lay off the given angle on coordinate axes. Drop a perpendicular from any point on the terminal side, and measure the distances x, y, and r, as in Fig. 7–3. Use these measured distances to write the six trigonometric ratios of the angle to two decimal places. Check your answer by calculator.

1. 30°   **2.** 55°
3. 14°   **4.** 37°

Plot the given point on coordinate axes, and connect it to the origin. Measure the angle formed with a protractor. Check by taking the tangent of your measured angle. It should equal the ratio y/x.

5. (3, 4)   **6.** (5, 3)   **7.** (1, 2)   **8.** (5, 4)
9. Write the six trigonometric ratios for angle θ in Fig. 7–7.

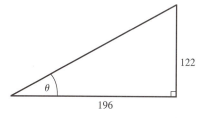

FIGURE 7–7

Each of the given points is on the terminal side of an angle. Compute the distance r from the origin to the point, and write the sin, cos, and tan of the angle.

10. (3, 5)   **11.** (4, 2)
12. (5, 3)   **13.** (2, 3)
14. (4, 5)   **15.** (4.75, 2.68)

Work to three significant digits on problems 10 through 15.

Evaluate. Work to four decimal places.

16. sin 47.3°   **17.** tan 44.4°   **18.** sec 29.5°
19. cot 18.2°   **20.** csc 37.5°   **21.** cos 21.27°
22. cos 86.75°   **23.** sec 12.3°   **24.** csc 12.67°

Write the sin, cos, and tan for each angle. Keep three decimal places.

25. 72.8°   **26.** 19.2°
27. 33.1°   **28.** 41.4°
29. 0.744 rad   **30.** 0.385 rad

Evaluate each trigonometric expression to three significant digits.

31. 5.27 sin 45.8° − 1.73
32. 2.84 cos 73.4° − 3.83 tan 36.2°
33. 3.72(sin 28.3° + cos 72.3°)
34. 11.2 tan 5 + 15.3 cos 3
35. 2.84(5.28 cos 2 − 2.82) + 3.35
36. 2.63 sin 2.4 + 1.36 cos 3.5 + 3.13 tan 2.5

7–3   Finding the Angle When the Trigonometric Function Is Given

The operation of finding the angle when the trigonometric function is given is the *inverse* of finding the function when the angle is given. There is special notation to indicate the inverse trigonometric function. If

$$\sin \theta = A$$

we write

$$\theta = \arcsin A$$

or

$$\theta = \sin^{-1} A$$

which is read "θ is the angle whose sine is A." Similarly, we use the symbols arccos A, $\cos^{-1} A$, arctan A, and so on.

	Do not confuse the **inverse** with the **reciprocal**.
Common Error	$$\frac{1}{\sin \theta} = (\sin \theta)^{-1} \neq \sin^{-1} \theta$$

Here we use the inverse trigonometric function keys, $\boxed{\sin^{-1}}$, $\boxed{\cos^{-1}}$, or $\boxed{\tan^{-1}}$, on the calculator. If you do not have such keys, look for an $\boxed{\text{arc}}$ or $\boxed{\text{inv}}$ key, and consult your manual on its use. See if you can duplicate the calculations in the following examples.

◆◆◆ **Example 14:** If $\sin \theta = 0.7337$, find θ in degrees to three significant digits.

Solution: Switch the calculator into the $\boxed{\text{Degree}}$ mode. Then

$$\sin^{-1} 0.7337 = 47.2°$$

Note that a calculator gives us acute angles only.   ◆◆◆

◆◆◆ **Example 15:** If $\tan x = 2.846$, find x in radians to four significant digits.

Solution: With the calculator in the $\boxed{\text{Radian}}$ mode,

$$\tan^{-1} 2.846 = 1.233$$ ◆◆◆

If the cotangent, secant, or cosecant is given, we first take the reciprocal of that value and then use the appropriate reciprocal relationship.

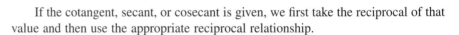
$$\sin \theta = \frac{1}{\csc \theta} \qquad \cos \theta = \frac{1}{\sec \theta} \qquad \tan \theta = \frac{1}{\cot \theta}$$

In this chapter we limit our work with inverse functions to *first-quadrant* angles only. In Chapter 15, we consider any angle.

◆◆◆ **Example 16:** If $\sec \theta = 1.573$, find θ in degrees to four significant digits.

Solution: Put the calculator into the $\boxed{\text{Degree}}$ mode. Then since

$$\cos \theta = \frac{1}{\sec \theta} = \frac{1}{1.573}$$

we first find the reciprocal of 1.573 and then take the inverse cosine.

$$\frac{1}{1.573} = 0.63573$$

Then

$$\cos^{-1} 0.63573 = 50.53$$ ◆◆◆

Exercise 3 ◆ Finding the Angle When the Trigonometric Function Is Given

Find the acute angle (in decimal degrees) whose trigonometric function is given. Keep three significant digits.

1. $\sin A = 0.500$ **2.** $\tan D = 1.53$ **3.** $\sin G = 0.528$
4. $\cot K = 1.77$ **5.** $\sin B = 0.483$ **6.** $\cot E = 0.847$

Without using tables or calculator, write the sin, cos, and tan of angle A. Leave your answer in fractional form.

7. $\sin A = \dfrac{3}{5}$ **8.** $\cot A = \dfrac{12}{5}$ **9.** $\cos A = \dfrac{12}{13}$

Evaluate the following, giving your answer in decimal degrees to three significant digits.

10. arcsin 0.635   **11.** arcsec 3.86   **12.** $\tan^{-1} 2.85$
13. $\cot^{-1} 1.17$ **14.** $\cos^{-1} 0.229$ **15.** arccsc 4.26

7–4 Solution of Right Triangles

Right Triangles

The right triangle (Fig. 7–8) was introduced in Sec. 6–2, where we also introduced the Pythagorean theorem.

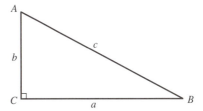

FIGURE 7–8 A right triangle. We will usually *label* a right triangle as shown here. We label the angles with capital letters A, B, and C, with C always the right angle. We label the sides with lowercase letters a, b, and c, with side a opposite angle A, side b opposite angle B, and side c (the hypotenuse) opposite angle C (the right angle).

Furthermore, since angle C is always 90°, Equation 139 becomes this formula for sum of the angles.

To these, we now add these trigonometric functions.

Pythagorean Theorem	$c^2 = a^2 + b^2$	145
Sum of the Angles	$A + B = 90°$	139
Trigonometric Functions	$\sin \theta = \dfrac{\text{side opposite to } \theta}{\text{hypotenuse}}$	146
	$\cos \theta = \dfrac{\text{side adjacent to } \theta}{\text{hypotenuse}}$	147
	$\tan \theta = \dfrac{\text{side opposite to } \theta}{\text{side adjacent to } \theta}$	148

These five equations will be our tools for solving any right triangle.

Solving Right Triangles When One Side and One Angle Are Known

To *solve* a triangle means to find all missing sides and angles (although in most practical problems we need find only one missing side or angle). We can solve any right triangle if we know one side and either another side or one angle.

To solve a right triangle when one side and one angle are known:

1. Make a sketch.
2. Find the missing angle by using Eq. 139.
3. Relate the known side to one of the missing sides by one of the trigonometric ratios. Solve for the missing side.
4. Repeat Step 3 to find the second missing side.
5. Check your work with the Pythagorean theorem.

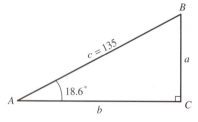

FIGURE 7–9

Realize that either of the two legs can be called *opposite* or *adjacent*, depending on which angle we are referring them to. Here b is adjacent to angle A but opposite to angle B.

◆◆◆ **Example 17:** Solve right triangle ABC if $A = 18.6°$ and $c = 135$.

Solution:

1. We make a sketch as shown in Figure 7–9.
2. Then, by Eq. 139,

$$B = 90° - A = 90° - 18.6° = 71.4°$$

3. Let us now find side a. We must use one of the trigonometric ratios. *But how do we know which one to use?* And further, *which of the two angles,* A *or* B, *should we write the trig ratio for?*
    It is simple. First, always work with the *given angle,* because if you made a mistake in finding angle B and then used it to find the sides, they would be wrong also. Then, to decide which trigonometric ratio to use, we note that side a is *opposite* to angle A and that the given side is the *hypotenuse.* Thus, our trig function must be one that relates the *opposite side* to the *hypotenuse.* Our obvious choice is Eq. 146.

$$\sin A = \frac{\text{opposite side}}{\text{hypotenuse}}$$

Substituting the given values, we obtain

$$\sin 18.6° = \frac{a}{135}$$

Solving for a yields

$$a = 135 \sin 18.6°$$
$$= 135(0.3190) = 43.1$$

4. We now find side b. Note that side b is *adjacent* to angle A. We therefore use Eq. 147.

$$\cos 18.6° = \frac{b}{135}$$

So

$$b = 135 \cos 18.6°$$
$$= 135(0.9478) = 128$$

We have thus found all missing parts of the triangle.

As a rough check of any triangle, see if the longest side is opposite the largest angle and if the shortest side is opposite the smallest angle. Also check that the hypotenuse is greater than either leg but less than their sum.

5. For a check, we see if the three sides will satisfy the Pythagorean theorem.

Check:

$$(43.1)^2 + (128)^2 \stackrel{?}{=} (135)^2$$
$$18{,}242 \stackrel{?}{=} 18{,}225$$

Since we are working to three significant digits, this is close enough for a check.   ◆◆◆

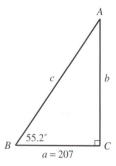

FIGURE 7–10

◆◆◆ **Example 18:** In right triangle ABC, angle $B = 55.2°$ and $a = 207$. Solve the triangle.

Solution:

1. We make a sketch as shown in Fig. 7–10.
2. By Eq. 139,

$$A = 90° - 55.2° = 34.8°$$

3. By Eq. 147,

$$\cos 55.2° = \frac{207}{c}$$

$$c = \frac{207}{\cos 55.2°} = \frac{207}{0.5707} = 363$$

4. Then, by Eq. 148,

$$\tan 55.2° = \frac{b}{207}$$

$$b = 207 \tan 55.2° = 207(1.439) = 298$$

5. Checking with the Pythagorean theorem, we have

$$(363)^2 \stackrel{?}{=} (207)^2 + (298)^2$$

$$131{,}769 \stackrel{?}{=} 131{,}653 \quad \text{(checks to within three significant digits)} \quad \blacklozenge\blacklozenge\blacklozenge$$

Tip	Whenever possible, use the *given information* for each computation, rather than some quantity previously calculated. This way, any errors in the early computation will not be carried along.

This, of course, is good advice for performing *any* computation, not just for solving right triangles.

Solving Right Triangles When Two Sides Are Known

1. Draw a diagram of the triangle.
2. Write the trigonometric ratio that relates one of the angles to the two given sides. Solve for the angle.
3. Subtract the angle just found from 90° to get the second angle.
4. Find the missing side by the Pythagorean theorem.
5. Check the computed side and angles with trigonometric ratios, as shown in the following example.

◆◆◆ **Example 19:** Solve right triangle *ABC* if $a = 1.48$ and $b = 2.25$.

Solution:

1. We sketch the triangle as shown in Fig. 7–11.
2. To find angle *A*, we note that the 1.48 side is *opposite* angle *A* and that the 2.25 side is *adjacent* to angle *A*. The trig ratio relating *opposite* and *adjacent* is the *tangent* (Eq. 148).

$$\tan A = \frac{1.48}{2.25} = 0.6578$$

from which

$$A = 33.3°$$

3. Solving for angle *B*, we have

$$B = 90° - A = 90 - 33.3 = 56.7°$$

4. We find side *c* by the Pythagorean theorem.

$$c^2 = (1.48)^2 + (2.25)^2 = 7.253$$

$$c = 2.69$$

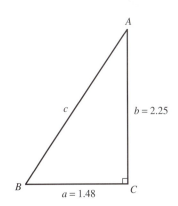

FIGURE 7–11

Check:

$$\sin 33.3° \stackrel{?}{=} \frac{1.48}{2.69}$$

$$0.549 \approx 0.550 \quad \text{(checks)}$$

and

$$\sin 56.7° \stackrel{?}{=} \frac{2.25}{2.69}$$

$$0.836 = 0.836 \quad \text{(checks)}$$

(Note that we could just as well have used another trigonometric function, such as the cosine, for our check.)   ◆◆◆

Cofunctions

The sine of angle A in Fig. 7–8 is

$$\sin A = \frac{a}{c}$$

But a/c is *also* the cosine of the complementary angle B. Thus we can write

$$\sin A = \cos B$$

Here $\sin A$ and $\cos B$ are called *cofunctions*. Similarly, $\cos A = \sin B$. These cofunctions and others in the same right triangle are given in the following boxes:

Cofunctions **Where** $A + B = 90°$ **or** $A + B = \dfrac{\pi}{2}$ **rad**	$\sin A = \cos B$	154a
	$\cos A = \sin B$	154b
	$\tan A = \cot B$	154c
	$\cot A = \tan B$	154d
	$\sec A = \csc B$	154e
	$\csc A = \sec B$	154f

In general, a trigonometric function of an acute angle is equal to the corresponding cofunction of the complementary angle.

◆◆◆ **Example 20:** If the cosine of the acute angle A in a right triangle ABC is equal to 0.725, find the sine of the other acute angle, B.

Solution: By Eq. 154b,

$$\sin B = \cos A = 0.725$$ ◆◆◆

Exercise 4 ◆ Solution of Right Triangles

Right Triangles with One Side and One Angle Given

Sketch each right triangle and find all of the missing parts. Assume the triangles to be labeled as in Fig. 7–8. Work to three significant digits.

1. $a = 155$ $A = 42.9°$ **2.** $b = 82.6$ $B = 61.4°$
3. $a = 1.74$ $B = 31.9°$ **4.** $b = 7.74$ $A = 22.5°$
5. $a = 284$ $A = 1.13$ rad   **6.** $b = 73.2$ $B = 0.655$ rad
7. $b = 9.26$ $B = 55.2°$ **8.** $a = 1.73$ $A = 39.3°$
9. $b = 82.4$ $A = 31.4°$ **10.** $a = 18.3$ $B = 44.1°$

Right Triangles with Two Sides Given

Sketch each right triangle and find all missing parts. Work to three significant digits and express the angles in decimal degrees.

11. $a = 382$ $b = 274$ **12.** $a = 3.88$ $c = 5.37$
13. $b = 3.97$ $c = 4.86$ **14.** $a = 63.9$ $b = 84.3$
15. $a = 27.4$ $c = 37.5$ **16.** $b = 746$ $c = 957$
17. $a = 41.3$ $c = 63.7$ **18.** $a = 4.82$ $b = 3.28$
19. $b = 228$ $c = 473$ **20.** $a = 274$ $c = 429$

Cofunctions

Express as a function of the complementary angle.

21. $\sin 38°$ **22.** $\cos 73°$ **23.** $\tan 19°$
24. $\sec 85.6°$ **25.** $\cot 63.2°$ **26.** $\csc 82.7°$
27. $\tan 35°14'$ **28.** $\cos 0.153$ rad   **29.** $\sin 0.475$ rad

Computer

30. Write a program or use a spreadsheet to compute the perimeter of a regular polygon of n sides inscribed in a circle having a diameter of 1 unit. Let n vary from 3 to 20, and for each n print **(a)** the number of sides, n; **(b)** the perimeter of the polygon; and **(c)** the difference between the circumference of the circle and the perimeter of the polygon.

31. Write a program to solve any right triangle. Have the program accept as input a known side and angle, or two known sides, and compute and print all of the sides and angles.

7–5   Applications of the Right Triangle

There are, of course, a huge number of applications for the right triangle, a few of which are given in the following examples and exercises. A typical application is that of finding a distance that cannot be measured directly, as shown in the following example.

◆◆◆ **Example 21:** To find the height of a flagpole (Fig. 7–12), a person measures 35.0 ft from the base of the pole and then measures an angle of 40.8° from a point 6.00 ft above the ground to the top of the pole. Find the height of the flagpole.

Estimate: If angle A were 45°, then BC would be the same length as AC, or 35 ft. But our angle is a bit less than 45°, so we expect BC to be less than 35 ft, say, 30 ft. Thus our guess for the entire height is about 36 ft.

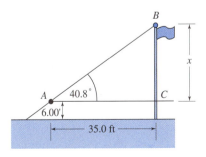

FIGURE 7–12

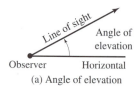

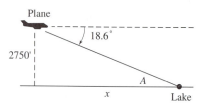

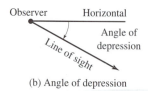

FIGURE 7–13 Angles of elevation and depression.

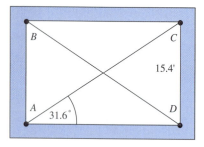

FIGURE 7–14

Solution: In right triangle ABC, BC is opposite the known angle, and AC is adjacent. Using the tangent, we get

$$\tan 40.8° = \frac{x}{35.0}$$

where x is the height of the pole above the observer. Then

$$x = 35.0 \tan 40.8° = 30.2 \text{ ft}$$

Adding 6.0 ft, we find that the total pole height is 36.2 ft, measured from the ground. ◆◆◆

◆◆◆ **Example 22:** From a plane at an altitude of 2750 ft, the pilot observes the angle of depression of a lake to be 18.6°. How far is the lake from a point on the ground directly beneath the plane?

Solution: We first note that an *angle of depression*, as well as an *angle of elevation,* is defined as the angle between a line of sight and the *horizontal,* as shown in Fig. 7–13. We then make a sketch for our problem (Fig. 7–14). Since the ground and the horizontal line drawn through the plane are parallel, angle A and the angle of depression (18.6°) are alternate interior angles. Therefore $A = 18.6°$. Then

$$\tan 18.6° = \frac{2750}{x}$$

$$x = \frac{2750}{\tan 18.6°} = 8170 \text{ ft}$$ ◆◆◆

◆◆◆ **Example 23:** A frame is braced by wires AC and BD, as shown in Fig. 7–15. Find the length of each wire.

Estimate: We can make a quick estimate if we remember that the sine of 30° is 0.5. Thus we expect wire AC to be a little shorter than twice side CD, or around 30 ft.

Solution: Noting that in right triangle ACD, CD is opposite the given angle and AC is the hypotenuse, we choose the sine.

$$\sin 31.6° = \frac{15.4}{AC}$$

$$AC = \frac{15.4}{\sin 31.6°} = 29.4 \text{ ft}$$ ◆◆◆

FIGURE 7–15

Exercise 5 ◆ Applications of the Right Triangle

Measuring Inaccessible Distances

1. From a point on the ground 255 m from the base of a tower, the angle of elevation to the top of the tower is 57.6°. Find the height of the tower.
2. A pilot 4220 m directly above the front of a straight train observes that the angle of depression of the end of the train is 68.2°. Find the length of the train.
3. From the top of a lighthouse 156 ft above the surface of the water, the angle of depression of a boat is observed to be 28.7°. Find the horizontal distance from the boat to the lighthouse.
4. An observer in an airplane 1520 ft above the surface of the ocean observes that the angle of depression of a ship is 28.8°. Find the straight-line distance from the plane to the ship.
5. The distance PQ across a swamp is desired. A line PR of length 59.3 m is laid off at right angles to PQ. Angle PRQ is measured at 47.6°. Find the distance PQ.

6. The angle of elevation of the top of a building from a point on the ground 275 ft from its base is 51.3°. Find the height of the building.

7. The angle of elevation of the top of a building from a point on the ground 75.0 yd from its base is 28.0°. How high is the building?

8. From the top of a hill 125 ft above a stream, the angles of depression of a point on the near shore and of a point on the opposite shore are 42.3° and 40.6°. Find the width of the stream between these two points.

9. From the top of a tree 15.0 m high on the shore of a pond, the angle of depression of a point on the other shore is 6.70°. What is the width of the pond?

Navigation

10. A passenger on a ship sailing due north at 8.20 km/h noticed that at 1 P.M., a buoy was due east of the ship. At 1:45 P.M., the bearing of the buoy from the ship was S 27°12′ E (Fig. 7–16). How far was the ship from the buoy at 1:00 P.M.?

11. An observer at a point P on a coast sights a ship in a direction N 43°15′ E. The ship is at the same time directly east of a point Q, 15.6 km due north of P. Find the distance of the ship from point P and from Q.

12. A ship sailing parallel to a straight coast is directly opposite one of two lights on the shore. The angle between the lines of sight from the ship to these lights is 27°50′, and it is known that the lights are 355 m apart. Find the perpendicular distance of the ship from the shore.

13. An airplane left an airport and flew 315 mi in the direction N 15°18′ E, then turned and flew 296 mi in the direction S 74°42′ E. How far, and in what direction, should it fly to return to the airport?

14. A ship is sailing due south at a speed of 7.50 km/h when a light is sighted with a bearing S 35°28′ E. One hour later the ship is due west of the light. Find the distance from the ship to the light at that instant.

15. After leaving port, a ship holds a course N 46°12′ E for 225 mi. Find how far north and how far east of the port the ship is now located.

Structures

16. A guy wire from the top of an antenna is anchored 53.5 ft from the base of the antenna and makes an angle of 85.2° with the ground. Find **(a)** the height h of the antenna and **(b)** the length L of the wire.

17. Find the angle θ and length AB in the truss shown in Fig. 7–17.

18. A house is 8.75 m wide, and its roof has an angle of inclination of 34.0° (Fig. 7–18). Find the length L of the rafters.

19. A guy wire 82.0 ft long is stretched from the ground to the top of a telephone pole 65.0 ft high. Find the angle between the wire and the pole.

Geometry

20. What is the angle, to the nearest tenth of a degree, between a diagonal AB of a cube and a diagonal AC of a face of that cube (Fig. 7–19)?

21. Two of the sides of an isosceles triangle have a length of $15\overline{0}$ units, and each of the base angles is 68.0°. Find the altitude and the base of the triangle.

22. The diagonal of a rectangle is 3 times the length of the shorter side. Find the angle, to the nearest tenth of a degree, between the diagonal and the longer side.

23. Find the angles between the diagonals of a rectangle whose dimensions are $58\overline{0}$ units × 940 units.

24. Find the area of a parallelogram if the lengths of the sides are 255 units and 482 units and if one angle is 83.2°.

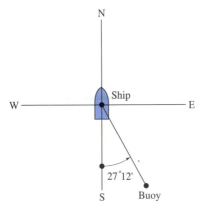

FIGURE 7–16 A compass direction of S 27°12′ E means an angle of 27°12′ measured from due south to the east. Also, remember that 1 minute of arc (1′) is equal to 1/60 degree, or

$$60' = 1°$$

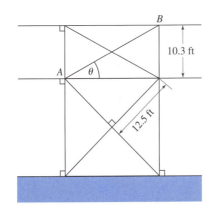

FIGURE 7–17

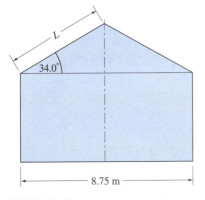

FIGURE 7–18

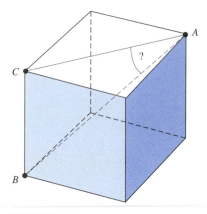

FIGURE 7–19

Trigonometry finds many uses in the machine shop. Given in problems 27 through 33 are some standard shop calculations.

25. Find the length of a side of a regular hexagon inscribed in a 125-cm-radius circle. *Hint*: Draw lines from the center of the hexagon to each vertex and to the midpoint of each side, to form right triangles.

26. Find the length of the side of a regular pentagon circumscribed about a circle of radius 244 in.

Shop Trigonometry

27. Find the dimension x in Fig. 7–20.

28. A bolt circle (Fig. 7–21) is to be made on a jig borer. Find the dimensions x_1, y_1, x_2, and y_2.

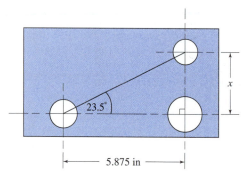

FIGURE 7–20

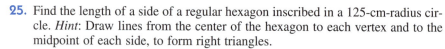

FIGURE 7–21   A bolt circle.

29. A bolt circle with a radius of 36.000 cm contains 24 equally spaced holes. Find the straight-line distance between the holes so that they may be stepped off with dividers.

30. A 9.000-in.-long shaft is to taper 3.500° and be 1.000 inch in diameter at the narrow end (Fig. 7–22). Find the diameter d of the larger end. *Hint*: Draw a line (shown dashed in the figure) from the small end to the large, to form a right triangle. Solve that triangle.

31. A measuring square having an included angle of 60.0° is placed over the pulley fragment of Fig. 7–23. The distance d from the corner of the square to the pulley rim is measured at 5.53 in. Find the pulley radius r.

32. A common way of measuring dovetails is with the aid of round plugs, as in Fig. 7–24. Find the distance x over the 0.5000-in.-diameter plugs. *Hint*: Find the length of side AB in the 30–60–90 triangle ABC, and then use AB to find x.

33. A bolt head (Fig. 7–25) measures 0.750 cm across the flats. Find the distance p across the corners and the width r of each flat.

FIGURE 7–22   Tapered shaft.

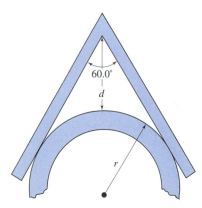

FIGURE 7–23   Measuring the radius of a broken pulley.

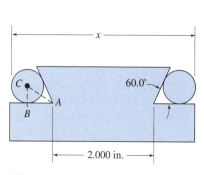

FIGURE 7–24   Measuring a dovetail.

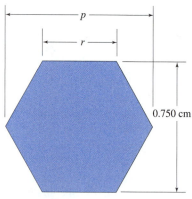

FIGURE 7–25

7–6   Vectors

Definition

A *vector quantity* is one that has *direction* as well as magnitude. For example, a velocity cannot be completely described without giving the direction of the motion, as well as the magnitude (the speed). Other vector quantities are force and acceleration.

Quantities that have magnitude but no direction are called *scalar* quantities. These include time, volume, mass, and so on.

Representation of Vectors

A vector is represented by a line segment whose length is proportional to the magnitude of the vector and whose direction is the same as the direction of the vector quantity (Fig. 7–26).

Vectors are represented differently in different textbooks, but they are usually written in boldface type. Here, we'll use the most common notation: **boldface** Roman capitals to represent vectors, and nonboldface *italic* capitals to represent scalar quantities. So in this textbook, **B** is understood to be a vector quantity, having both magnitude and direction, while *B* is understood to be a scalar quantity, having magnitude but no direction.

A vector can be designated by a single letter, or by two letters representing the endpoints, with the starting point given first. Thus the vector in Fig. 7–26 can be labeled

$$\mathbf{V} \quad \text{or} \quad \mathbf{AB}$$

or, when handwritten,

$$\vec{V} \quad \text{or} \quad \vec{AB}$$

The *magnitude* of a vector can be designated with absolute value symbols or by ordinary (nonboldface-italic) type. Thus the magnitude of vector **V** in Fig. 7–26 is

$$|\mathbf{V}| \quad \text{or} \quad V \quad \text{or} \quad |\mathbf{AB}| \quad \text{or} \quad AB$$

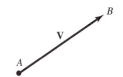

FIGURE 7–26   Representation of a vector.

Boldface letters are not practical in handwritten work. Instead, it is customary to place an *arrow* over vector quantities.

Polar Form and Rectangular Form

A vector is usually specified by giving its magnitude and direction. It is then said to be in *polar* form.

◆◆◆ **Example 24:** A certain vector has a magnitude of 5 units and makes an angle of 38° with the positive *x* axis.

We often write such a vector in the form

$$5\,\underline{/38°}$$

which is read "5 at an angle of 38°."   ◆◆◆

A vector drawn from the origin can also be designated by the *coordinates of its endpoint.*

◆◆◆ **Example 25:**

(a) The ordered pair (3, 5) is a vector. It designates a line segment drawn from the origin to the point (3, 5).
(b) The *ordered triple* of numbers (4, 6, 3) is a vector in three dimensions. It designates a point drawn from the origin to the point (4, 6, 3).

We cover vectors such as these in Chapter 15.

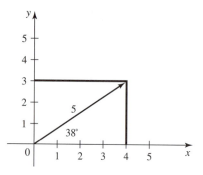

FIGURE 7–27

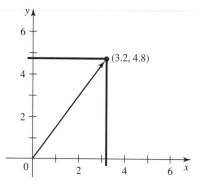

FIGURE 7–28

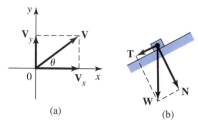

FIGURE 7–29 Rectangular components of a vector.

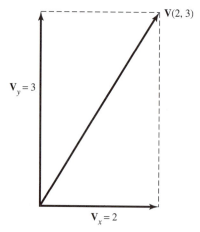

FIGURE 7–30

The component of a vector along an axis is also referred to as the *projection* of the vector onto that axis.

(c) The set of numbers (7, 3, 6, 1, 5) is a vector in five dimensions. We cannot draw it, but it is handled mathematically in the same way as the others.

A vector written this way is said to be in *rectangular* form.   ◆◆◆

Drawing a Vector on Coordinate Axes

When a vector is specified by its magnitude and direction, simply draw an arrow whose length is equal to its magnitude (using a suitable scale for the axes) in the given direction. The tail of the vector is placed at the origin, and the angle is measured, as usual, counterclockwise from the positive *x* direction.

◆◆◆ **Example 26:** Draw a vector whose magnitude is 5 and whose direction is 38°.

Solution: We measure 38° counterclockwise from the positive *x* axis and draw the vector from the origin, with length 5 units (Fig. 7–27).   ◆◆◆

When a vector is given by the coordinates of its endpoint, simply plot the point and connect it to the origin.

◆◆◆ **Example 27:** Draw the vector (3.2, 4.8).

Solution: We plot the point (3.2, 4.8) and connect it to the origin, as shown in Fig. 7–28. We draw the arrow at the given point, not at the origin.   ◆◆◆

Rectangular Components of a Vector

Any vector can be replaced by two (or more) vectors which, acting together, exactly duplicate the effect of the original vector. They are called the *components* of the vector. The components are usually chosen perpendicular to each other and are then called *rectangular components*. To *resolve* a vector means to replace it by its components.

◆◆◆ **Example 28:** A ball thrown from one person to another moves simultaneously in both the horizontal direction and the vertical direction. The ball thus has a horizontal and a vertical component of velocity.   ◆◆◆

◆◆◆ **Example 29:** Figure 7–29(a) shows a vector **V** resolved into its *x* component **V**$_x$ and its *y* component **V**$_y$.   ◆◆◆

◆◆◆ **Example 30:** Figure 7–29(b) shows a block on an inclined plane with its weight **W** resolved into a component **N** normal (perpendicular) to the plane and a component **T** tangential (parallel) to the plane.   ◆◆◆

◆◆◆ **Example 31:** Figure 7–30 shows the vector **V**(2, 3) resolved into its *x* and *y* components, **V**$_x$ and **V**$_y$ Figure 7–30. It is clear that the *x* component is 2 units, and the *y* component is 3 units. Note that the coordinates of the endpoint are also the magnitudes of the components.   ◆◆◆

Resolution of Vectors

We see in Fig. 7–29 that a vector and its rectangular components form a rectangle that has two *right triangles*. Thus we can resolve a vector into its rectangular components with the right-triangle trigonometry of this chapter. Since, in Fig. 7–29(a),

$$\sin \theta = \frac{V_y}{V}$$

Note that we have used nonboldface type to represent the magnitudes V and V_y of vectors $\mathbf{V}$ and $\mathbf{V}_y$. From this we get the magnitude V_y of the vector $\mathbf{V}_y$.

$$V_y = V \sin \theta$$

Similarly,

$$V_x = V \cos \theta$$

So

$$\mathbf{V}_y = V_y \underline{/90°} \quad \text{and} \quad \mathbf{V}_x = V_x \underline{/0°}$$

◆◆◆ **Example 32:** The vector $\mathbf{V}$ in Fig. 7–29(a) has a magnitude of 248 units and makes an angle θ of 38.2° with the x axis. Find the x and y components.

Solution:

$$\sin 38.2° = \frac{V_y}{248}$$

$$V_y = 248 \sin 38.2° = 153$$

Similarly,

$$V_x = 248 \cos 38.2° = 195$$

Thus we can write

$$\mathbf{V} = 248 \underline{/38.3°} = (195, 153)$$

in rectangular form.

◆◆◆

> Thus finding the rectangular components of a vector on coordinate axes is no different than converting the vector from polar to rectangular form. Some calculators have a key for converting between rectangular and polar coordinates. You can also use this key to find resultants and components of vectors. Check your manual.

◆◆◆ **Example 33:** A cable exerts a force of 558 N at an angle of 47.2° with the horizontal [Fig. 7–31(a)]. Resolve this force into vertical and horizontal components.

Solution: We draw a vector diagram as shown in Fig. 7–31(b). Then

$$\sin 47.2° = \frac{V_y}{558}$$

$$V_y = 558 \sin 47.2° = 409 \text{ N}$$

Similarly,

$$V_x = 558 \cos 47.2° = 379 \text{ N}$$

So

$$\mathbf{V}_y = 409\underline{/90°} \quad \text{and} \quad \mathbf{V}_x = 379\underline{/0°}$$

◆◆◆

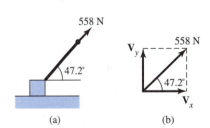

FIGURE 7–31

Resultant of Perpendicular Vectors

Just as any vector can be *resolved* into components, so can several vectors be *combined* into a single vector called the *resultant*, or *vector sum*. The process of combining vectors into a resultant is called *vector addition*.

When combining or adding two perpendicular vectors, we can use right-triangle trigonometry, just as we did for resolving a vector into rectangular components.

◆◆◆ **Example 34:** Find the resultant of two perpendicular vectors whose magnitudes are 485 and 627. Also find the angle that it makes with the 627-magnitude vector.

Solution: We draw a vector diagram as shown in Fig. 7–32. Then by the Pythagorean theorem,

$$\mathbf{R} = \sqrt{(485)^2 + (627)^2} = 793$$

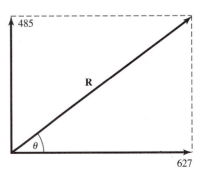

FIGURE 7–32 Resultant of two vectors.

and by Eq. 148,

$$\tan \theta = \frac{485}{627} = 0.774$$

$$\theta = 37.7°$$ ◆◆◆

◆◆◆ **Example 35:** If the components of vector **A** are

$$A_x = 735 \quad \text{and} \quad A_y = 593$$

find the magnitude of **A** and the angle θ that it makes with the x axis.

Solution: By the Pythagorean theorem,

$$A^2 = A_x^2 + A_y^2$$
$$= (735)^2 + (593)^2 = 891,900$$
$$A = 944$$

and

$$\tan \theta = \frac{A_y}{A_x}$$

$$= \frac{593}{735}$$

$$\theta = 38.9°$$

In other words,

$$\mathbf{A} = (735, 593) = 944\underline{/38.9°}$$ ◆◆◆

Resultant of Nonperpendicular Vectors

We can use right-triangle trigonometry to find the resultant of any number of non-perpendicular vectors, such as those shown in Fig. 7–33(a) as follows:

1.   Resolve each vector into its x and y components.
2.   Combine the x components into a single vector $\mathbf{R}_x$ in the x direction, and combine the y components into a single vector $\mathbf{R}_y$ in the y direction.
3.   Find the resultant **R** of $\mathbf{R}_x$ and $\mathbf{R}_y$.

◆◆◆ **Example 36:** Find the resultant of the three vectors in Fig. 7–33(a):

$$6.34\underline{/29.5°}, \quad 4.82\underline{/47.2°}, \quad \text{and} \quad 5.52\underline{/73.0°}$$

Solution: We first resolve each given vector into x and y components. We will arrange the computation in table form for convenience.

x components	y components
$6.34 \cos 29.5° = 5.52$	$6.34 \sin 29.5° = 3.12$
$4.82 \cos 47.2° = 3.27$	$4.82 \sin 47.2° = 3.54$
$5.52 \cos 73.0° = 1.61$	$5.52 \sin 73.0° = 5.28$
Sums: $R_x = 10.4$	$R_y = 11.9$

So

$$\mathbf{R}_x = 10.4\underline{/0°} \quad \text{and} \quad \mathbf{R}_y = 11.9\underline{/90°}$$

We now find the resultant of $\mathbf{R}_x$ and $\mathbf{R}_y$ [Fig. 7–33(b)]. The magnitude of the resultant is

$$R = \sqrt{R_x^2 + R_y^2} = \sqrt{(10.4)^2 + (11.9)^2} = 15.8$$

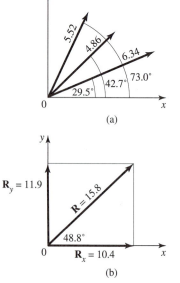

FIGURE 7–33

Now we find the angle.

$$\tan \theta = \frac{R_y}{R_x} = \frac{11.9}{10.4} = 1.14$$

So the resultant **R** is $15.8\,/48.8°$.   ◆◆◆

Exercise 6 ◆ Vectors

Drawing Vectors

Draw each vector on coordinate axes.

1. Magnitude = 3.84;
      angle = 72.4°
2. Magnitude = 9.46;
      angle = 28.4°
3. $36.2\,/44.8°$
4. $1.63\,/29.4°$
5. (2.84, 5.37)
6. (72.4, 58.3)

Resolution of Vectors

Given the magnitude of each vector and the angle θ that it makes with the x axis, find the x and y components.

7. Magnitude = 4.93 $\theta = 48.3°$
8. Magnitude = 835 $\theta = 25.8°$
9. Magnitude = 1.884 $\theta = 58.24°$
10. Magnitude = 362 $\theta = 13.8°$

Resultant of Perpendicular Vectors

In the following problems, the magnitudes of perpendicular vectors **A** and **B** are given. Find the resultant and the angle that it makes with vector **B**.

11. $A = 483$ $B = 382$
12. $A = 2.85$ $B = 4.82$
13. $A = 7364$ $B = 4837$
14. $A = 46.8$ $B = 38.6$
15. $A = 1.25$ $B = 2.07$
16. $A = 274$ $B = 529$

In the following problems, vector **A** has x and y components A_x and A_y. Find the magnitude of **A** and the angle that it makes with the x axis.

17. $A_x = 483$ $A_y = 382$
18. $A_x = 6.82$ $A_y = 4.83$
19. $A_x = 58.3$ $A_y = 37.2$
20. $A_x = 2.27$ $A_y = 3.97$

Resultant of Nonperpendicular Vectors

Find the resultant of each set of vectors.

21. $4.83\,/18.3°$ and $5.99\,/83.5°$
22. $13.5\,/29.3°$ and $27.8\,/77.2°$
23. $635\,/22.7°$ and $485\,/48.8°$
24. $83.2\,/49.7°$ and $52.5\,/66.3°$
25. $1.38\,/22.4°$, $2.74\,/49.5°$, and $3.32\,/77.3°$
26. $736\,/15.8°$, $586\,/69.2°$, and $826\,/81.5°$

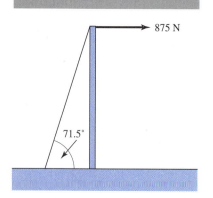

FIGURE 7–34

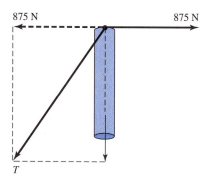

FIGURE 7–35

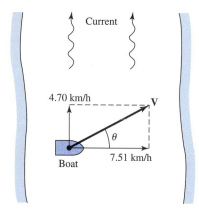

FIGURE 7–36

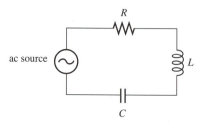

FIGURE 7–37

7–7 Applications of Vectors

Any vector quantity such as force, velocity, or impedance can be combined or resolved by the methods in the preceding section.

Force Vectors

◆◆◆ **Example 37:** A cable running from the top of a telephone pole creates a horizontal pull of 875 N, as shown in Fig. 7–34. A support cable running to the ground is inclined 71.5° from the horizontal. Find the tension in the support cable.

Solution: We draw the forces acting at the top of the pole as shown in Fig. 7–35. We see that the horizontal component of the tension T in the support cable must equal the horizontal pull of 875 N. So

$$\cos 71.5° = \frac{875}{T}$$

$$T = \frac{875}{\cos 71.5°} = 2760 \text{ N} \qquad ◆◆◆$$

Velocity Vectors

◆◆◆ **Example 38:** A river flows at the rate of 4.70 km/h. A rower, who can travel 7.51 km/h in still water, heads directly across the current. Find the rate and the direction of travel of the boat.

Solution: The boat is crossing the current and at the same time is being carried downstream (Fig. 7–36). Thus the velocity **V** has one component of 7.51 km/h across the current and another component of 4.70 km/h downstream. We find the resultant of these two components by the Pythagorean theorem.

$$V^2 = (7.51)^2 + (4.70)^2$$

from which

$$V = 8.86 \text{ km/h}$$

Now finding the angle θ yields

$$\tan \theta = \frac{4.70}{7.51}$$

from which

$$\theta = 32.0° \qquad ◆◆◆$$

Impedance Vectors

Vectors find extensive application in electrical technology, and one of the most common applications is in the calculation of impedances. Figure 7–37 shows a resistor, an inductor, and a capacitor, connected in series with an ac source. The *reactance X* is a measure of how much the capacitance and inductance retard the flow of current in such a circuit. It is the difference between the *capacitive reactance X_C* and the *inductive reactance X_L*.

| Reactance | $X = X_L - X_C$ | A98 |

The *impedance Z* is a measure of how much the flow of current in an ac circuit is retarded by all circuit elements, including the resistance. The magnitude of the impedance is related to the total resistance R and reactance X by the following formula:

Magnitude of Impedance	$|Z| = \sqrt{R^2 + X^2}$	A99

The impedance, resistance, and reactance form the three sides of a right triangle, the *vector impedance diagram* (Fig. 7–38). The angle ϕ between Z and R is called the *phase angle*.

Phase Angle	$\phi = \arctan \dfrac{X}{R}$	A100

◆◆◆ **Example 39:** The capacitive reactance of a certain circuit is 2720 Ω, the inductive reactance is 3260 Ω, and the resistance is 1150 Ω. Find the reactance and the magnitude of the impedance of the circuit; also find the phase angle.

Solution: By Eq. A98,

$$X = 3260 - 2720 = 540 \ \Omega$$

By Eq. A99,

$$|Z| = \sqrt{(1150)^2 + (540)^2} = 1270 \ \Omega$$

and by Eq. A100,

$$\phi = \arctan \frac{540}{1150} = 25.2° \qquad\qquad ◆◆◆$$

Exercise 7 ◆ Applications of Vectors

Force Vectors

1. What force, neglecting friction, must be exerted to drag a 56.5-N weight up a slope inclined 12.6° from the horizontal?
2. What is the largest weight that a tractor can drag up a slope that is inclined 21.2° from the horizontal if it is able to pull along the incline with a force of 2750 lb? Neglect the force due to friction.
3. Two ropes hold a crate as shown in Fig. 7–39. One is pulling with a force of 994 N in a direction 15.5° with the vertical, and the other is pulling with a force of 624 N in a direction 25.2° with the vertical. How much does the crate weigh?
4. A truck weighing 7280 lb moves up a bridge inclined 4.80° from the horizontal. Find the force of the truck normal (perpendicular) to the bridge.
5. A person has just enough strength to pull a 1270-N weight up a certain slope. Neglecting friction, find the angle at which the slope is inclined to the horizontal if the person is able to exert a pull of 551 N.
6. A person wishes to pull a 255-lb weight up an incline to the top of a wall 14.5 ft high. Neglecting friction, what is the length of the shortest incline (measured along the incline) that can be used if the person's pulling strength is 145 lb?

7. A truck weighing 18.6 tons stands on a hill inclined 15.4° from the horizontal. How large a force must be counteracted by brakes to prevent the truck from rolling downhill?

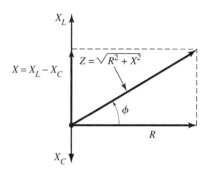

FIGURE 7–38 Vector impedance diagram.

We certainly are not attempting to teach ac circuits in a few paragraphs, but hope only to reinforce concepts learned in your other courses.

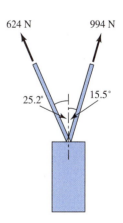

FIGURE 7–39

624 N   994 N   25.2°   15.5°

You will need the equations of equilibrium, Eqs. A13 and A14, for some of these problems.

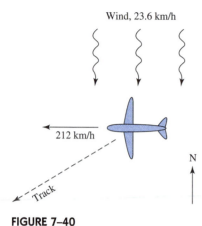

FIGURE 7–40

Velocity Vectors

8. A plane is headed due west with an air speed of 212 km/h (Fig. 7–40). It is driven from its course by a wind from due north blowing at 23.6 km/h. Find the ground speed of the plane and the actual direction of travel. Refer to Fig. 7–41 for definitions of flight terminology.

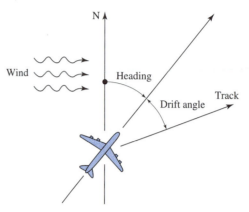

FIGURE 7–41   Flight terminology. The *heading* of an aircraft is the direction in which the craft is *pointed*. Due to air current, it usually will not travel in that direction but in an actual path called the *track*. The angle between the heading and the track is the *drift angle*. The *air speed* is the speed relative to the surrounding air, and the *ground speed* is the craft's speed relative to the earth.

9. At what air speed should an airplane head due west in order to follow a course S 80°15′ W if a 20.5-km/h wind is blowing from due north?

10. A pilot heading his plane due north finds the actual direction of travel to be N 5°12′ E. The plane's air speed is 315 mi/h, and the wind is from due west. Find the plane's ground speed and the velocity of the wind.

11. A certain escalator travels at a rate of 10.6 m/min, and its angle of inclination is 32.5°. What is the vertical component of the velocity? How long will it take a passenger to travel 10.0 m vertically?

12. A projectile is launched at an angle of 55.6° to the horizontal with a speed of 7550 m/min. Find the vertical and horizontal components of this velocity.

13. At what speed with respect to the water should a ship head due north in order to follow a course N 5°15′ E if a current is flowing due east at the rate of 10.6 mi/h?

See Fig. 7–16 if you've forgotten how compass directions are written.

Impedance Vectors

14. A circuit has a reactance of 2650 Ω and a phase angle of 44.6°. Find the resistance and the magnitude of the impedance.

15. A circuit has a resistance of 115 Ω and a phase angle of 72.0°. Find the reactance and the magnitude of the impedance.

16. A circuit has an impedance of 975 Ω and a phase angle of 28.0°. Find the resistance and the reactance.

17. A circuit has a reactance of 5.75 Ω and a resistance of 4.22 Ω. Find the magnitude of the impedance and the phase angle.

18. A circuit has a capacitive reactance of 1776 Ω, an inductive reactance of 5140 Ω, and a total impedance of 5560 Ω. Find the resistance and the phase angle.

••• CHAPTER 7 REVIEW PROBLEMS ••••••••••••••••••••••••••••••••••••••

Perform the angle conversions, and fill in the missing quantities.

	Decimal Degrees	Degrees–Minutes–Seconds	Radians	Revolutions
1.	38.2			
2.		25°28′45″		
3.			2.745	
4.				0.275

The following points are on the terminal side of an angle θ. Write the six trigonometric functions of the angle, and find the angle in decimal degrees, to three significant digits.

5. (3, 7)

6. (2, 5)

7. (4, 3)

8. (2.3, 3.1)

Write the six trigonometric functions of each angle. Keep four decimal places.

9. 72.9°

10. 35°13′33″

11. 1.05 rad

Find acute angle θ in decimal degrees.

12. $\sin \theta = 0.574$

13. $\cos \theta = 0.824$

14. $\tan \theta = 1.345$

Evaluate each expression. Give your answer as an acute angle in degrees.

15. arctan 2.86

16. $\cos^{-1} 0.385$

17. arcsec 2.447

Solve right triangle ABC.

18. $a = 746$ and $A = 37.2°$

19. $b = 3.72$ and $A = 28.5°$

20. $c = 45.9$ and $A = 61.4°$

21. Find the x and y components of a vector that has a magnitude of 885 and makes an angle of 66.3° with the x axis.

22. Find the magnitude of the resultant of two perpendicular vectors that have magnitudes of 54.8 and 39.4, and find the angle the resultant makes with the 54.8 vector.

23. A vector has x and y components of 385 and 275. Find the magnitude and the direction of that vector.

24. A telephone pole casts a shadow 13.5 m long when the angle of elevation of the sun is 15.4°. Find the height of the pole.

25. From a point 125 ft in front of a church, the angles of elevation of the top and base of its steeple are 22.5° and 19.6°, respectively. Find the height of the steeple.

26. A circuit has a resistance of 125 Ω, an impedance of 256 Ω, an inductive reactance of 312 Ω, and a positive phase angle. Find the capacitive reactance and the phase angle.

Evaluate to four decimal places.

27. cos 59.2°

28. csc 19.3°

29. tan 52.8°

30. sec 37.2°

31. cot 24.7°

32. sin 42.9°

Find, to the nearest tenth of a degree, the acute angle whose trigonometric function is given.

33. $\tan \theta = 1.7362$

34. $\sec \theta = 2.9914$

35. $\sin \theta = 0.7253$

36. $\csc \theta = 3.2746$

37. $\cot \theta = 0.9475$

38. $\cos \theta = 0.3645$

Evaluate each trigonometric expression to three significant digits.

39. 8.23 cos 15.8° + 8.73

40. 5.82 − 5.89 sin 16.2°

41. 9.26(cos 88.3° + sin 22.3°)

42. 52.2 sin 3 + 45.3 cos 2

Draw each vector on coordinate axes.

43. Magnitude = 726; angle = 32.4°

44. 12.6/74.8°

45. (7.84, 8.37)

Find the resultant of each set of vectors.

46. 633/68.3° and 483/23.5°

47. 13.2/22.7° and 22.6/76.8°

48. 8.28/82.3°, 9.14/15.5°, and 7.72/47.9°

49. Point *C* is due east of point *A* across a lake, as shown in Fig. 7–42. To measure the distance *AC*, a person walks 139.3 ft due south to point *B*, then walks 158.2 ft to *C*. Find the distance *AC* and the direction the person was walking when going from *B* to *C*.

Writing

50. Write a short paragraph, with examples, on how you might possibly use an idea from this chapter in a real-life situation: on the job, in a hobby, or around the house.

Team Project

51. Make a device for measuring angles in the horizontal plane, using a sighting tube of some sort and a protractor. Use this "transit" and a rope of known length to find some distance on your campus, such as the distance across a ball field, using the method of problem 5 in Exercise 5. Compare your results with the actual taped distance. Give a prize to the team that gets closest to the taped value.

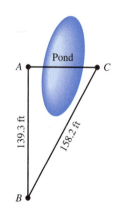

FIGURE 7–42

Factors and Factoring

In Chapter 2 we learned how to multiply expressions together—here we do the reverse. We find those quantities (*factors*) which, when multiplied together, give the original product.

Why bother with factoring? What is the point of changing an expression from one form to another? Because factoring, which may seem complicated at first, is actually essential to simplifying expressions and solving equations, especially quadratic equations which we discuss in Chapter 14. The value of factoring will become clear as we use it toward the end of this chapter to solve literal equations and to manipulate the formulas that are so important in technology. We also use factoring in Chapter 9 to simplify fractions.

The idea of a factor is not entirely new to us. You might glance back at Secs. 1–3 and 2–1 before starting this chapter.

8–1 Common Factors

Factors of an Expression

The *factors of an expression* are those quantities whose product is the original expression.

◆◆◆ **Example 1:** The factors of $x^2 - 9$ are $x + 3$ and $x - 3$, because

$$(x + 3)(x - 3) = x^2 + 3x - 3x - 9$$
$$= x^2 - 9$$

◆◆◆

Prime Factors

Many expressions have no factors other than 1 and themselves. Such expressions are called *prime*.

◆◆◆ **Example 2:** Two factors of 30 are 5 and 6. The 5 is prime because it cannot be factored. However, the 6 is not prime because it can be factored into 2 and 3. Thus the prime factors of 30 are 2, 3, and 5. ◆◆◆

◆◆◆ **Example 3:** The expressions x, $x - 3$, and $x^2 - 3x + 7$ are all prime. They cannot be factored further. ◆◆◆

◆◆◆ **Example 4:** The expressions $x^2 - 9$ and $x^2 + 5x + 6$ are not prime, because they can be factored. They are said to be *factorable*. ◆◆◆

Factoring

Factoring is the process of finding the factors of an expression. It is *the reverse of finding the product* of two or more quantities.

Multiplication (Finding the Product)	$x(x + 4) = x^2 + 4x$
Factoring (Finding the Factors)	$x^2 + 4x = x(x + 4)$

We factor an expression by *recognizing the form* of that expression and then applying standard rules for factoring. In the first type of factoring we will cover, we look for *common factors*.

Common Factors

If each term of an expression contains the same quantity (called the *common factor*), that quantity may be *factored out*.

Common Factor	$ab + ac = a(b + c)$	5

This is nothing but the *distributive law* that we studied earlier.

◆◆◆ **Example 5:** In the expression

$$2x + x^2$$

each term contains an x as a common factor. So we write

$$2x + x^2 = x(2 + x)$$

Most of the factoring we will do will be of this type.   ◆◆◆

◆◆◆ **Example 6:**

(a) $3x^3 - 2x + 5x^4 = x(3x^2 - 2 + 5x^3)$
(b) $3xy^2 - 9x^3y + 6x^2y^2 = 3xy(y - 3x^2 + 2xy)$
(c) $3x^3 - 6x^2y + 9x^4y^2 = 3x^2(x - 2y + 3x^2y^2)$
(d) $2x^2 + x = x(2x + 1)$ ◆◆◆

When factoring, don't confuse superscripts (powers) with subscripts.

◆◆◆ **Example 7:** The expression $x^3 + x^2$ is factorable.

$$x^3 + x^2 = x^2(x + 1)$$

But the expression

$$x_3 + x_2$$

is not factorable.   ◆◆◆

Common Error

Students are sometimes puzzled over the "1" in Example 7. Why should it be there? After all, when you remove a chair from a room, it is *gone;* there is nothing (zero) remaining where the chair used to be. If you remove an x by factoring, you might assume that nothing (zero) remains where the x used to be.

$$2x^2 + x = x(2x + 0)?$$

Prove to yourself that this is not correct by multiplying the factors to see if you get back the original expression.

Factors in the Denominator

Common factors may appear in the denominators of the terms as well as in the numerators.

◆◆◆ **Example 8:**

(a) $\dfrac{1}{x} + \dfrac{2}{x^2} = \dfrac{1}{x}\left(1 + \dfrac{2}{x}\right)$

(b) $\dfrac{x}{y^2} + \dfrac{x^2}{y} + \dfrac{2x}{3y} = \dfrac{x}{y}\left(\dfrac{1}{y} + x + \dfrac{2}{3}\right)$ ◆◆◆

Checking

To check if factoring has been done correctly, simply multiply your factors together and see if you get back the original expression.

◆◆◆ **Example 9:** Are $2xy$ and $x + 3 - y^2$ the factors of

$$2x^2y + 6xy - 2xy^3?$$

Solution: Multiplying the factors, we obtain

$$2xy(x + 3 - y^2) = 2x^2y + 6xy - 2xy^3 \quad \text{(checks)}$$ ◆◆◆

This check will tell us whether we have factored correctly but, of course, not whether we have factored *completely,* that is, found the prime factors.

Exercise 1 ◆ Common Factors

Factor each expression and check your result.

1. $3y^2 + y^3$
2. $6x - 3y$
3. $x^5 - 2x^4 + 3x^3$
4. $9y - 27xy$
5. $3a + a^2 - 3a^3$
6. $8xy^3 - 6x^2y^2 + 2x^3y$
7. $5(x + y) + 15(x + y)^2$
8. $\dfrac{u}{3} - \dfrac{u^7}{4} + \dfrac{u^3}{5}$
9. $\dfrac{3}{x} + \dfrac{2}{x^2} - \dfrac{5}{x^3}$
10. $\dfrac{3ab^2}{y^3} - \dfrac{6a^2b}{y^2} + \dfrac{12ab}{y}$
11. $\dfrac{5m}{2n} + \dfrac{15m^2}{4n^2} - \dfrac{25m^3}{8n}$
12. $\dfrac{16y^2}{9x^2} - \dfrac{8y^3}{3x^3} + \dfrac{24y^4}{9x}$
13. $5a^2b + 6a^2c$
14. $a^2c + b^2c + c^2d$
15. $4x^2y + cxy^2 + 3xy^3$
16. $4abx + 6a^2x^2 + 8ax$
17. $3a^3y - 6a^2y^2 + 9ay^3$
18. $2a^2c - 2a^2c^2 + 3ac$
19. $5acd - 2c^2d^2 + bcd$
20. $4b^2c^2 - 12abc - 9c^2$
21. $8x^2y^2 + 12x^2z^2$
22. $6xyz + 12x^2y^2z$
23. $3a^2b + abc - abd$
24. $5a^3x^2 - 5a^2x^3 + 10a^2x^2z$

Applications

25. When a bar of length L_0 is changed in temperature by an amount t, its new length L will be $L = L_0 + L_0\alpha t$, where α is the coefficient of thermal expansion. Factor the right side of this equation.

26. A sum of money a when invested for t years at an interest rate n will accumulate to an amount y, where $y = a + ant$. Factor the right side of this equation.

Compare your result for problem 27 with Eq. A70.

27. When a resistance R_1 is heated from a temperature t_1 to a new temperature t, it will increase in resistance by an amount $\alpha(t - t_1)R_1$, where α is the temperature coefficient of resistance. The final resistance will then be $R = R_1 + \alpha(t - t_1)R_1$. Factor the right side of this equation.

28. An item costing P dollars is reduced in price by 15%. The resulting price C is then $C = P - 0.15P$. Factor the right side of this equation.

29. The displacement of a uniformly accelerated body is given by

$$s = v_0t + \frac{a}{2}t^2$$

Factor the right side of this equation.

30. The sum of the voltage drops across the resistors in Fig. 8–1 must equal the battery voltage E.

$$E = iR_1 + iR_2 + iR_3$$

Factor the right side of this equation.

31. The mass of a spherical shell having an outside radius of r_2 and an inside radius r_1 is

$$\text{mass} = \tfrac{4}{3}\pi r_2^3 D - \tfrac{4}{3}\pi r_1^3 D$$

where D is the mass density of the material. Factor the right side of this equation.

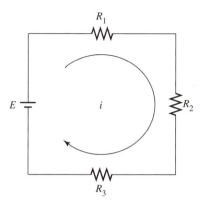

FIGURE 8–1

8–2 Difference of Two Squares

Form

An expression of the form

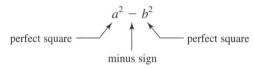

where one perfect square is subtracted from another, is called a *difference of two squares*. It arises when $(a - b)$ and $(a + b)$ are multiplied together.

Difference of Two Squares	$a^2 - b^2 = (a + b)(a - b)$	41

Remember this from Chapter 2?

Factoring the Difference of Two Squares

Once we recognize its form, the difference of two squares is easily factored.

◆◆◆ **Example 10:**

$$4x^2 - 9 = (2x)^2 - (3)^2$$

$$(2x)^2 - (3)^2 = (2x \quad)(2x \quad)$$

square root of the first term

$$(2x)^2 - (3)^2 = (2x \quad 3)(2x \quad 3)$$

square root of the last term

$$4x^2 - 9 = (2x + 3)(2x - 3)$$

opposite signs ◆◆◆

◆◆◆ **Example 11:**

(a) $y^2 - 1 = (y + 1)(y - 1)$
(b) $9a^2 - 16b^2 = (3a + 4b)(3a - 4b)$
(c) $49x^2 - 9a^2y^2 = (7x - 3ay)(7x + 3ay)$
(d) $1 - a^2b^2c^2 = (1 - abc)(1 + abc)$ ◆◆◆

Common Error	There is no similar rule for factoring the *sum* of two squares, such as $$a^2 + b^2$$

The difference of any two quantities that have *even powers* can be factored as the difference of two squares. *Express each quantity as a square* by means of the following equation:

Powers	$x^{ab} = (x^a)^b$	31

◆◆◆ **Example 12:**

(a) $a^4 - b^6 = (a^2)^2 - (b^3)^2$
$= (a^2 + b^3)(a^2 - b^3)$

(b) $x^{2a} - y^{6b} = (x^a)^2 - (y^{3b})^2$
$= (x^a + y^{3b})(x^a - y^{3b})$

(c) $4^{2p} - 9^{4p} = (4 - 9^2)^{2p}$
$= (2^2 - 9^2)^{2p}$
$= [(2 + 9)(2 - 9)]^{2p}$ ◆◆◆

Sometimes one or both terms in the given expression may be a fraction.

◆◆◆ **Example 13:**

$$\frac{1}{x^2} - \frac{1}{y^2} = \left(\frac{1}{x}\right)^2 - \left(\frac{1}{y}\right)^2$$

$$= \left(\frac{1}{x} + \frac{1}{y}\right)\left(\frac{1}{x} - \frac{1}{y}\right)$$ ◆◆◆

Sometimes one (or both) of the terms in an expression is itself a binomial.

◆◆◆ **Example 14:**

$$(a + b)^2 - x^2 = [(a + b) + x][(a + b) - x]$$ ◆◆◆

Factoring Completely

After factoring an expression, see if any of the factors themselves can be factored *again*.

◆◆◆ **Example 15:** Factor $a - ab^2$.

Solution: Always remove any common factors first. Factoring, we obtain

$$a - ab^2 = a(1 - b^2)$$

Now factoring the difference of two squares, we get

$$a - ab^2 = a(1 + b)(1 - b)$$ ◆◆◆

◆◆◆ **Example 16:**

$$x^4 - y^4 = (x^2 + y^2)(x^2 - y^2)$$

Now factoring the difference of two squares again, we obtain

$$x^4 - y^4 = (x^2 + y^2)(x + y)(x - y)$$ ◆◆◆

Exercise 2 ◆ Difference of Two Squares

Factor completely.

1. $4 - x^2$
2. $x^2 - 9$
3. $9a^2 - x^2$
4. $25 - x^2$
5. $4x^2 - 4y^2$
6. $9x^2 - y^2$
7. $x^2 - 9y^2$
8. $16x^2 - 16y^2$
9. $9c^2 - 16d^2$
10. $25a^2 - 9b^2$
11. $9y^2 - 1$
12. $4x^2 - 9y^2$

Expressions with Higher Powers or Literal Powers

13. $m^4 - n^4$
14. $a^8 - b^8$
15. $m^{2n} - n^{2m}$
16. $9^{2n} - 4b^{4n}$
17. $a^{16} - b^8$
18. $9a^2b^2 - 4c^4$
19. $25x^4 - 16y^6$
20. $36y^2 - 49z^6$
21. $16a^4 - 121$
22. $121a^4 - 16$
23. $25a^4b^4 - 9$
24. $121a^2 - 36b^4$

Applications

25. The thrust washer in Fig. 8–2 has a surface area of

$$\text{area} = \pi r_2^2 - \pi r_1^2$$

Factor this expression.

26. A flywheel of diameter d_1 (Fig. 8–3) has a balancing hole of diameter d_2 drilled through it. The mass M of the flywheel is then

$$\text{mass} = \frac{\pi d_1^2}{4} Dt - \frac{\pi d_2^2}{4} Dt$$

Factor this expression.

27. A spherical balloon shrinks from radius r_1 to radius r_2. The change in surface area is, from Eq. 129, $4\pi r_1^2 - 4\pi r_2^2$. Factor this expression.

28. When an object is released from rest, the distance fallen between time t_1 and time t_2 is $\frac{1}{2}gt_2^2 - \frac{1}{2}gt_1^2$, where g is the acceleration due to gravity. Factor this expression.

29. When a body of mass m slows down from velocity v_1 to velocity v_2, the decrease in kinetic energy is $\frac{1}{2}mv_1^2 - \frac{1}{2}mv_2^2$. Factor this expression.

30. The work required to stretch a spring from a length x_1 (measured from the unstretched position) to a new length x_2 is $\frac{1}{2}kx_2^2 - \frac{1}{2}kx_1^2$. Factor this expression.

31. The formula for the volume of a cylinder of radius r and height h is $v = \pi r^2 h$. Write an expression for the volume of a hollow cylinder that has an inside radius r, and outside radius R, and a height h. Then factor the expression completely.

32. A body at temperature T will radiate an amount of heat kT^4 to its surroundings and will absorb from the surroundings an amount of heat kT_s^4, where T_s is the temperature of the surroundings. Write an expression for the net heat transfer by radiation (amount radiated minus amount absorbed), and factor this expression completely.

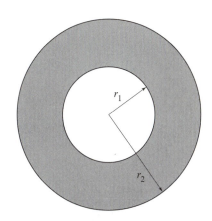

FIGURE 8–2 Circular washer.

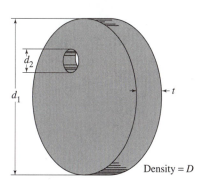

FIGURE 8–3

8–3   Factoring Trinomials

Trinomials

A *trinomial,* you recall, is a polynomial having *three terms.*

◆◆◆ **Example 17:** $2x^3 - 3x + 4$ is a trinomial.   ◆◆◆

A *quadratic trinomial* in x has an x^2 term, an x term, and a constant term.

◆◆◆ **Example 18:** $4x^2 + 3x - 5$ is a quadratic trinomial.   ◆◆◆

In this section we factor only quadratic trinomials.

Test for Factorability

Not all quadratic trinomials can be factored. We test for factorability as follows:

| Test for Factorability | The trinomial $ax^2 + bx + c$ (where a, b, and c are constants) is factorable if $b^2 - 4ac$ is a perfect square. | 44 |

In the margin: $b^2 - 4ac$ is called the *discriminant* (Eq. 101). We will use it in Chapter 14 to predict the nature of the roots of a quadratic equation.

◆◆◆ **Example 19:** Can the trinomial $6x^2 + x - 12$ be factored?

Solution: Using the test for factorability, Eq. 44, with $a = 6$, $b = 1$, and $c = -12$, we have

$$b^2 - 4ac = 1^2 - 4(6)(-12) = 289$$

By taking the square root of 289 on the calculator, we see that it is a perfect square ($289 = 17^2$), so the given trinomial is factorable. We will see later that its factors are $(2x + 3)$ and $(3x - 4)$.   ◆◆◆

The coefficient of the x^2 term in a quadratic trinomial is called the *leading co-efficient.* (It is the constant a in Eq. 44). We will first factor the easier type of trinomial, in which the leading coefficient is 1.

Trinomials with a Leading Coefficient of 1

We get such a trinomial when we multiply two binomials if the coefficients of the x terms in the binomials are also 1.

$$(x + a)(x + b) = x^2 + (a + b)x + ab$$

Note that the coefficient of the middle term of the trinomial equals the *sum* of a and b, and that the last term equals the *product* of a and b. Knowing this, we can reverse the process and find the factors when the trinomial is given.

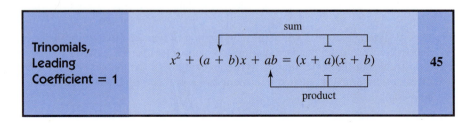

| Trinomials, Leading Coefficient = 1 | $x^2 + (a + b)x + ab = (x + a)(x + b)$ | 45 |

◆◆◆ **Example 20:** Factor the trinomial $x^2 + 8x + 15$.

Solution: From Eq. 45, this trinomial, if factorable, will factor as $(x + a)(x + b)$, where a and b have a sum of 8 and a product of 15. The integers 5 and 3 have a sum of 8 and a product of 15. Thus

$$x^2 + 8x + 15 = (x + 5)(x + 3)$$

◆◆◆

Using the Signs to Aid Factoring

The signs of the terms of the trinomial can tell you the signs of the factors.

1. If the sign of the last term is *positive,* both signs in the factors will be the same (both positive or both negative). The sign of the middle term of the trinomial will tell whether both signs in the factors are positive or negative.

◆◆◆ **Example 21:**

(a)
$$x^2 + 6x + 8 = (x + 4)(x + 2)$$

— positive, so signs in the factors are the same

positive, so both signs in the factors are positive

(b)
$$x^2 - 6x + 8 = (x - 4)(x - 2)$$

— positive, so signs in the factors are the same

negative, so both signs in the factors are negative

◆◆◆

2. If the sign of the last term of the trinomial is *negative,* the signs of a and b in the factors will differ (one positive, one negative). The sign of the middle term of the trinomial will tell which quantity, a or b, is larger (in absolute value), the positive one or the negative one.

◆◆◆ **Example 22:**

$$x^2 - 2x - 8 = (x - 4)(x + 2)$$

— negative, so the signs in the factors differ, one positive and one negative

— negative, so the larger number (4) has a negative sign

◆◆◆

◆◆◆ **Example 23:** Factor $x^2 + 2x - 15$.

Solution: We first look at the sign of the last term (-15). The negative sign tells us that the signs of the factors will differ.

$$x^2 + 2x - 15 = (x - \quad)(x + \quad)$$

We then find two numbers whose product is -15 and whose sum is 2. These two numbers must have opposite signs in order to have a negative product. Also, since the middle coefficient is $+2$, the positive number must be 2 greater than the negative number. Two numbers that meet these conditions are $+5$ and -3. So

$$x^2 + 2x - 15 = (x + 5)(x - 3)$$

◆◆◆

Remember to remove any *common factors* before factoring a trinomial.

◆◆◆ **Example 24:**

$$2x^3 + 4x^2 - 30x = 2x(x^2 + 2x - 15)$$
$$= 2x(x - 3)(x + 5)$$

◆◆◆

Trinomials Having More Than One Variable

◆◆◆ Example 25: Factor $x^2 + 5xy + 6y^2$.

Solution: One way to factor such a trinomial is, first, to temporarily drop the second variable (y in this example).

$$x^2 + 5x + 6$$

Then factor the resulting trinomial.

$$(x + 2)(x + 3)$$

Finally, note that if we tack y onto the second term of each factor, this will cause y to appear in the middle term of the trinomial and y^2 to appear in the last term of the trinomial. The required factors are then

$$(x + 2y)(x + 3y)$$ ◆◆◆

◆◆◆ Example 26: Factor $w^2x^2 + 3wxy - 4y^2$.

Solution: Temporarily dropping w and y yields

$$x^2 + 3x - 4$$

which factors into

$$(x - 1)(x + 4)$$

We then note that if a w were appended to each x term and a y to the second term in each binomial,

$$(wx - y)(wx + 4y)$$

the product of these factors will give the original expression.   ◆◆◆

Trinomials with Higher Powers

Trinomials with powers *greater than 2* may be factored if one exponent can be expressed as a *multiple of 2*, and if that exponent is *twice the other.*

◆◆◆ Example 27:

$$x^6 - x^3 - 6 = (x^3)^2 - (x^3) - 6$$
$$= (x^3 - 3)(x^3 + 2)$$ ◆◆◆

Trinomials with Variables in the Exponents

If one exponent is *twice the other,* you may be able to factor such a trinomial.

◆◆◆ Example 28:

$$x^{2n} + 5x^n + 6 = (x^n)^2 + 5(x^n) + 6$$
$$= (x^n + 3)(x^n + 2)$$ ◆◆◆

Substitution

You may prefer to handle some expressions by making a *substitution,* as in the following examples.

◆◆◆ Example 29: Factor $x^{6n} + 2x^{3n} - 3$.

Solution: We see that one exponent is twice the other. Let us substitute a new letter for x^{3n}. Let

$$w = x^{3n}$$

Side notes:

As with the difference of two squares, we can factor many trinomials that at first glance do not seem to be in the proper form. Examples 25 and 26 are some common types.

This technique will be valuable when we solve *equations of quadratic type* in Chapter 14.

Then

$$x^{6n} = (x^{3n})^2 = w^2$$

and our expression becomes

$$w^2 + 2w - 3$$

which factors into

$$(w + 3)(w - 1)$$

or, substituting back, we have

$$(x^{3n} + 3)(x^{3n} - 1)$$

◆◆◆

◆◆◆ **Example 30:** Factor $(x + y)^2 + 7(x + y) - 30$.

Solution: We might be tempted to expand this expression, but notice that if we make the substitution

$$w = x + y$$

we get

$$w^2 + 7w - 30$$

which factors into $(w + 10)(w - 3)$. Substituting back, we get the factors

$$(x + y + 10)(x + y - 3)$$

◆◆◆

Exercise 3 ◆ Factoring Trinomials

Test for Factorability

Test each trinomial for factorability.

1. $3x^2 - 5x - 9$
2. $3a^2 + 7a + 2$
3. $2x^2 - 12x + 18$
4. $5z^2 - 2z - 1$
5. $x^2 - 30x - 64$
6. $y^2 + 2y - 7$

Factor completely.

Trinomials with a Leading Coefficient of 1

7. $x^2 - 10x + 21$
8. $x^2 - 15x + 56$
9. $x^2 - 10x + 9$
10. $x^2 + 13x + 30$
11. $x^2 + 7x - 30$
12. $x^2 + 3x + 2$
13. $x^2 + 7x + 12$
14. $c^2 + 9c + 18$
15. $x^2 - 4x - 21$
16. $x^2 - x - 56$
17. $x^2 + 6x + 8$
18. $x^2 + 12x + 32$
19. $b^2 - 8b + 15$
20. $b^2 + b - 12$
21. $b^2 - b - 12$
22. $3b^2 - 30b + 63$
23. $2y^2 - 26y + 60$
24. $4z^2 - 16z - 84$
25. $3w^2 + 36w + 96$

Trinomials Having More Than One Variable

26. $x^2 - 13xy + 36y^2$
27. $x^2 + 19xy + 84y^2$
28. $a^2 - 9ab + 20b^2$
29. $a^2 + ab - 6b^2$
30. $a^2 + 3ab - 4b^2$
31. $a^2 + 2ax - 63x^2$
32. $x^2y^2 - 19xyz + 48z^2$

33. $a^2 - 20abc - 96b^2c^2$

34. $x^2 - 10xyz - 96y^2z^2$

35. $a^2 + 49abc + 48b^2c^2$

Trinomials with Higher Powers or with Variables in the Exponents

36. $a^{6p} + 13a^{3p} + 12$

37. $(a + b)^2 - 7(a + b) - 8$

38. $w(x + y)^2 + 5w(x + y) + 6w$

39. $x^4 - 9x^2y^2 + 20y^4$

40. $a^6b^6 - 23a^3b^3c^2 + 132c^4$

Applications

41. To find the thickness t of the angle iron in Fig. 8–4, it is necessary to solve the quadratic equation

$$t^2 - 14t + 24 = 0$$

Factor the left side of this equation. (*Extra:* Can you see where this equation comes from?)

42. To find the dimensions of a rectangular field having a perimeter of 70 ft and an area of 300 ft^2, we must solve the equation

$$x^2 - 35x + 300 = 0$$

Factor the left side of this equation.

43. To find the width of the frame in Fig. 8–5, we must solve the equation

$$x^2 - 11x + 10 = 0$$

Factor the left side of this equation.

44. To find two resistors that will give an equivalent resistance of 400 Ω when wired in series and 75 Ω when wired in parallel, we must solve the equation

$$R^2 - 400R + 30,000 = 0$$

Factor the left side of this equation.

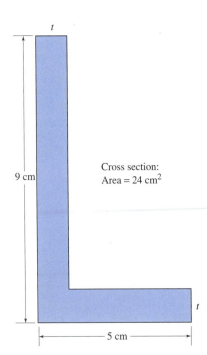

Cross section:
Area = 24 cm^2

9 cm

5 cm

FIGURE 8–4

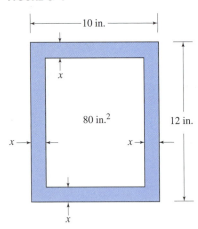

10 in.

x

80 in.^{2}

12 in.

x x

x

FIGURE 8–5

8–4   Factoring by Grouping

When the expression to be factored has four or more terms, these terms can sometimes be arranged in smaller groups that can be factored separately.

We'll use this same grouping idea later to factor certain trinomials.

◆◆◆ **Example 31:** Factor $ab + 4a + 3b + 12$.

Solution: Group the two terms containing the factor a, and factor the two containing the factor 3. Remove the common factor from each pair of terms.

$$(ab + 4a) + (3b + 12) = a(b + 4) + 3(b + 4)$$

Both terms now have the common factor $(b + 4)$, which we factor out.

$$(b + 4)(a + 3)$$ ◆◆◆

◆◆◆ **Example 32:** Factor $x^2 - y^2 + 2x + 1$.

Solution: Taking our cue from Example 31, we try grouping the x terms together.

$$(x^2 + 2x) + (-y^2 + 1)$$

But we are no better off than before. We then might notice that if the $+1$ term were grouped with the x terms, we would get a trinomial that could be factored. Thus

$$(x^2 + 2x + 1) - y^2$$
$$(x + 1)(x + 1) - y^2$$

or

$$(x + 1)^2 - y^2$$

We now have the difference of two squares. Factoring again gives

$$(x + 1 + y)(x + 1 - y) \qquad \blacklozenge\blacklozenge\blacklozenge$$

Exercise 4 ◆ Factoring by Grouping

Factor completely.

1. $a^3 + 3a^2 + 4a + 12$
2. $x^3 + x^2 + x + 1$
3. $x^3 - x^2 + x - 1$
4. $2x^3 - x^2 + 4x - 2$
5. $x^2 - bx + 3x - 3b$
6. $ab + a - b - 1$
7. $3x - 2y - 6 + xy$
8. $x^2y^2 - 3x^2 - 4y^2 + 12$
9. $x^2 + y^2 + 2xy - 4$
10. $x^2 - 6xy + 9y^2 - a^2$
11. $m^2 - n^2 - 4 + 4n$
12. $p^2 - r^2 - 6pq + 9q^2$

Expressions with Three or More Letters

13. $ax + bx + 3a + 3b$
14. $x^2 + xy + xz + yz$
15. $a^2 - ac + ab - bc$
16. $ay - by + ab - y^2$
17. $2a + bx^2 + 2b + ax^2$
18. $2xy + wy - wz - 2xz$
19. $6a^2 + 2ab - 3ac - bc$
20. $x^2 + bx + cx + bc$
21. $b^2 - bc + ab - ac$
22. $bx - cx + bc - x^2$
23. $x^2 - a^2 - 2ab - b^2$
24. $p^2 - y^2 - x^2 - 2xy$

8–5　The General Quadratic Trinomial

When we multiply the two binomials $(ax + b)$ and $(cx + d)$, we get a trinomial with a leading coefficient of ac, a middle coefficient of $(ad + bc)$, and a constant term of bd.

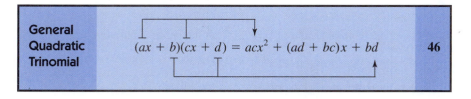

| General Quadratic Trinomial | $(ax + b)(cx + d) = acx^2 + (ad + bc)x + bd$ | 46 |

The general quadratic trinomial may be factored by trial and error or by the grouping method.

Trial and Error

To factor the general quadratic trinomial by trial and error, we look for four numbers, a, b, c, and d, such that

$$ac = \text{the leading coefficient}$$
$$ad + bc = \text{the middle coefficient}$$
$$bd = \text{the constant term}$$

Also, the *signs* in the factors are found in the same way as for trinomials with a leading coefficient of 1.

◆◆◆ **Example 33:** Factor $2x^2 + 5x + 3$.

Solution: The leading coefficient ac is 2, and the constant term bd is 3. Try

$$a = 1, \quad c = 2, \quad b = 3, \quad \text{and} \quad d = 1$$

Then $ad + bc = 1(1) + 3(2) = 7$. No good. It is supposed to be 5. We next try

$$a = 1, \quad c = 2, \quad b = 1, \quad \text{and} \quad d = 3$$

Then $ad + bc = 1(3) + 1(2) = 5$. This works. So

$$2x^2 + 5x + 3 = (x + 1)(2x + 3)$$

◆◆◆

Grouping Method

The grouping method eliminates the need for trial and error.

Some people have a knack for factoring and can quickly factor a trinomial by trial and error. Other rely on the longer but surer grouping method.

◆◆◆ **Example 34:** Factor $3x^2 - 16x - 12$.

Solution:

1. Multiply the leading coefficient and the constant term.

$$3(-12) = -36$$

2. Find two numbers whose product equals -36 and whose sum equals the middle coefficient, -16. Two such numbers are 2 and -18.
3. Rewrite the trinomial, splitting the middle term according to the selected factors $(-16x = 2x - 18x)$.

$$3x^2 + 2x - 18x - 12$$

Group the first two terms together and the last two terms together.

$$(3x^2 + 2x) + (-18x - 12)$$

4. Remove common factors from each grouping.

$$x(3x + 2) - 6(3x + 2)$$

5. Remove the common factor $(3x + 2)$ from the entire expression:

$$(3x + 2)(x - 6)$$

which are the required factors.   ◆◆◆

> **Common Error**
>
> It is easy to make a mistake when factoring out a negative quantity. Thus when we factored out a -6 in going from Step 3 to Step 4 in Example 34, we got
>
> $$(-18x - 12) = -6(3x + 2)$$
>
> but not
>
> $$-6(3x - 2)$$
>
> ↑
> └── incorrect!

Notice that if we had grouped the terms differently in Step 3 of Example 34, we would have arrived at the same factors as before.

$$3x^2 - 18x + 2x - 12$$
$$3x(x - 6) + 2(x - 6)$$
$$(3x + 2)(x - 6)$$

Sometimes an expression may be simplified before factoring, as we did in Sec. 8–3.

◆◆◆ **Example 35:** Factor $12(x + y)^{6n} - (x + y)^{3n}z - 6z^2$.

Solution: If we substitute $a = (x + y)^{3n}$, our expression becomes

$$12a^2 - az - 6z^2$$

Temporarily dropping the z's gives

$$12a^2 - a - 6$$

which factors into

$$(4a - 3)(3a + 2)$$

Replacing the z's, we get

$$12a^2 - az - 6z^2 = (4a - 3z)(3a + 2z)$$

Finally, substituting back $(x + y)^{3n}$ for a, we have

$$12(x + y)^{6n} - (x + y)^{3n}z - 6z^2 = [4(x + y)^{3n} - 3z][3(x + y)^{3n} + 2z] \quad ◆◆◆$$

Exercise 5 ◆ The General Quadratic Trinomial

Factor completely.

1. $4x^2 - 13x + 3$
2. $5a^2 - 8a + 3$
3. $5x^2 + 11x + 2$
4. $7x^2 + 23x + 6$
5. $12b^2 - b - 6$
6. $6x^2 - 7x + 2$
7. $2a^2 + a - 6$
8. $2x^2 + 3x - 2$
9. $5x^2 - 38x + 21$
10. $4x^2 + 7x - 15$
11. $3x^2 + 6x + 3$
12. $2x^2 + 11x + 12$
13. $3x^2 - x - 2$
14. $7x^2 + 123x - 54$
15. $4x^2 - 10x + 6$
16. $3x^2 + 11x - 20$
17. $4a^2 + 4a - 3$
18. $9x^2 - 27x + 18$
19. $9a^2 - 15a - 14$
20. $16c^2 - 48c + 35$

Expressions Reducible to Quadratic Trinomials

Factor completely.

21. $49x^6 + 14x^3y - 15y^2$
22. $-18y^2 + 42x^2 - 24xy$
23. $4x^6 + 13x^3 + 3$
24. $5a^{2n} - 8a^n + 3$
25. $5x^{4n} + 11x^{2n} + 2$
26. $12(a + b)^2 - (a + b) - 6$
27. $3(a + x)^{2n} - 3(a + x)^n - 6$

Applications

28. An object is thrown upward with an initial velocity of 32 ft/s from a building 128 ft above the ground. The height s of the object above the ground at any time t is given by

$$s = 128 + 32t - 16t^2$$

Factor the right side of this equation.

29. An object is thrown into the air with an initial velocity of 82 ft/s. To find the time it takes for the object to reach a height of 45 ft, we must solve the quadratic equation

$$16t^2 - 82t + 45 = 0$$

Factor the left side of this equation.

30. To find the depth of cut h needed to produce a flat of a certain width on a 1-in.-radius bar (Fig. 8–6), we must solve the equation

$$4h^2 - 8h + 3 = 0$$

Factor the left side of this equation.

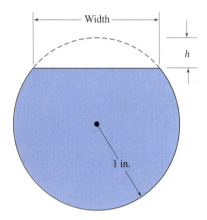

FIGURE 8–6

8–6   The Perfect Square Trinomial

The Square of a Binomial

In Chapter 2 we saw that the expression obtained when a binomial is squared is called a *perfect square trinomial*.

♦♦♦ **Example 36:** Square the binomial $(2x + 3)$.

Solution:

$$(2x + 3)^2 = 4x^2 + 6x + 6x + 9$$
$$= 4x^2 + 12x + 9$$ ♦♦♦

Note that in the perfect square trinomial obtained in Example 36, the first and last terms are the squares of the first and last terms of the binomial.

$$(2x + 3)^2$$

square ↙   ↘ square

$$4x^2 + 12x + 9$$

The middle term is twice the product of the terms of the binomial.

$$(2x + 3)^2$$

product ↓

$$6x$$

twice the product ↓

$$4x^2 + 12x + 9$$

Also, the constant term is always positive. In general, the following equations apply:

Perfect Square Trinomials	$(a + b)^2 = a^2 + 2ab + b^2$	47
	$(a - b)^2 = a^2 - 2ab + b^2$	48

Any quadratic trinomial can be manipulated into the form of a perfect square trinomial by a procedure called *completing the square*. We will use that method in Sec. 14–2 to derive the quadratic formula.

Factoring a Perfect Square Trinomial

We can factor a perfect square trinomial in the same way we factored the general quadratic trinomial in Sec. 8–5. However, the work is faster if we recognize that a trinomial is a perfect square. If it is, its factors will be the square of a binomial. The terms of that binomial are the square roots of the first and last terms of the trinomial. The sign in the binomial will be the same as the sign of the middle term of the trinomial.

◆◆◆ **Example 37:** Factor $a^2 - 4a + 4$.

Solution: The first and last terms are both perfect squares, and the middle term is twice the product of the square roots of the first and last terms. Thus the trinomial is a perfect square. Factoring, we obtain

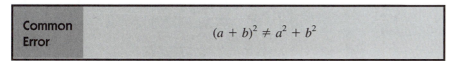

$$(a - 2)^2$$

◆◆◆

◆◆◆ **Example 38:**

$$9x^2y^2 - 6xy + 1 = (3xy - 1)^2$$

◆◆◆

Common Error	$(a + b)^2 \neq a^2 + b^2$

Exercise 6 ◆ The Perfect Square Trinomial

Factor completely.

1. $x^2 + 4x + 4$
2. $x^2 - 30x + 225$
3. $y^2 - 2y + 1$
4. $x^2 + 2x + 1$
5. $2y^2 - 12y + 18$
6. $9 - 12a + 4a^2$
7. $9 + 6x + x^2$
8. $4y^2 - 4y + 1$
9. $9x^2 + 6x + 1$
10. $16x^2 + 16x + 4$
11. $9y^2 - 18y + 9$
12. $16n^2 - 8n + 1$
13. $16 + 16a + 4a^2$
14. $1 + 20a + 100a^2$
15. $49a^2 - 28a + 4$

Perfect Square Trinomials with Two Variables

16. $a^2 - 2ab + b^2$
17. $x^2 + 2xy + y^2$
18. $a^2 - 14ab + 49b^2$
19. $a^2w^2 + 2abw + b^2$

20. $a^2 - 10ab + 25b^2$ **21.** $x^2 + 10ax + 25a^2$
22. $c^2 - 6cd + 9d^2$ **23.** $x^2 + 8xy + 16y^2$

Expressions That Can Be Reduced to Perfect Square Trinomials

24. $1 + 2x^2 + x^4$ **25.** $z^6 + 16z^3 + 64$
26. $36 + 12a^2 + a^4$ **27.** $49 - 14x^3 + x^6$
28. $a^2b^2 - 8ab^3 + 16b^4$ **29.** $a^4 - 2a^2b^2 + b^4$
30. $16a^2b^2 - 8ab^2c^2 + b^2c^4$ **31.** $4a^{2n} + 12a^nb^n + 9b^{2n}$

32. To find the width $2m$ of a road that will give a sight distance of 1000 ft on a curve of radius 500 ft, we must solve the equation

$$m^2 - 1000m + 250{,}000 = 0$$

Factor the left side of this equation.

8–7   Sum or Difference of Two Cubes

Definition

An expression such as

$$x^3 + 27$$

is called the *sum of two cubes* (x^3 and 3^3). In general, when we multiply the binomial ($a + b$) and the trinomial ($a^2 - ab + b^2$), we obtain

$$(a + b)(a^2 - ab + b^2) = a^3 - a^2b + ab^2 + a^2b - ab^2 + b^3$$
$$= a^3 + b^3$$

All but the cubed terms drop out, leaving the sum of two cubes.

Sum of Two Cubes	$a^3 + b^3 = (a + b)(a^2 - ab + b^2)$	42
Difference of Two Cubes	$a^3 - b^3 = (a - b)(a^2 + ab + b^2)$	43

When we recognize that an expression is the sum (or difference) of two cubes, we can write the factors immediately.

♦♦♦ **Example 39:** Factor $x^3 + 27$.

Solution: This expression is the sum of two cubes, $x^3 + 3^3$. Substituting into Eq. 42, with $a = x$ and $b = 3$, yields

$$x^3 + 27 = x^3 + 3^3 = (x + 3)(x^2 - 3x + 9)$$

same sign

opposite sign

always +

♦♦♦

◆◆◆ **Example 40:** Factor $27x^3 - 8y^3$.

Solution: This expression is the difference of two cubes, $(3x)^3 - (2y)^3$. Factoring gives us

$$27x^3 - 8y^3 = (3x)^3 - (2y)^3 = (3x - 2y)(9x^2 + 6xy + 4y^2)$$

same
sign

always +

opposite
sign

◆◆◆

Common Error	The middle term of the trinomials in Eqs. 42 and 43 is often mistaken as $2ab$. $$a^3 + b^3 \neq (a + b)(a^2 - 2ab + b^2)$$ no!

When the powers are multiples of 3, we may be able to factor the expression as the sum or difference of two cubes.

◆◆◆ **Example 41:**

$$a^6 - b^9 = (a^2)^3 - (b^3)^3$$

Factoring, we obtain

$$a^6 - b^9 = (a^2 - b^3)(a^4 + a^2b^3 + b^6)$$

◆◆◆

Factoring by Computer

A computer algebra system (CAS), such as *Derive, Maple,* or *Mathematica,* can, of course, factor very well. In fact, you might feel that they factor *too* well. For example, the expression

$$x^2 - 2$$

might be factored by CAS into

$$(x - \sqrt{2})(x + \sqrt{2})$$

Now this may be just what you want. However, if you want to avoid irrational factors like these, you can *adjust the level of factoring. Derive,* for example, has five levels of factoring: trivial, square-free, rational, radical, and complex, and *Mathematica* has similar commands.

Trivial factoring removes common factors. Thus $2x^3 - 4x^2 - 16x$ would be factored into

$$2x(x^2 - 2x - 8)$$

and *not* into $2x(x - 4)(x + 2)$.

Square-free factoring removes common factors and also factors trinomials, but it will not factor the difference of two squares. Thus

$$x^4 + 2x^3 - 3x^2 - 8x - 4$$

would factor into

$$(x + 1)^2 (x^2 - 4)$$

but not into

$$(x + 1)^2 (x - 2)(x + 2)$$

Rational factoring does what we normally refer to in this chapter as *complete* factoring. It will factor

$$2x^2 - 18 = 2(x - 3)(x + 3)$$

It will not factor $x^2 - 2$ into irrational factors.

Radical factoring will introduce radical factors, causing $x^2 - 2$ to be factored into $(x - \sqrt{2})(x + \sqrt{2})$.

Complex factoring may introduce complex factors. Thus $x^2 + 4$ would be factored into

$$(x - j2)(x + j2)$$

where j is equal to $\sqrt{-1}$, as explained in Chapter 21.

Exercise 7 ◆ Sum or Difference of Two Cubes

Factor completely.

1. $64 + x^3$	**2.** $1 - 64y^3$
3. $2a^3 - 16$	**4.** $a^3 - 27$
5. $x^3 - 1$	**6.** $x^3 - 64$
7. $x^3 + 1$	**8.** $a^3 - 343$
9. $a^3 + 64$	**10.** $x^3 + 343$
11. $x^3 + 125$	**12.** $64a^3 + 27$
13. $216 - 8a^3$	**14.** $343 - 27y^3$
15. $343 + 64x^3$	

More Difficult Types

16. $27x^9 + 512$
17. $y^9 + 64x^3$
18. $64a^{12} + x^{15}$
19. $27x^{15} + 8a^6$
20. $8x^{6p} + 8a^6$
21. $8a^{6x} - 125b^{3x}$
22. $64x^{12a} - 27y^{15a}$
23. $64x^{3n} - y^{9n}$
24. $x^3y^3 - z^3$

Applications

25. The volume of a hollow spherical shell having an inside radius of r_1 and an outside radius r_2 is $\frac{4}{3}\pi r_2^3 - \frac{4}{3}\pi r_1^3$. Factor this expression completely.

26. A cistern is in the shape of a hollow cube whose inside dimension is s and whose outside dimension is S. If it is made of concrete of density d, its weight is $dS^3 - ds^3$. Factor completely.

◆◆◆ CHAPTER 8 REVIEW PROBLEMS ◆◆◆◆◆◆◆◆◆◆◆◆◆◆◆◆◆◆◆◆◆◆◆◆◆◆◆◆

Factor completely.

1. $x^2 - 2x - 15$
2. $2a^2 + 3a - 2$
3. $x^6 - y^4$

4. $2ax^2 + 8ax + 8a$

5. $2x^2 + 3x - 2$

6. $a^2 + ab - 6b^2$

7. $(a - b)^3 + (c + d)^3$

8. $8x^3 - \dfrac{y^3}{27}$

9. $\dfrac{x^2}{y} - \dfrac{x}{y}$

10. $2ax^2y^2 - 18a$

11. $xy - 2y + 5x - 10$

12. $3a^2 - 2a - 8$

13. $x - bx - y + by$

14. $\dfrac{2a^2}{12} - \dfrac{8b^2}{27}$

15. $(y + 2)^2 - z^2$

16. $2x^2 - 20ax + 50a^2$

17. $x^2 - 7x + 12$

18. $4a^2 - (3a - 1)^2$

19. $16x^{4n} - 81y^{8n}$

20. $xy - y^2 + xz - yz$

21. $9a^2 + 12az^2 + 4z^4$

22. $4a^6 - 4b^6$

23. $x^{2m} + 2x^m + 1$

24. $4x^{6m} + 4x^{3m}y^m + y^{2m}$

25. $(x - y)^2 - z^2$

26. $x^2 - 2x - 3$

27. $1 - 16x^2$

28. $a^2 - 2a - 8$

29. $9x^4 - x^2$

30. $x^2 - 21x + 110$

31. $(p - q)^3 - 27$

32. $5a^2 - 20x^2$

33. $27a^3 - 8w^3$

34. $3x^2 - 6x - 45$

35. $16x^2 - 16xy + 4y^2$

36. $64m^3 - 27n^3$

37. $6ab + 2ay + 3bx + xy$

38. $15a^2 - 11a - 12$

39. $2y^4 - 18$

40. $ax - bx + ay - by$

41. The reduction in power in a resistance R by lowering the voltage across the resistor from V_2 to V_1 is

$$\frac{V_2^2}{R} - \frac{V_1^2}{R}$$

Factor this expression.

42. A conical tank of height h is filled to a depth d. The volume of liquid that can still be put into the tank is

$$V = \frac{\pi}{3}r_1^2 h - \frac{\pi}{3}r_2^2 d$$

where r_1 is the base radius of the tank and r_2 is the base radius of the liquid. Factor the right side of this equation.

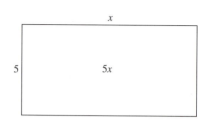

FIGURE 8–7

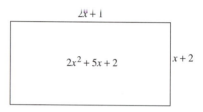

FIGURE 8–8

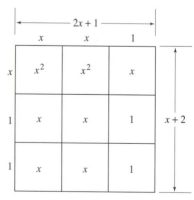

FIGURE 8–10

Computer

43. If you have access to a computer algebra system, use it to factor any of the expressions in this chapter. Be sure to set the level of factoring appropriately (**Rational** in *Derive,* for example).

Writing

44. We have studied the factoring of seven different types of expressions in this chapter. List them and give an example of each. State in words how to recognize each and how to tell one from the other. Also list at least four other expressions that are *not* one of the given seven, and state why each is different from those we have studied.

Team Project

45. The area of a rectangle is the *product* of its sides, so we can think of the product $5x$ as the area of a rectangle of sides 5 and x (Fig. 8–7).

Similarly, the trinomial $2x^2 + 5x + 2$ can be thought of as the area of a rectangle of sides $(2x + 1)$ and $(x + 2)$, because

$$2x^2 + 5x + 2 = (2x + 1)(x + 2)$$

as shown in Fig. 8–8.

To factor a trinomial using areas, cut rectangles of paper with areas equal to the terms in the trinomial. Thus the trinomial $2x^2 + 5x + 2$ can be represented by two squares of side x, five rectangles with sides 1 and x, and two squares of side 1 (Fig. 8–9). Choose any size for x, but avoid making it an integer value.

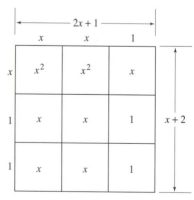

FIGURE 8–9

Arrange the individual squares to form a rectangle. There will be only one way to do this. The sides $(2x + 1)$ and $(x + 2)$ of this rectangle will be the factors of the trinomial, as shown in Fig. 8–10.

Use this method to factor the following trinomials:

$$x^2 + 3x + 2$$
$$2x^2 + 7x + 6$$
$$x^2 + 13x + 30$$

Expand the method to handle trinomials with *negative* terms. Use it to factor

$$x^2 + 2x - 3$$
$$x^2 + 7x - 30$$

Fractions and Fractional Equations

◆◆◆ **OBJECTIVES** ◆◆

When you have completed this chapter, you should be able to:

- Determine values of variables that result in undefined fractions.
- Convert common fractions to decimals, and vice versa.
- Reduce algebraic fractions.
- Add, subtract, and divide algebraic fractions.
- Simplify complex fractions.
- Solve fractional equations.
- Solve applied problems requiring fractional equations.
- Solve literal equations and formulas.

Although the calculator, the computer, and the metric system (all of which use decimal notation) have somewhat reduced the importance of common fractions, these fractions are still widely used. Algebraic fractions are, of course, as important as ever. We must be able to handle them in order to solve fractional equations and the word problems from which they arise.

Not all of this material is new to us. Some was covered in Chapter 2, and simple fractional equations were solved in Chapter 3. We also make heavy use of factoring in order to simplify algebraic fractions.

9–1   Simplification of Fractions

Parts of a Fraction

A fraction has a *numerator*, a *denominator*, and a *fraction line*.

$$\text{fraction line} \longrightarrow \frac{a}{b} \quad \begin{matrix} \longleftarrow \text{numerator} \\ \longleftarrow \text{denominator} \end{matrix}$$

Quotient

A fraction is a way of indicating a *quotient* of two quantities. The fraction *a/b* can be read "*a* divided by *b*."

Ratio

The quotient of two numbers or quantities is also spoken of as the *ratio* of those quantities. Thus the ratio of *x* to *y* is *x/y*.

Division by Zero

Since division by zero is not permitted, it should be understood in our work with fractions that *the denominator cannot be zero*.

◆◆◆ **Example 1:** What values of *x* are not permitted in the following fraction:

$$\frac{3x}{x^2 + x - 6}$$

Solution: Factoring the denominator, we get

$$\frac{3x}{x^2 + x - 6} = \frac{3x}{(x - 2)(x + 3)}$$

We see that an *x* equal to 2 or −3 will make $(x - 2)$ or $(x + 3)$ equal to zero. This will result in division by zero, so these values are not permitted.   ◆◆◆

Common Fractions

A fraction whose numerator and denominator are both integers is called a *common fraction*.

◆◆◆ **Example 2:** The following are common fractions:

$$\frac{2}{3}, \quad \frac{-9}{5}, \quad \frac{-124}{125} \quad \text{and} \quad \frac{18}{-11}$$ ◆◆◆

Algebraic Fractions

An *algebraic fraction* is one whose numerator and/or denominator contain *literal* quantities.

◆◆◆ **Example 3:** The following are algebraic fractions:

$$\frac{x}{y}, \quad \frac{\sqrt{x + 2}}{x}, \quad \frac{3}{y}, \quad \text{and} \quad \frac{x^2}{x - 3}$$ ◆◆◆

Rational Algebraic Fractions

Recall that a polynomial is an expression in which the exponents are nonnegative integers.

An algebraic fraction is called *rational* if the numerator and the denominator are both *polynomials*.

◆◆◆ **Example 4:** The following are rational fractions:

$$\frac{x}{y}, \quad \frac{3}{w^3}, \quad \text{and} \quad \frac{x^2}{x-3}$$

But $\dfrac{\sqrt{x+2}}{x}$ is not.   ◆◆◆

Proper and Improper Fractions

A *proper* common fraction is one whose numerator is smaller than its denominator.

◆◆◆ **Example 5:** $\frac{3}{5}$, $\frac{1}{3}$, and $\frac{9}{11}$ are proper fractions, whereas $\frac{8}{5}$, $\frac{3}{2}$, and $\frac{7}{4}$ are *improper* fractions.   ◆◆◆

A proper *algebraic* fraction is a rational fraction whose numerator is of *lower degree* than the denominator.

◆◆◆ **Example 6:** The following are proper fractions:

$$\frac{x}{x^2+2} \quad \text{and} \quad \frac{x^2+2x-3}{x^3+9}$$

However,

$$\frac{x^3-2}{x^2+x-3} \quad \text{and} \quad \frac{x^2}{y}$$

are improper fractions.   ◆◆◆

Mixed Form

A *mixed number* is the sum of an integer and a fraction.

◆◆◆ **Example 7:** The following are mixed numbers:

$$2\frac{1}{2}, \quad 5\frac{3}{4}, \quad \text{and} \quad 3\frac{1}{3}$$ ◆◆◆

A *mixed expression* is the sum or difference of a polynomial and a rational algebraic fraction.

◆◆◆ **Example 8:** The following are mixed expressions:

$$3x - 2 + \frac{1}{x} \quad \text{and} \quad y - \frac{y}{y^2+1}$$ ◆◆◆

Decimals and Fractions

To change a fraction to an equivalent decimal, simply divide the numerator by the denominator.

◆◆◆ **Example 9:** To write $\frac{9}{11}$ as a decimal, we divide 9 by 11.

$$\frac{9}{11} = 0.8181818181\overline{}\ldots$$

We get a *repeating decimal;* the dots following the number indicate that the digits continue indefinitely. ◆◆◆

To change a decimal number to a fraction, write a fraction with the decimal number in the numerator and 1 in the denominator. Multiply numerator and denominator by a multiple of 10 that will make the numerator a whole number. Finally, reduce to lowest terms.

◆◆◆ **Example 10:** Express 0.875 as a fraction.

Solution:

$$0.875 = \frac{0.875}{1} = \frac{875}{1000} = \frac{7}{8}$$

◆◆◆

To express a *repeating* decimal as a fraction, follow the steps in the next example.

◆◆◆ **Example 11:** Change the repeating decimal $0.81\overline{81}$ to a fraction.

Solution: Let $x = 0.81\overline{81}$. Multiplying by 100, we have

$$100x = 81.\overline{81}$$

Subtracting the first equation, $x = 0.81\overline{81}$, from the second gives us

$$99x = 81 \quad \text{(exactly)}$$

Dividing by 99 yields

$$x = \frac{81}{99} = \frac{9}{11} \quad \text{(reduced)}$$

◆◆◆

Simplifying a Fraction by Reducing to Lowest Terms

We reduce a fraction to lowest terms by dividing both numerator and denominator by any factor that is contained in both.

$$\frac{ad}{bd} = \frac{a}{b} \qquad \textbf{50}$$

◆◆◆ **Example 12:** Reduce the following to lowest terms. Write the answer without negative exponents.

(a) $\dfrac{9}{12} = \dfrac{3(3)}{4(3)} = \dfrac{3}{4}$

(b) $\dfrac{3x^2yz}{9xy^2z^3} = \dfrac{3}{9} \cdot \dfrac{x^2}{x} \cdot \dfrac{y}{y^2} \cdot \dfrac{z}{z^3} = \dfrac{x}{3yz^2}$ ◆◆◆

When possible, factor the numerator and the denominator. Then divide both numerator and denominator by any factors common to both.

◆◆◆ **Example 13:**

(a) $\dfrac{2x^2 + x}{3x} = \dfrac{(2x + 1)x}{3(x)} = \dfrac{2x + 1}{3}$

(b) $\dfrac{ab + bc}{bc + bd} = \dfrac{b(a + c)}{b(c + d)} = \dfrac{a + c}{c + d}$

(c) $\dfrac{2x^2 - 5x - 3}{4x^2 - 1} = \dfrac{(2x + 1)(x - 3)}{(2x + 1)(2x - 1)} = \dfrac{x - 3}{2x - 1}$

(d) $\dfrac{x^2 - ax + 2bx - 2ab}{2x^2 + ax - 3a^2} = \dfrac{x(x - a) + 2b(x - a)}{(x - a)(2x + 3a)}$

$\qquad\qquad = \dfrac{(x - a)(x + 2b)}{(x - a)(2x + 3a)} = \dfrac{x + 2b}{2x + 3a}$　　　◆◆◆

The process of striking out the same factors from numerator and denominator is called *canceling*.

	If a factor is missing from *even one term* in the numerator or denominator, that factor *cannot* be canceled.
	$$\dfrac{xy - z}{wx} \neq \dfrac{y - z}{w}$$
Common Errors	We may divide (or multiply) the numerator and denominator by the same quantity (Eq. 50), but we *may not add or subtract* the same quantity in the numerator and denominator, as this will change the value of the fraction. For example,
	$$\dfrac{3}{5} \neq \dfrac{3 + 1}{5 + 1} = \dfrac{4}{6} = \dfrac{2}{3}$$

Most students love canceling because they think they can cross out any term that stands in their way. If you use canceling, use it carefully!

Simplifying Fractions by Changing Signs

Recall from Chapter 2 that any two of the three signs of a fraction may be changed without changing the value of a fraction.

Rules of Signs	$\dfrac{+a}{+b} = \dfrac{-a}{-b} = -\dfrac{-a}{+b} = -\dfrac{+a}{-b} = \dfrac{a}{b}$	**10**
	$\dfrac{+a}{-b} = \dfrac{-a}{+b} = -\dfrac{-a}{-b} = -\dfrac{a}{b}$	**11**

We can sometimes use this idea to simplify a fraction, that is, to reduce it to lowest terms.

◆◆◆ **Example 14:** Simplify the fraction

$$-\frac{3x - 2}{2 - 3x}$$

Solution: We change the sign of the denominator *and* the sign of the entire fraction.

$$-\frac{3x-2}{2-3x} = +\frac{3x-2}{-(2-3x)} = \frac{3x-2}{-2+3x} = \frac{3x-2}{3x-2} = 1$$

changed

◆◆◆

Exercise 1 ◆ Simplification of Fractions

Hint: Factor the denominators in problems 4, 5, and 6.

In each fraction, what values of x, if any, are not permitted?

1. $\dfrac{12}{x}$

2. $\dfrac{x}{12}$

3. $\dfrac{18}{x-5}$

4. $\dfrac{5x}{x^2-49}$

5. $\dfrac{7}{x^2-3x+2}$

6. $\dfrac{3x}{8x^2-14x+3}$

Change each fraction to a decimal. Work to four decimal places.

7. $\dfrac{7}{12}$

8. $\dfrac{5}{9}$

9. $\dfrac{15}{16}$

10. $\dfrac{125}{155}$

11. $\dfrac{11}{3}$

12. $\dfrac{25}{9}$

Change each decimal to a fraction.

13. 0.4375

14. 0.390625

15. 0.6875

16. 0.28125

17. 0.7777 . . .

18. 0.636363 . . .

Simplify each fraction by manipulating the algebraic signs.

19. $\dfrac{a-b}{b-a}$

20. $-\dfrac{2x-y}{y-2x}$

21. $\dfrac{(a-b)(c-d)}{b-a}$

22. $\dfrac{w(x-y-z)}{y-x+z}$

Reduce to lowest terms. Write your answers without negative exponents.

23. $\dfrac{14}{21}$

24. $\dfrac{81}{18}$

25. $\dfrac{75}{35}$

26. $\dfrac{36}{44}$

27. $\dfrac{2ab}{6b}$

28. $\dfrac{12m^2n}{15mn^2}$

29. $\dfrac{21m^2p^2}{28mp^4}$

30. $\dfrac{abx-bx^2}{acx-cx^2}$

31. $\dfrac{4a^2-9b^2}{4a^2+6ab}$

32. $\dfrac{3a^2+6a}{a^2+4a+4}$

33. $\dfrac{x^2+5x}{x^2+4x-5}$

34. $\dfrac{xy-3y^2}{x^3-27y^3}$

35. $\dfrac{x^2 - 4}{x^3 - 8}$

36. $\dfrac{2a^3 + 6a^2 - 8a}{2a^3 + 2a^2 - 4a}$

37. $\dfrac{2m^3n - 2m^2n - 24mn}{6m^3 + 6m^2 - 36m}$

38. $\dfrac{9x^3 - 30x^2 + 25x}{3x^4 - 11x^3 + 10x^2}$

39. $\dfrac{2a^2 - 2}{a^2 - 2a + 1}$

40. $\dfrac{3a^2 - 4ab + b^2}{a^2 - ab}$

41. $\dfrac{x^2 - z^2}{x^3 - z^3}$

42. $\dfrac{2x^2}{6x - 4x^2}$

43. $\dfrac{2a^2 - 8}{2a^2 - 2a - 12}$

44. $\dfrac{2a^2 + ab - 3b^2}{a^2 - ab}$

45. $\dfrac{x^2 - 1}{2xy + 2y}$

46. $\dfrac{x^3 - a^2x}{x^2 - 2ax + a^2}$

47. $\dfrac{2x^4y^4 + 2}{3x^8y^8 - 3}$

48. $\dfrac{18a^2c - 6bc}{42a^2d - 14bd}$

49. $\dfrac{(x + y)^2}{x^2 - y^2}$

50. $\dfrac{mw + 3w - mz - 3z}{m^2 - m - 12}$

51. $\dfrac{b^2 - a^2 - 6b + 9}{5b - 15 - 5a}$

52. $\dfrac{3w - 3y - 3}{w^2 + y^2 - 2wy - 1}$

Computer

53. A computer algebra system (CAS) will simplify fractions, given the proper command. In *Derive, Mathematica,* and *Maple,* for example, the command *Simplify* will reduce a fraction to lowest terms. This same command will simplify nonfractional algebraic expressions as well. If you have access to a computer algebra system, use it to simplify any of the fractions in this exercise set.

9–2   Multiplication and Division of Fractions

Multiplication

We multiply a fraction *a/b* by another fraction *c/d* as follows:

Multiplying Fractions	$\dfrac{a}{b} \cdot \dfrac{c}{d} = \dfrac{ac}{bd}$	51

The product of two or more fractions is a fraction whose numerator is the product of the numerators of the original fractions and whose denominator is the product of the denominators of the original fractions.

◆◆◆ **Example 15:**

(a) $\dfrac{2}{3} \cdot \dfrac{5}{7} = \dfrac{2(5)}{3(7)} = \dfrac{10}{21}$

(b) $\dfrac{1}{2} \cdot \dfrac{2}{3} \cdot \dfrac{3}{5} = \dfrac{1(2)(3)}{2(3)(5)} = \dfrac{1}{5}$

(c) $5\dfrac{2}{3} \cdot 3\dfrac{1}{2} = \dfrac{17}{3} \cdot \dfrac{7}{2} = \dfrac{119}{6}$

◆◆◆

> **Common Error**
>
> When multiplying mixed numbers, students sometimes try to multiply the whole parts and the fractional parts separately.
>
> $$2\frac{2}{3} \times 4\frac{3}{5} \neq 8\frac{6}{15}$$
>
> The correct way is to write each mixed number as an improper fraction and to multiply as shown.
>
> $$2\frac{2}{3} \times 4\frac{3}{5} = \frac{8}{3} \times \frac{23}{5} = \frac{184}{15} = 12\frac{4}{15}$$

◆◆◆ **Example 16:**

(a) $\dfrac{2a}{3b} \cdot \dfrac{5c}{4a} = \dfrac{10ac}{12ab} = \dfrac{5c}{6b}$

Leave your product in factored form until after you simplify.

(b) $\dfrac{x}{x+2} \cdot \dfrac{x^2-4}{x^3} = \dfrac{x(x^2-4)}{(x+2)x^3}$

$$= \dfrac{x(x+2)(x-2)}{(x+2)x^3} = \dfrac{x-2}{x^2}$$

(c) $\dfrac{x^2+x-2}{x^2-4x+3} \cdot \dfrac{2x^2-3x-9}{2x^2+7x+6}$

$$= \dfrac{(x-1)(x+2)(x-3)(2x+3)}{(x-1)(x-3)(x+2)(2x+3)} = 1$$

◆◆◆

Division

To divide one fraction, a/b, by another fraction, c/d,

$$\dfrac{\dfrac{a}{b}}{\dfrac{c}{d}}$$

Can you say why it is permissible to multiply numerator and denominator by d/c?

we multiply numerator and denominator by d/c, as follows:

$$\dfrac{\dfrac{a}{b} \cdot \dfrac{d}{c}}{\dfrac{c}{d} \cdot \dfrac{d}{c}} = \dfrac{\dfrac{a}{b} \cdot \dfrac{d}{c}}{\dfrac{cd}{cd}} = \dfrac{\dfrac{a}{b} \cdot \dfrac{d}{c}}{1} = \dfrac{a}{b} \cdot \dfrac{d}{c} = \dfrac{ad}{bc}$$

We see that dividing by a fraction is the same as *multiplying by the reciprocal* of that fraction.

Division of Fractions	$\dfrac{a}{b} \div \dfrac{c}{d} = \dfrac{a}{b} \cdot \dfrac{d}{c} = \dfrac{ad}{bc}$	52

When dividing fractions, invert the divisor and multiply.

◆◆◆ **Example 17:**

(a) $\dfrac{2}{3} \div \dfrac{5}{7} = \dfrac{2}{3} \cdot \dfrac{7}{5} = \dfrac{14}{15}$

(b) $3\dfrac{2}{5} \div 2\dfrac{4}{15} = \dfrac{17}{5} \div \dfrac{34}{15}$

$\qquad\qquad = \dfrac{17}{5} \times \dfrac{15}{34} = \dfrac{3}{2}$

(c) $\dfrac{x}{y} \div \dfrac{x+2}{y-1} = \dfrac{x}{y} \cdot \dfrac{y-1}{x+2}$

$\qquad\qquad = \dfrac{x(y-1)}{y(x+2)}$

(d) $\dfrac{x^2 + x - 2}{x} \div \dfrac{x+2}{x^2} = \dfrac{x^2 + x - 2}{x} \cdot \dfrac{x^2}{x+2}$

$\qquad\qquad = \dfrac{(x+2)(x-1)x^2}{x(x+2)} = x(x-1)$

(e) $x \div \dfrac{\pi r^2 x}{4} = \dfrac{x}{1} \div \dfrac{\pi r^2 x}{4} = \dfrac{x}{1} \cdot \dfrac{4}{\pi r^2 x} = \dfrac{4}{\pi r^2}$ ◆◆◆

> Remember that all literal fractions have nonzero denominators, since division by zero is not permitted.

Common Error	We invert the *divisor* and multiply. Be sure not to invert the dividend.

Changing Improper Fractions to Mixed Form

To write an improper fraction in mixed form, divide the numerator by the denominator, and express the remainder as the numerator of a fraction whose denominator is the original denominator.

◆◆◆ **Example 18:** Write $\frac{45}{7}$ as a mixed number.

Solution: Dividing 45 by 7, we get 6 with a remainder of 3, so

$$\frac{45}{7} = 6\frac{3}{7}$$

◆◆◆

The procedure is the same for changing an *algebraic* fraction to mixed form. Divide the numerator by the denominator.

◆◆◆ **Example 19:**

$$\frac{x^2 + 3}{x} = \frac{x^2}{x} + \frac{3}{x}$$

$$= x + \frac{3}{x}$$

◆◆◆

Exercise 2 ◆ Multiplication and Division of Fractions

Multiply and reduce.

1. $\dfrac{1}{3} \times \dfrac{2}{5}$

2. $\dfrac{3}{7} \times \dfrac{21}{24}$

3. $\dfrac{2}{3} \times \dfrac{9}{7}$

4. $\dfrac{11}{3} \times 7$

5. $\dfrac{2}{3} \times 3\dfrac{1}{5}$

6. $3 \times 7\dfrac{2}{5}$

7. $3\dfrac{3}{4} \times 2\dfrac{1}{2}$

8. $\dfrac{3}{5} \times \dfrac{2}{7} \times \dfrac{5}{9}$

9. $3 \times \dfrac{5}{8} \times \dfrac{4}{5}$

10. $\dfrac{15a^2}{7b^2} \cdot \dfrac{28ab}{9a^3c}$

11. $\dfrac{a^4b^4}{2a^2y^n} \cdot \dfrac{a^2x}{xy^n}$

12. $\dfrac{3x^2y^2z^3}{4a^2b^2c^2} \cdot \dfrac{8a^3b^2c^2}{9x^2yz^3}$

13. $\dfrac{5m^2n^2p^4}{3x^2yz^3} \cdot \dfrac{21xyz^2}{20m^2n^2p^2}$

14. $\dfrac{x + y}{x - y} \cdot \dfrac{x^2 - y^2}{(x + y)^2}$

15. $\dfrac{x^2 - a^2}{xy} \cdot \dfrac{xy}{x + a}$

16. $\dfrac{x^4 - y^4}{a^2x^2} \cdot \dfrac{ax^3}{x^2 + y^2}$

17. $\dfrac{a}{x - y} \cdot \dfrac{b}{x + y}$

18. $\dfrac{x + y}{10} \cdot \dfrac{ax}{3(x + y)}$

19. $\dfrac{c}{x^2 - y^2} \cdot \dfrac{d}{x^2 - y^2}$

20. $\dfrac{x - 2}{2x + 3} \cdot \dfrac{2x^2 + x - 3}{x^2 - 4}$

21. $\dfrac{x^2 - 1}{x^2 + x - 6} \cdot \dfrac{x^2 + 2x - 8}{x^2 - 4x - 5}$

22. $\dfrac{5x - 1}{x^2 - x - 2} \cdot \dfrac{x - 2}{10x^2 + 13x - 3}$

23. $\dfrac{2x^2 + x - 1}{3x^2 - 11x - 4} \cdot \dfrac{3x^2 + 7x + 2}{2x^2 - 7x + 3}$

24. $\dfrac{x^3 + 1}{x + 1} \cdot \dfrac{x + 3}{x^2 - x + 1}$

25. $\dfrac{2x^2 + x - 6}{4x^2 - 3x - 1} \cdot \dfrac{4x^2 + 5x + 1}{x^2 - 4x - 12}$

Divide and reduce.

26. $\dfrac{7}{9} \div \dfrac{5}{3}$

27. $3\dfrac{7}{8} \div 2$

28. $\dfrac{7}{8} \div 4$

29. $\dfrac{9}{16} \div 8$

30. $24 \div \dfrac{5}{8}$

31. $\dfrac{5}{8} \div 2\dfrac{1}{4}$

32. $2\dfrac{7}{8} \div 1\dfrac{1}{2}$

33. $2\dfrac{5}{8} \div \dfrac{1}{2}$

34. $50 \div 2\dfrac{3}{5}$

35. $\dfrac{5abc^3}{3x^2} \div \dfrac{10ac^3}{6bx^2}$

36. $\dfrac{7x^2y}{3ad} \div \dfrac{2xy^2}{3a^2d}$

37. $\dfrac{4a^3x}{6dy^2} \div \dfrac{2a^2x^2}{8a^2y}$

38. $\dfrac{5x^2y^3z}{6a^2b^2c} \div \dfrac{10xy^3z^2}{8ab^2c^2}$

39. $\dfrac{3an + cm}{x^2 - y^2} \div (x^2 + y^2)$

40. $\dfrac{a^2 + 4a + 4}{d + c} \div (a + 2)$

41. $\dfrac{ac + ad + bc + bd}{c^2 - d^2} \div (a + b)$

42. $\dfrac{5(x + y)^2}{x - y} \div (x + y)$

43. $\dfrac{1}{x^2 + 17x + 30} \div \dfrac{1}{x + 15}$

44. $\dfrac{5xy}{a - x} \div \dfrac{10xy}{a^2 - x^2}$

45. $\dfrac{3x - 6}{2x + 3} \div \dfrac{x^2 - 4}{4x^2 + 2x - 6}$

46. $\dfrac{a^2 - a - 2}{5a - 1} \div \dfrac{a - 2}{10a^2 + 13a - 3}$

47. $\dfrac{2p^2 + p - 6}{4p^2 - 3p - 1} \div \dfrac{2p^2 - 5p + 3}{4p^2 + 9p + 2}$

48. $\dfrac{x^3 + 1}{x + 1} \div \dfrac{x^2 - x + 1}{3x + 9}$

49. $\dfrac{z^2 - 1}{z^2 + z - 6} \div \dfrac{z^2 - 4z - 5}{z^2 + 2z - 8}$

50. $\dfrac{x^2 + 5x + 6}{2x^2 - 3x + 1} \div \dfrac{x^2 - 9}{2x^2 - 7x + 3}$

Improper Fractions and Mixed Form

Change each improper fraction to a mixed number.

51. $\dfrac{5}{3}$

52. $\dfrac{11}{5}$

53. $\dfrac{29}{12}$

54. $\dfrac{17}{3}$

55. $\dfrac{47}{5}$

56. $\dfrac{125}{12}$

Change each improper algebraic fraction to a mixed expression.

57. $\dfrac{x^2 + 1}{x}$

58. $\dfrac{1 - 4x}{2x}$

59. $\dfrac{4x - x^2}{2x^2}$

Applications

60. For a thin lens, the relationship between the focal length f, the object distance p, and the image distance q is

$$f = \frac{pq}{p + q}$$

A second lens of focal length f_1 has the same object distance p but a different image distance q_1.

$$f_1 = \frac{pq_1}{p + q_1}$$

Find the ratio f/f_1 and simplify.

61. The mass density of an object is its mass divided by its volume. Write and simplify an expression for the density of a sphere having a mass m and a volume equal to $4\pi r^3/3$.

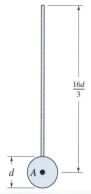

FIGURE 9–1

FIGURE 9–2

62. The pressure on a surface is equal to the total force divided by the area. Write and simplify an expression for the pressure on a circular surface of area $\pi d^2/4$ subjected to a distributed load F.

63. The stress on a bar in tension is equal to the load divided by the cross-sectional area. Write and simplify an expression for the stress in a bar having a trapezoidal cross section of area $(a + b)h/2$, subject to a load P.

64. The acceleration on a body is equal to the force on the body divided by its mass, and the mass equals the volume of the object times the density. Write and simplify an expression for the acceleration of a sphere having a volume $4\pi r^3/3$ and a density D, subjected to a force F.

65. To find the moment of the area A in Fig. 9–1, we must multiply the area $\pi d^2/4$ by the distance to the pivot, $16d/3$. Multiply and simplify.

66. The circle in Fig. 9–2 has an area $\pi d^2/4$, and the sector has an area equal to $ds/4$. Find the ratio of the area of the circle to the area of the sector.

Computer

67. If you have a computer algebra system, use it to do any of the multiplications or divisions in this exercise set. You need not tell the computer to multiply or divide; the instruction *Simplify* is all you need.

9–3   Addition and Subtraction of Fractions

Similar Fractions

Similar fractions (also called *like* fractions) are those having the same (common) denominator.

> To add or subtract similar fractions, *combine the numerators and place them over the common denominator.*

Addition and Subtraction of Fractions	$\dfrac{a}{b} \pm \dfrac{c}{b} = \dfrac{a \pm c}{b}$	53

◆◆◆ **Example 20:**

(a) $\dfrac{2}{3} + \dfrac{5}{3} = \dfrac{2 + 5}{3} = \dfrac{7}{3}$

(b) $\dfrac{1}{x} + \dfrac{3}{x} = \dfrac{1 + 3}{x} = \dfrac{4}{x}$

(c) $\dfrac{3x}{x + 1} - \dfrac{5}{x + 1} + \dfrac{x^2}{x + 1} = \dfrac{3x - 5 + x^2}{x + 1}$ ◆◆◆

Least Common Denominator

The least common denominator is also called the *least common multiple* (LCM) of the denominators.

The *least common denominator,* or LCD (also called the *lowest* common denominator), is the smallest expression that is exactly divisible by each of the denominators. Thus the LCD must contain all of the prime factors of each of the denominators. The common denominator of two or more fractions is simply the product of the denominators of those fractions. To find the *least* common denominator, before multiplying, drop any prime factor from one denominator that also appears in another denominator.

◆◆◆ **Example 21:** Find the LCD for the two fractions $\frac{3}{8}$ and $\frac{1}{18}$.

Solution: Factoring each denominator, we obtain

$$
\begin{array}{cc}
8 & 18 \\
(2)(2)(2) & (2)(3)(3)
\end{array}
$$

duplicates;
include only once
in LCD

Dropping one of the 2's that appear in both sets of factors, we find that our LCD is then the product of these factors.

$$\text{LCD} = (2)(2)(2)(3)(3) = 8(9) = 72 \qquad \text{◆◆◆}$$

For this simple problem, you probably found the LCD by inspection in less time than it took to read this example, and you are probably wondering what all the fuss is about. What we are really doing is developing a *method* that we can use on algebraic fractions when the LCD is *not* obvious.

◆◆◆ **Example 22:** Find the LCD for the fractions

$$\frac{5}{x^2 + x}, \qquad \frac{x}{x^2 - 1}, \quad \text{and} \quad \frac{9}{x^3 - x^2}$$

Solution: The denominator $x^2 + x$ has the prime factors x and $x + 1$; the denominator $x^2 - 1$ has the prime factors $x + 1$ and $x - 1$; and the denominator $x^3 - x^2$ has the prime factors x, x, and $x - 1$. Our LCD is thus

$$(x)(x)(x + 1)(x - 1) \quad \text{or} \quad x^2(x^2 - 1) \qquad \text{◆◆◆}$$

Combining Fractions with Different Denominators

To combine fractions with different denominators, first find the LCD. Then multiply numerator and denominator of each fraction by that quantity that will make the denominator equal to the LCD. Finally, combine as shown above, and simplify.

◆◆◆ **Example 23:** Add $\frac{1}{2}$ and $\frac{2}{3}$.

Solution: The LCD is 6, so

$$
\begin{aligned}
\frac{1}{2} + \frac{2}{3} &= \frac{1}{2}\left(\frac{3}{3}\right) + \frac{2}{3}\left(\frac{2}{2}\right) \\
&= \frac{3}{6} + \frac{4}{6} \\
&= \frac{7}{6} \qquad \text{◆◆◆}
\end{aligned}
$$

◆◆◆ **Example 24:** Add $\frac{3}{8}$ and $\frac{1}{18}$.

Solution: The LCD, from Example 21, is 72. So

$$
\begin{aligned}
\frac{3}{8} + \frac{1}{18} &= \frac{3}{8} \cdot \frac{9}{9} + \frac{1}{18} \cdot \frac{4}{4} \\
&= \frac{27}{72} + \frac{4}{72} \\
&= \frac{31}{72} \qquad \text{◆◆◆}
\end{aligned}
$$

It is not necessary to write the fractions with the *least* common denominator; any common denominator will work as well. But your final result will then have to be reduced to lowest terms.

The method for adding and subtracting unlike fractions can be summarized by the formula that follows.

| Combining Unlike Fractions | $\dfrac{a}{b} \pm \dfrac{c}{d} = \dfrac{ad}{bd} \pm \dfrac{bc}{bd} = \dfrac{ad \pm bc}{bd}$ | **54** |

◆◆◆ **Example 25:** Combine $x/2y - 5/x$.

Solution: The LCD will be the product of the two denominators, or $2xy$. So

$$\frac{x}{2y} - \frac{5}{x} = \frac{x}{2y}\left(\frac{x}{x}\right) - \frac{5}{x}\left(\frac{2y}{2y}\right)$$

$$= \frac{x^2}{2xy} - \frac{10y}{2xy}$$

$$= \frac{x^2 - 10y}{2xy}$$

◆◆◆

The procedure is the same, of course, even when the denominators are more complicated.

◆◆◆ **Example 26:** Combine the fractions

$$\frac{x + 2}{x - 3} + \frac{2x + 1}{3x - 2}$$

Solution: Our LCD is $(x - 3)(3x - 2)$, so

$$\frac{x + 2}{x - 3} + \frac{2x + 1}{3x - 2} = \frac{(x + 2)(3x - 2)}{(x - 3)(3x - 2)} + \frac{(2x + 1)(x - 3)}{(3x - 2)(x - 3)}$$

$$= \frac{(3x^2 + 4x - 4) + (2x^2 - 5x - 3)}{(x - 3)(3x - 2)}$$

$$= \frac{5x^2 - x - 7}{3x^2 - 11x + 6}$$

◆◆◆

Always factor the denominators completely. This will show whether the same factor appears in more than one denominator.

◆◆◆ **Example 27:** Combine and simplify:

$$\frac{1}{x - 3} + \frac{x - 1}{x^2 + 3x + 9} + \frac{x^2 + x - 3}{x^3 - 27}$$

Solution: Factoring gives

$$\frac{1}{x - 3} + \frac{x - 1}{x^2 + 3x + 9} + \frac{x^2 + x - 3}{(x - 3)(x^2 + 3x + 9)}$$

Our LCD is then $x^3 - 27$ or $(x - 3)(x^2 + 3x + 9)$. Multiplying and adding, we have

$$\frac{(x^2 + 3x + 9) + (x - 1)(x - 3) + (x^2 + x - 3)}{(x - 3)(x^2 + 3x + 9)}$$

$$= \frac{x^2 + 3x + 9 + x^2 - 4x + 3 + x^2 + x - 3}{(x - 3)(x^2 + 3x + 9)}$$

$$= \frac{3x^2 + 9}{x^3 - 27}$$

◆◆◆

Combining Integers and Fractions

To combine integers and fractions, treat the integer as a fraction having 1 as a denominator, and combine as shown above.

◆◆◆ **Example 28:**

$$3 + \frac{2}{9} = \frac{3}{1} + \frac{2}{9} = \frac{3}{1} \cdot \frac{9}{9} + \frac{2}{9}$$

$$= \frac{27}{9} + \frac{2}{9} = \frac{29}{9}$$

◆◆◆

The same procedure may be used to change a *mixed number* to an *improper fraction*.

◆◆◆ **Example 29:**

$$2\frac{1}{3} = 2 + \frac{1}{3} = \frac{6}{3} + \frac{1}{3}$$

$$= \frac{7}{3}$$

◆◆◆

Changing a Mixed Algebraic Expression to an Improper Fraction

This procedure is no different from that in the preceding section. Write the non-fractional expression as a fraction with 1 as the denominator, find the LCD, and combine, as shown in the following example.

◆◆◆ **Example 30:**

$$x^2 + \frac{5}{2x} = \frac{x^2}{1}\left(\frac{2x}{2x}\right) + \frac{5}{2x}$$

Adding, we obtain

$$x^2 + \frac{5}{2x} = \frac{2x^3 + 5}{2x}$$

◆◆◆

Partial Fractions

Those fractions whose sum is equal to a given fraction are called *partial fractions*.

◆◆◆ **Example 31:** If we add the two fractions $2/x$ and $3/x^2$, we get

$$\frac{2}{x} + \frac{3}{x^2} = \frac{2x}{x^2} + \frac{3}{x^2} = \frac{2x + 3}{x^2}$$

Turned around, we say that the fraction $(2x + 3)/x^2$ has the partial fractions $2/x$ and $3/x^2$.

◆◆◆

To be able to separate a fraction into its partial fractions is important in calculus. We mention them here but will not calculate any until later.

Exercise 3 ◆ Addition and Subtraction of Fractions

Don't use your calculator for these numerical problems. The practice you get working with common fractions will help you when doing algebraic fractions.

Common Fractions and Mixed Numbers

Combine and simplify.

1. $\dfrac{3}{5} + \dfrac{2}{5}$

2. $\dfrac{1}{8} - \dfrac{3}{8}$

3. $\dfrac{2}{7} + \dfrac{5}{7} - \dfrac{6}{7}$

4. $\dfrac{5}{9} + \dfrac{7}{9} - \dfrac{1}{9}$

5. $\dfrac{1}{3} - \dfrac{7}{3} + \dfrac{11}{3}$

6. $\dfrac{1}{5} - \dfrac{9}{5} + \dfrac{12}{5} - \dfrac{2}{5}$

7. $\dfrac{1}{2} + \dfrac{2}{3}$

8. $\dfrac{3}{5} - \dfrac{1}{3}$

9. $\dfrac{3}{4} + \dfrac{7}{16}$

10. $\dfrac{2}{3} + \dfrac{3}{7}$

11. $\dfrac{5}{9} - \dfrac{1}{3} + \dfrac{3}{18}$

12. $\dfrac{1}{2} + \dfrac{1}{3} + \dfrac{1}{5}$

13. $2 + \dfrac{3}{5}$

14. $3 - \dfrac{2}{3} + \dfrac{1}{6}$

15. $\dfrac{1}{5} - 7 + \dfrac{2}{3}$

16. $\dfrac{1}{5} + 2 - \dfrac{1}{3} + 5$

17. $\dfrac{1}{7} + 2 - \dfrac{3}{7} + \dfrac{1}{5}$

18. $5 - \dfrac{3}{5} + \dfrac{2}{15} - 2$

Rewrite each mixed number as an improper fraction.

19. $2\dfrac{2}{3}$

20. $1\dfrac{5}{8}$

21. $3\dfrac{1}{4}$

22. $2\dfrac{7}{8}$

23. $5\dfrac{11}{16}$

24. $9\dfrac{3}{4}$

Algebraic Fractions

Combine and simplify.

25. $\dfrac{1}{a} + \dfrac{5}{a}$

26. $\dfrac{3}{x} + \dfrac{2}{x} - \dfrac{1}{x}$

27. $\dfrac{2a}{y} + \dfrac{3}{y} - \dfrac{a}{y}$

28. $\dfrac{x}{3a} - \dfrac{y}{3a} + \dfrac{z}{3a}$

29. $\dfrac{5x}{2} - \dfrac{3x}{2}$

30. $\dfrac{7}{x + 2} - \dfrac{5}{x + 2}$

31. $\dfrac{3x}{a - b} + \dfrac{2x}{b - a}$

32. $\dfrac{a}{x^2} - \dfrac{b}{x^2} + \dfrac{c}{x^2}$

33. $\dfrac{7}{a + 1} + \dfrac{9}{1 + a} - \dfrac{3}{-a - 1}$

34. $\dfrac{2a}{x(x - 1)} - \dfrac{3a}{x^2 - x} + \dfrac{a}{x - x^2}$

35. $\dfrac{3a}{2x} + \dfrac{2a}{5x}$

36. $\dfrac{1}{x} + \dfrac{1}{y} + \dfrac{1}{z}$

37. $\dfrac{a}{3} + \dfrac{2}{x} - \dfrac{1}{2}$

38. $\dfrac{1}{x + 3} + \dfrac{1}{x - 2}$

39. $\dfrac{a + b}{3} - \dfrac{a - b}{2}$

40. $\dfrac{3}{a + b} + \dfrac{2}{a - b}$

41. $\dfrac{4}{x - 1} - \dfrac{5}{x + 1}$

42. $\dfrac{x + 1}{x - 1} - \dfrac{x - 1}{x + 1}$

43. $\dfrac{b}{a + b} - \dfrac{a}{b - a} - \dfrac{a - b}{a + b}$

44. $\dfrac{y^2 - 2xy + x^2}{x^2 - xy} + \dfrac{x}{x - y}$

45. $\dfrac{1}{2(x-1)} - \dfrac{1}{2(x+1)} + \dfrac{1}{x^2}$

46. $\dfrac{1+x}{1+x+x^2} + \dfrac{1-x}{1-x+x^2}$

47. $\dfrac{1}{x+1} + \dfrac{2}{x+2} + \dfrac{3}{x-3}$

48. $\dfrac{a-b}{ab} + \dfrac{b-c}{bc} + \dfrac{c-a}{ac}$

49. $\dfrac{x}{x^2-9} + \dfrac{x}{x+3}$

50. $\dfrac{x+3}{x^2-2x-8} - \dfrac{2x}{x-4}$

51. $\dfrac{5}{x-2} - \dfrac{3x+4}{x^3-8}$

52. $\dfrac{3x}{x-1} + \dfrac{5}{x^3+x^2-2x} + \dfrac{2}{x} + \dfrac{x-1}{x+2}$

53. $\dfrac{1}{x+d+1} - \dfrac{1}{x+1}$

54. $\dfrac{1}{(x+d)^2} - \dfrac{1}{x^2}$

55. $x+d+\dfrac{1}{x+d} - \left(x+\dfrac{1}{x}\right)$

56. $\dfrac{x+d}{(x+d-1)^2} - \dfrac{x}{(x-1)^2}$

Expressions of this form arise in calculus, when you are taking the derivative of a fraction by the delta method.

Mixed Form

Rewrite each mixed expression as an improper fraction.

57. $x + \dfrac{1}{x}$

58. $\dfrac{1}{2} - x$

59. $\dfrac{x}{x-1} - 1 - \dfrac{1}{x+1}$

60. $a + \dfrac{1}{a^2-b^2}$

61. $3 - a - \dfrac{2}{a}$

62. $\dfrac{5}{x} - 2x + \dfrac{3}{x^2}$

63. $\left(3b + \dfrac{1}{x}\right) - \left(2b - \dfrac{b}{ax}\right)$

64. $\left(5x + \dfrac{x+2}{3}\right) - \left(2x - \dfrac{2x-3}{4}\right)$

Applications

Treat the given numbers in these problems as exact, and leave your answers in fractional form.

65. A certain work crew can grade 7 mi of roadbed in 3 days, and another crew can do 9 mi in 4 days. How much can both crews together grade in 1 day?

Hint: In problem 65, first find the amount that each crew can do in one day (e.g., the first crew can grade 7/3 mi per day). Then add the separate amounts to get the daily total. You can use a similar approach to the other work problems in this group.

66. Liquid is running into a tank from a pipe that can fill four tanks in 3 days. Meanwhile, liquid is running out from a drain that can empty two tanks in 4 days. What will be the net change in volume in 1 day?

Do this group of problems without using your calculator, and leave your answers in fractional form.

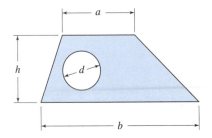

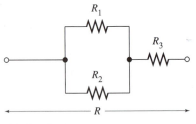

FIGURE 9–3

FIGURE 9–4

67. A planer makes a 1-m cutting stroke at a rate of 15 m/min and returns at 75 m/min. How long does it take for the cutting stroke and return? (See Eq. A17.)

68. Two resistors, 5 Ω and 15 Ω, are wired in parallel. What is the resistance of the parallel combination? (See Eq. A64.)

69. One crew can put together five machines in 8 days. Another crew can assemble three of these machines in 4 days. How much can both crews together assemble in 1 day?

70. One crew can assemble M machines in p days. Another crew can assemble N of these machines in q days. Write an expression for the number of machines that both crews together can assemble in 1 day, combine into a single term, and simplify.

71. A steel plate in the shape of a trapezoid (Fig. 9–3) has a hole of diameter d. The area of the plate, less the hole, is

$$\frac{(a + b)h}{2} - \frac{\pi d^2}{4}$$

Combine these two terms and simplify.

72. The total resistance R of Fig. 9–4 is

$$\frac{R_1 R_2}{R_1 + R_2} + R_3$$

Combine into a single term and simplify.

73. If a car travels a distance d at a rate V, the time required will be d/V. The car then continues for a distance d_1 at a rate V_1, and a third distance d_2 at rate V_2. Write an expression for the total travel time; then combine the three terms into a single term and simplify.

Computer

74. If you have a computer algebra system, use it to do any of the additions or subtractions in this exercise set.

9–4 Complex Fractions

Fractions that have only *one* fraction line are called *simple* fractions. Fractions with *more than one* fraction line are called *complex* fractions.

◆◆◆ **Example 32:** The following are complex fractions:

$$\frac{\dfrac{a}{b}}{c} \quad \text{and} \quad \frac{x - \dfrac{1}{x}}{\dfrac{x}{y} - x}$$

◆◆◆

We show how to simplify complex fractions in the following examples.

◆◆◆ **Example 33:** Simplify the complex fraction

$$\frac{\dfrac{1}{2} + \dfrac{2}{3}}{3 + \dfrac{1}{4}}$$

Solution: We can simplify this fraction by multiplying numerator and denominator by the least common denominator for all of the individual fractions. The denominators are 2, 3, and 4, so the LCD is 12. Multiplying, we obtain

$$\frac{\left(\dfrac{1}{2}+\dfrac{2}{3}\right)12}{\left(3+\dfrac{1}{4}\right)12} = \frac{6+8}{36+3} = \frac{14}{39}$$

◆◆◆

◆◆◆ **Example 34:** Simplify the complex fraction

$$\frac{1+\dfrac{a}{b}}{1-\dfrac{b}{a}}$$

Solution: The LCD for the two small fractions a/b and b/a is ab. Multiplying, we obtain

$$\frac{1+\dfrac{a}{b}}{1-\dfrac{b}{a}} \cdot \frac{ab}{ab} = \frac{ab+a^2}{ab-b^2}$$

or, in factored form,

$$\frac{a(b+a)}{b(a-b)}$$

◆◆◆

In the following example, we simplify a small section at a time and work outward. Try to follow the steps.

◆◆◆ **Example 35:**

$$\frac{a+\dfrac{1}{b}}{a+\dfrac{1}{b+\dfrac{1}{a}}} \cdot \frac{ab}{ab+1} = \frac{\dfrac{ab+1}{b}}{a+\dfrac{1}{\dfrac{ab+1}{a}}} \cdot \frac{ab}{ab+1}$$

$$= \frac{\dfrac{ab+1}{b}}{a+\dfrac{a}{ab+1}} \cdot \frac{ab}{ab+1}$$

$$= \frac{\dfrac{ab+1}{b}}{\dfrac{a(ab+1)+a}{ab+1}} \cdot \frac{ab}{ab+1}$$

$$= \frac{ab+1}{b} \cdot \frac{ab+1}{a(ab+1)+a} \cdot \frac{ab}{ab+1}$$

$$= \frac{a(ab+1)}{a(ab+1)+a} = \frac{ab+1}{ab+2}$$

◆◆◆

Exercise 4 ◆ Complex Fractions

Simplify. Leave your answers as improper fractions.

1. $\dfrac{\dfrac{2}{3} + \dfrac{3}{4}}{\dfrac{1}{5}}$

2. $\dfrac{\dfrac{3}{4} - \dfrac{1}{3}}{\dfrac{1}{2} + \dfrac{1}{6}}$

3. $\dfrac{\dfrac{1}{2} + \dfrac{1}{3} + \dfrac{1}{4}}{3 - \dfrac{4}{5}}$

4. $\dfrac{\dfrac{4}{5}}{\dfrac{1}{5} + \dfrac{2}{3}}$

5. $\dfrac{5 - \dfrac{2}{5}}{6 + \dfrac{1}{3}}$

6. $\dfrac{1}{2} + \dfrac{3}{\dfrac{2}{5} + \dfrac{1}{3}}$

7. $\dfrac{x + \dfrac{y}{4}}{x - \dfrac{y}{3}}$

8. $\dfrac{\dfrac{a}{b} + \dfrac{x}{y}}{\dfrac{a}{z} - \dfrac{x}{c}}$

9. $\dfrac{1 + \dfrac{x}{y}}{1 - \dfrac{x^2}{y^2}}$

10. $\dfrac{x + \dfrac{a}{c}}{x + \dfrac{b}{d}}$

11. $\dfrac{a^2 + \dfrac{x}{3}}{4 + \dfrac{x}{5}}$

12. $\dfrac{3a^2 - 3y^2}{\dfrac{a + y}{3}}$

13. $\dfrac{x + \dfrac{2d}{3ac}}{x + \dfrac{3d}{2ac}}$

14. $\dfrac{4a^2 - 4x^2}{\dfrac{a + x}{a - x}}$

15. $\dfrac{x^2 - \dfrac{y^2}{2}}{\dfrac{x - 3y}{2}}$

16. $\dfrac{\dfrac{ab}{7} - 3d}{3c - \dfrac{ab}{d}}$

17. $\dfrac{1 + \dfrac{1}{x + 1}}{1 - \dfrac{1}{x - 1}}$

18. $\dfrac{xy - \dfrac{3x}{ac}}{\dfrac{ac}{x} + 2c}$

19. $\dfrac{\dfrac{2m + n}{m + n} - 1}{1 - \dfrac{n}{m + n}}$

20. $\dfrac{\dfrac{x^3 + y^3}{x^2 - y^2}}{\dfrac{x^2 - xy + y^2}{x - y}}$

21. $\dfrac{1}{a + \dfrac{1}{1 + \dfrac{a + 1}{3 - a}}}$

22. $\dfrac{x + y}{x + y + \dfrac{1}{x - y + \dfrac{1}{x + y}}}$

23. $\dfrac{3abc}{bc + ac + ab} - \dfrac{\dfrac{a-1}{a} + \dfrac{b-1}{b} + \dfrac{c-1}{c}}{\dfrac{1}{a} + \dfrac{1}{b} + \dfrac{1}{c}}$

24. $\dfrac{\dfrac{1}{a} + \dfrac{1}{b+c}}{\dfrac{1}{a} - \dfrac{1}{b+c}}\left(1 + \dfrac{b^2 + c^2 - a^2}{2bc}\right)$

25. $\dfrac{\dfrac{x^2 - y^2 - z^2 - 2yz}{x^2 - y^2 - z^2 + 2yz}}{\dfrac{x - y - z}{x + y - z}}$

Applications

26. A car travels a distance d_1 at a rate V_1, then another distance d_2 at a rate V_2. The average speed for the entire trip is

$$\text{average speed} = \dfrac{d_1 + d_2}{\dfrac{d_1}{V_1} + \dfrac{d_2}{V_2}}$$

Simplify this complex fraction.

27. The equivalent resistance of two resistors in parallel is

$$\dfrac{R_1 R_2}{R_1 + R_2}$$

If each resistor is made of wire of resistivity ρ, with R_1 using a wire of length L_1 and cross-sectional area A_1, and R_2 having a length L_2 and area A_2, our expression becomes

See Eq. A71, $R = \rho L/A$.

$$\dfrac{\dfrac{\rho L_1}{A_1} \cdot \dfrac{\rho L_2}{A_2}}{\dfrac{\rho L_1}{A_1} + \dfrac{\rho L_2}{A_2}}$$

Simplify this complex fraction.

Computer

28. If you have a computer algebra system, use it to simplify any of the expressions in this exercise set.

9–5 Fractional Equations

Solving Fractional Equations

An equation in which one or more terms is a fraction is called a *fractional equation.* To solve a fractional equation, first eliminate the fractions by multiplying both sides of the equation by the least common denominator (LCD) of every term. We can do this because multiplying both sides of an equation by the same quantity (the LCD in this case) does not unbalance the equation. With the fractions thus eliminated, the equation is then solved as any nonfractional equation is.

◆◆◆ **Example 36:** Solve for x:

$$\frac{3x}{5} - \frac{x}{3} = \frac{2}{15}$$

Solution: Multiplying both sides of the equation by the LCD (15), we obtain

$$15\left(\frac{3x}{5} - \frac{x}{3}\right) = 15\left(\frac{2}{15}\right)$$

$$9x - 5x = 2$$

$$x = \frac{1}{2}$$ ◆◆◆

Equations with the Unknown in the Denominator

The procedure is the same when the unknown appears in the denominator of one or more terms. However, the LCD will now contain the unknown. Here it is understood that x cannot have a value that will make any of the denominators in the problem equal to zero. Such forbidden values are sometimes stated with the problem, but often they are not.

◆◆◆ **Example 37:** Solve for x:

$$\frac{2}{3x} = \frac{5}{x} + \frac{1}{2} \quad (x \neq 0)$$

Solution: The LCD is $6x$. Multiplying both sides of the equation yields

$$6x\left(\frac{2}{3x}\right) = 6x\left(\frac{5}{x} + \frac{1}{2}\right)$$

$$6x\left(\frac{2}{3x}\right) = 6x\left(\frac{5}{x}\right) + 6x\left(\frac{1}{2}\right)$$

$$4 = 30 + 3x$$

$$3x = -26$$

$$x = -\frac{26}{3}$$

We assume the integers in these equations to be exact numbers. Thus we leave our answers in fractional form rather than converting to (approximate) decimals.

Check:

$$\frac{2}{3\left(-\dfrac{26}{3}\right)} \overset{?}{=} \frac{5}{-\dfrac{26}{3}} + \frac{1}{2}$$

$$-\frac{2}{26} \overset{?}{=} -\frac{15}{26} + \frac{13}{26}$$

$$-\frac{2}{26} = -\frac{2}{26} \quad \text{(checks)}$$ ◆◆◆

◆◆◆ **Example 38:** Solve for x:

$$\frac{8x + 7}{5x + 4} = 2 - \frac{2x}{5x + 1}$$

Solution: Multiplying by the LCD $(5x + 4)(5x + 1)$ yields

$$(8x + 7)(5x + 1) = 2(5x + 4)(5x + 1) - 2x(5x + 4)$$
$$40x^2 + 35x + 8x + 7 = 2(25x^2 + 5x + 20x + 4) - 10x^2 - 8x$$
$$40x^2 + 43x + 7 = 50x^2 + 50x + 8 - 10x^2 - 8x$$
$$43x + 7 = 42x + 8$$
$$x = 1$$

Check:

$$\frac{8(1) + 7}{5(1) + 4} \overset{?}{=} 2 - \frac{2(1)}{5(1) + 1}$$

$$\frac{15}{9} \overset{?}{=} 2 - \frac{2}{6}$$

$$\frac{5}{3} = \frac{6}{3} - \frac{1}{3} \quad \text{(checks)}$$

◆◆◆

Common Error	The technique of multiplying by the LCD in order to eliminate the denominators is valid *only when we have an equation.* Do not multiply through by the LCD when there is no equation!

◆◆◆ **Example 39:** Solve for x:

$$\frac{3}{x - 3} = \frac{2}{x^2 - 2x - 3} + \frac{4}{x + 1}$$

Solution: The denominator of the second fraction factors into

$$x^2 - 2x - 3 = (x - 3)(x + 1)$$

The LCD is therefore $(x - 3)(x + 1)$. Multiplying both sides of the given equation by the LCD gives

$$3(x + 1) = 2 + 4(x - 3)$$
$$3x + 3 = 4x - 10$$
$$x = 13$$

◆◆◆

Exercise 5 ◆ Fractional Equations

Solve for x.

1. $2x + \dfrac{x}{3} = 28$

2. $4x + \dfrac{x}{5} = 42$

3. $x + \dfrac{x}{5} = 24$

4. $\dfrac{x}{6} + x = 21$

5. $3x - \dfrac{x}{7} = 40$

6. $x - \dfrac{x}{6} = 25$

7. $\dfrac{2x}{3} - x = -24$

8. $\dfrac{3x}{5} + 7x = 38$

9. $\dfrac{x}{2} + \dfrac{x}{3} + \dfrac{x}{4} = 26$

10. $\dfrac{x - 1}{2} = \dfrac{x + 1}{3}$

11. $\dfrac{3x - 1}{4} = \dfrac{2x + 1}{3}$

12. $x + \dfrac{2x}{3} + \dfrac{3x}{4} = 29$

13. $2x + \dfrac{x}{3} - \dfrac{x}{4} = 50$

14. $3x - \dfrac{2x}{3} - \dfrac{5x}{6} = 18$

15. $3x - \dfrac{x}{6} + \dfrac{x}{12} = 70$

16. $\dfrac{x}{4} + \dfrac{x}{6} + \dfrac{x}{8} = 26$

17. $\dfrac{6x - 19}{2} = \dfrac{2x - 11}{3}$

18. $\dfrac{7x - 40}{8} = \dfrac{9x - 80}{10}$

19. $\dfrac{3x - 116}{4} + \dfrac{180 - 5x}{6} = 0$

20. $\dfrac{3x - 4}{2} - \dfrac{3x - 1}{16} = \dfrac{6x - 5}{8}$

21. $\dfrac{x - 1}{8} - \dfrac{x + 1}{18} = 1$

22. $\dfrac{3x}{4} + \dfrac{180 - 5x}{6} = 29$

23. $\dfrac{15x}{4} = \dfrac{9}{4} - \dfrac{3 - x}{2}$

24. $\dfrac{x}{4} + \dfrac{x}{10} + \dfrac{x}{8} = 19$

Equations with Unknown in Denominator

25. $\dfrac{2}{3x} + 6 = 5$

26. $9 - \dfrac{4}{5x} = 7$

27. $4 + \dfrac{1}{x + 3} = 8$

28. $\dfrac{x + 5}{x - 2} = 5$

29. $\dfrac{x - 3}{x + 2} = \dfrac{x + 4}{x - 5}$

30. $\dfrac{3}{x + 2} = \dfrac{5}{x} + \dfrac{4}{x^2 + 2x}$

31. $\dfrac{9}{x^2 + x - 2} = \dfrac{7}{x - 1} - \dfrac{3}{x + 2}$

32. $\dfrac{3x + 5}{2x - 3} = \dfrac{3x - 3}{2x - 1}$

33. $\dfrac{4}{x^2 - 1} + \dfrac{1}{x - 1} + \dfrac{1}{x + 1} = 0$

34. $\dfrac{x}{3} - \dfrac{x^2 - 5x}{3x - 7} = \dfrac{2}{3}$

35. $\dfrac{3}{x + 1} - \dfrac{x + 1}{x - 1} = \dfrac{x^2}{1 - x^2}$

36. $\dfrac{x + 1}{2(x - 1)} - \dfrac{x - 1}{x + 1} = \dfrac{17 - x^2}{2(x^2 - 1)}$

37. $\dfrac{9x + 20}{36} - \dfrac{x}{4} = \dfrac{4x - 12}{5x - 4}$

38. $\dfrac{10x + 13}{18} - \dfrac{x + 2}{x - 3} = \dfrac{5x - 4}{9}$

39. $\dfrac{6x + 7}{10} - \dfrac{3x + 1}{5} = \dfrac{x - 1}{3x - 4}$

40. $\dfrac{5x}{1 - x} - \dfrac{7x}{3 + x} = \dfrac{12(1 - x^2)}{x^2 + 2x - 3}$

Computer

41. If you have a computer algebra system, use it to solve any of the equations in this exercise set.

9–6 Word Problems Leading to Fractional Equations

Solving Word Problems

The method for solving the word problems in this chapter (and in any other chapter) is no different from that given in Chapter 3. The resulting equations, however, will now contain fractions.

◆◆◆ **Example 40:** One-fifth of a certain number is 3 greater than one-sixth of that same number. Find the number.

Solution: Let x = the number. Then from the problem statement,

$$\frac{1}{5}x = \frac{1}{6}x + 3$$

Multiplying by the LCD (30), we obtain

$$6x = 5x + 90$$
$$x = 90$$

Check: Is one-fifth of 90 (which is 18) 3 greater than one-sixth of 90 (which is 15)? Yes.   ◆◆◆

Rate Problems

Fractional equations often arise when we solve problems involving *rates*. Problems of this type include:

Uniform motion: *amount traveled = rate of travel × time traveled*

Work: *amount of work done = rate of work × time worked*

Fluid flow: *amount of flow = flow rate × duration of flow*

Energy flow: *amount of energy = rate of energy flow × time*

Notice that these equations are mathematically identical. Word problems involving motion, flow of fluids or solids, flow of heat, electricity, solar energy, mechanical energy, and so on, can be handled in the same way.

To convince yourself of the similarity of all of these types of problems, consider the following work problem:

> If worker M can do a certain job in 5 h, and N can do the same amount in 8 h, how long will it take M and N together to do that same amount?

Are the following problems really any different?

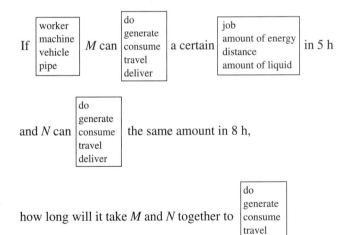

                                      that same amount?

We will first do a uniform motion problem and then proceed to work and flow problems.

Uniform Motion

Motion is called *uniform* when the speed does not change. The distance traveled at constant speed is related to the rate of travel and the elapsed time by the following equation:

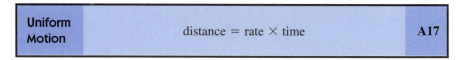

Uniform Motion	distance = rate $\times$ time	A17

Be careful not to use this formula for anything but *uniform* motion.

◆◆◆ **Example 41:** A train departs at noon traveling at a speed of 64 km/h. A car leaves the same station $\frac{1}{2}$ h later to overtake the train, traveling on a road parallel to the track. If the car's speed is 96 km/h, at what time and at what distance from the station will it overtake the train?

Estimate: The train gets a head start of (1/2)(64), or 32 km. But the car goes 32 km/h faster than the train, so every hour, the distance between car and train is reduced by 32 km. Thus it should take 1 hour for the car to overtake the train. The distance will be 1 hour's travel distance for the car, or 96 km.

Solution: Let x = time traveled by car (hours). Then $x + \frac{1}{2}$ = time traveled by train (hours). The distance traveled by the car is

$$\text{distance} = \text{rate} \times \text{time} = 96x$$

and for the train,

$$\text{distance} = 64\left(x + \frac{1}{2}\right)$$

But the distance is the same for car and train, so we can write the equation

$$96x = 64\left(x + \frac{1}{2}\right)$$

Removing parentheses yields

$$96x = 64x + 32$$
$$96x - 64x = 32$$
$$32x = 32$$
$$x = 1 \text{ h} = \text{time for car}$$
$$x + \frac{1}{2} = 1\tfrac{1}{2} \text{ h} = \text{time for train}$$

So the car overtakes the train at 1:30 P.M. at a distance of 96 km from the station. ◆◆◆

Work Problems

To tackle work problems, we need one simple idea.

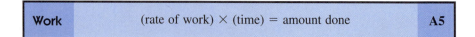

Work	(rate of work) $\times$ (time) = amount done	A5

We often use this equation to find the rate of work for a person or a machine.

In a typical work problem, there are two or more persons or machines doing work, each at a different rate. For each worker the work rate is the amount done by that worker divided by the time taken to do the work. That is, if a person can stamp 9 parts in 13 min, that person's work rate is $\frac{9}{13}$ part per minute.

••• Example 42: Crew A can assemble 2 cars in 5 days, and crew B can assemble 3 cars in 7 days. If both crews together assemble 100 cars, with crew B working 10 days longer than crew A, how many days (rounded to the nearest day) must each crew work?

Estimate: Working alone, crew A does 2 cars in 5 days, or 100 cars in 250 days. Similarly, crew B does 100 cars in about 231 days. Together they do 200 cars in $250 + 231$ or 481 days, or about 100 cars in 240 days. Thus we would expect each crew to work about half that, or about 120 days each. But crew B works 10 days longer than A, so we guess that they work about 5 days longer than 120 days, whereas crew A works about 5 days less than 120 days.

Solution: Let

$$x = \text{days worked by crew } A$$

and

$$x + 10 = \text{days worked by crew } B$$

The work rate of each crew is

$$\text{crew } A: \quad \text{rate} = \frac{2}{5} \text{ car per day}$$

$$\text{crew } B: \quad \text{rate} = \frac{3}{7} \text{ car per day}$$

The amount done by each crew equals their work rate times the number of days worked. The sum of the amounts done by the two crews must equal 100 cars.

$$\frac{2}{5}x + \frac{3}{7}(x + 10) = 100$$

We clear fractions by multiplying by the LCD, 35.

$$14x + 15(x + 10) = 3500$$
$$14x + 15x + 150 = 3500$$
$$29x = 3350$$
$$x = 116 \text{ days for crew } A$$
$$x + 10 = 126 \text{ days for crew } B \qquad \qquad \text{•••}$$

Fluid Flow and Energy Flow

For flow problems, we use the simple equation

$$\text{amount of flow} = \text{flow rate} \times \text{time}$$

Here we assume, of course, that the flow rate is constant.

••• Example 43: A certain small hydroelectric generating station can produce 17 megawatthours (MWh) of energy per year. After 4.0 months of operation, another generator is added which, by itself, can produce 11 MWh in 5.0 months. How many additional months are needed for a total of 25 MWh to be produced?

Solution: Let $x = $ additional months. The original generating station can produce 17/12 MWh per month for $4.0 + x$ months. The new generator can produce 11/5.0 MWh per month for x months. The problem states that the total amount produced (in MWh) is 25, so

$$\frac{17}{12}(4.0 + x) + \frac{11}{5.0}x = 25$$

Multiplying by the LCD, 60, we obtain

$$5.0(17)(4.0 + x) + 12(11)x = 25(60)$$
$$340 + 85x + 132x = 1500$$
$$217x = 1160$$
$$x = \frac{1160}{217} = 5.3 \text{ months}$$

◆◆◆

Exercise 6 ◆ Word Problems

Number Problems

Treat the numbers in Problems 1–5 as exact.

1. There is a number such that if $\frac{1}{5}$ of it is subtracted from 50 and this difference is multiplied by 4, the result will be 70 less than the number. What is the number?

2. The number 100 is separated into two parts such that if $\frac{1}{3}$ of one part is subtracted from $\frac{1}{4}$ of the other, the difference is 11. Find the two parts.

3. The second of two numbers is 1 greater than the first, and $\frac{1}{2}$ of the first plus $\frac{1}{5}$ of the first is equal to the sum of $\frac{1}{3}$ of the second and $\frac{1}{4}$ of the second. What are the numbers?

4. Find two consecutive numbers such that one-half of the larger exceeds one-third of the smaller by 9.

5. The difference between two positive numbers is 20, and $\frac{1}{7}$ of one is equal to $\frac{1}{3}$ of the other. What are the numbers?

Uniform Motion Problems

Depending on how you set them up, some of these problems may not result in fractional equations.

6. A person sets out from Boston and walks toward Portland at the rate of 3.0 mi/h. Three hours afterward, a second person sets out from the same place and walks in the same direction at the rate of 4.0 mi/h. How far from Boston will the second person overtake the first?

7. A courier who goes at the rate of 6.5 mi/h is followed, after 4.0 h, by another who goes at the rate of 7.5 mi/h. In how many hours will the second overtake the first?

8. A person walks to the top of a mountain at the rate of 2.0 mi/h and down the same way at the rate of 4.0 mi/h. If he walks for 6.0 h to reach the top of the mountain, how far is it to the top?

9. In going a certain distance, a train traveling at the rate of $\overline{40}$ mi/h takes 2.0 h less than a train traveling 30 mi/h. Find the distance.

10. Spacecraft A is over Houston at noon on a certain day and traveling at a rate of 300 km/h. Spacecraft B, attempting to overtake and dock with A, is over Houston at 1:15 P.M. and is traveling in the same direction as A, at 410 km/h. At what time will B overtake A? At what distance from Houston?

Work Problems

11. A laborer can do a certain job in 5 days, a second in 6 days, and a third in 8 days. In what time can the three together do the job?

12. Three masons build 318 m of wall. Mason A builds 7.0 m/day, B builds 6.0 m/day, and C builds 5.0 m/day. Mason B works twice as many days as A, and C works half as many days as A and B combined. How many days did each work?

13. If a carpenter can roof a house in 10 days and another can do the same in 14 days, how many days will it take if they work together?

14. A technician can assemble an instrument in 9.5 h. After working for 2.0 h, she is joined by another technician who, alone, could do the job in 7.5 h. How many additional hours are needed to finish the job?

15. A certain screw machine can produce a box of parts in 3.3 h. A new machine is to be ordered having a speed such that both machines working together would produce a box of parts in 1.4 h. How long would it take the new machine alone to produce a box of parts?

Fluid Flow Problems

16. A tank can be filled by a pipe in 3.0 h and emptied by another pipe in 4.0 h. How much time will be required to fill an empty tank if both are running?

17. Two pipes empty into a tank. One pipe can fill the tank in 8.0 h, and the other in 9.0 h. How long will it take both pipes together to fill the tank?

18. A tank has three pipes connected. The first, by itself, could fill the tank in $2\frac{1}{2}$ h, the second in 2 h, and the third in 1 h 40 min. In how many minutes will the tank be filled?

19. A tank can be filled by a certain pipe in 18.0 h. Five hours after this pipe is opened, it is supplemented by a smaller pipe which, by itself, could fill the tank in 24.0 h. Find the total time, measured from the opening of the larger pipe, to fill the tank.

20. At what rate must liquid be drained from a tank in order to empty it in 1.50 h if the tank takes 4.70 h to fill at the rate of 3.50 m³/min?

Energy Flow Problems

21. A certain power plant consumes 1500 tons of coal in 4.0 weeks. There is a stockpile of 10,000 tons of coal available when the plant starts operating. After 3.0 weeks in operation, an additional boiler, capable of using 2300 tons in 3.0 weeks, is put on line with the first boiler. In how many more weeks will the stockpile of coal be consumed?

22. A certain array of solar cells can generate 2.0 megawatthours (MWh) in 5.0 months (under standard conditions). After this array has been operating for 3.0 months, another array of cells is added which alone can generate 5.0 MWh in 7.0 months. How many additional months, after the new array has been added, is needed for the total energy generated from both arrays to be 10 MWh?

23. A landlord owns a house that consumes 2100 gal of heating oil in three winters. He buys another (insulated) house, and the two houses together use 1850 gal of oil in two winters. How many winters would it take the insulated house alone to use 1250 gal of oil?

24. A wind generator can charge 20 storage batteries in 24 h. After the generator has been charging for 6.0 h, another generator, which can charge the batteries in 36 h, is also connected to the batteries. How many additional hours are needed to charge the 20 batteries?

25. A certain solar collector can absorb 9000 Btu in 7.0 h. Another panel is added, and together they collect 35,000 Btu in 5.0 h. How long would it take the new panel alone to collect 35,000 Btu?

Financial Problems

26. After spending $\frac{1}{4}$ of my money, then $\frac{1}{5}$ of the remainder, I had $66 left. How many dollars did I start with?

27. A person who had inherited some money spent $\frac{3}{8}$ of it the first year and $\frac{4}{5}$ of the remainder the next year and had $1420 left. What was the inheritance?

28. Three children were left an inheritance of which the eldest received $\frac{2}{3}$, the second $\frac{1}{5}$, and the third the rest, which was $20,000. How much did each receive?

29. A's capital was $\frac{3}{4}$ of B's. If A's had been $500 less, it would have been only $\frac{1}{2}$ of B's. What was the capital of each?

The humorist Stephen Leacock, in his *A, B, and C—The Human Element in Mathematics*, tells of those word problem heroes: *A*, full-blooded and having great strength and endurance; *B*, quiet and easygoing; and *C*, the weakling. Here's a sample:

"The first time I ever saw these men was one evening after a regatta. They had all been rowing in it, and it had transpired that *A* could row as much in one hour as *B* in two or *C* in four. *B* and *C* had come in dead fagged and *C* was coughing badly. . . . Just then *A* came blustering in and shouted, "I say, you fellows, Smith has shown me three cisterns in his garden and he says we can pump them until tomorrow night. I bet I can beat you both . . . I heard *B* growl . . . but they went, and presently I could tell from the sound of the water that *A* was pumping four times as fast as *C*."

In this group of problems, as in the previous ones, we assume that the rates are constant.

Here are a few financial problems, similar to those we had in Sec. 3–3, except that these contain fractions. Give your answers to the nearest dollar. Glance back to that section if you need help in setting up these problems.

9–7   Literal Equations and Formulas

Literal Equations

A *literal equation* is one in which some or all of the constants are represented by letters.

◆◆◆ **Example 44:** The following is a literal equation:

$$a(x + b) = b(x + c)$$

However,

$$2(x + 5) = 3(x + 1)$$

is a numerical equation.   ◆◆◆

Formulas

A *formula* is a literal equation that relates two or more mathematical or physical quantities. These are the equations, mentioned in the introduction of this chapter, that describe the workings of the physical world. A listing of some of the common formulas used in technology is given in "Appendix A, Summary of Facts and Formulas."

◆◆◆ **Example 45:** Newton's second law is a formula relating the force acting on a body with its mass and acceleration. The equation is given in the following box.

Newton's Second Law	$F = ma$	A20

◆◆◆

Solving Literal Equations and Formulas

When we solve a literal equation or formula, we cannot, of course, get a *numerical* answer, as we could with equations that had only one unknown. Our object here is to *isolate* one of the letters on one side of the equal sign. We "solve for" one of the literal quantities.

◆◆◆ **Example 46:** Solve for x:

$$a(x + b) = b(x + c)$$

Solution: Our goal is to isolate x on one side of the equation. Removing parentheses, we obtain

$$ax + ab = bx + bc$$

Subtracting bx and then ab will place all of the x terms on one side of the equation.

$$ax - bx = bc - ab$$

Factoring to isolate x yields

$$x(a - b) = b(c - a)$$

Dividing by $(a - b)$, where $a \neq b$, gives us

$$x = \frac{b(c - a)}{a - b}$$

◆◆◆

◆◆◆ **Example 47:** Solve for x:

$$b\left(b + \frac{x}{a}\right) = d$$

Solution: Dividing both sides by b, we have

$$b + \frac{x}{a} = \frac{d}{b}$$

Subtracting b yields

$$\frac{x}{a} = \frac{d}{b} - b$$

Multiplying both sides by a gives us

$$x = a\left(\frac{d}{b} - b\right)$$

◆◆◆

◆◆◆ **Example 48:** The formula for the amount of heat q flowing by conduction through a wall of thickness L, conductivity k, and cross-sectional area A is

$$q = \frac{kA(t_1 - t_2)}{L}$$

where t_1 and t_2 are the temperatures of the warmer and cooler sides, respectively. Solve this equation for t_1.

Solution: Multiplying both sides by L/kA gives

$$\frac{qL}{kA} = t_1 - t_2$$

then adding t_2 to both sides, we get

$$t_1 = \frac{qL}{kA} + t_2$$

◆◆◆

Checking Literal Equations

As with numerical equations, we can substitute our expression for x back into the original equation and see if it checks.

◆◆◆ **Example 49:** We check the solution of Example 47 by substituting $a(d/b - b)$ for x in the original equation.

$$b\left(b + \frac{x}{a}\right) \overset{?}{=} d$$

$$b\left[b + \frac{a(d/b - b)}{a}\right] \overset{?}{=} d$$

$$b\left(b + \frac{d}{b} - b\right) \overset{?}{=} d$$

$$b\left(\frac{d}{b}\right) = d \quad \text{(checks)}$$

◆◆◆

Another simple "check" is to substitute numerical values. But avoid simple values such as 0 and 1 that may make an incorrect solution appear to check, and avoid any values that will make a denominator zero.

◆◆◆ Example 50: Check the results from Example 47.

Solution: Let us *choose* values for a, b, and d, say,

$$a = 2, \qquad b = 3, \quad \text{and} \quad d = 6$$

With these values,

$$x = a\left(\frac{d}{b} - b\right) = 2\left(\frac{6}{3} - 3\right) = -2$$

We now see if these values check in the original equation.

$$b\left(b + \frac{x}{a}\right) = d$$

$$3\left(3 + \frac{-2}{2}\right) \stackrel{?}{=} 6$$

$$3(3 - 1) \stackrel{?}{=} 6$$

$$3(2) = 6 \quad \text{(checks)}$$

Realize that this check is not a rigorous one, because we are testing the solution only for the specific values chosen. However, if it checks for those values, we can be *reasonably* sure that our solution is correct.   ◆◆◆

Literal Fractional Equations

To solve literal fractional equations, the procedure is the same as for other fractional equations: Multiply by the LCD to eliminate the fractions.

◆◆◆ Example 51: Solve the following for x in terms of a, b, and c:

$$\frac{x}{b} - \frac{a}{c} = \frac{x}{a}$$

Solution: Multiplying by the LCD (abc) yields

$$acx - a^2b = bcx$$

Rearranging so that all x terms are together on one side of the equation, we obtain

$$acx - bcx = a^2b$$

Factoring gives us

$$x(ac - bc) = a^2b$$

$$x = \frac{a^2b}{ac - bc}$$ ◆◆◆

◆◆◆ Example 52: Solve for x:

$$\frac{x}{x - a} - \frac{x + 2b}{x + a} = \frac{a^2 + b^2}{x^2 - a^2}$$

Solution: The LCD is $x^2 - a^2$, that is, $(x - a)(x + a)$. Multiplying each term by the LCD gives

$$x(x + a) - (x + 2b)(x - a) = a^2 + b^2$$

Removing parentheses and rearranging so that all of the x terms are together, we obtain

$$x^2 + ax - (x^2 + 2bx - ax - 2ab) = a^2 + b^2$$

$$x^2 + ax - x^2 - 2bx + ax + 2ab = a^2 + b^2$$

$$2ax - 2bx = a^2 - 2ab + b^2$$

Then we factor:

$$2x(a - b) = (a - b)^2$$

and divide by $2(a - b)$:

$$x = \frac{(a - b)^2}{2(a - b)} = \frac{a - b}{2}$$

◆◆◆

Sometimes a fractional equation will have some terms that are not fractions themselves, but the method of solution is the same.

◆◆◆ **Example 53:** Solve for x:

$$\frac{x}{b} + c = \frac{2x}{a + b} + b$$

Solution: We multiply both sides by the LCD, $b(a + b)$.

$$x(a + b) + c(b)(a + b) = 2x(b) + b(b)(a + b)$$

Removing parentheses and moving all of the x terms to one side gives

$$ax + bx + abc + b^2c = 2bx + ab^2 + b^3$$
$$ax + bx - 2bx = ab^2 + b^3 - abc - b^2c$$

Factoring yields

$$x(a - b) = b(ab + b^2 - ac - bc)$$
$$x = \frac{b(ab + b^2 - ac - bc)}{a - b}$$

◆◆◆

Exercise 7 ◆ Literal Equations and Formulas

Solve for x.

1. $2ax = bc$

2. $ax + dx = a - c$

3. $a(x + y) = b(x + z)$

4. $4x = 2x + ab$

5. $4acx - 3d^2 = a^2d - d^2x$

6. $a(2x - c) = a + c$

7. $a^2x - cd = b - ax + dx$

8. $3(x - r) = 2(x + p)$

9. $\frac{a}{2}(x - 3w) = z$

10. $cx - x = bc - b$
11. $3x + m = b$
12. $ax + m = cx + n$
13. $ax - bx = c + dx - m$
14. $3m + 2x - c = x + d$
15. $ax - ab = cx - bc$
16. $p(x - b) = qx + d$

17. $3(x - b) = 2(bx + c) - c(x - b)$

18. $a(bx + d) - cdx = c(dx + d)$

Literal Fractional Equations

19. $\dfrac{w + x}{x} = w(w + y)$

20. $\dfrac{ax + b}{c} = bx - a$

21. $\dfrac{p - q}{x} = 3p$

22. $\dfrac{bx - c}{ax - c} = 5$

23. $\dfrac{a - x}{5} = \dfrac{b - x}{2}$

24. $\dfrac{x}{a} - b = \dfrac{c}{d} - x$

25. $\dfrac{x}{a} - a = \dfrac{a}{c} - \dfrac{x}{c - a}$

26. $\dfrac{x}{a - 1} - \dfrac{x}{a + 1} = b$

27. $\dfrac{x - a}{x - b} = \left(\dfrac{2x - a}{2x - b}\right)^2$

28. $\dfrac{a - b}{bx + c} + \dfrac{a + b}{ax - c} = 0$

29. $\dfrac{x + a}{x - a} - \dfrac{x - b}{x + b} = \dfrac{2(a + b)}{x}$

30. $\dfrac{ax}{b} - \dfrac{b - x}{2c} + \dfrac{a(b - x)}{3d} = a$

Word Problems with Literal Solutions

31. Two persons are d miles apart. They set out at the same time and travel toward each other. One travels at the rate of m miles an hour, and the other at the rate of n miles an hour. How far will each have traveled when they meet?

32. If A can do a piece of work in p hours, B in q hours, C in r hours, and D in s hours, how many hours will it take to do the work if all work together?

33. After spending $1/n$ and $1/m$ of my money, I had a dollars left. How many dollars did I have at first?

Formulas

34. The correction C for the sag in a surveyor's tape weighing w lb/ft and pulled with a force of P lb is

$$C = \frac{w^2 L^3}{24 P^2} \text{ feet}$$

Solve this equation for the distance measured, L.

35. When a bar of length L_0 having a coefficient of linear thermal expansion α is increased in temperature by an amount Δt, it will expand to a new length L, where

$$L = L_0(1 + \alpha \Delta t)$$

Solve this equation for Δt.

36. Solve the equation given in problem 35 for the initial length L_0.

37. A rod of cross-sectional area a and length L will stretch by an amount e when subject to a tensile load of P. The modulus of elasticity is given by

$$E = \frac{PL}{ae}$$

Solve this equation for a.

38. The formula for the displacement s of a freely falling body having an initial velocity v_0 and acceleration a is

$$s = v_0 t + \tfrac{1}{2} a t^2$$

Solve this equation for a.

39. The formulas for the amount of heat flowing through a wall by conduction is

$$q = \frac{kA(t_1 - t_2)}{L}$$

Solve this equation for t_2.

40. An amount a invested at a simple interest rate n for t years will accumulate to an amount y, where $y = a + ant$. Solve for a.

41. The formula for the equivalent resistance R for the parallel combination of two resistors, R_1 and R_2, is

$$\frac{1}{R} = \frac{1}{R_1} + \frac{1}{R_2}$$

Solve this formula for R_2.

42. Applying Kirchhoff's law (Eq. A68) to loop 1 in Fig. 9–5, we get

$$E = I_1R_1 + I_1R_2 - I_2R_2$$

Solve for I_1.

43. If the resistance of a conductor is R_1 at temperature t_1, the resistance will change to a value R when the temperature changes to t, where

$$R = R_1[1 + \alpha(t - t_1)]$$

and α is the temperature coefficient of resistance at temperature t_1.

Solve this equation for t_1.

44. Solve the equation given in problem 43 for the initial resistance R_1.

45. Taking the moment M about point p in Fig. 9–6, we get

$$M = R_1L - F(L - x)$$

Solve for L.

46. Three masses, m_1, m_2, and m_3, are attached together and accelerated by means of a force F, where

$$F = m_1a + m_2a + m_3a$$

Solve for the acceleration a.

47. A bar of mass m_1 is attached to a sphere of mass m_2 (Fig. 9–7). The distance x to the center of gravity (C.G.) is

$$x = \frac{10m_1 + 25m_2}{m_1 + m_2}$$

Solve for m_1.

48. A ball of mass m is swung in a vertical circle (Fig. 9–8). At the top of its swing, the tension T in the cord plus the ball's weight mg is just balanced by the centrifugal force mv^2/R.

$$T + mg = \frac{mv^2}{R}$$

Solve for m.

49. The total energy of a body of mass m, moving with velocity v and located at a height y above some datum, is the sum of the potential energy mgy and the kinetic energy $\frac{1}{2}mv^2$.

$$E = mgy + \frac{1}{2}mv^2$$

Solve for m.

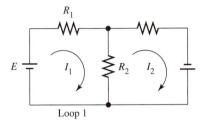

FIGURE 9–5

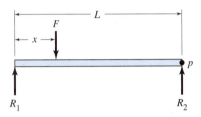

FIGURE 9–6

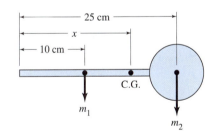

FIGURE 9–7

FIGURE 9–8

Computer

50. A computer algebra system is great for solving literal equations. If you have a CAS, use it to solve any of the equations in this exercise set. In *Derive,* you would highlight the equation to be solved and give the command *Solve.* You then have to tell the computer *which variable* to solve for.

••• CHAPTER 9 REVIEW PROBLEMS ••••••••••••••••••••••••••••••••••••

Perform the indicated operations and simplify.

1. $\dfrac{a^2 + b^2}{a - b} - a + b$

2. $\dfrac{3}{5x^2} - \dfrac{2}{15xy} + \dfrac{1}{6y^2}$

3. $\dfrac{3m + 1}{3m - 1} + \dfrac{m - 4}{5 - 2m} - \dfrac{3m^2 - 2m - 4}{6m^2 - 17m + 5}$

4. $\dfrac{\dfrac{1}{1 - x} - \dfrac{1}{1 + x}}{\dfrac{1}{1 - x^2} - \dfrac{1}{1 + x^2}}$

5. $(a^2 + 1 + a)\left(1 - \dfrac{1}{a} + \dfrac{1}{a^2}\right)$

6. $\dfrac{\dfrac{a - 1}{6} - \dfrac{2a - 7}{2}}{\dfrac{3a}{4} - 3}$

7. $\dfrac{1 + \dfrac{a - c}{a + c}}{1 - \dfrac{a - c}{a + c}}$

8. $\left(1 + \dfrac{x + y}{x - y}\right)\left(1 - \dfrac{x - y}{x + y}\right)$

9. $\dfrac{x^2 + 8x + 15}{x^2 + 7x + 10} - \dfrac{x - 1}{x - 2}$

10. $\dfrac{x^2 - 5ax + 6a^2}{x^2 - 8ax + 15a^2} - \dfrac{x - 7a}{x - 5a}$

11. $\dfrac{1}{a^2 - 7a + 12} + \dfrac{2}{a^2 - 4a + 3} - \dfrac{3}{a^2 - 5a + 4}$

12. $\dfrac{x^4 - y^4}{(x - y)^2} \times \dfrac{x - y}{x^2 + xy}$

13. $\dfrac{a^3 - b^3}{a^3 + b^3} \times \dfrac{a^2 - ab + b^2}{a - b}$

14. $\dfrac{3wx^2y^3}{7axyz} \times \dfrac{4a^3xz}{6aw^2y}$

15. $\dfrac{4a^2 - 9c^2}{4a^2 + 6ac}$

16. $\dfrac{b^2 - 5b}{b^2 - 4b - 5}$

17. $\dfrac{3a^2 + 6a}{a^2 + 4a + 4}$

18. $\dfrac{20(a^3 - c^3)}{4(a^2 + ac + c^2)}$

19. $\dfrac{x^2 - y^2 - 2yz - z^2}{x^2 + 2xy + y^2 - z^2}$

20. $\dfrac{2a^2 + 17a + 21}{3a^2 + 26a + 35}$

21. $\dfrac{(a + b)^2 - (c + d)^2}{(a + c)^2 - (b + d)^2}$

22. $\dfrac{1}{x - 2a} + \dfrac{a^2}{x^3 - 8a^3} - \dfrac{x + a}{x^2 + 2ax + 4a^2}$

Solve for x.

23. $cx - 5 = ax + b$

24. $a(x - 3) - b(x + 2) = c$

25. $\dfrac{5 - 3x}{4} + \dfrac{3 - 5x}{3} = \dfrac{3}{2} - \dfrac{5x}{3}$

26. $\dfrac{3x - 1}{11} - \dfrac{2 - x}{10} = \dfrac{6}{5}$

27. $\dfrac{x + 3}{2} + \dfrac{x + 4}{3} + \dfrac{x + 5}{4} = 16$

28. $\dfrac{2x + 1}{4} - \dfrac{4x - 1}{10} + \dfrac{5}{4} = 0$

29. $x^2 - (x - p)(x + q) = r$

30. $mx - n = \dfrac{nx - m}{p}$

31. $p = \dfrac{q - rx}{px - q}$

32. $m(x - a) + n(x - b) + p(x - c) = 0$

33. $\dfrac{x - 3}{4} - \dfrac{x - 1}{9} = \dfrac{x - 5}{6}$

34. $\dfrac{1}{x - 5} - \dfrac{1}{4} = \dfrac{1}{3}$

35. $\dfrac{2}{x - 4} + \dfrac{5}{2(x - 4)} + \dfrac{9}{2(x - 4)} = \dfrac{1}{2}$

36. $\dfrac{2}{x - 2} = \dfrac{5}{2(x - 1)}$

37. $\dfrac{6}{(2x + 1)} = \dfrac{4}{(x - 1)}$

38. $\dfrac{x + 2}{x - 2} - \dfrac{x - 2}{x + 2} = \dfrac{x + 7}{x^2 - 4}$

39. $\dfrac{10 - 7x}{6 - 7x} = \dfrac{5x - 4}{5x}$

40. $\dfrac{9x + 5}{6(x - 1)} + \dfrac{3x^2 - 51x - 71}{18(x^2 - 1)} = \dfrac{15x - 7}{9(x + 1)}$

41. It is estimated that a bulldozer can prepare a certain site in 15.0 days and that a larger bulldozer can prepare the same site in 11.0 days. If we assume that they do not get in each other's way, how long will it take the two machines, working together, to prepare the site?

42. The acceleration a_1 of a body of mass m_1 subjected to a force F_1 is, by Eq. A20, $a_1 = F_1/m_1$. Write an expression for the difference in acceleration between that body and another having a mass m^2 and force F^2. Combine the terms of that expression and simplify.

43. Multiply: $2\dfrac{3}{5} \times 7\dfrac{5}{9}$

44. Divide: $\dfrac{8x^3y^2z}{5a^4b^3c}$ by $\dfrac{4xy^5z^3}{15abc^3}$

45. Multiply: $\dfrac{3wx^2y^3}{7axyz}$ by $\dfrac{4a^3xz}{6aw^2y}$

46. Divide: $9\dfrac{5}{7} \div 3\dfrac{4}{9}$

47. Multiply: $5\dfrac{2}{7}$ by $2\dfrac{3}{4}$

48. Divide: $3\dfrac{3}{10}$ by $2\dfrac{2}{11}$

Writing

49. Suppose that a vocal member of your local school board says that the study of fractions is no longer important now that we have computers and insists that it be cut from the curriculum to save money.

Write a short letter to the editor of your local paper in which you agree or disagree. Give your reasons for retaining or eliminating the study of fractions.

Team Project

50. The ancient Egyptians wrote each fraction (except for $\frac{2}{3}$) as the sum of unit fractions (a fraction with a numerator of 1). Thus

$$\frac{3}{5} \text{ would be written as } \frac{1}{2} + \frac{1}{10}$$

and

$$\frac{5}{7} \text{ would be written as } \frac{1}{2} + \frac{1}{7} + \frac{1}{14}$$

This is called the *decomposition* of a fraction. Repetitions were not allowed. Thus $\frac{3}{5}$ would not be written as

$$\frac{1}{5} + \frac{1}{5} + \frac{1}{5}$$

Decompose each following proper fraction into unit fractions:

$$\frac{3}{4} \quad \frac{2}{5} \quad \frac{4}{5} \quad \frac{5}{6} \quad \frac{2}{7} \quad \frac{3}{7} \quad \frac{4}{7} \quad \frac{6}{7}$$

$$\frac{3}{8} \quad \frac{5}{8} \quad \frac{7}{8} \quad \frac{2}{9} \quad \frac{4}{9} \quad \frac{5}{9} \quad \frac{7}{9} \quad \frac{8}{9}$$

10

Systems of Linear Equations

OBJECTIVES •••

When you have completed this chapter, you should be able to:

- Find an approximate graphical solution to a system of two equations.
- Solve a system of two equations in two unknowns by the addition-subtraction method.
- Solve a system of two equations by substitution or by a computer technique.
- Solve a system of two equations having fractional coefficients or having the unknowns in the denominators.
- Solve a system of three or more equations by addition-subtraction or by substitution.
- Write a system of equations to describe a given word problem and to solve those equations.

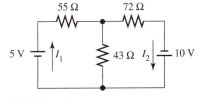

FIGURE 10–1

Usually, a physical situation can be described by a single equation. For example, the displacement of a freely falling body is described by Eq. A18, $s = v_0 t + \frac{1}{2}at^2$. Often, however, a situation can be described only by *more than one* equation. For example, to find the two currents I_1 and I_2 in the circuit of Fig. 10–1, we must solve the two equations

$$98I_1 - 43I_2 = 5$$
$$43I_1 - 115I_2 = 10$$

We must find values for I_1 and I_2 that satisfy *both* equations at the same time.

In this chapter we learn how to solve such *sets* of two or three linear equations, and we also get practice in writing *systems of equations* to describe a variety of technical problems. In Chapter 11 we will learn how to solve systems of equations by means of determinants, and in Chapter 12 we will solve sets of equations using matrices.

10–1   Systems of Two Linear Equations

Linear Equations

We have previously defined a *linear equation* as one of *first degree*.

◆◆◆ **Example 1:** The equation

$$3x + 5 = 20$$

is a linear equation in *one unknown*. We learned how to solve these in Chapter 3.
◆◆◆

◆◆◆ **Example 2:** The equation

$$2x - y = 3$$

is a linear equation in *two unknowns*. A linear equation with the terms written in this order is said to be in *general form*. If we graph this equation, we get a *straight line,* as shown in Fig. 10–2. Glance back at Sec. 5–4 if you've forgotten how to make such a graph.

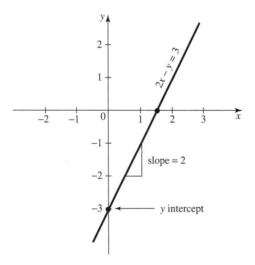

FIGURE 10–2   ◆◆◆

If a linear equation in two unknowns is solved for *y*, we get an equation in *slope-intercept* form, as explained in Sec. 5–4.

◆◆◆ **Example 3:** If we solve the equation from Example 2 for *y*, we get the same equation in slope-intercept form.

$$y = 2x - 3$$

Recall from Sec. 5–4 that the coefficient of the *x* term is the *slope* and the constant term is the *y intercept*. Thus in our example,

$$\text{slope} = 2 \qquad y \text{ intercept} = -3$$

The graph is, of course, the same as in Example 2.   ◆◆◆

A linear equation can have *any number* of unknowns.

◆◆◆ **Example 4:**

(a) $x - 3y + 2z = 5$ is a linear equation in three unknowns.
(b) $3x + 2y - 5z - w = 6$ is a linear equation in four unknowns.   ◆◆◆

Systems of Equations

A set of two or more equations that simultaneously impose conditions on all of the variables is called a *system of equations*.

◆◆◆ **Example 5:**

(a) $3x - 2y = 5$
     $x + 4y = 1$
     is a system of two linear equations in two unknowns.

(b) $x - 2y + 3z = 4$
     $3x + y - 2z = 1$
     $2x + 3y - z = 3$
     is a system of three linear equations in three unknowns.

(c) $2x - y = 5$
     $x - 2z = 3$
     $3y - z = 1$
     is also a system of three linear equations in three unknowns.

(d) $y = 2x^2 - 3$
     $y^2 = x + 5$
     is a system of two quadratic equations in two unknowns.   ◆◆◆

Solution to a System of Equations

The *solution to a system of equations* is a set of values of the unknowns that will *satisfy every equation* in the system.

◆◆◆ **Example 6:** The system of equations

$$x + y = 5$$
$$x - y = 3$$

is satisfied *only* by the values $x = 4$, $y = 1$, and by *no other* set of values. Thus the pair $(4, 1)$ is the solution to the system, and the equations are said to be *independent*.   ◆◆◆

To get a numerical solution for all of the unknowns in a system of linear equations, if one exists, *there must be as many independent equations as there are unknowns*. We first solve two equations in two unknowns; then later, three equations in three unknowns; and then larger systems. But for a solution to be possible, the number of equations must always equal the number of unknowns.

Approximate Graphical Solution to a Pair of Equations

Since any point on a curve satisfies the equation of that curve, the coordinates of the *points of intersection* of two curves will satisfy the equations of *both* curves. Thus we merely have to plot the two curves and find their point or points of intersection; these will be the solution to the pair of equations.

◆◆◆ **Example 7:** Graphically find the approximate solution to the pair of linear equations

$$3x - y = 1$$
$$x + y = 3$$

Solution: We plot the lines as in Chapter 5 and as shown in Fig. 10–3. Note that the lines intersect at the point $(1, 2)$. Our solution is then $x = 1$, $y = 2$.   ◆◆◆

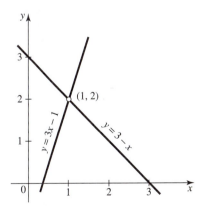

FIGURE 10–3 Graphical solution to a pair of equations. This is also a good way to get an approximate solution to a pair of *nonlinear* equations. See Exercise 2, problem 19.

A *linear* equation in two unknowns plots as a *straight line*. Thus two linear equations plot as two straight lines. If the lines intersect in a single point, the coordinates of that point will satisfy both equations, and hence that point is a solution to the set of equations. If the lines are *parallel,* there is no point whose coordinates satisfy both equations, and the set of equations is said to be *inconsistent*. If the two lines *coincide* at every point, the coordinates of any point on one line will satisfy both equations. Such a set of equations is called *dependent*.

Solution by Graphics Utility

The ability of the graphics calculator or graphing software on the computer to show the graphs of two equations on the screen at the same time makes it ideal for getting an approximate solution to a pair of equations. Further, the ability to magnify any portion of the display by *zooming in* or by changing the range, combined with the ⟨TRACE⟩ feature, enables us to find points of intersection to any desired accuracy.

◆◆◆ **Example 8:** Solve the pair of equations

$$1.53x + 3.35y = 7.62$$
$$2.84x - 1.94y = 4.82$$

to the nearest hundredth, using the graphics calculator.

Solution: The equations must be entered into the calculator in slope-intercept form, so we first solve each for y. Solving the first equation gives

$$3.35y = -1.53x + 7.62$$
$$y = -0.457x + 2.27 \tag{1}$$

From the second equation,

$$1.94y = 2.84x - 4.82$$
$$y = 1.46x - 2.48 \tag{2}$$

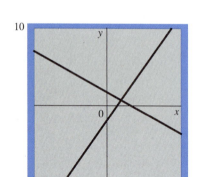

FIGURE 10–4

Next we enter Equations (1) and (2) into the graphics calculator and draw the graph. With both the x and y ranges set from -10 to $+10$, we get the graph shown in Fig. 10–4.

Turning on ⟨TRACE⟩, we move the cursor as close to the point of intersection as we can, and we read the x and y coordinates.

$$x \approx 2.421 \qquad y \approx 1.156$$

However, if we move the cursor one step, we see that the x and y values change more than 0.01 at this range setting. Thus we cannot get the required accuracy here. We must zoom in (see your calculator manual for how to do this) until the cursor step size is smaller than 0.01. This occurs when the x range is from 2 to 3 and the y range is from 0.5 to 2, approximately, depending on the calculator used. We then read

$$x = 2.48 \qquad y = 1.14$$

good to the nearest hundredth.　　　　　　　　　　　　　　　　　　　◆◆◆

Solving a Pair of Linear Equations by the Addition-Subtraction Method

The method of *addition-subtraction,* and the method of *substitution* that follows, both have the object of *eliminating* one of the unknowns.

Some graphics calculators will automatically find the point of intersections of two plotted curves. Consult your user's manual to see if your calculator has this feature.

In Chapter 12 we will use the matrix features of the graphics calculator to solve sets of equations.

In the addition-subtraction method, we eliminate one of the unknowns by first (if necessary) multiplying each equation by such numbers that will make the coefficients of one unknown in both equations equal in absolute value. The two equations are then added or subtracted so as to eliminate that variable.

Let us first use this method for the same system that we solved graphically in Example 7.

◆◆◆ **Example 9:** Solve by the addition-subtraction method:

$$3x - y = 1$$
$$x + y = 3$$

Solution: Simply adding the two equations causes y to drop out.

$$
\begin{aligned}
3x - y &= 1 \\
\underline{x + y} &= \underline{3} \\
4x &= 4 \\
x &= 1
\end{aligned}
$$

Substituting into the second given equation yields

$$1 + y = 3$$
$$y = 2$$

Our solution is then $x = 1$, $y = 2$, as found graphically in Example 7.   ◆◆◆

In the next example we must multiply one equation by a constant before adding.

◆◆◆ **Example 10:** Solve by the addition-subtraction method:

$$2x - 3y = -4$$
$$x + y = 3$$

Solution: Multiply the second equation by 3.

$$
\begin{aligned}
2x - 3y &= -4 \\
\underline{3x + 3y} &= \underline{9} \\
\text{Add:}\quad 5x &= 5
\end{aligned}
$$

We have thus reduced our two original equations to a single equation in one unknown. Solving for x gives

$$x = 1$$

Substituting into the second original equation, we have

$$1 + y = 3$$
$$y = 2$$

So the solution is $x = 1$, $y = 2$.

Check: Substitute into the first original equation.

$$2(1) - 3(2) \overset{?}{=} -4$$
$$2 - 6 = -4 \quad \text{(checks)}$$

Also substitute into the second original equation.

$$1 + 2 = 3 \quad \text{(checks)}$$ ◆◆◆

Often it is necessary to multiply *both* given equations by suitable factors, as shown in the following example.

◆◆◆ **Example 11:** Solve by addition or subtraction:

$$5x - 3y = 19$$
$$7x + 4y = 2$$

Solution: Multiply the first equation by 4 and the second by 3.

$$
\begin{aligned}
20x - 12y &= 76 \\
21x + 12y &= 6 \\
\hline
\text{Add:} \quad 41x \phantom{{}- 12y} &= 82 \\
x &= 2
\end{aligned}
$$

Substituting $x = 2$ into the first equation gives

$$
\begin{aligned}
5(2) - 3y &= 19 \\
3y &= -9 \\
y &= -3
\end{aligned}
$$

So the solution is $x = 2$, $y = -3$.

These values check when substituted into each of the original equations (work not shown). Notice that we could have eliminated the x terms by multiplying the first equation by 7 and the second by -5, and adding. The results, of course, would have been the same: $x = 2$, $y = -3$.   ◆◆◆

The coefficients in the preceding examples were integers, but in applications they will usually be approximate numbers. If so, we must retain the proper number of significant digits, as in the following example.

◆◆◆ **Example 12:** Solve for x and y:

$$2.64x + 8.47y = 3.72$$
$$1.93x + 2.61y = 8.25$$

We are less likely to make a mistake adding rather than subtracting, so it is safer to multiply by a negative number and add the resulting equations, as we do here.

Solution: Let us eliminate y. We multiply the first equation by 2.61 and the second equation by -8.47. Let us carry at least three digits and/or two decimal places in our calculation, and round our answer to three digits at the end.

$$
\begin{aligned}
6.89x + 22.11y &= 9.71 \\
-16.35x - 22.11y &= -69.88 \\
\hline
-9.46x \phantom{{}- 22.11y} &= -60.17 \\
x &= 6.36
\end{aligned}
$$

Another approach is to divide each equation by the coefficient of its x term, thus making each x coefficient equal to 1. Then subtract one equation from the other.

Substituting into the first equation yields

$$
\begin{aligned}
2.64(6.36) + 8.47y &= 3.72 \\
8.47y = 3.72 - 16.79 &= -13.07 \\
y &= -1.54
\end{aligned}
$$
        ◆◆◆

Substitution Method

To use the *substitution method* to solve a pair of linear equations, first solve either original equation for one unknown in terms of the other unknown. Then substitute this expression into the other equation, thereby eliminating one unknown.

◆◆◆ **Example 13:** Solve by substitution:

$$7x - 9y = 1 \tag{1}$$
$$y = 5x - 17 \tag{2}$$

Solution: We substitute $(5x - 17)$ for y in the first equation and get

$$7x - 9(5x - 17) = 1$$
$$7x - 45x + 153 = 1$$
$$-38x = -152$$
$$x = 4$$

Substituting $x = 4$ into Equation (2) gives

$$y = 5(4) - 17$$
$$y = 3$$

So our solution is $x = 4$, $y = 3$. ◆◆◆

If, as in this example, one or both of the given equations are already in explicit form, then the substitution method is probably the easiest. Otherwise, many students find the addition-subtraction method easier.

Systems Having No Solution

Certain systems of equations have no unique solution. If you try to solve either of these types, *both variables will vanish.*

◆◆◆ **Example 14:** Solve the system

$$2x + 3y = 5 \qquad (1)$$
$$6x + 9y = 2 \qquad (2)$$

Solution: Multiply the first equation by 3.

$$6x + 9y = 15$$
$$6x + 9y = \ \ 2$$

Subtract: $\quad 0 = 13 \quad$ (no solution) ◆◆◆

If both variables vanish and an *inequality* results, as in Example 14, the system is called *inconsistent.* The equations would plot as two *parallel lines.* There is no point of intersection and hence no solution.

If both variables vanish and an *equality* results (such as $4 = 4$), the system is called *dependent.* The two equations would plot as a *single line,* indicating that there are infinitely many solutions.

It does not matter much whether a system is inconsistent or dependent; in either case we get no useful solution. But the practical problems that we solve here will always have numerical solutions, so if your variables vanish, go back and check your work.

A Computer Technique: The Gauss-Seidel Method

The *Gauss-Seidel method* is a simple computer technique for solving systems of equations. Suppose that we wish to solve the set of equations

$$x - 2y + 2 = 0 \qquad (1)$$
$$3x - 2y - 6 = 0 \qquad (2)$$

We first solve the first equation for x in terms of y, and the second equation for y in terms of x.

$$x = 2y - 2 \qquad (3)$$
$$y = \frac{3x - 6}{2} \qquad (4)$$

We then *guess* at the value of y, substitute this value into Equation (3), and obtain a value for x. This x is substituted into (4), and a value of y is obtained. These values for x and y will not be the correct values; they are only our *first approximation* to the true values.

The latest value for y is then put into (3), producing a new x; which is then put into (4), producing a new y; which is then put into (3); and so on. We repeat the computation until the values no longer change (we say that they *converge* on the true values). If the values of x and y get very large (*diverge* instead of converge), we solve the first equation for y and the second equation for x, and the computation will converge.

This type of repetitive process is called *iteration,* or the *method of successive approximations.*

◆◆◆ **Example 15:** Calculate x and y for the equations above using the Gauss-Seidel method. Take $y = 0$ for the first guess.

Solution: From Equation (3) we get

$$x = 2(0) - 2 = -2$$

Substituting $x = -2$ into (4) yields

$$y = \frac{3(-2) - 6}{2} = -6$$

With y equal to -6, we then use (3) to find a new x, and so on. We get the following values:

x	y
-2	-6
-14	-24
-50	-78
-158	-240
-482	-726
$-1,454$	$-2,184$
$-4,370$	$-6,558$
$-13,118$	$-19,680$
$-39,362$	$-59,046$
.	.
.	.
.	.

Note that the values are *diverging*. We try again, this time solving Equation (1) for y,

$$y = \frac{x + 2}{2}$$

and solving (2) for x,

$$x = \frac{2y + 6}{3}$$

Repeating the computation, starting with $x = 0$, gives the following values:

x	y
2.666667	1
3.555556	2.333334
3.851852	2.777778
3.950617	2.925926
3.983539	2.975309
3.994513	2.991769
3.998171	2.997256
3.99939	2.999086
3.999797	2.999695
3.999932	2.999899
3.999977	2.999966
3.999992	2.999989
3.999998	2.999996
3.999999	2.999999
4	3
4	3
.	.
.	.
.	.

which converge on $x = 4$ and $y = 3$. ◆◆◆

How do we know what value to take for our first guess? Choose whatever values seem "reasonable." If you have no idea, then choose any values, such as the zeros in this example. Actually, it does not matter much what values we start with. If the computation is going to converge, it will converge regardless of the initial values chosen. However, the closer your guess, the *faster* the convergence.

Exercise 1 ◆ Systems of Two Linear Equations

Graphical Solution

Graphically find the approximate solution to each system of equations. If you have a graphics calculator, use the $\boxed{\text{ZOOM}}$ and $\boxed{\text{TRACE}}$ features to find the solution.

1. $2x - y = 5$
$x - 3y = 5$

2. $x + 2y = -7$
$5x - y = 9$

3. $x - 2y = -3$
$3x + y = 5$

4. $4x + y = 8$
$2x - y = 7$

5. $2x + 5y = 4$
$5x - 2y = -3$

6. $x - 2y + 2 = 0$
$3x - 6y + 2 = 0$

No applications are given here; they are located in Exercise 3.

Algebraic Solution

Solve each system of equations by addition-subtraction, by substitution, or by computer.

7. $2x + y = 11$
$3x - y = 4$

8. $x - y = 7$
$3x + y = 5$

9. $4x + 2y = 3$
$-4x + y = 6$

10. $2x - 3y = 5$
$3x + 3y = 10$

11. $3x - 2y = -15$
$5x + 6y = 3$

12. $7x + 6y = 20$
$2x + 5y = 9$

13. $x + 5y = 11$
$3x + 2y = 7$

14. $4x - 5y = -34$
$2x - 3y = -22$

15. $x = 11 - 4y$
$5x - 2y = 11$

16. $2x - 3y = 3$
$4x + 5y = 39$

17. $7x - 4y = 81$
$5x - 3y = 57$

18. $3x + 4y = 85$
$5x + 4y = 107$

19. $3x - 2y = 1$
$2x + y = 10$

20. $5x - 2y = 3$
$2x + 3y = 5$

21. $y = 9 - 3x$
$x = 8 - 2y$

22. $y = 2x - 3$
$x = 19 - 3y$

23. $29.1x - 47.6y = 42.8$
$11.5x + 72.7y = 25.8$

24. $4.92x - 8.27y = 2.58$
$6.93x + 2.84y = 8.36$

25. $4n = 18 - 3m$
$m = 8 - 2n$

26. $5p + 4q - 14 = 0$
$17p = 31 + 3q$

27. $3w = 13 + 5z$
$4w - 7z - 17 = 0$

28. $3u = 5 + 2v$
$5v + 2u = 16$

29. $3.62x = 11.7 + 4.73y$
$4.95x - 7.15y - 12.8 = 0$

30. $3.03a = 5.16 + 2.11b$
$5.63b + 2.26a = 18.8$

31. $4.17w = 14.7 - 3.72v$
$v = 8.11 - 2.73w$

32. $5.66p + 4.17q - 16.9 = 0$
$13.7p = 32.2 + 3.61q$

Graphics Calculator

33. Use your graphics calculator to solve any of the systems in this exercise set. Work to three significant digits.

34. If your calculator has a $\boxed{\text{TABLE}}$ feature, you can use it to solve a set of two equations. Solve each equation for y, calling the y values Y1 and Y2. Then enter each equation, and display X, Y1, and Y2 in a table. Scroll down the table to find the value of X at which Y1 and Y2 are nearly equal. Then reduce the range and the step size to get a solution to whatever precision you want.

Computer

35. Write a program or use a spreadsheet for the Gauss-Seidel method described in Sec. 10–1. Use it to solve any of the sets of equations in this exercise set.

36. A computer algebra system (CAS) will easily solve a set of equations. First enter the equations in the format required by your system. Then simply highlight the equations and select the *Solve* command. Use this method to solve any of the sets of equations in this exercise set.

10–2   Other Systems of Equations

Systems with Fractional Coefficients

When one or more of the equations in our system have fractional coefficients, simply multiply the entire equation by the LCD, and proceed as before.

> The techniques in this section apply not only to two equations in two unknowns, but to larger systems as well.

◆◆◆ **Example 16:** Solve for x and y:

$$\frac{x}{2} + \frac{y}{3} = \frac{5}{6} \tag{1}$$

$$\frac{x}{4} - \frac{y}{2} = \frac{7}{4} \tag{2}$$

Solution: Multiply Equation (1) by 6 and (2) by 4.

$$3x + 2y = 5 \tag{3}$$

$$x - 2y = 7 \tag{4}$$

Add (3) and (4): $4x \quad\quad = 12$

$$x = 3$$

Having x, we can now substitute back to get y. It is not necessary to substitute back into one of the original equations. We choose instead the easiest place to substitute, such as Equation (4). (However, when *checking,* be sure to substitute your answers into both of the *original* equations.) Substituting $x = 3$ into (4) gives

$$3 - 2y = 7$$
$$2y = 3 - 7 = -4$$
$$y = -2 \qquad\qquad ◆◆◆$$

Fractional Equations with Unknowns in the Denominator

> To save space, we will not show the check in many of these examples.

The same method (multiplying by the LCD) can be used to clear fractions when the unknowns appear in the denominators. Note that such equations are not linear (that is, not of first degree) as were the equations that we have solved so far, but we are able to solve them by the same methods.

> Of course, neither x nor y can equal zero in this system.

◆◆◆ **Example 17:** Solve the system

$$\frac{10}{x} - \frac{9}{2y} = 6 \tag{1}$$

$$\frac{20}{3x} + \frac{15}{y} = 1 \tag{2}$$

Solution: We multiply Equation (1) by $2xy$ and (2) by $3xy$.

$$20y - 9x = 12xy \tag{3}$$

$$20y + 45x = 3xy \tag{4}$$

Subtracting (4) from (3) gives

$$-54x = 9xy$$

$$y = -6$$

Substituting $y = -6$ back into (1), we have

$$\frac{10}{x} - \frac{9}{2(-6)} = 6$$

$$\frac{10}{x} = 6 - \frac{3}{4} = \frac{21}{4}$$

$$\frac{x}{10} = \frac{4}{21}$$

$$x = \frac{40}{21}$$

So our solution is $x = 40/21$, $y = -6$. ◆◆◆

A convenient way to solve nonlinear systems such as these is to *substitute new variables* that will make the equations linear. Solve in the usual way and then substitute back.

◆◆◆ **Example 18:** Solve the nonlinear system of equations

$$\frac{2}{3x} + \frac{3}{5y} = 17 \tag{1}$$

$$\frac{3}{4x} + \frac{2}{3y} = 19 \tag{2}$$

Solution: We substitute $m = 1/x$ and $n = 1/y$ and get the linear system

$$\frac{2m}{3} + \frac{3n}{5} = 17 \tag{3}$$

$$\frac{3m}{4} + \frac{2n}{3} = 19 \tag{4}$$

> This technique of substitution is very useful, and we will use it again later to reduce certain equations to quadratic form. Do not confuse this kind of substitution with the method of substitution that we studied in Sec. 10–1.

Again, we clear fractions by multiplying each equation by its LCD. Multiplying Equation (3) by 15 gives

$$10m + 9n = 255 \tag{5}$$

and multiplying (4) by 12 gives

$$9m + 8n = 228 \tag{6}$$

Using the addition-subtraction method, we multiply (5) by 8 and multiply (6) by -9.

$$80m + 72n = 2040$$
$$\underline{-81m - 72n = -2052}$$
$$\text{Add:} \quad -m = -12$$
$$m = 12$$

Substituting $m = 12$ into (5) yields

$$120 + 9n = 255$$

$$n = 15$$

Finally, we substitute back to get x and y.

$$x = \frac{1}{m} = \frac{1}{12} \quad \text{and} \quad y = \frac{1}{n} = \frac{1}{15}$$ ◆◆◆

Common Error	Students often forget that last step. We are solving not for m and n, but for x and y.

Literal Equations

We use the method of addition-subtraction or the method of substitution for solving systems of equations with literal coefficients, treating the literals as if they were numbers.

◆◆◆ **Example 19:** Solve for x and y in terms of m and n:

$$2mx + ny = 3 \tag{1}$$
$$mx + 3ny = 2 \tag{2}$$

Solution: We will use the addition-subtraction method. Multiply the second equation by -2.

$$
\begin{array}{l}
2mx + ny = 3 \\
-2mx - 6ny = -4 \\
\hline
\text{Add:} \quad -5ny = -1 \\
y = \dfrac{1}{5n}
\end{array}
$$

Substituting back into the second original equation, we obtain

$$mx + 3n\left(\frac{1}{5n}\right) = 2$$

$$mx + \frac{3}{5} = 2$$

$$mx = 2 - \frac{3}{5} = \frac{7}{5}$$

$$x = \frac{7}{5m}$$ ◆◆◆

◆◆◆ **Example 20:** Solve for x and y by the addition-subtraction method:

$$a_1x + b_1y = c_1 \tag{1}$$
$$a_2x + b_2y = c_2 \tag{2}$$

Solution: Multiplying the first equation by b_2 and the second equation by b_1, we obtain

$$
\begin{array}{l}
a_1b_2x + b_1b_2y = b_2c_1 \\
a_2b_1x + b_1b_2y = b_1c_2 \\
\hline
\text{Subtract:} \quad (a_1b_2 - a_2b_1)x = b_2c_1 - b_1c_2
\end{array}
$$

Dividing by $a_1b_2 - a_2b_1$ gives

$$x = \frac{b_2c_1 - b_1c_2}{a_1b_2 - a_2b_1}$$

Now solving for y, we multiply the first equation by a_2 and the second equation by a_1. Writing the second equation above the first, we get

$$a_1 a_2 x + a_1 b_2 y = a_1 c_2$$

$$a_1 a_2 x + a_2 b_1 y = a_2 c_1$$

Subtract: $\overline{(a_1 b_2 - a_2 b_1)y = a_1 c_2 - a_2 c_1}$

Dividing by $a_1 b_2 - a_2 b_1$ yields

$$y = \frac{a_1 c_2 - a_2 c_1}{a_1 b_2 - a_2 b_1}$$

◆◆◆

This result may be summarized as follows:

Two Linear Equations in Two Unknowns	The solution to the set of equations $$a_1 x + b_1 y = c_1$$ $$a_2 x + b_2 y = c_2$$ is $$x = \frac{b_2 c_1 - b_1 c_2}{a_1 b_2 - a_2 b_1} \qquad y = \frac{a_1 c_2 - a_2 c_1}{a_1 b_2 - a_2 b_1}$$ $$(a_1 b_2 - a_2 b_1 \ne 0)$$ 63

Thus Equation 63 is a formula for solving a pair of linear equations. We simply have to identify the six numbers a_1, b_1, c_1, a_2, b_2, and c_2 and substitute them into the equations. But be sure to put the equations into standard form first.

◆◆◆ **Example 21:** Solve using Eq. 63:

$$9y = 7x - 15$$

$$-17 + 5x = 8y$$

Solution: Rewrite the equations in the form given in Eq. 63.

$$7x - 9y = 15$$

$$5x - 8y = 17$$

Then substitute, with

$$a_1 = 7 \qquad b_1 = -9 \qquad c_1 = 15$$

$$a_2 = 5 \qquad b_2 = -8 \qquad c_2 = 17$$

$$x = \frac{b_2 c_1 - b_1 c_2}{a_1 b_2 - a_2 b_1} = \frac{-8(15) - (-9)(17)}{7(-8) - 5(-9)}$$

$$= -3$$

$$y = \frac{a_1 c_2 - a_2 c_1}{a_1 b_2 - a_2 b_1} = \frac{7(17) - (5)(15)}{-11}$$

$$= -4$$

◆◆◆

Exercise 2 ◆ Other Systems of Equations

Fractional Coefficients

Solve simultaneously.

1. $\dfrac{x}{5} + \dfrac{y}{6} = 18$

$\dfrac{x}{2} - \dfrac{y}{4} = 21$

2. $\dfrac{x}{2} + \dfrac{y}{3} = 7$

$\dfrac{x}{3} + \dfrac{y}{4} = 5$

3. $\dfrac{x}{3} + \dfrac{y}{4} = 8$

$x - y = -3$

4. $\dfrac{x}{2} + \dfrac{y}{3} = 5$

$\dfrac{x}{3} + \dfrac{y}{2} = 5$

5. $\dfrac{3x}{5} + \dfrac{2y}{3} = 17$

$\dfrac{2x}{3} + \dfrac{3y}{4} = 19$

6. $\dfrac{x}{7} + 7y = 251$

$\dfrac{y}{7} + 7x = 299$

7. $\dfrac{m}{2} + \dfrac{n}{3} - 3 = 0$

$\dfrac{n}{2} + \dfrac{m}{5} = \dfrac{23}{10}$

8. $\dfrac{p}{6} - \dfrac{q}{3} + \dfrac{1}{3} = 0$

$\dfrac{2p}{3} - \dfrac{3q}{4} - 1 = 0$

9. $\dfrac{r}{6.20} - \dfrac{s}{4.30} = \dfrac{1}{3.10}$

$\dfrac{r}{4.60} - \dfrac{s}{2.30} = \dfrac{1}{3.50}$

Unknowns in the Denominator

Solve simultaneously.

10. $\dfrac{8}{x} + \dfrac{6}{y} = 3$

$\dfrac{6}{x} + \dfrac{15}{y} = 4$

11. $\dfrac{1}{x} + \dfrac{3}{y} = 11$

$\dfrac{5}{x} + \dfrac{4}{y} = 22$

12. $\dfrac{5}{x} + \dfrac{6}{y} = 7$

$\dfrac{7}{x} + \dfrac{9}{y} = 10$

13. $\dfrac{2}{x} + \dfrac{4}{y} = 14$

$\dfrac{6}{x} - \dfrac{2}{y} = 14$

14. $\dfrac{6}{x} + \dfrac{8}{y} = 1$

$\dfrac{7}{x} - \dfrac{11}{y} = -9$

15. $\dfrac{2}{5x} + \dfrac{5}{6y} = 14$

$\dfrac{2}{5x} - \dfrac{3}{4y} = -5$

16. $\dfrac{5}{3a} + \dfrac{2}{5b} = 7$

$\dfrac{7}{6a} - 3 = \dfrac{1}{10b}$

17. $\dfrac{1}{5z} + \dfrac{1}{6w} = 18$

$\dfrac{1}{4w} - \dfrac{1}{2z} + 21 = 0$

18. $\dfrac{8.10}{5.10t} + \dfrac{1.40}{3.60s} = 1.80$

$\dfrac{2.10}{1.40s} - \dfrac{1.30}{5.20t} + 3.10 = 0$

Graphics Calculator

19. We can use a graphics calculator to solve approximately a pair of nonlinear equations just as we solved linear equations in Sec. 10–1. We graph the two equations and use the $\boxed{\text{ZOOM}}$ and $\boxed{\text{TRACE}}$ features to find the point of intersection to any desired accuracy. Remember that you must solve the equations for y before entering them into the calculator. Use the graphics calculator to solve any of the preceding pairs of equations in this exercise set.

Use the graphics calculator to solve the following sets of nonlinear equations.

20. $y = 4.83x + 7.82$
$y = 3.44x^2 + 2.08x$

21. $y = 7.14 - 8.83x^2$
$y = 1.89x^3 + 2.87x$

22. $y = 3.75x^4 + 4.75$
$y = 5.82 - 3.35x$

23. $y = 4.82x - 3.82x^3$
$y = 2.84x + 6.27$

Literal Equations

Solve for x and y in terms of the other literal quantities.

24. $3x - 2y = a$

$2x + \ y = b$

25. $ax + by = r$

$ax + cy = s$

26. $ax - dy = c$

$mx - ny = c$

27. $px - qy + pq = 0$

$2px - 3qy = 0$

Computer

28. Program Eqs. 63 into your computer so that x and y will be computed and printed whenever the six coefficients

$$a_1, \quad b_1, \quad c_1 \quad \text{and} \quad a_2, \quad b_2, \quad c_2$$

are entered into the program. Test the program on a system of equations that you have already solved algebraically.

29. A CAS will solve a pair of equations with literal coefficients, except now you will not get numerical answers but will get expressions containing the literal quantities.

Enter the two literal equations just as you do for equations with numerical coefficients. Now, however, the program will prompt you to enter the solution variables (usually x and y).

Use your CAS to solve any of the literal equations in this exercise set.

10–3   Word Problems with Two Unknowns

Many problems contain two or more unknowns that must be found. To solve such problems, we must write *as many independent equations as there are unknowns*. Otherwise, it is not possible to obtain numerical answers, although one unknown could be expressed in terms of another.

Set up these problems as we did in Chapters 3 and 9, and solve the resulting system of equations by any of the methods of this chapter.

We will start with a financial problem. We did financial problems in one unknown in Sec. 3–3. You may want to flip through that section before starting here.

◆◆◆ **Example 22:** A certain investment in bonds had a value of $248,000 after 4 years, and of $260,000 after 5 years, at simple interest (Fig. 10–5). Find the amount invested and the interest rate. [The formula $y = P(1 + nt)$ gives the amount y obtained by investing an amount P for t years at an interest rate n.]

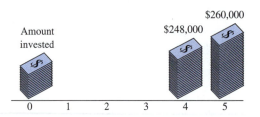

FIGURE 10–5

Estimate: We know that the original investment has to be less than $248,000. We might also reason that the investment cannot be too small, if it is to grow to $248,000 in only 4 years. Also some familiarity with bonds (or a quick inquiry of someone who is familiar with them) reveals that any kind of bond historically has yields somewhere between a few percent and about 15%.

Solution: For the 4-year investment, $t = 4$ years and $y = \$248,000$. Substituting into the given formula yields

$$\$248,000 = P(1 + 4n)$$

Then substitute again, with $t = 5$ years and $y = \$260,000$.

$$\$260,000 = P(1 + 5n)$$

Thus we get two equations in two unknowns: P (the amount invested) and n (the interest rate). Next we remove parentheses.

$$248,000 = P + 4Pn \tag{1}$$

$$260,000 = P + 5Pn \tag{2}$$

At this point we may be tempted to subtract one of these equations from the other to eliminate P. But this will not work because P remains in the term containing Pn. Instead, we multiply Equation (1) by 5 and (2) by -4 to eliminate the Pn term.

$$\begin{aligned} 1,240,000 &= 5P + 20Pn \\ -1,040,000 &= -4P - 20Pn \end{aligned}$$

Add: $\$200,000 = P$

Substituting back into (1) gives us

$$248,000 = 200,000 + 4(200,000)n$$

from which

$$n = 0.06$$

Thus a sum of $200,000 was invested at 6%.

Check: For the 4-year period,

$$y = 200,000[1 + 4(0.06)] = \$248,000 \quad \text{(checks)}$$

and for the 5-year period,

$$y = 200,000[1 + 5(0.06)] = \$260,000 \quad \text{(checks)} \qquad ◆◆◆$$

We'll now do a **mixture** problem. We did mixture problems with one unknown in Sec. 3–4.

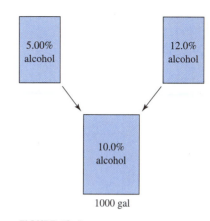

FIGURE 10–6

◆◆◆ **Example 23:** Two different gasohol mixtures are available, one containing 5.00% alcohol, and the other 12.0% alcohol. How much of each, to the nearest gallon, should be mixed to obtain 1000 gal of gasohol containing 10.0% alcohol (Fig. 10–6)?

Estimate: Note that using 500 gallons of each original mixture would give the 1000 gallons needed, but with a percent alcohol midway between 5% and 12%, or around 8.5%. This is lower than needed, so we reason that we need more than 500 gallons of the 12% mixture and less than 500 gallons of the 5% mixture.

Solution: We let

$$x = \text{gal of 5.00\% gasohol needed}$$
$$y = \text{gal of 12.0\% gasohol needed}$$

So

$$x + y = 1000$$

and

$$0.05x + 0.12y = 0.10(1000)$$

are our two equations in two unknowns. Multiplying the first equation by 5 and the second equation by 100, we have

$$5x + 5y = 5{,}000$$
$$5.00x + 12.0y = 10{,}000$$

Subtracting the first equation from the second yields

$$7.0y = 5000$$
$$y = 714 \text{ gal}$$

So

$$x = 1000 - y = 286 \text{ gal} \qquad ◆◆◆$$

Our next application is a work problem. Glance back at Sec. 9–6 to see how we set up work problems in one unknown.

◆◆◆ **Example 24:** During a certain day, two computer printers are observed to process 1705 form letters, with the slower printer in use for 5.5 h and the faster for 4.0 h (Fig. 10–7). On another day the slower printer works for 6.0 h and the faster for 6.5 h, and together they print 2330 form letters. How many letters can each print in an hour, working alone?

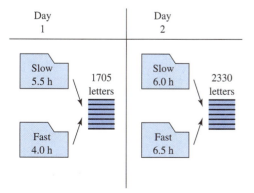

FIGURE 10–7

Estimate: Assume for now that both printers work at the same rate. On the first day they work a total of 9.5 h and print 1705 letters, or $1705 \div 9.5 \approx 180$ letters per hour, and on the second day, $2330 \div 121.5 \approx 186$ letters per hour. But we expect more from the fast printer, say, around 200 letters/h, and less from the slow printer, say, around 150 letters/h.

Solution: We let

$$x = \text{rate of slow printer, letters/h}$$
$$y = \text{rate of fast printer, letters/h}$$

We write two equations to express, for each day, the total amount of work produced, remembering from Chapter 9 that

$$\text{amount of work} = \text{work rate} \times \text{time}$$

On the first day, the slow printer produces $5.5x$ letters while the fast printer produces $4.0y$ letters. Together they produce

$$5.5x + 4.0y = 1705$$

Similarly, for the second day,

$$6.0x + 6.5y = 2330$$

Using the addition-subtraction method, we multiply the first equation by 6.5 and the second by -4.0.

$$
\begin{array}{rl}
35.75x + 26y =& 11{,}083 \\
-24x - 26y =& -9{,}320 \\
\hline
\text{Add:} \quad 11.75x =& 1{,}763 \\
x =& 150 \text{ letters/h}
\end{array}
$$

Substituting back yields

$$5.5(150) + 4y = 1705$$
$$y = 220 \text{ letters/h}$$ ◆◆◆

Our next example will be from statics. Refer to Sec. 3–5 for the basic formulas and to Sec. 7–6 for the resolution of vectors.

◆◆◆ **Example 25:** Find the forces F_1 and F_2 in Fig. 10–8.

Solution: We resolve each vector into its x and y components. We will take components to the right and upward as positive. As in Sec. 7–6, we will arrange the values in a table, for convenience.

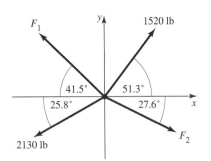

FIGURE 10–8

Force	x component	y component
F_1	$-F_1 \cos 41.5° = -0.749F_1$	$F_1 \sin 41.5° = 0.663F_1$
F_2	$F_2 \cos 27.6° = 0.886F_2$	$-F_2 \sin 27.6° = -0.463F_2$
1520 lb	$1520 \cos 51.3° = 950$	$1520 \sin 51.3° = 1186$
2130 lb	$-2130 \cos 25.8° = -1918$	$-2130 \sin 25.8° = -927$

We then set the sum of the x components to zero,

$$\Sigma F_x = -0.749F_1 + 0.886F_2 + 950 - 1918 = 0$$

or

$$-0.749F_1 + 0.886F_2 = 968 \tag{1}$$

and set the sum of the y components to zero,

$$\Sigma F_y = 0.663F_1 - 0.463F_2 + 1186 - 927 = 0$$

or

$$0.663F_1 - 0.463F_2 = -259 \qquad (2)$$

Now we divide Equation (1) by 0.749 and divide (2) by 0.663 and get

$$
\begin{aligned}
-F_1 + 1.183F_2 &= 1292 \\
F_1 - 0.698F_2 &= -391
\end{aligned} \qquad (3)
$$

Add: $\quad\overline{0.485F_2 = 901}$

$$F_2 = 1860 \text{ lb}$$

Substituting back into (3) gives

$$
\begin{aligned}
F_1 &= 1.183F_2 - 1292 \\
&= 1.183(1860) - 1292 \\
&= 908 \text{ lb}
\end{aligned} \qquad \blacklozenge\blacklozenge\blacklozenge
$$

Exercise 3 ◆ Word Problems with Two Unknowns

Number Problems

1. The sum of two numbers is 24 and their difference is 8. What are the numbers?

2. The sum of two numbers is 29 and their difference is 5. What are the numbers?

3. There is a fraction such that if 3 is added to the numerator, its value will be $\frac{1}{3}$, and if 1 is subtracted from the denominator, its value will be $\frac{1}{5}$. What is the fraction?

4. The sum of the two digits of a certain number is 9. If 9 is added to the original number, the new number will have the original digits reversed. Find the number. (*Hint:* Let x = the tens digit and y = the ones digit, so that the value of the number is $10x + y$.)

5. A number consists of two digits. The number is 2 more than 8 times the sum of the digits, and if 54 is subtracted from the number, the digits will be reversed. Find the number.

6. If the larger of two numbers is divided by the smaller, the quotient is 7 and the remainder is 4. But if 3 times the greater is divided by twice the smaller, the quotient is 11 and the remainder is 4. What are the numbers?

7. The sum of two numbers divided by 2 gives a quotient of 24, and their difference divided by 2 gives a quotient of 17. What are the numbers?

Some of these problems can be set up using just one unknown, but for now use two unknowns for the practice.

Geometry Problems

8. If the width of a certain rectangle is increased by 3 and the length decreased by 3, the area is seen to increase by 6 [Fig. 10–9(a)]. But if the width is reduced by 5 and the length increased by 3, the area decreases by 90 [Fig. 10–9(b)]. Find the original dimensions.

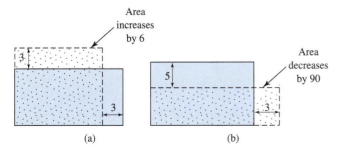

(a)   (b)

FIGURE 10–9

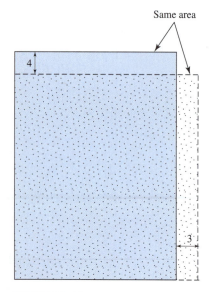

FIGURE 10–10

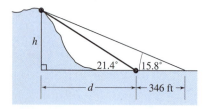

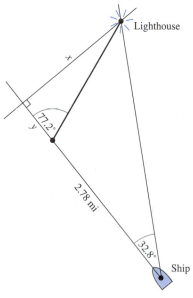

FIGURE 10–11

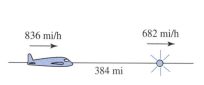

FIGURE 10–12

9. If the width of a certain rectangle is increased by 3 and the length reduced by 4, we get a square with the same area as the original rectangle (Fig. 10–10). Find the length and the width of the original rectangle.

Trigonometry Problems

10. A surveyor measures the angle of elevation of a hill at 15.8° (Fig. 10–11). She then moves 346 ft closer, on level ground, and measures the angle of elevation at 21.4°. Find the height h of the hill and the distance d. (*Hint:* Write the expression for the tangent of the angle of elevation, at each location, and solve the two resulting equations simultaneously.)

11. A ship traveling a straight course sights a lighthouse at an angle of 32.8° (Fig. 10–12). After the ship sails another 2.78 mi, the lighthouse is at an angle of 77.2°. Find the distances x and y.

Uniform Motion Problems

12. In a first race, A gives B a headstart of $10\overline{0}$ ft and overtakes him in 4.00 min. In a second race he gains 750 ft on B when B runs 9000 ft. Find the rate at which each runs.

13. A and B run two races of $28\overline{0}$ ft. In the first race, A gives B a start of 70.0 ft, and neither wins the race. In the second race, A gives B a start of 35.0 ft and beats him by $6\frac{2}{3}$ s. How many feet can each run in a second?

14. A certain river has a speed of 2.50 mi/h. A rower travels downstream for 1.50 h and returns in 4.50 h. Find his rate in still water, and find the one-way distance traveled.

15. A canoeist can paddle 20.0 mi down a certain river and back in 8.0 h (40.0 mi round trip). She can also paddle 5.0 mi down river in the same time as she paddles 3.0 mi up river. Find her rate in still water, and find the rate of the current.

16. A space shuttle and a damaged satellite are 384 mi apart (Fig. 10–13). The shuttle travels at 836 mi/h, and the satellite at 682 mi/h, in the same direction. How long will it take the shuttle to overtake the satellite, and in what distance?

Nonuniform Motion Problems

17. The arm of an industrial robot starts at a speed v_0 and drops 34.8 cm in 4.28 s, at constant acceleration a. Its motion is described by

$$s = v_0 t + a t^2 / 2$$

or

$$34.8 = 4.28 v_0 + a(4.28)^2 / 2$$

In another trial, the arm is found to drop 58.3 cm in 5.57 s, with the same initial speed and acceleration. Find v_0 and a.

18. Crates start at the top of an inclined roller ramp with a speed of v_0 (Fig. 10–14). They roll down with constant acceleration, reaching a speed of 16.3 ft/s after 5.58 s. The motion is described by Eq. A19,

$$v = v_0 + at$$

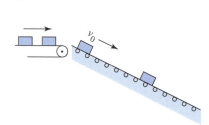

FIGURE 10–13   **FIGURE 10–14**

or

$$16.3 = v_0 + 5.58a$$

The crates are also seen to reach a speed of 18.5 ft/s after 7.03 s. Find v_0 and a.

Financial Problems

19. A certain investment, at simple interest, amounted in 5 years to $3000 and in 6 years to $3100. Find the amount invested, to the nearest dollar, and the rate of interest (see Eq. A9).

20. A shipment of 21 computer keyboards and 33 monitors cost $35,564.25. Another shipment of 41 keyboards and 36 monitors cost $49,172.50. Find the cost of each keyboard and each monitor.

21. A person invested $4400, part of it in railroad bonds bearing 6.2% interest and the remainder in state bonds bearing 9.7% interest (Fig. 10–15), and she received the same income from each. How much, to the nearest dollar, was invested in each?

22. A farmer bought $\overline{100}$ acres of land, part at $370 an acre and part at $450 (Fig. 10–16), paying for the whole $42,200. How much land was there in each part?

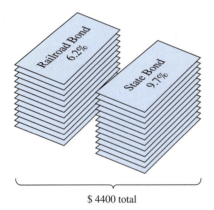

FIGURE 10–15

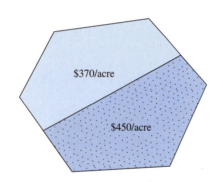

FIGURE 10–16 One hundred acres for $42,200.

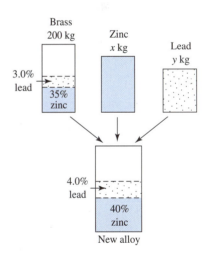

FIGURE 10–17

23. If I lend my money at 6% simple interest for a given time, I shall receive $720 interest; but if I lend it for 3 years longer, I shall receive $1800. Find the amount of money and the time.

Mixture Problems

24. A certain brass alloy contains 35% zinc and 3.0% lead (Fig. 10–17). Then x kg of zinc and y kg of lead are added to 200 kg of the original alloy to make a new alloy that is 40% zinc and 4.0% lead.

    (a) Show that the amount of zinc is given by

$$0.35(200) + x = 0.40(200 + x + y)$$

and that the amount of lead is given by

$$0.03(200) + y = 0.04(200 + x + y)$$

    (b) Solve for x and y.

25. A certain concrete mixture contains 5.00% cement and 8.00% sand (Fig. 10–18). How many pounds of this mixture and how many pounds of sand should be combined with 255 lb of cement to make a batch that is 12.0% cement and 15.0% sand?

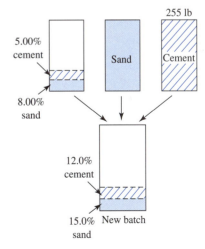

FIGURE 10–18

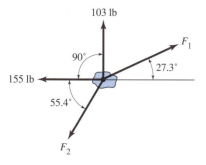

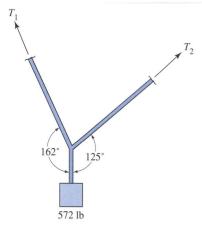

FIGURE 10–19

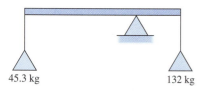

FIGURE 10–20

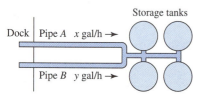

FIGURE 10–21

Storage tanks

Dock | Pipe A x gal/h →

Pipe B y gal/h →

FIGURE 10–22

26. A distributor has two gasohol blends: one that contains 5.00% alcohol and another with 11.0% alcohol. How many gallons of each must be mixed to make 500 gal of gasohol containing 9.50% alcohol?

27. A potting mixture contains 12.0% peat moss and 6.00% vermiculite. How much peat and how much vermiculite must be added to 100 lb of this mixture to produce a new mixture having 15.0% peat and 15.0% vermiculite?

Statics

28. Find the forces F_1 and F_2 in Fig. 10–19.

29. Find the tensions in the ropes in Fig. 10–20.

30. When the 45.3-kg mass in Fig. 10–21 is increased to 100 kg, the balance point shifts 15.4 cm. Find the length of the bar and the original distance from the balance point to the 45.3-kg mass.

Work Problems

31. A carpenter and a helper can do a certain job in 15.0 days. If the carpenter works 1.50 times as fast as the helper, how long would it take each, working alone, to do the job?

32. During one week, two machines produce a total of 27,210 parts, with the faster machine working for 37.5 h and the slower for 28.2 h. During another week, they produce 59,830 parts, with the faster machine working 66.5 h and the slower machine working 88.6 h. How many parts can each, working alone, produce in an hour?

Flow Problems

33. Two different-sized pipes lead from a dockside to a group of oil storage tanks (Fig. 10–22). On one day the two pipes are seen to deliver 117,000 gal, with pipe A in use for 3.5 h and pipe B for 4.5 h. On another day the two pipes deliver 151,200 gal, with pipe A operating for 5.2 h and pipe B for 4.8 h. Assume that pipe A can deliver x gal/h and pipe B can deliver y gal/h.

(a) Show that the total gallons of oil delivered for the two days are given by the equations

$$3.5x + 4.5y = 117,000$$
$$5.2x + 4.8y = 151,200$$

(b) Solve for x and y.

34. Working together, two conveyors can fill a certain bin in 6.00 h. If one conveyor works 1.80 times as fast as the other, how long would it take each to fill the bin working alone?

Energy Flow

35. A hydroelectric generating plant and a coal-fired generating plant together supply a city of 255,000 people, with the hydro plant producing 1.75 times the power of the coal plant. How many people could each service alone?

36. During a certain week, a small wind generator and a small hydro unit together produce 5880 kWh, with the wind generator operating only 85.0% of the time. During another week, the two units produce 6240 kWh, with the wind generator working 95.0% of the time and the hydro unit down 7.50 h for repairs. Assuming that each unit has a constant output when operating, find the number of kilowatts produced by each in 1.00 h.

Electrical

37. To find the currents I_1 and I_2 in Fig. 10–23, we use Eq. A68 in each loop and get the following pair of equations:

$$6.00 - R_1I_1 - R_2I_1 + R_2I_2 = 0$$
$$12.0 - R_3I_2 - R_2I_2 + R_2I_1 = 0$$

Solve for I_1 and I_2 if $R_1 = 736\ \Omega$, $R_2 = 386\ \Omega$, and $R_3 = 375\ \Omega$.

38. Use Eq. A68 to write a pair of equations for the circuit of Fig. 10–24 as in problem 37. Solve these equations for I_1 and I_2.

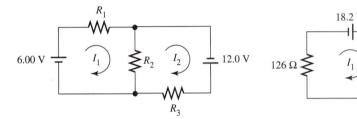

FIGURE 10–23   **FIGURE 10–24**

39. The resistance R of a conductor at temperature t is given by Eq. A70.

$$R = R_1(1 + \alpha t)$$

where R_1 is the temperature at 0°C and α is the temperature coefficient of resistance. A coil of this conductor is found to have a resistance of 31.2 Ω at 25.4°C and a resistance of 35.7 Ω at 57.3°C. Find R_1 and α.

40. The resistance R of two resistors in parallel (Fig. 10–25) is given by Eq. A64.

$$1/R = 1/R_1 + 1/R_2$$

R is found to be 283.0 Ω, but if R_1 is doubled and R_2 halved, then R is found to be 291.0 Ω. Find R_1 and R_2.

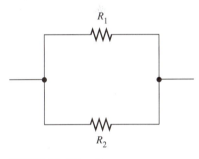

FIGURE 10–25

10–4   Systems of Three Linear Equations

Our strategy here is to reduce a given system of three equations in three unknowns to a system of two equations in two unknowns, which we already know how to solve.

Addition-Subtraction Method

We take any two of the given equations and, by addition-subtraction or substitution, eliminate one variable, obtaining a single equation in two unknowns. We then take another pair of equations (which must include the one not yet used, as well as one of those already used) and similarly obtain a second equation in the *same two* unknowns. This pair of equations can then be solved simultaneously, and the values obtained are substituted back to obtain the third variable.

In the following chapter we will use determinants to solve sets of three or more equations.

◆◆◆ **Example 26:** Solve the following:

$$6x - 4y - \ 7z = 17 \qquad (1)$$
$$9x - 7y - 16z = 29 \qquad (2)$$
$$10x - 5y - \ 3z = 23 \qquad (3)$$

It is a good idea to number your equations, as in this example, to help keep track of your work.

Solution: Let us start by eliminating x from Equations (1) and (2).

$$\begin{array}{llrl}
\text{Multiply Equation (1) by 3:} & 18x - 12y - 21z = & 51 & (4) \\
\text{Multiply (2) by } -2: & -18x + 14y + 32z = & -58 & (5) \\
\text{Add:} & 2y + 11z = & -7 & (6)
\end{array}$$

We now eliminate the same variable, x, from Equations (1) and (3).

$$\begin{array}{llrl}
\text{Multiply (1) by } -5: & -30x + 20y + 35z = & -85 & (7) \\
\text{Multiply (3) by 3:} & 30x - 15y - 9z = & 69 & (8) \\
\text{Add:} & 5y + 26z = & -16 & (9)
\end{array}$$

Now we solve (6) and (9) simultaneously.

$$\begin{array}{llr}
\text{Multiply (6) by 5:} & 10y + 55z = & -35 \\
\text{Multiply (9) by } -2: & -10y - 52z = & 32 \\
\text{Add:} & 3z = & -3 \\
& z = & -1
\end{array}$$

Substituting $z = -1$ into (6) gives us

$$2y + 11(-1) = -7$$
$$y = 2$$

Substituting $y = 2$ and $z = -1$ into (1) yields

$$6x - 4(2) - 7(-1) = 17$$
$$x = 3$$

Our solution is then $x = 3$, $y = 2$, and $z = -1$.

Check: We check a system of three (or more) equations in the same way that we checked a system of two equations: by substituting back into the original equations.

Substituting into (1) gives

$$6(3) - 4(2) - 7(-1) = 17$$
$$18 - 8 + 7 = 17 \quad \text{(checks)}$$

Substituting into (2), we get

$$9(3) - 7(2) - 16(-1) = 29$$
$$27 - 14 + 16 = 29 \quad \text{(checks)}$$

Finally we substitute into (3).

$$10(3) - 5(2) - 3(-1) = 23$$
$$30 - 10 + 3 = 23 \quad \text{(checks)} \qquad \blacklozenge\blacklozenge\blacklozenge$$

Substitution Method

A *sparse* system (one in which many terms are missing) is often best solved by substitution, as in the following example.

◆◆◆ **Example 27:** Solve by substituting:

$$\begin{array}{rl}
x - y = & 4 \qquad (1) \\
x + z = & 8 \qquad (2) \\
x - y + z = & 10 \qquad (3)
\end{array}$$

Solution: From the first two equations, we can write both y and z in terms of x.

$$y = x - 4$$

and

$$z = 8 - x$$

Substituting these back into the third equation yields

$$x - (x - 4) + (8 - x) = 10$$

from which

$$x = 2$$

Substituting back gives

$$y = x - 4 = 2 - 4 = -2$$

and

$$z = 8 - x = 8 - 2 = 6$$

Our solution is then $x = 2$, $y = -2$, and $z = 6$.　　　◆◆◆

Fractional Equations

We use the same techniques for solving a set of three fractional equations that we did for two equations:

1. Multiply each equation by its LCD to eliminate fractions.
2. If the unknowns are in the denominators, substitute new variables that are the reciprocals of the originals.

◆◆◆ **Example 28:** Solve for x, y, and z:

$$\frac{4}{x} + \frac{9}{y} - \frac{8}{z} = 3 \tag{1}$$

$$\frac{8}{x} - \frac{6}{y} + \frac{4}{z} = 3 \tag{2}$$

$$\frac{5}{3x} + \frac{7}{2y} - \frac{2}{z} = \frac{3}{2} \tag{3}$$

Solution: We make the substitution

$$p = \frac{1}{x} \qquad q = \frac{1}{y} \quad \text{and} \quad r = \frac{1}{z}$$

and also multiply Equation (3) by its LCD, 6, to clear fractions.

$$4p + 9q - 8r = 3 \tag{4}$$
$$8p - 6q + 4r = 3 \tag{5}$$
$$10p + 21q - 12r = 9 \tag{6}$$

We multiply (5) by 2 and add it to (4), eliminating the r terms.

$$20p - 3q = 9 \tag{7}$$

Then we multiply (5) by 3 and add it to (6).

$$34p + 3q = 18 \tag{8}$$

Adding (7) and (8) gives

$$54p = 27$$

$$p = \frac{1}{2}$$

Then from (8),

$$3q = 18 - 17 = 1$$

$$q = \frac{1}{3}$$

and from (5),

$$4r = 3 - 4 + 2 = 1$$

$$r = \frac{1}{4}$$

Returning to our original variables, we have $x = 2$, $y = 3$, and $z = 4$.   ◆◆◆

Literal Equations

The same techniques apply to literal equations, as shown in the following example.

◆◆◆ **Example 29:** Solve for x, y, and z:

$$x + 3y = 2c \tag{1}$$
$$y + 2z = a \tag{2}$$
$$2z - x = 3 \tag{3}$$

Solution: We first notice that adding Equations (1) and (3) will eliminate x.

$$\begin{array}{lr} x + 3y = 2c & (1) \\ \underline{-x + 2z = + 3} & (3) \\ \text{Add (1) + (3):} \quad 3y + 2z = 2c + 3 & (1) + (3) \end{array}$$

From this equation we subtract (2), eliminating z:

$$\begin{array}{lr} 3y + 2z = 2c + 3 & (1) + (3) \\ \underline{y + 2z = + a} & (2) \\ \text{Subtract (2):} \quad 2y = 2c + 3 - a \end{array}$$

from which we obtain

$$y = \frac{2c + 3 - a}{2}$$

Substituting back into (1), we have

$$x = 2c - 3\left(\frac{2c + 3 - a}{2}\right)$$

which simplifies to

$$x = \frac{3a - 2c - 9}{2}$$

Substituting this expression for x back into (3) gives

$$2z = 3 + \left(\frac{3a - 2c - 9}{2}\right)$$

which simplifies to

$$z = \frac{3a - 2c - 3}{4} \qquad \blacklozenge\blacklozenge\blacklozenge$$

♦♦♦ **Example 30:** Find the forces F_1, F_2, and F_3 in Fig. 10–26.

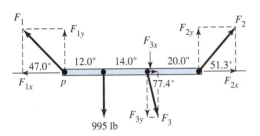

FIGURE 10–26

Solution: We resolve each force into its x and y components (see Sec. 7–6 to refresh your memory). By Eq. A13, the sum of the vertical forces must be zero, so

$$F_{1y} + F_{2y} - F_{3y} - 995 = 0 \qquad (1)$$

By Eq. A14, the sum of the horizontal forces must be zero, giving

$$F_{1x} - F_{2x} - F_{3x} = 0 \qquad (2)$$

By Eq. A15, the sum of the moments about any point, p in this example, must be zero, so we get

$$12.0(995) + 26.0F_{3y} - 46.0F_{2y} = 0 \qquad (3)$$

Using the trigonometric functions, we change these equations as follows:

$$F_1 \sin 47.0° + F_2 \sin 51.3° - F_3 \sin 77.4° - 995 = 0 \qquad (4)$$
$$F_1 \cos 47.0° - F_2 \cos 51.3° - F_3 \cos 77.4° = 0 \qquad (5)$$
$$12.0(995) + 26.0 F_3 \sin 77.4° - 46.0 F_2 \sin 51.3° = 0 \qquad (6)$$

After evaluating the trigonometric functions and simplifying, we have

$$0.731F_1 + 0.780F_2 - 0.976F_3 = 995 \qquad (7)$$
$$0.682F_1 - 0.625F_2 - 0.218F_3 = 0 \qquad (8)$$
$$-35.9F_2 + 25.4F_3 = -11{,}940 \qquad (9)$$

Solving Equation (9) for F_3 gives

$$F_3 = 1.41F_2 - 471 \qquad (10)$$

Substituting this into (7) and (8) and simplifying, we get

$$F_1 - 0.818F_2 = \quad 733 \qquad (11)$$
$$F_1 - 1.369F_2 = -150$$

Subtracting, we get $0.551F_2 = 883$, from which

$$F_2 = 16\overline{0}0 \text{ lb}$$

Substituting back into (11) gives

$$F_1 = 733 + 0.818(1600) = 2040 \text{ lb}$$

and substituting into (10) we get

$$F_3 = 1.41(1600) - 471 = 1790 \text{ lb} \qquad \blacklozenge\blacklozenge\blacklozenge$$

Exercise 4 ◆ Systems of Three Linear Equations

Solve each system of equations.

1. $x + y = 35$
$x + z = 40$
$y + z = 45$

2. $x + y + z = 12$
$x - y = 2$
$x - z = 4$

3. $3x + y = 5$
$2y - 3z = -5$
$x + 2z = 7$

4. $x - y = 5$
$y - z = -6$
$2x - z = 2$

5. $x + y + z = 18$
$x - y + z = 6$
$x + y - z = 4$

6. $x + y + z = 90$
$2x - 3y = -20$
$2x + 3z = 145$

7. $x + 2y + 3z = 14$
$2x + y + 2z = 10$
$3x + 4y - 3z = 2$

8. $x + y + z = 35$
$x - 2y + 3z = 15$
$y - x + z = -5$

9. $x - 2y + 2z = 5$
$5x + 3y + 6z = 57$
$x + 2y + 2z = 21$

10. $1.21x + 1.48y + 1.63z = 6.83$
$4.94x + 4.27y + 3.63z = 21.7$
$2.88x + 4.15y - 2.79z = 2.76$

11. $2.51x - 4.48y + 3.13z = 10.8$
$2.84x + 1.37y - 1.66z = 6.27$
$1.58x + 2.85y - 1.19z = 27.3$

12. $5a + b - 4c = -5$
$3a - 5b - 6c = -20$
$a - 3b + 8c = -27$

13. $p + 3q - r = 10$
$5p - 2q + 2r = 6$
$3p + 2q + r = 13$

Fractional Equations

14. $x + \dfrac{y}{3} = 5$

$x + \dfrac{z}{3} = 6$

$y + \dfrac{z}{3} = 9$

15. $\dfrac{1}{x} + \dfrac{1}{y} = 5$

$\dfrac{1}{y} + \dfrac{1}{z} = 7$

$\dfrac{1}{x} + \dfrac{1}{z} = 6$

16. $\dfrac{x}{10} + \dfrac{y}{5} + \dfrac{z}{20} = \dfrac{1}{4}$

$x + y + z = 6$

$\dfrac{x}{3} + \dfrac{y}{2} + \dfrac{z}{6} = 1$

17. $\dfrac{1}{x} + \dfrac{2}{y} - \dfrac{1}{z} = -3$

$\dfrac{3}{x} + \dfrac{1}{y} + \dfrac{1}{z} = 4$

$\dfrac{1}{x} - \dfrac{1}{y} + \dfrac{2}{z} = 6$

Literal Equations

Solve for x, y, and z.

18. $x - y = a$
$y + z = 3a$
$5z - x = 2a$

19. $x + a = y + z$
$y + a = 2x + 2z$
$z + a = 3x + 3y$

20. $ax + by = (a + b)c$
$by + cz = (c + a)b$
$ax + cz = (b + c)a$

21. $x + y + 2z = 2(b + c)$
$x + 2y + z = 2(a + c)$
$2x + y + z = 2(a + b)$

22. $a_1x + b_1y + c_1z = k_1$
$a_2x + b_2y + c_2z = k_2$
$a_3x + b_3y + c_3z = k_3$

Solving problem 22 gives us formulas for solving a set of three equations. We'll find them useful in the next chapter.

Number Problems

23. The sum of the digits in a certain three-digit number is 21. The sum of the first and third digits is twice the middle digit. If the hundreds and tens digits are interchanged, the number is reduced by 90. Find the number.

24. Two fractions have the same denominator. If 1 is subtracted from the numerator of the smaller fraction, its value will be $\frac{1}{3}$ of the larger fraction; but if 1 is subtracted from the numerator of the larger, its value will be twice that of the smaller fraction. The difference between the fractions is $\frac{1}{3}$. What are the fractions?

25. A certain number is expressed by three digits whose sum is 10. The sum of the first and last digits is $\frac{2}{3}$ of the second digit; and if 198 is subtracted from the number, the digits will be reversed. What is the number?

Electrical Problems

26. When writing Kirchhoff's law (Eq. A68) for a certain three-loop network, we get the set of equations

$$3I_1 + 2I_2 - 4I_3 = 4$$
$$I_1 - 3I_2 + 2I_3 = -5$$
$$2I_1 + I_2 - I_3 = 3$$

where I_1, I_2, and I_3 are the loop currents in amperes. Solve for these currents.

27. For the three-loop network of Fig. 10–27:
   (a) Use Kirchhoff's law to show that the currents may be found from

$$159I_1 - 50.9I_2 \qquad\qquad = 1$$
$$407I_1 - 2400I_2 + 370I_3 = 1$$
$$142I_2 - 380I_3 = 1$$

   (b) Solve this set of equations for the three currents.

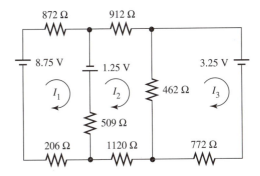

FIGURE 10–27

Statics Problems

28. Find the forces F_1, F_2, and F_3 in Fig. 10–28.

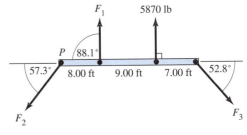

FIGURE 10–28

29. Find the forces F_1, F_2, and F_3 in Fig. 10–29.

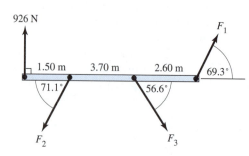

FIGURE 10–29

Computer

30. The Gauss-Seidel method described in Sec. 10–1 works for any number of equations. For three equations in three unknowns, for example, write the equations in the form

$$x = f(y, z)$$
$$y = g(x, z)$$
$$z = h(x, y)$$

and proceed as before. Try it on one of the systems given in this exercise set. If the computation diverges, then solve each equation for a different variable than before, and try again. It may take two or three tries to find a combination that works.

31. We used a CAS to solve a set of two equations in problems 35 and 36 in Exercise 1. Such programs can, of course, solve larger systems. The equations are entered in the same way.

   Use a CAS to solve any of the equations in this exercise set.

◆◆◆ CHAPTER 10 REVIEW PROBLEMS ◆◆◆◆◆◆◆◆◆◆◆◆◆◆◆◆◆◆◆◆◆◆◆◆◆◆◆◆◆◆

Solve each system of equations by any method.

1. $4x + 3y = 27$
   $2x - 5y = -19$

2. $\dfrac{x}{3} + \dfrac{y}{2} = \dfrac{4}{3}$

   $\dfrac{x}{2} + \dfrac{y}{3} = \dfrac{7}{6}$

3. $\dfrac{15}{x} + \dfrac{4}{y} = 1$

   $\dfrac{5}{x} - \dfrac{12}{y} = 7$

4. $2x + 4y - 3z = 22$
   $4x - 2y + 5z = 18$
   $6x + 7y - z = 63$

5. $5x + 3y - 2z = 5$
   $3x - 4y + 3z = 13$
   $x + 6y - 4z = -8$

6. $3x - 5y - 2z = 14$
   $5x - 8y - z = 12$
   $x - 3y - 3z = 1$

7. $\dfrac{3}{x+y} + \dfrac{4}{x-z} = 2$

$\dfrac{6}{x+y} + \dfrac{5}{y-z} = 1$

$\dfrac{4}{x-z} + \dfrac{5}{y-z} = 2$

8. $4x + 2y - 26 = 0$
$3x + 4y = 39$

9. $2x - 3y + 14 = 0$
$3x + 2y = 44$

10. $\dfrac{2}{x} + \dfrac{1}{y} = \dfrac{4}{3}$

$\dfrac{3}{x} + \dfrac{5}{y} = \dfrac{19}{6}$

11. $\dfrac{9}{x} + \dfrac{8}{y} = \dfrac{43}{6}$

$\dfrac{3}{x} + \dfrac{10}{y} = \dfrac{29}{6}$

12. $\dfrac{x}{a} + \dfrac{y}{b} = p$

$\dfrac{x}{b} + \dfrac{y}{a} = q$

13. $x + y = a$
$x + z = b$
$y + z = c$

14. $\dfrac{5x}{6} + \dfrac{2y}{5} = 14$

$\dfrac{3x}{4} - \dfrac{2y}{5} = 5$

15. $\dfrac{2x}{7} + \dfrac{2y}{3} = \dfrac{16}{3}$

$x + y = 12$

16. $x + y + z = 35$
$5x + 4y + 3z = 22$
$3x + 4y - 3z = 2$

17. $2x - 4y + 3z = 10$
$3x + y - 2z = 6$
$x - 3y - z = 20$

18. $5x + y - 4z = -5$
$3x - 5y - 6z = -20$
$x - 3y + 8z = -27$

19. $x + 3y - z = 10$
$5x - 2y + 2z = 6$
$3x + 2y + z = 13$

20. $x + 21y = 2$
$2x + 27y = 19$

21. $2x - y = 9$
$5x - 3y = 14$

22. $6x - 2y + 5z = 53$
$5x + 3y + 7z = 33$
$x + y + z = 5$

23. $9x + 4y = 54$
$4x + 9y = 89$

24. A sum of money was divided between A and B so that A's share was to B's share as 5 is to 3. Also, A's share exceeded $\frac{5}{9}$ of the whole sum by \$50. What was each share?

25. A and B together can do a job in 12 days, and B and C together can do the same job in 16 days. How long would it take them all working together to do the job if A does $1\frac{1}{2}$ times as much as C?

26. If the numerator of a certain fraction is increased by 2 and its denominator diminished by 2, its value will be 1. If the numerator is increased by the denominator and the denominator is diminished by 5, its value will be 5. Find the fraction.

Writing

27. Suppose that you have a number of pairs of equations to solve in order to find the loop currents in a circuit that you are designing. Each pair of equations has different values of the variables. You must leave for a week but want your assistant to solve the equations in your absence. Write step-by-step instructions to be followed, using the addition-subtraction method, including instructions as to the number of digits to be retained.

Team Project

28. A certain nickel silver alloy contains the following:

Copper	55.9%	Lead	0.10%
Zinc	31.25%	Nickel	12.00%
Tin	0.50%	Manganese	0.25%

How many pounds of zinc, tin, lead, nickel, and manganese must be added to 400 lb of this alloy to make a new "leaded nickel silver" with the following composition:

Copper	44.50%	Lead	1.00%
Zinc	42.00%	Nickel	10.00%
Tin	0.50%	Manganese	2.00%

11

Determinants

OBJECTIVES

When you have completed this chapter, you should be able to:

- Evaluate second-order determinants.
- Evaluate third-order determinants by minors.
- Use the properties of determinants to reduce the order of a determinant.
- Evaluate determinants by a systematic method suitable for the computer.
- Solve a system of any number of linear equations by determinants.
- Evaluate a determinant using a calculator or a computer.

In Chapter 10, we learned how to solve sets of two or three equations by the graphing, addition-subtraction, and substitution methods, and by computer using the Gauss-Seidel method. These methods, however, become unwieldy when the number of equations to be solved is greater than three.

In this chapter we introduce the idea of a *determinant*. We learn how to evaluate determinants and then how to use them to solve a set of equations. We start with sets of two equations and then expand our use of determinants so that we can solve systems with any number of equations.

We show how to solve these systems by hand, using determinants, as well as with a calculator or a computer. That way, you should obtain both an understanding of the mathematics, as well as fast and practical tools for solving such sets.

11–1   Second-Order Determinants

Definitions

In Chapter 10, we studied the set of equations

$$a_1x + b_1y = c_1$$
$$a_2x + b_2y = c_2$$

Now let us refer to the values

$$\begin{vmatrix} a_1 \\ a_2 \end{vmatrix} \quad \text{as the column of } x \text{ coefficients,}$$

$$\begin{vmatrix} b_1 \\ b_2 \end{vmatrix} \quad \text{as the column of } y \text{ coefficients, and}$$

$$\begin{vmatrix} c_1 \\ c_2 \end{vmatrix} \quad \text{as the column of constants.}$$

When we solved this set of equations, we got the solution

$$x = \frac{b_2c_1 - b_1c_2}{a_1b_2 - a_2b_1} \quad \text{and} \quad y = \frac{a_1c_2 - a_2c_1}{a_1b_2 - a_2b_1}$$

> It is common practice to use the same word *determinant* to refer to the *symbol*
> $$\begin{vmatrix} a_1 & b_1 \\ a_2 & b_2 \end{vmatrix}$$
> as well as to the value $(a_1b_2 - a_2b_1)$. We'll follow this practice.

The denominator $(a_1b_2 - a_2b_1)$ of each of these fractions is called the *determinant* of the coefficients a_1, b_1, a_2, and b_2. It is commonly expressed by the symbol

$$\begin{vmatrix} a_1 & b_1 \\ a_2 & b_2 \end{vmatrix}$$

with it being understood that this signifies the product of the upper-left and lower-right numbers, minus the product of the upper-right and lower-left numbers. Thus

$$\begin{vmatrix} a_1 & b_1 \\ a_2 & b_2 \end{vmatrix} = a_1b_2 - a_2b_1$$

◆◆◆ **Example 1:** In the determinant

$$\begin{vmatrix} 2 & 5 \\ 6 & 1 \end{vmatrix}$$

each of the numbers 2, 5, 6, and 1 is called an *element*. There are two *rows,* the first row containing the elements 2 and 5, and the second row having the elements 6 and 1. There are also two *columns,* the first with elements 2 and 6, and the second with elements 5 and 1. The *value* of this determinant is

$$2(1) - 5(6) = 2 - 30 = -28$$

◆◆◆

Thus a determinant is written as a *square array,* enclosed between vertical bars. The number of rows equals the number of columns. The *order* of a determinant is equal to the number of rows or columns that it contains. Thus the determinants above are of second order.

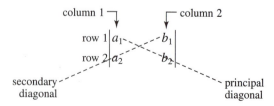

The diagonal running from upper left to lower right is called the *principal diagonal.* The other is called the *secondary diagonal.*

Second-Order Determinant	$\begin{vmatrix} a_1 & b_1 \\ a_2 & b_2 \end{vmatrix} = a_1b_2 - a_2b_1$	65

The value of a second-order determinant is equal to the product of the elements on the principal diagonal minus the product of the elements on the secondary diagonal.

••• Example 2:

$$\begin{vmatrix} 2 & -1 \\ 3 & 5 \end{vmatrix} = 2(5) - 3(-1) = 13$$

•••

Common Errors	Students sometimes add the numbers on a diagonal instead of *multiplying*. Be careful not to do that.
	Don't just ignore a zero element. It causes the product along its diagonal to be zero.

••• Example 3:

$$\begin{vmatrix} 4 & a \\ x & 3b \end{vmatrix} = 4(3b) - x(a) = 12b - ax$$

•••

••• Example 4:

$$\begin{vmatrix} 0 & 3 \\ 5 & 2 \end{vmatrix} = 0(2) - 3(5) = -15$$

•••

Evaluating Determinants by Calculator

Most graphics calculators can evaluate determinants if all of the elements are numbers.

••• Example 5: Evaluate the following determinant using a graphics calculator:

$$\begin{vmatrix} 3 & 2 \\ 5 & 1 \end{vmatrix}$$

Solution: Locate the matrix editor on your calculator. Then follow the instructions for entering each element in the proper order. Finally, choose the command to evaluate the determinant, which is probably labeled $\boxed{\text{DET}}$. The value of the determinant will then be displayed, in this case, a value of -7.   •••

Unfortunately, calculators and computer algebra systems vary so widely that we cannot give more specific instructions than these. As before, you must consult your user's manual.

Evaluating Determinants on a Computer Algebra System (CAS)

A computer algebra system (CAS) can evaluate determinants, even when the elements contain *algebraic symbols*.

••• Example 6: Evaluate the following determinant using a CAS:

$$\begin{vmatrix} x & 3 \\ 2y & w \end{vmatrix}$$

Solution: We locate the matrix editor on our CAS and enter 2 for the number of rows and of columns where requested. We then enter the four elements, x, 3, $2y$, and w, as directed. Following the instructions for evaluating the determinant of this matrix gives

$$wx - 6y$$

 •••

Solving a System of Equations by Determinants

We now return to the set of equations

$$a_1 x + b_1 y = c_1 \tag{1}$$
$$a_2 x + b_2 y = c_2$$

If, in the solution for x,

$$x = \frac{b_2 c_1 - b_1 c_2}{a_1 b_2 - a_2 b_1} \tag{2}$$

we replace the denominator by the square array

This denominator is called the *determinant of the coefficients*, or sometimes the *determinant of the system*. It is usually given the Greek capital letter delta (Δ). If this determinant equals zero, there is no solution to the set of equations.

$$\begin{vmatrix} a_1 & b_1 \\ a_2 & b_2 \end{vmatrix} = a_1 b_2 - a_2 b_1$$

we get

$$x = \frac{b_2 c_1 - b_1 c_2}{\begin{vmatrix} a_1 & b_1 \\ a_2 & b_2 \end{vmatrix}} \tag{3}$$

Now look at the numerator of Equation (3), which is $b_2 c_1 - b_1 c_2$. If this expression were the value of some determinant, it is clear that the elements on the principal diagonal must be b_2 and c_1 and that the elements on the secondary diagonal must be b_1 and c_2. One such determinant is

column of constants

$$\begin{vmatrix} c_1 & b_1 \\ c_2 & b_2 \end{vmatrix} = b_2 c_1 - b_1 c_2$$

Our solution for x can then be expressed as

$$x = \frac{\begin{vmatrix} c_1 & b_1 \\ c_2 & b_2 \end{vmatrix}}{\begin{vmatrix} a_1 & b_1 \\ a_2 & b_2 \end{vmatrix}}$$

An expression for y can be developed in a similar way. Thus the solution to the system of Equations (1), in terms of determinants, is given by the following rule:

Named after the Swiss mathematician Gabriel Cramer (1704–52), Cramer's rule, as we will see, is valid for systems with any number of linear equations.

| Cramer's Rule | $x = \dfrac{\begin{vmatrix} c_1 & b_1 \\ c_2 & b_2 \end{vmatrix}}{\begin{vmatrix} a_1 & b_1 \\ a_2 & b_2 \end{vmatrix}}$ and $y = \dfrac{\begin{vmatrix} a_1 & c_1 \\ a_2 & c_2 \end{vmatrix}}{\begin{vmatrix} a_1 & b_1 \\ a_2 & b_2 \end{vmatrix}}$ | 70 |

The solution for any variable is a fraction whose denominator is the determinant of the coefficients and whose numerator is that same determinant, except that the column of coefficients for the variable for which we are solving is replaced by the column of constants.

Of course, the denominator Δ cannot be zero, or we get division by zero, indicating that the set of equations has no unique solution.

◆◆◆ **Example 7:** Solve by determinants:

$$2x - 3y = 1$$
$$x + 4y = -5$$

Solution: We first evaluate the determinant of the coefficients, because if $\Delta = 0$, there is no unique solution, and we have no need to proceed further.

$$\Delta = \begin{vmatrix} 2 & -3 \\ 1 & 4 \end{vmatrix} = 2(4) - 1(-3) = 8 + 3 = 11$$

Solving for x, we have

column of constants

$$x = \frac{\begin{vmatrix} 1 & -3 \\ -5 & 4 \end{vmatrix}}{\Delta} = \frac{1(4) - (-5)(-3)}{11} = \frac{4 - 15}{11} = -1$$

Solving for y yields

column of constants

$$y = \frac{\begin{vmatrix} 2 & 1 \\ 1 & -5 \end{vmatrix}}{\Delta} = \frac{2(-5) - 1(1)}{11} = \frac{-10 - 1}{11} = -1$$

It's easier to find y by substituting x back into one of the previous equations, but we use determinants instead, to show how it is done.

We would, of course, check our answers by substituting back into the original equations, as we did in Chapter 10.

Common Error

Don't forget to first arrange the given equations into the form

$$a_1x + b_1y = c_1$$
$$a_2x + b_2y = c_2$$

before writing the determinants.

◆◆◆ **Example 8:** Solve for x and y by determinants:

$$2ax - by + 3a = 0$$
$$4y = 5x - a$$

Solution: We first rearrange our equations.

$$2ax - by = -3a$$
$$5x - 4y = \quad a$$

The determinant of the coefficients is

$$\Delta = \begin{vmatrix} 2a & -b \\ 5 & -4 \end{vmatrix} = 2a(-4) - 5(-b) = -8a + 5b$$

So

$$x = \frac{\begin{vmatrix} -3a & -b \\ a & -4 \end{vmatrix}}{\Delta} = \frac{(-3a)(-4) - a(-b)}{5b - 8a}$$
$$= \frac{12a + ab}{5b - 8a}$$

and

$$y = \frac{\begin{vmatrix} 2a & -3a \\ 5 & a \end{vmatrix}}{\Delta} = \frac{2a(a) - 5(-3a)}{5b - 8a}$$
$$= \frac{2a^2 + 15a}{5b - 8a}$$

◆◆◆

Exercise 1 ◆ Second-Order Determinants

Find the value of each second-order determinant.

1. $\begin{vmatrix} 3 & 2 \\ 1 & -4 \end{vmatrix}$

2. $\begin{vmatrix} -2 & 5 \\ 3 & -3 \end{vmatrix}$

3. $\begin{vmatrix} 0 & 5 \\ -3 & -4 \end{vmatrix}$

4. $\begin{vmatrix} 8 & 3 \\ -1 & 2 \end{vmatrix}$

5. $\begin{vmatrix} -4 & 5 \\ 7 & -2 \end{vmatrix}$

6. $\begin{vmatrix} 7 & -6 \\ -5 & 4 \end{vmatrix}$

7. $\begin{vmatrix} 4.82 & 2.73 \\ 2.97 & 5.28 \end{vmatrix}$

8. $\begin{vmatrix} 48.7 & -53.6 \\ 4.93 & 9.27 \end{vmatrix}$

9. $\begin{vmatrix} -2/3 & 2/5 \\ -1/3 & 4/5 \end{vmatrix}$

10. $\begin{vmatrix} 1/2 & 1/4 \\ -1/2 & -1/4 \end{vmatrix}$

11. $\begin{vmatrix} a & b \\ c & d \end{vmatrix}$

12. $\begin{vmatrix} 2m & 3n \\ -m & 4n \end{vmatrix}$

13. $\begin{vmatrix} \sin\theta & 3 \\ \tan\theta & \sin\theta \end{vmatrix}$

14. $\begin{vmatrix} 2i_1 & i_2 \\ -3i_1 & 4i_2 \end{vmatrix}$

15. $\begin{vmatrix} 3i_1 & i_2 \\ \sin\theta & \cos\theta \end{vmatrix}$

16. $\begin{vmatrix} 3x & 2 \\ x & 5 \end{vmatrix}$

Solve each pair of equations by determinants.

17. $2x + y = 11$
$3x - y = 4$

18. $5x + 7y = 101$
$7x - y = 55$

19. $3x - 2y = -15$
$5x + 6y = 3$

20. $7x + 6y = 20$
$2x + 5y = 9$

21. $x + 5y = 11$
$3x + 2y = 7$

22. $4x - 5y = -34$
$2x - 3y = -22$

23. $x + 4y = 11$
$5x - 2y = 11$

24. $2x - 3y = 3$
$4x + 5y = 39$

25. $7x - 4y = 81$
$5x - 3y = 57$

26. $3x + 4y = 85$
$5x + 4y = 107$

27. $3x - 2y = 1$
$2x + y = 10$

28. $5x - 2y = 3$
$2x + 3y = 5$

29. $y = 9 - 3x$
$x = 8 - 2y$

30. $y = 2x - 3$
$x = 19 - 3y$

31. $29.1x - 47.6y = 42.8$
$11.5x + 72.7y = 25.8$

32. $4.92x - 8.27y = 2.58$
$6.93x + 2.84y = 8.36$

33. $4n = 18 - 3m$
$m = 8 - 2n$

34. $5p + 4q - 14 = 0$
$17p = 31 + 3q$

35. $3w = 13 + 5z$
$4w - 7z - 17 = 0$

36. $3u = 55 + 2v$
$5v + 2u = 16$

Fractional and Literal Equations

Solve for x and y by determinants.

37. $\dfrac{x}{3} + \dfrac{y}{2} = 1$

$\dfrac{x}{2} - \dfrac{y}{3} = -1$

38. $px - qy + pq = 0$
$2px - 3qy = 0$

39. $ax + by = p$
$cx + dy = q$

40. $\dfrac{x}{2} - \dfrac{3y}{4} = 0$

$\dfrac{x}{3} + \dfrac{y}{4} = 0$

41. $\dfrac{3}{x} + \dfrac{8}{y} = 3$

$\dfrac{15}{x} - \dfrac{4}{y} = 4$

42. $4mx - 3ny = 6$

$3mx + 2ny = -7$

Applications

43. A shipment of 4 cars and 2 trucks cost $172,172. Another shipment of 3 cars and 5 trucks cost $209,580. Find the cost of each car and each truck.

44. An airplane and a helicopter are 125 mi apart. The airplane is traveling at 226 mi/h, and the helicopter at 85.0 mi/h, both in the same direction. How long will it take the airplane to overtake the helicopter, and in what distance from the initial position of the airplane?

45. A certain alloy contains 31.0% zinc and 3.50% lead. Then x kg of zinc and y kg of lead are added to 325 kg of the original alloy to make a new alloy that is 45.0% zinc and 4.80% lead. The amount of zinc is given by

$$0.310(325) + x = 0.450(325 + x + y)$$

and the amount of lead is given by

$$0.0350(325) + y = 0.0480(325 + x + y)$$

Solve for x and y.

Graphing Calculator

46. Use your graphing calculator to evaluate any of the determinants in problems 1 through 10.

47. Use your graphing calculator to solve any of the sets of equations in problems 17 through 32. Remember, the value of each variable is the quotient of *two* determinants.

Computer

48. Using a CAS, evaluate any of the determinants in this exercise set having numerical *or* literal elements.

49. Use a CAS to solve any of the sets of equations in this exercise set. Remember, the value of each variable is the quotient of *two* determinants.

11–2 Third-Order Determinants

The technique we used for finding the value of a second-order determinant *cannot* be used for higher-order determinants. We will now show two other methods—first, what we'll call the *diagonal* method, which works *only for third-order determinants;* and then *development by minors,* which works for *any* determinant.

The Diagonal Method

The value of a third-order determinant can be found as follows. To the right of the determinant, rewrite the first two columns.

$$\begin{vmatrix} a_1 & b_1 & c_1 \\ a_2 & b_2 & c_2 \\ a_3 & b_3 & c_3 \end{vmatrix} \begin{matrix} a_1 & b_1 \\ a_2 & b_2 \\ a_3 & b_3 \end{matrix}$$

Then multiply the elements along each of the six diagonals. The value of the determinant is the sum of the products along those diagonals running downward to the right, *minus* the products along those diagonals running downward to the left.

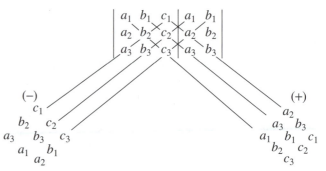

We will not emphasize this method, because it works only for third-order determinants and also because it is not suitable for computer solution.

The value of the determinant is then given by the following equation:

Third-Order Determinant	$\begin{vmatrix} a_1 & b_1 & c_1 \\ a_2 & b_2 & c_2 \\ a_3 & b_3 & c_3 \end{vmatrix} = \begin{aligned} & a_1b_2c_3 + a_3b_1c_2 + a_2b_3c_1 \\ & - a_3b_2c_1 - a_1b_3c_2 - a_2b_1c_3 \end{aligned}$	**66**

◆◆◆ **Example 9:** Evaluate the following third-order determinant:

$$\begin{vmatrix} 2 & 1 & 0 \\ 3 & -2 & 1 \\ 5 & 1 & 0 \end{vmatrix}$$

Solution: Rewriting the first two columns to the right of the determinant, we have

$$\begin{vmatrix} 2 & 1 & 0 \\ 3 & -2 & 1 \\ 5 & 1 & 0 \end{vmatrix}\begin{matrix} 2 & 1 \\ 3 & -2 \\ 5 & 1 \end{matrix}$$

Multiplying along the diagonals, we obtain

$$2(-2)(0) + 1(1)(5) + 0(3)(1) - [0(-2)(5) + 2(1)(1) + 1(3)(0)]$$
$$= \quad 0 \quad + \quad 5 \quad + \quad 0 \quad - [\quad 0 \quad + \quad 2 \quad + \quad 0]$$
$$= 5 - 2 = 3$$

◆◆◆

Minors

The *minor of an element in a determinant* is a determinant of next lower order, obtained by deleting the row and the column in which that element lies.

◆◆◆ **Example 10:** We find the minor of element c in the determinant

$$\begin{vmatrix} a & b & c \\ d & e & f \\ g & h & i \end{vmatrix}$$

by striking out the first row and the third column.

$$\begin{vmatrix} a & b & c \\ d & e & f \\ g & h & i \end{vmatrix} \quad \text{or} \quad \begin{vmatrix} d & e \\ g & h \end{vmatrix}$$

◆◆◆

Sign Factor

The *sign factor* for an element depends upon the position of the element in the determinant. To find the sign factor, add the row number and the column number of the element. If the sum is even, the sign factor is $+1$; if the sum is odd, the sign factor is -1.

A minor with the sign factor attached is called a *signed minor*. It is also called a *cofactor* of the given element.

◆◆◆ Example 11:

(a) For the determinant in Example 10, element c is in the first row and the third column. The sum of the row and column numbers

$$1 + 3 = 4$$

is even, so the sign factor for the element c is $+1$.

(b) The sign factor for the element b in Example 10 is -1. **◆◆◆**

The sign factor for the element in the upper left-hand corner of the determinant is $+1$, and the signs alternate according to the pattern

$$\begin{vmatrix} + & - & + & - & + & - & \cdot \\ - & + & - & + & - & \cdot & \cdot \\ + & - & + & - & \cdot & \cdot & \cdot \\ - & + & - & \cdot & \cdot & \cdot & \cdot \\ + & - & \cdot & \cdot & \cdot & \cdot & \cdot \\ - & \cdot & \cdot & \cdot & \cdot & \cdot & \cdot \\ \cdot & \cdot & \cdot & \cdot & \cdot & \cdot & \cdot \end{vmatrix}$$

so that the sign factor may be found simply by *counting off* from the upper left corner.

Thus for any element in a determinant, we can write three quantities:

1. The element itself.
2. The sign factor of the element.
3. The minor of the element.

◆◆◆ Example 12: For the element in the second row and the first column in the following determinant:

$$\begin{vmatrix} -2 & 5 & 1 \\ 3 & -2 & 4 \\ 1 & -4 & 2 \end{vmatrix}$$

1. the element itself is 3;
2. the sign factor of that element is -1; and
3. the minor of that element is

$$\begin{vmatrix} 5 & 1 \\ -4 & 2 \end{vmatrix}$$

◆◆◆

Evaluating a Determinant by Minors

We now define the *value of a determinant* as follows:

Value of a Determinant	To find the value of a determinant of order n:	
	1. Multiply each of the n elements in any row (or column) by the sign factor of that element and the minor of that element.	68
	2. Take the algebraic sum of those n products.	

◆◆◆ **Example 13:** Evaluate by minors:

$$\begin{vmatrix} 3 & 2 & 4 \\ 1 & 6 & -2 \\ -1 & 5 & -3 \end{vmatrix}$$

Solution: We first choose a row or a column for development. The work of expansion is greatly reduced if that row or column contains zeros. Our given determinant has no zeros, so let us choose column 1, which at least contains some 1's.

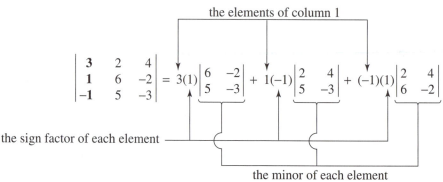

$$= 3[(6)(-3) - (-2)(5)] - 1[(2)(-3) - (4)(5)]$$
$$\quad - 1[(2)(-2) - 4(6)]$$
$$= -24 + 26 + 28$$
$$= 30$$

◆◆◆

◆◆◆ **Example 14:** Evaluate by minors:

$$\begin{vmatrix} a_1 & b_1 & c_1 \\ a_2 & b_2 & c_2 \\ a_3 & b_3 & c_3 \end{vmatrix}$$

We would get the same result if we chose another row or column for development.

Solution: Choosing, say, the first row for development, we get

$$a_1 \begin{vmatrix} b_2 & c_2 \\ b_3 & c_3 \end{vmatrix} - b_1 \begin{vmatrix} a_2 & c_2 \\ a_3 & c_3 \end{vmatrix} + c_1 \begin{vmatrix} a_2 & b_2 \\ a_3 & b_3 \end{vmatrix}$$

$$= a_1(b_2c_3 - b_3c_2) - b_1(a_2c_3 - a_3c_2) + c_1(a_2b_3 - a_3b_2)$$
$$= a_1b_2c_3 - a_1b_3c_2 - a_2b_1c_3 + a_3b_1c_2 + a_2b_3c_1 - a_3b_2c_1$$
$$= a_1b_2c_3 + a_3b_1c_2 + a_2b_3c_1 - a_3b_2c_1 - a_1b_3c_2 - a_2b_1c_3$$

◆◆◆

The value of the determinant is thus given by the following formula:

Third-Order Determinant	$\begin{vmatrix} a_1 & b_1 & c_1 \\ a_2 & b_2 & c_2 \\ a_3 & b_3 & c_3 \end{vmatrix} = \begin{matrix} a_1b_2c_3 + a_3b_1c_2 + a_2b_3c_1 \\ - a_3b_2c_1 - a_1b_3c_2 - a_2b_1c_3 \end{matrix}$	66

Solving a System of Three Equations by Determinants

If we were to *algebraically* solve the system of equations

$$a_1x + b_1y + c_1z = k_1$$
$$a_2x + b_2y + c_2z = k_2$$
$$a_3x + b_3y + c_3z = k_3$$

The solution of this system was given as a problem in Exercise 4 in Chapter 10.

we would get the following solution:

<table>
<tr><td>**Three Equations in Three Unknowns**</td><td>$$x = \frac{b_2c_3k_1 + b_1c_2k_3 + b_3c_1k_2 - b_2c_1k_3 - b_3c_2k_1 - b_1c_3k_2}{a_1b_2c_3 + a_3b_1c_2 + a_2b_3c_1 - a_3b_2c_1 - a_1b_3c_2 - a_2b_1c_3}$$ $$y = \frac{a_1c_3k_2 + a_3c_2k_1 + a_2c_1k_3 - a_3c_1k_2 - a_1c_2k_3 - a_2c_3k_1}{a_1b_2c_3 + a_3b_1c_2 + a_2b_3c_1 - a_3b_2c_1 - a_1b_3c_2 - a_2b_1c_3}$$ $$z = \frac{a_1b_2k_3 + a_3b_1k_2 + a_2b_3k_1 - a_3b_2k_1 - a_1b_3k_2 - a_2b_1k_3}{a_1b_2c_3 + a_3b_1c_2 + a_2b_3c_1 - a_3b_2c_1 - a_1b_3c_2 - a_2b_1c_3}$$</td><td>64</td></tr>
</table>

Notice that the three denominators are identical and are equal to the value of the determinant formed from the coefficients of the unknowns (Eq. 66). As before, we call it the *determinant of the coefficients,* Δ.

Furthermore, we can obtain the numerator of each from the determinant of the coefficients by replacing the coefficients of the variable in question with the constants k_1, k_2, and k_3. The solution to our set of equations

$$a_1x + b_1y + c_1z = k_1$$

$$a_2x + b_2y + c_2z = k_2$$

$$a_3x + b_3y + c_3z = k_3$$

is then given by Cramer's rule:

<table>
<tr><td>**Cramer's Rule**</td><td>$$x = \frac{\begin{vmatrix} k_1 & b_1 & c_1 \\ k_2 & b_2 & c_2 \\ k_3 & b_3 & c_3 \end{vmatrix}}{\Delta} \qquad y = \frac{\begin{vmatrix} a_1 & k_1 & c_1 \\ a_2 & k_2 & c_2 \\ a_3 & k_3 & c_3 \end{vmatrix}}{\Delta} \qquad z = \frac{\begin{vmatrix} a_1 & b_1 & k_1 \\ a_2 & b_2 & k_2 \\ a_3 & b_3 & k_3 \end{vmatrix}}{\Delta}$$ where $$\Delta = \begin{vmatrix} a_1 & b_1 & c_1 \\ a_2 & b_2 & c_2 \\ a_3 & b_3 & c_3 \end{vmatrix} \neq 0$$</td><td>71</td></tr>
</table>

Although we do not prove it, Cramer's rule works for higher-order systems as well. We restate it now in words.

<table>
<tr><td>**Cramer's Rule**</td><td>The solution for any variable is a fraction whose denominator is the determinant of the coefficients and whose numerator is the same determinant, except that the column of coefficients for the variable for which we are solving is replaced by the column of constants.</td><td>69</td></tr>
</table>

◆◆◆ **Example 15:** Solve by determinants:

$$x + 2y + 3z = 14$$
$$2x + y + 2z = 10$$
$$3x + 4y - 3z = 2$$

Solution: We first write the determinant of the coefficients

$$\Delta = \begin{vmatrix} 1 & 2 & 3 \\ 2 & 1 & 2 \\ 3 & 4 & -3 \end{vmatrix}$$

and expand it by minors. Choosing the first row for expansion gives

$$\Delta = \begin{vmatrix} 1 & 2 & 3 \\ 2 & 1 & 2 \\ 3 & 4 & -3 \end{vmatrix} = 1\begin{vmatrix} 1 & 2 \\ 4 & -3 \end{vmatrix} - 2\begin{vmatrix} 2 & 2 \\ 3 & -3 \end{vmatrix} + 3\begin{vmatrix} 2 & 1 \\ 3 & 4 \end{vmatrix}$$

$$= 1(-3 - 8) - 2(-6 - 6) + 3(8 - 3)$$

$$= -11 + 24 + 15 = 28$$

As when we are solving a set of two equations, a Δ of zero would mean that the system had no unique solution, and we would stop here.

We now make a new determinant by replacing the coefficients of x, $\begin{vmatrix} 1 \\ 2 \\ 3 \end{vmatrix}$, by the column of constants $\begin{vmatrix} 14 \\ 10 \\ 2 \end{vmatrix}$, and expand it by, say, the second column.

$$\begin{vmatrix} 14 & 2 & 3 \\ 10 & 1 & 2 \\ 2 & 4 & -3 \end{vmatrix} = -2\begin{vmatrix} 10 & 2 \\ 2 & -3 \end{vmatrix} + 1\begin{vmatrix} 14 & 3 \\ 2 & -3 \end{vmatrix} - 4\begin{vmatrix} 14 & 3 \\ 10 & 2 \end{vmatrix}$$

$$= -2(-30 - 4) + 1(-42 - 6) - 4(28 - 30)$$

$$= 68 - 48 + 8 = 28$$

Dividing this value by Δ gives x.

$$x = \frac{28}{\Delta} = \frac{28}{28} = 1$$

Next we replace the coefficients of y with the column of constants and expand the first column by minors.

$$\begin{vmatrix} 1 & 14 & 3 \\ 2 & 10 & 2 \\ 3 & 2 & -3 \end{vmatrix} = 1\begin{vmatrix} 10 & 2 \\ 2 & -3 \end{vmatrix} - 2\begin{vmatrix} 14 & 3 \\ 2 & -3 \end{vmatrix} + 3\begin{vmatrix} 14 & 3 \\ 10 & 2 \end{vmatrix}$$

$$= 1(-30 - 4) - 2(-42 - 6) + 3(28 - 30)$$

$$= -34 + 96 - 6 = 56$$

Dividing by Δ gives the value of y.

$$y = \frac{56}{\Delta} = \frac{56}{28} = 2$$

We can get z also by determinants or, more easily, by substituting back. Substituting $x = 1$ and $y = 2$ into the first equation, we obtain

$$1 + 2(2) + 3z = 14$$
$$3z = 9$$
$$z = 3$$

The solution is thus $(1, 2, 3)$.   ◆◆◆

Often, some terms will have zero coefficients or decimal coefficients. Further, the terms may be out of order, as in the following example.

◆◆◆ **Example 16:** Solve for x, y, and z:

$$23.7y + 72.4x = 82.4 - 11.3x$$
$$25.5x - 28.4z + 19.3 = 48.2y$$
$$13.4 + 66.3z = 39.2x - 10.5$$

Solution: We rewrite each equation in the form $ax + by + cz = k$, combining like terms as we go and putting in the missing terms with zero coefficients. We do this *before* writing the determinant.

$$83.7x + 23.7y + \quad 0z = \quad 82.4 \qquad (1)$$
$$25.5x - 48.2y - 28.4z = -19.3$$
$$-39.2x + \quad 0y + 66.3z = -23.9 \qquad (2)$$

The determinant of the system is then

$$\Delta = \begin{vmatrix} 83.7 & 23.7 & 0 \\ 25.5 & -48.2 & -28.4 \\ -39.2 & 0 & 66.3 \end{vmatrix}$$

Let us develop the first row by minors.

$$\Delta = 83.7 \begin{vmatrix} -48.2 & -28.4 \\ 0 & 66.3 \end{vmatrix} - 23.7 \begin{vmatrix} 25.5 & -28.4 \\ -39.2 & 66.3 \end{vmatrix} + 0$$
$$= 83.7(-48.2)(66.3) - 23.7[25.5(66.3) - (-28.4)(-39.2)]$$
$$= -281,000 \quad \text{(to three significant digits)}$$

Now solving for x gives us

$$x = \frac{\begin{vmatrix} 82.4 & 23.7 & 0 \\ -19.3 & -48.2 & -28.4 \\ -23.9 & 0 & 66.3 \end{vmatrix}}{\Delta}$$

Let us develop the first row of the determinant by minors. We get

$$82.4 \begin{vmatrix} -48.2 & -28.4 \\ 0 & 66.3 \end{vmatrix} - 23.7 \begin{vmatrix} -19.3 & -28.4 \\ -23.9 & 66.3 \end{vmatrix} + 0$$
$$= 82.4(-48.2)(66.3) - 23.7[(-19.3)(66.3) - (-28.4)(-23.9)]$$
$$= -217,000$$

Dividing by Δ yields

$$x = \frac{-217,000}{-281,000} = 0.772$$

We solve for y and z by substituting back. From Equation (1), we get

$$23.7y = 82.4 - 83.7(0.772) = 17.8$$
$$y = 0.751$$

and from (2)

$$66.3z = -23.9 + 39.2(0.772) = 6.36$$
$$z = 0.0959$$

The solution is then $x = 0.772$, $y = 0.751$, and $z = 0.0959$. ◆◆◆

Exercise 2 ◆ Third-Order Determinants

Evaluate each determinant.

1. $\begin{vmatrix} 1 & 0 & 2 \\ 3 & 1 & 0 \\ 1 & 2 & 1 \end{vmatrix}$

2. $\begin{vmatrix} 2 & -1 & 3 \\ 0 & 2 & 1 \\ 3 & -2 & 4 \end{vmatrix}$

3. $\begin{vmatrix} -3 & 1 & 2 \\ 0 & -1 & 5 \\ 6 & 0 & 1 \end{vmatrix}$

4. $\begin{vmatrix} -1 & 0 & 3 \\ 2 & 0 & -2 \\ 1 & -3 & 4 \end{vmatrix}$

5. $\begin{vmatrix} 5 & 1 & 2 \\ -3 & 2 & -1 \\ 4 & -3 & 5 \end{vmatrix}$

6. $\begin{vmatrix} 1.0 & 2.4 & -1.5 \\ -2.6 & 0 & 3.2 \\ -2.9 & 1.0 & 4.1 \end{vmatrix}$

7. $\begin{vmatrix} 2 & 1 & 3 \\ 0 & -2 & 4 \\ 0 & 1 & 5 \end{vmatrix}$

8. $\begin{vmatrix} 1 & 5 & 4 \\ -3 & 6 & -2 \\ -1 & 5 & 3 \end{vmatrix}$

Solve by determinants.

Some of these problems are identical to those given in Exercise 4 in Chapter 10.

9. $x + y + z = 18$
$x - y + z = 6$
$x + y - z = 4$

10. $x + y + z = 12$
$x - y = 2$
$x - z = 4$

11. $x + y = 35$
$x + z = 40$
$y + z = 45$

12. $x + y + z = 35$
$x - 2y + 3z = 15$
$y - x + z = -5$

13. $x + 2y + 3z = 14$
$2x + y + 2z = 10$
$3x + 4y - 3z = 2$

14. $x + y + z = 90$
$2x - 3y = -20$
$2x + 3z = 145$

15. $x - 2y + 2z = 5$
$5x + 3y + 6z = 57$
$x + 2y + 2z = 21$

16. $3x + y = 5$
$2y - 3z = -5$
$x + 2z = 7$

17. $2x - 4y + 3z = 10$
$3x + y - 2z = 6$
$x + 3y - z = 20$

18. $x - y = 5$
$y - z = -6$
$2x - z = 2$

19. $1.15x + 1.95y + 2.78z = 15.3$
$2.41x + 1.16y + 3.12z = 9.66$
$3.11x + 3.83y - 2.93z = 2.15$

20. $1.52x - 2.26y + 1.83z = 4.75$
$4.72x + 3.52y + 5.83z = 45.2$
$1.33x + 2.61y + 3.02z = 18.5$

Fractional Equations

Solve by determinants.

21. $x + \dfrac{y}{3} = 5$

$x + \dfrac{z}{3} = 6$

$y + \dfrac{z}{3} = 9$

22. $\dfrac{x}{10} + \dfrac{y}{5} + \dfrac{z}{20} = \dfrac{1}{4}$

$x + y + z = 6$

$\dfrac{x}{3} + \dfrac{y}{2} + \dfrac{z}{6} = 1$

23. $\dfrac{1}{x} + \dfrac{1}{y} = 5$

$\dfrac{1}{y} + \dfrac{1}{z} = 7$

$\dfrac{1}{x} + \dfrac{1}{z} = 6$

24. $\dfrac{1}{x} + \dfrac{2}{y} - \dfrac{1}{z} = -3$

$\dfrac{3}{x} + \dfrac{1}{y} + \dfrac{1}{z} = 4$

$\dfrac{1}{x} - \dfrac{1}{y} + \dfrac{2}{z} = 6$

Applications

25. Applying Kirchhoff's law to a certain three-loop network gives

$$283I_1 - 274I_2 + 163I_3 = 352$$
$$428I_1 + 163I_2 + 373I_3 = 169$$
$$338I_1 - 112I_2 - 227I_3 = 825$$

Solve this set of equations for the three currents

26. Find the forces F_1, F_2, and F_3 in Fig. 11–1. Neglect the weight of the beam.

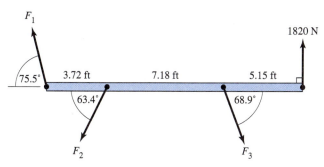

FIGURE 11–1

Graphics Calculator

27. We use a graphics calculator to evaluate third-order determinants in the same way as for second-order determinants, as we showed in Sec. 11–1. Be sure to change the dimensions from 2 × 2 to 3 × 3. Use the graphics calculator to evaluate any of the determinants in this exercise set.

28. Use a graphics calculator to solve any of the systems of equations in this exercise set. Evaluate the determinant of the system and the determinant for x, and divide to get x. Repeat for the other unknowns.

Computer

29. To evaluate a determinant on a CAS, enter the elements in the same form as we showed in Sec. 11–1, and give the command Simplify. Use a CAS to evaluate any determinant in this exercise set.

30. Use a CAS to solve any system of equations in this exercise set by evaluating and then dividing the appropriate determinants.

11–3   Higher-Order Determinants

We can evaluate a determinant of any order by repeated use of the method of minors. Thus any row or column of a fifth-order determinant can be developed, thereby reducing the determinant to five fourth-order determinants. Each of these can then be developed into four third-order determinants, and so on, until only second-order determinants remain.

Obviously, this method involves lots of work. To avoid it, we first *simplify* the determinant, using the *properties of determinants* given in the following section, and then apply a systematic method for reducing the determinants by minors shown after that.

Properties of Determinants

A determinant larger than third order can usually be evaluated faster if we first reduce its order before expanding by minors. We do this by applying the various *properties of determinants* as shown in the following examples.

Zero Row or Column	If all elements in a row (or column) are zero, then the value of the determinant is zero.	**72**

◆◆◆ **Example 17:**

$$\begin{vmatrix} 4 & 8 & 0 \\ 5 & 2 & 0 \\ 3 & 9 & 0 \end{vmatrix} = 0$$

◆◆◆

Identical Rows or Columns	If two rows (or columns) are identical, then the value of a determinant is zero.	**73**

◆◆◆ **Example 18:** We find the value of the determinant

$$\begin{vmatrix} 2 & 1 & 3 \\ 4 & 2 & 5 \\ 4 & 2 & 5 \end{vmatrix}$$

by expanding by minors using, say, the first row.

$$2\begin{vmatrix} 2 & 5 \\ 2 & 5 \end{vmatrix} - 1\begin{vmatrix} 4 & 5 \\ 4 & 5 \end{vmatrix} + 3\begin{vmatrix} 4 & 2 \\ 4 & 2 \end{vmatrix}$$
$$= 2(10 - 10) - 1(20 - 20) + 3(8 - 8)$$
$$= 0$$

◆◆◆

Zeros below the Principal Diagonal	If all elements below the principal diagonal are zeros, then the value of the determinant is the product of the elements along the principal diagonal.	**74**

◆◆◆ **Example 19:** The determinant

Expand this determinant by minors and see for yourself that its value is indeed 6.

$$\begin{vmatrix} 2 & 9 & 7 & 8 \\ 0 & 1 & 5 & 9 \\ 0 & 0 & 3 & 4 \\ 0 & 0 & 0 & 1 \end{vmatrix}$$

has the value

$$(2)(1)(3)(1) = 6$$

◆◆◆

Interchanging Rows with Columns	If we change the rows to columns and the columns to rows, the value of a determinant will be unchanged.	**75**

◆◆◆ **Example 20:** The value of the determinant

$$\begin{vmatrix} 5 & 2 \\ 4 & 3 \end{vmatrix}$$

is

$$(5)(3) - (2)(4) = 7$$

If the first column now becomes the first row, and the second column becomes the second row, we get the determinant

$$\begin{vmatrix} 5 & 4 \\ 2 & 3 \end{vmatrix}$$

which has the value

$$(5)(3) - (4)(2) = 7$$

as before. ◆◆◆

Interchange of Two Rows (or Columns)	A determinant will change sign when we interchange two rows (or columns).	76

◆◆◆ **Example 21:** We find the value of the determinant

$$\begin{vmatrix} 3 & 0 & 2 \\ 1 & 4 & 3 \\ 2 & 1 & 2 \end{vmatrix}$$

by expanding by minors using the first row.

$$3\begin{vmatrix} 4 & 3 \\ 1 & 2 \end{vmatrix} - 0 + 2\begin{vmatrix} 1 & 4 \\ 2 & 1 \end{vmatrix} = 3(8 - 3) + 2(1 - 8) = 1$$

Let us now interchange, say, the first and second columns.

$$\begin{vmatrix} 0 & 3 & 2 \\ 4 & 1 & 3 \\ 1 & 2 & 2 \end{vmatrix}$$

We find its value again by expanding using the first row.

$$0 - 3\begin{vmatrix} 4 & 3 \\ 1 & 2 \end{vmatrix} + 2\begin{vmatrix} 4 & 1 \\ 1 & 2 \end{vmatrix} = -3(8 - 3) + 2(8 - 1) = -1$$

So interchanging two columns has reversed the sign of the determinant. ◆◆◆

Multiplying by a Constant	If each element in a row (or column) is multiplied by some constant, then the value of the determinant is multiplied by that constant.	77

◆◆◆ **Example 22:** We saw that the value of the determinant in Example 21 was equal to 1.

$$\begin{vmatrix} 3 & 0 & 2 \\ 1 & 4 & 3 \\ 2 & 1 & 2 \end{vmatrix} = 1$$

Let us now multiply the elements in the third row by some constant, say, 3.

$$\begin{vmatrix} 3 & 0 & 2 \\ 1 & 4 & 3 \\ 6 & 3 & 6 \end{vmatrix}$$

This new determinant has the value

$$3(24 - 9) + 2(3 - 24) = 45 - 42 = 3$$

or three times its previous value. ◆◆◆

We can also use this rule *in reverse,* to *remove a factor* from a row or a column.

◆◆◆ **Example 23:** The determinant

$$\begin{vmatrix} 3 & 1 & 0 \\ 30 & 10 & 15 \\ 3 & 2 & 1 \end{vmatrix}$$

could be evaluated as it is, but let us first factor a 3 from the first column.

$$\begin{vmatrix} 3 & 1 & 0 \\ 30 & 10 & 15 \\ 3 & 2 & 1 \end{vmatrix} = 3 \begin{vmatrix} 1 & 1 & 0 \\ 10 & 10 & 15 \\ 1 & 2 & 1 \end{vmatrix}$$

Then we factor a 5 from the second row.

$$3 \begin{vmatrix} 1 & 1 & 0 \\ 10 & 10 & 15 \\ 1 & 2 & 1 \end{vmatrix} = 3(5) \begin{vmatrix} 1 & 1 & 0 \\ 2 & 2 & 3 \\ 1 & 2 & 1 \end{vmatrix}$$

$$= 15[(2-6)-(2-3)] = -45 \qquad ◆◆◆$$

Multiples of One Row (or Column) Added to Another	If the elements of a row (or column) are multiplied by some factor and are then added to the corresponding elements of another row or column, the value of a determinant is unchanged.	**78**

◆◆◆ **Example 24:** Again using the determinant from Example 21,

$$\begin{vmatrix} 3 & 0 & 2 \\ 1 & 4 & 3 \\ 2 & 1 & 2 \end{vmatrix} = 1$$

let us get a new second row by multiplying the third row by -4 and adding those elements to the second row. The new elements in the second row are then

$$1 + 2(-4) = -7, \qquad 4 + 1(-4) = 0, \quad \text{and} \quad 3 + 2(-4) = -5$$

giving a new determinant

$$\begin{vmatrix} 3 & 0 & 2 \\ -7 & 0 & -5 \\ 2 & 1 & 2 \end{vmatrix}$$

whose value we find by developing the second column by minors.

$$-1 \begin{vmatrix} 3 & 2 \\ -7 & -5 \end{vmatrix} = -1[3(-5) - 2(-7)] = -[-15 + 14] = 1 \quad \text{(as before)}$$

Notice, however, that we have *introduced another zero* into the determinant, making it easier to evaluate. This, of course, is the point of the whole operation. ◆◆◆

Exercise 3 ◆ Higher-Order Determinants

Evaluate each determinant.

1. $\begin{vmatrix} 4 & 3 & 1 & 0 \\ -1 & 2 & -3 & 5 \\ 0 & 1 & -1 & 2 \\ 0 & 2 & -3 & 5 \end{vmatrix}$

2. $\begin{vmatrix} -1 & 3 & 0 & 2 \\ 2 & -1 & 1 & 0 \\ 5 & 2 & -2 & 0 \\ 1 & -1 & 3 & 1 \end{vmatrix}$

3. $\begin{vmatrix} 2 & 0 & -1 & 0 \\ 0 & 0 & 2 & -1 \\ 1 & 3 & 2 & 1 \\ 3 & 1 & 1 & -2 \end{vmatrix}$

4. $\begin{vmatrix} 1 & 2 & -1 & 1 \\ -1 & 1 & 2 & 3 \\ 3 & -1 & 1 & 2 \\ 1 & 2 & -1 & 1 \end{vmatrix}$

5. $\begin{vmatrix} 3 & 1 & 0 & 2 & 4 \\ 1 & 2 & 4 & 0 & 1 \\ 2 & 3 & 1 & 4 & 2 \\ 1 & 2 & 0 & 2 & 1 \\ 3 & 4 & 1 & 3 & 1 \end{vmatrix}$

6. $\begin{vmatrix} 2 & 1 & 5 & 3 & 6 \\ 1 & 4 & 2 & 4 & 3 \\ 3 & 1 & 2 & 4 & 1 \\ 5 & 2 & 3 & 1 & 4 \\ 4 & 5 & 2 & 3 & 1 \end{vmatrix}$

Four Equations in Four Unknowns

Solve by determinants.

7. $\begin{aligned} x + y + 2z + w &= 18 \\ x + 2y + z + w &= 17 \\ x + y + z + 2w &= 19 \\ 2x + y + z + w &= 16 \end{aligned}$

8. $\begin{aligned} 2x - y - z - w &= 0 \\ x - 3y + z + w &= 0 \\ x + y - 4z + w &= 0 \\ x + y \qquad + w &= 36 \end{aligned}$

9. $\begin{aligned} 3x - 2y - z + w &= -3 \\ -x - y + 3z + 2w &= 23 \\ x + 3y - 2z + w &= -12 \\ 2x - y - z - 3w &= -22 \end{aligned}$

10. $\begin{aligned} x + 2y &= 5 \\ y + 2z &= 8 \\ z + 2u &= 11 \\ u + 2x &= 6 \end{aligned}$

11. $\begin{aligned} x + y &= a + b \\ y + z &= b + c \\ z + w &= a - b \\ w - x &= c - b \end{aligned}$

12. $\begin{aligned} 2x - 3y + z - w &= -6 \\ x + 2y - z \qquad &= 8 \\ 3y + z + 3w &= 0 \\ 3x - y \qquad + w &= 0 \end{aligned}$

Five Equations in Five Unknowns

Solve by determinants.

13. $\begin{aligned} x + y &= 9 \\ y + z &= 11 \\ z + w &= 13 \\ w + u &= 15 \\ u + x &= 12 \end{aligned}$

14. $\begin{aligned} 3x + 4y + z &= 35 \\ 3z + 2y - 3w &= 4 \\ 2x - y + 2w &= 17 \\ 3z - 2w + v &= 9 \\ w + y &= 13 \end{aligned}$

15. $\begin{aligned} w + v + x + y &= 14 \\ w + v + x + z &= 15 \\ w + v + y + z &= 16 \\ w + x + y + z &= 17 \\ v + x + y + z &= 18 \end{aligned}$

Graphics Calculator

16. We use a graphics calculator to evaluate any order of determinant in the same way as for second-order determinants, as we showed in Sec. 11–1. Be sure to enter the dimensions of the determinant to be evaluated. Use the graphics calculator to evaluate any of the determinants in this exercise set.

17. Use a graphics calculator to solve any of the systems of equations in this exercise set. Evaluate the determinant of the system and the determinant for x, and divide to get x. Repeat for the other unknowns.

Computer

18. To evaluate a determinant on a CAS, enter the elements in the same form as we showed in Sec. 11–1, and give the command *Simplify*. Use a CAS to evaluate any determinant in this exercise.

19. Use a CAS to solve any system of equations in this exercise set by evaluating and then dividing the appropriate determinants.

20. Applying Kirchhoff's law to a certain four-loop network gives the following equations:

$$57.2I_1 + 92.5I_2 - 23.0I_3 - 11.4I_4 = 38.2$$
$$95.3I_1 - 14.9I_2 + 39.0I_3 + 59.9I_4 = 29.3$$
$$66.3I_1 + 81.4I_2 - 91.5I_3 + 33.4I_4 = -73.6$$
$$38.2I_1 - 46.6I_2 + 30.1I_3 + 93.2I_4 = 55.7$$

Solve for the four loop currents by calculator or computer.

21. The following equations result when Kirchhoff's law is applied to a certain four-loop network:

$$1.27I_1 - 5.27I_2 + 4.27I_3 + 9.63I_4 = 6.82$$
$$7.92I_1 + 9.36I_2 - 9.72I_3 + 4.14I_4 = -8.83$$
$$8.36I_1 - 2.27I_2 + 4.77I_3 + 7.33I_4 = 3.93$$
$$7.37I_1 + 9.36I_2 - 3.82I_3 - 2.73I_4 = 5.04$$

Find the four currents by calculator or computer.

22. Four types of computers are made by a company, each requiring the following numbers of hours for four steps:

	Assembly	Burn-In	Inspection	Testing
Model A	4.50 h	12 h	1.25 h	2.75 h
Model B	5.25 h	12 h	1.75 h	3.00 h
Model C	6.55 h	24 h	2.25 h	3.75 h
Model D	7.75 h	36 h	3.75 h	4.25 h
Available	15,157 h	43,680 h	4824 h	8928 h

Also shown in the table are the monthly production hours available for each step. How many of each type of computer, rounded to the nearest unit, can be made each month?

23. We have available four bronze alloys containing the following percentages of copper, zinc, and lead:

	Alloy 1	Alloy 2	Alloy 3	Alloy 4
Copper	52.0	53.0	54.0	55.0
Zinc	30.0	38.0	20.0	38.0
Lead	3.00	2.00	4.00	3.00

How many kilograms of each alloy should be taken to produce $60\overline{0}$ kg of a new alloy that is 53.8% copper, 30.1% zinc, and 3.20% lead?

••• CHAPTER 11 REVIEW PROBLEMS ••••••••••••••••••••••••••••••

Evaluate.

1. $\begin{vmatrix} 6 & 1/2 & -2 \\ 3 & 1/4 & 4 \\ 2 & -1/2 & 3 \end{vmatrix}$

2. $\begin{vmatrix} 2 & 7 & -2 & 8 \\ 4 & 1 & 1 & -3 \\ 0 & 3 & -1 & 4 \\ 6 & 4 & 2 & -8 \end{vmatrix}$

3. $\begin{vmatrix} 0 & n & m \\ -n & 0 & l \\ -m & -l & 0 \end{vmatrix}$

4. $\begin{vmatrix} 8 & 2 & 0 & 1 & 4 \\ 0 & 1 & 4 & 2 & 7 \\ 2 & 6 & 3 & 8 & 0 \\ 1 & 4 & 2 & 6 & 5 \\ 4 & 6 & 8 & 3 & 5 \end{vmatrix}$

5. $\begin{vmatrix} 25 & 23 & 19 \\ 14 & 11 & 9 \\ 21 & 17 & 14 \end{vmatrix}$

6. $\begin{vmatrix} x+y & x-y \\ x-y & x+y \end{vmatrix}$

7. $\begin{vmatrix} 9 & 13 & 17 \\ 11 & 15 & 19 \\ 17 & 21 & 25 \end{vmatrix}$

8. $\begin{vmatrix} 5 & -3 & -2 & 0 \\ 4 & 1 & -6 & 2 \\ -1 & 4 & 3 & -5 \\ 0 & 6 & -4 & 2 \end{vmatrix}$

9. $\begin{vmatrix} 1 & 2 & 3 \\ 3 & 1 & 2 \\ 2 & 3 & 1 \end{vmatrix}$

10. $\begin{vmatrix} 2 & -3 & 1 \\ -2 & 4 & 5 \\ 3 & -1 & -4 \end{vmatrix}$

11. $\begin{vmatrix} 1 & 2 & 2 & 4 \\ 1 & 4 & 4 & 1 \\ 1 & 1 & 2 & 2 \\ 4 & 8 & 11 & 13 \end{vmatrix}$

12. $\begin{vmatrix} a-b & -2a \\ 2b & a-b \end{vmatrix}$

13. $\begin{vmatrix} 3 & 1 & 5 & 2 \\ 4 & 10 & 14 & 6 \\ 8 & 9 & 1 & 4 \\ 6 & 15 & 21 & 9 \end{vmatrix}$

14. $\begin{vmatrix} 7 & 8 & 9 \\ 28 & 35 & 40 \\ 21 & 26 & 30 \end{vmatrix}$

15. $\begin{vmatrix} 1 & 5 & 2 \\ 4 & 7 & 3 \\ 9 & 8 & 6 \end{vmatrix}$

16. $\begin{vmatrix} 3 & -6 \\ -5 & 4 \end{vmatrix}$

17. $\begin{vmatrix} 6 & 4 & 7 \\ 9 & 0 & 8 \\ 5 & 3 & 2 \end{vmatrix}$

18. $\begin{vmatrix} 3 & 2 & 1 & 3 \\ 4 & 2 & 2 & 4 \\ 2 & 3 & 1 & 6 \\ 10 & 4 & 5 & 8 \end{vmatrix}$

Solve by determinants.

19. $x + y + z + w = -4$
$x + 2y + 3z + 4w = 0$
$x + 3y + 6z + 10w = 9$
$x + 4y + 10z + 20w = 24$

20. $x + 2y + 3z = 4$
$3x + 5y + z = 18$
$4x + y + 2z = 12$

21. $4x + 3y = 27$
$2x - 5y = -19$

22. $8x - 3y - 7z = 85$
$x + 6y - 4z = 12$
$2x - 5y + z = 33$

23. $x + 2y = 10$
$2x - 3y = -1$

24. $x + y + z + u = 1$
$2x + 3y - 4z + 5u = -31$
$3x - 4y + 5z + 6u = -22$
$4x + 5y - 6z - u = -13$

25. $2x - 3y = 7$
$5x + 2y = 27$

26. $6x + y = 60$
$3x + 2y = 39$

27. $2x + 5y = 29$
$2x - 5y = -21$

28. $4x + 3y = 7$
$2x - 3y = -1$

29. $4x - 5y = 3$
$3x + 5y = 11$

30. $x + 5y = 41$
$3x - 2y = 21$

31. $x + 2y = 7$
$x + y = 5$

32. $8x - 3y = 22$
$4x + 5y = 18$

33. $3x + 4y = 25$
$4x + 3y = 21$

34. $37.7x = 59.2 + 24.6y$
$28.3x - 39.8y - 62.5 = 0$

35. $5.03a = 8.16 + 5.11b$
$3.63b + 7.26a = 28.8$

36. $541x + 216y + 412z = 866$
$211x + 483y - 793z = 315$
$215x + 495y + 378z = 253$

37. $72.3x + 54.2y + 83.3z = 52.5$
$52.2x - 26.6y + 83.7z = 75.2$
$33.4x + 61.6y + 30.2z = 58.5$

38. A link in a certain mechanism starts at a speed v_0 and travels 27.5 inches in 8.25 s, at constant acceleration a. Its motion is described by Eq. A18,

$$s = v_0 t + \tfrac{1}{2}at^2$$

or

$$27.5 = 8.25v_0 + \tfrac{1}{2}a(8.25)^2$$

In another trial, the link is found to travel 34.6 inches in 11.2 s, with the same initial speed and acceleration. Find v_0 and a.

39. The following equations result when Kirchhoff's law is applied to a certain four-loop network:

$$14.7I_1 - 25.7I_2 + 14.7I_3 + 19.3I_4 = 26.2$$
$$17.2I_1 + 19.6I_2 - 19.2I_3 + 24.4I_4 = -28.3$$
$$18.6I_1 - 22.7I_2 + 24.7I_3 + 17.3I_4 = 23.3$$
$$27.7I_1 + 19.6I_2 - 33.2I_3 - 42.3I_4 = 25.4$$

Find the four currents by calculator or computer.

Writing

40. Suppose that you often solve sets of three or more equations on the job. Write a memo to your boss asking for a desktop computer to do the work. It should include a description of the method to be used which can be understood by an engineer who does not have extensive computer experience.

Team Project

41. Solve the following by any method:

$$28.3x + 29.2y - 33.1z + 72.4u + 29.4v = 39.5$$
$$73.2x - 28.4y + 59.3z - 27.4u + 49.2v = 82.3$$
$$33.7x + 10.3y + 72.3z + 29.3u - 21.2v = 28.4$$
$$92.3x - 39.5y + 29.5z - 10.3u + 82.2v = 73.4$$
$$88.3x + 29.3y + 10.3z + 84.2u + 29.3v = 39.4$$

12

Matrices

◆◆◆ **OBJECTIVES** ◆◆◆

When you have completed this chapter, you should be able to:

- Identify various types of arrays, matrices, and vectors, and give their dimensions.
- Write the transpose of a matrix.
- Add, subtract, and multiply matrices.
- Solve systems of equations using matrix inversion.

◆◆◆

The method of determinants is fine for solving a set of equations by hand, but it is not easy to program for the computer. One popular method of solving sets of equations by computer and graphics calculator is called *matrix inversion*. We will learn that method later in this chapter, but first we must learn about *matrices* and how to manipulate them.

12–1   Definitions

Arrays

A set of numbers, called *elements,* arranged in a pattern, is called an *array.* Arrays are named for the shape of the pattern made by the elements.

♦♦♦ **Example 1:** The array

$$\begin{pmatrix} 3 & 6 & 4 \\ 1 & 3 & 6 \\ 9 & 2 & 4 \end{pmatrix}$$

is called a *square array.*
   The array

$$\begin{pmatrix} 6 & 2 & 8 \\ 3 & 2 & 7 \end{pmatrix}$$

is called a *rectangular array.*
   The array

$$\begin{pmatrix} 7 & 3 & 9 & 1 \\ & 8 & 3 & 2 \\ & & 9 & 1 \\ & & & 5 \end{pmatrix}$$

is called a *triangular array.*   ♦♦♦

Vectors

A *vector* is an array consisting of a single row or a single column.

♦♦♦ **Example 2:** The array $(2, 6, -2, 8)$ is called a *row vector,* and the array

$$\begin{pmatrix} 7 \\ 3 \\ 9 \\ 2 \end{pmatrix}$$

is called a *column vector.*   ♦♦♦

   What do these vectors have to do with the vectors we defined in Chapter 7, where we used vectors to represent directed quantities such as force and velocity?
   The two apparently different definitions agree, if we think of a vector as the *coordinates of the endpoint* of a line segment drawn from the origin.

♦♦♦ **Example 3:**

(a) The vector $(4, 7)$ represents a line segment in a plane drawn from the origin to the point $(4, 7)$, as shown in Fig. 12–1(a).
(b) The vector $(3, 2, 4)$ represents a line segment in three dimensions, drawn from the origin to the point $(3, 2, 4)$, as shown in Fig. 12–1(b).
(c) The vector $(5, 8, 1, 3, 6)$ can be thought of as representing a line segment in five dimensions, although we cannot draw it. However, we can work mathematically with this vector using the same rules as for the others.   ♦♦♦

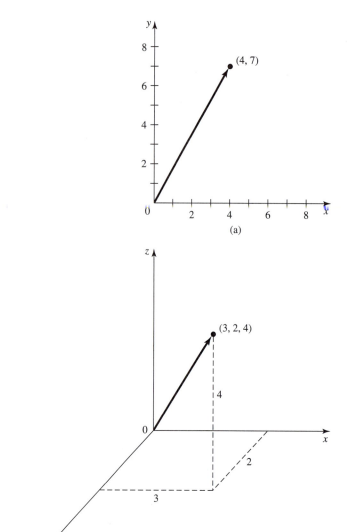

FIGURE 12–1

(b)

Matrices

A *matrix* is a *rectangular* array.

◆◆◆ **Example 4:** The following rectangular arrays are matrices:

A matrix is sometimes referred to as a *vector of vectors*.

$$\begin{pmatrix} 2 & 5 & 7 \\ 6 & 3 & 1 \end{pmatrix} \tag{a}$$

$$\begin{pmatrix} 7 & 3 & 5 \\ 1 & 9 & 3 \\ 8 & 3 & 2 \end{pmatrix} \tag{b}$$

Further, (b) is a *square* matrix. For a square matrix, the diagonal running from the upper-left element to the lower-right element is called the *main diagonal*. In (b), the elements on the main diagonal are 7, 9, and 2. The diagonal from upper right to lower left is called the *secondary* diagonal. In our example, the elements 5, 9, and 8 are on the secondary diagonal.   ◆◆◆

In everyday language, an array is a *table* and a vector is a *list*.

Subscripts

Each element in an array is located in a horizontal *row* and a vertical *column*. We indicate the row and column by means of *subscripts*.

◆◆◆ **Example 5:** The element a_{25} is located in row 2 and column 5.　　　　◆◆◆

Thus an element in an array needs *double subscripts* to give its location. An element in a list, such as b_7, needs only a *single subscript*.

Dimensions

A matrix will have, in general, m rows and n columns. The numbers m and n are the *dimensions* of the matrix, as, for example, a 4×5 matrix.

◆◆◆ **Example 6:** The matrix

$$\begin{pmatrix} 2 & 3 & 5 & 9 \\ 1 & 0 & 8 & 3 \end{pmatrix}$$

has the dimensions 2×4, and the matrix

$$\begin{pmatrix} 5 & 3 & 8 \\ 1 & 0 & 3 \\ 2 & 5 & 2 \\ 1 & 0 & 7 \end{pmatrix}$$

has the dimensions 4×3.　　　　◆◆◆

Scalars

A single number, as opposed to an array of numbers, is called a *scalar*.

◆◆◆ **Example 7:** Some scalars are 5, 693.6, and -24.3.　　　　◆◆◆

A scalar can also be thought of as an array having just one row and one column.

Naming a Matrix

We will often denote or *name* a matrix with a single letter.

◆◆◆ **Example 8:** We can let

$$\mathbf{A} = \begin{pmatrix} 2 & 5 & 1 \\ 0 & 4 & 3 \end{pmatrix}$$

Thus we can represent an entire array by a single symbol.　　　　◆◆◆

The Null or Zero Matrix

A matrix in which all elements are zero is called the *null* or *zero* matrix.

◆◆◆ **Example 9:** The matrix

$$\mathbf{O} = \begin{pmatrix} 0 & 0 & 0 \\ 0 & 0 & 0 \end{pmatrix}$$

is a 2×3 zero matrix. We will denote the zero matrix by the boldface letter $\mathbf{O}$.

◆◆◆

The Unit Matrix

If all of the elements *not* on the main diagonal of a square matrix are 0, we have a *diagonal matrix.* If, in addition, the elements on the main diagonal are equal, we have a *scalar matrix.* If the elements on the main diagonal of a scalar matrix are 1's, we have a *unit matrix* or an *identity matrix.*

◆◆◆ **Example 10:** The following matrices are all diagonal matrices:

$$\mathbf{A} = \begin{pmatrix} 3 & 0 & 0 \\ 0 & 5 & 0 \\ 0 & 0 & 7 \end{pmatrix} \qquad \mathbf{B} = \begin{pmatrix} 6 & 0 & 0 \\ 0 & 6 & 0 \\ 0 & 0 & 6 \end{pmatrix} \qquad \mathbf{I} = \begin{pmatrix} 1 & 0 & 0 \\ 0 & 1 & 0 \\ 0 & 0 & 1 \end{pmatrix}$$

B and **I** are, in addition, scalar matrices, and **I** is a unit matrix. We usually denote the unit matrix by the letter **I**.   ◆◆◆

Equality of Matrices

When we are comparing two matrices *of the same dimension,* elements in each matrix having the same row and column subscripts are called *corresponding elements.* Two matrices of the same dimension are equal if their corresponding elements are equal.

◆◆◆ **Example 11:** The matrices

$$\begin{pmatrix} x & 2 & 5 \\ 3 & 1 & 7 \end{pmatrix} \quad \text{and} \quad \begin{pmatrix} 4 & 2 & 5 \\ 3 & 1 & y \end{pmatrix}$$

and equal only if $x = 4$ and $y = 7$.   ◆◆◆

Transpose of a Matrix

The *transpose* $\mathbf{A}'$ of a matrix $\mathbf{A}$ is obtained by changing the rows to columns and changing the columns to rows.

◆◆◆ **Example 12:** The transpose of the matrix

$$\begin{pmatrix} 1 & 2 & 3 & 4 \\ 5 & 6 & 7 & 8 \end{pmatrix}$$

is

$$\begin{pmatrix} 1 & 5 \\ 2 & 6 \\ 3 & 7 \\ 4 & 8 \end{pmatrix}$$ ◆◆◆

Determinant of a Square Matrix

We form the determinant of a square matrix by writing the same elements enclosed between vertical bars.

◆◆◆ **Example 13:** The determinant of the square matrix

$$\begin{pmatrix} a & b & c & d \\ e & f & g & h \\ i & j & k & l \\ m & n & o & p \end{pmatrix}$$

is

$$\begin{vmatrix} a & b & c & d \\ e & f & g & h \\ i & j & k & l \\ m & n & o & p \end{vmatrix}$$

◆◆◆

In Chapter 11 we learned how to evaluate determinants, so we'll not repeat that material here. We do, however, need a definition that will be useful later in this chapter.

A square matrix is called *singular* if its determinant is zero; it is called *nonsingular* if its determinant is not zero.

We'll see later when we want to find the *inverse* of a square matrix that the inverse exists only for a nonsingular matrix.

Exercise 1 ◆ Definitions

Given the following arrays:

$$A = \begin{pmatrix} 2 & 5 & 1 \\ 6 & 3 & 7 \\ 1 & 6 & 9 \\ 7 & 4 & 2 \end{pmatrix} \qquad B = \begin{pmatrix} 7 \\ 3 \\ 9 \\ 2 \end{pmatrix} \qquad C = \begin{pmatrix} f & i & q & w \\ & g & w & k \\ & & c & z \\ & & & b \end{pmatrix}$$

$$D = \begin{pmatrix} 6 & 2 & 0 & 1 \\ 2 & 8 & 3 & 9 \end{pmatrix} \qquad E = \begin{pmatrix} x & y \\ z & w \end{pmatrix} \qquad F = \begin{pmatrix} 0 & 0 & 0 & 0 \\ 0 & 0 & 0 & 0 \end{pmatrix}$$

$$G = 7 \qquad H = \begin{pmatrix} 3 & 8 \\ & 5 \end{pmatrix} \qquad I = (3 \quad 8 \quad 4 \quad 6)$$

Which of the nine arrays shown above is:

1. a rectangular array?
2. a square array?
3. a triangular array?
4. a column vector?
5. a row vector?
6. a table?
7. a list?
8. a scalar?
9. a null matrix?
10. a matrix?

For matrix **A** above, find each of the following elements:

11. a_{32}
12. a_{41}

Give the dimensions of:

13. matrix **A**.
14. matrix **B**.
15. matrix **D**.
16. matrix **E**.

Write the transpose of:

17. matrix **F**.
18. matrix **D**.

19. Under what conditions will the matrices

$$\begin{pmatrix} 6 & x \\ w & 8 \end{pmatrix} \quad \text{and} \quad \begin{pmatrix} y & 2 \\ 7 & z \end{pmatrix}$$

be equal?

Graphics Calculator

20. Most graphics calculators can do matrix manipulations. First locate the matrix editor on your calculator, probably labeled something such as $\boxed{\text{MAT}}$ or

MATRX . In the screen that appears, you will be asked to name the matrix and enter the matrix dimensions, that is, the number of rows and of columns. Then you will be prompted to enter the elements, one at a time.

Check your calculator manual to learn how to enter a matrix. Then locate the keystrokes for transposing a matrix. Use your calculator to find the transpose of any matrix in this exercise set.

Computer

21. A computer algebra system (CAS) is ideal for manipulating matrices. In *Derive,* for example, one way to enter a matrix or vector is as follows. From the main menu, choose *Author, Matrix.* A box will appear in which you enter the number of rows and of columns. Following that, another dialog box will appear into which you enter the elements of the matrix. Once the matrix is entered, you may perform various operations on it.

Study the manual for your CAS to learn how to enter a matrix. Then use it to find the transpose of any matrix in this exercise set.

12–2 Operations with Matrices

In earlier chapters we learned how to add, subtract, and multiply numbers and algebraic quantities. We now perform these operations on matrices.

Addition and Subtraction of Matrices

To add two matrices of the same dimensions, simply combine corresponding elements. We first show the addition of two vectors.

◆◆◆ Example 14:

$$(3 \quad 2 \quad -1 \quad 5) + (2 \quad 4 \quad 5 \quad -2) = (5 \quad 6 \quad 4 \quad 3)$$ ◆◆◆

Geometric Interpretation of the Sum of Two Vectors

We have given a geometric interpretation to a vector as the coordinates of the endpoint of a line segment drawn from the origin, as in Fig. 12–1. Thus the vector $(3 \quad 2)$ can represent the vector **A** shown in Fig. 12–2. Let us plot another vector **B**$(1 \quad 4)$ on the same graph as **A** and find the sum **A** + **B**.

$$\mathbf{A} + \mathbf{B} = (3 \quad 2) + (1 \quad 4) = (4 \quad 6)$$

Thus the geometric interpretation of the sum **A** + **B** is the *resultant* of **A** and **B**.

◆◆◆ Example 15: Find the resultant of the vectors $(3 \quad 2)$ and $(1 \quad 4)$.

Solution: Adding the vectors gives the resultant.

$$(3 \quad 2) + (1 \quad 4) = (4 \quad 6)$$ ◆◆◆

We'll now show the addition of matrices.

◆◆◆ Example 16:

$$\begin{pmatrix} 2 & -3 & 4 & 1 \\ 0 & 5 & 6 & -8 \end{pmatrix} + \begin{pmatrix} 5 & 8 & 1 & 2 \\ 9 & -3 & 0 & 1 \end{pmatrix} = \begin{pmatrix} 7 & 5 & 5 & 3 \\ 9 & 2 & 6 & -7 \end{pmatrix}$$ ◆◆◆

Subtraction is done in a similar way.

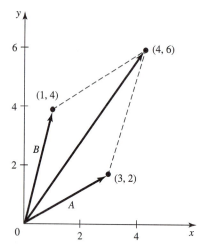

FIGURE 12–2 The resultant of two vectors. In Sec. 15–5 we'll show that this is an illustration of the *parallelogram method* for finding the resultant of two vectors.

Addition and subtraction are not *defined* for two matrices of *different* dimensions.

◆◆◆ **Example 17:**

$$\begin{pmatrix} 2 & -3 & 4 & 1 \\ 0 & 5 & 6 & -8 \end{pmatrix} - \begin{pmatrix} 5 & 8 & 1 & 2 \\ 9 & -3 & 0 & 1 \end{pmatrix} = \begin{pmatrix} -3 & -11 & 3 & -1 \\ -9 & 8 & 6 & -9 \end{pmatrix}$$ ◆◆◆

As with ordinary addition, the *commutative law* (Eq. 1) and the *associative law* (Eq. 3) apply.

Commutative Law	$A + B = B + A$	79

Associative Law	$A + (B + C) = (A + B) + C$ $= (A + C) + B$	80

Product of a Scalar and a Matrix

To multiply a matrix by a scalar, multiply each element in the matrix by the scalar. Let's start by multiplying a vector by a scalar.

◆◆◆ **Example 18:**

$$3(2 \quad 4 \quad -3 \quad 1) = (6 \quad 12 \quad -9 \quad 3)$$ ◆◆◆

The operation of multiplying a vector by a scalar can be interpreted geometrically. For example, if we multiply vector **A** by 2, we get

$$2A = 2(3 \quad 2) = (6 \quad 4)$$

Notice that this vector is collinear with **A** but of twice the length. So multiplying a vector by a scalar merely changes the length of the vector.

Now let's multiply a matrix by a scalar.

◆◆◆ **Example 19:**

$$3\begin{pmatrix} 2 & 4 & 1 \\ 5 & 7 & 3 \end{pmatrix} = \begin{pmatrix} 6 & 12 & 3 \\ 15 & 21 & 9 \end{pmatrix}$$ ◆◆◆

Working in reverse, we can *factor* a scalar from a matrix. Thus in Example 19, we can think of the number 3 as being *factored out*.

In symbols, the formula is expressed as follows:

Product of a Scalar and a Matrix	$k\begin{pmatrix} a & b \\ c & d \end{pmatrix} = \begin{pmatrix} ka & kb \\ kc & kd \end{pmatrix}$	87

Multiplication of Vectors and Matrices

Not all matrices can be multiplied. Further, for matrices **A** and **B**, it may be possible to find the product **AB** but *not* the product **BA**.

Conformable Matrices	The product **AB** of two matrices **A** and **B** is defined only when the number of columns in **A** equals the number of rows in **B**.	82

Two matrices for which multiplication is defined are called *conformable matrices*.

◆◆◆ **Example 20:** If

$$A = \begin{pmatrix} 3 & 5 \\ 2 & 1 \\ 7 & 4 \\ 9 & 0 \end{pmatrix} \quad \text{and} \quad B = \begin{pmatrix} 7 & 3 & 8 \\ 4 & 7 & 1 \end{pmatrix}$$

we can find the product **AB**, because the number of columns in **A** (2 columns) equals the number of rows in **B** (2 rows). However, we *cannot* find the product **BA** because the number of columns in **B** (3 columns) does not equal the number of rows in **A** (4 rows). ◆◆◆

For two or more conformable matrices, the *associative* law and the *distributive* law for multiplication apply.

Associative Law	$A(BC) = (AB)C = ABC$	84

Distributive Law	$A(B + C) = AB + AC$	85

These equations are similar to Eqs. 4 and 5 for ordinary multiplication.

The *dimensions* of the matrix obtained by multiplying an $m \times p$ matrix with a $p \times n$ matrix are equal to $m \times n$. In other words, the product matrix **AB** will have the same number of *rows* as **A** and the same number of *columns* as **B**.

Dimensions of the Product	$(m \times p)(p \times n) = (m \times n)$	86

◆◆◆ **Example 21:** The product **AB** of a 3×5 matrix **A** and a 5×7 matrix **B** is a 3×7 matrix. ◆◆◆

Scalar Product of Two Vectors

Whatever we say about matrices applies, of course, to vectors as well.

We will show matrix multiplication first with vectors and then extend it to other matrices.

The *scalar product* (also called the *dot product* or *inner product*) of a row vector **A** and a column vector **B** is defined only when the number of columns in **A** is equal to the number of rows in **B**.

If each vector contains m elements, the dimensions of the product will be, by Eq. 86,

$$(1 \times m)(m \times 1) = 1 \times 1$$

Thus the product is a scalar, hence the name *scalar product*.

To multiply a row vector by a column vector, multiply the first elements of each, then the second, and so on, and add the products obtained.

◆◆◆ **Example 22:**

$$(2 \quad 6 \quad 4 \quad 8)\begin{pmatrix} 3 \\ 5 \\ 1 \\ 7 \end{pmatrix} = [(2)(3) + (6)(5) + (4)(1) + (8)(7)]$$

$$= (6 + 30 + 4 + 56) = 96 \qquad ◆◆◆$$

We state this operation as the following formula:

$$(a \quad b \quad \ldots)\begin{pmatrix} x \\ y \\ \cdot \\ \cdot \\ \cdot \end{pmatrix} = (ax + by + \cdot \cdot \cdot) \qquad \textbf{88}$$

◆◆◆ **Example 23:** A certain store has four types of radios, priced at $71, $62, $83, and $49. On a particular day, the store sells quantities of 8, 3, 7, and 6, respectively. If we represent the prices by a row vector and the quantities by a column vector, and multiply, we get

$$(71 \quad 62 \quad 83 \quad 49)\begin{pmatrix} 8 \\ 3 \\ 7 \\ 6 \end{pmatrix} = [71(8) + 62(3) + 83(7) + 49(6)]$$

$$= 1629$$

The interpretation of this number is obviously the total dollar income from the sale of the four types of radios on that day. ◆◆◆

Product of a Row Vector and a Matrix

By Eq. 86, the product of a row vector and a matrix having n columns will be a row vector having n elements.

$$(1 \times p)(p \times n) = (1 \times n)$$

We saw earlier that the product of a row vector and a column vector is a single number, or scalar. Thus the product of the given row vector and the first column of the matrix will give a scalar. This scalar will be *the first element in the product vector.*

We find the *second* element in the product vector by multiplying the second column in the matrix by the row vector, and so on.

◆◆◆ **Example 24:** Multiply

$$(1 \quad 3 \quad 0)\begin{pmatrix} 7 & 5 \\ 1 & 0 \\ 4 & 9 \end{pmatrix}$$

Solution: The dimensions of the product will be

$$(1 \times 3)(3 \times 2) = (1 \times 2)$$

Multiplying the vector and the first column of the matrix gives

$$(1 \quad 3 \quad 0)\begin{pmatrix} \mathbf{7} & 5 \\ \mathbf{1} & 0 \\ \mathbf{4} & 9 \end{pmatrix} = [\mathbf{1(7)} + \mathbf{3(1)} + \mathbf{0(4)} \qquad * \qquad]$$

$$= (\mathbf{10} \qquad *)$$

Multiplying the vector and the second column of the matrix yields

$$(1 \quad 3 \quad 0)\begin{pmatrix} 7 & \mathbf{5} \\ 1 & \mathbf{0} \\ 4 & \mathbf{9} \end{pmatrix} = [10 \qquad \mathbf{1(5)} + \mathbf{3(0)} + \mathbf{0(9)}]$$

$$= (10 \quad 5)$$

◆◆◆

Here we use the asterisk (*) as a placeholder to denote the other element still to be computed.

◆◆◆ **Example 25:** Continuing Example 23, let us suppose that on the second day, the store sells quantities of 5, 9, 1, and 4, respectively, of the four radios. If we again represent the prices by a row vector, but now represent the quantities by a matrix and multiply, we get

$$(71 \quad 62 \quad 83 \quad 49)\begin{pmatrix} 8 & 5 \\ 3 & 9 \\ 7 & 1 \\ 6 & 4 \end{pmatrix} = \begin{bmatrix} 71(8) + 62(3) + 83(7) + 49(6) \\ 71(5) + 62(9) + 83(1) + 49(4) \end{bmatrix}$$
$$= (1629 \qquad 1192)$$

We interpret these numbers as the total dollar income from the sale of the four radios on each of the two days.   ◆◆◆

Product of Two Matrices

We now multiply two matrices **A** and **B**. We already know how to multiply a row vector and a matrix. So let us think of the matrix **A** as several row vectors. Each of these row vectors, multiplied by the matrix **B**, will produce a row vector in the product **AB**.

◆◆◆ **Example 26:** Find the product **AB** of the matrices

$$\mathbf{A} = \begin{pmatrix} 1 & 0 \\ 3 & 2 \end{pmatrix} \qquad \mathbf{B} = \begin{pmatrix} 5 & 4 & 8 \\ 6 & 7 & 9 \end{pmatrix}$$

Solution: The dimensions of the product will be

$$(2 \times 2)(2 \times 3) = (2 \times 3)$$

We think of the matrix **A** as being two row vectors, each of which, when multiplied by matrix **B**, will produce a row vector in the product matrix.

$$\underset{\mathbf{A}}{\begin{pmatrix} 1 & 0 \\ 3 & 2 \end{pmatrix}} \underset{\mathbf{B}}{\begin{pmatrix} 5 & 4 & 8 \\ 6 & 7 & 9 \end{pmatrix}} = \underset{\mathbf{AB}}{\begin{pmatrix} * & * & * \\ * & * & * \end{pmatrix}}$$

Multiplying by the first row vector in **A**, we have

$$\underset{\mathbf{A}}{\begin{pmatrix} \mathbf{1} & \mathbf{0} \\ 3 & 2 \end{pmatrix}} \underset{\mathbf{B}}{\begin{pmatrix} 5 & 4 & 8 \\ 6 & 7 & 9 \end{pmatrix}} = \underset{\mathbf{AB}}{\begin{bmatrix} 1(5) + 0(6) & 1(4) + 0(7) & 1(8) + 0(9) \\ * & * & * \end{bmatrix}}$$

$$= \begin{bmatrix} 5 & 4 & 8 \\ * & * & * \end{bmatrix}$$

Then multiplying by the second row vector in **A** yields

$$\underset{\mathbf{A}}{\begin{pmatrix} 1 & 0 \\ \mathbf{3} & \mathbf{2} \end{pmatrix}} \underset{\mathbf{B}}{\begin{pmatrix} 5 & 4 & 8 \\ 6 & 7 & 9 \end{pmatrix}} = \underset{\mathbf{AB}}{\begin{bmatrix} 5 & 4 & 8 \\ 3(5) + 2(6) & 3(4) + 2(7) & 3(8) + 2(9) \end{bmatrix}}$$

giving the final product

$$\underset{\mathbf{A}}{\begin{pmatrix} 1 & 0 \\ 3 & 2 \end{pmatrix}} \underset{\mathbf{B}}{\begin{pmatrix} 5 & 4 & 8 \\ 6 & 7 & 9 \end{pmatrix}} = \underset{\mathbf{AB}}{\begin{pmatrix} 5 & 4 & 8 \\ 27 & 26 & 42 \end{pmatrix}}$$ ◆◆◆

◆◆◆ **Example 27:** Find the product **AB** of the matrices

$$\underset{\mathbf{A}}{\begin{pmatrix} 1 & 0 & 2 \\ 0 & 1 & 1 \\ 3 & 0 & 2 \\ 1 & 1 & 0 \\ 2 & 1 & 1 \end{pmatrix}} \underset{\mathbf{B}}{\begin{pmatrix} 3 & 0 & 1 & 1 \\ 1 & 2 & 3 & 0 \\ 0 & 1 & 2 & 1 \end{pmatrix}}$$

Solution: We think of the matrix **A** as five row vectors. Each of these five row vectors, multiplied by the matrix **B**, will produce a row vector in the product **AB**.

$$
\begin{matrix}
(1 & 0 & 2) \\
(0 & 1 & 1) \\
(3 & 0 & 2) \\
(1 & 1 & 0) \\
(2 & 1 & 1)
\end{matrix}
\begin{pmatrix}
3 & 0 & 1 & 1 \\
1 & 2 & 3 & 0 \\
0 & 1 & 2 & 1
\end{pmatrix}
=
\begin{matrix}
(& & &) \\
(& & &) \\
(& & &) \\
(& & &) \\
(& & &)
\end{matrix}
$$

$$
\begin{matrix}
\mathbf{A} & \mathbf{B} & \mathbf{AB} \\
(5 \times 3) & (3 \times 4) & (5 \times 4)
\end{matrix}
$$

Multiplying the *first* row vector $(1 \quad 0 \quad 2)$ in **A** by the matrix **B** gives the first row vector in **AB**.

$$
\begin{matrix}
\mathbf{(1} & \mathbf{0} & \mathbf{2)} \\
(0 & 1 & 1) \\
(3 & 0 & 2) \\
(1 & 1 & 0) \\
(2 & 1 & 1)
\end{matrix}
\begin{pmatrix}
3 & 0 & 1 & 1 \\
1 & 2 & 3 & 0 \\
0 & 1 & 2 & 1
\end{pmatrix}
=
\begin{matrix}
\mathbf{(3} & \mathbf{2} & \mathbf{5} & \mathbf{3)} \\
(& & &) \\
(& & &) \\
(& & &) \\
(& & &)
\end{matrix}
$$

$$
\begin{matrix}
\mathbf{A} & \mathbf{B} & \mathbf{AB}
\end{matrix}
$$

Multiplying the *second* row vector in **A** and the matrix **B** gives a second row vector in **AB**.

$$
\begin{matrix}
(1 & 0 & 2) \\
\mathbf{(0} & \mathbf{1} & \mathbf{1)} \\
(3 & 0 & 2) \\
(1 & 1 & 0) \\
(2 & 1 & 1)
\end{matrix}
\begin{pmatrix}
3 & 0 & 1 & 1 \\
1 & 2 & 3 & 0 \\
0 & 1 & 2 & 1
\end{pmatrix}
=
\begin{matrix}
\mathbf{(3} & \mathbf{2} & \mathbf{5} & \mathbf{3)} \\
\mathbf{(1} & \mathbf{3} & \mathbf{5} & \mathbf{1)} \\
(& & &) \\
(& & &) \\
(& & &)
\end{matrix}
$$

$$
\begin{matrix}
\mathbf{A} & \mathbf{B} & \mathbf{AB}
\end{matrix}
$$

The *third* row vector in **A** produces a third row vector in **AB**.

$$
\begin{matrix}
(1 & 0 & 2) \\
(0 & 1 & 1) \\
\mathbf{(3} & \mathbf{0} & \mathbf{2)} \\
(1 & 1 & 0) \\
(2 & 1 & 1)
\end{matrix}
\begin{pmatrix}
3 & 0 & 1 & 1 \\
1 & 2 & 3 & 0 \\
0 & 1 & 2 & 1
\end{pmatrix}
=
\begin{matrix}
\mathbf{(3} & \mathbf{2} & \mathbf{5} & \mathbf{3)} \\
\mathbf{(1} & \mathbf{3} & \mathbf{5} & \mathbf{1)} \\
\mathbf{(9} & \mathbf{2} & \mathbf{7} & \mathbf{5)} \\
(& & &) \\
(& & &)
\end{matrix}
$$

$$
\begin{matrix}
\mathbf{A} & \mathbf{B} & \mathbf{AB}
\end{matrix}
$$

And so on, until we get the final product.

$$
\begin{pmatrix}
1 & 0 & 2 \\
0 & 1 & 1 \\
3 & 0 & 2 \\
1 & 1 & 0 \\
2 & 1 & 1
\end{pmatrix}
\begin{pmatrix}
3 & 0 & 1 & 1 \\
1 & 2 & 3 & 0 \\
0 & 1 & 2 & 1
\end{pmatrix}
=
\begin{pmatrix}
3 & 2 & 5 & 3 \\
1 & 3 & 5 & 1 \\
9 & 2 & 7 & 5 \\
4 & 2 & 4 & 1 \\
7 & 3 & 7 & 3
\end{pmatrix}
$$

$$
\begin{matrix}
\mathbf{A} & \mathbf{B} & \mathbf{AB}
\end{matrix}
$$
◆◆◆

We can state the product of two matrices as a formula as follows:

| Product of Two Matrices | $\begin{pmatrix} a & b & c \\ d & e & f \end{pmatrix}\begin{pmatrix} u & x \\ v & y \\ w & z \end{pmatrix} = \begin{pmatrix} au + bv + cw & ax + by + cz \\ du + ev + fw & dx + ey + fz \end{pmatrix}$ | **92** |

This formula works, of course, for conformable matrices with any number of elements.

◆◆◆ **Example 28:** Continuing Examples 23 and 25, how much would the income from the four radios be for each of the two days if the prices were discounted to

$$\$69 \quad \$57 \quad \$75 \quad \text{and} \quad \$36$$

and if they were marked up to

$$\$75 \quad \$65 \quad \$90 \quad \text{and} \quad \$57$$

Solution: Our prices can be represented by the matrix

$$\begin{pmatrix} 71 & 62 & 83 & 49 \\ 69 & 57 & 75 & 36 \\ 75 & 65 & 90 & 57 \end{pmatrix}$$

where the four columns represent the four different types of radios, as before, and the three rows represent the regular price, the discounted price, and the marked-up price. The quantities of each radio sold on each of the two days are represented by the same matrix as in Example 25. To find the income from the four radios for each of the three pieces for each on each of the two days, we multiply the two matrices.

$$\begin{pmatrix} 71 & 62 & 83 & 49 \\ 69 & 57 & 75 & 36 \\ 75 & 65 & 90 & 57 \end{pmatrix} \begin{pmatrix} 8 & 5 \\ 3 & 9 \\ 7 & 1 \\ 6 & 4 \end{pmatrix} = \begin{pmatrix} 1629 & 1192 \\ 1464 & 1077 \\ 1767 & 1278 \end{pmatrix}$$

The figures in the first row, \$1629 and \$1192, are the amounts from the sale of radios on the first and second days, as in Example 25. The other two rows give the incomes with the discounted, and then the marked-up, prices.   ◆◆◆

Product of Square Matrices

We multiply square matrices **A** and **B** just as we do other matrices. But unlike for rectangular matrices, both the products **AB** and **BA** will be defined, although they usually will not be equal.

◆◆◆ **Example 29:** Find the products **AB** and **BA** for the square matrices

$$\mathbf{A} = \begin{pmatrix} 0 & 1 & 2 \\ 1 & 1 & 0 \\ 2 & 3 & 1 \end{pmatrix} \qquad \mathbf{B} = \begin{pmatrix} 2 & 1 & 0 \\ 1 & 0 & 3 \\ 3 & 1 & 1 \end{pmatrix}$$

and show that the products are not equal.

Solution: Multiplying as shown earlier, we have

$$\mathbf{AB} = \begin{pmatrix} 0 & 1 & 2 \\ 1 & 1 & 0 \\ 2 & 3 & 1 \end{pmatrix} \begin{pmatrix} 2 & 1 & 0 \\ 1 & 0 & 3 \\ 3 & 1 & 1 \end{pmatrix} = \begin{pmatrix} 7 & 2 & 5 \\ 3 & 1 & 3 \\ 10 & 3 & 10 \end{pmatrix}$$

and

$$\mathbf{BA} = \begin{pmatrix} 2 & 1 & 0 \\ 1 & 0 & 3 \\ 3 & 1 & 1 \end{pmatrix} \begin{pmatrix} 0 & 1 & 2 \\ 1 & 1 & 0 \\ 2 & 3 & 1 \end{pmatrix} = \begin{pmatrix} 1 & 3 & 4 \\ 6 & 10 & 5 \\ 3 & 7 & 7 \end{pmatrix}$$ ◆◆◆

   We see that multiplying the matrices in the reverse order does *not* give the same result.

Commutative Law	$\mathbf{AB} \neq \mathbf{BA}$ Matrix multiplication is *not* commutative.	83

Product of a Matrix and a Column Vector

We have already multiplied a row vector **A** by a matrix **B**. Now we multiply a matrix **A** by a column vector **B**.

••• Example 30: Multiply

$$\begin{pmatrix} 3 & 1 & 2 & 0 \\ 1 & 3 & 5 & 2 \end{pmatrix}\begin{pmatrix} 2 \\ 1 \\ 3 \\ 4 \end{pmatrix}$$

Solution: The dimensions of the product will be

Here, of course, we use the same rules as for finding the product of any two matrices.

$$(2 \times 4)(4 \times 1) = (2 \times 1)$$

Multiplying the column vector by the first row in the matrix yields

$$\begin{pmatrix} \mathbf{3} & \mathbf{1} & \mathbf{2} & \mathbf{0} \\ 1 & 3 & 5 & 2 \end{pmatrix}\begin{pmatrix} 2 \\ 1 \\ 3 \\ 4 \end{pmatrix} = \begin{pmatrix} \mathbf{3(2)} + \mathbf{1(1)} + \mathbf{2(3)} + \mathbf{0(4)} \end{pmatrix} = \begin{pmatrix} 13 \end{pmatrix}$$

Then multiplying by the second row, we have

$$\begin{pmatrix} 3 & 1 & 2 & 0 \\ \mathbf{1} & \mathbf{3} & \mathbf{5} & \mathbf{2} \end{pmatrix}\begin{pmatrix} 2 \\ 1 \\ 3 \\ 4 \end{pmatrix} = \begin{pmatrix} 13 \\ \mathbf{1(2)} + \mathbf{3(1)} + \mathbf{5(3)} + \mathbf{2(4)} \end{pmatrix} = \begin{pmatrix} 13 \\ 28 \end{pmatrix}$$

•••

Note that the product of a matrix and a column vector is a column vector.

Most of the matrix multiplication that we do later will be of a matrix and a column vector.

••• Example 31: Five students decided to raise money by selling T-shirts for three weeks. The following table shows the number of shirts sold by each student during each week:

		Week		
		1	*2*	*3*
	1	8	2	6
	2	3	5	2
Student	*3*	6	5	4
	4	7	3	1
	5	2	3	4

They charged $7.50 per shirt the first week and then lowered the price to $6.00 the second week and to $5.50 the third week. Find the total amount raised by each student for the three weeks.

Solution: We know that to get the total amounts, we must multiply the matrix **B** representing the number of T-shirts sold,

$$\mathbf{B} = \begin{pmatrix} 8 & 2 & 6 \\ 3 & 5 & 2 \\ 6 & 5 & 4 \\ 7 & 3 & 1 \\ 2 & 3 & 4 \end{pmatrix}$$

by a vector **C** representing the price per shirt. But should **C** be a row vector or a column vector? And should we find the product **BC** or the product **CB**?

If we let **C** be a 1×3 row vector and try to multiply it by the 5×3 matrix **B**, we see that we cannot, because these matrices are not comfortable for either product **BC** or **CB**.

On the other hand, if we let **C** be a 3×1 column vector,

$$\mathbf{C} = \begin{pmatrix} 7.50 \\ 6.00 \\ 5.50 \end{pmatrix}$$

the product **CB** is not defined, but the product **BC** gives a column vector

$$\mathbf{BC} = \begin{pmatrix} 8 & 2 & 6 \\ 3 & 5 & 2 \\ 6 & 5 & 4 \\ 7 & 3 & 1 \\ 2 & 3 & 4 \end{pmatrix} \begin{pmatrix} 7.50 \\ 6.00 \\ 5.50 \end{pmatrix} = \begin{pmatrix} 105.00 \\ 63.50 \\ 97.00 \\ 76.00 \\ 55.00 \end{pmatrix}$$

which represents the total earned by each student.   ◆◆◆

Exercise 2 ◆ Operations with Matrices

Addition and Subtraction of Matrices

Combine the following vectors and matrices.

1. $(3 \quad 8 \quad 2 \quad 1) + (9 \quad 5 \quad 3 \quad 7)$

2. $\begin{pmatrix} 6 & 3 & 9 \\ 2 & 1 & 3 \end{pmatrix} + \begin{pmatrix} 8 & 2 & 7 \\ 1 & 6 & 3 \end{pmatrix}$

3. $\begin{pmatrix} 5 \\ -3 \\ 5 \\ 8 \end{pmatrix} + \begin{pmatrix} 7 \\ 0 \\ -4 \\ 8 \end{pmatrix} - \begin{pmatrix} -2 \\ 7 \\ -5 \\ 2 \end{pmatrix}$

4. $\begin{pmatrix} 6 & 2 & -7 \\ -3 & 5 & 9 \\ 2 & 5 & 3 \end{pmatrix} - \begin{pmatrix} 5 & 2 & 9 \\ 1 & 4 & 8 \\ 7 & -3 & 4 \end{pmatrix} + \begin{pmatrix} 1 & -6 & 7 \\ 2 & 4 & -9 \\ -6 & 2 & 7 \end{pmatrix}$

5. $\begin{pmatrix} 1 & 5 & -2 & 1 \\ 0 & -2 & 3 & 5 \\ -2 & 9 & 3 & -2 \\ 5 & -1 & 2 & 6 \end{pmatrix} - \begin{pmatrix} -6 & 4 & 2 & 8 \\ 9 & 2 & -6 & 3 \\ -5 & 2 & 6 & 7 \\ 6 & 3 & 7 & 0 \end{pmatrix}$

Product of a Scalar and a Matrix

Multiply each scalar and vector.

6. $6\begin{pmatrix} 9 \\ -2 \\ 6 \\ -3 \end{pmatrix}$

7. $-3(5 \quad 9 \quad -2 \quad 7 \quad 3)$

Multiply each scalar and matrix.

8. $7\begin{pmatrix} 7 & -3 & 7 \\ -2 & 9 & 5 \\ 9 & 1 & 0 \end{pmatrix}$

9. $-3\begin{pmatrix} 4 & 2 & -3 & 0 \\ -1 & 6 & 3 & -4 \end{pmatrix}$

Remove any factors from the vector or matrix.

10. $\begin{pmatrix} 25 \\ -15 \\ 30 \\ -5 \end{pmatrix}$

11. $\begin{pmatrix} 21 & 7 & -14 \\ 49 & 63 & 28 \\ 14 & -21 & 56 \\ -42 & 70 & 84 \end{pmatrix}$

If

$$\mathbf{A} = \begin{pmatrix} 2 & 5 & -1 \\ -3 & 2 & 6 \\ 9 & -4 & 5 \end{pmatrix} \quad \text{and} \quad \mathbf{B} = \begin{pmatrix} 0 & 1 & 6 \\ -7 & 2 & 1 \\ 4 & -2 & 0 \end{pmatrix}$$

find the following.

12. A − 4B   **13. 3A + B**   **14. 2A + 2B**

15. Show that $2(\mathbf{A} + \mathbf{B}) = 2\mathbf{A} + 2\mathbf{B}$, thus demonstrating the distributive law (Eq. 85).

Scalar Product of Two Vectors

Multiply each pair of vectors.

16. $(6 \quad -2)\begin{pmatrix} -1 \\ 7 \end{pmatrix}$

17. $(5 \quad 3 \quad 7)\begin{pmatrix} 2 \\ 6 \\ 1 \end{pmatrix}$

18. $(-3 \quad 6 \quad -9)\begin{pmatrix} 8 \\ -3 \\ 4 \end{pmatrix}$

19. $(-4 \quad 3 \quad 8 \quad -5)\begin{pmatrix} 7 \\ -6 \\ 0 \\ 3 \end{pmatrix}$

Product of a Row Vector and a Matrix

Multiply each row vector and matrix.

20. $(4 \quad -2)\begin{pmatrix} 3 & 6 \\ -1 & 5 \end{pmatrix}$

21. $(3 \quad -1)\begin{pmatrix} 4 & 7 & -2 & 4 & 0 \\ 2 & -6 & 8 & -3 & 7 \end{pmatrix}$

22. $(2 \quad 8 \quad 1)\begin{pmatrix} 5 & 6 \\ 3 & 1 \\ 1 & 0 \end{pmatrix}$

23. $(-3 \quad 6 \quad -1)\begin{pmatrix} 5 & -2 & 7 \\ -6 & 2 & 5 \\ 9 & 0 & -1 \end{pmatrix}$

24. $(9 \quad 1 \quad 4)\begin{pmatrix} 6 & 2 & 1 & 5 & -8 \\ 7 & 1 & -3 & 0 & 3 \\ 1 & -5 & 0 & 3 & 2 \end{pmatrix}$

25. $(2 \quad 4 \quad 3 \quad -1 \quad 0)\begin{pmatrix} 4 & 3 & 5 & -1 & 0 \\ 1 & 3 & 5 & -2 & 6 \\ 2 & -4 & 3 & 6 & 5 \\ 3 & 4 & -1 & 6 & 5 \\ 2 & 3 & 1 & 4 & -6 \end{pmatrix}$

Product of Two Matrices

Multiply each pair of matrices.

26. $\begin{pmatrix} 7 & 2 \\ 1 & 3 \end{pmatrix}\begin{pmatrix} 4 & 1 & 2 \\ 3 & 2 & 5 \end{pmatrix}$

27. $\begin{pmatrix} 3 & 1 & 4 \\ 1 & -3 & 5 \end{pmatrix}\begin{pmatrix} 4 & 2 \\ -1 & 5 \\ 2 & -6 \end{pmatrix}$

28. $\begin{pmatrix} 4 & 2 & -1 \\ -2 & 1 & 6 \end{pmatrix} \begin{pmatrix} 5 & -1 & 2 & 3 \\ 1 & 3 & 5 & 2 \\ 0 & 1 & -2 & 4 \end{pmatrix}$

29. $\begin{pmatrix} 5 & 1 \\ -2 & 0 \\ 1 & 3 \end{pmatrix} \begin{pmatrix} -4 & 1 \\ 0 & 5 \end{pmatrix}$

30. $\begin{pmatrix} 2 & 4 \\ 0 & -1 \\ 2 & 5 \end{pmatrix} \begin{pmatrix} 0 & -2 & 3 & -2 \\ 1 & 3 & -4 & 1 \end{pmatrix}$

31. $\begin{pmatrix} 4 & 1 & -2 & 3 \\ 0 & 2 & 1 & 5 \end{pmatrix} \begin{pmatrix} 3 & -1 \\ 0 & 2 \\ 3 & 1 \\ 0 & -2 \end{pmatrix}$

32. $\begin{pmatrix} -1 & 0 & 3 & 1 \\ 2 & 1 & 0 & -3 \end{pmatrix} \begin{pmatrix} 1 & 2 & 3 & 0 \\ 3 & 0 & -1 & 2 \\ -1 & 4 & 3 & 0 \\ 2 & 0 & -2 & 1 \end{pmatrix}$

33. $\begin{pmatrix} 3 & 1 \\ -2 & 0 \\ 1 & -3 \\ 5 & 0 \end{pmatrix} \begin{pmatrix} -1 & 0 \\ 3 & 2 \end{pmatrix}$

34. $\begin{pmatrix} 1 & 0 & -2 \\ 2 & 4 & 1 \\ 0 & -1 & 3 \end{pmatrix} \begin{pmatrix} 4 & 0 & 1 & -2 \\ 0 & 2 & 1 & 3 \\ 1 & 0 & -3 & 2 \end{pmatrix}$

35. $\begin{pmatrix} 2 & 1 & 0 & -1 \\ 1 & 2 & -1 & 1 \\ 0 & -2 & 4 & 0 \end{pmatrix} \begin{pmatrix} 1 & 2 & 0 & -1 \\ 3 & 1 & -1 & 3 \\ 2 & 1 & 0 & 3 \\ 3 & 5 & 1 & 0 \end{pmatrix}$

Product of Square Matrices

Find the product **AB** and the product **BA** for each pair of square matrices.

36. $\Lambda = \begin{pmatrix} 2 & 3 \\ 5 & 1 \end{pmatrix}$ $B = \begin{pmatrix} 4 & 6 \\ 7 & 9 \end{pmatrix}$

37. $A = \begin{pmatrix} 2 & -1 & 0 \\ 1 & 0 & 2 \\ -2 & 1 & 0 \end{pmatrix}$ $B = \begin{pmatrix} 2 & 1 & 0 \\ 3 & 0 & 1 \\ -2 & 4 & 1 \end{pmatrix}$

Using the matrices in problem 36, find the following.

38. A^2 **39.** B^3

Product of a Matrix and a Column Vector

Multiply each matrix and column vector.

40. $\begin{pmatrix} 2 & 1 \\ 0 & 3 \end{pmatrix} \begin{pmatrix} 3 \\ 1 \end{pmatrix}$

41. $\begin{pmatrix} 2 & 0 & 1 \\ 1 & 5 & 2 \end{pmatrix} \begin{pmatrix} 3 \\ 0 \\ -1 \end{pmatrix}$

42. $\begin{pmatrix} 2 & 1 & 0 \\ 1 & 0 & -2 \\ 3 & 5 & 0 \\ 0 & -1 & 3 \end{pmatrix} \begin{pmatrix} 3 \\ 0 \\ -1 \end{pmatrix}$

43. $\begin{pmatrix} 0 & 2 & -1 & 4 \\ 1 & 5 & 3 & 0 \\ -2 & 0 & 1 & 5 \end{pmatrix} \begin{pmatrix} 2 \\ 0 \\ -1 \\ 3 \end{pmatrix}$

44. The unit matrix **I** is a square matrix that has 1's on the main diagonal and 0's elsewhere. Show that the product of **I** and another square matrix **A** is simply **A**.

Multiplying by the Unit Matrix	$\mathbf{AI = IA = A}$	94

Applications

45. A certain store has four brands of athletic shoes, priced at $81, $72, $93, and $69. On a particular day, the store sells quantities of 2, 3, 7, and 4, respectively. Represent the prices and the quantities by vectors, and multiply them to get the total dollar income from the sale of the four brands of shoes on that day.

46. A certain store has computers, monitors, and keyboards priced at $2171, $625, and $149. On a particular day, the store sells quantities of 2, 3, and 5, respectively, and on the next day 4, 4, and 3. Represent the prices by a row vector and the quantities by a matrix. Multiply to get the receipts per day.

47. A building contractor has orders for 2 one-story houses, 4 two-story houses, and 3 three-story houses. He estimates that the amounts needed (in units) for each are as given in the following table:

	Lumber	Labor	Plumbing	Electrical
One-story	2.0	6.2	1.6	1.4
Two-story	3.1	7.5	2.5	2.1
Three-story	4.2	9.3	3.1	2.9

Represent the number of houses ordered by a row vector and the costs by a matrix. Multiply the two to get a list of needed amounts of the four items for the three kinds of houses combined.

48. In problem 47, suppose that the cost per unit for the four items are $10,400; $3620; $11,800; and $9850, respectively. Represent these costs by a column vector, and multiply it by the matrix to get a vector giving the total costs for each kind of house.

49. Two pizza shops sell the following numbers of pizzas:

On Monday			
	Cheese	**Ham**	**Veggie**
Shop *A*	32	26	19
Shop *B*	41	33	43

On Tuesday			
	Cheese	**Ham**	**Veggie**
Shop *A*	35	18	22
Shop *B*	33	26	32

The prices of the pizzas are $9.93, $10.45, and $11.20, respectively. Represent the sales for Monday by a matrix **M**, the sales for Tuesday by a matrix **T**, and the prices by a vector **P**.

(a) Add **M** and **T** to get the total sales per pizza per shop for Monday and Tuesday.
(b) Subtract **M** from **T** to get the change in sales from Monday to Tuesday.
(c) Multiply **M** and **P** to get the total income per shop for Monday, and multiply **T** and **P** to get the total income per shop for Tuesday.

Graphics Calculator

50. To add, subtract, or multiply two matrices on a calculator, first name and enter each of the matrices as in Exercise 1. Then simply type the names of two matrices to be added or multiplied, with the proper sign of operation between them. Thus to add matrices **A** and **B**, type

$$[A] + [B]$$

Pressing ⎣ENTER⎦ will display the sum of the two matrices. Use your graphics calculator to do any of the matrix operations in this exercise set.

Computer

51. A spreadsheet is ideal for matrix operations because the data are already in the form of lists or tables. Each cell in the spreadsheet corresponds to an element in the matrix.

To multiply two matrices that have already been entered in the spreadsheet, locate the command to do matrix multiplication. In *Lotus*, for example, it is

Data/Matrix/Multiply

and in *Quattro Pro* it is

Tools/Numeric/Multiply

Your spreadsheet will have something similar. You will then be prompted to identify the first matrix, then the second matrix, and finally the location where you want the product matrix.

Try this procedure for any of the multiplications in this exercise set.

52. Matrix operations are easy to do on a CAS. First you enter each of the two matrices, as in Exercise 1, each on its own line. If the first matrix is on line #12, say, and the second is on line #13, you would multiply them simply by writing a new line saying

#12 · #13

Use your CAS to perform any of the matrix operations in this exercise set.

12–3   The Inverse of a Matrix

Inverse of a Square Matrix

Recall from Sec. 12–1 that the unit matrix **I** is a square matrix that has 1's along its main diagonal and 0's elsewhere, such as

$$\mathbf{I} = \begin{pmatrix} 1 & 0 & 0 & 0 \\ 0 & 1 & 0 & 0 \\ 0 & 0 & 1 & 0 \\ 0 & 0 & 0 & 1 \end{pmatrix}$$

We now define the *inverse* of any given matrix **A** as another matrix $\mathbf{A}^{-1}$ such that the product of the matrix and its inverse is equal to the unit matrix.

| Product of a Matrix and Its Inverse | $\mathbf{AA}^{-1} = \mathbf{A}^{-1}\mathbf{A} = \mathbf{I}$ | 93 |

The product of a matrix and its inverse is the unit matrix.

In *ordinary* algebra, if b is some nonzero quantity, then

$$b\left(\frac{1}{b}\right) = bb^{-1} = 1$$

In *matrix* algebra, the unit matrix $\mathbf{I}$ has properties similar to those of the number 1, and the inverse of a matrix has properties similar to the *reciprocal* in ordinary algebra.

◆◆◆ **Example 32:** The inverse of the matrix

$$\mathbf{A} = \begin{pmatrix} 1 & 0 \\ 2 & 0.5 \end{pmatrix} \quad \text{is} \quad \mathbf{A}^{-1} = \begin{pmatrix} 1 & 0 \\ -4 & 2 \end{pmatrix}$$

because

$$\mathbf{AA}^{-1} = \begin{pmatrix} 1 & 0 \\ 2 & 0.5 \end{pmatrix}\begin{pmatrix} 1 & 0 \\ -4 & 2 \end{pmatrix} = \begin{pmatrix} 1(1) + 0(-4) & 1(0) + 0(2) \\ 2(1) + 0.5(-4) & 2(0) + 0.5(2) \end{pmatrix}$$

$$= \begin{pmatrix} 1 & 0 \\ 0 & 1 \end{pmatrix}$$

Not every matrix has an inverse. The inverse exists only for a *nonsingular* matrix, that is, one whose determinant is not zero. Such a matrix is said to be *invertible*.

and

$$\mathbf{A}^{-1}\mathbf{A} = \begin{pmatrix} 1 & 0 \\ -4 & 2 \end{pmatrix}\begin{pmatrix} 1 & 0 \\ 2 & 0.5 \end{pmatrix} = \begin{pmatrix} 1(1) + 0(2) & 1(0) + 0(0.5) \\ -4(1) + 2(2) & -4(0) + 2(0.5) \end{pmatrix}$$

$$= \begin{pmatrix} 1 & 0 \\ 0 & 1 \end{pmatrix}$$

◆◆◆

Finding the Inverse of a Matrix

When we solved sets of linear equations in Chapter 10, we were able to:

1. Interchange two equations.
2. Multiply an equation by a nonzero constant.
3. Add a constant multiple of one equation to another equation.

Thus for a matrix that represents such a system of equations, we may perform the following transformations without altering the meaning of a matrix:

| Elementary Transformations of a Matrix | 1. Interchange any rows.
2. Multiply a row by a nonzero constant.
3. Add a constant multiple of one row to another row. | 96 |

One method to find the inverse of a matrix is to apply the rules of matrix transformation to transform the given matrix into a unit matrix, while at the same time performing the same operations on a unit matrix. The unit matrix, after the transformations, becomes the inverse of the original matrix.

$$\begin{array}{cc} \mathbf{A} & \mathbf{I} \\ \downarrow & \downarrow \\ \mathbf{I} & \mathbf{A}^{-1} \end{array}$$

We will not try to prove the method here, but only show its use.

◆◆◆ **Example 33:** Find the inverse of the matrix of Example 32.

$$\mathbf{A} = \begin{pmatrix} 1 & 0 \\ 2 & 0.5 \end{pmatrix}$$

Solution: To the given matrix **A**, we append a unit matrix,

$$\left(\begin{array}{cc|cc} 1 & 0 & 1 & 0 \\ 2 & 0.5 & 0 & 1 \end{array}\right)$$

obtaining a *double-square* matrix. We then double the first row and subtract it from the second.

$$\left(\begin{array}{cc|cc} 1 & 0 & 1 & 0 \\ 0 & 0.5 & -2 & 1 \end{array}\right)$$

We now double the second row.

$$\left(\begin{array}{cc|cc} 1 & 0 & 1 & 0 \\ 0 & 1 & -4 & 2 \end{array}\right)$$

*Our original matrix **A** is now the unit matrix, and the original unit matrix is now the inverse of **A**.*

$$\mathbf{A}^{-1} = \begin{pmatrix} 1 & 0 \\ -4 & 2 \end{pmatrix}$$

◆◆◆

◆◆◆ **Example 34:** Find the inverse of the matrix

$$\mathbf{A} = \begin{pmatrix} a_1 & b_1 \\ a_2 & b_2 \end{pmatrix}$$

Solution: We append the unit matrix as follows:

$$\left(\begin{array}{cc|cc} a_1 & b_1 & 1 & 0 \\ a_2 & b_2 & 0 & 1 \end{array}\right)$$

Then we multiply the first row by b_2 and the second by b_1, and subtract the second from the first.

$$\left(\begin{array}{cc|cc} a_1b_2 - a_2b_1 & 0 & b_2 & -b_1 \\ a_2b_1 & b_1b_2 & 0 & b_1 \end{array}\right)$$

We now divide the first row by $a_1b_2 - a_2b_1$.

$$\left(\begin{array}{cc|cc} 1 & 0 & \dfrac{b_2}{a_1b_2 - a_2b_1} & \dfrac{-b_1}{a_1b_2 - a_2b_1} \\ a_2b_1 & b_1b_2 & 0 & b_1 \end{array}\right)$$

Then we multiply the first row by a_2b_1 and subtract it from the second. After combining fractions, we get

$$\left(\begin{array}{cc|cc} 1 & 0 & \dfrac{b_2}{a_1b_2 - a_2b_1} & \dfrac{-b_1}{a_1b_2 - a_2b_1} \\ 0 & b_1b_2 & \dfrac{-a_2b_1b_2}{a_1b_2 - a_2b_1} & \dfrac{a_1b_1b_2}{a_1b_2 - a_2b_1} \end{array}\right)$$

We then divide the second row by b_1b_2.

$$\left(\begin{array}{cc|cc} 1 & 0 & \dfrac{b_2}{a_1b_2 - a_2b_1} & \dfrac{-b_1}{a_1b_2 - a_2b_1} \\ 0 & 1 & \dfrac{-a_2}{a_1b_2 - a_2b_1} & \dfrac{a_1}{a_1b_2 - a_2b_1} \end{array}\right)$$

The inverse of our matrix is then

$$\mathbf{A}^{-1} = \frac{1}{a_1 b_2 - a_2 b_1} \begin{pmatrix} b_2 & -b_1 \\ -a_2 & a_1 \end{pmatrix}$$

after $1/(a_1 b_2 - a_2 b_1)$ has been factored out.

Comparing this result with the original matrix, we see that:

1. The elements on the main diagonal are interchanged.
2. The elements on the secondary diagonal have reversed sign.
3. It is divided by the determinant of the original matrix. ◆◆◆

Shortcut for Finding the Inverse of a 2 × 2 Matrix

The results of Example 34 provide us with a way to quickly write the inverse of a 2 × 2 matrix. We obtain the inverse of a 2 × 2 matrix by:

1. interchanging the elements on the main diagonal;
2. reversing the signs of the elements on the secondary diagonal; and
3. dividing by the determinant of the original matrix.

◆◆◆ **Example 35:** Find the inverse of

$$\mathbf{A} = \begin{pmatrix} 5 & -2 \\ 1 & 3 \end{pmatrix}$$

Solution: We interchange the 5 and the 3 and reverse the signs of the -2 and the 1.

$$\begin{pmatrix} 3 & 2 \\ -1 & 5 \end{pmatrix}$$

Then we divide by the determinant of the original matrix.

$$5(3) - (-2)(1) = 17$$

So

$$\mathbf{A}^{-1} = \frac{1}{17} \begin{pmatrix} 3 & 2 \\ -1 & 5 \end{pmatrix}$$ ◆◆◆

Exercise 3 ◆ The Inverse of a Matrix

Find the inverse of each matrix. Work to at least three decimal places.

1. $\begin{pmatrix} 4 & 8 \\ -5 & 0 \end{pmatrix}$

2. $\begin{pmatrix} -2 & 4 \\ 3 & 1 \end{pmatrix}$

3. $\begin{pmatrix} 8 & -3 \\ 1 & 6 \end{pmatrix}$

4. $\begin{pmatrix} 1 & 5 & 3 \\ 2 & 3 & 0 \\ 6 & 8 & 1 \end{pmatrix}$

5. $\begin{pmatrix} -1 & 3 & 7 \\ 4 & -2 & 0 \\ 7 & 2 & -9 \end{pmatrix}$

6. $\begin{pmatrix} 18 & 23 & 71 \\ 82 & 49 & 28 \\ 36 & 19 & 84 \end{pmatrix}$

7. $\begin{pmatrix} 9 & 1 & 0 & 3 \\ 4 & 3 & 5 & 2 \\ 7 & 1 & 0 & 3 \\ 1 & 4 & 2 & 7 \end{pmatrix}$

8. $\begin{pmatrix} -2 & 1 & 5 & -6 \\ 7 & -3 & 2 & -6 \\ -3 & 2 & 6 & 1 \\ 5 & 2 & 7 & 5 \end{pmatrix}$

Graphics Calculator

9. To invert a square matrix on a graphics calculator, first name and enter the matrix as you did in Exercise 1. Then locate the instruction on your calculator for inverting a matrix. On many calculators it is the $\boxed{x^{-1}}$ key. If so, to invert a matrix named **A**, you enter

$$\mathbf{A}^{-1}$$

and the inverse of matrix **A** will be displayed. Try this method on any of the matrices above.

Computer

10. Most spreadsheets have a command to invert a matrix. In *Lotus,* for example, the command is

Data/Matrix/Invert

and in *Quattro Pro,* the command is

Tools/Numeric/Invert

You will be prompted to highlight the matrix to be inverted and to indicate where the inverted matrix is to be placed. Try this procedure on any of the matrices in this exercise set.

11. A CAS can be used to invert a matrix. Enter the matrix as you did in Exercise 1. Then locate the command to invert the matrix. In *Derive,* you simply raise the matrix to the -1 power. Try this method on any of the matrices above.

12–4   Solving a System of Linear Equations by Matrix Inversion

Representing a System of Equations by Matrices

Suppose that we have the set of equations

$$2x + y + 2z = 10$$
$$x + 2y + 3z = 14$$
$$3x + 4y - 3z = 2$$

The left sides of these equations can be represented by the product of a square matrix **A** made up of the coefficients, called the *coefficient matrix,* and a column vector **X** made up of the unknowns. If

$$\mathbf{A} = \begin{pmatrix} 2 & 1 & 2 \\ 1 & 2 & 3 \\ 3 & 4 & -3 \end{pmatrix} \quad \text{and} \quad \mathbf{X} = \begin{pmatrix} x \\ y \\ z \end{pmatrix}$$

then

$$\mathbf{AX} = \begin{pmatrix} 2 & 1 & 2 \\ 1 & 2 & 3 \\ 3 & 4 & -3 \end{pmatrix} \begin{pmatrix} x \\ y \\ z \end{pmatrix} = \begin{pmatrix} 2x + y + 2z \\ x + 2y + 3z \\ 3x + 4y - 3z \end{pmatrix}$$

Note that our product is a column vector having *three* elements. The expression $(2x + y + 2z)$ is a *single element* in that vector.

If we now represent the constants in our set of equations by the column vector

$$\mathbf{B} = \begin{pmatrix} 10 \\ 14 \\ 2 \end{pmatrix}$$

our system of equations can be written, using matrices, as

$$\begin{pmatrix} 2 & 1 & 2 \\ 1 & 2 & 3 \\ 3 & 4 & -3 \end{pmatrix} \begin{pmatrix} x \\ y \\ z \end{pmatrix} = \begin{pmatrix} 10 \\ 14 \\ 2 \end{pmatrix}$$

or in the following more compact form:

Matrix Form for a System of Equations	$\mathbf{AX} = \mathbf{B}$	95

Solving Systems of Equations by Matrix Inversion

Suppose that we have a system of equations, which we represent in matrix form as

$$\mathbf{AX} = \mathbf{B}$$

where $\mathbf{A}$ is the coefficient matrix, $\mathbf{X}$ is the column vector of the unknowns, and $\mathbf{B}$ is the column of constants. Multiplying both sides by $\mathbf{A}^{-1}$, the inverse of $\mathbf{A}$, we have

$$\mathbf{A}^{-1}\mathbf{AX} = \mathbf{A}^{-1}\mathbf{B}$$

or

$$\mathbf{IX} = \mathbf{A}^{-1}\mathbf{B}$$

since $\mathbf{A}^{-1}\mathbf{A} = \mathbf{I}$. But, by Eq. 94, $\mathbf{IX} = \mathbf{X}$, so we have the following formula:

Solving a Set of Equations Using the Inverse	$\mathbf{X} = \mathbf{A}^{-1}\mathbf{B}$	98

The solutions to a set of equations can be obtained by multiplying the column of constants by the inverse of the coefficient matrix.

This is called the *matrix inversion* method. To use this method, set up the equations in standard form. Write $\mathbf{A}$, the matrix of the coefficients. Then find $\mathbf{A}^{-1}$ by matrix inversion or by the shortcut method (for a 2×2 matrix). Finally, multiply $\mathbf{A}^{-1}$ by the column of constants $\mathbf{B}$ to get the solution.

◆◆◆ **Example 36:** Solve the system of equations

$$\begin{aligned} x - 2y &= -3 \\ 3x + y &= 5 \end{aligned}$$

by the matrix inversion method.

Solution: The coefficient matrix $\mathbf{A}$ is

$$\mathbf{A} = \begin{pmatrix} 1 & -2 \\ 3 & 1 \end{pmatrix}$$

We now find the inverse, $\mathbf{A}^{-1}$, of $\mathbf{A}$. We append the unit matrix,

$$\left(\begin{array}{cc|cc} 1 & -2 & 1 & 0 \\ 3 & 1 & 0 & 1 \end{array} \right)$$

and then multiply the first row by -3 and add it to the second row.

$$\left(\begin{array}{cc|cc} 1 & -2 & 1 & 0 \\ 0 & 7 & -3 & 1 \end{array} \right)$$

There are several other methods for using matrices to solve a system of equations, including the *Gauss elimination* and the *unit matrix* method.

The matrix inversion method is built into many calculators and computer programs.

We could have used the shortcut method here, since this is a 2×2 matrix. We have, however, used the longer general method to give another example of its use.

Then we multiply the second row by 2/7 and add it to the first row.

$$\begin{pmatrix} 1 & 0 & | & 1/7 & 2/7 \\ 0 & 7 & | & -3 & 1 \end{pmatrix}$$

Finally, we divide the second row by 7.

$$\begin{pmatrix} 1 & 0 & | & 1/7 & 2/7 \\ 0 & 1 & | & -3/7 & 1/7 \end{pmatrix}$$

Thus the inverse of our coefficient matrix $\mathbf{A}$ is

$$\mathbf{A}^{-1} = \begin{pmatrix} 1/7 & 2/7 \\ -3/7 & 1/7 \end{pmatrix}$$

We then get our solution by multiplying the inverse $\mathbf{A}^{-1}$ by the column of constants, $\mathbf{B}$.

$$\begin{array}{ccc} \mathbf{A}^{-1} & \mathbf{B} & = & \mathbf{X} \end{array}$$

$$\begin{pmatrix} 1/7 & 2/7 \\ -3/7 & 1/7 \end{pmatrix} \begin{pmatrix} -3 \\ 5 \end{pmatrix} = \begin{pmatrix} -3/7 + 10/7 \\ 9/7 + 5/7 \end{pmatrix} = \begin{pmatrix} 1 \\ 2 \end{pmatrix}$$

so $x = 1$ and $y = 2$. ◆◆◆

We see that even for a system of only two equations, manual solution by matrix inversion is a lot of work. Our main reason for covering it is for its wide use in computer solutions of sets of equations.

◆◆◆ **Example 37:** Solve the general system of two equations

$$a_1 x + b_1 y = c_1$$
$$a_2 x + b_2 y = c_2$$

by matrix inversion.

Solution: We write the coefficient matrix

$$\mathbf{A} = \begin{pmatrix} a_1 & b_1 \\ a_2 & b_2 \end{pmatrix}$$

The inverse of our coefficient matrix has already been found in Example 34. It is

$$\mathbf{A}^{-1} = \frac{1}{a_1 b_2 - a_2 b_1} \begin{pmatrix} b_2 & -b_1 \\ -a_2 & a_1 \end{pmatrix}$$

By Eq. 98 we now get the column of unknowns by multiplying the inverse by the column of coefficients.

$$\mathbf{X} = \mathbf{A}^{-1}\mathbf{B}$$

$$= \frac{1}{a_1 b_2 - a_2 b_1} \begin{pmatrix} b_2 & -b_1 \\ -a_2 & a_1 \end{pmatrix} \begin{pmatrix} c_1 \\ c_2 \end{pmatrix}$$

$$= \begin{pmatrix} \dfrac{b_2 c_1 - b_1 c_2}{a_1 b_2 - a_2 b_1} \\[2mm] \dfrac{a_1 c_2 - a_2 c_1}{a_1 b_2 - a_2 b_1} \end{pmatrix}$$

Thus we again get the following, now-familiar, equations:

$$x = \frac{b_2 c_1 - b_1 c_2}{a_1 b_2 - a_2 b_1} \qquad y = \frac{a_1 c_2 - a_2 c_1}{a_1 b_2 - a_2 b_1} \qquad \textbf{63}$$

◆◆◆

Exercise 4 ◆ Solving a System of Linear Equations by Matrix Inversion

1. Solve any of the sets of equations in Chapter 10 or 11 by matrix inversion.

Graphics Calculator

2. You learned how to invert a square matrix in Exercise 3 and how to multiply matrices in Exercise 2. Now to solve a system of equations, simply enter the coefficient matrix, invert it, and multiply it by the column of constants. If the coefficient matrix is **A** and the column of constants is **B,** we would enter

$$\mathbf{A^{-1}B}$$

Try this on any system of equations.

Computer

3. You learned how to invert a square matrix in your CAS in Exercise 3 and how to multiply matrices in Exercise 2. Now use those same commands to solve a system of equations. Enter the coefficient matrix, invert it, and multiply it by the column of constants. Try this procedure on any system of equations.

4. You learned how to invert a square matrix with your spreadsheet in Exercise 3 and how to multiply matrices in Exercise 2. Now combine the two operations to solve a system of equations by entering the coefficient matrix, inverting it, and multiplying it by the column of constants. Try this method on any system of equations.

◆◆◆ CHAPTER 12 REVIEW PROBLEMS ◆◆◆◆◆◆◆◆◆◆◆◆◆◆◆◆◆◆◆◆◆◆◆◆◆◆◆◆◆◆◆◆

1. Combine:

$$(4 \quad 9 \quad -2 \quad 6) + (2 \quad 4 \quad 9 \quad -1) - (4 \quad 1 \quad -3 \quad 7)$$

2. Multiply:

$$\begin{pmatrix} 8 & 4 \\ 9 & -5 \end{pmatrix} \begin{pmatrix} 3 & 7 \\ -3 & 6 \end{pmatrix}$$

3. Multiply:

$$(7 \quad -2 \quad 5) \begin{pmatrix} 6 \\ -2 \\ 8 \end{pmatrix}$$

4. Find the inverse:

$$\begin{pmatrix} 5 & -2 & 5 & 1 \\ 2 & 8 & -3 & -1 \\ -4 & 2 & 7 & 5 \\ 9 & 2 & 5 & 3 \end{pmatrix}$$

5. Find the inverse:

$$\begin{pmatrix} 4 & -4 \\ 3 & 1 \end{pmatrix}$$

6. Combine:

$$\begin{pmatrix} 5 & 9 \\ -4 & 7 \end{pmatrix} + \begin{pmatrix} -2 & 6 \\ 3 & 8 \end{pmatrix}$$

7. Multiply:

$$\begin{pmatrix} 1 & 2 \\ 0 & 3 \end{pmatrix} \begin{pmatrix} 1 & 2 & 4 \\ 3 & 0 & 1 \end{pmatrix}$$

8. Multiply

$$\begin{pmatrix} 8 & 2 & 6 & 1 \\ 9 & 0 & 1 & 4 \\ 1 & 6 & 9 & 3 \\ 6 & 9 & 3 & 5 \end{pmatrix} \begin{pmatrix} 5 \\ 3 \\ 2 \\ 8 \end{pmatrix}$$

9. Solve:

$$\begin{aligned} x + y + z + w &= -4 \\ x + 2y + 3z + 4w &= 0 \\ x + 3y + 6z + 10w &= 9 \\ x + 4y + 10z + 20w &= 24 \end{aligned}$$

10. Multiply:

$$7\begin{pmatrix} 5 & -3 \\ -4 & 8 \end{pmatrix}$$

11. Solve:

$$\begin{aligned} 5x + 3y - 2z &= 5 \\ 3x - 4y + 3z &= 13 \\ x + 6y - 4z &= -8 \end{aligned}$$

12. Find the inverse:

$$\begin{pmatrix} 5 & 1 & 4 \\ -2 & 6 & 9 \\ 4 & -1 & 0 \end{pmatrix}$$

13. Solve:

$$\begin{aligned} 4x + 3y &= 27 \\ 2x - 5y &= -19 \end{aligned}$$

14. Multiply:

$$9(6 \quad 3 \quad -2 \quad 8)$$

15. The four loop currents (mA) in a certain circuit satisfy the equations

$$\begin{aligned} 376I_1 + 374I_2 - 364I_3 - 255I_4 &= 284 \\ 898I_1 - 263I_2 + 229I_3 + 274I_4 &= -847 \\ 364I_1 + 285I_2 - 857I_3 - 226I_4 &= 746 \\ 937I_1 + 645I_2 - 387I_3 + 847I_4 &= -254 \end{aligned}$$

Solve for the four currents.

16. A factory makes TVs, stereos, CD players, and tape recorders, which it sells for $181, $62, $33, and $49, respectively. On a particular day, the factory makes quantities of 232, 373, 737, and 244, respectively. Represent the prices and the quantities by vectors, and multiply them to get the total value of the four items made on that day.

17. On the following day, the factory in problem 16 makes quantities of the given items of 273, 836, 361, and 227. Represent the prices by a row vector and the quantities made for both days by a matrix. Multiply to get a table giving the total values per day.

18. A company has two plants that make the following numbers of items:

	In April		
	Skis	**Snowboards**	**Surfboards**
Plant *A*	132	126	129
Plant *B*	241	133	143

	In May		
	Skis	**Snowboards**	**Surfboards**
Plant *A*	135	218	122
Plant *B*	133	126	232

The prices of the items are $129.53, $110.65, and $411.60, respectively. Represent the sales for April by a matrix **A**, the sales for May by a matrix **M**, and the prices by a vector **P**.

(a) Add **A** and **M** to get the total production per item per plant for April and May.

(b) Subtract **A** from **M** to get the change in production from April to May.

(c) Multiply **A** and **P** to get the total income for April, and multiply **M** and **P** to get the total income for May.

Writing

19. Suppose that you have submitted a lab report in which you stated that you have used matrix inversion to solve some sets of equations. The report was rejected because you failed to say anything about that method.

   Write one paragraph explaining matrix inversion to be inserted into the next draft of your report. Start by telling what matrix inversion is, and then go on to say how it works.

Team Project

20. In Chapter 10 we learned how to solve a system of equations by classical methods. In Chapter 11 we used determinants, and in this chapter we learned matrix inversion. For each method we could do the solution manually, and usually by calculator, by a CAS, or with a spreadsheet.

   Given a system of equations by your instructor, solve that system as many ways as you can. See which team can get the correct solution by the greatest number of means.

13

Exponents and Radicals

◆◆◆ **OBJECTIVES** ◆◆

When you have completed this chapter, you should be able to:

- Use the laws of exponents to simplify and combine expressions having integral exponents.
- Simplify radicals by removing perfect powers, by rationalizing the denominator, and by reducing the index.
- Add, subtract, multiply, and divide radicals.
- Solve radical equations.

◆◆

In Chapter 2, "Introduction to Algebra," we studied expressions with integral exponents and derived some laws of exponents. We expand upon that work in this chapter, using the laws of exponents and our additional skills of factoring and manipulation of fractions, to simplify more complex expressions.

We also define *fractional exponent,* which we'll see is another way of writing a radical. We had a brief look at *radicals* in Chapter 1, where we computed roots by calculator. There, all of the problems were numerical, but here we will have letter quantities under the radical sign. We will learn each operation first with numbers, because they are familiar, and then go on to letters. However, if you simply want the decimal value of a radical containing only numbers, you use your calculator as was shown in Chapter 1.

Here, we'll simplify, add, subtract, multiply, and divide radicals and solve some simple radical equations. We must postpone the harder radical equations until we can solve quadratic equations (Chapter 14).

13–1 Integral Exponents

In this section, we continue the study of exponents that we started in Sec. 2–3. We repeat the laws of exponents derived there and use them to simplify harder expressions than before. You should glance back at that section before starting here.

Negative Exponents

The law that we derived in Sec. 2–3 for negative exponents is repeated here.

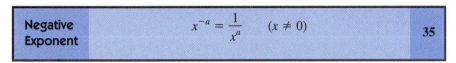

| Negative Exponent | $x^{-a} = \dfrac{1}{x^a}$ $(x \neq 0)$ | 35 |

When taking the reciprocal of a base raised to a power, **change the sign of the exponent.**

We'll use the laws of exponents mainly to simplify expressions, to make them easier to work with in later computations, such as solving equations containing exponents. For example, we use the law for negative exponents to rewrite an expression so that it does not contain a negative exponent.

◆◆◆ **Example 1:**

(a) $7^{-1} = \dfrac{1}{7}$

(b) $x^{-1} = \dfrac{1}{x}$

(c) $z^{-3} = \dfrac{1}{z^3}$

(d) $xy^{-1} = \dfrac{x}{y}$

(e) $\dfrac{ab^{-2}}{c^{-3}d} = \dfrac{ac^3}{b^2 d}$

◆◆◆

Another use for negative exponents is to rewrite an expression so that it does not contain fractions.

◆◆◆ **Example 2:**

(a) $\dfrac{1}{w} = w^{-1}$

(b) $\dfrac{x}{y} = xy^{-1}$

(c) $\dfrac{p^3}{q^2} = p^3 q^{-2}$

(d) $\dfrac{ab^2 c}{w^3 x^2 y} = ab^2 cw^{-3} x^{-2} y^{-1}$

◆◆◆

Zero Exponents

The law that we derived in Sec. 2–3 for zero exponents is repeated here.

| Zero Exponent | $x^0 = 1$ $(x \neq 0)$ | 34 |

Any quantity (except 0) raised to the zero power equals 1.

We use this law to rewrite an expression so that it contains no zero exponents.

◆◆◆ **Example 3:**

(a) $367^0 = 1$

(b) $(\sin 37.2°)^0 = 1$

(c) $x^2 y^0 z = x^2 z$

(d) $(abc)^0 = 1$

◆◆◆

We'll use these laws frequently in the examples to come.

Common Error	$x^{-a} \neq x^{1/a}$

Power Raised to a Power

Another law that we derived in Chapter 1 is for raising a power to a power.

Powers	$(x^a)^b = x^{ab} = (x^b)^a$	31

When raising a power to a power, keep the same base and **multiply the exponents.**

◆◆◆ **Example 4:**

(a) $(2^2)^2 = 2^4 = 16$

(b) $(3^{-1})^4 = 3^{-4} = \dfrac{1}{3^4} = \dfrac{1}{81}$

(c) $(a^2)^3 = a^6$

(d) $(z^{-3})^{-2} = z^6$

(e) $(x^5)^{-2} = x^{-10} = \dfrac{1}{x^{10}}$

◆◆◆

Products

The two laws of exponents that apply to products are given here again.

Products	$x^a \cdot x^b = x^{a+b}$	29

When multiplying powers of the same base, keep the same base and **add the exponents.**

Product Raised to a Power	$(xy)^n = x^n \cdot y^n$	32

When a product is raised to a power, each factor may be **separately** raised to the power.

◆◆◆ **Example 5:**

(a) $2axy^3z(3a^2xyz^3) = 6a^3x^2y^4z^4$

(b) $(3bx^3y^2)^3 = 27b^3x^9y^6$

(c) $\left(\dfrac{xy^2}{3}\right)^2 = \dfrac{x^2y^4}{9}$

◆◆◆

Common Error	Parentheses are important when we are raising products to a power. $(2x)^0 = 1$ but $2x^0 = 2(1) = 2$ $(4x)^{-1} = \dfrac{1}{4x}$ but $4x^{-1} = \dfrac{4}{x}$

We try to leave our expressions without zero or negative exponents.

◆◆◆ **Example 6:**

(a) $3p^3q^2r^{-4}(4p^{-5}qr^4) = 12p^{-2}q^3r^0 = \dfrac{12q^3}{p^2}$

(b) $(2m^2n^3x^{-2})^{-3} = 2^{-3}m^{-6}n^{-9}x^6 = \dfrac{x^6}{8m^6n^9}$

◆◆◆

Of course, our old rules for multiplying multinomials still apply.

◆◆◆ **Example 7:**

(a) $3x^{-2}y^3(2ax^3 - x^2y + ax^4y^{-3})$
$= 6axy^3 - 3x^0y^4 + 3ax^2y^0 = 6axy^3 - 3y^4 + 3ax^2$

(b) $(p^2x - q^2y^{-2})(p^{-2} - q^{-2}y^2)$
$= p^0x - p^2q^{-2}xy^2 - p^{-2}q^2y^{-2} + q^0y^0$

$= x - \dfrac{p^2xy^2}{q^2} - \dfrac{q^2}{p^2y^2} + 1$

◆◆◆

Letters in the exponents are handled just as if they were numbers.

◆◆◆ **Example 8:**

$$x^{n+1}y^{1-n}(x^{2n-3}y^{n-1}) = x^{n+1+2n-3}y^{1-n+n-1}$$
$$= x^{3n-2}y^0 = x^{3n-2}$$

◆◆◆

Quotients

Our two laws of exponents that apply to quotients are repeated here.

Quotients	$\dfrac{x^a}{x^b} = x^{a-b} \qquad (x \neq 0)$	30

When dividing powers of the same base, keep the same base and **subtract the exponents.**

Quotient Raised to a Power	$\left(\dfrac{x}{y}\right)^n = \dfrac{x^n}{y^n} \qquad (y \neq 0)$	33

When a quotient is raised to a power, the numerator and the denominator may be **separately** raised to the power.

◆◆◆ **Example 9:**

(a) $\dfrac{x^5}{x^3} = x^2$

(b) $\dfrac{16p^5q^{-3}}{8p^2q^{-5}} = 2p^3q^2$

(c) $\dfrac{x^{2p-1}y^2}{x^{p+4}y^{1-p}} = x^{2p-1-p-4}y^{2-1+p}$
$= x^{p-5}y^{1+p}$

(d) $\left(\dfrac{x^3y^2}{p^2q}\right)^3 = \dfrac{x^9y^6}{p^6q^3}$

◆◆◆

As before, we try to leave our expressions without zero or negative exponents.

••• Example 10:

(a) $\dfrac{8a^3b^{-5}c^3}{4a^5b^{-1}c^3} = 2a^{-2}b^{-4}c^0 = \dfrac{2}{a^2b^4}$ (b) $\left(\dfrac{a}{b}\right)^{-2} = \dfrac{a^{-2}}{b^{-2}} = \dfrac{b^2}{a^2} = \left(\dfrac{b}{a}\right)^2$ •••

Note in Example 10(b) that the negative exponent had the effect of inverting the fraction.

Don't forget our hard-won skills of factoring, of combining fractions over a common denominator, and of simplifying complex fractions. Use them where needed.

••• Example 11:

(a) $\dfrac{1}{(5a)^2} + \dfrac{1}{(2a^2b^{-4})^2} = \dfrac{1}{25a^2} + \dfrac{1}{\dfrac{4a^4}{b^8}}$

We simplify the compound fraction and combine the two fractions over a common denominator, getting

$$\dfrac{1}{25a^2} + \dfrac{b^8}{4a^4} = \dfrac{4a^2}{100a^4} + \dfrac{25b^8}{100a^4}$$

$$= \dfrac{4a^2 + 25b^8}{100a^4}$$

(b) $(x^{-2} - 3y)^{-2} = \dfrac{1}{(x^{-2} - 3y)^2} = \dfrac{1}{\left(\dfrac{1}{x^2} - 3y\right)^2}$

$$= \dfrac{1}{\left(\dfrac{1 - 3x^2y}{x^2}\right)^2}$$

$$= \dfrac{x^4}{(1 - 3x^2y)^2}$$

(c) $\dfrac{p^{4m} - q^{2n}}{p^{2m} + q^n} = \dfrac{(p^{2m})^2 - (q^n)^2}{p^{2m} + q^n}$

Factoring the difference of two squares in the numerator and canceling gives us

$$\dfrac{(p^{2m} + q^n)(p^{2m} - q^n)}{p^{2m} + q^n} = p^{2m} - q^n$$

•••

Exercise 1 ◆ Integral Exponents

Simplify, and write without negative exponents.

1. $3x^{-1}$ **2.** $4a^{-2}$

3. $2p^0$ **4.** $(3m)^{-1}$

5. $a(2b)^{-2}$ **6.** x^2y^{-1}

7. a^3b^{-2} **8.** m^2n^{-3}

9. p^3q^{-1} **10.** $(2x^2y^3z^4)^{-1}$

11. $(3m^3n^2p)^0$ **12.** $(5p^2q^2r^2)^3$

13. $(4a^3b^2c^6)^{-2}$ **14.** $(a^p - b^q)^2$

15. $(x + y)^{-1}$

16. $(2a + 3b)^{-1}$

17. $(m^{-2} - 6n)^{-2}$

18. $(4p^2 + 5q^{-4})^{-2}$

19. $2x^{-1} + y^{-2}$

20. $p^{-1} - 3q^{-2}$

21. $(3m)^{-3} - 2n^{-2}$

22. $(5a)^{-2} + (2a^2b^{-4})^{-2}$

23. $(x^n + y^m)^2$

24. $(x^{a-1} + y^{a-2})(x^a + y^{a-1})$

25. $(16x^6y^0 \div 8x^4y) \div 4xy^6$

26. $(a^{n-1} + b^{n-2})(a^n + b^{n-1})$

27. $(72p^6q^7 \div 9p^4q) \div 8pq^6$

28. $\dfrac{(p^2 - pq)^8}{(p - q)^4}$

29. $\left(\dfrac{x^{m+n}}{x^n}\right)^m$

30. $\dfrac{27x^{3n} + 8y^{3n}}{3x^n + 2y^n}$

31. $\left(\dfrac{3}{b}\right)^{-1}$

32. $\left(\dfrac{x}{4}\right)^{-1}$

33. $\left(\dfrac{2x}{3y}\right)^{-2}$

34. $\left(\dfrac{3p}{4y}\right)^{-1}$

35. $\left(\dfrac{2px}{3qy}\right)^{-3}$

36. $\left(\dfrac{p^2}{q^0}\right)^{-1}$

37. $\left(\dfrac{3a^4b^3}{5x^2y}\right)^2$

38. $\left(\dfrac{x^2}{y}\right)^{-3}$

39. $\left(\dfrac{-2a^3x^3}{3b^2y}\right)^{2n}$

40. $\left(\dfrac{3p^2y^3}{4qx^4}\right)^{-2}$

41. $\left(\dfrac{2p^{-2}z^3}{3q^{-2}x^{-4}}\right)^{-1}$

42. $\left(\dfrac{9m^4n^3}{3p^3q}\right)^3$

43. $\left(\dfrac{5w^2}{2z}\right)^p$

44. $\left(\dfrac{5a^5b^3}{2x^2y}\right)^{3n}$

45. $\left(\dfrac{3n^{-2}y^3}{5m^{-3}x^4}\right)^{-2}$

46. $\left(\dfrac{z^0z^{2n-2}}{z^n}\right)^3$

47. $\left(\dfrac{1 + a}{1 + b}\right)^3 \div \dfrac{1 + a^3}{b^2 - 1}$

48. $\left(\dfrac{p^2 - q^2}{p + q}\right)^2 \cdot \dfrac{(p^3 + q^3)^2}{(p + q)^2}$

49. $\dfrac{(x^2 - xy)^7}{(x - y)^5}$

50. $\dfrac{(a^n)^{2n} - (b^{2n})^n}{(a^n)^n + (b^n)^n}$

51. $\dfrac{27a^{3n} - 8b^{3n}}{3a^n - 2b^n}$

52. $\left(\dfrac{a^2a^{2n-2}}{a^n}\right)^2$

53. $\left(\dfrac{x + 1}{y + 1}\right)^3 \div \dfrac{x^3 + 1}{y^2 - 1}$

54. $\left(\dfrac{a^2 - b^2}{x + y}\right)^2 \cdot \dfrac{(x^3 + y^3)^2}{(a + b)^2}$

55. $\left(\dfrac{7x^3}{15y^3a^{m-1}} \div \dfrac{x^{n+1}}{a^{m+n}}\right)\dfrac{y^n}{a^{m-1}}$

56. $\dfrac{a^7b^8c^9 + a^6b^7c^8 - 3a^5b^6c^7}{a^5b^6c^7}$

57. $\dfrac{128a^4b^3 - 48a^5b^2 - 40a^6b + 15a^7}{3a^2 - 8ab}$

58. $\left(\dfrac{x^{m+n}}{x^n}\right)^m \div \left(\dfrac{x^n}{x^{m+n}}\right)^{n+m}$

59. $\left(\dfrac{5a^3}{25b^3a^{m-1}} \div \dfrac{a^{n+1}}{a^{m+n}}\right)\dfrac{b^n}{a^{m-1}}$

60. $\dfrac{p^4q^8r^9 - p^6q^4r^8 + 3p^5q^6r^4}{p^3q^2r^4}$

61. $\dfrac{3x^4y^3 + 4x^5y^2 + 3x^6y + 4x^7}{4x^2 + 3xy}$

62. $\left(\dfrac{p^n}{p^{m+n}}\right)^{n-m} \div \left(\dfrac{p^{m+n}}{p^n}\right)^m$

63. $\dfrac{a^{-1}+b^{-1}}{a^{-1}-b^{-1}} \cdot \dfrac{a^{-2}-b^{-2}}{a^{-2}+b^{-2}} \cdot \dfrac{1}{\left(1+\dfrac{b}{a}\right)^{-2} + \left(1+\dfrac{a}{b}\right)^{-2}}$

Applications

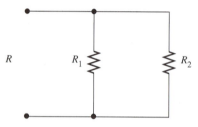

FIGURE 13–1

64. The resistance R of two resistors R_1 and R_2 wired in parallel, as shown in Fig. 13–1, is given by Eq. A64.

$$\frac{1}{R} = \frac{1}{R_1} + \frac{1}{R_2}$$

Write this equation without fractions.

65. When a current I causes a power P in a resistance R, that resistance is given by Eq. A67.

$$R = \frac{P}{I^2}$$

Write this equation without fractions.

66. The volume of a cube of side a is a^3. If we double the length of the side, the volume becomes

$$(2a)^3$$

Simplify this expression.

67. The power to a resistor is given by Eq. A67. If the current is then doubled and the resistance is halved, we have

$$(2I)^2\left(\frac{R}{2}\right)$$

Simplify this expression.

Computer

68. Using your CAS, enter an expression, highlight it, and give the command to simplify. Use it to simplify any of the expressions in this exercise, but realize that you may not get the result in exactly the same form given in the answer key.

13–2 Simplification of Radicals

Relation between Fractional Exponents and Radicals

A quantity raised to a *fractional exponent* can also be written as a *radical*. A radical consists of a *radical sign*, a quantity under the radical sign called the *radicand*, and the *index* of the radical. A radical is written as follows:

radical sign

index

radicand

where n is an integer. An index of 2, for a square root, is usually not written.

Unless otherwise stated, we'll assume that when the index is even, the radicand is *positive*. We'll consider square roots and fourth roots of negative numbers when we discuss imaginary numbers in Chapter 21.

◆◆◆ **Example 12:**

(a) $4^{1/2} = \sqrt{4} = 2$

(b) $8^{1/3} = \sqrt[3]{8} = 2$

(c) $x^{1/2} = \sqrt{x}$

(d) $y^{1/4} = \sqrt[4]{y}$

(e) $w^{-1/2} = \dfrac{1}{\sqrt{w}}$

◆◆◆

Recall that we used this relation in Chapter 1 to take any root of a number on a calculator that could do only square or cube roots.

◆◆◆ **Example 13:** Simplify $\sqrt[5]{382}$.

Solution:

$$\sqrt[5]{382} = 382^{1/5} = 3.28$$

by calculator, rounded to three significant digits.

◆◆◆

In general, the following equation applies:

Relation between Exponents and Radicals	$a^{1/n} = \sqrt[n]{a}$	36

If a quantity is raised to a fractional exponent m/n (where m and n are integers), we have

$$a^{m/n} = (a^m)^{1/n} = \sqrt[n]{a^m}$$

Further,

$$a^{m/n} = (a^{1/n})^m = (\sqrt[n]{a})^m$$

So we have the following formula:

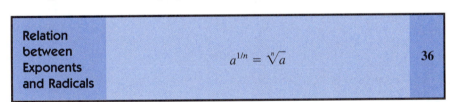

	$a^{m/n} = \sqrt[n]{a^m} = (\sqrt[n]{a})^m$	37

◆◆◆ **Example 14:**

(a) $8^{2/3} = (\sqrt[3]{8})^2 = (2)^2 = 4$

(b) $\sqrt[5]{(68.3)^3} = (68.3)^{3/5} = 12.6$, by calculator

(c) $\sqrt[3]{x^4} = x^{4/3}$

(d) $(\sqrt[5]{y^2})^3 = (y^{2/5})^3 = y^{6/5}$ ◆◆◆

We use these definitions to switch between *exponential form* and *radical form.*

◆◆◆ **Example 15:** Express $x^{1/2}y^{-1/3}$ in radical form.

Solution:

$$x^{1/2}y^{-1/3} = \frac{x^{1/2}}{y^{1/3}} = \frac{\sqrt{x}}{\sqrt[3]{y}}$$

◆◆◆

◆◆◆ **Example 16:** Express $\sqrt[3]{y^2}$ in exponential form.

Solution:

$$\sqrt[3]{y^2} = (y^2)^{1/3} = y^{2/3}$$

◆◆◆

◆◆◆ **Example 17:** Find the value of $\sqrt[4]{(81)^3}$ without using a calculator.

Solution: Instead of cubing 81 and then taking the fourth root, let us take the fourth root first.

$$\sqrt[4]{(81)^3} = (\sqrt[4]{81})^3 = (3)^3 = 27$$

◆◆◆

Common Error	Don't confuse the **coefficient** of a radical with the **index** of a radical. $$3\sqrt{x} \neq \sqrt[3]{x}$$

Root of a Product

We have several *rules of radicals,* which are similar to the laws of exponents and, in fact, are derived from them. The first rule is for products. By our definition of a radical,

$$\sqrt[n]{ab} = (ab)^{1/n}$$

Using the law of exponents for a product and then returning to radical form,

$$(ab)^{1/n} = a^{1/n}b^{1/n} = \sqrt[n]{a}\,\sqrt[n]{b}$$

So our first rule of radicals is as follows:

Root of a Product	$\sqrt[n]{ab} = \sqrt[n]{a}\,\sqrt[n]{b}$	**38**

The root of a product equals the product of the roots of the factors.

◆◆◆ **Example 18:** We may split the radical $\sqrt{9x}$ into two radicals, as follows:

$$\sqrt{9x} = \sqrt{9}\,\sqrt{x} = 3\sqrt{x}$$

since $\sqrt{9} = 3$.   ◆◆◆

◆◆◆ **Example 19:** Write as a single radical $\sqrt{7}\,\sqrt{2}\,\sqrt{x}$.

Solution: By Eq. 38,

$$\sqrt{7}\,\sqrt{2}\,\sqrt{x} = \sqrt{7(2)x} = \sqrt{14x}$$

◆◆◆

Common Error	There is no similar rule for the root of a sum. $$\sqrt[n]{a + b} \neq \sqrt[n]{a} + \sqrt[n]{b}$$

◆◆◆ **Example 20:** $\sqrt{9} + \sqrt{16}$ does *not* equal $\sqrt{25}$. ◆◆◆

Common Error	Equation 38 does not hold when a and b are both *negative* and the index is even.

◆◆◆ **Example 21:**

$$(\sqrt{-4})^2 = \sqrt{-4}\,\sqrt{-4} \neq \sqrt{(-4)(-4)}$$
$$\neq \sqrt{16} = +4$$

Instead, we convert to *imaginary numbers,* as we will show in Sec. 21–1. ◆◆◆

Root of a Quotient

Just as the root of a product can be split up into the roots of the factors, the root of a quotient can be expressed as the root of the numerator divided by the root of the denominator. We first write the quotient in exponential form.

$$\sqrt[n]{\frac{a}{b}} = \left(\frac{a}{b}\right)^{1/n} = \frac{a^{1/n}}{b^{1/n}}$$

by the law of exponents for quotients. Returning to radical form, we have the following:

Root of a Quotient	$$\sqrt[n]{\frac{a}{b}} = \frac{\sqrt[n]{a}}{\sqrt[n]{b}}$$	39

The root of a quotient equals the quotient of the roots of numerator and denominator.

◆◆◆ **Example 22:** The radical $\sqrt{\dfrac{w}{25}}$ can be written $\dfrac{\sqrt{w}}{\sqrt{25}}$ or $\dfrac{\sqrt{w}}{5}$. ◆◆◆

Simplest Form for a Radical

A radical is said to be in *simplest form* when:

1. The radicand has been reduced as much as possible.
2. There are no radicals in the denominator and no fractional radicands.
3. The index has been made as small as possible.

Rather than "simplifying," we are actually putting radicals into a *standard form,* so that they may be combined or compared.

Removing Factors from the Radicand

Using the fact that

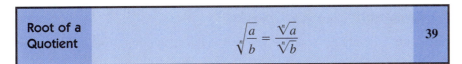

$$\sqrt{x^2} = x, \qquad \sqrt[3]{x^3} = x, \ldots, \qquad \sqrt[n]{x^n} = x$$

try to factor the radicand so that one or more of the factors is a perfect nth power (where n is the index of the radical). Then use Eq. 38 to split the radical into two or more radicals, some of which can then be reduced.

Remember that we are doing numerical problems as a way of learning the rules. If you simply want the decimal value of a radical expression containing only numbers, use your calculator.

◆◆◆ Example 23:

$$\sqrt{50} = \sqrt{(25)(2)} = \sqrt{25}\,\sqrt{2} = 5\sqrt{2}$$ ◆◆◆

◆◆◆ Example 24:

$$\sqrt{x^3} = \sqrt{x^2 x} = \sqrt{x^2}\,\sqrt{x} = x\sqrt{x}$$ ◆◆◆

◆◆◆ Example 25: Simplify $\sqrt{50x^3}$.

Solution: We factor the radicand so that some factors are perfect squares.

$$\sqrt{50x^3} = \sqrt{(25)(2)x^2 x}$$

Then, by Eq. 38,

$$= \sqrt{25}\,\sqrt{x^2}\,\sqrt{2x} = 5x\sqrt{2x}$$ ◆◆◆

◆◆◆ Example 26: Simplify $\sqrt{24y^5}$.

Solution: We look for factors of the radicand that are perfect squares.

$$\sqrt{24y^5} = \sqrt{4(6)y^4 y} = 2y^2\sqrt{6y}$$ ◆◆◆

◆◆◆ Example 27:

$$\sqrt[4]{24y^5} = \sqrt[4]{24y^4 y} = y\sqrt[4]{24y}$$ ◆◆◆

◆◆◆ Example 28: Simplify $\sqrt[3]{24y^5}$.

Solution: We look for factors of the radicand that are perfect cubes.

$$\sqrt[3]{24y^5} = \sqrt[3]{8(3)y^3 y^2}$$
$$= \sqrt[3]{8y^3}\,\sqrt[3]{3y^2} = 2y\sqrt[3]{3y^2}$$ ◆◆◆

When the radicand contains more than one term, try to *factor out* a perfect n^{th} power (where n is the index).

◆◆◆ Example 29: Simplify $\sqrt{4x^2 y + 12x^4 z}$.

Solution: We factor $4x^2$ from the radicand and then remove it from under the radical sign.

$$\sqrt{4x^2 y + 12x^4 z} = \sqrt{4x^2(y + 3x^2 z)}$$
$$= 2x\sqrt{y + 3x^2 z}$$ ◆◆◆

Rationalizing the Denominator

An expression is considered in simpler form when its denominators contain no radicals. To put it into this form is called *rationalizing* the denominator. We will show how to rationalize the denominator when it is a square root, a cube root, or a root with any index and when it has more than one term.

If the denominator is a square root, multiply numerator and denominator of the fraction by a quantity that will make the radicand in the denominator a perfect square. Note that we are eliminating radicals from the *denominator* and that the numerator may still contain radicals. Further, even though we call this process *simplifying*, the resulting radical may look more complicated than the original.

◆◆◆ Example 30:

$$\frac{5}{\sqrt{2}} = \frac{5}{\sqrt{2}} \cdot \frac{\sqrt{2}}{\sqrt{2}} = \frac{5\sqrt{2}}{\sqrt{4}} = \frac{5\sqrt{2}}{2}$$ ◆◆◆

When the *entire* fraction is under the radical sign, we make the denominator of that fraction a perfect square and remove it from the radical sign.

◆◆◆ **Example 31:**

$$\sqrt{\frac{3x}{2y}} = \sqrt{\frac{3x(2y)}{2y(2y)}} = \sqrt{\frac{6xy}{4y^2}} = \frac{\sqrt{6xy}}{2y}$$

◆◆◆

If the denominator is a *cube* root, we must multiply numerator and denominator by a quantity that will make the quantity under the radical sign a perfect cube.

◆◆◆ **Example 32:** Simplify

$$\frac{7}{\sqrt[3]{4}}$$

Solution: To rationalize the denominator $\sqrt[3]{4}$, we multiply numerator and denominator of the fraction by $\sqrt[3]{2}$. We chose this as the factor because it results in a perfect cube (8) under the radical sign.

$$\frac{7}{\sqrt[3]{4}} = \frac{7}{\sqrt[3]{4}} \cdot \frac{\sqrt[3]{2}}{\sqrt[3]{2}} = \frac{7\sqrt[3]{2}}{\sqrt[3]{8}} = \frac{7\sqrt[3]{2}}{2}$$

◆◆◆

The same principle applies regardless of the index. In general, if the index is n, we must make the quantity under the radical sign (in the denominator) a perfect nth power.

◆◆◆ **Example 33:**

$$\frac{2y}{3\sqrt[5]{x}} = \frac{2y}{3\sqrt[5]{x}} \cdot \frac{\sqrt[5]{x^4}}{\sqrt[5]{x^4}} = \frac{2y\sqrt[5]{x^4}}{3\sqrt[5]{x^5}} = \frac{2y\sqrt[5]{x^4}}{3x}$$

◆◆◆

Sometimes the denominator will have more than one term.

◆◆◆ **Example 34:**

$$\sqrt{\frac{a}{a^2 + b^2}} = \sqrt{\frac{a}{a^2 + b^2} \cdot \frac{a^2 + b^2}{a^2 + b^2}} = \frac{\sqrt{a(a^2 + b^2)}}{a^2 + b^2}$$

◆◆◆

Reducing the Index

We can sometimes reduce the index by writing the radical in exponential form and then reducing the fractional exponent, as in the next example.

◆◆◆ **Example 35:**

(a) $\sqrt[6]{x^3} = x^{3/6} = x^{1/2} = \sqrt{x}$

(b) $\sqrt[4]{4x^2y^2} = \sqrt[4]{(2xy)^2}$
$$= (2xy)^{2/4} = (2xy)^{1/2}$$
$$= \sqrt{2xy}$$

◆◆◆

Exercise 2 ◆ Simplification of Radicals

Exponential and Radical Forms

Express in radical form.

1. $a^{1/4}$

2. $x^{1/2}$

3. $z^{3/4}$

4. $a^{1/2}b^{1/4}$

5. $(m - n)^{1/2}$

6. $(x^2y)^{-1/2}$

7. $\left(\dfrac{x}{y}\right)^{-1/3}$

8. $a^0b^{-3/4}$

Express in exponential form.

9. $\sqrt{b}$ **10.** $\sqrt[3]{x}$ **11.** $\sqrt{y^2}$

12. $4\sqrt[3]{xy}$ **13.** $\sqrt[3]{a+b}$ **14.** $\sqrt[n]{x^m}$

15. $\sqrt{x^2 y^2}$ **16.** $\sqrt[n]{a^n b^{3n}}$

Simplifying Radicals

Write in simplest form.

> Do not use your calculator for any of these numerical problems. Leave the answers in radical form.

17. $\sqrt{18}$ **18.** $\sqrt{75}$

19. $\sqrt{63}$ **20.** $\sqrt[3]{16}$

21. $\sqrt[3]{-30}$ **22.** $\sqrt{48}$

23. $\sqrt{a^3}$ **24.** $3\sqrt{50x^5}$

25. $\sqrt{36x^2 y}$ **26.** $\sqrt[3]{x^2 y^5}$

27. $x\sqrt[3]{16x^3 y}$ **28.** $\sqrt[4]{64m^2 n^4}$

29. $3\sqrt[5]{32xy^{11}}$ **30.** $6\sqrt[3]{16x^4}$

31. $\sqrt{a^3 - a^2 b}$ **32.** $x\sqrt{x^4 - x^3 y^2}$

33. $\sqrt{9m^3 + 18n}$ **34.** $\sqrt{2x^3 + x^4 y}$

35. $\sqrt[3]{x^4 - a^2 x^3}$ **36.** $x\sqrt{a^3 + 2a^2 b + ab^2}$

37. $(a+b)\sqrt{a^3 - 2a^2 b + ab^2}$ **38.** $\sqrt{\dfrac{2}{5}}$

39. $\sqrt{\dfrac{3}{7}}$ **40.** $\sqrt{\dfrac{2}{3}}$

41. $\sqrt[3]{\dfrac{1}{4}}$ **42.** $\sqrt{\dfrac{5}{8}}$

43. $\sqrt[3]{\dfrac{2}{9}}$ **44.** $\sqrt[4]{\dfrac{7}{8}}$

45. $\sqrt{\dfrac{1}{2x}}$ **46.** $\sqrt{\dfrac{5m}{7n}}$

47. $\sqrt{\dfrac{3a^3}{5b}}$ **48.** $\sqrt{\dfrac{5ab}{6xy}}$

49. $\sqrt[3]{\dfrac{1}{x^2}}$ **50.** $\sqrt[3]{\dfrac{81x^4}{16yz^2}}$

51. $\sqrt[6]{\dfrac{4x^6}{9}}$ **52.** $\sqrt{x^2 - \left(\dfrac{x}{2}\right)^2}$

> In these more complicated types, you should start by simplifying the expression under the radical sign.

53. $(x^2 - y^2)\sqrt{\dfrac{x}{x+y}}$ **54.** $\sqrt{a^3 + \left(\dfrac{a^3}{2}\right)^3}$

55. $(m+n)\sqrt{\dfrac{m}{m-n}}$ **56.** $\sqrt{\left(\dfrac{x+1}{2}\right)^2 - x}$

57. $\sqrt{\dfrac{8a^2 - 48a + 72}{3a}}$

Applications

58. The period ω_n for simple harmonic motion is given by Eq. A37.

$$\omega_n = \sqrt{\dfrac{kg}{W}}$$

Write this equation in exponential form.

59. The impedance Z of a series RLC circuit is given by Eq. A99.

$$Z = \sqrt{R^2 + X^2}$$

Write this equation in exponential form.

60. Given Eq. A99 from problem 59, write the expression for Z when $X = 2R$, and simplify.

61. The hypotenuse in right triangle ABC, shown in Fig. 13–2, is given by the Pythagorean theorem.

$$c = \sqrt{a^2 + b^2}$$

Write an expression for c when $b = 3a$, and simplify.

62. In problem 58, rationalize the denominator in Eq. A37.

63. A stone is thrown upward with a horizontal velocity of 40 ft/s and an upward velocity of 60 ft /s. At t seconds it will have a horizontal displacement H equal to $40t$ and a vertical displacement V equal to $60t - 16t^2$. The straight-line distance S from the stone to the launch point is found by the Pythagorean theorem. Write an equation for S in terms of t, and simplify.

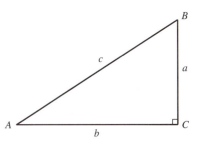

FIGURE 13–2

Computer

64. You can, of course, use a CAS to simplify radicals. Just type in the radical expression and use the command to simplify. Try it with any of the expressions in this exercise set.

13–3 Operations with Radicals

Adding and Subtracting Radicals

Radicals are called *similar* if they have the same index and the same radicand, such as $5\sqrt[3]{2x}$ and $3\sqrt[3]{2x}$. We add and subtract radicals by *combining similar radicals*.

> One reason we learned to put radicals into a standard form is to combine them.

◆◆◆ Example 36:

$$5\sqrt{y} + 2\sqrt{y} - 4\sqrt{y} = 3\sqrt{y} \qquad \text{◆◆◆}$$

Radicals that may look similar at first glance may not actually be similar.

◆◆◆ Example 37: The radicals

$$\sqrt{2x} \quad \text{and} \quad \sqrt{3x}$$

are *not* similar. ◆◆◆

Common Error	Do not try to combine radicals that are not similar. $$\sqrt{2x} + \sqrt{3x} \neq \sqrt{5x}$$

Radicals that do not appear to be similar at first may turn out to be so after simplification.

◆◆◆ Example 38:

(a) $\sqrt{18x} - \sqrt{8x} = 3\sqrt{2x} - 2\sqrt{2x} = \sqrt{2x}$

(b) $\sqrt[3]{24y^4} + \sqrt[3]{81x^3y} = 2y\sqrt[3]{3y} + 3x\sqrt[3]{3y}$

  Factor: $= (2y + 3x)\sqrt[3]{3y}$

(c) $5y\sqrt{\dfrac{x}{y}} - 2\sqrt{xy} + x\sqrt{\dfrac{y}{x}}$

$$= 5y\sqrt{\dfrac{xy}{y^2}} - 2\sqrt{xy} + x\sqrt{\dfrac{yx}{x^2}}$$

$$= 5\sqrt{xy} - 2\sqrt{xy} + \sqrt{xy} = 4\sqrt{xy}$$

◆◆◆

Multiplying Radicals

Radicals having the *same index* can be multiplied by using Eq. 38.

$$\sqrt[n]{a}\sqrt[n]{b} = \sqrt[n]{ab}$$

◆◆◆ **Example 39:**

(a) $\sqrt{x}\sqrt{2y} = \sqrt{2xy}$

(b) $(5\sqrt[3]{3a})(2\sqrt[3]{4b}) = 10\sqrt[3]{12ab}$

(c) $(2\sqrt{m})(5\sqrt{n})(3\sqrt{mn}) = 30\sqrt{m^2n^2} = 30mn$

(d) $\sqrt{2x}\sqrt{3x} = \sqrt{6x^2} = x\sqrt{6}$

(e) $\sqrt{3x}\sqrt{3x} = 3x$

(f) $\sqrt[3]{2x}\sqrt[3]{4x^2} = \sqrt[3]{8x^3} = 2x$

◆◆◆

We multiply radicals having *different indices* by first going to exponential form, then multiplying using Eq. 29 ($x^a x^b = x^{a+b}$), and finally returning to radical form.

◆◆◆ **Example 40:**

$$\sqrt{a}\sqrt[4]{b} = a^{1/2}b^{1/4}$$
$$= a^{2/4}b^{1/4} = (a^2b)^{1/4}$$

Or, in radical form,

$$= \sqrt[4]{a^2b}$$

◆◆◆

When the radicands are the same, the work is even easier.

◆◆◆ **Example 41:** Multiply $\sqrt[3]{x}$ by $\sqrt[6]{x}$.

Solution:

$$\sqrt[3]{x}\sqrt[6]{x} = x^{1/3}x^{1/6} = x^{2/6}x^{1/6}$$

By Eq. 29,

$$= x^{3/6} = x^{1/2}$$
$$= \sqrt{x}$$

◆◆◆

Multinomials containing radicals are multiplied in the way we learned in Sec. 2–4.

◆◆◆ **Example 42:**

(a) $(3 + \sqrt{x})(2\sqrt{x} - 4\sqrt{y}) = 3(2\sqrt{x}) - 3(4\sqrt{y}) + 2\sqrt{x}\sqrt{x} - 4\sqrt{x}\sqrt{y}$
$$= 6\sqrt{x} - 12\sqrt{y} + 2x - 4\sqrt{xy}$$

(b) $\sqrt{50x}(\sqrt{2x} - \sqrt{xy}) = \sqrt{100x^2} - \sqrt{50x^2y}$
$$= 10x - 5x\sqrt{2y}$$

(c) $(\sqrt{x} - \sqrt{y})(\sqrt{x} + \sqrt{y}) = \sqrt{x^2} + \sqrt{xy} - \sqrt{xy} - \sqrt{y^2}$
$$= \sqrt{x^2} - \sqrt{y^2}$$
$$= x - y$$

◆◆◆

To raise a radical to a *power,* we simply multiply the radical by itself the proper number of times.

◆◆◆ **Example 43:** Cube the expression

$$2x\sqrt{y}$$

Solution:

$$(2x\sqrt{y})^3 = (2x\sqrt{y})(2x\sqrt{y})(2x\sqrt{y})$$
$$= 2^3x^3(\sqrt{y})^3$$
$$= 8x^3\sqrt{y^3} = 8x^3y\sqrt{y} \qquad ◆◆◆$$

Dividing Radicals

Radicals having the same indices can be divided using Eq. 39.

$$\frac{\sqrt[n]{a}}{\sqrt[n]{b}} = \sqrt[n]{\frac{a}{b}}$$

◆◆◆ **Example 44:**

(a) $$\frac{\sqrt{4x^5}}{\sqrt{2x}} = \sqrt{\frac{4x^5}{2x}} = \sqrt{2x^4} = \sqrt{2}x^2$$

(b) $$\frac{\sqrt{3a^7} + \sqrt{12a^5} - \sqrt{6a^3}}{\sqrt{3a}} = \sqrt{\frac{3a^7}{3a}} + \sqrt{\frac{12a^5}{3a}} - \sqrt{\frac{6a^3}{3a}}$$
$$= \sqrt{a^6} + \sqrt{4a^4} - \sqrt{2a^2}$$
$$= a^3 + 2a^2 - \sqrt{2}a \qquad ◆◆◆$$

It is a common practice to rationalize the denominator after division.

◆◆◆ **Example 45:**

$$\frac{\sqrt{10x}}{\sqrt{5y}} = \sqrt{\frac{10x}{5y}} = \sqrt{\frac{2x}{y}}$$

Rationalizing the denominator, we obtain

$$\sqrt{\frac{2x}{y}} = \sqrt{\frac{2x}{y} \cdot \frac{y}{y}} = \frac{\sqrt{2xy}}{y} \qquad ◆◆◆$$

If the indices are different, we go to exponential form, as we did for multiplication. Divide using Eq. 30 and then return to radical form.

◆◆◆ **Example 46:**

$$\frac{\sqrt[3]{a}}{\sqrt[4]{b}} = \frac{a^{1/3}}{b^{1/4}} = \frac{a^{4/12}}{b^{3/12}} = \left(\frac{a^4}{b^3}\right)^{1/12}$$

Returning to radical form, we obtain

$$\left(\frac{a^4}{b^3}\right)^{1/12} = \sqrt[12]{\frac{a^4}{b^3}}$$
$$= \sqrt[12]{\frac{a^4}{b^3} \cdot \frac{b^9}{b^9}}$$
$$= \sqrt[12]{\frac{a^4b^9}{b^{12}}}$$
$$= \frac{1}{b}\sqrt[12]{a^4b^9} \qquad ◆◆◆$$

When dividing by a binomial containing *square* roots, multiply the divisor and the dividend by the *conjugate* of that binomial (a binomial that differs from the original dividend only in the sign of one term).

Remember that when we multiply a binomial, for example, $(a + b)$, by its conjugate $(a - b)$,

$$(a + b)(a - b) = a^2 - ab + ab - b^2$$
$$= a^2 - b^2$$

we get an expression where the cross-product terms ab and $-ab$ drop out, and the remaining terms are both squares. Thus if a or b were square roots, our final expression would have no square roots.

We use this operation to remove square roots from the denominator, as shown in the following example.

◆◆◆ **Example 47:** Divide $(3 + \sqrt{x})$ by $(2 - \sqrt{x})$.

Solution: The conjugate of the divisor $2 - \sqrt{x}$ is $2 + \sqrt{x}$. Multiplying divisor and dividend by $2 + \sqrt{x}$, we get

$$\frac{3 + \sqrt{x}}{2 - \sqrt{x}} = \frac{3 + \sqrt{x}}{2 - \sqrt{x}} \cdot \frac{2 + \sqrt{x}}{2 + \sqrt{x}}$$

$$= \frac{6 + 3\sqrt{x} + 2\sqrt{x} + \sqrt{x}\sqrt{x}}{4 + 2\sqrt{x} - 2\sqrt{x} - \sqrt{x}\sqrt{x}}$$

$$= \frac{6 + 5\sqrt{x} + x}{4 - x}$$

◆◆◆

Exercise 3 ◆ Operations with Radicals

Addition and Subtraction of Radicals

Combine as indicated

1. $2\sqrt{24} - \sqrt{54}$
2. $\sqrt{300} + \sqrt{108} - \sqrt{243}$
3. $\sqrt{24} - \sqrt{96} + \sqrt{54}$
4. $\sqrt{128} - \sqrt{18} + \sqrt{32}$
5. $2\sqrt{50} + \sqrt{72} + 3\sqrt{18}$
6. $\sqrt[3]{384} - \sqrt[3]{162} + \sqrt[3]{750}$
7. $2\sqrt[3]{2} - 3\sqrt[3]{16} + \sqrt[3]{-54}$
8. $3\sqrt[3]{108} + 2\sqrt[3]{32} - \sqrt[3]{256}$
9. $\sqrt[3]{625} - 2\sqrt[3]{135} - \sqrt[3]{320}$
10. $5\sqrt[3]{320} + 2\sqrt[3]{40} - 4\sqrt[3]{135}$
11. $\sqrt[4]{768} - \sqrt[4]{48} - \sqrt[4]{243}$
12. $3\sqrt{\dfrac{5}{4}} + 2\sqrt{45}$
13. $\sqrt{128x^2y} - \sqrt{98x^2y} + \sqrt{162x^2y}$
14. $3\sqrt{\dfrac{1}{3}} - 2\sqrt{\dfrac{3}{4}} + 4\sqrt{3}$
15. $7\sqrt{\dfrac{27}{50}} - 3\sqrt{\dfrac{2}{3}}$
16. $4\sqrt{50} - 2\sqrt{72} + \dfrac{3}{\sqrt{2}}$
17. $\sqrt{a^2x} + \sqrt{b^2x}$
18. $\sqrt{2b^2xy} - \sqrt{2a^2xy}$
19. $\sqrt{x^2y} - \sqrt{4a^2y}$
20. $\sqrt{80a^3} - 3\sqrt{20a^3} - 2\sqrt{45a^3}$
21. $4\sqrt{3a^2x} - 2a\sqrt{48x}$
22. $\sqrt[3]{125x^2} - 2\sqrt[3]{8x^2}$
23. $\sqrt[3]{1250a^3b} + \sqrt[3]{270c^3b}$
24. $2\sqrt[3]{ab^7} + 3\sqrt[3]{a^7b} + 2\sqrt[3]{8a^4b^4}$
25. $\sqrt[5]{a^{13}b^{11}c^{12}} - 2\sqrt[5]{a^8bc^2} + \sqrt[5]{a^3b^6c^7}$
26. $\sqrt{\dfrac{9x}{16}} - \dfrac{3\sqrt{x}}{4}$
27. $\sqrt[6]{8x^3} - 2\sqrt[4]{\dfrac{x^2}{4}} + 3\sqrt{8x}$
28. $\sqrt{\dfrac{x}{a^2}} + 2\sqrt{\dfrac{x}{b^2}} - 3\sqrt{\dfrac{x}{c^2}}$

29. $3\sqrt{\dfrac{3x}{4y^2}} + 2\sqrt{\dfrac{x}{27y^2}} - \sqrt{\dfrac{x}{3y^2}}$

30. $\sqrt{(a+b)^2 x} + \sqrt{(a-b)^2 x}$

Multiplication of Radicals

Multiply.

31. $3\sqrt{3}$ by $5\sqrt{3}$

32. $2\sqrt{3}$ by $3\sqrt{8}$

33. $\sqrt{8}$ by $\sqrt{160}$

34. $\sqrt{\dfrac{5}{8}}$ by $\sqrt{\dfrac{3}{4}}$

35. $4\sqrt[3]{45}$ by $2\sqrt[3]{3}$

36. $3\sqrt{3}$ by $2\sqrt[3]{2}$

37. $3\sqrt{2}$ by $2\sqrt[3]{3}$

38. $2\sqrt{3}$ by $\sqrt[4]{5}$

39. $2\sqrt[3]{3}$ by $5\sqrt[4]{4}$

40. $2\sqrt[3]{24}$ by $\sqrt[9]{\dfrac{8}{27}}$

41. $2x\sqrt{3a}$ by $3\sqrt{y}$

42. $3\sqrt[3]{9a^2}$ by $\sqrt[3]{3abc}$

43. $\sqrt[5]{4xy^2}$ by $\sqrt[5]{8x^2y}$

44. $\sqrt{\dfrac{a}{b}}$ by $\sqrt{\dfrac{c}{d}}$

45. $\sqrt{a}$ by $\sqrt[4]{b}$

46. $\sqrt[3]{x}$ by $\sqrt{y}$

47. $\sqrt{xy}$ by $2\sqrt{xz}$ and $\sqrt[3]{x^2y^2}$

48. $\sqrt[3]{a^2b}$ by $\sqrt[3]{2a^2b^2}$ and $\sqrt{3a^3b^2}$

49. $(\sqrt{5} - \sqrt{3})$ by $2\sqrt{3}$

50. $(x + \sqrt{y})$ by $\sqrt{y}$

51. $\sqrt{x^3 - x^4y}$ by $\sqrt{x}$

52. $(\sqrt{x} - \sqrt{2})$ by $(2\sqrt{x} + \sqrt{2})$

53. $(a + \sqrt{b})$ by $(a - \sqrt{b})$

54. $(\sqrt{x} + \sqrt{y})$ by $(\sqrt{x} + \sqrt{y})$

55. $(4\sqrt{x} + 2\sqrt{y})$ by $(4\sqrt{x} - 5\sqrt{y})$

56. $(x + \sqrt{xy})$ by $(\sqrt{x} - \sqrt{y})$

Powers

Square the following expressions.

57. $3\sqrt{y}$

58. $4\sqrt[3]{4x^2}$

59. $3x\sqrt[3]{2x^2}$

60. $5 + 4\sqrt{x}$

61. $3 - 5\sqrt{a}$

62. $\sqrt{a} + 5a\sqrt{b}$

Cube the following radicals.

63. $5\sqrt{2x}$

64. $2x\sqrt{3x}$

65. $5\sqrt[3]{2ax}$

Division of Radicals

Divide.

66. $8 \div 3\sqrt{2}$

67. $6\sqrt{72} \div 12\sqrt{32}$

68. $(4\sqrt[3]{3} - 3\sqrt[3]{2}) \div \sqrt[3]{2}$

69. $5\sqrt[3]{12} \div 10\sqrt{8}$

70. $9\sqrt[3]{18} \div 3\sqrt{6}$

71. $12\sqrt[3]{4a^3} \div 4\sqrt[3]{2a^5}$

72. $12\sqrt{256} \div 18\sqrt{2}$

73. $(4\sqrt[3]{4} + 2\sqrt[3]{3} + 3\sqrt[3]{6}) \div \sqrt[3]{6}$

74. $2 \div \sqrt[3]{6}$

75. $\sqrt{72} \div 2\sqrt[4]{64}$

76. $8 \div 2\sqrt[3]{4}$

77. $8\sqrt[3]{ab} \div 4\sqrt{ac}$

78. $\sqrt[3]{4ab} \div \sqrt[4]{2ab}$

79. $2\sqrt[4]{a^2b^3c^2} \div 4\sqrt[3]{ab^2c^2}$

80. $6\sqrt[3]{x^4y^7z} \div \sqrt[4]{xy}$

81. $(3 + \sqrt{2}) \div (2 - \sqrt{2})$

82. $5 \div \sqrt{3x}$

83. $4\sqrt{x} \div \sqrt{a}$

84. $10 \div \sqrt[3]{9x^2}$

85. $12 \div \sqrt[3]{4x^2}$

86. $5\sqrt{a^2 b} \div 3\sqrt[3]{a^2 b^2}$

87. $(10\sqrt{6} - 3\sqrt{5}) \div (2\sqrt{6} - 4\sqrt{5})$

88. $\sqrt{x} \div (\sqrt{x} + \sqrt{y})$

89. $a \div (a + \sqrt{b})$

90. $(a + \sqrt{b}) \div (a - \sqrt{b})$

91. $(\sqrt{x} - \sqrt{y}) \div (\sqrt{x} + \sqrt{y})$

92. $(3\sqrt{m} - \sqrt{2n}) \div (\sqrt{3n} + \sqrt{m})$

93. $(5\sqrt{x} + \sqrt{2y}) \div (2\sqrt{3x} + 2\sqrt{y})$

94. $(5\sqrt{2p} - 3\sqrt{3q}) \div (3\sqrt{3p} + 2\sqrt{5q})$

95. $(2\sqrt{3a} + 4\sqrt{5b}) \div (3\sqrt{2a} + 4\sqrt{b})$

Computer

96. Use a CAS for any of the problems in this exercise set. Simply enter the entire expression and give the command to simplify.

13–4   Radical Equations

An equation in which the unknown is under a radical sign is called a *radical equation*. To solve a radical equation, it is necessary to isolate the radical term on one side of the equal sign and then raise both sides to whatever power will eliminate the radical.

◆◆◆ **Example 48:** Solve for x:

$$\sqrt{x - 5} - 4 = 0$$

Solution: Rearranging yields

$$\sqrt{x - 5} = 4$$

Square both sides: $x - 5 = 16$

$$x = 21$$

We will solve more difficult radical equations in Chapter 14.

Check:

$$\sqrt{21 - 5} - 4 \stackrel{?}{=} 0$$
$$\sqrt{16} - 4 \stackrel{?}{=} 0$$
$$4 - 4 = 0 \quad \text{(checks)}$$

◆◆◆

◆◆◆ **Example 49:** Solve for x:

$$\sqrt{x^2 - 6x} = x - 9$$

Solution: Squaring, we obtain

$$x^2 - 6x = x^2 - 18x + 81$$
$$12x = 81$$
$$x = \frac{81}{12} = \frac{27}{4}$$

Remember that we take only the principal (positive) root.

Check:

$$\sqrt{\left(\frac{27}{4}\right)^2 - 6\left(\frac{27}{4}\right)} \stackrel{?}{=} \frac{27}{4} - 9$$

$$\sqrt{\frac{729}{16} - \frac{162}{4}} \stackrel{?}{=} \frac{27}{4} - \frac{36}{4}$$

$$\sqrt{\frac{81}{16}} = \frac{9}{4} \neq -\frac{9}{4} \quad \text{(does not check)}$$

Thus the given equation has no solution. ◆◆◆

Common Error	The squaring process often introduces *extraneous roots*. These are discarded because they do not satisfy the original equation. Check your answer in the original equation.

If the equation has more than one radical, isolate one at a time and square both sides, as in the following example.

◆◆◆ **Example 50:** Solve for x:

$$\sqrt{x - 32} + \sqrt{x} = 16$$

Solution: Rearranging gives

$$\sqrt{x - 32} = 16 - \sqrt{x}$$

Squaring yields

$$x - 32 = (16)^2 - 32\sqrt{x} + x$$

We rearrange again to isolate the radical.

$$32\sqrt{x} = 256 + 32 = 288$$
$$\sqrt{x} = 9$$

Squaring again, we obtain

$$x = 81$$

Check:

$$\sqrt{81 - 32} + \sqrt{81} \stackrel{?}{=} 16$$
$$\sqrt{49} + 9 \stackrel{?}{=} 16$$
$$7 + 9 = 16 \quad \text{(checks)} \qquad ◆◆◆$$

Tip: It is usually better to isolate and square the *most complicated* radical first.

If a radical equation contains a fraction, we proceed as we did with other fractional equations and multiply both sides by the least common denominator.

◆◆◆ **Example 51:** Solve the following:

$$\sqrt{x - 3} + \frac{1}{\sqrt{x - 3}} = \sqrt{x}$$

Solution: Multiplying both sides by $\sqrt{x - 3}$ gives

$$x - 3 + 1 = \sqrt{x}\sqrt{x - 3}$$
$$x - 2 = \sqrt{x^2 - 3x}$$

Squaring both sides yields

$$x^2 - 4x + 4 = x^2 - 3x$$
$$x = 4$$

Check:

$$\sqrt{4-3} + \frac{1}{\sqrt{4-3}} \overset{?}{=} \sqrt{4}$$

$$1 + \frac{1}{1} = 2 \quad \text{(checks)} \qquad \qquad \text{◆◆◆}$$

Radical equations having indices other than 2 are solved in a similar way.

◆◆◆ **Example 52:** Solve for x:

$$\sqrt[3]{x-5} = 2$$

Solution: Cubing both sides, we obtain

$$x - 5 = 8$$
$$x = 13$$

Check:

$$\sqrt[3]{13-5} \overset{?}{=} 2$$
$$\sqrt[3]{8} = 2 \quad \text{(checks)} \qquad \qquad \text{◆◆◆}$$

Approximate Solution of Radical Equations by Graphing

We learned how to find the approximate solution of equations by graphics utility in Sec. 5–6. Recall that we moved all terms of the given equation to one side and graphed that expression. Then by zooming in and using $\boxed{\text{TRACE}}$, we were able to find the value of x at which the curve crossed the x axis to any precision we wanted. That method works for any equation, and we use it here to solve a radical equation.

◆◆◆ **Example 53:** Find an approximate solution to the equation given in Example 52.

Solution: Rearranging the equation gives

$$\sqrt[3]{x-5} - 2 = 0$$

So we graph the function

$$y = \sqrt[3]{x-5} - 2$$

as shown in Fig. 13–3. Using $\boxed{\text{TRACE}}$ and $\boxed{\text{ZOOM}}$, we verify that the root is at $x = 13$. ◆◆◆

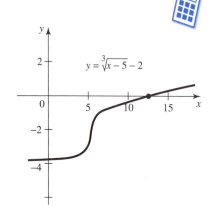

FIGURE 13–3

Exercise 4 ◆ Radical Equations

Radical Equations

Solve for x and check.

1. $\sqrt{x} = 6$ **2.** $\sqrt{x} + 5 = 9$

3. $\sqrt{7x + 8} = 6$ **4.** $\sqrt{3x - 2} = 5$

5. $\sqrt{2.95x - 1.84} = 6.23$ **6.** $\sqrt{5.88x + 4.92} = 7.72$

7. $\sqrt{x + 1} = \sqrt{2x - 7}$ **8.** $2 = \sqrt[4]{1 + 3x}$

9. $\sqrt{3x + 1} = 5$ **10.** $\sqrt[3]{2x} = 4$

11. $\sqrt{x - 3} = \dfrac{4}{\sqrt{x - 3}}$ **12.** $x + 2 = \sqrt{x^2 + 6}$

13. $\sqrt{x^2 - 7.25} = 8.75 - x$

14. $\sqrt{x - 15.5} = 5.85 - \sqrt{x}$

15. $\dfrac{6}{\sqrt{3 + x}} = \sqrt{x + 3}$

16. $\sqrt{12 + x} = 2 + \sqrt{x}$

17. $\sqrt[5]{x - 7} = 1$

18. $\sqrt{x^2 - 2x} - \sqrt{1 - x} = 1 - x$

19. $\sqrt{1.25x} - \sqrt{1.25x + 14.3} + 3.18 = 0$

20. $2.10x + \sqrt{4.41x^2 - 8.36} = 9.37$

21. $\sqrt{x - 2} - \sqrt{x} = \dfrac{1}{\sqrt{x - 2}}$

22. $\sqrt{\dfrac{1}{x}} - \sqrt{x} = \sqrt{1 + x}$

23. $\sqrt{5x - 19} - \sqrt{5x + 14} = -3$

24. $\sqrt[3]{x - 3} = 2$

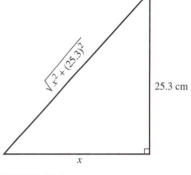

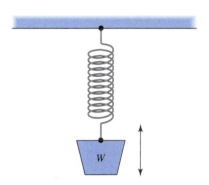

FIGURE 13–4

Applications

25. The right triangle in Fig. 13–4 has one side equal to 25.3 cm and a perimeter of 68.4 cm, so

$$x + 25.3 + \sqrt{x^2 + (25.3)^2} = 68.4$$

Solve this equation for x and for the hypotenuse.

26. Find the missing side and the hypotenuse of a right triangle that has one side equal to 293 in. and a perimeter of 994 in.

27. Find the side and the hypotenuse of a right triangle that has a side of 2.73 m and a perimeter of 11.4 m.

28. The natural frequency f_n of a body under simple harmonic motion (Fig. 13–5) is found from Eqs. A37 and A38.

$$f_n = \frac{1}{2\pi}\sqrt{\frac{kg}{W}}$$

FIGURE 13–5

where g is the gravitational constant, k is a constant, and W is the weight.

Solve for W.

29. The magnitude Z of the impedance in an RLC circuit (Fig. 13–6) having a resistance R, an inductance L, and a capacitance C, to which is applied a voltage of frequency ω, is given by Eq. A99.

$$Z = \sqrt{R^2 + \left(\omega L - \frac{1}{\omega C}\right)^2}$$

Solve for C.

30. The resonant frequency ω_n for the circuit shown in Fig. 13–6 is given by Eq. A91.

$$\omega_n = \sqrt{\frac{1}{LC}}$$

Solve for L.

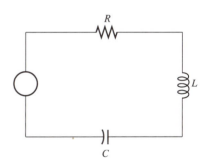

FIGURE 13–6

Graphics Calculator

31. Use your graphics calculator to solve approximately any of the above equations, to three significant digits.

Computer

32. Use your CAS to solve any of the equations above. Enter the equation and give the command *Solve.*

33. If you have written a program to solve equations by the *midpoint method,* as suggested in Sec. 5–6, use your program to solve any of the radical equations above. If not, you may want to write that program now, for it will be very useful in later chapters as well.

••• CHAPTER 13 REVIEW PROBLEMS ••••••••••••••••••••••••••••••••••

Simplify.

1. $\sqrt{52}$

2. $\sqrt{108}$

3. $\sqrt[3]{162}$

4. $\sqrt[4]{9}$

5. $\sqrt[6]{4}$

6. $\sqrt{81a^2x^3y}$

7. $3\sqrt[4]{81x^5}$

8. $\sqrt{ab^2 - b^3}$

9. $\sqrt[3]{(a-b)^5x^4}$

10. $5\sqrt{\dfrac{7x}{12y}}$

11. $\sqrt{\dfrac{a-2}{a+2}}$

Perform the indicated operations and simplify. Don't use a calculator.

12. $4b\sqrt{3y} \cdot 5\sqrt{x}$

13. $\sqrt{x} \cdot \sqrt{x^3 - x^4y}$

14. $(\sqrt{x} + \sqrt{y})(\sqrt{x} - \sqrt{y})$

15. $\dfrac{3 - 2\sqrt{3}}{2 - 5\sqrt{2}}$

16. $\dfrac{\sqrt{x} - \sqrt{y}}{\sqrt{x} + \sqrt{y}}$

17. $\dfrac{x + \sqrt{x^2 - y^2}}{x - \sqrt{x^2 - y^2}}$

18. $\sqrt{98x^2y^2} - \sqrt{128x^2y^2}$

19. $4\sqrt[3]{125x^2} + 3\sqrt[3]{8x^2}$

20. $3\sqrt{x^2 - y^2} - 2\sqrt{\dfrac{x+y}{x-y}}$

21. $\sqrt{a}\sqrt[3]{b}$

22. $\sqrt[3]{2abc^2} \cdot \sqrt{abc}$

23. $(3 + 2\sqrt{x})^2$

24. $(4x\sqrt{2x})^3$

25. $3\sqrt{50} - 2\sqrt{32}$

26. $3\sqrt{72} - 4\sqrt{8} + \sqrt{128}$

27. $5\sqrt{\dfrac{9}{8}} - 2\sqrt{\dfrac{25}{18}}$

28. $2\sqrt{2} \div \sqrt[3]{2}$

29. $\sqrt{2ab} \div \sqrt{4ab^2}$

30. $9 \div \sqrt[3]{7x^2}$

31. $3\sqrt{9} \cdot 4\sqrt{8}$

Solve for x and check:

32. $\dfrac{\sqrt{x} - 8}{\sqrt{x} - 6} = \dfrac{\sqrt{x} - 4}{\sqrt{x} + 2}$

33. $\sqrt{x + 6} = 4$

34. $\sqrt{2x - 7} = \sqrt{x - 3}$

35. $\sqrt[5]{2x + 4} = 2$

36. $\sqrt{4x^2 - 3} = 2x - 1$

37. $\sqrt{x} + \sqrt{x - 9.75} = 6.23$

38. $\sqrt[3]{21.5x} = 2.33$

Simplify, and write without negative exponents.

39. $(x^{n-1} + y^{n-2})(x^n + y^{n-1})$

40. $(72a^6b^7 \div 9a^4b) \div 8ab^6$

41. $\left(\dfrac{9x^4y^3}{6x^3y}\right)^3$

42. $\left(\dfrac{5w^2}{2x}\right)^n$

43. $\left(\dfrac{2x^5y^3}{x^2y}\right)^3$

44. $5a^{-1}$

45. $3w^{-2}$

46. $2r^0$

47. $(3x)^{-1}$

48. $2a^{-1} + b^{-2}$

49. $x^{-1} - 2y^{-2}$

50. $(5a)^{-3} - 3b^{-2}$

51. $(3x)^{-2} + (2x^2y^{-4})^{-2}$

52. $(a^n + b^m)^2$

53. $(p^{a-1} + q^{a-2})(p^a + q^{a-1})$

54. $(16a^6b^0 \div 8a^4b) \div 4ab^6$

55. p^2q^{-1}

56. x^3y^{-2}

57. r^2s^{-3}

58. a^3b^{-1}

59. $(3x^3y^2z)^0$

60. $(5a^2b^2c^2)^3$

61. The volume V of a sphere of radius r is given by Eq. 128.

$$V = \frac{4}{3}\pi r^3$$

If the radius is tripled, the volume is then

$$V = \frac{4}{3}\pi (3r)^3$$

Simplify this expression.

62. Find the missing side and the hypotenuse of a right triangle that has one side equal to 154 in. and a perimeter of 558 in.

63. The geometric mean B between two numbers A and C is

$$B = \sqrt{AC}$$

Use the rules of radicals to write this equation in a different form.

64. Equation A40 gives the damped angular velocity ω_d of a suspended weight as

$$\omega_d = \sqrt{\omega_n^2 - \frac{c^2 g^2}{\omega^2}}$$

Rationalize the denominator and simplify.

Writing

65. Explain how exponents and radicals are really two different ways of writing the same expression. Also explain why, if they are the same, we need both.

Team Project

66. Use Hero's formula, Eq. 138, to write an expression for the area of an equilateral triangle of side a, and simplify. Compare your result with what you get when multiplying half the base times the altitude, (Eq. 137).

14

Quadratic Equations

••• **OBJECTIVES** ••

When you have completed this chapter, you should be able to:

- Solve quadratic equations by factoring, by completing the square, and by using the quadratic formula.
- Solve literal and fractional quadratic equations.
- Solve radical equations that lead to quadratics.
- Solve word problems that result in quadratic equations.
- Graph quadratic equations, and solve them graphically.
- Solve polynomial equations of higher degree by factoring.
- Solve equations by the numerical method of simple iteration.
- Solve systems of two quadratic equations.

••

So far we have learned how to solve linear (first-degree) equations and sets of first-degree equations. But many technical problems require us to solve more complicated equations than that. In this chapter we study *second-degree equations (quadratics),* and later we cover *logarithmic, exponential,* and *trigonometric equations.* Those equation types do not include every type that exists, but does include many of the common types.

Quadratic equations arise in many technical problems, such as the simple falling-body problem: "An object is thrown downward with a speed of 15 ft/s. How long will it take to fall 100 ft?"

The displacement s of a freely falling body is given by $s = v_0 t + \frac{1}{2}at^2$, and if we substitute $s = 100$, $v_0 = 15$, and $a = 32$, we get

$$100 = 15t + 16t^2$$

Now how do we solve that equation? How shall we find the values of t that will make the equation balance? Even this simple problem requires us to solve a quadratic equation.

Let us leave this problem and return to it after we have learned how to solve quadratics. Before we start, you might want to leaf through Chapter 8 if you have gotten rusty on factoring. Look especially at Secs. 8–1, 8–3, and 8–6 on common factors, trinomials, and the perfect square trinomial.

14–1   Solving Quadratics by Factoring

Terminology

Recall that in a polynomial, all powers of *x* are positive integers.

A polynomial equation of second degree is called a *quadratic equation*. It is common practice to refer to it simply as a *quadratic*.

◆◆◆ **Example 1:** The following equations are quadratic equations:

(a) $4x^2 - 5x + 2 = 0$ (b) $x^2 = 58$
(c) $9x^2 - 5x = 0$ (d) $2x^2 - 7 = 0$

Equation (a) is called a *complete* quadratic; (b) and (d), which have no *x* terms, are called *pure* quadratics; and (c), which has no constant term, is called an *incomplete* quadratic.   ◆◆◆

A quadratic is in *general form* when it is written in the following form, where *a*, *b*, and *c* are constants:

General Form of a Quadratic	$ax^2 + bx + c = 0$	99

◆◆◆ **Example 2:** Write the quadratic equation

$$7 - 4x = \frac{5x^2}{3}$$

in general form, and identify *a*, *b*, and *c*.

Solution: Subtracting $5x^2/3$ from both sides and writing the terms in descending order of the exponents, we obtain

$$-\frac{5x^2}{3} - 4x + 7 = 0$$

Quadratics in general form are usually written without fractions and with the first term positive. Multiplying by -3, we get

$$5x^2 + 12x - 21 = 0$$

The equation is now in general form, with $a = 5$, $b = 12$, and $c = -21$.   ◆◆◆

Number of Roots

We'll see later in this chapter that the maximum number of roots that certain types of equations may have is equal to the degree of the equation. Thus a quadratic equation, being of degree 2, has *two solutions* or *roots*. The two roots are sometimes equal, or they may be imaginary or complex numbers. However, in applications, we'll see that one of the two roots must sometimes be discarded.

◆◆◆ **Example 3:** The quadratic equation

$$x^2 = 4$$

has two roots, $x = 2$ and $x = -2$.   ◆◆◆

Solving Pure Quadratics

To solve a *pure* quadratic, we simply isolate the x^2 term and then take the square root of both sides, as in the following example.

◆◆◆ Example 4: Solve $3x^2 - 75 = 0$.

Solution: Adding 75 to both sides and dividing by 3, we obtain

$$3x^2 = 75$$
$$x^2 = 25$$

Taking the square root yields

$$x = \pm\sqrt{25} = \pm 5$$

Check: We check our solution the same way as for other equations, by substituting back into the original equation. Now, however, we must check two solutions.

$$\text{Substitute } +5: \quad 3(5)^2 - 75 = 0$$
$$75 - 75 = 0 \quad \text{(checks)}$$
$$\text{Substitute } -5: \quad 3(-5)^2 - 75 = 0$$
$$75 - 75 = 0 \quad \text{(checks)} \qquad \text{◆◆◆}$$

> When taking the square root, be sure to keep both the plus *and* the minus values. *Both* will satisfy the equation.

At this point students usually grumble: "First we're told that $\sqrt{4} = +2$ only, not ± 2 (Sec. 1–5). *Now* we're told to keep both plus and minus. What's going on here?" Here's the difference. When we solve a quadratic, we know that it must have two roots, so we keep both values. When we evaluate a square root, such as $\sqrt{4}$, which is not the solution of a quadratic, we agree to take only the positive value to avoid ambiguity.

The roots might sometimes be irrational, as in the following example.

◆◆◆ Example 5: Solve $4x^2 - 15 = 0$.

Solution: Following the same steps as before, we get

$$4x^2 = 15$$
$$x^2 = \frac{15}{4}$$
$$x = \pm\sqrt{\frac{15}{4}} = \pm\frac{\sqrt{15}}{2} \qquad \text{◆◆◆}$$

Solving Incomplete Quadratics

To solve an *incomplete* quadratic, remove the common factor x from each term, and set each factor equal to zero.

◆◆◆ Example 6: Solve

$$x^2 + 5x = 0$$

Solution: Factoring yields

$$x(x + 5) = 0$$

We use this idea often in this chapter. If we have the product of two quantities a and b set equal to zero, $ab = 0$, this equation will be true if $a = 0$ ($0 \cdot b = 0$) or if $b = 0$ ($a \cdot 0 = 0$), or if both are zero ($0 \cdot 0 = 0$).

Note that this expression will be true if either or both of the two factors equal zero. We therefore set each factor in turn equal to zero.

$$x = 0 \qquad x + 5 = 0$$
$$x = -5$$

The two solutions are thus $x = 0$ and $x = -5$. ◆◆◆

Common Error	Do not cancel an x from the terms of an incomplete quadratic. That will cause a root to be lost. In the last example, if we had said $$x^2 = -5x$$ and had divided by x, $$x = -5$$ we would have obtained the correct root $x = -5$ but would have lost the root $x = 0$.

Solving Complete Quadratics

We now consider a quadratic that has all of its terms in place: the *complete* quadratic.

General Form	$ax^2 + bx + c = 0$	**99**

First write the quadratic in general form, as given in Eq. 99. Factor the trinomial (if possible) by the methods of Chapter 8, and set each factor equal to zero.

◆◆◆ **Example 7:** Solve by factoring:

$$x^2 - x - 6 = 0$$

Solution: Factoring gives

$$(x - 3)(x + 2) = 0$$

This equation will be satisfied if either or both of the two factors $(x - 3)$ and $(x + 2)$ are zero. We therefore set each factor in turn to equal zero.

$$x - 3 = 0 \qquad x + 2 = 0$$

so the roots are

$$x = 3 \quad \text{and} \quad x = -2$$ ◆◆◆

Common Error	When the product of two quantities is zero, as in $$(x - 3)(x + 2) = 0$$ we can set each factor equal to zero, getting $x - 3 = 0$ and $x + 2 = 0$. But *this is valid only when the product is zero.* Thus if $$(x - 3)(x + 2) = 5$$ we *cannot* say that $x - 3 = 5$ and $x + 2 = 5$.

Often an equation must first be simplified before factoring.

◆◆◆ **Example 8:** Solve for x:

$$x(x - 8) = 2x(x - 1) + 9$$

Solution: Removing parentheses gives

$$x^2 - 8x = 2x^2 - 2x + 9$$

Collecting terms, we get

$$x^2 + 6x + 9 = 0$$

Factoring yields

$$(x + 3)(x + 3) = 0$$

which gives the double root,

$$x = -3 \qquad\qquad ◆◆◆$$

Sometimes at first glance an equation will not look like a quadratic. The following example shows a fractional equation which, after simplification, turns out to be a quadratic.

◆◆◆ **Example 9:** Solve for x:

$$\frac{3x - 1}{4x + 7} = \frac{x + 1}{x + 7}$$

Solution: We start by multiplying both sides by the LCD, $(4x + 7)(x + 7)$. We get

$$(3x - 1)(x + 7) = (x + 1)(4x + 7)$$

or

$$3x^2 + 20x - 7 = 4x^2 + 11x + 7$$

Collecting terms gives

$$x^2 - 9x + 14 = 0$$

Factoring yields

$$(x - 7)(x - 2) = 0$$

so $x = 7$ and $x = 2$. ◆◆◆

Writing the Equation When the Roots Are Known

Given the roots, we simply reverse the process to find the equation.

◆◆◆ **Example 10:** Write the quadratic equation that has the roots $x = 2$ and $x = -5$.

Solution: If the roots are 2 and -5, we know that the factors of the equation must be $(x - 2)$ and $(x + 5)$. So

$$(x - 2)(x + 5) = 0$$

Multiplying gives us

$$x^2 + 5x - 2x - 10 = 0$$

So

$$x^2 + 3x - 10 = 0$$

is the original equation. ◆◆◆

Solving Radical Equations

In Chapter 13 we solved *simple* radical equations. We isolated a radical on one side of the equation and then squared both sides. Here we solve equations in which this squaring operation results in a quadratic equation.

◆◆◆ **Example 11:** Solve for x:

$$3\sqrt{x-1} - \frac{4}{\sqrt{x-1}} = 4$$

Solution: We clear fractions by multiplying through by $\sqrt{x-1}$.

$$3(x-1) - 4 = 4\sqrt{x-1}$$
$$3x - 7 = 4\sqrt{x-1}$$

Squaring both sides yields

$$9x^2 - 42x + 49 = 16(x-1)$$

Removing parentheses and collecting terms gives

$$9x^2 - 58x + 65 = 0$$

Factoring gives

$$(x-5)(9x-13) = 0$$
$$x = 5 \quad \text{and} \quad x = \frac{13}{9}$$

Check: When $x = 5$,

$$3\sqrt{5-1} - \frac{4}{\sqrt{5-1}} \stackrel{?}{=} 4$$

$$3(2) - \frac{4}{2} = 4 \quad \text{(checks)}$$

When $x = \frac{13}{9}$,

$$\sqrt{\frac{13}{9} - 1} = \sqrt{\frac{4}{9}} = \frac{2}{3}$$

So

$$3 \cdot \frac{2}{3} - \frac{4}{\frac{2}{3}} \stackrel{?}{=} 4$$

$$2 - 6 \neq 4 \quad \text{(does not check)}$$

Our solution is then $x = 5$.   ◆◆◆

Remember that the squaring operation sometimes gives an extraneous root that will not check.

Exercise 1 ◆ Solving Quadratics by Factoring

Pure and Incomplete Quadratics

Solve for x.

1. $2x = 5x^2$
2. $2x - 40x^2 = 0$
3. $3x(x-2) = x(x+3)$
4. $2x^2 - 6 = 66$
5. $5x^2 - 3 = 2x^2 + 24$
6. $7x^2 + 4 = 3x^2 + 40$
7. $(x+2)^2 = 4x + 5$
8. $5x^2 - 2 = 3x^2 + 6$
9. $8.25x^2 - 2.93x = 0$
10. $284x = 827x^2$

Complete Quadratics

Solve for x.

11. $x^2 + 2x - 15 = 0$

12. $x^2 + 6x - 16 = 0$

13. $x^2 - x - 20 = 0$

14. $x^2 + 13x + 42 = 0$

15. $x^2 - x - 2 = 0$

16. $x^2 + 7x + 12 = 0$

17. $2x^2 - 3x - 5 = 0$

18. $4x^2 - 10x + 6 = 0$

19. $2x^2 + 5x - 12 = 0$

20. $3x^2 - x - 2 = 0$

21. $5x^2 + 14x - 3 = 0$

22. $5x^2 + 3x - 2 = 0$

23. $\dfrac{2 - x}{3} = 2x^2$

24. $(x - 6)(x + 6) = 5x$

25. $x(x - 5) = 36$

26. $(2x - 3)^2 = 2x - x^2$

27. $x^2 - \dfrac{5}{6}x = \dfrac{1}{6}$

28. $\dfrac{7}{x + 4} - \dfrac{1}{4 - x} = \dfrac{2}{3}$

> There are no applications in the first few exercise sets. These will come a little later.

Literal Equations

Solve for x.

29. $4x^2 + 16ax + 12a^2 = 0$

30. $14x^2 - 23ax + 3a^2 = 0$

31. $9x^2 + 30bx + 24b^2 = 0$

32. $24x^2 - 17xy + 3y^2 = 0$

Writing the Equation from the Roots

Write the quadratic equations that have the following roots.

33. $x = 4$ and $x = 7$

34. $x = -3$ and $x = -5$

35. $x = \frac{2}{3}$ and $x = \frac{3}{5}$

36. $x = p$ and $x = q$

Radical Equations

Solve each radical equation for x, and check.

37. $\sqrt{5x^2 - 3x - 41} = 3x - 7$

38. $3\sqrt{x - 1} - \dfrac{4}{\sqrt{x - 1}} = 4$

39. $\sqrt{x + 5} = \dfrac{12}{\sqrt{x + 12}}$

40. $\sqrt{7x + 8} - \sqrt{5x - 4} = 2$

Computer

41. Use your program for solving equations by the midpoint method (Sec. 5–6) to find the roots of any of the quadratics in this exercise set.

42. If you have a CAS, use it to solve any of the equations in this exercise set.

14–2 Solving Quadratics by Completing the Square

If a quadratic equation is not factorable, it is possible to manipulate it into factorable form by a procedure called *completing the square.* The form into which we shall put our expression is the *perfect square trinomial,* Eqs. 47 and 48, which we studied in Sec. 8–6.

The method of completing the square is really too cumbersome to be a practical tool for solving quadratics. The main reason we learn it is to derive the quadratic formula. Furthermore, the method of *completing the square* is a useful technique that we'll use again in later chapters.

Some computer algebra systems can automatically complete the square. *Maple,* for example, uses the *Completesquare()* command.

In the perfect square trinomial:

1.   The first and last terms are perfect squares.
2.   The middle term is twice the product of the square roots of the outer terms.

To complete the square, we manipulate our given expression so that these two conditions are met. This is best shown by an example.

♦♦♦ **Example 12:** Solve the quadratic $x^2 - 8x + 6 = 0$ by completing the square.

Solution: Subtracting 6 from both sides, we obtain

$$x^2 - 8x = -6$$

We complete the square by *adding the square of half the coefficient of the x term* to both sides. The coefficient of x is -8. We take half of -8 and square it, getting $(-4)^2$ or 16. Adding 16 to both sides yields

$$x^2 - 8x + 16 = -6 + 16 = 10$$

Factoring, we have

$$(x - 4)^2 = 10$$

Taking the square root of both sides, we obtain

$$x - 4 = \pm \sqrt{10}$$

Finally, we add 4 to both sides.

$$x = 4 \pm \sqrt{10}$$

$$\cong 7.16 \quad \text{or} \quad 0.838 \qquad \text{♦♦♦}$$

Common Error

When you are adding the quantity needed to complete the square to the left-hand side, it is easy to forget to add the same quantity to the right-hand side.

$$x^2 - 8x \boxed{+16} = -6 \boxed{+16}$$

don't forget

If the x^2 term has a coefficient other than 1, divide through by this coefficient *before* completing the square.

♦♦♦ **Example 13:** Solve:

$$2x^2 + 4x - 3 = 0$$

Solution: Rearranging and dividing by 2 gives

$$x^2 + 2x = \frac{3}{2}$$

Completing the square by adding 1 (half the coefficient of the x term, squared) to both sides of the equation, we have

$$x^2 + 2x + 1 = \frac{3}{2} + 1$$

$$(x + 1)^2 = \frac{5}{2}$$

$$x + 1 = \pm \sqrt{\frac{5}{2}} = \frac{\pm\sqrt{10}}{2}$$

$$x = -1 \pm \frac{1}{2}\sqrt{10}$$

♦♦♦

Common Error	Don't forget to divide through by the coefficient of the x^2 term *before* completing the square.

Exercise 2 ◆ Solving Quadratics by Completing the Square

Solve each quadratic by completing the square.

1. $x^2 - 8x + 2 = 0$
2. $x^2 + 4x - 9 = 0$
3. $x^2 + 7x - 3 = 0$
4. $x^2 - 3x - 5 = 0$
5. $4x^2 - 3x - 5 = 0$
6. $2x^2 + 3x - 3 = 0$
7. $3x^2 + 6x - 4 = 0$
8. $5x^2 - 2x - 1 = 0$
9. $4x^2 + 7x - 5 = 0$
10. $8x^2 - 4x - 3 = 0$
11. $x^2 + 6.25x - 3.27 = 0$
12. $2.45x^2 + 6.88x - 4.17 = 0$

Quadratics having noninteger coefficients can also be solved by completing the square. The method is exactly the same, although the work is harder.

Computer

13. If you have a CAS, use it to solve any of the equations in this exercise set.

14–3 Solving Quadratics by Formula

Of the several methods we have for solving quadratics, the most useful is the quadratic formula. It will work for any quadratic, regardless of the type of roots, and it can easily be programmed for the computer.

Derivation of the Quadratic Formula

We wish to find the roots of the equation

$$ax^2 + bx + c = 0$$

by completing the square. We start by subtracting c from both sides and dividing by a.

$$x^2 + \frac{b}{a}x = -\frac{c}{a}$$

We're assuming here that the coefficient a is not zero.

Completing the square gives

$$x^2 + \frac{b}{a}x + \left(\frac{b}{2a}\right)^2 = \frac{b^2}{4a^2} - \frac{c}{a}$$

Factoring, we obtain

$$\left(x + \frac{b}{2a}\right)^2 = \frac{b^2 - 4ac}{4a^2}$$

Taking the square root yields

$$x + \frac{b}{2a} = \pm\frac{\sqrt{b^2 - 4ac}}{2a}$$

Rearranging, we get the following well-known formula for finding the roots of the quadratic equation, $ax^2 + bx + c = 0$:

Quadratic Formula	$x = \dfrac{-b \pm \sqrt{b^2 - 4ac}}{2a}$	**100**

Using the Quadratic Formula

Simply put the given equation into general form (Eq. 99); list a, b, and c; and substitute them into the formula.

◆◆◆ **Example 14:** Solve $2x^2 - 5x - 3 = 0$ by the quadratic formula.

Solution: The equation is already in general form, with

$$a = 2 \qquad b = -5 \qquad c = -3$$

Substituting into Eq. 100, we obtain

$$x = \frac{-(-5) \pm \sqrt{(-5)^2 - 4(2)(-3)}}{2(2)}$$

$$= \frac{5 \pm \sqrt{25 + 24}}{4}$$

$$= \frac{5 \pm \sqrt{49}}{4} = \frac{5 \pm 7}{4}$$

$$= 3 \quad \text{and} \quad -\frac{1}{2}$$

◆◆◆

Common Error	Always rewrite a quadratic in general form before trying to use the quadratic formula.

◆◆◆ **Example 15:** Solve the equation $5.25x - 2.94x^2 + 6.13 = 0$ by the quadratic formula.

Solution: The constants a, b, and c are *not* 5.25, -2.94, and 6.13, as you might think at first glance. We must rearrange the terms.

$$-2.94x^2 + 5.25x + 6.13 = 0$$

Although not a necessary step, dividing through by the coefficient of x^2 simplifies the work a bit.

$$x^2 - 1.79x - 2.09 = 0$$

Substituting into the quadratic formula with $a = 1$, $b = -1.79$, and $c = -2.09$, we have

$$x = \frac{1.79 \pm \sqrt{(-1.79)^2 - 4(1)(-2.09)}}{2(1)}$$

$$= 2.59 \quad \text{and} \quad -0.805$$

◆◆◆

We now have the tools to finish the falling-body problem started in the introduction to this chapter.

◆◆◆ **Example 16:** Solve the equation

$$100 = 15t + 16t^2$$

(where t is in seconds) to three significant digits.

Solution: We first go to general form.

$$16t^2 + 15t - 100 = 0$$

Substituting into the quadratic formula gives us

$$t = \frac{-15 \pm \sqrt{225 - 4(16)(-100)}}{2(16)}$$

$$= \frac{-15 \pm \sqrt{6625}}{32} = \frac{-15 \pm 81.4}{32}$$

$$\simeq 2.08 \text{ s} \quad \text{and} \quad -3.01 \text{ s}$$

In this problem, a negative elapsed time makes no sense, so we discard the negative root. Thus it will take 2.08 s for an object to fall 100 ft when thrown downward with a speed of 15 ft/s. ◆◆◆

Nonreal Roots

Quadratics resulting from practical problems will usually yield real roots if the equation is properly set up. Other quadratics, however, could give *nonreal roots,* as shown in the next example.

◆◆◆ **Example 17:** Solve:

$$x^2 + 2x + 5 = 0$$

Solution: From Eq. 100,

$$x = \frac{-2 \pm \sqrt{4 - 4(1)(5)}}{2} = \frac{-2 \pm \sqrt{-16}}{2}$$

The expression $\sqrt{-16}$ can be written $\sqrt{-1}\sqrt{16}$ or $j4$, where $j = \sqrt{-1}$. So

$$x = \frac{-2 \pm j4}{2} = -1 \pm j2$$

◆◆◆

See Chapter 21 for a complete discussion of imaginary and complex numbers.

Predicting the Nature of the Roots

As shown by Example 17, when the quantity in Eq. 100 under the radical sign ($b^2 - 4ac$) is negative, the roots are not real. It should also be evident that if $b^2 - 4ac$ is zero, the roots will be equal. Thus we can use the quantity $b^2 - 4ac$, called the *discriminant,* to predict what kind of roots an equation will give, without having to actually find the roots.

The Discriminant	If a, b, and c are real, and if $b^2 - 4ac > 0$, the roots are real and unequal. if $b^2 - 4ac = 0$, the roots are real and equal. if $b^2 - 4ac < 0$, the roots are not real.	**101**

◆◆◆ **Example 18:** What sort of roots can you expect from the equation

$$3x^2 - 5x + 7 = 0?$$

Solution: Computing the discriminant, we obtain

$$b^2 - 4ac = (-5)^2 - 4(3)(7) = 25 - 84 = -59$$

A negative discriminant tells us that the roots are not real. ◆◆◆

Do not lose too much sleep over the discriminant. If you need to know the nature of the roots of a quadratic and cannot remember Eq. 101, all you have to do is calculate the roots themselves. This is not much more work than finding the discriminant.

Exercise 3 ◆ Solving Quadratics by Formula

Solve by quadratic formula to three significant digits.

1. $x^2 - 12x + 28 = 0$
2. $x^2 - 6x + 7 = 0$
3. $x^2 + x - 19 = 0$
4. $x^2 - x - 13 = 0$
5. $3x^2 + 12x - 35 = 0$
6. $29.4x^2 - 48.2x - 17.4 = 0$
7. $36x^2 + 3x - 7 = 0$
8. $28x^2 + 29x + 7 = 0$
9. $49x^2 + 21x - 5 = 0$
10. $16x^2 - 16x + 1 = 0$
11. $3x^2 - 10x + 4 = 0$
12. $x^2 - 34x + 22 = 0$
13. $3x^2 + 5x = 7$
14. $4x + 5 = x^2 + 2x$
15. $x^2 - 4 = 4x + 7$
16. $x^2 - 6x - 14 = 3$
17. $6x - 300 = 205 - 3x^2$
18. $3x^2 - 25x = 5x - 73$
19. $2x^2 + 100 = 32x - 11$
20. $33 - 3x^2 - 10x = 0$
21. $4.26x + 5.74 = 1.27x^2 + 2.73x$
22. $1.83x^2 - 4.26 = 4.82x + 7.28$
23. $x^2 - 6.27x - 14.4 = 3.17$
24. $6.47x - 338 = 205 - 3.73x^2$
25. $x(2x - 3) = 3x(x + 4) - 2$
26. $2.95(x^2 + 8.27x) = 7.24x(4.82x - 2.47) + 8.73$
27. $(4.20x - 5.80)(7.20x - 9.20) = 8.20x + 9.90$
28. $(2x - 1)^2 + 6 = 6(2x - 1)$

Nature of the Roots

Determine whether the roots of each quadratic are real or nonreal and whether they are equal or unequal.

29. $x^2 - 5x - 11 = 0$
30. $x^2 - 6x + 15 = 0$
31. $2x^2 - 4x + 7 = 0$
32. $2x^2 + 3x - 2 = 0$
33. $3x^2 - 3x + 5 = 0$
34. $4x^2 + 4x + 1 = 0$

Computer

35. Write a program or use a spreadsheet to compute and print the roots of any quadratic once the constants a, b, and c are entered. Have the computer evaluate the discriminant before computing the roots. If the discriminant is negative, have the run stop. This will prevent the program from crashing when the computer tries to take the square root of a negative number.

36. Use your program for solving equations by the midpoint method (Sec. 5–6) to find the roots of any of the quadratics in this exercise set.

37. If you have a CAS, use it to solve any of the equations in this exercise set.

14–4 Applications and Word Problems

Now that we have the tools to solve any quadratic, let us go on to problems from technology that require us to solve these equations. At this point you might want to take a quick look at Chapter 3 and review some of the suggestions for setting up and solving word problems. You should set up these problems just as you did then.

When you solve the resulting quadratic, you will get two roots, of course. If one of the roots does not make sense in the physical problem (such as a beam having a length of −2000 ft), throw it away. But do not be too hasty. Often a second root will give an unexpected but good answer.

◆◆◆ **Example 19:** The angle iron in Fig. 14–1 has a cross-sectional area of 53.4 cm². Find the thickness x.

Estimate: Let's assume that we have two rectangles of width x, with lengths of 15.6 cm and 10.4 cm. Setting their combined area equal to 53.3 cm² gives

$$x(15.6 + 10.4) = 53.4$$

Thus we get the approximation $x \approx 2.05$ cm. But since our assumed lengths were too great, because we have counted the small square in the corner twice, our estimated value of x must be too small. We thus conclude that $x > 2.05$ cm.

Solution: We divide the area into two rectangles, as shown by the dashed line at the bottom of Fig. 14–1. One rectangle has an area of $10.4x$, and the other has an area of $(15.6 - x)x$. Since the sum of these areas must be 53.4,

$$10.4x + (15.6 - x)x = 53.4$$

Putting this equation into standard form, we get

$$10.4x + 15.6x - x^2 = 53.4$$
$$x^2 - 26.0x + 53.4 = 0$$

By the quadratic formula,

$$x = \frac{26.0 \pm \sqrt{676 - 4(53.4)}}{2} = \frac{26.0 \pm 21.5}{2}$$

So

$$x = \frac{26.0 - 21.5}{2} = 2.25 \text{ cm}$$

and

$$x = \frac{26.0 + 21.5}{2} = 23.8 \text{ cm}$$

We discard 23.8 cm because it is an impossible solution in this problem.

Check: Does our answer meet the requirements of the original statement? Let us compute the area of the angle iron using our value of 2.25 cm for the thickness. We get

$$\text{area} = 2.25(13.35) + 2.25(10.4)$$
$$= 30.0 + 23.4 = 53.4 \text{ cm}^2$$

which is the required area. Our answer is also a little bigger than our 2.05 cm estimate, as expected. ◆◆◆

◆◆◆ **Example 20:** A certain train is to be replaced with a "bullet" train that goes 40.0 mi/h faster than the old train and that will make its regular 850-mi run in 3.00 h less time. Find the speed of each train.

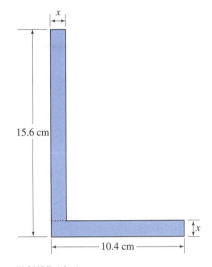

15.6 cm

10.4 cm

FIGURE 14–1

Solution: Let

$$x = \text{rate of old train} \quad \text{(mi/h)}$$

Then

$$x + 40.0 = \text{rate of bullet train} \quad \text{(mi/h)}$$

The time it takes the old train to travel 850 mi at x mi/h is, by Eq. A17,

$$\text{time} = \frac{\text{distance}}{\text{rate}} = \frac{850}{x} \quad \text{(h)}$$

The time for the bullet train is then $(850/x - 3.00)$ h. Applying Eq. A17 for the bullet train gives us

$$\text{rate} \times \text{time} = \text{distance}$$

$$(x + 40.0)\left(\frac{850}{x} - 3.00\right) = 850$$

Removing parentheses, we have

$$850 - 3.00x + \frac{34,000}{x} - 120 = 850$$

Collecting terms and multiplying through by x gives

$$-3.00x^2 + 34,000 - 120x = 0$$

or

$$x^2 + 40.0x - 11,330 = 0$$

Solving for x by the quadratic formula yields

$$x = \frac{-40.0 \pm \sqrt{1600 - 4(-11,330)}}{2}$$

If we drop the negative root, we get

$$x = \frac{-40.0 + 217}{2} = 88.3 \text{ mi/h} = \text{speed of old train}$$

and

$$x + 40.0 = 128.3 \text{ mi/h} = \text{speed of bullet train} \qquad \text{◆◆◆}$$

Exercise 4 ◆ Applications and Word Problems

Number Problems

1. What fraction added to its reciprocal gives $2\frac{1}{6}$?
2. Find three consecutive numbers such that the sum of their squares will be 434.
3. Find two numbers whose difference is 7 and the difference of whose cubes is 1267.
4. Find two numbers whose sum is 11 and whose product is 30.
5. Find two numbers whose difference is 10 and the sum of whose squares is 250.
6. A number increased by its square is equal to 9 times the next higher number. Find the number.

The "numbers" in problems 2 through 6 are all positive integers.

Geometry Problems

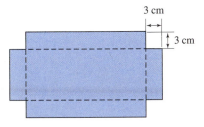

7. A rectangle is to be 2 m longer than it is wide and have an area of 24 m². Find its dimensions.

8. One leg of a right triangle is 3 cm greater than the other leg, and the hypotenuse is 15 cm. Find the legs of the triangle.

9. A rectangular sheet of brass is twice as long as it is wide. Squares, 3 cm × 3 cm, are cut from each corner (Fig. 14–2), and the ends are turned up to form an open box having a volume of 648 cm³. What are the dimensions of the original sheet of brass?

FIGURE 14–2

10. The length, width, and height of a cubical shipping container are all decreased by 1.0 ft, thereby decreasing the volume of the cube by 37 ft³. What was the volume of the original container?

11. Find the dimensions of a rectangular field that has a perimeter of 724 m and an area of 32,400 m².

12. A flat of width w is to be cut on a bar of radius r (Fig. 14–3). Show that the required depth of cut x is given by the formula

$$x = r \pm \sqrt{r^2 - \frac{w^2}{4}}$$

13. A casting in the shape of a cube is seen to shrink 0.175 in. on a side, with a reduction in volume of 2.48 in.³. Find the original dimensions of the cube.

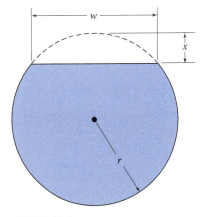

14. The cylinder in Fig. 14–4 has a surface area of 846 cm², including the ends. Find its radius.

FIGURE 14–3

15. A cylindrical tank having a diameter of 75.5 in. is placed so that it touches a wall, as in Fig. 14–5. Find the radius x of the largest pipe that can fit into the space between the tank, the wall, and the floor. *Hint:* First use the Pythagorean theorem to show that $OC = 53.4$ in. and that $OP = 15.6$ in. Then in triangle OQR, $(OP - x)^2 = x^2 + x^2$.

Uniform Motion

16. A truck travels $35\overline{0}$ mi to a delivery point, unloads, and, now empty, returns to the starting point at a speed 8.00 mi/h greater than on the outward trip. What was the speed of the outward trip if the total round-trip driving time was 14.4 h?

17. An airplane flies 355 mi to city A. Then, with better winds, it continues on to city B, 448 mi from A, at a speed 15.8 mi/h greater than on the first leg of the trip. The total flying time was 5.20 h. Find the speed at which the plane traveled to city A.

12.0 cm

18. An express bus travels a certain 250-mi route in 1.0 h less time than it takes a local bus to travel a 240-mi route. Find the speed of each bus if the speed of the local is 10 mi/h less than that of the express.

19. A trucker calculates that if he increased his average speed by $2\overline{0}$ km/h, he could travel his $80\overline{0}$-km route in 2.0 h less time than usual. Find his usual speed.

FIGURE 14–4

20. A boat sails 30 km at a uniform rate. If the rate had been 1 km/h more, the time of the sailing would have been 1 h less. Find the rate of travel.

Work Problems

21. A certain punch press requires 3 h longer to stamp a box of parts than does a newer-model punch press. After the older press has been punching a box of parts for 5 h, it is joined by the newer machine. Together, they finish the box of parts in 3 additional hours. How long does it take each machine, working alone, to punch a box of parts?

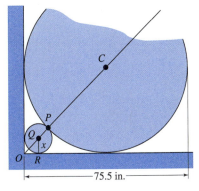

FIGURE 14–5

22. Two water pipes together can fill a certain tank in 8.40 h. The smaller pipe alone takes 2.50 h longer than the larger pipe to fill that same tank. How long would it take the larger pipe alone to fill the tank?

23. A laborer built 35 m of stone wall. If she had built 2 m less each day, it would have taken her 2 days longer. How many meters did she build each day, working at her usual rate?

24. A woman worked part-time a certain number of days, receiving for her pay $1800. If she had received $10 per day less than she did, she would have had to work 3 days longer to earn the same sum. How many days did she work?

Simply Supported Beam

25. For a simply supported beam of length l having a distributed load of w lb/ft (Fig. 14–6), the bending moment M at any distance x from one end is given by

$$M = \frac{1}{2}wlx - \frac{1}{2}wx^2$$

Find the locations on the beam where the bending moment is zero.

26. A simply supported beam, 25.0 ft long, carries a distributed load of 1550 lb/ft. At what distances from an end of the beam will the bending moment be 112,000 ft·lb? (Use the equation from problem 25.)

Freely Falling Body

27. An object is thrown upward with a velocity of 145 ft/s. When will it be 85.0 ft above its initial position?

28. An object is thrown upward with an initial speed of $12\overline{0}$ m/s. Find the time for it to return to its starting point. (In the metric system, $g = 980$ cm/s².)

Electrical Problems

29. Referring to Fig. 14–7, determine what two resistances will give a total resistance of $78\overline{0}$ Ω when wired in series and 105 Ω when wired in parallel. (See Eqs. A63 and A64.)

30. Find two resistances that will give an equivalent resistance of 9070 Ω in series and 1070 Ω in parallel.

31. In the circuit of Fig. 14–8, the power P dissipated in the load resistor R_1 is

$$P = EI - I^2R$$

If the voltage E is 115 V, and if $R = 10\overline{0}$ Ω, find the current I needed to produce a power of 29.3 W in the load.

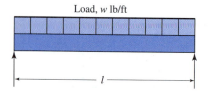

Load, w lb/ft

FIGURE 14–6 Simply supported beam with a uniformly distributed load.

Use $s = v_0t + \frac{1}{2}gt^2$, Eq. A18, for these falling-body problems, but be careful of the signs. If you take the upward direction as positive, g will be *negative*.

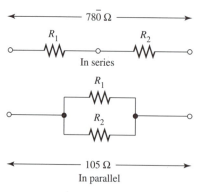

$78\overline{0}$ Ω

R_1 R_2

In series

R_1

R_2

105 Ω

In parallel

FIGURE 14–7

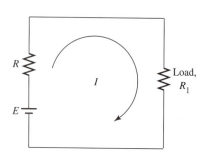

R

E

I

Load, R_1

FIGURE 14–8

32. The reactance X of a capacitance C and an inductance L in series (Fig. 14–9) is

$$X = \omega L - \frac{1}{\omega C}$$

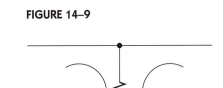

FIGURE 14–9

where ω is the angular frequency in rad/s, L is in henries, and C is in farads. Find the angular frequency needed to make the reactance equal to 1500 Ω, if $L = 0.5$ henry and $C = 0.2 \times 10^{-6}$ farad (0.2 microfarad).

33. Figure 14–10 shows two currents flowing in a single resistor R. The total current in the resistor will be $I_1 + I_2$, so the power dissipated is

$$P = (I_1 + I_2)^2 R$$

If $R = 100\ \Omega$ and $I_2 = 0.2$ A, find the current I_1 needed to produce a power of 9.0 W.

FIGURE 14–10

34. A *square-law device* is one whose output is proportional to the square of the input. A junction field-effect transistor (JFET) (Fig. 14–11) is such a device. The current I that will flow through an n-channel JFET when a voltage V is applied is

$$I = A\left(1 - \frac{V}{B}\right)^2$$

where A is the drain saturation current and B is the gate source pinch-off voltage. Solve this equation for V.

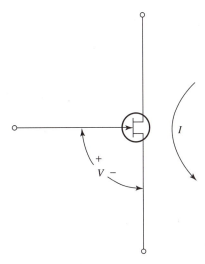

FIGURE 14–11

35. A certain JFET has a drain saturation current of 4.8 mA and a gate source pinch-off voltage of -2.5 V. What input voltage is needed to produce a current of 1.5 mA?

14–5 Graphing the Quadratic Function

Quadratic Functions

A *quadratic function* is one whose highest-degree term is of second degree.

◆◆◆ **Example 21:** The following functions are quadratic functions:

(a) $f(x) = 5x^2 - 3x + 2$

(b) $f(x) = 9 - 3x^2$

(c) $f(x) = x(x + 7)$

(d) $f(x) = x - 4 - 3x^2$ ◆◆◆

The parabola is one of the four *conic sections* (the curves obtained when a cone is intersected by a plane at various angles), and it has many interesting properties and applications. We take only a very brief look at the parabola here, with the main treatment saved for our study of the conic sections.

The Parabola

When we plot a quadratic function, we get the well-known and extremely useful curve called the *parabola*.

◆◆◆ **Example 22:** Plot the quadratic function $y = x^2 + x - 3$ for $x = -3$ to $x = 3$.

Solution: We can graph the function using a graphics calculator or graphing software on a computer, or by computing a table of point pairs as was shown in Sec. 5–2.

x	−3	−2	−1	0	1	2
y	3	−1	−3	−3	−1	3

We plot these points and connect them with a smooth curve, a parabola, as shown in Fig. 14–12.

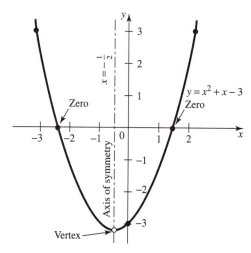

FIGURE 14–12

Note that the curve in this example is symmetrical about the line $x = -\frac{1}{2}$. This line is called the *axis of symmetry*. The point where the parabola crosses the axis of symmetry is called the *vertex*.

Recall that in Chapter 5 we said that a low point in a curve is called a *minimum point*, and that a high point is called a *maximum point*. Thus the vertex of this parabola is also a minimum point.   ◆◆◆

Solving Quadratics Graphically

In Sec. 4–5 we saw that a graphical solution to the equation

$$f(x) = 0$$

could be obtained by plotting the function

$$y = f(x)$$

and locating the points where the curve crosses the x axis. Those points are called the *zeros* of the function. Those points are then the *solution, or roots,* of the equation. Notice that in Fig. 14–12 there are two zeros, corresponding to the two roots of the equation.

This, of course, is not a practical way to solve a quadratic. It is much faster and more accurate to use the formula. However, the same method is valuable for finding the roots of other equations that cannot be as easily solved algebraically.

◆◆◆ **Example 23:** Graphically find the approximate roots of the equation

$$x^2 + x - 3 = 0$$

Solution: The function

$$y = x^2 + x - 3$$

is already plotted in Fig. 14–12. Reading the x intercepts as accurately as possible, we get for the roots

$$x \simeq -2.3 \quad \text{and} \quad x \simeq 1.3$$

If we were using a graphics utility, we could use $\boxed{\text{TRACE}}$ and $\boxed{\text{ZOOM}}$ to get the roots to whatever accuracy we wanted. ◆◆◆

Exercise 5 ◆ Graphing the Quadratic Function

Plot the following quadratic functions. Use a graphics utility or a point-by-point plot as directed by your instructor. Label the vertex, any zeros, and the axis of symmetry.

1. $y = x^2 + 3x - 15$
2. $y = 5x^2 + 14x - 3$
3. $y = 8x^2 - 6x + 1$
4. $y = 2.74x^2 - 3.12x + 5.38$

Computer

5. If you have graphing software, use it to make any of the graphs in this exercise set. If you had written a program for generating a table of point pairs (see Chapter 5, Exercise 2), use it to compute the points needed to graph any of these equations.

Parabolic Arch

6. The parabolic arch is often used in construction because of its great strength. The equation of the bridge arch in Fig. 14–13 is $y = 0.0625x^2 - 5.00x + 100$. Find the distances a, b, c, and d.
7. Plot the parabolic arch of problem 6, taking values of x every 10 ft.

Parabolic Reflector

8. The parabolic solar collector in Fig. 14–14 has the equation $x^2 = 125y$. Plot the curve, taking values of x every 10 cm.
9. Find the depth d of the collector of problem 8 at its center.

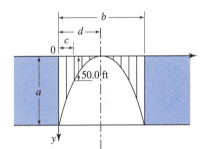

FIGURE 14–13   A parabolic bridge arch. Note that the y axis is positive downward.

Vertical Highway Curves

Where there is a change in the slope of a highway surface, such as at the top of a hill or the bottom of a dip, the roadway is often made parabolic in shape to provide a smooth transition between the two different grades.

To construct a vertical highway curve (Fig. 14–15), we use the following property of the parabola:

The offset C from a tangent to a parabola is proportional to the square of the distance x from the point of tangency,

or $C = kx^2$, where k is a constant of proportionality, found from a known offset.

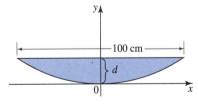

FIGURE 14–14

A parabolic surface has the property that all of the incoming rays that are parallel to the axis of symmetry, after being reflected from the surface of the parabola, will converge to a single point called the *focus*. This property is used in optical and radio telescopes, solar collectors, searchlights, microwave and radar antennas, and other devices.

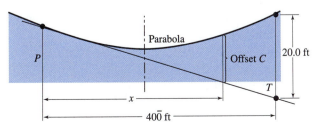

FIGURE 14–15   A dip in a highway.

10. If the offset is 20.0 ft at a point $40\overline{0}$ ft from the point of tangency, find the constant of proportionality k.

11. For the highway curve of problem 10, find the offset at a point 225 ft from the point of tangency.

Square-Law Devices

12. For the n-channel JFET of problem 35 in Exercise 4, plot the curve of current I versus applied voltage V, using the vertical axis for current and the horizontal axis for voltage. Compute I for values of V from -2.5 V to 0 V. (The curve obtained is a parabola and is called the *transconductance curve.*) Graphically determine the voltage needed to produce a current of 1.00 mA.

14–6   Equations of Quadratic Type

Recognition of Form

The general quadratic formula, Eq. 99, can be written

$$a(x)^2 + b(x) + c = 0$$

The form of this equation obviously does not change if we put a different symbol in the parentheses, as long as we put the same one in both places. The equations

$$a(Q)^2 + b(Q) + c = 0$$

and

$$a(\$)^2 + b(\$) + c = 0$$

are still quadratics. The form still does not change if we put more complicated expressions in the parentheses, such as

$$a(x^2)^2 + b(x^2) + c = 0$$

or

$$a(\sqrt{x})^2 + b(\sqrt{x}) + c = 0$$

or

$$a(x^{-1/3})^2 + b(x^{-1/3}) + c = 0$$

The preceding three equations can no longer be called quadratics, however, because they are no longer of second degree. They are called *equations of quadratic type.* They can be easily recognized because the power of x in one term is always twice the power of x in another term (and there are no other terms containing powers of x).

Solving Equations of Quadratic Type

The solution of equations of quadratic type is best shown by examples.

◆◆◆ **Example 24:** Solve:

$$x^4 - 5.12x^2 + 4.08 = 0$$

Solution: We see that the power of x in the first term is twice the power of x in the second term. We make the substitution and let

$$w = x^2$$

Our original equation becomes the quadratic

$$w^2 - 5.12w + 4.08 = 0$$

Its roots are

$$w = \frac{-(-5.12) \pm \sqrt{(-5.12)^2 - 4(1)(4.08)}}{2(1)} = 4.132 \quad \text{and} \quad 0.9872$$

Substituting back, we have

$$w = x^2 = 4.132 \quad \text{and} \quad w = x^2 = 0.9872$$

from which we obtain

$$x = \pm 2.03 \quad \text{and} \quad x = \pm 0.994$$

Check: These four values of x each satisfy the original equation (work is not shown). ◆◆◆

◆◆◆ **Example 25:** Solve the equation

$$2x^{-1/3} + x^{-2/3} + 1 = 0$$

Solution: Inspecting the powers of x, we see that one of them, $-\frac{2}{3}$, is twice the other, $-\frac{1}{3}$. We make the substitution and let

$$w = x^{-1/3}$$

Our equation becomes

$$2w + w^2 + 1 = 0$$

or

$$w^2 + 2w + 1 = 0$$

which factors into

$$(w + 1)(w + 1) = 0$$

Setting each factor equal to zero yields

$$w + 1 = 0 \quad \text{and} \quad w + 1 = 0$$

So

$$w = -1 = x^{-1/3}$$

Cubing both sides, we get

$$x^{-1} = (-1)^3 = -1$$
$$= \frac{1}{x} = -1$$

So

$$x = \frac{1}{-1} = -1$$

Check:
$$2(-1)^{-1/3} + (-1)^{-2/3} + 1 \stackrel{?}{=} 0$$
$$2(-1) + (1) + 1 = 0$$
$$0 = 0 \quad \text{(checks)} \qquad ◆◆◆$$

Exercise 6 ◆ Equations of Quadratic Type

Solve for all real values of x.

1. $x^6 - 6x^3 + 8 = 0$ **2.** $x^4 - 10x^2 + 9 = 0$

3. $x^{-1} - 2x^{-1/2} + 1 = 0$ **4.** $9x^4 - 6x^2 + 1 = 0$

5. $x^{-2/3} - x^{-1/3} = 0$

6. $x^{-6} + 19x^{-3} = 216$

7. $x^{-10/3} + 244x^{-5/3} + 243 = 0$

8. $x^6 - 7x^3 = 8$

9. $x^{2/3} + 3x^{1/3} = 4$

10. $64 + 63x^{-3/2} - x^{-3} = 0$

11. $81x^{-3/4} - 308 - 64x^{3/4} = 0$

12. $27x^6 + 46x^3 = 16$

13. $x^{1/2} - x^{1/4} = 20$

14. $3x^{-2} + 14x^{-1} = 5$

15. $2.35x^{1/2} - 1.82x^{1/4} = 43.8$

16. $1.72x^{-2} + 4.75x^{-1} = 2.24$

Computer

17. Use your program for finding roots of equations from Sec. 5–6 to solve any of the equations in this exercise set.

18. A computer is especially good for solving the complex equations in this section and in the one to follow. A CAS will give exact roots when it can. Otherwise, you can find the roots approximately. When asking for an approximate root, we must enter the range of x values between which we want the computer to search for a root. If there is more than one root within the selected range, the computer will give just one of these, and you will have to change the range to get another root. The simplest way to get these ranges is to have the computer graph the equation and to note where the graph crosses the x axis.

Use a computer to solve any of the equations in this exercise set.

14–7   Simple Equations of Higher Degree

Polynomial of Degree n

In Sec. 13–1, we used factoring to find the roots of a quadratic equation. We now expand this idea to include polynomials of higher degree. A polynomial, remember, has exponents that are all positive integers.

Polynomial of Degree n	$a_0x^n + a_1x^{n-1} + \cdots + a_{n-1}x + a_n$	102

The Factor Theorem

In Sec. 14–1, the expression

$$x^2 - x - 6 = 0$$

factored into

$$(x - 3)(x + 2) = 0$$

and hence had the roots

$$x = 3 \quad \text{and} \quad x = -2$$

In a similar way, the cubic equation

$$x^3 - 7x - 6 = 0$$

has the factors

$$(x + 1)(x - 3)(x + 2) = 0$$

and hence has the roots

$$x = -1 \qquad x = 3 \qquad x = -2$$

In general, the following equation applies:

Factor Theorem	If a polynomial equation $f(x) = 0$ has a root r, then $(x - r)$ is a factor of the polynomial $f(x)$. Conversely, if $(x - r)$ is a factor of a polynomial $f(x)$, then r is a root of $f(x) = 0$.	**103**

Number of Roots

A polynomial equation of degree n has n and only n roots.

Two or more of the roots may be equal, however, and some may be nonreal, giving the appearance of fewer than n roots; but there can never be *more* than n roots.

Solving Polynomial Equations

If an expression cannot be factored directly, we proceed largely by trial and error to obtain one root. Once a root r is obtained (and verified by substituting in the equation), we obtain a *reduced* or *depressed* equation by dividing the original equation by $(x - r)$. The process is then repeated.

> A graph of the function is very helpful in finding the roots.

◆◆◆ **Example 26:** The graph of Fig. 14–16 shows that one root of the cubic equation $16x^3 - 13x + 3 = 0$ is $x = -1$. Find the other roots.

Solution: For a cubic, we expect a maximum of three roots. By the factor theorem, since $x = -1$ is a root, then $(x + 1)$ must be a factor of the given equation.

$$(x + 1)(\text{other factor}) = 0$$

We obtain the "other factor" by dividing the given equation by $(x + 1)$, getting $16x^2 - 16x + 3$ (work is not shown). The "other factor," or reduced equation, is then the quadratic,

$$16x^2 - 16x + 3 = 0$$

By the quadratic formula,

$$x = \frac{16 \pm \sqrt{16^2 - 4(16)(3)}}{2(16)} = \frac{1}{4} \text{ and } \frac{3}{4}$$

Our three roots are therefore $x = -1$, $x = \frac{1}{4}$, and $x = \frac{3}{4}$. ◆◆◆

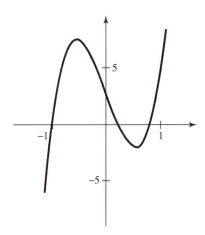

FIGURE 14–16 Graph of $y = 16x^3 - 13x + 3$.

Solution of Equations by Approximate Methods

For us to solve an equation by the factor theorem, the equation must, of course, be factorable. This requirement severely limits the use of this method for practical problems.

Fortunately, we have several ways to solve nonfactorable equations approximately. That the solution is "approximate" is usually not a difficulty, because these methods allow us to obtain as many significant digits as we want.

We have already described:

1. Graphical solution of equations, with or without the graphics calculator (Sec. 5–5).
2. Computer solution of equations by the midpoint method (Sec. 5–5).
3. The use of commercially available computer software, such as *MathCAD, Derive, Maple, Mathematica,* and so forth.

Here we give another computer method.

Simple Iteration: A Computer Method

We already have the midpoint method for computer solution of equations, and now we learn the method of *simple iteration*.

This is also known as the *method of Picard*, after Charles Émile Picard, 1856–1941.

With this method, we split the given equation into two or more terms, one of which is the unknown itself, and we move the unknown to the left side of the equation. We then guess the value of x and substitute it into the right side of the equation to obtain another value of x, which is then substituted into the equation, and so on, until our values converge on the correct value of x.

◆◆◆ **Example 27:** Find one root of the equation from Example 26,

$$16x^3 - 13x + 3 = 0$$

by simple iteration. Assume that there is a root somewhere near $x = 1$.

Solution: The equation contains an x in both the first and second terms. We can choose either to isolate on the left side of the equation. Let us choose the first term. Rearranging yields

$$16x^3 = 13x - 3$$

Dividing gives

$$x^3 = 0.8125x - 0.1875$$

if we work to four significant digits. Taking the cube root of both sides, we get

$$x = (0.8125x - 0.1875)^{1/3} \tag{1}$$

Note that we have not *solved* the equation, because there is still an x on the right side of the equation.

For a first approximation, let $x = 1$. Substituting into the right of Equation (1) gives

$$x = (0.8125 - 0.1875)^{1/3}$$
$$= 0.8550$$

This, of course, is not a root, but only a *second approximation* to a root. We now take the value 0.8550 and substitute it into the right side of Equation (1).

$$x = [0.8125(0.8550) - 0.1875]^{1/3}$$
$$= 0.7975$$

This third approximation is substituted into (1), giving

$$x = [0.8125(0.7975) - 0.1875]^{1/3}$$
$$= 0.7722$$

We continue, getting the following table of values:

$$0.8550$$
$$0.7975$$
$$0.7722$$
$$0.7605$$
$$0.7550$$
$$0.7524$$
$$0.7512$$
$$0.7506$$
$$0.7503$$

If the computation diverges instead of converges, we split the equation in a different way and try again. In the example here, we would isolate the x in the middle term, rather than the one in the first term.

Notice that we are converging on the value

$$x = 0.750$$

which is one of the roots found in Example 26. With a different first guess, we can usually make the computation converge on one of the other roots. ◆◆◆

Exercise 7 ◆ Simple Equations of Higher Degree

Find the remaining roots of each polynomial, given one root.

1. $x^3 + x^2 - 17x + 15 = 0,$ $x = 1$
2. $x^3 + 9x^2 + 2x - 48 = 0,$ $x = 2$
3. $x^3 - 2x^2 - x + 2 = 0,$ $x = 1$
4. $x^3 + 9x^2 + 26x + 24 = 0,$ $x = -2$
5. $2x^3 + 7x^2 - 7x - 12 = 0,$ $x = -1$
6. $3x^3 - 7x^2 + 4 = 0,$ $x = 1$
7. $9x^3 + 21x^2 - 20x - 32 = 0,$ $x = -1$
8. $16x^4 - 32x^3 - 13x^2 + 29x - 6 = 0,$ $x = -1$

Graphics Calculator

9. Use a graphics calculator to find the roots of the above equations to three significant digits, using the $\boxed{\text{TRACE}}$ and $\boxed{\text{ZOOM}}$ features.

Computer

10. Use mathematical computation software or a CAS to find the roots of any of the above equations to three significant digits. You will have to be in $\boxed{\text{APPROXIMATE}}$ mode and will need to enter the values of x between which the root is located.
11. Write a program or use a spreadsheet for solving equations by simple iteration. Use it to find the roots of any of the above equations to three significant digits.
12. Use your program for finding roots by the midpoint method to find the roots of any of the above equations to three significant digits.

Find the roots of the following equations to three significant digits. Use a graphics calculator, mathematical software, simple iteration, or the midpoint method.

13. $15.2x^3 - 11.3x + 1.72 = 0$
14. $1.26x^3 - 1.83x + 1.03 = 0$
15. $8.62x^3 - 6.27x + 1.62 = 0$
16. $3.92x^4 - 3.84x - 6.26 = 0$
17. $4.82x^4 - 16.1x^3 + 3.25 = 0$
18. $8.33x^4 + 3.95x^3 - 2.25x - 9.02 = 0$

14–8   Systems of Quadratic Equations

In Chapter 10 we solved systems of *linear* equations. We now use the same techniques, substitution or addition-subtraction, to solve sets of two equations in two unknowns where one or both equations are of second degree.

One Equation Linear and One Quadratic

These systems are best solved by substitution.

◆◆◆ **Example 28:** Solve for x and y to two decimal places.

$$2x + y = 5$$
$$2x^2 + y^2 = 17$$

Solution: From the first equation,

$$x = \frac{5 - y}{2}$$

Squaring gives

$$x^2 = \frac{25 - 10y + y^2}{4}$$

Substituting into the second equation yields

$$\frac{2(25 - 10y + y^2)}{4} + y^2 = 17$$

$$25 - 10y + y^2 + 2y^2 = 34$$

We get the quadratic,

$$3y^2 - 10y - 9 = 0$$

which, by quadratic formula, gives

$$y = 4.07 \quad \text{and} \quad y = -0.74$$

Substituting back for x gives

$$x = \frac{5 - 4.07}{2} = 0.47 \quad \text{and} \quad x = \frac{5 - (-0.74)}{2} = 2.87$$

So our two solutions are (0.47, 4.07) and (2.87, −0.74). The plots of the two given equations, shown in Fig. 14–17, show points of intersection at those coordinates.

♦♦♦

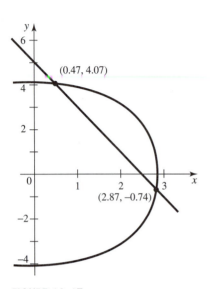

FIGURE 14–17

♦♦♦ Example 29: Solve for x and y:

$$x + y = 2$$
$$xy + 15 = 0$$

Solution: From the first equation,

$$y = 2 - x$$

Substituting into the second equation, we have

$$x(2 - x) + 15 = 0$$

from which

$$x^2 - 2x - 15 = 0$$

Factoring gives

$$(x - 5)(x + 3) = 0$$
$$x = 5 \quad \text{and} \quad x = -3$$

Substituting back yields

$$y = 2 - 5 = -3$$

and

$$y = 2 - (-3) = 5$$

So the solutions are

$$x = 5, y = -3 \quad \text{and} \quad x = -3, y = 5$$

as shown in Fig. 14–18.

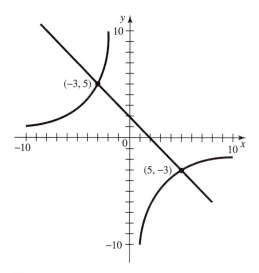

FIGURE 14–18   ◆◆◆

Both Equations Quadratic

These systems may be solved either by substitution or by addition-subtraction.

◆◆◆ **Example 30:** Solve for x and y:

$$3x^2 - 4y = 47$$
$$7x^2 + 6y = 33$$

Solution: We multiply the first equation by 3 and the second by 2.

$$9x^2 - 12y = 141$$
$$\underline{14x^2 + 12y = 66}$$

$$\text{Add:} \quad 23x^2 = 207$$
$$x^2 = 9$$
$$x = \pm 3$$

Substituting back, we get $y = -5$ for both $x = 3$ and $x = -3$. The plot of the given curves, given in Fig. 14–19, clearly shows the points of intersections $(3, -5)$ and $(-3, -5)$.

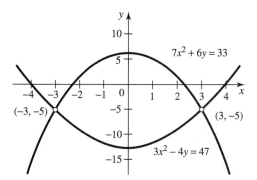

FIGURE 14–19   Intersection of two parabolas. This is a good way to get an approximate solution to any pair of equations that can readily be graphed.   ◆◆◆

The graphs of the equations in the preceding examples intersected in two points. However, the graphs of two second-degree equations can sometimes intersect in as many as four points.

◆◆◆ **Example 31:** Solve for x and y:

$$3x^2 + 4y^2 = 76$$
$$-11x^2 + 3y^2 = 4$$

Solution: We multiply the first equation by 3 and the second by -4.

$$9x^2 + 12y^2 = 228$$
$$\underline{44x^2 - 12y^2 = -16}$$
$$\text{Add: } 53x^2 \qquad = 212$$
$$x^2 = 4$$
$$x = \pm 2$$

From the first equation, when x is either $+2$ or -2,

$$4y^2 = 76 - 12 = 64$$
$$y = \pm 4$$

So our solutions are

$$\begin{array}{cccc} x = 2 & x = 2 & x = -2 & x = -2 \\ y = 4 & y = -4 & y = 4 & y = -4 \end{array}$$ ◆◆◆

Sometimes we can eliminate the squared terms from a pair of equations and obtain a linear equation. This linear equation can then be solved simultaneously with one of the original equations.

◆◆◆ **Example 32:** Solve for x and y:

$$x^2 + y^2 - 3y = 4$$
$$x^2 + y^2 + 2x - 5y = 2$$

Solution: Subtracting the second equation from the first gives

$$-2x + 2y = 2$$

or

$$y = x + 1$$

Substituting $x + 1$ for y in the first equation gives

$$x^2 + (x + 1)^2 - 3(x + 1) = 4$$
$$2x^2 - x - 6 = 0$$

Factoring yields

$$(x - 2)(2x + 3) = 0$$

from which

$$x = 2 \quad \text{and} \quad x = -3/2$$

Substituting $x = 2$ into both original equations gives

$$y = 0 \qquad y = 2 \qquad y = 3$$

Substituting $x = -3/2$ into both original equations gives

$$y = 7/2 \qquad y = 11/2 \qquad y = -1/2$$

giving solutions of $(2, 3)$, $(-3/2, -1/2)$, $(2, 0)$, $(2, 2)$, $(-3/2, 7/2)$, and $(-3/2, 11/2)$. However, only two of these,

$$(2, 3) \quad \text{and} \quad (-3/2, -1/2)$$

check in *both* original equations. Figure 14–20 shows plots of the two original equations, which graph as circles. The points found are the coordinates of the points of intersection of the two circles.   ◆◆◆

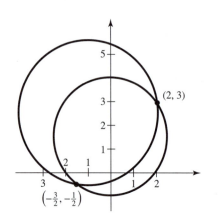

FIGURE 14–20

Solution by Graphics Calculator or Computer

In Sec. 10–1 we showed how to use a graphics calculator or graphing software on a computer to solve a system of linear equations. We simply graphed the two lines and found the coordinates of the point of intersection. Then using the TRACE and ZOOM features, we could obtain the coordinates with any degree of accuracy that we wanted.

   The procedure here is the same, except now there will usually be more than one point of intersection. Also, since the calculator or computer requires the equations to be in *explicit form,* we must solve each equation for *y*.

◆◆◆ **Example 33:** Use a graphics utility to solve the following system of equations to three significant digits:

$$2.74x + 1.04y = 4.88$$
$$2.03x^2 + 1.63y^2 = 15.3$$

Solution: We first solve each equation for *y*. The first equation gives

$$y = \frac{4.88 - 2.74x}{1.04}$$

and from the second original equation we get

$$y = \pm \sqrt{\frac{15.3 - 2.03x^2}{1.63}}$$

We need both the plus and the minus in this equation to get both the upper and the lower portions of the curve. So in our graphics calculator or computer program we enter the equations

$$Y_1 = (4.88 - 2.74x)/1.04$$
$$Y_2 = ((15.3 - 2.03x{}^{\wedge}2)/1.63){}^{\wedge}(1/2)$$
$$Y_3 = -((15.3 - 2.03x{}^{\wedge}2)/1.63){}^{\wedge}(1/2)$$

We get the graph shown in Fig. 14–21. Using TRACE, we read the coordinates of one of the points of intersection. We then zoom in repeatedly until our reading of the coordinates no longer changes in the third digit. We then repeat this operation for the other point. Doing this, we get the points

$$(0.651, 2.98) \quad \text{and} \quad (2.37, -1.55)$$

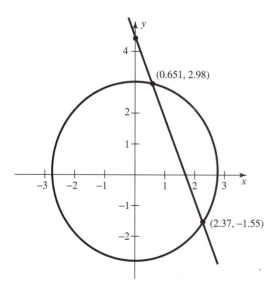

FIGURE 14–21

The approximate solution to our set of equations, to three significant digits, is then

$$x = 0.651, \; y = 2.98 \quad \text{and} \quad x = 2.37, \; y = -1.55 \qquad \text{◆◆◆}$$

Exercise 8 ◆ Systems of Quadratic Equations

Solve for x and y.

1. $x - y = 4$
   $x^2 - y^2 = 32$

2. $x + y = 5$
   $x^2 + y^2 = 13$

3. $x + 2y = 7$
   $2x^2 - y^2 = 14$

4. $x - y = 15$
   $x = 2y^2$

5. $x + 4y = 14$
   $y^2 - 2y + 4x = 11$

6. $x^2 + y = 3$
   $x^2 - y = 4$

7. $2x^2 + 3y = 10$
   $x^2 - y = 8$

8. $2x^2 - 3y = 15$
   $5x^2 + 2y = 12$

9. $x + 3y^2 = 7$
   $2x - y^2 = 9$

10. $x^2 - y^2 = 3$
    $x^2 + 3y^2 = 7$

11. $x - y = 11$
    $xy + 28 = 0$

12. $x^2 + y^2 - 8x - 4y + 16 = 0$
    $x^2 + y^2 - 4x - 8y + 16 = 0$

Graphics Calculator and Computer

Use a graphics calculator or graphing software on a computer to find an approximate graphical solution to any of the above problems. Also use either device to solve the following to three significant digits.

13. $3.52x^2 + 4.82y^2 = 34.2$
    $5.92x - 3.72y = 4.58$

14. $15.1x^2 + 8.23y^2 = 23.2$
    $9.22x - 5.47y = 2.84$

15. $7.15x^2 - 6.68y = 4.27$
    $4.83x - 3.75y^2 = 1.83$

16. $11.2x - 14.2y^2 + 13.6 = 0$
    $15.9x^2 - 13.7y = 14.8$

◆◆◆ CHAPTER 14 REVIEW PROBLEMS ◆◆◆◆◆◆◆◆◆◆◆◆◆◆◆◆◆◆◆◆◆◆◆◆◆◆◆◆◆◆◆

Solve each equation for any real roots. Give any approximate answers to three significant digits.

1. $y^2 - 5y - 6 = 0$

2. $2x^2 - 5x = 2$

3. $x^4 - 13x^2 + 36 = 0$

4. $(a + b)x^2 + 2(a + b)x = -a$

5. $w^2 - 5w = 0$

6. $x(x - 2) = 2(-2 - x + x^2)$

7. $\dfrac{r}{3} = \dfrac{r}{r + 5}$

8. $6y^2 + y - 2 = 0$

9. $3t^2 - 10 = 13t$

10. $2.73x^2 + 1.47x - 5.72 = 0$

11. $\dfrac{1}{t^2} + 2 = \dfrac{3}{t}$

12. $3w^2 + 2w - 11 = 0$

13. $9 - x^2 = 0$

14. $\dfrac{2y}{3} = \dfrac{3}{5} + \dfrac{2}{y}$

15. $2x(x + 2) = x(x + 3) + 5$

16. $\dfrac{ax^2}{b} + \dfrac{bx}{c} + \dfrac{c}{a} = 0$

17. $y + 6 = 5y^{1/2}$

18. $18 + w^2 + 11w = 0$

19. $9y^2 + y = 5$

20. $\dfrac{5}{w} - \dfrac{4}{w^2} = \dfrac{3}{5}$

21. $\dfrac{z}{2} = \dfrac{5}{z}$

22. $3x - 6 = \dfrac{5x + 2}{4x}$

23. $2x^2 + 3x = 2$

24. $x^2 + Rx - R^2 = 0$

25. $27x = 3x^2$

26. $\dfrac{n}{n + 1} = \dfrac{2}{n + 3} - 1$

27. $z^{2/3} + 8 = 9z^{1/3}$

28. $6w^2 + 13w + 6 = 0$

29. $5y^2 = 125$

30. $\dfrac{t + 2}{t} = \dfrac{4t}{3}$

31. $1.26x^2 - 11.8 = 1.13x$

32. $2.62z^2 + 4.73z - 5.82 = 0$

33. $5.12y^2 + 8.76y - 9.89 = 0$

34. $3w^2 + 2w^4 - 2 = 0$

35. $27.2w^2 + 43.6w = 45.2$

36. $2.35x^{-2} + 4.26x^{-1} - 5.57 = 0$

37. $3.21z^6 - 21.3 = 4.23z^3$

38. If one root is $x = 1$, find the other roots of the following equation:
$$x^3 + x^2 - 5x + 3$$

39. Solve for x and y.
$$x^2 + y^2 = 55$$
$$x + y = 7$$

40. Solve for x and y.
$$x - y^2 = 1$$
$$x^2 + 3y^2 = 7$$

41. If one root is $x = -1$, find the other roots of the following equation:
$$x^4 - 2x^3 - 13x^2 + 14x + 24 = 0$$

42. Plot the parabola $y = 3x^2 + 2x - 6$. Label the vertex, the axis of symmetry, and any zeros.

43. Solve for x and y.
$$x + 2y = 4$$
$$2x^2 - 4y = 1$$

44. Solve for x and y.

$$2x + y = 5$$
$$2x^2 + y = 8$$

45. A person purchased some bags of insulation for $1000. If she had purchased 5 more bags for the same sum, they would have cost 12 cents less per bag. How many did she buy?

46. The perimeter of a rectangular field is 184 ft, and its area 1920 ft². Find its dimensions.

47. The rectangular yard of Fig. 14–22 is to be enclosed by fence on three sides, and an existing wall is to form the fourth side. The area of the yard is to be 450 m², and its length is to be twice its width. Find the dimensions of the yard.

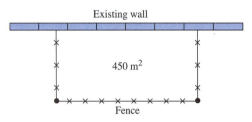

Existing wall

450 m²

Fence

FIGURE 14–22

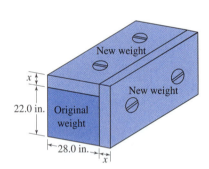

New weight

New weight

x

22.0 in.   Original weight

28.0 in. x

FIGURE 14–23

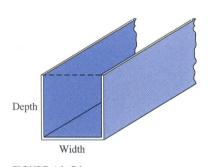

Depth

Width

FIGURE 14–24

48. A fast train runs 8.0 mi/h faster than a slow train and takes 3.0 h less to travel 288 mi. Find the rates of the trains.

49. A man started to walk 3 mi, intending to arrive at a certain time. After walking 1 mi, he was detained 10 min and had to walk the rest of the way 1 mi/h faster in order to arrive at the intended time. What was his original speed?

50. A rectangular field is 12 m longer than it is wide and contains 448 m². What are the lengths of its sides?

51. A tractor wheel, 15 ft in circumference, revolves in a certain number of seconds. If it revolved in a time longer by 1.0 s, the tractor would travel 14,400 ft less in 1.0 h. In how many seconds does it revolve?

52. The iron counterweight in Fig. 14–23 is to have its weight increased by 50% by plates of iron bolted along the top and side (but not at the ends). The top plate and the side plate have the same thickness. Find their thickness.

53. A 26-in.-wide strip of steel is to have its edges bent up at right angles to form an open trough, as in Fig. 14–24. The cross-sectional area is to be 80 in.². Disregarding the thickness of the steel sheet, find the width and the depth of the trough.

54. A boat sails 30 mi at a uniform rate. If the rate had been 1 mi/h less, the time of the sailing would have been 1 h more. Find the rate of travel.

55. In a certain number of hours a woman traveled 36.0 km. If she had traveled 1.50 km more per hour, it would have taken her 3.00 h less to make the journey. How many kilometers did she travel per hour?

56. The length of a rectangle court exceeds its width by 2 m. If the length and the width were each increased by 3 m, the area of the court would be 80 m². Find the dimensions of the court.

57. The area of a certain square will double if its dimensions are increased by 6.0 ft and 4.0 ft, respectively. Find its dimensions.

58. A mirror 18 in. by 12 in. is to be set in a frame of uniform width, and the area of the frame is to be equal to that of the glass. Find the width of the frame.

Writing

59. As in Chapter 3, our writing assignment is to make up a word problem. But now we want a word problem that leads to a quadratic equation. As before, swap with a classmate; solve each other's problem; note anything unclear, unrealistic, or ambiguous; and then rewrite your problem if needed.

Team Project

60. The bending moment for a simply supported beam carrying a distributed load w (Fig. 14–6) is $M = \frac{1}{2}wlx - \frac{1}{2}wx^2$. The bending moment curve is therefore a parabola. Plot the curve of M vs. x for a 10-ft-long beam carrying a load of 1000 lb/ft. Take 1-ft intervals along the beam. Graphically locate **(a)** the points of zero bending moment and **(b)** the point of maximum bending moment.

15

Oblique Triangles and Vectors

In Chapter 7 we defined the trigonometric functions for angles of any size, but we used them only for acute angles. Here we learn how to find the trigonometric functions of obtuse angles, negative angles, angles greater than one revolution, and angles with terminal sides on the coordinate axes. We need the trigonometric functions of obtuse angles to solve oblique triangles and the trigonometric functions of angles larger than 180° for vectors and other applications.

An *oblique triangle* is one that does not contain a right angle. In this chapter we derive two new formulas, the *law of sines* and the *law of cosines,* to enable us to solve oblique triangles quickly. We cannot, of course, use the methods we derived for solving right triangles (the Pythagorean theorem or the six trigonometric functions) to solve oblique triangles, although we will use these relationships to derive the *law of sines* and the *law of cosines.*

We also continue our study of *vectors* in this chapter. In Chapter 7 we dealt with vectors at right angles to each other; here we consider vectors at any angle.

15–1 Trigonometric Functions of Any Angle

Definition of the Trigonometric Functions

We defined the trigonometric functions of any angle in Sec. 7–2 but have so far done problems only with acute angles. We turn now to larger angles.

Figures 15–1(a)–(c) show angles in the second, third, and fourth quadrants, and Figure 15–1(d) shows an angle greater than 360°. The trigonometric functions of any of these angles are defined exactly as for an acute angle in quadrant I. From any point P on the terminal side of the angle we drop a perpendicular to the x axis, forming a right triangle with legs x and y and with a hypotenuse r. The six trigonometric ratios are then given by Eqs. 146 through 151, just as before, except that some of the ratios might now be negative.

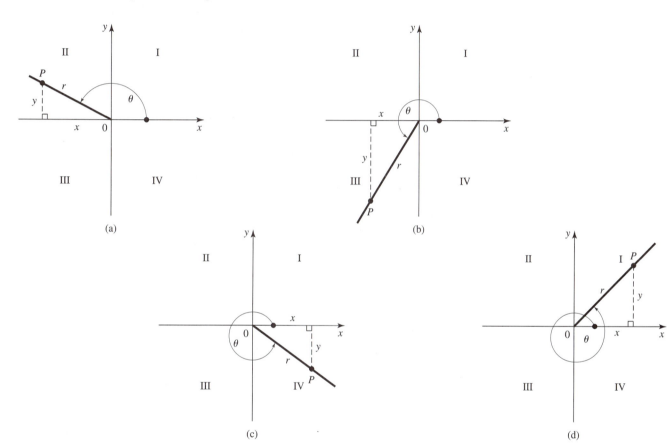

(a)

(b)

(c)

(d)

FIGURE 15–1

◆◆◆ **Example 1:** A point on the terminal side of angle θ has the coordinates $(-3, -5)$. Write the six trigonometric functions of θ to three significant digits.

Solution: We sketch the angle as shown in Fig. 15–2 and see that it lies in the third quadrant. We find distance r by the Pythagorean theorem.

$$r^2 = (-3)^2 + (-5)^2 = 9 + 25 = 34$$

$$r = 5.83$$

Then, by Eqs. 146 through 151, with $x = -3$, $y = -5$, and $r = 5.83$,

$$\sin \theta = \frac{y}{r} = \frac{-5}{5.83} = -0.858$$

$$\cos \theta = \frac{x}{r} = \frac{-3}{5.83} = -0.515$$

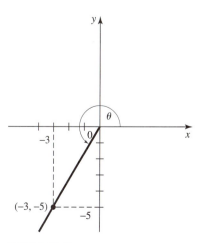

FIGURE 15–2

$$\tan \theta = \frac{y}{x} = \frac{-5}{-3} = \quad 1.67$$

$$\cot \theta = \frac{x}{y} = \frac{-3}{-5} = \quad 0.600$$

$$\sec \theta = \frac{r}{x} = \frac{5.83}{-3} = -1.94$$

$$\csc \theta = \frac{r}{y} = \frac{5.83}{-5} = -1.17$$ ◆◆◆

Algebraic Signs of the Trigonometric Functions

We saw in Chapter 7 that the trigonometric functions of first-quadrant angles were always positive. From the preceding example, it is clear that some of the trigonometric functions of angles in the second, third, and fourth quadrants are negative, because x or y can be negative (r is always positive). Figure 15–3 shows the signs of the trigonometric functions in each quadrant.

Instead of trying to remember which trigonometric functions are negative in which quadrants, just sketch the angle and note whether x or y is negative. From this information you can figure out whether the function you want is positive or negative.

Why bother learning the signs when a calculator gives them to us automatically? One reason is that you will need them when using a calculator to find the *inverse* of a function, as discussed later.

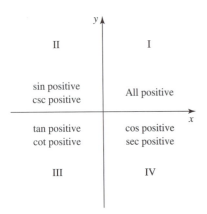

FIGURE 15–3

◆◆◆ **Example 2:** What is the algebraic sign of csc 315°?

Solution: We make a sketch, such as in Fig. 15–4. It is not necessary to draw the angle accurately, but it must be shown in the proper quadrant, quadrant IV in this case. We note that y is negative and that r is (always) positive. So

$$\csc 315° = \frac{r}{y} = \frac{(+)}{(-)} = \text{negative}$$ ◆◆◆

Trigonometric Functions of Any Angle by Calculator

In Sec. 7–2 we learned how to find the six trigonometric functions of an acute angle. Now we will find the trigonometric functions of any angle, acute or obtuse, positive or negative, or less than or greater than 360°.

The keystrokes are exactly the same as those we used for acute angles, and the calculator will automatically give the correct algebraic sign. On some calculators we enter the angle first and then press the proper trigonometric key. On other calculators we press the trigonometric key first and then enter the angle. Refer to your manual to see which way your calculator works.

But for either type of calculator, you must put it into the proper mode, DEGREE or RADIAN, before doing the calculation.

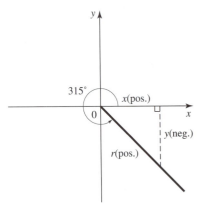

FIGURE 15–4

◆◆◆ **Example 3:** Use your calculator to verify the following calculations to four significant digits:

(a) sin 212° = −0.5299
(b) cos 163° = −0.9563
(c) tan 314° = −1.036   ◆◆◆

As before, to find the cotangent, secant, and cosecant, we use the reciprocal relations.

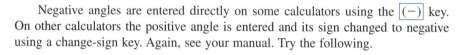

$$\tan \theta = \frac{1}{\cot \theta}$$

Reciprocal Relationships

$$\cos \theta = \frac{1}{\sec \theta}$$

152

$$\sin \theta = \frac{1}{\csc \theta}$$

◆◆◆ **Example 4:** Verify the following calculations to four significant digits:

(a) $\cot 101° = -0.1944$
(b) $\sec 213° = -1.192$
(c) $\csc 294° = -1.095$ ◆◆◆

Negative angles are entered directly on some calculators using the $\boxed{(-)}$ key. On other calculators the positive angle is entered and its sign changed to negative using a change-sign key. Again, see your manual. Try the following.

◆◆◆ **Example 5:** The following calculations are worked to four significant digits:

(a) $\tan(-35.0°) = -0.7002$
(b) $\sec(-125°) = -1.743$ ◆◆◆

Angles greater than $360°$ are handled in the same manner as any other angle.

◆◆◆ **Example 6:**

(a) $\cos 412° = 0.6157$
(b) $\csc 555° = -3.864$ ◆◆◆

Evaluating Trigonometric Expressions

In Sec. 7–2 we evaluated some simple trigonometric expressions. Here we will evaluate some that are slightly more complicated and some that contain obtuse angles.

◆◆◆ **Example 7:** Evaluate the expression

$$(\sin^2 48° + \cos 62°)^3$$

to four significant digits.

Solution: The notation $\sin^2 48°$ is the same as $(\sin 48°)^2$. Let us carry five digits and round to four in the last step.

$$(\sin^2 48° + \cos 62°)^3 = [(0.74314)^2 + 0.46947]^3$$
$$= (1.0217)^3 = 1.067 \qquad ◆◆◆$$

◆◆◆ **Example 8:** Evaluate the expression

$$\cos 123.5° - \sin^2 242.7°$$

to four significant digits.

Solution: Since

$$\cos 123.5° = -0.55194$$

and

$$\sin 242.7° = -0.88862$$

we get

$$\cos 123.5° - \sin^2 242.7° = -0.55194 - (-0.88862)^2$$
$$= -1.342$$

rounded to four digits.   ◆◆◆

Common Error	Do not confuse an exponent that is on the angle with one that is on the entire function. $$(\sin \theta)^2 = \sin^2 \theta \neq \sin \theta^2$$

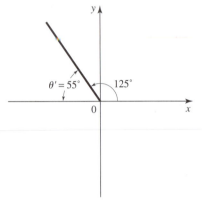

FIGURE 15–5   Reference angle θ'. It is also called the *working* angle.

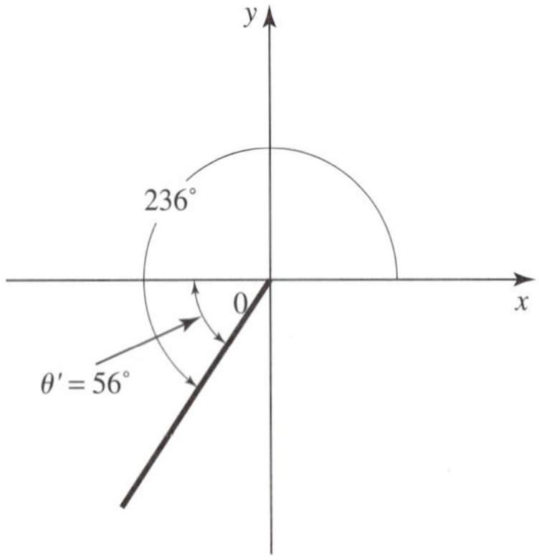

Reference Angle

Finding the trigonometric function of any angle by calculator is no problem—the calculator does all of the thinking. This is not the case, however, when we are given the function and are asked to find the angle. For this operation we will use the *reference angle*.

For an angle in standard position on coordinate axes, the *acute* angle that its terminal side makes with the x axis is called the *reference angle, θ'*.

◆◆◆ **Example 9:** The reference angle θ' for an angle of 125° is

$$\theta' = 180° - 125° = 55°$$

as shown in Fig. 15–5.   ◆◆◆

◆◆◆ **Example 10:** The reference angle θ' for an angle of 236° is

$$\theta' = 236° - 180° = 56°$$

as in Fig. 15–6.   ◆◆◆

FIGURE 15–6

We treat the quadrantal angles, 180° and 360°, as *exact* numbers. Thus the rounding of our answer is determined by the significant digits in the given angle.

◆◆◆ **Example 11:** The reference angle θ' for an angle of 331.6° (Fig. 15–7) is

$$\theta' = 360° - 331.6° = 28.4°$$

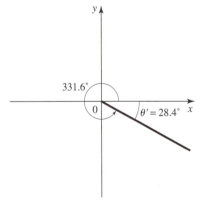

FIGURE 15–7   ◆◆◆

◆◆◆ **Example 12:** The reference angle θ' for an angle of $375°15'$ is

$$\theta' = 375°15' - 360° = 15°15'$$

as in Fig. 15–8. ◆◆◆

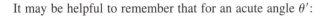

Common Error	The reference angle is measured always from the x axis, never from the y axis.

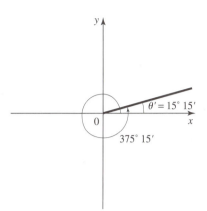

FIGURE 15–8

It may be helpful to remember that for an acute angle θ':

$180° - \theta'$ is in quadrant II.
$180° + \theta'$ is in quadrant III.
$360° - \theta'$ is in quadrant IV.

Finding the Angle When the Function Is Given

We use the same procedure as was given for acute angles. However, a calculator gives us just *one* angle when we enter the value of a trigonometric function. But we know that there are infinitely many angles that have the same value of that trigonometric function. We must use the angle given by calculator to determine the reference angle, which we then use to compute as many larger angles as we wish having the same trigonometric ratio. Usually, we want only the two positive angles less than $360°$ that have the required trigonometric function.

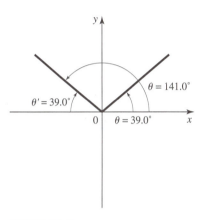

FIGURE 15–9

◆◆◆ **Example 13:** If $\sin \theta = 0.6293$, find two positive values of θ less than $360°$. Work to the nearest tenth of a degree.

Solution: We use the $\boxed{\sin^{-1}}$ key (or the $\boxed{\text{arc}}\ \boxed{\text{sin}}$ or $\boxed{\text{inv}}\ \boxed{\text{sin}}$ keys) and get

$$\theta' = 39.0°$$

The sine is positive in the first and second quadrants. As shown in Fig. 15–9, our first-quadrant angle is $39.0°$, and our second-quadrant angle is, taking $39.0°$ as the reference angle,

$$\theta = 180° - 39.0° = 141.0° \qquad ◆◆◆$$

Check your work by taking the sine. In this case, $\sin 141.0° = 0.6293$, which checks.

◆◆◆ **Example 14:** Find, to the nearest tenth of a degree, the two positive angles less than $360°$ that have a tangent of -2.25.

Solution: From the calculator, $\tan^{-1}(-2.25) = -66.0°$, so our reference angle is $66°$. The tangent is negative in the second and fourth quadrants. As shown in Fig. 15–10, our second-quadrant angle is

$$180° - 66.0° = 114.0°$$

and our fourth-quadrant angle is

$$360° - 66.0° = 294.0° \qquad ◆◆◆$$

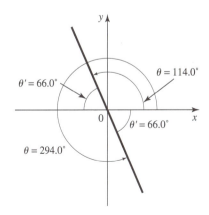

◆◆◆ **Example 15:** Find $\cos^{-1} 0.575$ to the nearest tenth of a degree.

Solution: From the calculator,

$$\cos^{-1} 0.575 = 54.9°$$

FIGURE 15–10

The cosine is positive in the first and fourth quadrants. Our fourth-quadrant angle is

$$360° - 54.9° = 305.1°$$

◆◆◆

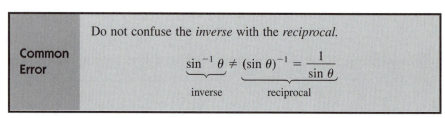

	Do not confuse the *inverse* with the *reciprocal*.
Common Error	$$\underbrace{\sin^{-1} \theta}_{\text{inverse}} \neq \underbrace{(\sin \theta)^{-1}}_{\text{reciprocal}} = \frac{1}{\sin \theta}$$

When the cotangent, secant, or cosecant is given, we use the reciprocal relationships (Eqs. 152) as in the following example.

◆◆◆ **Example 16:** Find two positive angles less than 360° that have a secant of -4.22. Work to the nearest tenth of a degree.

Solution: If we let the angle be θ, then, by Eq. 152b,

$$\cos \theta = \frac{1}{\sec \theta} = \frac{1}{-4.22} = -0.237$$

By calculator,

$$\theta = 103.7°$$

◆◆◆

The secant is also negative in the third quadrant. Our reference angle (Fig. 15–11) is

$$\theta' = 180° - 103.7° = 76.3°$$

so the third-quadrant angle is

$$\theta = 180° + 76.3° = 256.3°$$

◆◆◆

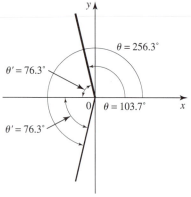

FIGURE 15–11

◆◆◆ **Example 17:** Evaluate $\arcsin(-0.528)$ to the nearest tenth of a degree.

Solution: As before, we seek only two positive angles less than 360°. By calculator,

$$\arcsin(-0.528) = -31.9°$$

which is a fourth-quadrant angle. As a positive angle, it is

$$\theta = 360° - 31.9° = 328.1°$$

The sine is also negative in the third quadrant. Using 31.9° as our reference angle, we have

$$\theta = 180° + 31.9° = 211.9°$$

◆◆◆

Special Angles

The angles in Table 15–1 appear so frequently in problems that it is convenient to be able to write their trigonometric functions from memory.

TABLE 15–1

Angle	Sin	Cos	Tan
0°	0	1	0
30°	$\frac{1}{2}$	$\sqrt{3}/2$ (or 0.8660)	$\sqrt{3}/3$ (or 0.5774)
45°	$\sqrt{2}/2$ (or 0.7071)	$\sqrt{2}/2$ (or 0.7071)	1
60°	$\sqrt{3}/2$ (or 0.8660)	$\frac{1}{2}$	$\sqrt{3}$ (or 1.732)
90°	1	0	undefined
180°	0	-1	0
270°	-1	0	undefined
360°	0	1	0

The angles $0°$, $90°$, $180°$, and $360°$ are called *quadrantal* angles because the terminal side of each of them lies along one of the coordinate axes. Notice that the tangent is undefined for $90°$ and $270°$, angles whose terminal side is on the y axis. The reason is that for any point (x, y) on the y axis, the value of x is zero. Since the tangent is equal to y/x, we have division by zero, which is not defined.

Exercise 1 ◆ Trigonometric Functions of Any Angle

Signs of the Trigonometric Functions

State in what quadrant or quadrants the terminal side of θ can lie if:

Assume all angles in this exercise set to be in standard position.

1. $\theta = 123°$.
2. $\theta = 272°$.
3. $\theta = -47°$.
4. $\theta = -216°$.
5. $\theta = 415°$.
6. $\theta = -415°$.
7. $\theta = 845°$.
8. $\sin \theta$ is positive.
9. $\cos \theta$ is negative.
10. $\sec \theta$ is positive.
11. $\cos \theta$ is positive and $\sin \theta$ is negative.
12. $\sin \theta$ and $\cos \theta$ are both negative.
13. $\tan \theta$ is positive and $\csc \theta$ is negative.

State whether the following expressions are positive or negative. Do not use your calculator, and try not to refer to your book.

14. $\sin 174°$
15. $\cos 329°$
16. $\tan 227°$
17. $\sec 332°$
18. $\cot 206°$
19. $\csc 125°$
20. $\sin(-47°)$
21. $\tan(-200°)$
22. $\cos 400°$

Give the algebraic signs of the sine, cosine, and tangent of the following.

23. $110°$
24. $206°$
25. $335°$
26. $-48°$
27. $500°$

Trigonometric Functions

Sketch the angle and write the six trigonometric functions if the terminal side of the angle passes through the given point.

28. $(3, 5)$
29. $(-4, 12)$
30. $(24, -7)$
31. $(-15, -8)$
32. $(0, -3)$
33. $(5, 0)$

Leave your answers for problems 34 through 40 in fractional or radical form.

Write the sine, cosine, and tangent of θ if:

34. $\sin \theta = -\frac{3}{5}$ and $\tan \theta$ is positive.
35. $\cos \theta = -\frac{4}{5}$ and θ is in the third quadrant.
36. $\csc \theta = -\frac{25}{7}$ and $\cos \theta$ is negative.
37. $\tan \theta = 2$ and θ is not in the first quadrant.
38. $\cot \theta = -\frac{4}{3}$ and $\sin \theta$ is positive.
39. $\sin \theta = \frac{2}{3}$ and θ is not in the first quadrant.
40. $\cos \theta = -\frac{7}{9}$ and $\tan \theta$ is negative.

Write, to four significant digits, the sine, cosine, and tangent of each angle.

41. 101°   **42.** 216°   **43.** 331°
44. 125.8°   **45.** −62.85°   **46.** −227.4°
47. 486°   **48.** −527°   **49.** 114°23′
50. 264°15′45″   **51.** −166°55′   **52.** 1.15°
53. −11°18′   **54.** 412°   **55.** 238°

Evaluating Trigonometric Expressions

Find the numerical value of each expression to four significant digits.

56. sin 35° + cos 35°
57. sin 125° tan 225°
58. cos 270° cos 150° + sin 270° sin 150°

59. $\dfrac{\sin^2 155°}{1 + \cos 155°}$

60. sin² 75°
61. tan² 125° − cos² 125°
62. (cos² 206° + sin 206°)²
63. $\sqrt{\sin^2 112° - \cos 112°}$

Inverse Trigonometric Functions

Find all positive angles less than 360° whose trigonometric function is given. Give any approximate answers to the nearest tenth of a degree.

64. sin θ = ½.   **65.** tan θ = −1.
66. cot θ = −√3.   **67.** cos θ = 0.8372.
68. csc θ = −3.85.   **69.** tan θ = 6.372.
70. cos θ = −½.   **71.** cot θ = 0.
72. cos θ = −1.   **73.** sin θ = −0.6358.
74. cot θ = −2.8458.   **75.** tan θ = 1.7361.
76. cos θ = 0.3759.

Evaluate to the nearest tenth of a degree. Find only positive angles less than 360°.

77. arcsin(−0.736)   **78.** arcsec 2.85
79. cos⁻¹ 0.827   **80.** arccot 5.22
81. arctan(−4.48)   **82.** csc⁻¹ 5.02

Evaluate from memory. Do not use a calculator.

83. sin 30°   **84.** cos 45°   **85.** tan 180°
86. sin 360°   **87.** tan 45°   **88.** cos 30°
89. sin 180°   **90.** cos 360°   **91.** tan 0°
92. sin 90°   **93.** cos 180°   **94.** sin 45°
95. tan 360°   **96.** cos 270°   **97.** cos 60°
98. sin 60°   **99.** cos 0°   **100.** tan 90°
101. cos 90°

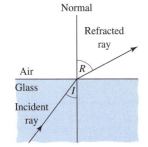

FIGURE 15–12   Refraction of light.

Computer

102. Figure 15–12 shows a ray of light passing from glass to air. The angle of refraction *R* is related to the angle of incidence *I* by Snell's law.

$$\frac{\sin R}{\sin I} = \text{constant}$$

Using a value for the constant (called the *index of refraction*) of 1.5, write a program or use a spreadsheet that will compute and print angle R for values of I from 0 to 90°. Have the computer inspect $\sin R$ each time it is computed, and when its value exceeds 1.00, stop the computation and print "TOTAL INTERNAL REFLECTION."

15–2 Law of Sines

Derivation

We first derive the *law of sines* for an oblique triangle in which all three angles are acute, such as in Fig. 15–13. We start by breaking the given triangle into two right triangles by drawing altitude h to side AB.

In right triangle ACD,

$$\sin A = \frac{h}{b} \quad \text{or} \quad h = b \sin A$$

In right triangle BCD,

$$\sin B = \frac{h}{a} \quad \text{or} \quad h = a \sin B$$

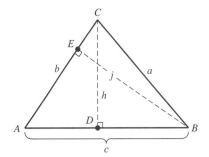

FIGURE 15–13 Derivation of the law of sines.

So

$$b \sin A = a \sin B$$

Dividing by $\sin A \sin B$, we have

$$\frac{a}{\sin A} = \frac{b}{\sin B}$$

Similarly, drawing altitude j to side AC, and using triangles BEC and AEB, we get

$$j = a \sin C = c \sin A$$

or

$$\frac{a}{\sin A} = \frac{c}{\sin C}$$

Combining this with the previous result, we obtain the following equation:

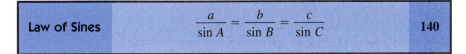

| Law of Sines | $\dfrac{a}{\sin A} = \dfrac{b}{\sin B} = \dfrac{c}{\sin C}$ | 140 |

The sides of a triangle are proportional to the sines of the opposite angles.

When one of the angles of the triangle is obtuse, as in Fig. 15–14, the derivation is nearly the same. We draw an altitude h to the extension of side AB. Then:

In right triangle ACD: $\sin A = \dfrac{h}{b}$ or $h = b \sin A$

In right triangle BCD: $\sin(180 - B) = \dfrac{h}{a}$

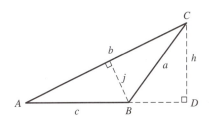

FIGURE 15–14

But $\sin(180 - B) = \sin B$, so

$$h = a \sin B$$

The derivation then proceeds exactly as before.

Solving a Triangle When Two Angles and One Side Are Given (AAS or ASA)

Recall that "solving a triangle" means to find all missing sides and angles. Here we use the law of sines to solve an oblique triangle. To do this, we must have *a known side opposite to a known angle,* as well as another angle. We abbreviate these given conditions as AAS (angle–angle–side) or ASA (angle–sine–angle).

◆◆◆ **Example 18:** Solve triangle *ABC* where $A = 32.5°$, $B = 49.7°$, and $a = 226$.

Solution: We want to find the unknown angle *C* and sides *b* and *c*. We sketch the triangle as shown in Fig. 15–15. The missing angle is found by Eq. 139.

$$C = 180° - 32.5° - 49.7° = 97.8°$$

Then, by the law of sines,

$$\frac{a}{\sin A} = \frac{b}{\sin B}$$

$$\frac{226}{\sin 32.5°} = \frac{b}{\sin 49.7°}$$

Solving for *b*, we get

$$b = \frac{226 \sin 49.7°}{\sin 32.5°} = 321$$

Again using the law of sines, we have

$$\frac{a}{\sin A} = \frac{c}{\sin C}$$

$$\frac{226}{\sin 32.5°} = \frac{c}{\sin 97.8°}$$

So

$$c = \frac{226 \sin 97.8°}{\sin 32.5°} = 417$$

◆◆◆

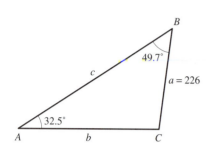

FIGURE 15–15

A diagram drawn more or less to scale can serve as a good check of your work and reveal inconsistencies in the given data. As another rough check, see that the longest side is opposite the largest angle and that the shortest side is opposite the smallest angle.

Notice that if two angles are given, the third can easily be found, since all angles of a triangle must have a sum of 180°. This means that if two angles and an included side are given (ASA), the problem can be solved as shown above for AAS.

Solving a Triangle When Two Sides and One Angle Are Given (SSA): The Ambiguous Case

In Example 18, *two angles and one side* were given. We can also use the law of sines when *one angle and two sides* are given, provided that the given angle is opposite one of the given sides. But we must be careful here. Sometimes we will get no solution, one solution, or two solutions, depending on the given information. The possibilities are given in Table 15–2.

We see that we will sometimes get *two* correct solutions, both of which may be reasonable in a given application.

Common Error	If it appears that you will get no solution when doing an application of oblique triangles, it probably means that the data are incorrect or that the problem is not properly set up. Don't give up at that point, but go back and check your work.

TABLE 15–2

When angle A is obtuse, and:	Then:	Example:
1. $a \leq c$	Side a is too short to intersect side b, so there is *no solution*.	
2. $a > c$	Side a can intersect side b in only one place, so we get *one solution*.	

When angle A is acute, and:	Then:	Example:
1. $a \geq c$	Side a is too long to intersect side b in more than one place, so there is *one solution*.	
2. $a < h < c$ where the altitude h is $h = c \sin A$	Side a is too short to touch side b, so we get *no solution*.	
3. $a = h < c$ where $h = c \sin A$	Side a just reaches side b, so we get a right triangle having *one solution*.	
4. $a > h < c$ where $h = c \sin A$	Side a can intersect side b in two places, giving *two solutions*. This is the **ambiguous case**.	

Another simple way to check for the number of solutions is to make a sketch. But the sketch must be fairly accurate, as in the following example.

◆◆◆ **Example 19:** Solve triangle ABC where $A = 27.6°$, $a = 112$, and $c = 165$.

Solution: Let's first calculate the altitude h to find out how many solutions we may have.

$$h = c \sin A$$
$$= 165 \sin 27.6° = 77.4$$

We see that side a (112) is greater than h (77.4) but less than c (165), so we have the ambiguous case with two solutions, as verified by Fig. 15–16. We will solve for both possible triangles. By the law of sines,

$$\frac{\sin C}{165} = \frac{\sin 27.6°}{112}$$

$$\sin C = \frac{165 \sin 27.6°}{112} = 0.6825$$

$$C = 43.0°$$

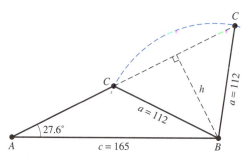

FIGURE 15–16   The ambiguous case.

Recall from Sec. 15–1 that there are two angles less than 180° for which the sine is positive. One of them, θ, is in the first quadrant, and the other, $180° - \theta$, is in the second quadrant.

This is *one* of the possible values for C. The other is

$$C = 180° - 43.0° = 137.0°$$

We now find the two corresponding values for side b and angle B.
When $C = 43.0°$,

$$B = 180° - 27.6° - 43.0° = 109.4°$$

So

$$\frac{b}{\sin 109.4°} = \frac{112}{\sin 27.6°}$$

from which $b = 228$.
When $C = 137.0°$,

$$B = 180° - 27.6° - 137.0° = 15.4°$$

So

$$\frac{b}{\sin 15.4°} = \frac{112}{\sin 27.6°}$$

from which $b = 64.2$. So our two solutions are given in the following table:

	A	B	C	a	b	c
1.	27.6°	109.4°	43.0°	112	228	165
2.	27.6°	15.4°	137.0°	112	64.2	165

◆◆◆

Common Error	In a problem such as the preceding one, it is easy to forget the second possible solution ($C = 137.0°$), especially since a calculator will give only the acute angle when computing the arc sine.

Not every SSA problem has two solutions, as we'll see in the following example.

◆◆◆ **Example 20:** Solve triangle ABC if $A = 35.2°$, $a = 525$, and $c = 412$.

Solution: When we sketch the triangle (Fig. 15–17), we see that the given information allows us to draw the triangle in only one way. We will get a unique solution. By the law of sines,

$$\frac{\sin C}{412} = \frac{\sin 35.2°}{525}$$

$$\sin C = \frac{412 \sin 35.2°}{525} = 0.4524$$

Thus the two possible values for C are

$$C = 26.9° \quad \text{and} \quad C = 180° - 26.9° = 153.1°$$

Our sketch, even if crudely drawn, shows that C cannot be obtuse, so we discard the 153.1° value. Then, by Eq. 139,

$$B = 180° - 35.2° - 26.9° = 117.9°$$

Using the law of sines once again, we get

$$\frac{b}{\sin 117.9°} = \frac{525}{\sin 35.2°}$$

$$b = \frac{525 \sin 117.9°}{\sin 35.2°} = 805$$

◆◆◆

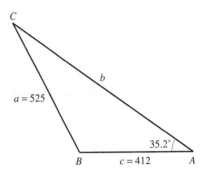

FIGURE 15–17

It is sometimes convenient, as in this example, to use the *reciprocals* of the expressions in the law of sines.

Exercise 2 ◆ Law of Sines

Data for triangle ABC are given in the following table. Solve for the missing parts.

	Angles			Sides		
A	**B**	**C**	**a**	**b**	**c**
1.	46.3°		228	304	
2.	65.9°			1.59	1.46
3.		43.0°		15.0	21.0
4. 21.45°	35.17°		276.0		
5. 61.9°		47.0°	7.65		
6. 126°	27.0°			119	
7.	31.6°	44.8°		11.7	
8. 15.0°		72.0°			375
9.	125°	32.0°			58.0
10. 24.14°	38.27°		5562		
11.	55.38°	18.20°		77.85	
12. 44.47°		63.88°			1.065
13. 18.0°	12.0°		50.7		
14.		45.55°		1137	1010

Computer

15. Write a program to solve a triangle by the law of sines. Have it *menu-driven* with the following choices:

1. Two angles and an opposite side given (AAS)
2. Two sides and one angle given (SSA)
3. Two angles and the included side given (ASA)

These triangles are shown in Fig. 15–20. For each, have the program ask for the given sides and angles, and then compute and print the missing sides and angles.

15–3   Law of Cosines

Derivation

Consider an oblique triangle ABC as shown in Fig. 15–18. As we did for the law of sines, we start by dividing the triangle into two right triangles by drawing an altitude h to side AC.

In right triangle ABD,

$$c^2 = h^2 + (AD)^2$$

But $AD = b - CD$. Substituting, we get

$$c^2 = h^2 + (b - CD)^2 \qquad (1)$$

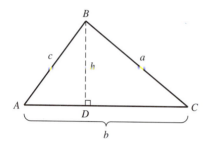

FIGURE 15–18   Derivation of the law of cosines.

Now, in right triangle BCD, by Eq. 147,

$$\frac{CD}{a} = \cos C$$

or

$$CD = a \cos C$$

Substituting $a \cos C$ for CD in Equation (1) yields

$$c^2 = h^2 + (b - a \cos C)^2$$

Squaring, we have

$$c^2 = h^2 + b^2 - 2ab \cos C + a^2 \cos^2 C \qquad (2)$$

Let us leave this expression for the moment and write the Pythagorean theorem for the same triangle BCD.

$$h^2 = a^2 - (CD)^2$$

Again substituting $a \cos C$ for CD, we obtain

$$h^2 = a^2 - (a \cos C)^2$$
$$= a^2 - a^2 \cos^2 C$$

Substituting this expression for h^2 back into (2), we get

$$c^2 = a^2 - a^2 \cos^2 C + b^2 - 2ab \cos C + a^2 \cos^2 C$$

Collecting terms, we get the law of cosines.

$$c^2 = a^2 + b^2 - 2ab \cos C$$

We can repeat the derivation, with perpendiculars drawn to side AB and to side BC, and get two more forms of the law of cosines.

Most students find it easier to think of the law of cosines in terms of the parts of the triangle, that is, *the given angle, the side opposite to the given angle, and the sides adjacent to the given angle*, rather than by letters of the alphabet.

Law of Cosines	$a^2 = b^2 + c^2 - 2bc \cos A$	
	$b^2 = a^2 + c^2 - 2ac \cos B$	**141**
	$c^2 = a^2 + b^2 - 2ab \cos C$	

Notice that when the angle in these equations is 90°, the law of cosines reduces to the Pythagorean theorem.

The square of any side equals the sum of the squares of the other two sides minus twice the product of the other sides and the cosine of the opposite angle.

◆◆◆ Example 21: Find side x in Fig. 15–19.

Solution: By the law of cosines,

$$x^2 = (1.24)^2 + (1.87)^2 - 2(1.24)(1.87) \cos 42.8°$$
$$= 5.03 - 4.64(0.7337) = 1.63$$
$$x = 1.28 \text{ m} \qquad \text{◆◆◆}$$

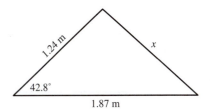

FIGURE 15–19

When to Use the Law of Sines or the Law of Cosines

It is sometimes not clear whether to use the law of sines or the law of cosines to solve a triangle. We use the law of sines when we have a *known side opposite a known angle*. We use the law of cosines only when the law of sines does not work, that is, for all other cases. In Fig. 15–20, the heavy lines indicate the known information and may help in choosing the proper law.

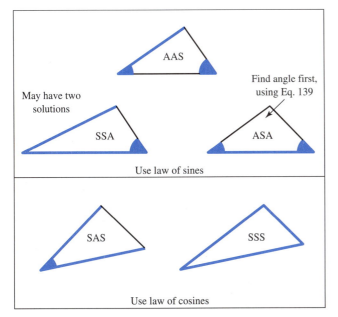

FIGURE 15–20 When to use the law of sines or the law of cosines.

Using the Cosine Law When Two Sides and the Included Angle Are Known

We can solve triangles by the law of cosines if we know *two sides and the angle between them,* or if we know *three sides.* We consider the first of these cases in the following example.

◆◆◆ Example 22: Solve triangle ABC where $a = 184$, $b = 125$, and $C = 27.2°$.

Solution: We make a sketch as shown in Fig. 15–21. Then, by the law of cosines,

$$c^2 = a^2 + b^2 - 2ab \cos C$$
$$= (184)^2 + (125)^2 - 2(184)(125) \cos 27.2°$$
$$c = \sqrt{(184)^2 + (125)^2 - 2(184)(125) \cos 27.2°}$$
$$= 92.6$$

Now that we have a known side opposite a known angle, we can use the law of sines to find angle A or angle B. Which shall we find first?

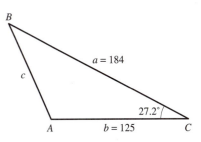

FIGURE 15–21

Notice that we cannot initially use the law of sines because we do not have a known side opposite a known angle.

Use the law of sines to *find the acute angle first* (angle B in this example). If, instead, you solve for the obtuse angle first, you may forget to subtract the angle obtained from the calculator from 180°. Further, if one of the angles is so close to 90° that you cannot tell from your sketch if it is acute or obtuse, find the other angle first, and then subtract the two known angles from 180° to obtain the third angle.

By the law of sines,

$$\frac{\sin B}{125} = \frac{\sin 27.2°}{92.6}$$

$$\sin B = \frac{125 \sin 27.2°}{92.6} = 0.617$$

$$B = 38.1° \quad \text{and} \quad B = 180 - 38.1 = 141.9°$$

We drop the larger value because our sketch shows us that B must be acute. Then, by Eq. 139,

$$A = 180° - 27.2° - 38.1° = 114.7° \qquad \qquad \text{◆◆◆}$$

In our next example, the given angle is *obtuse*.

◆◆◆ Example 23: Solve triangle ABC where $b = 16.4$, $c = 10.6$, and $A = 128.5°$.

Solution: We make a sketch as shown in Fig. 15–22. Then, by the law of cosines,

$$a^2 = (16.4)^2 + (10.6)^2 - 2(16.4)(10.6) \cos 128.5°$$

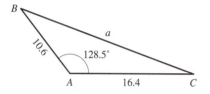

FIGURE 15–22

Common Error	The cosine of an obtuse angle is *negative*. Be sure to use the proper algebraic sign when applying the law of cosines to an obtuse angle.

In our example,

$$\cos 128.5° = -0.6225$$

So

$$a = \sqrt{(16.4)^2 + (10.6)^2 - 2(16.4)(10.6)(-0.6225)} = 24.4$$

By the law of sines,

$$\frac{\sin B}{16.4} = \frac{\sin 128.5°}{24.4}$$

$$\sin B = \frac{16.4 \sin 128.5°}{24.4} = 0.526$$

$$B = 31.7° \quad \text{and} \quad B = 180 - 31.7 = 148.3°$$

We drop the larger value because our sketch shows us that B must be acute. Then, by Eq. 139,

$$C = 180° - 31.7° - 128.5° = 19.8° \qquad \qquad \text{◆◆◆}$$

Using the Cosine Law When Three Sides Are Known

When three sides of an oblique triangle are known, we can use the law of cosines to solve for one of the angles. A second angle is found using the law of sines, and the third angle is found by subtracting the other two from 180°.

◆◆◆ **Example 24:** Solve triangle ABC in Fig. 15–23, where $a = 128$, $b = 146$, and $c = 222$.

Solution: We start by writing the law of cosines for any of the three angles. A good way to avoid ambiguity is to find the largest angle first (the law of cosines will tell us if it is acute or obtuse), and then we are sure that the other two angles are acute.

Writing the law of cosines for angle C gives

$$(222)^2 = (128)^2 + (146)^2 - 2(128)(146) \cos C$$

Solving for $\cos C$ gives

$$\cos C = -0.3099$$

Since the cosine is negative, C must be obtuse, so

$$C = 108.1°$$

Then by the law of sines

$$\frac{\sin A}{128} = \frac{\sin 108.1°}{222}$$

from which $\sin A = 0.5482$. Since we know that A is acute, we get

$$A = 33.2°$$

Finally,

$$B = 180° - 108.1° - 33.2° = 38.7° \qquad \text{◆◆◆}$$

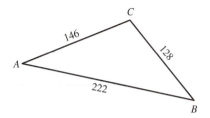

FIGURE 15–23

Exercise 3 ◆ Law of Cosines

Data for triangle ABC are given in the following table. Solve for the missing parts.

	Angles			Sides		
A	**B**	**C**	**a**	**b**	**c**
1.		106.0°	15.7	11.2	
2.	51.4°		1.95		1.46
3. 68.3°				18.3	21.7
4.			728	906	663
5.		27.3°	128	152	
6.	128°		1.16		1.95
7. 135°				275	214
8.		35.2°	77.3	81.4	
9.			11.3	15.6	12.8
10.	41.7°		199		202
11. 115°				46.8	51.3
12.			1.475	1.836	2.017
13.	67.0°		9.08	6.75	
14.	129°		186		179
15. 158°				1.77	1.99
16.			41.8	57.2	36.7
17.	41.77°		1445	1502	
18.	108.8°		7.286		6.187
19.			97.3	81.4	88.5
20. 36.29°				47.28	51.36

Computer

21. Write a program to solve a triangle by the law of cosines. Have it *menu-driven* with the following choices:

1. Two sides and the included angle given (SAS)
2. Three sides given (SSS)

These triangles are shown in Fig. 15–20. For each, have the program ask for the given sides and angles, and then compute and print the missing sides and angles.

15–4   Applications

As with right triangles, oblique triangles have many applications in technology, as you will see in the exercises for this section. Follow the same procedures for setting up these problems as we used for other word problems, and solve the resulting triangle by the law of sines or the law of cosines, or both.

If an *area* of an oblique triangle is needed, either compute all the sides and use Hero's formula (Eq. 138), or find an altitude with right-triangle trigonometry and use Eq. 137.

◆◆◆ **Example 25:** Find the area of the gusset in Fig. 15–24(a).

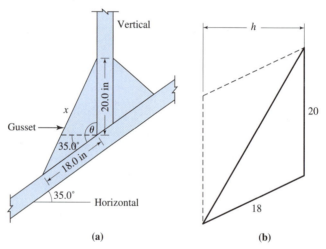

FIGURE 15–24

Solution: We first find θ.

$$\theta = 90° + 35.0° = 125.0°$$

We now know two sides and the angle between them, so can find side x by the law of cosines.

$$x^2 = (18.0)^2 + (20.0)^2 - 2(18.0)(20.0) \cos 125° = 1137$$
$$x = 33.7 \text{ in.}$$

We find the area of the gusset by Hero's formula (Eq. 138).

$$s = \tfrac{1}{2}(18.0 + 20.0 + 33.7) = 35.9$$
$$\text{area} = \sqrt{35.9(35.9 - 18.0)(35.9 - 20.0)(35.9 - 33.7)}$$
$$= 150 \text{ in.}^2$$

Check: Does the answer look reasonable? Let's place the gusset inside the parallelogram, as shown in Fig. 15–24(b), and estimate the height h by eye at about 15 in. Thus the area of the parallelogram would be 15×20, or 300 in.^{2}, just double that found for the triangle.   ◆◆◆

◆◆◆ **Example 26:** A ship takes a sighting on two buoys. At a certain instant, the bearing of buoy A is N 44.23° W, and that of buoy B is N 62.17° E. The distance between the buoys is 3.60 km, and the bearing of B from A is N 87.87° E. Find the distance of the ship from each buoy.

Estimate: Let us draw the figure with a ruler and protractor as shown in Fig. 15–25. Notice how the compass directions are laid out, starting from the north and turning in the indicated direction. Measuring, we get $SA = 1.7$ units and $SB = 2.8$ units. If you try it, you will probably get slightly different values.

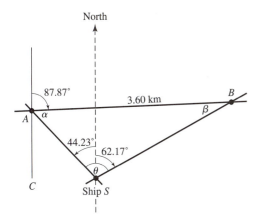

FIGURE 15–25

Solution: Calculating the angles of triangle ABS gives

$$\theta = 44.23° + 62.17° = 106.40°$$

Since angle SAC is 44.23°,

$$\alpha = 180° - 87.87° - 44.23° = 47.90°$$

and

$$\beta = 180° - 106.40° - 47.90° = 25.70°$$

From the law of sines,

$$\frac{SA}{\sin 25.70°} = \frac{SB}{\sin 47.90°} = \frac{3.60}{\sin 106.40°}$$

from which

$$SA = \frac{3.60 \sin 25.70°}{\sin 106.40°} = 1.63 \text{ km}$$

and

$$SB = \frac{3.60 \sin 47.90°}{\sin 106.40°} = 2.78 \text{ km}$$

which agree with our estimated values.　　　　　　　　　　　　◆◆◆

Exercise 4 ◆ Applications

Determining Inaccessible Distances

1. Two stakes, A and B, are 88.6 m apart. From a third stake C, the angle ACB is 85.4°, and from A, the angle BAC is 74.3°. Find the distance from C to each of the other stakes.

This set of problems contains no vectors, which are covered in the next section.

Remember that angles of elevation or depression are always measured from the *horizontal*.

2. From a point on level ground, the angles of elevation of the top and the bottom of an antenna standing on top of a building are 32.6° and 27.8°, respectively. If the building is 125 ft high, how tall is the antenna?

3. A triangular lot measures 115 m, 187 m, and 215 m along its sides. Find the angles between the sides.

4. Two boats are 45.5 km apart. Both are traveling toward the same point, which is 87.6 km from one of them and 77.8 km from the other. Find the angle at which their paths intersect.

Navigation

Hint: Draw your diagram after *t* hours have elapsed.

5. A ship is moving at 15.0 km/h in the direction N 15.0° W. A helicopter with a speed of 22.0 km/h is due east of the ship. In what direction should the helicopter travel if it is to meet the ship?

6. City *A* is 215 miles N 12.0° E from city *B*. The bearing of city *C* from *B* is S 55.0° E. The bearing of *C* from *A* is S 15.0° E. How far is *C* from *A*? From *B*?

7. A ship is moving in a direction S 24.25° W at a rate of 8.60 mi/h. If a launch that travels at 15.4 mi/h is due west of the ship, in what direction should it travel in order to meet the ship?

8. A ship is 9.50 km directly east of a port. If the ship sails southeast for 2.50 km, how far will it be from the port?

9. From a plane flying due east, the bearing of a radio station is S 31.0° E at 1 P.M. and S 11.0° E at 1:20 P.M. The ground speed of the plane is 625 km/h. Find the distance of the plane from the station at 1 P.M.

Structures

10. A tower stands vertically on sloping ground whose inclination with the horizontal is 11.6°. From a point 42.0 m downhill from the tower (measured along the slope), the angle of elevation of the top of the tower is 18.8°. How tall is the tower?

11. A vertical antenna stands on a slope that makes an angle of 8.70° with the horizontal. From a point directly uphill from the antenna, the angle of elevation of its top is 61.0°. From a point 16.0 m farther up the slope (measured along the slope), the angle of elevation of its top is 38.0°. How tall is the antenna?

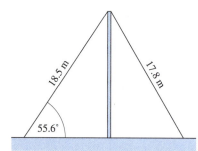

FIGURE 15–26

12. A power pole on level ground is supported by two wires that run from the top of the pole to the ground (Fig. 15–26). One wire is 18.5 m long and makes an angle of 55.6° with the ground, and the other wire is 17.8 m long. Find the angle that the second wire makes with the ground.

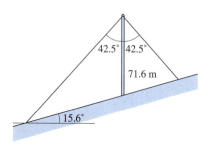

FIGURE 15–27

13. A 71.6-m-high antenna mast is to be placed on sloping ground, with the cables making an angle of 42.5° with the top of the mast (Fig. 15–27). Find the lengths of the two cables.

14. In the roof truss in Fig. 15–28, find the lengths of members *AB*, *BD*, *AC*, and *AD*.

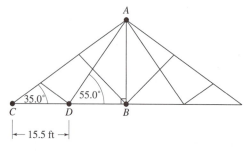

FIGURE 15–28   Roof truss.

15. From a point on level ground between two power poles of the same height, cables are stretched to the top of each pole. One cable is 52.6 ft long, the other is 67.5 ft long, and the angle of intersection between the two cables is 125°. Find the distance between the poles.

16. A pole standing on level ground makes an angle of 85.8° with the horizontal. The pole is supported by a 22.0-ft prop whose base is 12.5 ft from the base of the pole. Find the angle made by the prop with the horizontal.

Mechanisms

17. In the slider crank mechanism of Fig. 15–29, find the distance x between the wrist pin W and the crank center C when $\theta = 35.7°$.

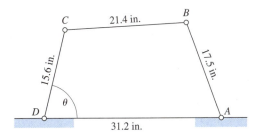

FIGURE 15–29

18. In the four-bar linkage of Fig. 15–30, find angle θ when angle BAD is 41.5°.

FIGURE 15–30

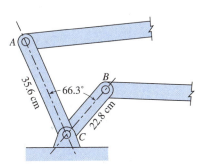

FIGURE 15–31

19. Two links, AC and BC, are pivoted at C, as shown in Fig. 15–31. How far apart are A and B when angle ACB is 66.3°?

Geometry

20. Find angles A, B, and C in the quadrilateral in Fig. 15–32.
21. Find side AB in the quadrilateral in Fig. 15–33.
22. Two sides of a parallelogram are 22.8 and 37.8 m, and one of the diagonals is 42.7 m. Find the angles of the parallelogram.
23. Find the lengths of the sides of a parallelogram if its diagonal, which is 125 mm long, makes angles with the sides of 22.7° and 15.4°.

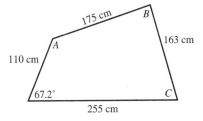

FIGURE 15–32

24. Find the lengths of the diagonals of a parallelogram, two of whose sides are 3.75 m and 1.26 m; their included angle is 68.4°.
25. A *median* of a triangle is a line joining a vertex to the midpoint of the opposite side. In triangle ABC, $A = 62.3°$, $b = 112$, and the median from C to the midpoint of c is 186. Find c.
26. The sides of a triangle are 124, 175, and 208. Find the length of the median drawn to the longest side.
27. The angles of a triangle are in the ratio 3:4:5, and the shortest side is 994. Solve the triangle.
28. The sides of a triangle are in the ratio 2:3:4. Find the cosine of the largest angle.

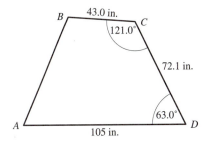

FIGURE 15–33

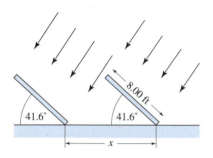

FIGURE 15–34   Solar panels.

29. Two solar panels are to be placed as shown in Fig. 15–34. Find the minimum distance x so that the first panel will not cast a shadow on the second when the angle of elevation of the sun is 18.5°.

30. Find the overhang x so that the window in Fig. 15–35 will be in complete shade when the sun is 60° above the horizontal.

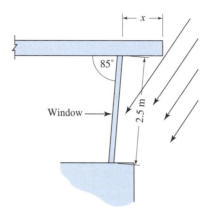

FIGURE 15–35

Computer

31. Combine your triangle-solving programs from Exercises 2 and 3 into a single program to solve any triangle. The menu should contain the five options shown in Fig. 15–20. Also compute the area of the triangle by Hero's formula.

32. For the slider crank mechanism of Fig. 15–29, compute and print the position x of the wrist pin for values of θ from 0 to 180° in 15° steps.

15–5   Addition of Vectors

In Sec. 7–6 we added two or more nonperpendicular vectors by first resolving each into components, then adding the components, and finally resolving the components into a single resultant. Now, by using the law of sines and the law of cosines, we can combine two nonperpendicular vectors directly, with much less time and effort. However, when more than two vectors must be added, it is faster to resolve each into its x and y components, combine the x components and the y components, and then find the resultant of those two perpendicular vectors.

In this section we show both methods.

Vector Diagram

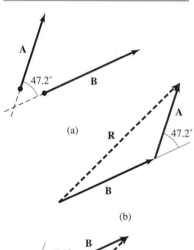

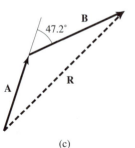

FIGURE 15–36   Addition of vectors.

We can illustrate the resultant, or vector sum, of two vectors by means of a diagram. Suppose that we wish to add vectors **A** and **B** in Fig. 15–36(a). If we draw the two vectors *tip to tail*, as in Fig. 15–36(b), the resultant **R** will be the vector that will complete the triangle when drawn from the tail of the first vector to the tip of the second vector. It does not matter whether vector **A** or vector **B** is drawn first; the same resultant will be obtained either way, as shown in Fig. 15–36(c).

The *parallelogram method* will give the same result. To add the same two vectors **A** and **B** as before, we first draw the given vectors *tail to tail* (Fig. 15–37) and complete a parallelogram by drawing lines from the tips of the given vectors, parallel to the given vectors. The resultant **R** is then the diagonal of the parallelogram drawn from the intersections of the tails of the original vectors.

Finding the Resultant of Two Nonperpendicular Vectors

Whichever method we use for drawing two vectors, *the resultant is one side of an oblique triangle.* To find the length of the resultant and the angle that it makes with one of the original vectors, we simply solve the oblique triangle by the methods we learned earlier in this chapter.

◆◆◆ **Example 27:** Two vectors, **A** and **B**, make an angle of 47.2° with each other as shown in Fig. 15–37. If their magnitudes are $A = 125$ and $B = 146$, find the magnitude of the resultant **R** and the angle that **R** makes with vector **B**.

Solution: We make a vector diagram, either tip to tail or by the parallelogram method. Either way, we must solve the oblique triangle in Fig. 15–38 for R and ϕ. Finding θ yields

$$\theta = 180° - 47.2° = 132.8°$$

By the law of cosines,

$$R^2 = (125)^2 + (146)^2 - 2(125)(146) \cos 132.8° = 61{,}740$$
$$R = 248$$

Then, by the law of sines,

$$\frac{\sin \phi}{125} = \frac{\sin 132.8}{248}$$

$$\sin \phi = \frac{125 \sin 132.8}{248} = 0.3698$$

$$\phi = 21.7° \qquad\qquad ◆◆◆$$

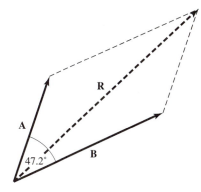

FIGURE 15–37 Parallelogram method.

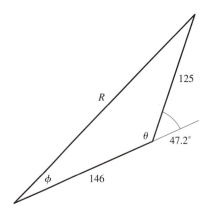

FIGURE 15–38

Addition of Several Vectors

The law of sines and the law of cosines are good for adding *two* nonperpendicular vectors. However, when *several* vectors are to be added, we usually break each into its x and y components and combine them, as in the following example.

◆◆◆ **Example 28:** Find the resultant of the vectors shown in Fig. 15–39(a).

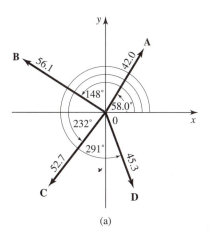

(a)

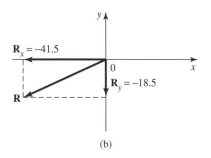

(b)

FIGURE 15–39

Solution: The x component of a vector of magnitude V at any angle θ is

$$V \cos \theta$$

and the y component is

$$V \sin \theta$$

These equations apply for an angle in any quadrant. We compute and tabulate the x and y components of each original vector and find the sums of each as shown in the following table.

Vector	x Component	y Component
A	$42.0 \cos 58.0° = 22.3$	$42.0 \sin 58.0° = 35.6$
B	$56.1 \cos 148° = -47.6$	$56.1 \sin 148° = 29.7$
C	$52.7 \cos 232° = -32.4$	$52.7 \sin 232° = -41.5$
D	$45.3 \cos 291° = 16.2$	$45.3 \sin 291° = 42.3$
R	$R_x = -41.5$	$R_y = -18.5$

The two vectors $\mathbf{R}_x$ and $\mathbf{R}_y$ are shown in Fig. 15–39(b). We find their resultant $\mathbf{R}$ by the Pythagorean theorem.

$$R^2 = (-41.5)^2 + (-18.5)^2 = 2065$$
$$R = 45.4$$

We find the angle θ by

$$\theta = \arctan \frac{R_y}{R_x}$$

$$= \arctan \frac{-18.5}{-41.5}$$

$$= 24.0° \quad \text{or} \quad 204°$$

Since our resultant is in the third quadrant, we drop the 24.0° value. Thus the resultant has a magnitude of 45.4 and a direction of 204°. This is often written in the form

$$\mathbf{R} = 45.4\underline{/204°} \qquad\qquad ◆◆◆$$

Exercise 5 ◆ Addition of Vectors

The magnitudes of vectors $\mathbf{A}$ and $\mathbf{B}$ are given in the following table, as well as the angle between the vectors. For each, find the magnitude R of the resultant and the angle that the resultant makes with vector $\mathbf{B}$.

Magnitudes		
A	*B*	*Angle*
1. 244	287	21.8°
2. 1.85	2.06	136°
3. 55.9	42.3	55.5°
4. 1.006	1.745	148.4°
5. 4483	5829	100.0°
6. 35.2	23.8	146°

This is called *polar form*, which we'll cover in Chapter 17.

In Sec. 7–7 we gave examples of applications of force vectors, velocity vectors, and impedance vectors. You might want to review that section before trying the applications here. These problems are set up in exactly the same way, only now they will require the solution of an oblique triangle rather than a right triangle.

Force Vectors

7. Two forces of 18.6 N and 21.7 N are applied to a point on a body. The angle between the forces is 44.6°. Find the magnitude of the resultant and the angle that it makes with the larger force.

8. Two forces whose magnitudes are 187 lb and 206 lb act on an object. The angle between the forces is 88.4°. Find the magnitude of the resultant force.

9. A force of 125 N pulls due west on a body, and a second force pulls N 28.7° W. The resultant force is 212 N. Find the second force and the direction of the resultant.

10. Forces of 675 lb and 828 lb act on a body. The smaller force acts due north; the larger force acts N 52.3° E. Find the direction and the magnitude of the resultant.

11. Two forces of 925 N and 1130 N act on an object. Their lines of action make an angle of 67.2° with each other. Find the magnitude and the direction of their resultant.

12. Two forces of 136 lb and 251 lb act on an object with an angle of 53.9° between their lines of action. Find the magnitude of their resultant and its direction.

13. The resultant of two forces of 1120 N and 2210 N is 2870 N. What angle does the resultant make with each of the two forces?

14. Three forces are in equilibrium: 212 N, 325 N, and 408 N. Find the angles between their lines of action.

Velocity Vectors

15. As an airplane heads west with an air speed of 325 mi/h, a wind with a speed of 35.0 mi/h causes the plane to travel slightly south of west with a ground speed of 305 mi/h. In what direction is the wind blowing? In what direction does the plane travel?

16. A boat heads S 15.0° E on a river that flows due west. The boat travels S 11.0° W with a speed of 25.0 km/h. Find the speed of the current and the speed of the boat in still water.

17. A pilot wishes to fly in the direction N 45.0° E. The wind is from the west at 36.0 km/h, and the plane's speed in still air is 388 km/h. Find the heading and the ground speed.

18. The heading of a plane is N 27.7° E, and its air speed is 255 mi/h. If the wind is blowing from the south with a velocity of 42.0 mi/h, find the actual direction of travel of the plane and its ground speed.

19. A plane flies with a heading of N 48.0° W and an air speed of 584 km/h. It is driven from its course by a wind of 58.0 km/h from S 12.0° E. Find the ground speed and the drift angle of the plane.

See Fig. 7–42 for definitions of the flight terms used in these problems.

Current and Voltage Vectors

20. We will see later that it is possible to represent an alternating current or voltage by a vector whose length is equal to the maximum amplitude of the current or voltage, placed at an angle that we later define as the *phase angle*. Then to add two alternating currents or voltages, we *add the vectors* representing those voltages or currents in the same way that we add force or velocity vectors.

   A current I_1 is represented by a vector of magnitude 12.5 A at an angle of 15.6°, and a second current I_2 is represented by a vector of magnitude 7.38 A at an angle of 132°, as shown in Fig. 15–40. Find the magnitude and the direction of the sum of these currents, represented by the vector I.

21. Figure 15–41 shows two impedances in parallel, with the currents in each represented by

$$I_1 = 18.4 \text{ A} \quad \text{at } 51.5°$$

and

$$I_2 = 11.3 \text{ A} \quad \text{at } 0°$$

The current I will be the vector sum of I_1 and I_2. Find the magnitude and the direction of the vector representing I.

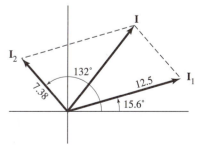

FIGURE 15–40

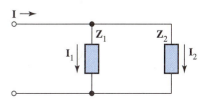

FIGURE 15–41

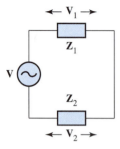

FIGURE 15–42

22. Figure 15–42 shows two impedances in series, with the voltage drop V_1 equal to 92.4 V at 71.5° and V_2 equal to 44.2 V at −53.8°. Find the magnitude and the direction of the vector representing the total drop **V**.

Addition of Several Vectors

Find the resultant of each of the following sets of vectors.

23. $273\underline{/34.0°}$, $179\underline{/143°}$, $203\underline{/225°}$, $138\underline{/314°}$
24. $72.5\underline{/284°}$, $28.5\underline{/331°}$, $88.2\underline{/104°}$, $38.9\underline{/146°}$

Computer

25. Write a program or use a spreadsheet that will accept as input the magnitude and the direction of any number of vectors. Have the computer resolve each vector into x and y components, combine these components into a single x component and y component, and compute and print the magnitude and direction of the resultant. Try your program on problems 23 and 24.

26. A certain airplane has an air speed of 225 mi/h, and the wind is from the northeast at 45 mi/h. Write a program or use a spreadsheet to compute and print the ground speed and the actual direction of travel of the plane, for headings that vary from due north completely around the compass, in steps of 15°.

••• CHAPTER 15 REVIEW PROBLEMS •••••••••••••••••••••••••••••••••••

Solve oblique triangle ABC if:

1. $C = 135°$ $a = 44.9$ $b = 39.1$
2. $A = 92.4°$ $a = 129$ $c = 83.6$
3. $B = 38.4°$ $a = 1.84$ $c = 2.06$
4. $B = 22.6°$ $a = 2840$ $b = 1170$

5. $A = 132°$ $b = 38.2$ $c = 51.8$

In what quadrant(s) will the terminal side of θ lie if:

6. $\theta = 227°$ 7. $\theta = -45°$ 8. $\theta = 126°$ 9. $\theta = 170°$
10. $\tan \theta$ is negative

Without using book or calculator, state the algebraic sign of:

11. $\tan 275°$ 12. $\sec(-58°)$
13. $\cos 183°$ 14. $\cos 45°$
15. $\sin 300°$

Write the sin, cos, and tan, to three significant digits, for the angle whose terminal side passes through the given point.

16. $(-2, 5)$ 17. $(-3, -4)$ 18. $(5, -1)$

Two vectors of magnitudes A and B are separated by an angle θ. Find the resultant and the angle that the resultant makes with vector **B**.

19. $A = 837$ $B = 527$ $\theta = 58.2°$
20. $A = 2.58$ $B = 4.82$ $\theta = 82.7°$
21. $A = 44.9$ $B = 29.4$ $\theta = 155°$
22. $A = 8374$ $B = 6926$ $\theta = 115.4°$

23. From a ship sailing north at the rate of 18.0 km/h, the bearing of a lighthouse is N 18°15′ E. Ten minutes later the bearing is N 75°46′ E. How far is the ship from the lighthouse at the time of the second observation?

Write the sin, cos, and tan, to four decimal places, of:

24. 273°   **25.** 175°   **26.** 334°36'

27. 127°22'   **28.** 114°

Evaluate each expression to four significant digits.

29. $\sin 35° \cos 35°$

30. $\tan^2 68°$

31. $(\cos 14° + \sin 14°)^2$

32. What angle does the slope of a hill make with the horizontal if a vertical tower 18.5 m tall, located on the slope of the hill, is found to subtend an angle of 25.5° from a point 35.0 m directly downhill from the foot of the tower, measured along the slope?

33. Three forces are in equilibrium. One force of 457 lb acts in the direction N 28.0° W. The second force acts N 37.0° E. Find the direction of the third force of magnitude 638 lb.

Find to the nearest tenth of a degree all nonnegative values of θ less than 360°.

34. $\cos \theta = 0.736$

35. $\tan \theta = -1.16$

36. $\sin \theta = 0.774$

Evaluate to the nearest tenth of a degree.

37. arcsin 0.737

38. $\tan^{-1} 4.37$

39. $\cos^{-1} 0.174$

40. A ship wishes to travel in the direction N 38.0° W. The current is from due east at 4.20 mi/h, and the speed of the ship in still water is 18.5 mi/h. Find the direction in which the ship should head and the speed of the ship in the actual direction of travel.

41. Two forces of 483 lb and 273 lb act on a body. The angle between the lines of action of these forces is 48.2°. Find the magnitude of the resultant and the angle that it makes with the 483-lb force.

42. Find the vector sum of two voltages, $V_1 = 134\underline{/24.5°}$ and $V_2 = 204\underline{/85.7°}$.

43. A point on a rotating wheel has a tangential velocity of 523 cm/s. Find the x and y components of the velocity when the point is in the position shown in Fig. 15–43.

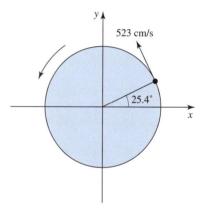

FIGURE 15–43

Writing

44. Suppose that you have analyzed a complex bridge truss made up of many triangles, sometimes using the law of sines and sometimes the more time-consuming law of cosines. Your client's accountant, angry over the size of your bill, has attacked your report for often using the longer law when it is clear to him that the shorter law of sines is also "good for solving triangles." Write a letter to your client explaining why you sometimes had to use one law and sometimes the other.

Team Project

45. Four mutually tangent circles are shown in Fig. 15–44. Find the radius of the shaded circle.

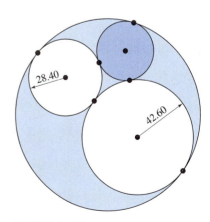

FIGURE 15–44

16

Radian Measure, Arc Length, and Rotation

◆◆◆ **OBJECTIVES** ◆◆◆

When you have completed this chapter, you should be able to:

- Convert angles between radians, degrees, and revolutions.
- Write the trigonometric functions of an angle given in radians.
- Compute arc length, radius, or central angle.
- Compute the angular velocity of a rotating body.
- Compute the linear speed of a point on a rotating body.
- Solve applied problems involving arc length or rotation.

◆◆◆

Until now we have usually been expressing the measures of angles in degrees or in revolutions. Another angular measure, much more useful in many cases, is the *radian*. The radian is not an arbitrarily chosen unit like the degree or grad. Instead, it uses a part of the circle itself (the radius) as a unit.

Here we review how to convert between radians, degrees, and revolutions. We go on to study two important uses for the radian. We find *arc lengths* and then study the *rotation of rigid bodies*.

16–1   Radian Measure

A *central angle* is one whose vertex is at the center of the circle (Fig. 16–1). An *arc* is a portion of the circle. In Chapter 6 we introduced the *radian:* If an arc is laid off along a circle with a length equal to the radius of the circle, the central angle subtended by this arc is defined as *one radian,* as shown in Fig. 16–2.

Angle Conversion

An arc having a length equal to the radius of the circle subtends a central angle of one radian; an arc having a length of twice the radius subtends a central angle of two radians; and so on. Thus an arc with a length of 2π times the radius (the entire circumference) subtends a central angle of 2π radians. Therefore 2π radians is equal to 1 revolution, or 360°. This gives us the following conversions for angular measure:

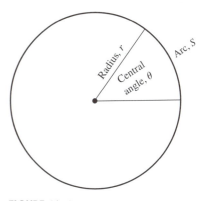

FIGURE 16–1

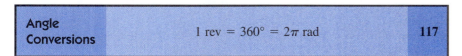

| Angle Conversions | $1 \text{ rev} = 360° = 2\pi \text{ rad}$ | **117** |

We convert angular units in the same way that we converted other units in Chapter 1.

◆◆◆ **Example 1:** Convert 47.6° to radians and revolutions.

Solution: By Eq. 117,

$$47.6°\left(\frac{2\pi \text{ rad}}{360°}\right) = 0.831 \text{ rad}$$

Note that 2π and 360° are exact numbers, so we keep the same number of digits in our answer as in the given angle. Now converting to revolutions, we obtain

$$47.6°\left(\frac{1 \text{ rev}}{360°}\right) = 0.132 \text{ rev}$$

◆◆◆

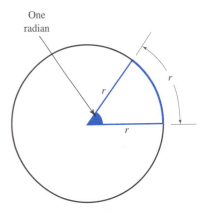

FIGURE 16–2

If we divide 360° by 2π we get

$$1 \text{ rad} \approx 57.3°$$

Some prefer to use Eq. 117 in this approximate form instead of the one given.

◆◆◆ **Example 2:** Convert 1.8473 rad to degrees and revolutions.

Solution: By Eq. 117,

$$1.8473 \text{ rad}\left(\frac{360°}{2\pi \text{ rad}}\right) = 105.84°$$

and

$$1.8473 \text{ rad}\left(\frac{1 \text{ rev}}{2\pi \text{ rad}}\right) = 0.29401 \text{ rev}$$

◆◆◆

◆◆◆ **Example 3:** Convert 1.837520 rad to degrees, minutes, and seconds.

Solution: We first convert to decimal degrees.

$$1.837520 \text{ rad}\left(\frac{360°}{2\pi \text{ rad}}\right) = 105.2821°$$

Converting the decimal part (0.2821°) to minutes, we obtain

$$0.2821°\left(\frac{60'}{1°}\right) = 16.93'$$

Converting the decimal part of 16.93′ to seconds yields

$$0.93'\left(\frac{60''}{1\ \text{min}}\right) = 56''$$

So 1.837520 rad = 105°16′56″.   ◆◆◆

Radian Measure in Terms of π

Not only can we express radians in decimal form, but it is also very common to express radian measure in terms of π. We know that 180° equals π radians, so

$$90° = \frac{\pi}{2}\ \text{rad}$$

$$45° = \frac{\pi}{4}\ \text{rad}$$

$$15° = \frac{\pi}{12}\ \text{rad}$$

and so on. Thus to convert an angle from degrees to radians in terms of π, multiply the angle by (π rad/180°), and reduce to lowest terms.

◆◆◆ **Example 4:** Express 135° in radian measure in terms of π.

Solution:

$$135°\left(\frac{\pi\ \text{rad}}{180°}\right) = \frac{135\pi}{180}\ \text{rad} = \frac{3\pi}{4}\ \text{rad}$$

     ◆◆◆

Common Error

Students sometimes confuse the decimal value of π with the degree equivalent of π radians.

What is the value of π? 180 or 3.1416 . . . ?

Remember that the approximate decimal value of π is always

$$\pi \simeq 3.1416$$

but that π radians converted to degrees equals

$$\pi\ \text{radians} = 180°$$

Note that we write "radians" or "rad" after π when referring to the angle, but not when referring to the decimal value.

To convert an angle from radians to degrees, multiply the angle by (180°/π rad). Cancel the π in numerator and denominator, and reduce.

◆◆◆ **Example 5:** Convert $7\pi/9$ rad to degrees.

Solution:

$$\frac{7\pi}{9}\ \text{rad}\left(\frac{180°}{\pi\ \text{rad}}\right) = \frac{7(180)}{9}\ \text{deg} = 140°$$

     ◆◆◆

Trigonometric Functions of Angles in Radians

We use a calculator to find the trigonometric functions of angles in radians just as we did for angles in degrees. However, we first *switch the calculator into radian mode*. Consult your calculator manual for how to switch to this mode. Many calculators will display an $\boxed{R}$ or the word $\boxed{\text{RAD}}$ when in radian mode.

On some calculators you enter the angle and then press the required trigono-metric function key. On other calculators you must press the trigonometric function key *first* and then enter the angle. Be sure you know which way your own calcula-tor works.

◆◆◆ Example 6: Use your calculator to verify the following to four decimal places:

(a) sin 2.83 rad = 0.3066
(b) cos 1.52 rad = 0.0508
(c) tan 0.463 rad = 0.4992   ◆◆◆

To find the cotangent, secant, or cosecant, we use the reciprocal relations, just as when working in degrees.

◆◆◆ Example 7: Find sec 0.733 rad to three decimal places.

Solution: We put the calculator into radian mode. The reciprocal of the secant is the cosine, so we take the cosine of 0.733 rad and get

$$\cos 0.733 \text{ rad} = 0.7432$$

Then taking the reciprocal of 0.7432 gives

$$\sec 0.733 = \frac{1}{\cos 0.733} = \frac{1}{0.7432} = 1.346$$ ◆◆◆

◆◆◆ Example 8: Use your calculator to verify the following to four decimal places:

(a) csc 1.33 rad = 1.0297
(b) cot 1.22 rad = 0.3659
(c) sec 0.726 rad = 1.3372   ◆◆◆

The Inverse Trigonometric Functions

The inverse trigonometric functions are found the same way as when working in de-grees. Just be sure that your calculator is in radian mode.

As with the trigonometric functions, some calculators require that you press the function key before entering the number, and some require that you press it after en-tering the number.

◆◆◆ Example 9: Use your calculator to verify the following, in radians, to four dec-imal places:

(a) arcsin 0.2373 = 0.2396 rad
(b) $\cos^{-1} 0.5152 = 1.0296$ rad
(c) arctan 3.246 = 1.2720 rad   ◆◆◆

To find the arccot, arcsec, and arccsc, we first take the reciprocal of the given function and then find the inverse function, as shown in the following example.

◆◆◆ Example 10: If $\theta = \cot^{-1} 2.745$, find θ in radians to four decimal places.

Solution: If the cotangent of θ is 2.745, then

$$\cot \theta = 2.745$$

so

$$\frac{1}{\tan \theta} = 2.745$$

Remember that the inverse trigonomet-ric function can be written in two differ-ent ways. Thus the inverse sine can be written

$$\text{arcsin } \theta \quad \text{or} \quad \sin^{-1} \theta$$

Also recall that there are infinitely many angles that have a particular value of a trigonometric function. Of these, we are finding just the smallest positive angle.

Taking reciprocals of both sides gives

$$\tan \theta = \frac{1}{2.745} = 0.3643$$

So

$$\theta = \tan^{-1} 0.3643 = 0.3494 \qquad \text{◆◆◆}$$

To find the trigonometric function of an angle in radians expressed in terms of π using the calculator, it is necessary first to convert the angle to decimal form.

◆◆◆ Example 11: Find $\cos(5\pi/12)$ to four significant digits.

Solution: Converting $5\pi/12$ to decimal form, we have

$$5(\pi)/12 = 1.3090$$

With our calculator in radian mode, we then take the cosine.

$$\cos 1.3090 \text{ rad} = 0.2588 \qquad \text{◆◆◆}$$

◆◆◆ Example 12: Evaluate to four significant digits:

$$5 \cos(2\pi/5) + 4 \sin^2(3\pi/7)$$

Solution: From the calculator,

$$\cos \frac{2\pi}{5} = 0.30902 \quad \text{and} \quad \sin \frac{3\pi}{7} = 0.97493$$

so

$$5 \cos \frac{2\pi}{5} + 4 \sin^2 \frac{3\pi}{7} = 5(0.30902) + 4(0.97493)^2$$

$$= 5.347 \qquad \text{◆◆◆}$$

Areas of Sectors and Segments

Sectors and segments of a circle (Fig. 16–3) were defined in Sec. 6–4. Now we will compute their areas.

The area of a circle of radius r is given by πr^2, so the area of a semicircle, of course, is $\pi r^2/2$; the area of a quarter circle is $\pi r^2/4$; and so on. The segment area is the same fractional part of the whole area as the central angle is of the whole.

Thus if the central angle is 1/4 revolution, the sector area is also 1/4 of the total circle area.

If the central angle (in radians) is $\theta/2\pi$ revolution, the sector area is also $\theta/2\pi$ the total circle area. So

$$\text{area of sector} = \pi r^2 \left(\frac{\theta}{2\pi} \right)$$

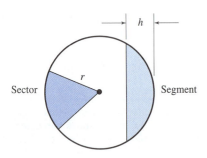

FIGURE 16–3 Segment and sector of a circle.

Area of a Sector	$A = \dfrac{r^2 \theta}{2}$ where θ is in radians	**116**

◆◆◆ **Example 13:** Find the area of a sector having a radius of 8.25 m and a central angle of 46.8°.

Solution: We first convert the central angle to radians.

$$46.8°(\pi \text{ rad}/180°) = 0.8168 \text{ rad}$$

Then, by Eq. 116,

$$\text{area} = \frac{(8.25)^2(0.8168)}{2} = 27.8 \text{ m}^2$$

◆◆◆

The area of a *segment* of a circle (Fig. 16–3) is

Area of a Segment	$r^2 \arccos \dfrac{r-h}{r} - (r-h)\sqrt{2rh-h^2}$
	where $\arccos \dfrac{r-h}{r}$ is in radians

◆◆◆ **Example 14:** Compute the area of a segment having a height of 10.0 cm in a circle of radius 25.0 cm.

Solution: Substituting into the given equation gives

$$\text{area} = (25.0)^2 \arccos \frac{25.0-10.0}{25.0} - (25.0-10.0)\sqrt{2(25.0)(10.0)-(10.0)^2}$$

$$= 625 \arccos 0.600 - 15.0\sqrt{500-100}$$

$$= 625(0.9273) - 15.0(20.0) = 280 \text{ cm}^2$$

◆◆◆

Exercise 1 ◆ Radian Measure

Convert to radians.

1. 47.8° **2.** 18.7° **3.** 35°15′
4. 0.370 rev **5.** 1.55 rev **6.** 1.27 rev

Convert to revolutions.

7. 1.75 rad **8.** 2.30 rad **9.** 3.12 rad
10. 0.0633 rad **11.** 1.12 rad **12.** 0.766 rad

Convert to degrees (decimal).

13. 2.83 rad **14.** 4.275 rad **15.** 0.372 rad
16. 0.236 rad **17.** 1.14 rad **18.** 0.116 rad

Convert each angle given in degrees to radian measure in terms of π.

19. 60° **20.** 130° **21.** 66°
22. 240° **23.** 126° **24.** 105°
25. 78° **26.** 305° **27.** 400°
28. 150° **29.** 81° **30.** 189°

Convert each angle given in radian measure to degrees.

31. $\dfrac{\pi}{8}$ **32.** $\dfrac{2\pi}{3}$ **33.** $\dfrac{9\pi}{11}$

34. $\dfrac{3\pi}{5}$ **35.** $\dfrac{\pi}{9}$ **36.** $\dfrac{4\pi}{5}$

37. $\dfrac{7\pi}{8}$ **38.** $\dfrac{5\pi}{9}$ **39.** $\dfrac{2\pi}{15}$

40. $\dfrac{6\pi}{7}$ **41.** $\dfrac{\pi}{12}$ **42.** $\dfrac{8\pi}{9}$

Calculator

Use a calculator to evaluate to four significant digits.

43. $\sin \dfrac{\pi}{3}$ **44.** $\tan 0.442$ **45.** $\cos 1.063$

46. $\tan(-\dfrac{2\pi}{3})$ **47.** $\cos \dfrac{3\pi}{5}$ **48.** $\sin(-\dfrac{7\pi}{8})$

49. $\sec 0.355$ **50.** $\csc \dfrac{4\pi}{3}$ **51.** $\cot \dfrac{8\pi}{9}$

52. $\tan \dfrac{9\pi}{11}$ **53.** $\cos(-\dfrac{6\pi}{5})$ **54.** $\sin 1.075$

55. $\cos 1.832$ **56.** $\cot 2.846$ **57.** $\sin 0.6254$

58. $\csc 0.8163$ **59.** $\arcsin 0.7263$ **60.** $\arccos 0.6243$

61. $\cos^{-1} 0.2320$ **62.** $\text{arccot } 1.546$ **63.** $\sin^{-1} 0.2649$

64. $\csc^{-1} 2.6263$ **65.** $\arctan 3.7253$ **66.** $\text{arcsec } 2.8463$

67. $\sin^2 \dfrac{\pi}{6} + \cos \dfrac{\pi}{6}$ **68.** $7 \tan^2 \dfrac{\pi}{9}$ **69.** $\cos^2 \dfrac{3\pi}{4}$

70. $\dfrac{\pi}{6} \sin \dfrac{\pi}{6}$ **71.** $\sin \dfrac{\pi}{8} \tan \dfrac{\pi}{8}$ **72.** $3 \sin \dfrac{\pi}{9} \cos^2 \dfrac{\pi}{9}$

73. Find the area of a sector having a radius of 5.92 in. and a central angle of 62.5°.

74. Find the area of a sector having a radius of 3.15 m and a central angle of 28.3°.

75. Find the area of a segment of height 12.4 cm in a circle of radius 38.4 cm.

76. Find the area of a segment of height 55.4 inches in a circle of radius 122.6 in.

77. A weight bouncing on the end of a spring moves with *simple harmonic motion* according to the equation $y = 4 \cos 25t$, where y is in inches. Find the displacement y when $t = 2.00$ s. (In this equation, the angle $25t$ must be in radians.)

78. The angle D (measured at the earth's center) between two points on the earth's surface is found by

$$\cos D = \sin L_1 \sin L_2 + \cos L_1 \cos L_2 \cos(M_1 - M_2)$$

where L_1 and M_1 are the latitude and longitude, respectively, of one point, and L_2 and M_2 are the latitude and longitude of the second point. Find the angle between Pittsburgh (latitude 40° N, longitude 81° W) and Houston (latitude 30° N, longitude 95° W) by substituting into this equation.

79. A grinding machine for granite uses a grinding disk that has four abrasive pads with the dimensions shown in Fig. 16–4. Find the area of each pad.

80. A partial pulley is in the form of a sector with a cylindrical hub, as shown in Fig. 16–5. Using the given dimensions, find the volume of the pulley.

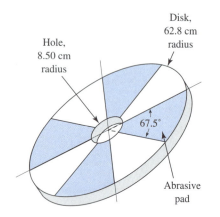

Hole, 8.50 cm radius

Disk, 62.8 cm radius

67.5°

Abrasive pad

FIGURE 16–4

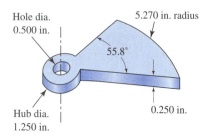

Hole dia. 0.500 in.

5.270 in. radius

55.8°

0.250 in.

Hub dia. 1.250 in.

FIGURE 16–5

16–2 Arc Length

We have seen that arc length, radius, and central angle are related to each other. In Fig. 16–6, if the arc length s is equal to the radius r, we have θ equal to 1 radian. For other lengths s, the angle θ is equal to the *ratio* of s to r.

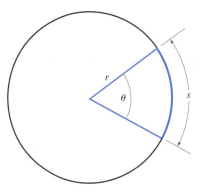

| Central Angle | $$\theta = \frac{s}{r}$$ where θ must be in radians | 115 |

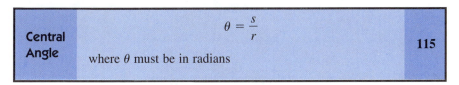

FIGURE 16–6 Relationship between arc length, radius, and central angle.

The central angle in a circle (in radians) is equal to the ratio of the intercepted arc and the radius of the circle.

When you are dividing s by r to obtain θ, s and r must have the same units. The units cancel, *leaving θ as a dimensionless ratio* of two lengths. Thus the radian is not a unit of measure like the degree or inch, although we usually carry it along as if it were.

We can use Eq. 115 to find any of the quantities θ, r, or s when the other two are known.

◆◆◆ **Example 15:** Find the angle that would intercept an arc of 27.0 ft in a circle of radius 21.0 ft.

Solution: From Eq. 115,

$$\theta = \frac{s}{r} = \frac{27.0 \text{ ft}}{21.0 \text{ ft}} = 1.29 \text{ rad}$$

◆◆◆

◆◆◆ **Example 16:** Find the arc length intercepted by a central angle of 62.5° in a 10.4-cm-radius circle.

Solution: Converting the angle to radians, we get

$$62.5°\left(\frac{\pi \text{ rad}}{180°}\right) = 1.09 \text{ rad}$$

By Eq. 115,

$$s = r\theta = 10.4 \text{ cm}(1.09) = 11.3 \text{ cm}$$

◆◆◆

◆◆◆ **Example 17:** Find the radius of a circle in which an angle of 2.06 rad intercepts an arc of 115 ft.

Solution: By Eq. 115,

$$r = \frac{s}{\theta} = \frac{115 \text{ ft}}{2.06} = 55.8 \text{ ft}$$

◆◆◆

| Common Error | Be sure that r and s have the same units. Convert if necessary. |

◆◆◆ **Example 18:** Find the angle that intercepts a 35.8-in. arc in a circle of radius 49.2 cm.

Solution: We divide s by r and convert units at the same time.

$$\theta = \frac{s}{r} = \frac{35.8 \text{ in.}}{49.2 \text{ cm}} \cdot \frac{2.54 \text{ cm}}{1 \text{ in.}} = 1.85 \text{ rad}$$

◆◆◆

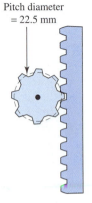

Pitch diameter = 22.5 mm

FIGURE 16–7   Rack and pinion.

♦♦♦ **Example 19:** How far will the rack in Fig. 16–7 move when the pinion rotates 300.0°?

Solution: Converting to radians gives us

$$\theta = 300.0° \left(\frac{\pi \text{ rad}}{180°} \right) = 5.236 \text{ rad}$$

Then, by Eq. 115,

$$s = r\theta = 11.25 \text{ mm}(5.236) = 58.9 \text{ mm}$$

As a rough check, we note that the pinion rotates less than 1 revolution, so the rack will travel a distance less than the circumference of the pinion. This circumference is 22.5π or about 71 mm. So our answer of 58.91 mm seems reasonable.   ♦♦♦

♦♦♦ **Example 20:** How many miles north of the equator is a town of latitude 43.6° N. Assume that the earth is a sphere of radius 3960 mi, and refer to the definitions of latitude and longitude shown in Fig. 16–8.

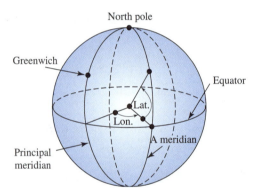

FIGURE 16–8   *Latitude* is the angle (measured at the earth's center) between a point on the earth and the equator. *Longitude* is the angle between the meridian passing through a point on the earth and the principal (or prime) meridian passing through Greenwich, England.

Estimate: Our given latitude angle is about 1/8 of a circle, so the required distance must be about 1/8 of the earth's circumference (8000π or 25,000 mi), or about 3120 mi.

Solution: The latitude angle, in radians, is

$$\theta = 43.6° \left(\frac{\pi \text{ rad}}{180°} \right) = 0.761 \text{ rad}$$

Then, by Eq. 115,

$$s = r\theta = 3960 \text{ mi}(0.761) = 3010 \text{ mi}$$ ♦♦♦

Exercise 2 ♦ Arc Length

In the following table, s is the length of arc subtended by a central angle θ in a circle of radius r. Fill in the missing values.

	r	θ	s
1.	4.83 in.	$2\pi/5$	
2.	11.5 cm	1.36 rad	
3.	284 ft	46°24′	
4.	2.87 m	1.55 rad	
5.	64.8 in.	38.5°	
6.	28.3 ft		32.5 ft
7.	263 mm		582 mm
8.	21.5 ft		18.2 ft
9.	3.87 m		15.8 ft
10.		$\pi/12$	88.1 in.
11.		77.2°	1.11 cm
12.		2.08 rad	3.84 m
13.		12°55′	28.2 ft
14.		$5\pi/6$	125 mm

Applications

15. A certain town is at a latitude of 35.2°N. Find the distance in miles from the town to the north pole.

16. The hour hand of a clock is 85.5 mm long. How far does the tip of the hand travel between 1:00 A.M. and 11:00 A.M.?

17. Find the radius of a circular railroad track that will cause a train to change direction by 17.5° in a distance of 180 m.

18. The pulley attached to the tuning knob of a radio (Fig. 16–9) has a radius of 35 mm. How far will the needle move if the knob is turned a quarter of a revolution?

19. Find the length of contact ABC between the belt and pulley in Fig. 16–10.

Assume the earth to be a sphere with a radius of 3960 mi. Actually, the distance from pole to pole is about 27 mi less than the diameter at the equator.

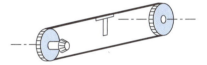

FIGURE 16–9

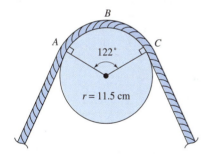

FIGURE 16–10 Belt and pulley.

20. One circular "track" on a magnetic disk used for computer data storage is located at a radius of 155 mm from the center of the disk. If 10 "bits" of data can be stored in 1 mm of track, how many bits can be stored in the length of track subtending an angle of $\pi/12$ rad?

21. If we assume the earth's orbit around the sun to be circular, with a radius of 93 million mi, how many miles does the earth travel (around the sun) in 125 days?

22. Find the latitude of a city that is 1265 mi from the equator.

23. A 1.25-m-long pendulum swings 5.75° on each side of the vertical. Find the length of arc traveled by the end of the pendulum.

24. A brake band is wrapped around a drum (Fig. 16–11). If the band has a width of 92.0 mm, find the area of contact between the band and the drum.

25. Sheet metal is to be cut from the pattern of Fig. 16–12(a) and bent to form the frustum of a cone [Fig. 16–12(b)], with top and bottom open. Find the dimensions r and R and the angle θ in degrees.

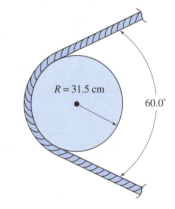

FIGURE 16–11 Brake drum.

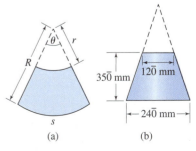

FIGURE 16–12

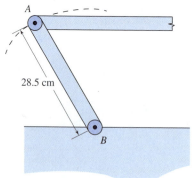

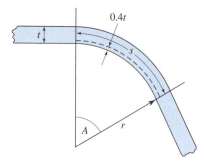

28.5 cm

FIGURE 16–13

26. An isosceles triangle is to be inscribed in a circle of radius 1.000. Find the angles of the triangle if its base subtends an arc of length 1.437.

27. A satellite is in a circular orbit 225 mi above the equator of the earth. How many miles must it travel for its longitude to change by 85.0°?

28. City *B* is due north of city *A*. City *A* has a latitude of 14°37′ N, and city *B* has a latitude of 47°12′ N. Find the distance in kilometers between the cities.

29. The link *AB* in the mechanism of Fig. 16–13 rotates through an angle of 28.3°. Find the distance traveled by point *A*.

30. Find the radius *R* of the sector gear of Fig. 16–14.

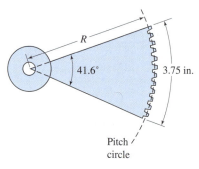

FIGURE 16–14   Sector gear.

31. A circular highway curve has a radius of 325.500 ft and a central angle of 15°25′15″ measured to the centerline of the road. Find the length of the curve.

32. When metal is in the process of being bent, the amount *s* that must be allowed for the bend is called the *bending allowance,* as shown in Fig. 16–15. Assume that the neutral axis (the line at which there is no stretching or compression of the metal) is at a distance from the inside of the bend equal to 0.4 of the metal thickness *t*.

(a) Show that the bending allowance is

$$s = \frac{A(r + 0.4t)\pi}{180}$$

where *A* is the angle of bend in degrees.

(b) Find the bending allowance for a 60° bend in $\frac{1}{4}$-in.-thick steel with a radius of 1.50 in.

33. In Sec. 16–1 we gave the formula for the area of a circular sector of radius *r* and central angle θ: area = $r^2\theta/2$ (Eq. 116). Using Eq. 115, $\theta = s/r$, show that the area of a sector is also equal to $rs/2$, where *s* is the length of the arc intercepted by the central angle.

34. Find the area of a sector having a radius of 34.8 cm and an arc length of 14.7 cm.

FIGURE 16–15   Bending allowance.

Computer

35. For the angles from 0 to 10°, with steps every $\frac{1}{2}$ degree, compute and print the angle in radians, the sine of the angle, and the tangent of the angle. What do you notice about these three columns of figures? What is the largest angle for which the sine and the tangent are equal to the angle in radians to three significant digits?

16–3 Uniform Circular Motion

Angular Velocity

Let us consider a rigid body that is rotating about a point O, as shown in Fig. 16–16. The *angular velocity* ω is a measure of the *rate* at which the object rotates. The motion is called *uniform* when the angular velocity is constant. The units of angular velocity are degrees, radians, or revolutions, per unit time.

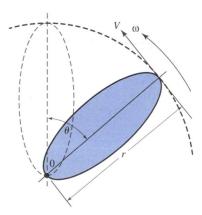

FIGURE 16–16 Rotating body. The symbol ω is lowercase Greek omega, not the letter w.

Angular Displacement

The angle θ through which a body rotates in time t is called the *angular displacement*. It is related to ω and t by the following equation:

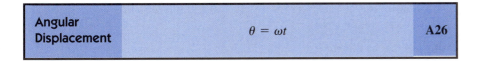

Angular Displacement	$\theta = \omega t$	A26

The angular displacement is the product of the angular velocity and the elapsed time.

Equation A26 is similar to our old formula (distance = rate × time) for linear motion.

◆◆◆ **Example 21:** A wheel is rotating with an angular velocity of 1800 rev/min. How many revolutions does the wheel make in 1.5 s?

Solution: We first make the units of time consistent. Converting yields

$$\omega = \frac{1800 \text{ rev}}{\text{min}} \cdot \frac{1 \text{ min}}{60 \text{ s}} = 30 \text{ rev/s}$$

Then, by Eq. A26,

$$\theta = \omega t = \frac{30 \text{ rev}}{\text{s}}(1.5 \text{ s}) = 45 \text{ rev}$$ ◆◆◆

◆◆◆ **Example 22:** Find the angular velocity in revolutions per minute of a pulley that rotates 275° in 0.750 s.

Solution: By Eq. A26,

$$\omega = \frac{\theta}{t} = \frac{275°}{0.750 \text{ s}} = 367 \text{ deg/s}$$

Converting to rev/min, we obtain

$$\omega = \frac{367°}{\text{s}} \cdot \frac{60 \text{ s}}{\text{min}} \cdot \frac{1 \text{ rev}}{360°} = 61.2 \text{ rev/min}$$ ◆◆◆

◆◆◆ **Example 23:** How long will it take a spindle rotating at 3.55 rad/s to make 1000 revolutions?

Solution: Converting revolutions to radians, we get

$$\theta = 1000 \text{ rev} \cdot \frac{2\pi \text{ rad}}{1 \text{ rev}} = 6280 \text{ rad}$$

Then, by Eq. A26,

$$t = \frac{\theta}{\omega} = \frac{6280 \text{ rad}}{3.55 \text{ rad/s}} = 1770 \text{ s} = 29.5 \text{ min}$$ ◆◆◆

Linear Speed

For any point on a rotating body, the linear displacement per unit time along the circular path is called the *linear speed*. The linear speed is zero for a point at the center of rotation and is directly proportional to the distance r from the point to the center of rotation. If ω is expressed in radians per unit time, the linear speed v is given by the following equation:

Linear Speed	$v = \omega r$	A27

The linear speed of a point on a rotating body is proportional to the distance from the center of rotation.

◆◆◆ **Example 24:** A wheel is rotating at 2450 rev/min. Find the linear speed of a point 35.0 cm from the center.

Solution: We first express the angular velocity in terms of radians.

$$\omega = \frac{2450 \text{ rev}}{\text{min}} \cdot \frac{2\pi \text{ rad}}{\text{rev}} = 15,400 \text{ rad/min}$$

Then, by Eq. A27,

$$v = \omega r = \frac{15,400 \text{ rad}}{\text{min}}(35.0 \text{ cm}) = 539,000 \text{ cm/min}$$

$$= 89.8 \text{ m/s}$$ ◆◆◆

What became of "radians" in our answer? Shouldn't the final units be rad cm/min? No. Remember that radians is a dimensionless ratio; it is the ratio of two lengths (arc length and radius) whose units cancel.

Common Error	Remember when using Eq. A27 that the angular velocity must be expressed in *radians* per unit time.

◆◆◆ **Example 25:** A belt having a speed of 885 in./min turns a 12.5-in.-radius pulley. Find the angular velocity of the pulley in rev/min.

Solution: By Eq. A27,

$$\omega = \frac{v}{r} = \frac{885 \text{ in.}}{\text{min}} \div 12.5 \text{ in.} = 70.8 \text{ rad/min}$$

Converting yields

$$\omega = \frac{70.8 \text{ rad}}{\text{min}} \cdot \frac{1 \text{ rev}}{2\pi \text{ rad}} = 11.3 \text{ rev/min}$$ ◆◆◆

Exercise 3 ◆ Uniform Circular Motion

Angular Velocity

Fill in the missing values.

rev/min	*rad/s*	*deg/s*
1. 1850		
2.	5.85	
3.		77.2
4.	$3\pi/5$	
5.		48.1
6. 22,600		

7. A flywheel makes 725 revolutions in a minute. How many degrees does it rotate in 1.00 s?

8. A propeller on a wind generator rotates 60.0° in 1.00 s. Find the angular velocity of the propeller in revolutions per minute.

9. A gear is rotating at 2550 rev/min. How many seconds will it take to rotate through an angle of 2.00 rad?

Linear Speed

10. A milling machine cutter has a diameter of 75.0 mm and is rotating at 56.5 rev/min. What is the linear speed at the edge of the cutter?

11. A sprocket 3.00 inches in diameter is driven by a chain that moves at a speed of 55.5 in./s. Find the angular velocity of the sprocket in rev/min.

12. A capstan on a magnetic tape drive rotates at $\overline{3600}$ rad/min and drives the tape at a speed of 45.0 m/min. Find the diameter of the capstan in millimeters.

13. A blade on a water turbine turns 155° in 1.25 s. Find the linear speed of a point on the tip of the blade 0.750 m from the axis of rotation.

14. A steel bar 6.50 inches in diameter is being ground in a lathe. The surface speed of the bar is 55.0 ft/min. How many revolutions will the bar make in 10.0 s?

15. Assuming the earth to be a sphere 7920 mi in diameter, calculate the linear speed in miles per hour of a point on the equator due to the rotation of the earth about its axis.

16. Assuming the earth's orbit about the sun to be a circle with a radius of 93.0×10^6 mi, calculate the linear speed of the earth around the sun.

17. A car is traveling at a rate of 65.5 km/h and has tires that have a radius of 31.6 cm. Find the angular velocity of the wheels of the car.

18. A wind generator has a propeller 21.7 ft in diameter, and the gearbox between the propeller and the generator has a gear ratio of 1:44 (with the generator shaft rotating faster than the propeller). Find the tip speed of the propeller when the generator is rotating at $\overline{1800}$ rev/min.

••• **CHAPTER 16 REVIEW PROBLEMS** ••••••••••••••••••••••••••••••••••••

Convert to degrees.

1. $\dfrac{3\pi}{7}$

2. $\dfrac{9\pi}{4}$

3. $\dfrac{\pi}{9}$

4. $\dfrac{2\pi}{5}$

5. $\dfrac{11\pi}{12}$

6. Find the central angle in radians that would intercept an arc of 5.83 m in a circle of radius 7.29 m.

7. Find the angular velocity in rad/s of a wheel that rotates 33.5 revolutions in 1.45 min.

8. A wind generator has blades 3.50 m long. Find the tip speed when the blades are rotating at 35 rev/min.

Convert to radians in terms of π.

9. $300°$

10. $150°$

11. $230°$

12. $145°$

Evaluate to four significant digits.

13. $\sin \dfrac{\pi}{9}$

14. $\cos^2 \dfrac{\pi}{8}$

15. $\sin\left(\dfrac{2\pi}{5}\right)^2$

16. $3 \tan \dfrac{3\pi}{7} - 2 \sin \dfrac{3\pi}{7}$

17. $4 \sin^2 \dfrac{\pi}{9} - 4 \sin\left(\dfrac{\pi}{9}\right)^2$

18. What arc length is subtended by a central angle of $63.4°$ in a circle of radius 4.85 in.?

19. A winch has a drum 30.0 cm in diameter. The steel cable wrapped around the drum is to be pulled in at a rate of 8.00 ft/s. Ignoring the thickness of the cable, find the angular speed of the drum in revolutions per minute.

20. Find the linear velocity of the tip of a 5.50-in.-long minute hand of a clock.

21. A satellite has a circular orbit around the earth with a radius of 4250 mi. The satellite makes a complete orbit of the earth in 3 days, 7 h, and 35 min. Find the linear speed of the satellite.

Evaluate to four decimal places. (Angles are in radians.)

22. $\cos 1.83$

23. $\tan 0.837$

24. $\csc 2.94$

25. $\sin 4.22$

26. $\cot 0.371$

27. $\sec 3.38$

28. $\cos^2 1.74$

29. $\sin(2.84)^2$

30. $(\tan 0.475)^2$

31. $\sin^2(2.24)^2$

32. Find the area of a sector with a central angle of $29.3°$ and a radius of 37.2 in.

33. Find the radius of a circle in which an arc of length 384 mm intercepts a central angle of 1.73 rad.

34. Find the arc length intercepted in a circle of radius 3.85 m by a central angle of 1.84 rad.

35. A parts-storage tray, shown in Fig. 16–17, is in the form of a partial sector of a circle. Using the dimensions shown, find the inside volume of the tray in cubic centimeters.

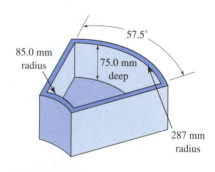

85.0 mm radius 57.5° 75.0 mm deep 287 mm radius

FIGURE 16–17

Writing

36. When we multiply an angular velocity in radians per second by a length in feet, we get a linear speed in feet per second. Explain why radians do not appear in the units for linear speed.

Team Projects

37. The kinetic energy of a rotating body is given by

$$E = \tfrac{1}{2}Mk^2\omega^2 \qquad \text{joules (J)}$$

where M is the mass in kilograms, ω is the angular velocity in rad/s, and k is the radius of gyration in meters, which for a sphere of radius r is $\sqrt{2/5}r$.

Your team should weigh and measure a golf ball, a billiard ball, a baseball, a softball, and a basketball. Then compute the kinetic energy of rotation for each, assuming that it is spinning at a rate of 3.0 revolutions per second.

38. Would you like to tell your friends that you measured the circumference of the earth with just a yardstick? In the third century B.C., the Greek scientific writer, astronomer, mathematician, and poet Eratosthenes (276?–194? B.C.) did just that. At Syene (now Aswan), he noted that the sun was directly overhead at the solstice by observing that the sun's rays shone directly into a deep well. He also learned that at that instant, the sun cast a shadow of 7.5° at Alexandria 500 mi to the north. He used these data to calculate the circumference of the earth.

(a) Calculate the circumference of the earth, using Eratosthenes' data.

(b) Repeat Eratosthenes' experiment. First devise a way to measure the angle of elevation of the sun when it is on your meridian of longitude. *Caution:* Use the *shadow* cast by a vertical stick for this. Never look directly at the sun, especially with a telescope!

Then enlist an assistant who is located at least 100 mi north or south of your location (the farther the better), and determine the north-south distance between the two locations from a map. Have this person determine the angle of the sun, when it is on the meridian, using your method. That same day, do the same at your location. Use the data from both locations to calculate the circumference of the earth.

17

Graphs of the Trigonometric Functions

••• **OBJECTIVES** •••

When you have completed this chapter, you should be able to:

- Find the amplitude, period, frequency, and phase shift for a sine wave or a cosine wave.
- Graph the sine function, cosine function, or tangent function.
- Write a sine function or cosine function, given the amplitude, period, and phase shift.
- Graph composite functions.
- Graph the sine or cosine wave as a function of time.
- Convert between polar and rectangular coordinates.
- Graph points and equations in polar coordinates.
- Graph parametric equations in polar coordinates.

Here we apply the graphing techniques of Chapter 5 to graph some *periodic* functions—the sine, cosine, and tangent—and we'll see how the constants in the equation for each affect the shape and position of the curves. We define some terms that will be applicable to the study of periodic functions in general, such as *cycle, period,* and *frequency.* This material has applications to such fields as alternating current, sound waves, mechanical vibrations, reciprocating machinery such as a gasoline or steam engine, and so on.

Finally, we introduce a second coordinate system (we have already studied rectangular coordinates) called the *polar coordinate system,* and we learn how to make transformations from one system to the other.

17–1 The Sine Curve

Periodic Functions

A curve that *repeats* its shape over and over is called a *periodic* curve or *periodic waveform,* as shown in Fig. 17–1. The function that has this graph is thus called a *periodic function.* Each repeated portion of the curve is called a *cycle.*

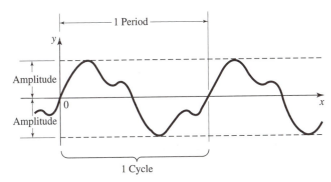

FIGURE 17–1   A periodic waveform.

The *x* axis represents either *an angle* (in degrees or radians) or *time* (usually seconds or milliseconds).

The *period* is the *x* distance taken for one cycle. Depending on the units on the *x* axis, the period is expressed in degrees per cycle, radians per cycle, or seconds per cycle.

The *frequency* of a periodic waveform is the number of cycles that will fit into 1 unit (degree, radian, or second) along the *x* axis. It is the reciprocal of the period.

$$\text{frequency} = \frac{1}{\text{period}}$$

Units of frequency are cycles/degree, cycles/radian, or, most commonly, cycles/second (called *hertz*), depending on the units of the period.

The *amplitude* is half the difference between the greatest and least value of a waveform (Fig. 17–1). The peak-to-peak value of a curve is equal to twice the amplitude.

The *instantaneous value* of a waveform is its *y* value at any given *x*.

◆◆◆ **Example 1:** For the voltage curve in Fig. 17–2, find the period, frequency, amplitude, peak-to-peak value, and the instantaneous value of *y* when *x* is 50 ms.

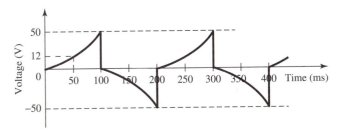

FIGURE 17–2

> Recall that a *complete graph* of a function is one that contains all of the features of interest: intercepts, maximum and minimum points, and so forth. Thus a complete graph of a periodic waveform must contain at least one cycle.

Solution: By observation of Fig. 17–2,

$$\text{period} = 200 \text{ ms/cycle}$$

$$\text{frequency} = \frac{1 \text{ cycle}}{200 \text{ ms}}$$

$$= 5 \text{ cycles/s} = 5 \text{ hertz (Hz)}$$

$$\text{amplitude} = 50 \text{ V}$$

$$\text{peak-to-peak value} = 100 \text{ V}$$

$$\text{instantaneous value when } x \text{ is } 50 \text{ ms} = 12 \text{ V}$$ ◆◆◆

Graph of the Sine Function $y = \sin x$

> The sine curve or sine wave is also called a *sinusoid*.

We will graph the sine curve just as we graphed other functions in Sec. 5–2. We choose values of x and substitute into the equation to get corresponding values of y, and then we plot the resulting table of point pairs. A little later we will learn a faster method.

The units of angular measure used for the x axis are sometimes degrees and sometimes radians. We will show examples of each.

◆◆◆ **Example 2:** Plot the sine curve $y = \sin x$ for values of x from 0 to 360°.

Solution: Let us choose 30° intervals and make a table of point pairs.

x	0	30°	60°	90°	120°	150°	180°	210°	240°	270°	300°	330°	360°
y	0	0.5	0.866	1	0.866	0.5	0	−0.5	−0.866	−1	−0.866	−0.5	0

Plotting these points gives the periodic curve shown in Fig. 17–3. If we had plotted points from $x = 360°$ to 720°, we would have gotten another cycle of the sine curve, identical to the one just plotted. We see then that the sine curve is periodic, requiring 360° for each cycle. Thus the period is equal to 360°.

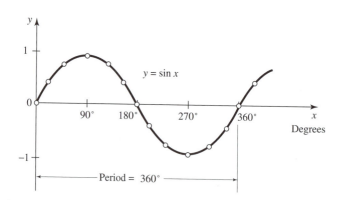

FIGURE 17–3 $y = \sin x$ (x is in degrees).   ◆◆◆

We'll now do an example in radians.

> It is not easy to work in radian measure when we are used to thinking of angles in degrees. You might want to glance back at Sec. 16–1.

◆◆◆ **Example 3:** Plot $y = \sin x$ for values of x from 0 to 2π rad.

Solution: We could choose integer values for x, such as 1, 2, 3, . . . , or follow the usual practice of using multiples and fractions of π. This will give us points at the peaks and valleys of the curve and the point where the curve crosses the x axis. We take intervals of $\pi/6$ and make the following table of point pairs:

x (rad)	0	$\dfrac{\pi}{6}$	$\dfrac{\pi}{3}$	$\dfrac{\pi}{2}$	$\dfrac{2\pi}{3}$	$\dfrac{5\pi}{6}$	π	$\dfrac{7\pi}{6}$	$\dfrac{4\pi}{3}$	$\dfrac{3\pi}{2}$	$\dfrac{5\pi}{3}$	$\dfrac{11\pi}{6}$	2π
y	0	0.5	0.866	1	0.866	0.5	0	-0.5	-0.866	-1	-0.866	-0.5	0

Our curve, shown in Fig. 17–4, is the same as before, except for the scale on the x axis. There, we show both integer values of x and multiples of π.

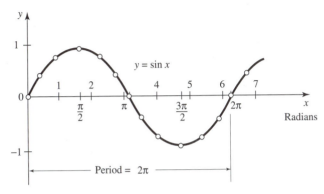

FIGURE 17–4 $y = \sin x$ (x is in radians).

Figure 17–5 shows the same curve with both radian and degree scales on the x axis. Notice that one cycle of the curve fits into a rectangle whose height is twice the amplitude of the sine curve and whose width is equal to the period. We use this later for rapid sketching of periodic curves.

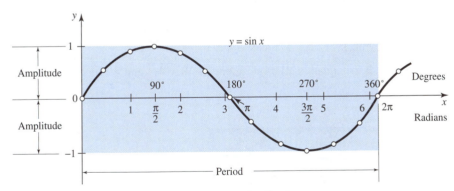

FIGURE 17–5 The sine curve (x is in degrees and radians).

Graphing the Sine Function with a Graphics Calculator

The graphics calculator makes it easy both to graph the trigonometric functions and to explore how the constants in the equation affect the graph. We will do the simplest case here and more complicated ones later.

◆◆◆ **Example 4:** Use the graphics calculator to graph $y = \sin x$ for x from $-30°$ to $400°$.

Solution: We put the calculator in $\boxed{\text{DEGREE}}$ mode and set the range.

$$\text{Xmin} = -30 \qquad \text{Xmax} = 400 \qquad \text{Xscl} = 30$$

Once you have the graph, use the TRACE feature to find the coordinates of zeros and of the maximum and minimum points.

You may recall that the sine of an angle varies between 1 and −1. So let us set the y range as follows:

$$\text{Ymin} = -1.5 \qquad \text{Ymax} = 1.5 \qquad \text{Yscl} = 1$$

We enter the equation Y1 = sin x, press GRAPH , and get the display shown in Fig. 17–6.

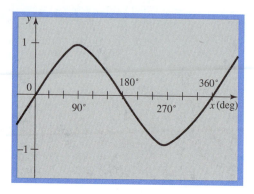

FIGURE 17–6   Graph of y = sin x.

The graphics calculator will not show labels on the axes. We include them for clarity.
◆◆◆

We'll now do an example in radians.

◆◆◆ **Example 5:** Use the graphics calculator to graph y = sin x for x from −4 radians to 4 radians.

Solution: We put the calculator into RADIAN mode and set the range.

$$\text{Xmin} = -4 \qquad \text{Xmax} = 4 \qquad \text{Xscl} = 0.7854 \quad (\pi/4)$$
$$\text{Ymin} = -1.5 \qquad \text{Ymax} = 1.5 \qquad \text{Yscl} = 1$$

Again we enter the equation Y1 = sin x, press GRAPH , and get the display shown in Fig. 17–7.

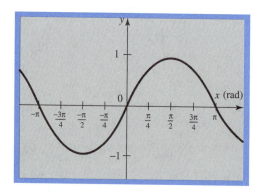

FIGURE 17–7   Graph of y = sin x.

Graph of y = a sin x

When the sine function has a coefficient a, then the absolute value of a

is the *amplitude* of the sine wave. In the preceding examples, the constant *a* was equal to 1, and we got sine waves with amplitudes of 1. In examples that follow, *a* will have other values. For now, we still let $b = 1$ and $c = 0$.

◆◆◆ **Example 6:** Plot the sine curve $y = 3 \sin x$ for $x = 0$ to 2π.

Solution: When we make a table of point pairs, it becomes apparent that all the *y* values will be three times as large as the ones in the preceding example. These points are plotted in Fig. 17–8.

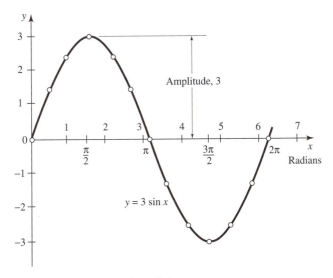

FIGURE 17–8 Graph of $y = 3 \sin x$.

Use your graphics calculator to plot

$$y = \sin x$$

and

$$y = 3 \sin x$$

in the same viewing rectangle. Work in either degrees or radians. How do the two graphs differ?

◆◆◆

Exercise 1 ◆ The Sine Curve

Periodic Waveforms

Find the period, frequency, and amplitude for each periodic waveform.

1. Figure 17–9(a).
2. Figure 17–9(b).

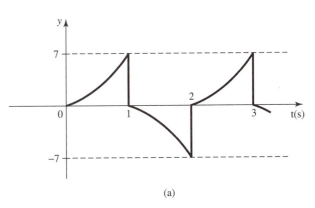

(a)

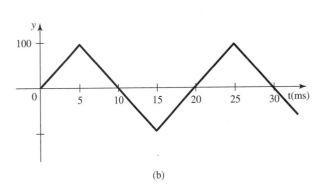

(b)

FIGURE 17–9

The Sine Function

Graph each function. Find the amplitude and period.

3. $y = 2 \sin x$ **4.** $y = 1.5 \sin x$
5. $y = -3 \sin x$ **6.** $y = 4.25 \sin x$

Graph each function and find the amplitude.

7. $y = -2.5 \sin x$ **8.** $y = -3.75 \sin x$

9. As the ladder in Fig. 17–10 is raised, the height h increases as θ increases.
(a) Write the function $h = f(\theta)$.
(b) Graph $h = f(\theta)$ for $\theta = 0$ to 90°.

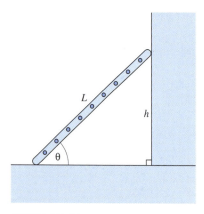

FIGURE 17–10   A ladder.

10. Figure 17–11 shows a Scotch yoke mechanism, or Scotch crosshead.
(a) Write an expression for the displacement y of the rod PQ as a function of θ and of the length R of the rotating arm.
(b) Graph $y = f(\theta)$ for $\theta = 0$ to 360°.

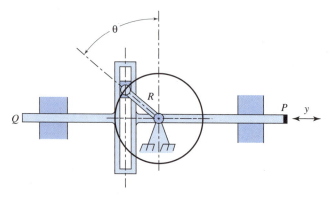

FIGURE 17–11   Scotch crosshead.

11. For the pendulum of Fig. 17–12:

   (**a**) Write an equation for the horizontal displacement x as a function of θ.

   (**b**) Graph $x = f(\theta)$ for $\theta = 0$ to $30°$.

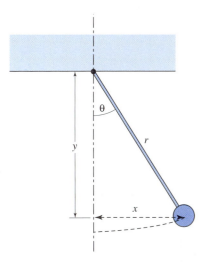

FIGURE 17–12   A pendulum.

Graphics Calculator or Computer

12. Use your graphics calculator or graphing utility on a computer to graph any of the sine waves in this exercise set.

17–2   The General Sine Wave

In the preceding section, we considered only sine waves that had a period of $360°$ and those that passed through the origin. Here we will graph sine waves of any period and those that are *shifted* in the x direction. We'll start with sine waves of any period.

Graph of $y = a \sin bx$

For the sine function $y = a \sin x$, we saw that the curve repeated in an interval of $360°$ (or 2π radians). We called this interval the *period*. Now we will see how the constant b affects the graph of the equation.

◆◆◆ **Example 7:** Plot $y = \sin 2x$ for $x = 0$ to $360°$.

Solution: As before, we make a table of point pairs, taking $30°$ intervals.

x	0	30°	60°	90°	120°	150°	180°	210°	240°	270°	300°	330°	360°
y	0	0.866	0.866	0	−0.866	−0.866	0	0.866	0.866	0	−0.866	−0.866	0

Plotting these points and connecting them with a smooth curve gives the graph shown in Fig. 17–13. Notice that the curve repeats twice in the $360°$ interval instead of just once as it did before. In other words, the period is half that of $y = \sin x$.

The sine wave $y = \sin x$ is sometimes called the *fundamental*, and the sine waves

$$y = \sin bx, \quad \text{where } b = 1, 2, 3, \ldots$$

are called the *harmonics*. Thus $y = \sin 2x$ is the *second harmonic*.

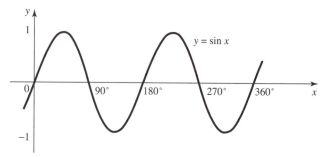

FIGURE 17–13

In general,

$$y = a \sin bx$$

will have a period of $360°/b$, if x is in degrees, or $2\pi/b$ radians, if x is in radians.

Period	$$P = \dfrac{360}{b} \text{ degrees/cycle}$$ $$P = \dfrac{2\pi}{b} \text{ radians/cycle}$$	211

◆◆◆ **Example 8:** Find the period for the sine curve $y = \sin 4x$ in degrees and also in radians per cycle.

Solution:

$$P = \frac{360°}{4} = 90°\text{/cycle}$$

or

$$P = \frac{2\pi}{4} = \frac{\pi}{2} \text{ radians/cycle}$$

◆◆◆

The *frequency f* is the reciprocal of the period.

Frequency	$$f = \dfrac{b}{360} \text{ cycles/degree}$$ $$f = \dfrac{b}{2\pi} \text{ cycles/radian}$$	212

◆◆◆ **Example 9:** Determine the amplitude, period, and frequency of the sine curve $y = 2 \sin(\pi x/2)$, where x is in radians, and graph one cycle of the curve.

Solution: Here, $a = 2$ and $b = \pi/2$, so the amplitude is 2, and the period is

$$P = \frac{2\pi}{\pi/2} = 4 \text{ rad/cycle}$$

or, in degrees,

$$P = \frac{360°}{\pi/2} = 229.2°\text{/cycle}$$

We make a table of point pairs for $x = 0$ to 4 rad.

x (rad)	0	$\frac{1}{2}$	1	$\frac{3}{2}$	2	$\frac{5}{2}$	3	$\frac{7}{2}$	4
y	0	1.41	2	1.41	0	-1.41	-2	-1.41	0

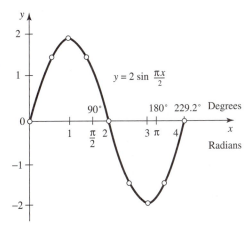

FIGURE 17–14 Graph of $y = 2 \sin(\pi x/2)$.

These points are plotted in Fig. 17–14. The frequency is

$$f = \frac{1}{P} = \frac{1}{4} \text{ cycle/rad}$$

In degrees,

$$f = \frac{1}{229.2°} = 0.00436 \text{ cycle/deg}$$

◆◆◆

◆◆◆ **Example 10:** Use the graphics calculator to graph the following in the same viewing window:

$$y = \sin x, \quad y = \sin 2x, \quad \text{and} \quad y = \sin 0.5x$$

for x from $-30°$ to $400°$.

Solution: We put the calculator in $\boxed{\text{DEGREE}}$ mode and set the range.

$$\text{Xmin} = -3.0 \qquad \text{Xmax} = 400 \qquad \text{Xscl} = 45$$
$$\text{Ymin} = -1.5 \qquad \text{Ymax} = 1.5 \qquad \text{Yscl} = 1$$

We enter the equations, press $\boxed{\text{GRAPH}}$, and get the display shown in Fig. 17–15.

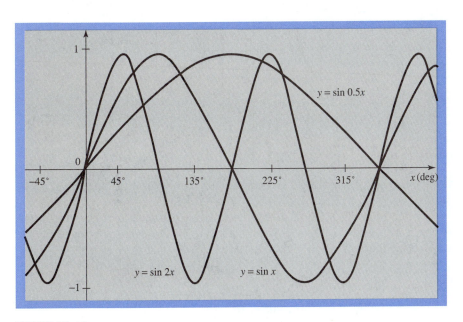

FIGURE 17–15

Notice that the graph of $y = \sin 2x$ completes two cycles in the same interval that $y = \sin x$ completes in one cycle. This is not surprising: Because $2x$ is twice as large as x, $2x$ will reach the value of 360° (a full cycle) twice as soon as x.

Similarly, $0.5x$ will reach 360° half as soon as x will, so $y = \sin 0.5x$ completes only half a cycle in the same interval that $y = \sin x$ completes a full cycle.   ◆◆◆

Graph of the General Sine Function $y = a \sin(bx + c)$

The graph of $y = a \sin bx$, we have seen, passes through the origin. However, the graph of the general sine function, given in the following box, will be shifted in the x direction by an amount called the *phase displacement,* or *phase shift.*

General Sine Curve	$y = a \sin(bx + c)$	210

◆◆◆ **Example 11:** Graph $y = \sin(3x - 90°)$.

Solution: Since $a = 1$ and $b = 3$, we expect an amplitude of 1 and a period of $360/b = 120°$. The effect that the constant c has on the curve will be clearer if we put the given expression into the form

$$y = a \sin b(x + c/b)$$

by factoring. Thus our equation becomes

$$y = \sin 3(x - 30°)$$

We make a table of point pairs and plot each point as shown in Fig. 17–16.

x	0	15°	30°	45°	60°	75°	90°	105°	120°	135°	150°
y	−1	−0.707	0	0.707	1	0.707	0	−0.707	−1	−0.707	0

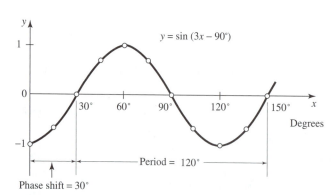

FIGURE 17–16   Sine curve with phase shift.

Our curve, unlike those plotted earlier, does not pass through the origin.   ◆◆◆

We see from Fig. 17–16 that the phase shift is that x distance from the origin at which the positive half-cycle of the sine wave starts. To find this value for the general sine function $y = a \sin(bx + c)$, we set $y = 0$ and solve for x.

$$y = a \sin(bx + c) = 0$$

$$\sin(bx + c) = 0$$

$$bx + c = 0°, \ 180°, \ . \ . \ .$$

We discard 180° because that is the start of the negative half-cycle. So $bx + c = 0$, and

$$x = -\frac{c}{b}$$

The formula for phase shift is summarized in the following box:

Phase Shift	phase shift $= -\dfrac{c}{b}$	213
	When the phase shift is *positive*, shift the curve to the *right* of its unshifted position.	
	When the phase shift is *negative*, shift the curve to the *left* of its unshifted position.	

The phase shift will have the same units, radians or degrees, as the constant c. Where no units are indicated for c, assume it to be in radians. If the phase shift is *negative*, it means that the positive half-cycle starts at a point to the left of the origin, so *the curve is shifted to the left.* Conversely, a *positive* phase shift means that the curve is shifted to the *right*.

◆◆◆ **Example 12:** In the equation $y = \sin(x + 45°)$,

$$a = 1, \qquad b = 1, \quad \text{and} \quad c = 45°$$

$$\text{phase shift} = -\frac{c}{b} = -45°$$

so the curve is shifted 45° to the left as shown in Fig. 17–17(b). For comparison, Figure 17–17 also shows (a) the unshifted sine wave and (c) the sine wave shifted 45° to the right.

When you study *translation of axes* in analytic geometry, you will see that replacing x in a function by $(x − h)$, where h is a constant, causes the graph of the function to be shifted h units in the x direction.

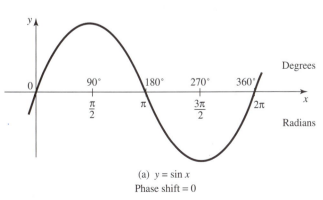

(a) $y = \sin x$
Phase shift $= 0$

FIGURE 17–17

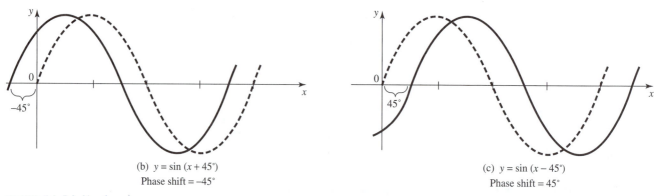

(b) $y = \sin(x + 45°)$
Phase shift $= -45°$

(c) $y = \sin(x - 45°)$
Phase shift $= 45°$

FIGURE 17–17 *Continued*   ◆◆◆

◆◆◆ **Example 13:** For the sine function $y = 3\sin(4x - 2)$, we have

$$a = 3, \qquad b = 4, \quad \text{and} \quad c = -2$$

so

$$\text{phase shift} = -\frac{c}{b} = \frac{1}{2}\text{ rad}$$

The curve is thus shifted $\frac{1}{2}$ rad to the right as shown in Fig. 17–18.

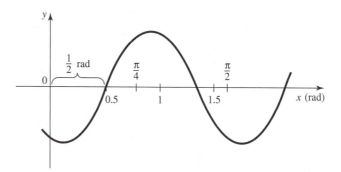

FIGURE 17–18   ◆◆◆

The formulas are summarized as follows:

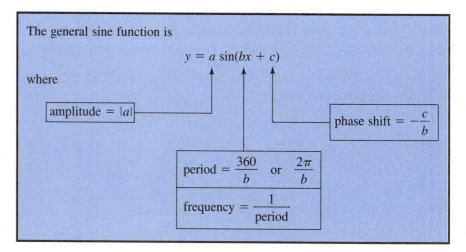

The general sine function is

$$y = a\sin(bx + c)$$

where

amplitude $= |a|$

phase shift $= -\dfrac{c}{b}$

$$\text{period} = \frac{360}{b} \quad \text{or} \quad \frac{2\pi}{b}$$

$$\text{frequency} = \frac{1}{\text{period}}$$

In our next example we reverse the procedure. We show how to write the equation of the curve, given the amplitude, period, and phase shift.

◆◆◆ **Example 14:** Write the equation of a sine curve in which $a = 5$, with a period of π, and a phase shift of -0.5 rad.

Solution: The period is given as π, so

$$P = \pi = \frac{2\pi}{b}$$

so

$$b = \frac{2\pi}{\pi} = 2$$

Then

$$\text{phase shift} = -0.5 = -\frac{c}{b}$$

Since $b = 2$, we get

$$c = -2(-0.5) = 1$$

Substituting into the general equation of the sine function

$$y = a \sin(bx + c)$$

with $a = 5$, $b = 2$, and $c = 1$ gives

$$y = 5 \sin(2x + 1) \qquad \text{◆◆◆}$$

The graphics calculator can, of course, easily graph the general sine function. It can also be used to illustrate the effect of phase shift, as in the following example.

◆◆◆ **Example 15:** Use the graphics calculator to graph the following in the same viewing window:

$$y = 2 \sin 3x \quad \text{and} \quad y = 2 \sin(3x + 45°)$$

for x from $-30°$ to $135°$.

Solution: We put the calculator in $\boxed{\text{DEGREE}}$ mode and set the range.

$$\text{Xmin} = -30 \qquad \text{Xmax} = 135 \qquad \text{Xscl} = 45$$
$$\text{Ymin} = -2.5 \qquad \text{Ymax} = 2.5 \qquad \text{Yscl} = 1$$

We enter the equations, press $\boxed{\text{GRAPH}}$, and get the display shown in Figure 17–19. Notice that $-c/b = -45/3 = -15°$, so that the curve $y = 2 \sin(3x + 45°)$ is shifted $15°$ to the left.

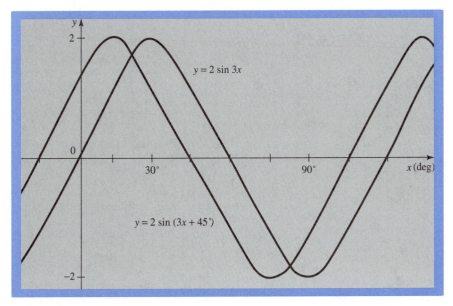

FIGURE 17–19   ◆◆◆

The graphing calculator makes it easy to graph sine waves of various frequencies and phase shifts when the constants a, b, and c may be approximate numbers, as in the following example.

◆◆◆ **Example 16:** Use the graphics calculator to make a complete graph of

$$y = -3.24 \sin(1.22x - 0.526)$$

in radians. Then use the $\boxed{\text{TRACE}}$ and $\boxed{\text{ZOOM}}$ features to find the coordinates of the intercepts and the maximum and minimum points to three significant digits.

Solution: We put the calculator in $\boxed{\text{RADIAN}}$ mode and enter the equation. We note that the amplitude is 3.24, so let us set the Ymax to 4 and Ymin to -4, with Yscl equal to 1. We can set the X range either by first calculating the period or by using trial and error with the $\boxed{\text{ZOOM}}$ feature. Either way we get a complete graph, with a little to spare, with the X range set approximately to the following values:

$$\text{Xmin} = -3 \qquad \text{Xmax} = 4 \qquad \text{Xscl} = 1$$
$$\text{Ymin} = -4 \qquad \text{Ymax} = 4 \qquad \text{Yscl} = 1$$

The coordinates of the points of interest are shown in Fig. 17–20.

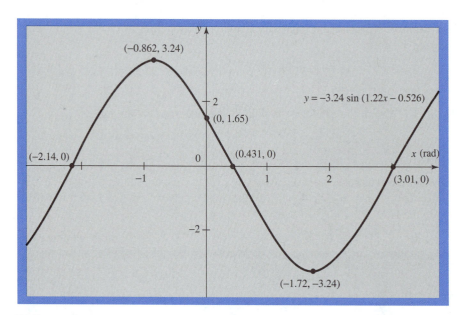

FIGURE 17–20

◆◆◆

Exercise 2 ◆ The General Sine Wave

Graph each sine function. Find the amplitude, period, and phase shift.

1. $y = \sin 2x$

2. $y = \sin \dfrac{x}{3}$

3. $y = 0.5 \sin 2x$

4. $y = 100 \sin \dfrac{3x}{2}$

5. $y = 2 \sin 3x$

6. $y = \sin \pi x$

7. $y = \sin 3\pi x$

8. $y = -3 \sin \pi x$

9. $y = 4 \sin(x - 180°)$

10. $y = -2 \sin(x + 90°)$

11. $y = \sin(x + 90°)$

12. $y = 20 \sin(2x - 45°)$

13. $y = \sin(x - 1)$

14. $y = 0.5 \sin\left(\dfrac{x}{2} - 1\right)$

15. $y = \sin\left(x + \dfrac{\pi}{2}\right)$

16. $y = \sin\left(x + \dfrac{\pi}{8}\right)$

17. Write the equations of the sine curves in Fig. 17–21.

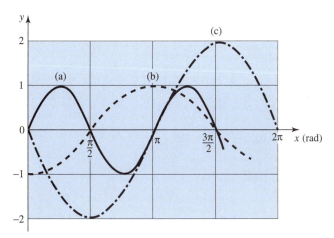

FIGURE 17–21

18. Write the equation of a sine curve with $a = 3$, a period of 4π, and a phase shift of zero.

19. Write the equation of a sine curve with $a = -2$, a period of 6π, and a phase shift of $-\pi/4$.

Graphics Calculator or Computer

Use a graphics calculator or a graphics program on the computer to plot any of the equations in this exercise. Also use it to graph the following:

20. $y = 2.63 \sin(1.63x - 0.823)$

21. $y = 45.3 \sin(4.22x + 0.372)$

22. $y = -5.21 \sin(0.234x + 1.35)$

23. $y = 3.82 \sin(2.33x + 24.6°)$

24. $y = 25.6 \sin(4.42x - 45.2°)$

25. $y = -61.3 \sin(3.27x - 58.2°)$

26. The projectile shown in Fig. 17–22 is fired with an initial velocity v at an angle θ with the horizontal. Its range is given by

$$x = \frac{v^2}{g} \sin 2\theta$$

where g is the acceleration due to gravity.

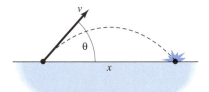

FIGURE 17–22 A projectile.

(a) Graph the range of the projectile for $\theta = 0$ to $90°$. Take $v = 100$ ft/s and $g = 32.2$ ft/s^2.

(b) What angle gives a maximum range?

27. For the pendulum of Fig. 17–12:

   (a) Write an equation for the horizontal displacement x as a function of θ, taking $\theta = 0$ when the pendulum is 15° from the vertical.

   (b) Graph $x = f(\theta)$ for $\theta = 0$ to 30°. Take θ as increasing in the counterclockwise direction.

28. The vertical distance x for cog A in Fig. 17–23 is given by $x = r \sin \theta$. Write an expression for the vertical distance to cog B.

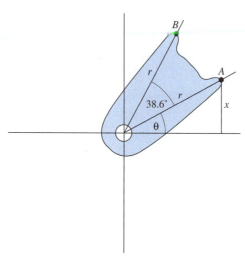

FIGURE 17–23

29. In the mechanism shown in Fig. 17–24, the rotating arm A lifts the follower F vertically.

   (a) Write an expression for the vertical displacement y as a function of θ. Take $\theta = 0$ in the extreme clockwise position shown.

   (b) Graph $y = f(\theta)$ for $\theta = 0$ to 90°.

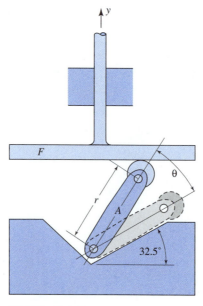

FIGURE 17–24

17-3 Quick Sketching of the Sine Curve

As shown in the preceding examples, we can plot any curve by computing and plotting a set of point pairs. But a faster way to get a sketch is to draw a rectangle of width P and height $2a$, and then sketch the sine curve (whose *shape* does not vary) within that box. First determine the amplitude, period, and phase shift. Then do the following:

1. Draw two horizontal lines each at a distance $|a|$ from the x axis.
2. Draw a vertical line at a distance P from the origin.
3. Subdivide the period P into four equal parts. Label the x axis at these points, and draw vertical lines through them.
4. Lightly sketch in the sine curve.
5. Shift the curve by the amount of the phase shift.

◆◆◆ **Example 17:** Make a quick sketch of $y = 2 \sin(3x + 60°)$.

Solution: We have $a = 2$, $b = 3$, and $c = 60°$. From Eq. 211, the period is $P = 360°/3 = 120°$ and, by Eq. 213, phase shift $= -60°/3 = -20°$. The steps for sketching the curve are shown in Fig. 17–25.

(a) Draw horizontals 2 units above and below x axis

(b) Draw a vertical at a distance P (120°) from the origin

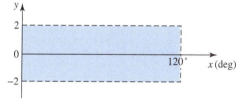

(c) Divide the period into 4 equal parts

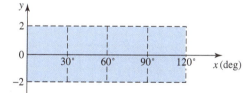

(d) Sketch the curve lightly

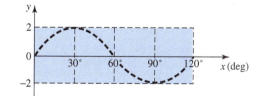

(e) Shift the curve by an amount of the phase shift (20°), right for a positive phase shift and left for a negative one

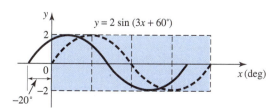

FIGURE 17–25 Quick sketching of the sine curve.

We now do an example where the units are radians.

◆◆◆ **Example 18:** Make a quick sketch of

$$y = 1.5 \sin(5x - 2)$$

Solution: From the given equation,

$$a = 1.5, \qquad b = 5, \quad \text{and} \quad c = -2 \text{ rad}$$

◆◆◆

so

$$P = \frac{2\pi}{5} = 1.26 \text{ rad}$$

and

$$\text{phase shift} = -\left(\frac{-2}{5}\right) = 0.4 \text{ rad}$$

As shown in Fig. 17–26(a), we draw a rectangle whose height is $2 \times 1.5 = 3$ units and whose width is 1.26 rad; then we draw three verticals spaced apart by $1.26/4 = 0.315$ rad. We sketch a sine wave [shown dashed in Fig. 17–26(b)] into this rectangle, and then shift it 0.4 rad to the right to get the final curve as shown in Fig. 17–26(b).

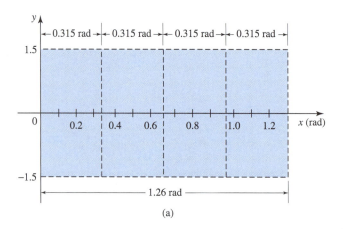

(a)

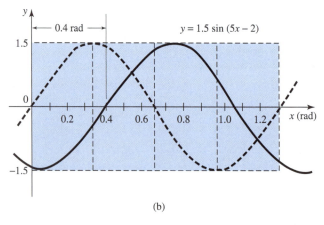

(b)

FIGURE 17–26

◆◆◆

Exercise 3 ◆ Quick Sketching of the Sine Curve

Make a quick sketch of any of the functions in Exercise 1 or 2, as directed by your instructor.

17–4 Graphs of More Trigonometric Functions

Graph of the Cosine Function $y = \cos x$

Let us graph one cycle of the function $y = \cos x$. We make a table of point pairs and plot each point. The graph of the cosine function is shown in Fig. 17–27.

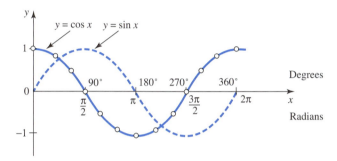

x	0	$\frac{\pi}{6}$	$\frac{\pi}{3}$	$\frac{\pi}{2}$	$\frac{2\pi}{3}$	$\frac{5\pi}{6}$	π	$\frac{7\pi}{6}$	$\frac{4\pi}{3}$	$\frac{3\pi}{2}$	$\frac{5\pi}{3}$	$\frac{11\pi}{6}$	2π
y	1	0.866	0.5	0	−0.5	−0.866	−1	−0.866	−0.5	0	0.5	0.866	1

FIGURE 17–27 $y = \cos x$.

Cosine and Sine Curves Related

Note in Fig. 17–27 that the cosine curve and the sine curve, shown dashed, have the same shape. In fact, the cosine curve appears to be identical to a sine curve shifted 90° to the left, or

$$\cos \theta = \sin(\theta + 90°) \qquad (1)$$

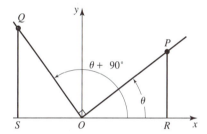

FIGURE 17–28

We can show that Equation (1) is true. We lay out the two angles θ and $\theta + 90°$ (Fig. 17–28), choose points P and Q so that $OP = OQ$, and drop perpendiculars PR and QS to the x axis. Since triangles OPR and OQS are congruent, we have $OR = QS$. The cosine of θ is then

$$\cos \theta = \frac{OR}{OP} = \frac{QS}{OQ} = \sin(\theta + 90°)$$

which verifies (1).

> Since the sine and cosine curves are identical except for phase shift, we can use either one to describe periodically varying quantities.

Graph of the General Cosine Function $y = a \cos(bx + c)$

The equation for the general cosine function is

$$y = a \cos(bx + c)$$

The amplitude, period, frequency, and phase shift are found the same way as for the sine curve, and, of course, the quick plotting method works here, too.

◆◆◆ **Example 19:** Find the amplitude, period, frequency, and phase shift for the curve $y = 3 \cos(4x − 120°)$, and make a graph.

Solution: From the equation, a = amplitude = 3. Since $b = 4$, the period is

$$\text{period} = \frac{360°}{4} = 90°/\text{cycle}$$

and

$$\text{frequency} = \frac{1}{P} = 0.0111 \text{ cycle/degree}$$

Since $c = -120°$,

$$\text{phase shift} = -\frac{c}{b} = -\frac{-120°}{4} = 30°$$

So we expect the curve to be shifted 30° to the right. For a quick plot we draw a rectangle with height twice the amplitude and a length of one period as shown in Fig. 17–29. We sketch in the cosine curve, shown dashed, and then shift it 30° to the right to get the final curve, shown solid.

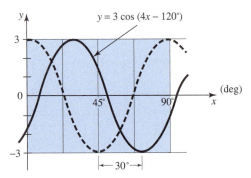

FIGURE 17–29   ◆◆◆

Graph of the Tangent Function $y = \tan x$

We now plot the function $y = \tan x$. As usual, we make a table of point pairs.

x	0	30°	60°	90°	120°	150°	180°	210°	240°	270°	300°	330°	360°
y	0	0.577	1.73		−1.73	−0.577	0	0.577	1.73		−1.73	−0.577	0

The tangent is not defined at 90° (or 270°), as we saw in Sec. 15–1. But for angles slightly less than 90° (or 270°), the tangent is large and grows larger the closer the angle gets to 90° (for example, $\tan 89.9° = 573$, $\tan 89.99° = 5730$, etc.). The same thing happens for angles slightly larger than 90° (or 270°) except that the tangent is now negative ($\tan 90.1° = -573$, $\tan 90.01° = -5730$, etc.). With this information, we graph the tangent curve in Fig. 17–30.

The terms *period* and *frequency* have the same meaning as before, but *amplitude* is not used in connection with the tangent function, as the curve extends to infinity in the positive and negative directions.

The vertical lines at $x = 90°$ and 270° that the tangent graph approaches but never touches are called *asymptotes*.

As with the sine and cosine functions, the tangent function may also take the form

$$y = a \tan(bx + c)$$

where the constants, a, b, and c affect the shape, period, and phase angle of the function.

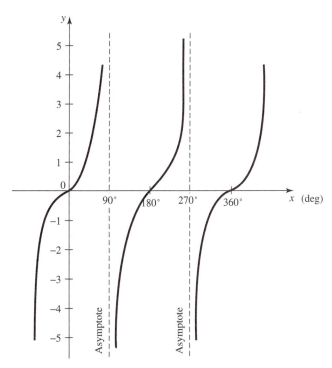

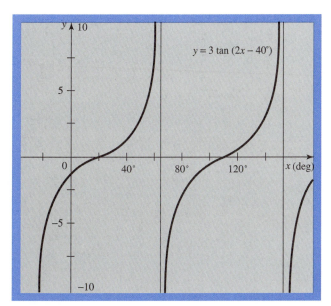

FIGURE 17–30

◆◆◆ **Example 20:** Use a graphics calculator to graph two cycles of the function

$$y = 3 \tan(2x - 40°)$$

Solution: We put the calculator into DEGREE mode and enter the equation. By trial and error we quickly find that the range settings that give us two cycles are approximately

$$\text{Xmin} = -30 \qquad \text{Xmax} = 170 \qquad \text{Xscl} = 20$$
$$\text{Ymin} = -10 \qquad \text{Ymax} = 10 \qquad \text{Yscl} = 1$$

We get the graph shown in Fig. 17–31.

FIGURE 17–31 ◆◆◆

Graphs of Cotangent, Secant, and Cosecant Functions

For completeness, the graphs of all six trigonometric functions are shown in Fig. 17–32. To make them easier to compare, the same horizontal scale is used for each curve. Only the sine and cosine curves find much use in technology. Notice that of the six, they are the only curves that have no "gaps" or "breaks." They are called *continuous* curves, while the others are called *discontinuous*.

If the need should arise to graph a cotangent, secant, or cosecant function, simply make a table of point pairs and plot them, or use a graphics calculator or graphing utility on a computer. With the calculator or the computer, you will need to use the reciprocal relations.

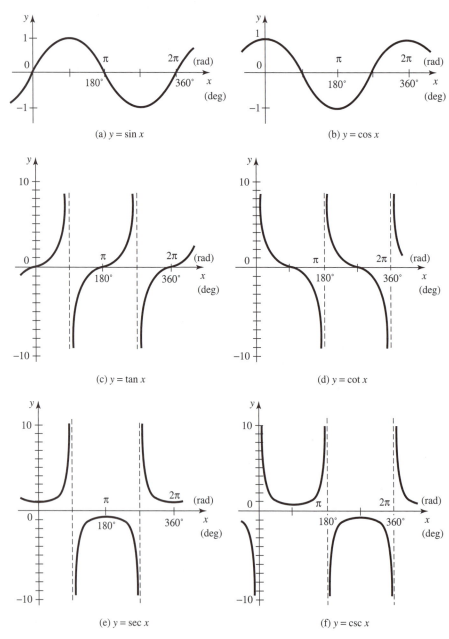

FIGURE 17–32   Graphs of the six trigonometric functions.

♦♦♦ **Example 21:** To graph the function

$$y = 3 \sec(2x - 1)$$

for $x = -2$ to 3 radians, we set the calculator in $\boxed{\text{RADIAN}}$ mode and set the x range.

$$\text{Xmin} = -2 \qquad \text{Xmax} = 3 \qquad \text{Xscl} = .5$$

We enter the function

$$\text{Y1} = 3/\cos(2x - 1)$$

and press $\boxed{\text{GRAPH}}$. We find a suitable y scale by trial and error. Some values that work well are

$$\text{Ymin} = -10 \qquad \text{Ymax} = 10 \qquad \text{Yscl} = 2.5$$

The resulting graph is shown in Fig. 17–33.

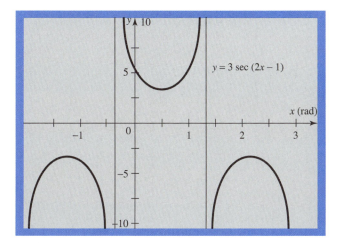

FIGURE 17–33   ◆◆◆

Composite Curves

One way to graph a function containing several terms is to graph each term separately and then add (or subtract) them on the graph paper. This method, called the method of *addition of ordinates,* is especially useful when one or more terms of the expression to be graphed are trigonometric functions.

◆◆◆ **Example 22:** Graph the function $y = 8/x + \cos x$, where x is in radians, from $x = 0$ to 12 rad.

Solution: We separately graph $y_1 = 8/x$ and $y_2 = \cos x$, shown dashed in Fig. 17–34. We add the ordinates graphically to get the composite curve, shown as a solid line.

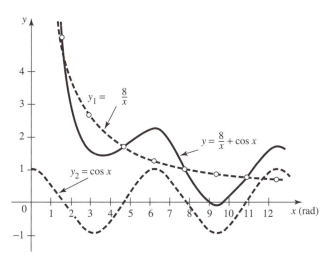

FIGURE 17–34   Graphing by addition of ordinates.   ◆◆◆

Exercise 4 ◆ Graphs of More Trigonometric Functions

The Cosine Curve

Make a complete graph of each function. Find the amplitude, period, and phase shift.

1. $y = 3 \cos x$ **2.** $y = -2 \cos x$ **3.** $y = \cos 3x$

4. $y = \cos 2x$ **5.** $y = 2 \cos 3x$ **6.** $y = 3 \cos 2x$

7. $y = \cos(x - 1)$ **8.** $y = \cos(x + 2)$ **9.** $y = 3 \cos\left(x - \dfrac{\pi}{4}\right)$

10. $y = 2 \cos(3x + 1)$

The Tangent Curve

Make a complete graph of each function.

11. $y = 2 \tan x$ **12.** $y = \tan 4x$ **13.** $y = 3 \tan 2x$

14. $y = \tan\left(x - \dfrac{\pi}{2}\right)$ **15.** $y = 2 \tan(3x - 2)$ **16.** $y = 4 \tan\left(x + \dfrac{\pi}{6}\right)$

Cotangent, Secant, and Cosecant Curves

Make a complete graph of each function.

17. $y = 2 \cot 2x$ **18.** $y = 3 \sec 4x$ **19.** $y = \csc 3x$

20. $y = \cot(x - 1)$ **21.** $y = 2 \sec(3x + 1)$ **22.** $y = \csc(2x + 0.5)$

Composite Curves

Graph by addition of ordinates (x is in radians).

23. $y = x + \sin x$ **24.** $y = \cos x - x$

25. $y = \dfrac{2}{x} - \sin x$ **26.** $y = 2x + 5 \cos x$

27. $y = \sin x + \cos 2x$ **28.** $y = \sin x + \sqrt{x}$

29. $y = 2 \sin 2x - \cos 3x$ **30.** $y = 3 \cos 3x + 2 \sin 4x$

31. For the pendulum of Fig. 17–12:
  (a) Write an equation for the vertical distance y as a function of θ.
  (b) Graph $y = f(\theta)$ for $\theta = 0$ to 30°.
32. Repeat problem 31 taking $\theta = 0$ when the pendulum is 20° from the vertical and increasing counterclockwise.
33. A rocket is rising vertically, as shown in Fig. 17–35, and is tracked by a radar dish R.
  (a) Write an expression for the altitude y of the rocket as a function of θ.
  (b) Graph $y = f(\theta)$ for $\theta = 0$ to 60°.
34. The ship in Fig. 17–36 travels in a straight line while keeping its searchlight trained on the reef.
  (a) Write an expression for the distance d as a function of the angle θ.
  (b) Graph $d = f(\theta)$ for $\theta = 0$ to 90°.

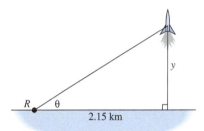

FIGURE 17–35   Tracking a rocket.

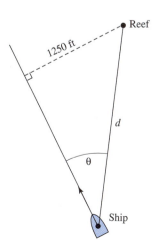

FIGURE 17–36

35. If a weight W is dragged along a surface with a coefficient of friction f, as shown in Fig. 17–37, the force needed is

$$F = \frac{fW}{f \sin \theta + \cos \theta}$$

Graph F as a function of θ for $\theta = 0$ to $90°$. Take $f = 0.55$ and $W = 5.35$ kg.

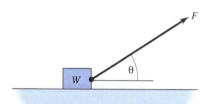

FIGURE 17–37   A weight dragged along a surface.

Graphics Calculator or Computer

Use a graphics calculator or a graphing program on a computer to graph any of the above functions. Also use it to make a complete graph of the following.

36. $y = 3.72 \cos(1.39x + 0.836)$
37. $y = 23.1 \tan(2.24x + 35.2°)$
38. $y = 0.723 \cot(3.82x + 12.2°)$
39. $y = 4.82 \sec(4.19x + 1.36)$

17–5   The Sine Wave as a Function of Time

The sine curve can be generated in a simple geometric way. Figure 17–38 shows a vector OP rotating with a constant angular velocity ω. A rotating vector is called a *phasor*. Its angular velocity ω is almost always given in radians per second (rad/s).

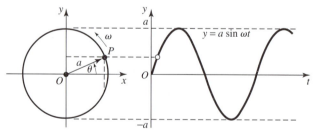

FIGURE 17–38   The sine curve generated by a rotating vector.

If the length of the phasor is a, then its projection on the y axis is

$$y = a \sin \theta$$

But since by Eq. A26 the angle θ at any instant t is equal to ωt,

$$y = a \sin \omega t$$

Further, if the phasor does not start from the x axis but has some *phase angle* ϕ, we get

$$y = a \sin(\omega t + \phi)$$

Notice in this equation that y *is a function of time,* rather than of an angle.

We'll soon see that a sine wave that is a function of time is more appropriate for representing alternating current than a sine wave that is a function of θ.

◆◆◆ **Example 23:** Write the equation for a sine wave generated by a phasor of length 8 rotating with an angular velocity of 300 rad/s and with a phase angle of 0.

Solution: Substituting, with $a = 8$, $\omega = 300$ rad/s, and $\phi = 0$, we have

$$y = 8 \sin 300t$$

◆◆◆

Period

Recall that the *period* was defined as the distance along the x axis taken by one cycle of the waveform. When the units on the x axis represented an angle, the period was in radians or degrees. Now that the x axis shows time, the period will be in seconds. From Eq. 211,

$$P = \frac{2\pi}{b}$$

But in the equation $y = a \sin \omega t$, $b = \omega$, so we have the following formula:

Period of a Sine Wave	$P = \dfrac{2\pi}{\omega}$	A77

where P is in seconds and ω is in rad/s.

◆◆◆

◆◆◆ **Example 24:** Find the period and the amplitude of the sine wave $y = 6 \sin 10t$, and make a graph with time t shown on the horizontal axis.

Solution: The period, from Eq. A77, is

$$P = \frac{2\pi}{\omega} = \frac{2(3.142)}{10} = 0.628 \text{ s}$$

Thus it takes 628 ms for a full cycle, 314 ms for a half-cycle, 157 ms for a quarter-cycle, and so forth. This sine wave, with amplitude 6, is plotted in Fig. 17–39. ◆◆◆

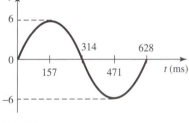

FIGURE 17–39

Common Error	Don't confuse the function $$y = \sin bx$$ with the function $$y = \sin \omega t$$ In the first case, y is a function of an angle x, and b is a coefficient that has no units. In the second case, y is a function of time t, and ω is an angular velocity in radians per second.

Frequency

From Eq. 212 we see that the frequency f of a periodic waveform is equal to the reciprocal of the period P. So the frequency of a sine wave is given by the following equation:

Frequency of a Sine Wave	$f = \dfrac{1}{P} = \dfrac{\omega}{2\pi}$ (hertz)	A78

where P is in seconds and ω is in rad/s.

The unit of frequency is therefore cycles/s, or *hertz* (Hz).

$$1 \text{ Hz} = 1 \text{ cycle/s}$$

Higher frequencies are often expressed in kilohertz (kHz), where

$$1 \text{ kHz} = 10^3 \text{ Hz}$$

or in megahertz (MHz), where

$$1 \text{ MHz} = 10^6 \text{ Hz}$$

◆◆◆ **Example 25:** The frequency of the sine wave of Example 24 is

$$f = \frac{1}{P} = \frac{1}{0.628} = 1.59 \text{ Hz}$$

◆◆◆

When the period P is not wanted, the angular velocity can be obtained directly from Eq. A78 by noting that $\omega = 2\pi f$.

◆◆◆ **Example 26:** Find the angular velocity of a 1000-Hz source.

Solution: From Eq. A78,

$$\omega = 2\pi f = 2(3.14)(1000) = 6280 \text{ rad/s}$$

◆◆◆

A sine wave as a function of time can also have a phase shift, expressed in either degrees or radians, as in the following example.

◆◆◆ **Example 27:** Given the sine wave $y = 5.83 \sin(114t + 15°)$, find (a) the amplitude, (b) the angular velocity, (c) the period, (d) the frequency, and (e) the phase angle; and (f) make a graph.

Solution:
(a) The amplitude is 5.83 units.
(b) The angular velocity is $\omega = 114$ rad/s.
(c) From Eq. A77, the period is

$$P = \frac{2\pi}{\omega} = \frac{2\pi}{114} = 0.0551 \text{ s} = 55.1 \text{ ms}$$

(d) From Eq. A78, the frequency is

$$f = \frac{1}{P} = \frac{1}{0.0551} = 18.1 \text{ Hz}$$

(e) The phase angle is 15°. It is not unusual to see a sine wave given with ωt in radians and the phase angle in degrees. We find the phase shift, in units of time, the same way that we found the phase shift for the general sine function in Sec. 17–1. We find the value of t at which the positive half-cycle starts. As before, we set y equal to zero and solve for t.

Even though it is often done, some people object to mixing degrees and radians in the same equation. If you do it, work with extra care.

$$y = \sin(\omega t + \phi) = 0$$
$$\omega t = -\phi$$
$$t = -\frac{\phi}{\omega} = \text{phase shift}$$

Substituting our values for ω and ϕ (which we first convert to 0.262 radian) gives

$$\text{phase shift} = -\frac{0.262 \text{ rad}}{114 \text{ rad/s}} = -0.00230 \text{ s}$$

So our curve is shifted 2.30 milliseconds to the left.

(f) This sine wave is graphed in Fig. 17–40. We draw a rectangle of height 2(5.83) and width 55.1 ms. We subdivide the rectangle into four rectangles of equal width and sketch in the sine wave, shown dashed. We then shift the sine wave 2.30 ms to the left to get the final (solid) curve.

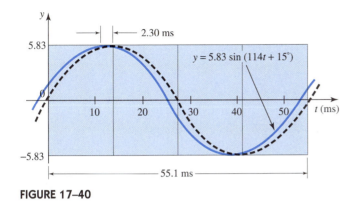

FIGURE 17–40   ◆◆◆

Common Error	It's easy to get confused as to which direction to shift the curve. Check your work by substituting the value of an x intercept (such as -0.00230 s in Example 27) back into the original equation. You should get a value of zero for y.

Adding Sine Waves of the Same Frequency

We have seen that $A \sin \omega t$ is the y component of a phasor of magnitude A rotating at angular velocity ω. Similarly, $B \sin(\omega t + \theta)$ is the y component of a phasor of magnitude B rotating at the same angular velocity ω, but with a phase angle ϕ between A and B. Since each is the y component of a phasor, their sum is equal to the sum of the y components of the two phasors, in other words, simply the y component of the resultant of those phasors.

Thus to add two sine waves of the same frequency, we simply *find the resultant of the phasors representing those sine waves.*

◆◆◆ **Example 28:** Express $y = 2.00 \sin \omega t + 3.00 \sin(\omega t + 60°)$ as a single sine wave.

Solution: We represent $2.00 \sin \omega t$ by a phasor of length 2.00, and we represent $3.00 \sin(\omega t + 60°)$ as a phasor of length 3.00, at an angle of 60°. These phasors are shown in Fig. 17–41. We wish to add $y = 2.00 \sin \omega t$ and $y = 3.00 \sin(\omega t + 60°)$. By the law of cosines,

$$R = \sqrt{2.00^2 + 3.00^2 - 2(2.00)(3.00) \cos 120°} = 4.36$$

Then, by the law of sines,

$$\frac{\sin \phi}{3.00} = \frac{\sin 120°}{4.36}$$

$$\sin \phi = 0.596$$

$$\phi = 36.6°$$

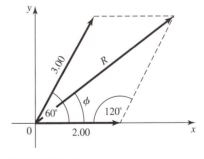

FIGURE 17–41

Thus

$$y = 2.00 \sin \omega t + 3.00 \sin(\omega t + 60°)$$

$$= 4.36 \sin(\omega t + 36.6°)$$

Figure 17–42 shows the two original sine waves as well as a graph of 4.36 sin(ωt + 36.6°). Note that the sum of the ordinates of the original waves, at any value of ωt, is equal to the ordinate of the final wave.

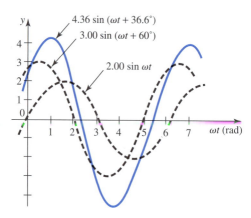

FIGURE 17–42   ◆◆◆

Use a graphics calculator or a computer to graph the original function and the final one in Example 28 on the same axes. The two curves should be identical.

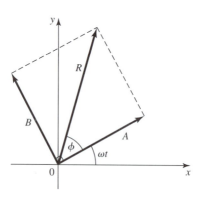

FIGURE 17–43

In Sec. 18–2 we will derive this same equation using the trigonometric identities.

Adding Sine and Cosine Waves of the Same Frequency

In Sec. 17–2 we verified that

$$\cos \omega t = \sin(\omega t + 90°)$$

and saw that a cosine wave $A \cos \omega t$ is identical to the sine wave $A \sin \omega t$, except for a phase difference of 90° between the two curves (see Fig. 17–27). Thus we can find the sum ($A \sin \omega t + B \cos \omega t$) of a sine and a cosine wave the same way that we added two sine waves in the preceding section: by finding the resultant of the phasors representing each sine wave. Since the angle between the two phasors is 90° (Fig. 17–43), the resultant has a magnitude

$$R = \sqrt{A^2 + B^2}$$

and is at an angle

$$\phi = \arctan \frac{B}{A}$$

So we have the following equation:

Addition of a Sine Wave and a Cosine Wave	$A \sin \omega t + B \cos \omega t = R \sin(\omega t + \phi)$ where $R = \sqrt{A^2 + B^2}$ and $\phi = \arctan \dfrac{B}{A}$	214

Thus we can represent the sum of a sine and a cosine function by a single sine function.

◆◆◆ **Example 29:** Express $y = 274 \sin \omega t + 371 \cos \omega t$ as a single sine function.

Solution: The magnitude of the resultant is

$$R = \sqrt{(274)^2 + (371)^2} = 461$$

and the phase angle is

$$\phi = \arctan \frac{371}{274} = \arctan 1.354 = 53.6°$$

Note that you can use the rectangular-to-polar conversion keys on your calculator to do this calculation.

So

$$y = 274 \sin \omega t + 371 \cos \omega t$$
$$= 461 \sin(\omega t + 53.6°)$$

Figure 17–44 shows the original sine and cosine functions and a graph of $y = 461 \sin(\omega t + 53.6°)$. Note that the sum of the ordinates of the original waves equals the ordinate of the computed wave at any abscissa.

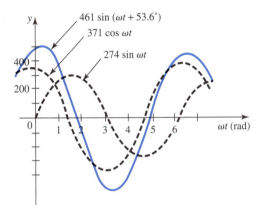

FIGURE 17–44   ◆◆◆

Alternating Current

When a loop of wire rotates in a magnetic field, any portion of the wire cuts the field while traveling first in one direction and then in the other direction. Since the polarity of the voltage induced in the wire depends on the direction in which the field is cut, the induced current will travel first in one direction and then in the other: It *alternates*.

The same is true with an armature more complex than a simple loop. We get an *alternating current*. Just as the rotating point P in Fig. 17–38 generated a sine wave, the current induced in the rotating armature will be sinusoidal.

The sinusoidal voltage or current can be described by the equation

$$y = a \sin(\omega t + \phi)$$

or, in electrical terms,

$$i = I_m \sin(\omega t + \phi_2) \tag{A76}$$

where i is the current at time t, I_m the maximum current (the amplitude), and ϕ_2 the phase angle. Also, if we let V_m stand for the amplitude of the voltage wave, the instantaneous voltage v becomes

$$v = V_m \sin(\omega t + \phi_1) \tag{A75}$$

We have given the phase angle ϕ subscripts because, in a given circuit, the current and voltage waves will usually have different phase angles.

Equations A77 and A78 still apply here.

$$\text{period} = \frac{2\pi}{\omega} \quad \text{(s)}$$

$$\text{frequency} = \frac{\omega}{2\pi} \quad \text{(Hz)}$$

◆◆◆

♦♦♦ **Example 30:** Utilities in the United States supply alternating current at a frequency of 60 Hz. Find the angular velocity ω and the period P.

Solution: By Eq. A78,

$$P = \frac{1}{f} = \frac{1}{60} = 0.0167 \text{ s}$$

and, by Eq. A77,

$$\omega = \frac{2\pi}{P} = \frac{2\pi}{0.0167} = 377 \text{ rad/s}$$

These are good numbers to remember.   ♦♦♦

♦♦♦ **Example 31:** A certain alternating current has an amplitude of 1.5 A and a frequency of 60 Hz (cycles/s). Taking the phase angle as zero, write the equation for the current as a function of time, find the period, and find the current at $t = 0.01$ s.

Solution: From Example 30,

$$P = 0.0167 \text{ s} \quad \text{and} \quad \omega = 377 \text{ rad/s}$$

so the equation is

$$i = 1.5 \sin(377t)$$

When $t = 0.01$ s,

$$i = 1.5 \sin(377)(0.01) = -0.882 \text{ A}$$

as shown in Fig. 17–45.

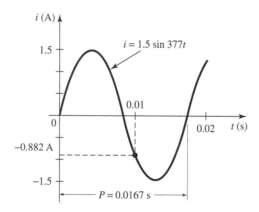

FIGURE 17–45   ♦♦♦

Phase Shift

When writing the equation of a single alternating voltage or current, we are usually free to choose the origin anywhere along the time axis. So we can place it to make the phase angle equal to zero. However, when there are *two* curves on the same graph that are *out of phase,* we usually locate the origin so that the phase angle of one curve is zero.

If the difference in phase between two sine waves is t seconds, then we often say that one curve *leads* the other by t seconds. Conversely, we could say that one curve *lags* the other by t seconds.

◆◆◆ **Example 32:** Figure 17–46 shows voltage and current waves, with the voltage wave leading a current wave by 5 ms. Write the equations for the two waves.

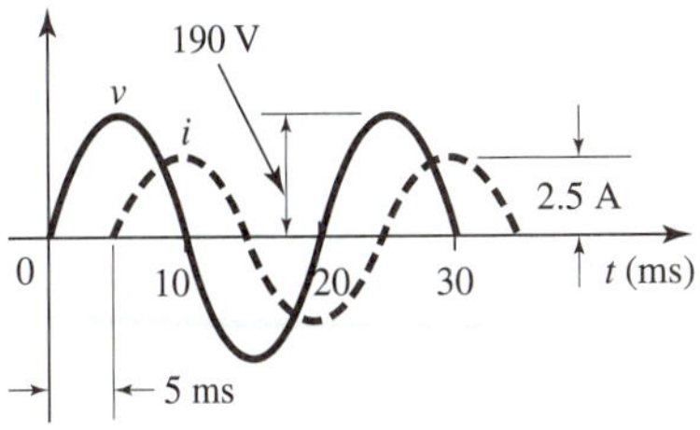

Solution: For the voltage wave,

$$V_m = 190, \qquad \phi = 0, \quad \text{and} \quad P = 0.02 \text{ s}$$

By Eq. A77,

$$\omega = \frac{2\pi}{0.02} = 100\pi = 314 \text{ rad/s}$$

so

$$v = 190 \sin 314t$$

where v is expressed in volts. For the current wave, $I_m = 2.5$ A. Since the curve is shifted to the right by 5 ms,

$$\text{phase shift} = 0.005 \text{ s} = -\frac{\phi}{\omega}$$

so the phase angle ϕ is

$$\phi = -(0.005 \text{ s})(100\pi \text{ rad/s}) = -\frac{\pi}{2} \text{ rad} = -90°$$

The angular frequency is the same as before, so

$$i = 2.5 \sin(314t - 90°)$$

where i is expressed in amperes. With ϕ in radians,

$$i = 2.5 \sin\left(314t - \frac{\pi}{2}\right)$$

where i is expressed in amperes. ◆◆◆

FIGURE 17–46

Exercise 5 ◆ The Sine Wave as a Function of Time

Find the period and the angular velocity of a repeating waveform that has a frequency of:

1. 68 Hz.   **2.** 10 Hz.   **3.** 5000 Hz.

Find the frequency (in hertz) and the angular velocity of a repeating waveform whose period is:

4. 1 s.   **5.** $\frac{1}{8}$ s.   **6.** 95 ms.

7. If a periodic waveform has a frequency of 60 Hz, how many seconds will it take to complete 200 cycles?

8. Find the frequency in Hz for a wave that completes 150 cycles in 10 s.

Find the period and the frequency of a sine wave that has an angular velocity of:

9. 455 rad/s.   **10.** 2.58 rad/s.   **11.** 500 rad/s.

Find the period, amplitude, and phase angle for:

12. the sine wave shown in Fig. 17–47(a).

13. the sine wave shown in Fig. 17–47(b).

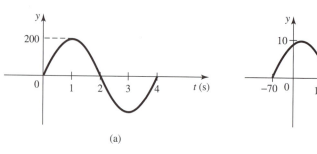

(a)   (b)

FIGURE 17–47

14. Write an equation for a sine wave generated by a phasor of length 5 rotating with an angular velocity of 750 rad/s and with a phase angle of 0°.

15. Repeat problem 14 with a phase angle of 15°.

Plot each sine wave.

16. $y = \sin t$

17. $y = 3 \sin 377t$

18. $y = 54 \sin(83t - 20°)$

19. $y = 375 \sin\left(55t + \dfrac{\pi}{4}\right)$

Adding Sine Waves and Cosine Waves

Express as a single sine wave.

20. $y = 3.00 \sin \omega t + 5.50 \sin(\omega t + 30°)$
21. $y = 18.3 \sin \omega t + 26.3 \sin(\omega t + 75°)$
22. $y = 2.47 \sin \omega t + 1.83 \sin(\omega t - 24°)$
23. $y = 384 \sin \omega t + 536 \sin(\omega t - 48°)$
24. $y = 8370 \sin \omega t + 7570 \cos \omega t$
25. $y = 7.37 \cos \omega t + 5.83 \sin \omega t$
26. $y = 74.2 \sin \omega t + 69.3 \cos \omega t$
27. $y = 364 \sin \omega t + 550 \cos \omega t$
28. The weight Fig. 17–48 is pulled down 2.50 in. and released. The distance x is given by

$$x = 2.50 \cos \omega t$$

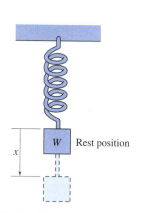

FIGURE 17–48 Weight hanging from a spring.

Graph x as a function of t for two complete cycles. Take $\omega = 42.5$ rad/s.

29. The arm in the Scotch crosshead of Fig. 17–11 is rotating at a rate of 2.55 rev/s.
(a) Write an equation for the displacement y as a function of time.
(b) Graph that equation for two complete cycles.

Alternating Current

30. An alternating current has the equation

$$i = 25 \sin(635t - 18°)$$

where i is given in amperes, A. Find the maximum current, period, frequency, phase angle, and the instantaneous current at $t = 0.01$ s.

31. Given an alternating voltage $v = 4.27 \sin(463t + 27°)$, find the maximum voltage, period, frequency, and phase angle, and the instantaneous voltage at $t = 0.12$ s.

32. Write an equation of an alternating voltage that has a peak value of 155 V, a frequency of 60 Hz, and a phase angle of 22°.

33. Write an equation for an alternating current that has a peak amplitude of 49.2 mA, a frequency of 35 Hz, and a phase angle of 63.2°.

17–6 Polar Coordinates

The Polar Coordinate System

The *polar coordinate system* (Fig. 17–49) consists of a *polar axis*, passing through point O, which is called the *pole*. The location of a point P is given by its distance r from the pole, called the *radius vector*, and by the angle θ, called the *polar angle* (sometimes called the *vectorial* angle or *reference* angle). The polar angle is called *positive* when measured counterclockwise from the polar axis, and *negative* when measured clockwise.

The *polar coordinates* of a point P are thus r and θ, usually written in the form $P(r, \theta)$, or as r/θ (read "r at an angle of θ").

♦♦♦ **Example 33:** A point at a distance 5 from the origin with a polar angle of 28° can be written

$$P(5, 28°) \quad \text{or} \quad 5\underline{/28°} \qquad ♦♦♦$$

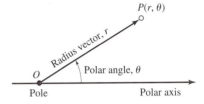

FIGURE 17–49 Polar coordinates.

Most of our graphing will continue to be in rectangular coordinates, but in some cases polar coordinates will be more convenient.

Plotting Points in Polar Coordinates

For plotting in polar coordinates, it is convenient, although not essential, to have *polar coordinate paper* (Fig. 17–50). This paper has concentric circles, equally spaced, and an angular scale in degrees or radians.

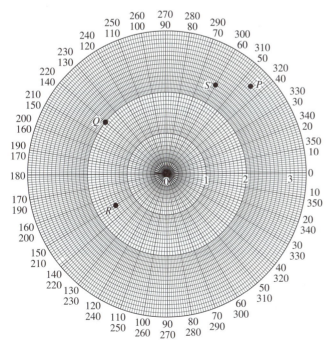

FIGURE 17–50

◆◆◆ **Example 34:** The points $P(3, 45°)$, $Q(-2, 320°)$, $R(1.5, 7\pi/6)$, and $S(2.5, -5\pi/3)$ are plotted in Fig. 17–50.

Notice that to plot $Q(-2, 320°)$, which has a negative value for r, we first locate $(2, 320°)$ and then find the corresponding point along the diameter and in the opposite quadrant. ◆◆◆

Graphing Equations in Polar Coordinates

To graph a function $r = f(\theta)$, simply assign convenient values to θ and compute the corresponding value for r. Then plot the resulting table of point pairs.

◆◆◆ **Example 35:** Graph the function $r = \cos \theta$.

Solution: Let us take values for θ every 30° and make a table.

θ	0	30°	60°	90°	120°	150°	180°	210°	240°	270°	300°	330°	360°
r	1	0.87	0.5	0	−0.5	−0.87	−1	−0.87	−0.5	0	0.5	0.87	1

Plotting these points, we get a circle as shown in Fig. 17–51.

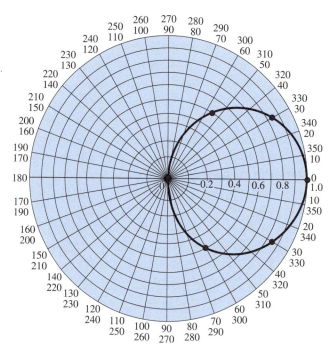

FIGURE 17–51 Graph of $r = \cos \theta$. The polar equation of a circle that passes through the pole and has its center at (c, α) is $r = 2c \cos(\theta - \alpha)$. ◆◆◆

In Chapters 5 and 14, we graphed a parabola in rectangular coordinates. We now graph an equation in polar coordinates that also results in a parabola.

◆◆◆ **Example 36:** Graph the parabola

$$r = \frac{p}{1 - \cos \theta}$$

with $p = 1$.

Solution: As before, we compute r for selected values of θ.

θ	0°	30°	60°	90°	120°	150°	180°	210°	240°	270°	300°	330°	360°
r		7.46	2.00	1.00	0.667	0.536	0.500	0.536	0.667	1.00	2.00	7.46	

Note that we get division by zero at $\theta = 0°$ and 360°, so that the curve does not exist there. Plotting these points (except for 7.46, which is off the graph), we get the parabola shown in Fig. 17–52. ◆◆◆

Some of the more interesting curves in polar coordinates are shown in Fig. 17–53.

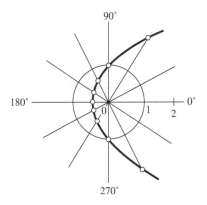

FIGURE 17–52

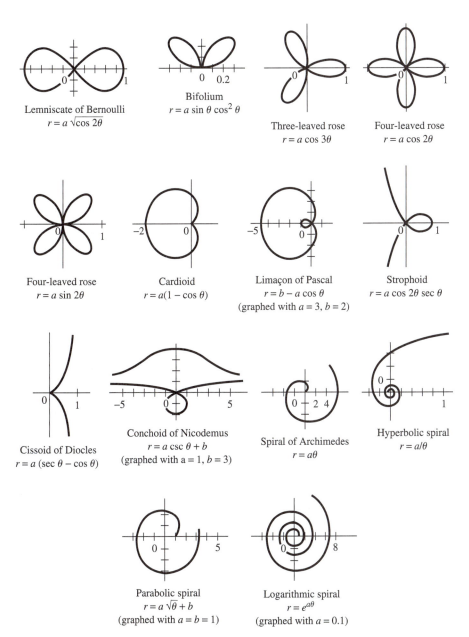

Lemniscate of Bernoulli
$r = a\sqrt{\cos 2\theta}$

Bifolium
$r = a \sin \theta \cos^2 \theta$

Three-leaved rose
$r = a \cos 3\theta$

Four-leaved rose
$r = a \cos 2\theta$

Four-leaved rose
$r = a \sin 2\theta$

Cardioid
$r = a(1 - \cos \theta)$

Limaçon of Pascal
$r = b - a \cos \theta$
(graphed with $a = 3$, $b = 2$)

Strophoid
$r = a \cos 2\theta \sec \theta$

Cissoid of Diocles
$r = a (\sec \theta - \cos \theta)$

Conchoid of Nicodemus
$r = a \csc \theta + b$
(graphed with a = 1, b = 3)

Spiral of Archimedes
$r = a\theta$

Hyperbolic spiral
$r = a/\theta$

Parabolic spiral
$r = a \sqrt{\theta} + b$
(graphed with $a = b = 1$)

Logarithmic spiral
$r = e^{a\theta}$
(graphed with $a = 0.1$)

FIGURE 17–53 Some curves in polar coordinates. Unless otherwise noted, the curves were graphed with $a = 1$.

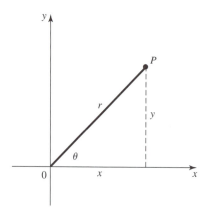

FIGURE 17–54   Rectangular and polar coordinates of a point.

Some calculators have special keys for converting between polar and rectangular coordinates. Check your manual.

Transforming between Polar and Rectangular Coordinates

We can easily see the relationships between rectangular coordinates and polar coordinates when we put both systems on a single diagram (Fig. 17–54). Using the trigonometric functions and the Pythagorean theorem, we get the following equations:

Rectangular	$x = r \cos \theta$	158
	$y = r \sin \theta$	159
Polar	$r = \sqrt{x^2 + y^2}$	160
	$\theta = \arctan \dfrac{y}{x}$	161

◆◆◆ Example 37: What are the polar coordinates of $P(3, 4)$?

Solution:

$$r = \sqrt{9 + 16} = 5$$

$$\theta = \arctan \frac{4}{3} = 53.1°$$

So the polar coordinates of P are $(5, 53.1°)$.   ◆◆◆

◆◆◆ Example 38: The polar coordinates of a point are $(8, 125°)$. What are the rectangular coordinates?

Solution:

$$x = 8 \cos 125° = -4.59$$
$$y = 8 \sin 125° = 6.55$$

So the rectangular coordinates are $(-4.59, 6.55)$.   ◆◆◆

We may also use Eqs. 158 through 161 to transform *equations* from one system of coordinates to the other.

◆◆◆ Example 39: Transform the polar equation $r = \cos \theta$ to rectangular coordinates.

Solution: Multiplying both sides by r yields

$$r^2 = r \cos \theta$$

Then, by Eqs. 158 and 160,

$$x^2 + y^2 = x$$ ◆◆◆

◆◆◆ Example 40: Transform the rectangular equation $2x + 3y = 5$ into polar form.

Solution: By Eqs. 158 and 160,

$$2r \cos \theta + 3r \sin \theta = 5$$

or

$$r(2 \cos \theta + 3 \sin \theta) = 5$$

$$r = \frac{5}{2 \cos \theta + 3 \sin \theta}$$

Exercise 6 ◆ Polar Coordinates

Plot each point in polar coordinates.

1. $(4, 35°)$ **2.** $(3, 120°)$ **3.** $(2.5, 215°)$

4. $(3.8, 345°)$ **5.** $\left(2.7, \dfrac{\pi}{6}\right)$ **6.** $\left(3.9, \dfrac{7\pi}{8}\right)$

7. $\left(-3, \dfrac{\pi}{2}\right)$ **8.** $\left(4.2, \dfrac{2\pi}{5}\right)$ **9.** $(3.6, -20°)$

10. $(-2.5, -35°)$ **11.** $\left(-1.8, -\dfrac{\pi}{6}\right)$ **12.** $\left(-3.7, -\dfrac{3\pi}{5}\right)$

Graph the following curves from Fig. 17–53.

13. The lemniscate of Bernoulli   **14.** The bifolium
15. The three-leaved rose   **16.** The four-leaved rose, $r = a \cos 2\theta$
17. The cardioid   **18.** The limaçon of Pascal
19. The strophoid   **20.** The cissoid of Diocles
21. The conchoid of Nicodemus   **22.** The spiral of Archimedes
23. The hyperbolic spiral   **24.** The parabolic spiral
25. The logarithmic spiral

Graph each function in polar coordinates.

26. $r = 2 \cos \theta$ **27.** $r = 3 \sin \theta$
28. $r = 3 \sin \theta + 3$ **29.** $r = 2 \cos \theta - 1$
30. $r = 3 \cos 2\theta$ **31.** $r = 2 \sin 3\theta$
32. $r = \sin 2\theta - 1$ **33.** $r = 2 + \cos 3\theta$

Write the polar coordinates of each point.

34. $(2.00, 5.00)$ **35.** $(3.00, 6.00)$
36. $(1.00, 4.00)$ **37.** $(4.00, 3.00)$
38. $(2.70, -1.80)$ **39.** $(-4.80, -5.90)$
40. $(207, 186)$ **41.** $(-312, -509)$
42. $(1.08, -2.15)$

Write the rectangular coordinates of each point.

43. $(5.00, 47.0°)$ **44.** $(6.30, 227°)$

45. $(445, 312°)$ **46.** $\left(3.60, \dfrac{\pi}{5}\right)$

47. $\left(-4.00, \dfrac{3\pi}{4}\right)$

48. $\left(18.3, \dfrac{2\pi}{3}\right)$

49. $(15.0, -35.0°)$

50. $(-12.0, -48.0°)$

51. $\left(-9.80, -\dfrac{\pi}{5}\right)$

Write each polar equation in rectangular form.

52. $r = 6$ **53.** $r = 2 \sin\theta$
54. $r = \sec\theta$ **55.** $r^2 = 1 - \tan\theta$
56. $r(1 - \cos\theta) = 1$ **57.** $r^2 = 4 - r\cos\theta$

Write each rectangular equation in polar form.

58. $x = 2$ **59.** $y = -3$
60. $x = 3 - 4y$ **61.** $x^2 + y^2 = 1$
62. $3x - 2y = 1$ **63.** $y = x^2$

Graphing Calculator and Computer

Some graphics calculators, and some computer software such as *Derive,* allow the plotting of polar equations. If you are able, use your calculator or computer to plot any of the polar equations in this exercise set. Also plot the following, taking a large enough range for θ to give a complete graph.

Some calculators require that a polar equation first be transformed into a pair of parametric equations before they can be graphed. We show how to do this in the next section.

64. $r = 1.42 \cos 2\theta$
65. $r = 2.53 \sin 2\theta + 4.23$
66. $r = 1.55 \sin^2 2\theta$
67. $r = 2.35 \sin^2\theta + 2.94 \cos^2\theta$

17–7   Graphing Parametric Equations

Parametric Equations with Trigonometric Functions

In Chapter 5 we discussed parametric equations of the form

$$x = g(t)$$

and

$$y = h(t)$$

where x and y are each expressed in terms of a third variable t, called the *parameter.* There we graphed parametric equations in which the $h(t)$ and $g(t)$ were *algebraic* functions; here we graph some parametric equations in which the functions are *trigonometric.*

The procedure is the same. We assign values to the parameter (which will now be θ) and compute x and y for each θ. The table of (x, y) pairs is then plotted in rectangular coordinates.

◆◆◆ **Example 41:** Plot the parametric equations

$$x = 3 \cos \theta$$

$$y = 2 \sin 2\theta$$

Solution: We select values for θ and compute x and y.

θ	0	$\dfrac{\pi}{4}$	$\dfrac{\pi}{2}$	$\dfrac{3\pi}{4}$	π	$\dfrac{5\pi}{4}$	$\dfrac{3\pi}{2}$	$\dfrac{7\pi}{4}$	2π
x	3	2.12	0	-2.12	-3	-2.12	0	2.12	3
y	0	2	0	-2	0	2	0	-2	0

Each (x, y) pair is plotted in Fig. 17–55.

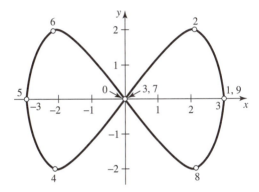

FIGURE 17–55   Patterns of this sort can be obtained by applying ac voltages to both the horizontal and the vertical deflection plates of an oscilloscope. Called *Lissajous figures,* they can indicate the relative amplitudes, frequencies, and phase angles of the two applied voltages.   ◆◆◆

The graphs of some common parametric equations in which the functions are trigonometric are given in Fig. 17–56.

Note that these parametric equations are graphed in *rectangular coordinates,* not in polar coordinates.

Converting a Polar Equation to Parametric Form

Some graphics calculators cannot graph a polar equation directly. It is first necessary to convert the polar equation to a pair of parametric equations and then plot those.

To convert the polar equation

$$r = f(\theta)$$

to the parametric equations

$$x = g(\theta) \quad \text{and} \quad y = h(\theta)$$

we use Eqs. 158 and 159. Thus

$$x = r \cos \theta = f(\theta) \cos \theta$$

and

$$y = r \sin \theta = f(\theta) \sin \theta$$

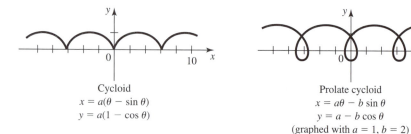

Cycloid
$$x = a(\theta - \sin \theta)$$
$$y = a(1 - \cos \theta)$$

Prolate cycloid
$$x = a\theta - b \sin \theta$$
$$y = a - b \cos \theta$$
(graphed with $a = 1, b = 2$)

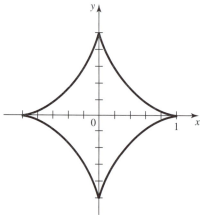

Hypocycloid of four cusps
or astroid
$$x = a \cos^3 \theta$$
$$y = a \sin^3 \theta$$

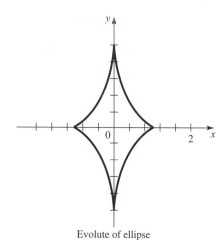

Evolute of ellipse
$$x = a \cos^3 \theta$$
$$y = b \sin^3 \theta$$
(graphed with $a = 1, b = 2$)

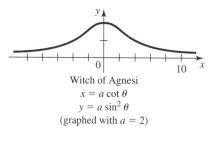

Witch of Agnesi
$$x = a \cot \theta$$
$$y = a \sin^2 \theta$$
(graphed with $a = 2$)

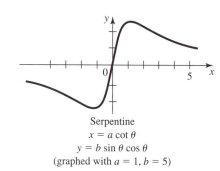

Serpentine
$$x = a \cot \theta$$
$$y = b \sin \theta \cos \theta$$
(graphed with $a = 1, b = 5$)

FIGURE 17–56   Graphs of some parametric equations. Unless otherwise noted, the curves were graphed with $a = 1$.

◆◆◆ **Example 42:** Write the polar equation $r = 3 \sin 2\theta$ in parametric form, and plot using a graphics calculator.

Solution: Here, $f(\theta) = 3 \sin 2\theta$, so

$$x = f(\theta) \cos \theta = 3 \sin 2\theta \cos \theta$$

and

$$y = f(\theta) \sin \theta = 3 \sin 2\theta \sin \theta$$

We put the calculator into $\boxed{\text{RADIAN}}$ and $\boxed{\text{PARAMETRIC}}$ modes. We then enter the equations (using T instead of θ).

$$\text{Xlt} = 3 \sin 2\text{T} \cos \text{T}$$
$$\text{Ylt} = 3 \sin 2\text{T} \sin \text{T}$$

We set the range.

$$\text{T:} \quad 0 \text{ to } 2\pi$$
$$\text{X:} \quad -4 \text{ to } 4$$
$$\text{Y:} \quad -4 \text{ to } 4$$

and get the four-leaved rose shown in Fig. 17–53.   ◆◆◆

Exercise 7 ◆ Graphing Parametric Equations

Graph each curve from Fig. 17–56.

1. The cycloid
2. The prolate cycloid
3. The hypocycloid of four cusps
4. The evolute of ellipse
5. The witch of Agnesi
6. The serpentine

Graph each pair of parametric equations.

7. $x = \sin \theta$
$y = \sin \theta$

8. $x = \sin \theta$
$y = 2 \sin \theta$

9. $x = \sin \theta$
$y = \sin 2\theta$

10. $x = \sin 2\theta$
$y = \sin 3\theta$

11. $x = \sin \theta$
$y = \sin(\theta + \pi/4)$

12. $x = \sin \theta$
$y = \sin(2\theta - \pi/6)$

Rewrite each polar equation in parametric form, and graph.

13. $r = 3 \cos \theta$
14. $r = 2 \sin \theta$
15. $r = 2 \cos 3\theta$
16. $r = 3 \sin 4\theta$
17. $r = 3 \sin 3\theta - 2$
18. $r = 4 + \cos 2\theta$

Graphing Calculator or Computer

Use your graphing calculator or graphing program on a computer to graph any of the parametric equations in this exercise set. Also use it to graph the following. Take a large enough range for θ to get a complete graph.

19. $x = \sin 2\theta$
$y = \sin 3\theta$

20. $x = \sin \theta$
$y = 1.32 \sin 2\theta$

21. $x = 5.83 \sin 2\theta$
$y = 4.24 \sin \theta$

22. $x = 6.27 \sin \theta$
$y = 4.83 \sin(\theta + 0.527)$

◆◆◆ CHAPTER 17 REVIEW PROBLEMS ◆◆◆◆◆◆◆◆◆◆◆◆◆◆◆◆◆◆◆◆◆◆◆◆◆◆◆◆

Graph one cycle of each curve and find the period, amplitude, and phase shift.

1. $y = 3 \sin 2x$

2. $y = 5 \cos 3x$

3. $y = 1.5 \sin\left(3x + \dfrac{\pi}{2}\right)$

4. $y = -5 \cos\left(x - \dfrac{\pi}{6}\right)$

5. $y = 2.5 \sin\left(4x + \dfrac{2\pi}{9}\right)$

6. $y = 4 \tan x$

7. Write the equation of a sine curve with $a = 5$, a period of 3π, and a phase shift of $-\frac{\pi}{6}$.

Graph by addition of ordinates.

8. $y = \sin x + \cos 2x$

9. $y = x + \cos x$

10. $y = 2 \sin x - \dfrac{1}{x}$

Plot each point in polar coordinates.

11. $(3.4, 125°)$

12. $(-2.5, 228°)$

13. $\left(1.7, -\dfrac{\pi}{6}\right)$

Graph each equation in polar coordinates.

14. $r^2 = \cos 2\theta$

15. $r + 2 \cos \theta = 1$

Write the polar coordinates of each point.

16. $(7, 3)$

17. $(-5.30, 3.80)$

18. $(-24, -52)$

Write the rectangular coordinates of each point.

19. $(3.80, 48.0°)$

20. $\left(-65, \dfrac{\pi}{9}\right)$

21. $(3.80, -44.0°)$

Transform into rectangular form.

22. $r(1 - \cos \theta) = 2$

23. $r = 2 \cos \theta$

Transform into polar form.

24. $x - 3y = 2$

25. $5x + 2y = 1$

26. Find the frequency and the angular velocity of a sine wave that has a period of 2.5 s.

27. Find the period and the angular velocity of a cosine wave that has a frequency of 120 Hz.

28. Find the period and the frequency of a sine wave that has an angular velocity of 44.8 rad/s.

29. If a sine wave has a frequency of 30 Hz, how many seconds will it take to complete 100 cycles?

30. What frequency must a sine wave have in order to complete 500 cycles in 2 s?

31. Graph the parametric equations $x = 2 \sin \theta$, $y = 3 \sin 4\theta$.

32. Write an equation for an alternating current that has a peak amplitude of 92.6 mA, a frequency of 82 Hz, and a phase angle of 28.3°.

33. Write each of the following as a single sine wave:
   (a) $y = 63.7 \sin \omega t + 42.9 \sin(\omega t - 38°)$
   (b) $y = 1.73 \sin \omega t + 2.64 \cos \omega t$

34. Given an alternating voltage $v = 27.4 \sin(736t + 37°)$, find the maximum voltage, period, frequency, phase angle, and the instantaneous voltage at $t = 0.25$ s.

Convert to parametric form, and plot.

35. $r = 2 \cos 3\theta$ **36.** $r = 3 \sin 4\theta$
37. Graph: $y = 2.5 \sin(2x + 32°)$ **38.** Graph: $y = 1.5 \sin(3x - 41°)$

39. Graph: $y = -3.6 \sin\left(x + \dfrac{\pi}{3}\right)$ **40.** Graph: $y = 12.4 \sin\left(2x - \dfrac{\pi}{4}\right)$

Writing

41. Given the general sine wave $y = a \sin(bx + c)$, explain in your own words how a change in one of the constants (a, b, or c) will affect the appearance of the sine wave.

42. *Doppler effect:* As a train approaches, the sound of its whistle seems higher than when the train recedes. This is called the *Doppler effect.* Read about and write a short paper on the Doppler effect, including an explanation of why it occurs.

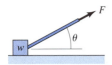

FIGURE 17–57

Team Projects

43. A force F (Fig. 17–57) pulls the weight along a horizontal surface. If f is the coefficient of friction, then

$$F = \frac{fW}{f \sin \theta + \cos \theta}$$

Graphically find the value of θ, to two significant digits, so that the least force is required. Do this for values of f of 0.50, 0.60, and 0.70.

44. *Beats:* Beats occur when two sine waves of slightly different frequencies are combined.
(a) Read about beats, and write a short paper on the subject.
(b) Graph two sine waves, having frequencies of 24 Hz and 30 Hz, for about 20 cycles. Then graph the sum of those two sine waves.
(c) From the above graphs, what can you deduce about the frequency of the beats?
(d) Try to produce beats, using a musical instrument, tuning forks, or some other device.
(e) Name one practical use for beats.

45. Graph the function that it is the sum of the following four functions:

$$y = \sin x$$
$$y = (1/3) \sin 3x$$
$$y = (1/5) \sin 5x$$
$$y = (1/7) \sin 7x$$

Use a graphics calculator or computer or manual addition of ordinates on graph paper.
    What can you say about the shape of the composite curve? What do you think would happen if you added in more sine waves [$y = (1/9) \sin 9x$ and so on]?

18

Trigonometric Identities and Equations

◆◆◆ OBJECTIVES ◆◆

When you have completed this chapter, you should be able to:

- Write a trigonometric expression in terms of the sine and cosine.
- Simplify a trigonometric expression using the fundamental identities.
- Prove trigonometric identities using the fundamental identities.
- Simplify expressions or prove identities using the sum or difference formulas, the double-angle formulas, or the half-angle formulas.
- Solve trigonometric equations.
- Add a sine wave and a cosine wave of the same frequency.
- Evaluate inverse trigonometric functions.

◆◆

In Sec. 3–1 we said that an *identity* is an equation that is true for *all* meaningful values of the variables, as contrasted with a *conditional* equation, which is true only for certain values of the variable. Thus $(x + 2)(x - 2) = x^2 - 4$ is true for every value of x. A *trigonometric identity* is one that contains one or more trigonometric functions. We have already seen that $\sin \theta = 1/\csc \theta$ for any angle θ (except, of course, for values of θ for which $\csc \theta = 0$).

We derive some new identities in this chapter, which we can then use to simplify trigonometric expressions, to verify other identities, and to aid in the solution of trigonometric equations.

18–1   Fundamental Identities

Reciprocal Relations

We have already encountered the reciprocal relations in Sec. 7–2, and we repeat them here.

Reciprocal Relations		
$\sin \theta = \dfrac{1}{\csc \theta}$ or $\csc \theta = \dfrac{1}{\sin \theta}$ or $\sin \theta \csc \theta = 1$	**152a**
$\cos \theta = \dfrac{1}{\sec \theta}$ or $\sec \theta = \dfrac{1}{\cos \theta}$ or $\cos \theta \sec \theta = 1$	**152b**
$\tan \theta = \dfrac{1}{\cot \theta}$ or $\cot \theta = \dfrac{1}{\tan \theta}$ or $\tan \theta \cot \theta = 1$	**152c**

◆◆◆ **Example 1:** Simplify $\dfrac{\cos \theta}{\sec^2 \theta}$.

Solution: Using Eq. 152b gives us

$$\frac{\cos \theta}{\sec^2 \theta} = \cos \theta (\cos^2 \theta) = \cos^3 \theta$$

◆◆◆

Quotient Relations

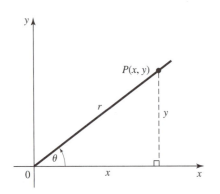

FIGURE 18–1   Angle in standard position.

Figure 18–1 shows an angle θ in standard position, as when we first defined the trigonometric functions in Sec. 7–2. We see that

$$\sin \theta = \frac{y}{r} \quad \text{and} \quad \cos \theta = \frac{x}{r}$$

Dividing yields

$$\frac{\sin \theta}{\cos \theta} = \frac{\dfrac{y}{r}}{\dfrac{x}{r}} = \frac{y}{x}$$

But $y/x = \tan \theta$, so we have the following equations:

Quotient Relations		
$\tan \theta = \dfrac{\sin \theta}{\cos \theta}$	**162**
$\cot \theta = \dfrac{\cos \theta}{\sin \theta}$	**163**

where $\cot \theta$ is found by taking the reciprocal of $\tan \theta$.

◆◆◆ **Example 2:** Rewrite the expression

$$\frac{\cot \theta}{\csc \theta} - \frac{\tan \theta}{\sec \theta}$$

so that it contains only the sine and cosine, and simplify.

Solution:

$$\frac{\cot \theta}{\csc \theta} - \frac{\tan \theta}{\sec \theta} = \frac{\cos \theta}{\sin \theta} \cdot \frac{\sin \theta}{1} - \frac{\sin \theta}{\cos \theta} \cdot \frac{\cos \theta}{1}$$

$$= \cos \theta - \sin \theta \qquad \text{◆◆◆}$$

Pythagorean Relations

We can get three more relations by applying the Pythagorean theorem to the triangle in Figure 18–1.

$$x^2 + y^2 = r^2$$

Dividing through by r^2, we get

$$\frac{x^2}{r^2} + \frac{y^2}{r^2} = 1$$

or

$$\left(\frac{x}{r}\right)^2 + \left(\frac{y}{r}\right)^2 = 1$$

But $x/r = \cos \theta$ and $y/r = \sin \theta$, so the following equation applies:

Pythagorean Relation	$\sin^2 \theta + \cos^2 \theta = 1$	164

Recall that $\sin^2 \theta$ is a short way of writing $(\sin \theta)^2$.

Returning to equation $x^2 + y^2 = r^2$, we now divide through by x^2.

$$1 + \frac{y^2}{x^2} = \frac{r^2}{x^2}$$

But $y/x = \tan \theta$ and $r/x = \sec \theta$, so we have the following:

Pythagorean Relation	$1 + \tan^2 \theta = \sec^2 \theta$	165

Now dividing $x^2 + y^2 = r^2$ by y^2,

$$\frac{x^2}{y^2} + 1 = \frac{r^2}{y^2}$$

But $x/y = \cot \theta$ and $r/y = \csc \theta$. Thus:

Pythagorean Relation	$1 + \cot^2 \theta = \csc^2 \theta$	166

◆◆◆ Example 3: Simplify

$$\sin^2 \theta - \csc^2 \theta - \tan^2 \theta + \cot^2 \theta + \cos^2 \theta + \sec^2 \theta$$

Solution: By the Pythagorean relations,

$$\sin^2 \theta - \csc^2 \theta - \tan^2 \theta + \cot^2 \theta + \cos^2 \theta + \sec^2 \theta$$

$$= (\sin^2 \theta + \cos^2 \theta) - (\tan^2 \theta - \sec^2 \theta) + (\cot^2 \theta - \csc^2 \theta)$$

$$= 1 - (-1) + (-1) = 1 \qquad \text{◆◆◆}$$

Simplification of Trigonometric Expressions

One use of the trigonometric identities is the simplification of expressions, as in the preceding example. We now give a few more examples.

◆◆◆ **Example 4:** Simplify $(\cot^2 \theta + 1)(\sec^2 \theta - 1)$.

Solution: By Eqs. 165 and 166,

$$(\cot^2 \theta + 1)(\sec^2 \theta - 1) = \csc^2 \theta \tan^2 \theta$$

By Eqs. 152a and 162,

$$= \frac{1}{\sin^2 \theta} \cdot \frac{\sin^2 \theta}{\cos^2 \theta}$$

$$= \frac{1}{\cos^2 \theta}$$

and, by Eq. 152b,

$$= \sec^2 \theta \qquad\qquad ◆◆◆$$

◆◆◆ **Example 5:** Simplify $\dfrac{1 - \sin^2 \theta}{\sin \theta + 1}$.

Solution: Factoring the difference of two squares in the numerator gives

$$\frac{1 - \sin^2 \theta}{\sin \theta + 1} = \frac{(1 - \sin \theta)(1 + \sin \theta)}{\sin \theta + 1} = 1 - \sin \theta \qquad\qquad ◆◆◆$$

◆◆◆ **Example 6:** Simplify $\dfrac{\cos \theta}{1 - \sin \theta} + \dfrac{\sin \theta - 1}{\cos \theta}$.

Solution: Combining the two fractions over a common denominator, we have

$$\frac{\cos \theta}{1 - \sin \theta} + \frac{\sin \theta - 1}{\cos \theta} = \frac{\cos^2 \theta + (\sin \theta - 1)(1 - \sin \theta)}{(1 - \sin \theta) \cos \theta}$$

$$= \frac{\cos^2 \theta + \sin \theta - \sin^2 \theta - 1 + \sin \theta}{(1 - \sin \theta) \cos \theta}$$

Students tend to forget the rules of *algebraic operations* when working trigonometric expressions. We still need to factor, to combine fractions over a common denominator, and so on.

From Eq. 164,

$$= \frac{-\sin^2 \theta - \sin^2 \theta + 2 \sin \theta}{(1 - \sin \theta) \cos \theta}$$

$$= \frac{-2 \sin^2 \theta + 2 \sin \theta}{(1 - \sin \theta) \cos \theta}$$

Factor the numerator:

$$= \frac{2 \sin \theta(-\sin \theta + 1)}{(1 - \sin \theta) \cos \theta}$$

$$= \frac{2 \sin \theta}{\cos \theta}$$

By Eq. 162,

$$= 2 \tan \theta \qquad\qquad ◆◆◆$$

Trigonometric Identities

We know that an *identity* is an equation that is true for any values of the unknowns, and we have presented eight trigonometric identities so far in this chapter. We now use these eight identities to *verify* or *prove* whether a given identity is true. We do this by trying to transform one side of the identity (usually the more complicated side) until it is identical to the other side.

A good way to start is to rewrite each side so that it contains only sines and cosines.

◆◆◆ **Example 7:** Prove the identity $\sec\theta\csc\theta = \tan\theta + \cot\theta$.

Solution: Expressing the given identity in terms of sines and cosines gives

$$\sec\theta\csc\theta = \tan\theta + \cot\theta$$

$$\frac{1}{\cos\theta}\cdot\frac{1}{\sin\theta} = \frac{\sin\theta}{\cos\theta} + \frac{\cos\theta}{\sin\theta}$$

If we are to make one side identical to the other, each must finally have the same number of terms. But the left side now has one term and the right has two, so we combine the two fractions on the right over a common denominator.

$$\frac{1}{\sin\theta\cos\theta} = \frac{\sin^2\theta + \cos^2\theta}{\sin\theta\cos\theta}$$

By Eq. 164,

$$\frac{1}{\sin\theta\cos\theta} = \frac{1}{\sin\theta\cos\theta}$$

◆◆◆

> Proving identities is *not easy.* It takes a good knowledge of the fundamental identities, lots of practice, and often several false starts. Do not be discouraged if you don't get them right away.

Common Error	Each side of an identity must be worked *separately;* we cannot transpose, or multiply both sides by the same thing, the way we do with equations. The reason is that we do not yet know if the two sides are in fact equal; that is what we are trying to prove.

◆◆◆ **Example 8:** Prove the identity $\csc x + \tan x = \dfrac{\cot x + \sin x}{\cos x}$.

Solution: As in Example 7, we have a single term on one side of the equals sign and two terms on the other. Now, however, it may be easier to split the single term into two, rather than combine the two terms into one, as we did before.

$$\csc x + \tan x = \frac{\cot x}{\cos x} + \frac{\sin x}{\cos x}$$

Switching to sines and cosines yields

> Some computer algebra systems such as *Derive* are able to simplify, expand, and otherwise manipulate trigonometric expressions.

$$\frac{1}{\sin x} + \frac{\sin x}{\cos x} = \frac{\dfrac{\cos x}{\sin x}}{\cos x} + \frac{\sin x}{\cos x}$$

$$= \frac{1}{\sin x} + \frac{\sin x}{\cos x}$$

◆◆◆

Using the Graphics Calculator for Trigonometric Identities

We can use a graphics calculator or a graphing utility on the computer to plot each side of a trigonometric identity. If the two graphs are not identical, we can be sure that we do not have an identity. If the two graphs appear to be identical, we probably have an identity, although this does not constitute a proof.

◆◆◆ **Example 9:** Graphically verify the identity

$$\tan x \sec x = \sin x$$

Solution: In the same viewing window we graph

$$Y_1 = \tan x (1/\cos x)$$

and

$$Y_2 = \sin x$$

The two graphs, shown in Fig. 18–2, are different, so we do not have an identity.

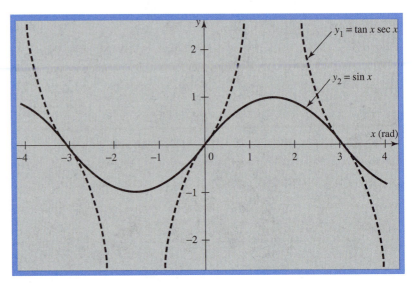

FIGURE 18–2

◆◆◆

◆◆◆ **Example 10:** Graphically verify the identity

$$\sin^2 x + \cos^2 x = 1$$

Solution: We graph

$$y = \sin^2 x + \cos^2 x$$

As shown in Fig. 18–3, we get a horizontal line with a slope of 0 and a y-intercept of 1, in other words, the graph of $y = 1$. Therefore, we have strongly confirmed, but not proven, that our given relation is an identity.

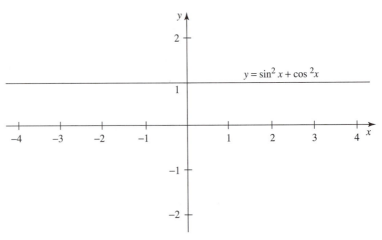

FIGURE 18–3   ◆◆◆

◆◆◆ **Example 11:** Graphically verify the identity

$$\cot x = \cot x \sec^2 x - \tan x$$

Solution: We graph

$$Y_1 = 1/\tan x$$

and

$$Y_2 = (1/\tan x)(1/\cos x)^2 - \tan x$$

in the same viewing window, as shown in Fig. 18–4. We get a single curve again, strongly indicating, but not proving, that our given relation is an identity.

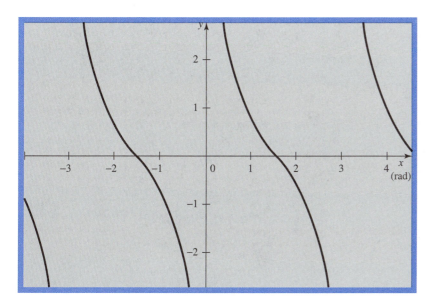

FIGURE 18–4   ◆◆◆

Exercise 1 ◆ Fundamental Identities

Change to an expression containing only sin and cos.

1. $\tan x - \sec x$

2. $\cot x + \csc x$

3. $\tan \theta \csc \theta$

4. $\sec \theta - \tan \theta \sin \theta$

5. $\dfrac{\tan \theta}{\csc \theta} + \dfrac{\sin \theta}{\tan \theta}$

6. $\cot x + \tan x$

Simplify.

7. $1 - \sec^2 x$

8. $\dfrac{\csc \theta}{\sin \theta}$

9. $\dfrac{\cos \theta}{\cot \theta}$

10. $\sin \theta \csc \theta$

11. $\tan \theta \csc \theta$

12. $\dfrac{\sin \theta}{\csc \theta}$

13. $\sec x \sin x$

14. $\sec x \sin x \cot x$

15. $\csc \theta \tan \theta - \tan \theta \sin \theta$

16. $\dfrac{\cos x}{\cot x \sin x}$

17. $\cot \theta \tan^2 \theta \cos \theta$

18. $\dfrac{\tan x(\csc^2 x - 1)}{\sin x + \cot x \cos x}$

19. $\dfrac{\sin \theta}{\cos \theta \tan \theta}$

20. $\dfrac{\sin^2 x + \cos^2 x}{1 - \cos^2 x}$

21. $\dfrac{1}{\sec^2 x} + \dfrac{1}{\csc^2 x}$

22. $\sin \theta(\csc \theta + \cot \theta)$

23. $\csc x - \cot x \cos x$

24. $1 + \dfrac{\tan^2 \theta}{1 + \sec \theta}$

25. $\dfrac{\sec x - \csc x}{1 - \cot x}$

26. $\dfrac{1}{1 + \sin x} + \dfrac{1}{1 - \sin x}$

27. $\sec^2 x(1 - \cos^2 x)$

28. $\tan x + \dfrac{\cos x}{\sin x + 1}$

29. $\cos \theta \sec \theta - \dfrac{\sec \theta}{\cos \theta}$

30. $\cot^2 x \sin^2 x + \tan^2 x \cos^2 x$

Prove each identity.

All identities in this chapter *can* be proven.

31. $\tan x \cos x = \sin x$

32. $\tan x = \dfrac{\sec x}{\csc x}$

33. $\dfrac{\sin x}{\csc x} + \dfrac{\cos x}{\sec x} = 1$

34. $\sin \theta = \dfrac{1}{\cot \theta \sec \theta}$

35. $(\cos^2 \theta + \sin^2 \theta)^2 = 1$

36. $\tan x = \dfrac{\tan x - 1}{1 - \cot x}$

37. $\dfrac{\csc \theta}{\sec \theta} = \cot \theta$

38. $\cot^2 x = \dfrac{\cos x}{\tan x \sin x}$

39. $\cos x + 1 = \dfrac{\sin^2 x}{1 - \cos x}$

40. $\csc x - \sin x = \cot x \cos x$

41. $\cot^2 x - \cos^2 x = \cos^2 x \cot^2 x$

42. $\csc x = \cos x \cot x + \sin x$

43. $1 = (\csc x - \cot x)(\csc x + \cot x)$

44. $\tan x = \dfrac{\tan x + \sin x}{1 + \cos x}$

45. $\dfrac{\tan x + 1}{1 - \tan x} = \dfrac{\sin x + \cos x}{\cos x - \sin x}$

46. $\cot x = \cot x \sec^2 x - \tan x$

47. $\dfrac{\sin \theta + 1}{1 - \sin \theta} = (\tan \theta + \sec \theta)^2$

48. $\dfrac{1 + \sin \theta}{1 - \sin \theta} = \dfrac{1 + \csc \theta}{\csc \theta - 1}$

49. $(\sec \theta - \tan \theta)(\tan \theta + \sec \theta) = 1$

50. $\dfrac{1 + \cot \theta}{\csc \theta} = \dfrac{\tan \theta + 1}{\sec \theta}$

Computer and Graphics Calculator

51. Use a graphics calculator or a graphing utility on the computer to graphically verify any of the identities in this exercise set. You may also use the same method to check your answers in problems 1 through 30.

52. Graph the expression

$$\sin^2 x + \cos^2 x$$

How do you interpret your resulting graph? Also graph

$$\sec^2 x - \tan^2 x$$

and

$$\csc^2 x - \cot^2 x$$

and interpret the results.

53. Use a CAS to simplify any of the above expressions or to aid in proving the above identities.

18–2   Sum or Difference of Two Angles

Sine of the Sum of Two Angles

We wish now to derive a formula for the sine of the sum of two angles, say, α and β. For example, is it true that

$$\sin 20° + \sin 30° = \sin 50°?$$

Try it on your calculator—you will see that it is not true.

We start by drawing two positive acute angles, α and β (Fig. 18–5), small enough so that their sum ($\alpha + \beta$) is also acute. From any point P on the terminal side of β we draw perpendicular AP to the x axis, and draw perpendicular BP to line OB. Since the angle between two lines equals the angle between the perpendiculars to those two lines, we note that angle APB is equal to α.

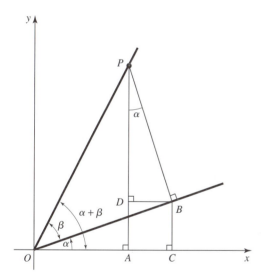

FIGURE 18–5

Then

$$\sin(\alpha + \beta) = \frac{AP}{OP} = \frac{AD + PD}{OP} = \frac{BC + PD}{OP}$$

$$= \frac{BC}{OP} + \frac{PD}{OP}$$

But in triangle OBC,

$$BC = OB \sin \alpha$$

and in triangle PBD,

$$PD = PB \cos \alpha$$

Substituting, we obtain

$$\sin(\alpha + \beta) = \frac{OB \sin \alpha}{OP} + \frac{PB \cos \alpha}{OP}$$

But in triangle OPB,

$$\frac{OB}{OP} = \cos \beta$$

and

$$\frac{PB}{OP} = \sin \beta$$

Thus:

$$\sin(\alpha + \beta) = \sin \alpha \cos \beta + \cos \alpha \sin \beta$$

We have proven this true for acute angles whose sum is also acute. These identities are, in fact, true for any size angles, positive or negative, although we will not take the space to prove this.

Common Error

The sine of the sum of two angles *is not* the sum of the sines of each angle.

$$\sin(\alpha + \beta) \neq \sin \alpha + \sin \beta$$

Cosine of the Sum of Two Angles

Again using Fig. 8–5, we can derive an expression for $\cos(\alpha + \beta)$. From Eq. 147,

$$\cos(\alpha + \beta) = \frac{OA}{OP} = \frac{OC - AC}{OP} = \frac{OC - BD}{OP}$$

$$= \frac{OC}{OP} - \frac{BD}{OP}$$

Now, in triangle OBC,

$$OC = OB \cos \alpha$$

and in triangle PDB,

$$BD = PB \sin \alpha$$

Substituting, we obtain

$$\cos(\alpha + \beta) = \frac{OB \cos \alpha}{OP} - \frac{PB \sin \alpha}{OP}$$

As before,

$$\frac{OB}{OP} = \cos \beta \quad \text{and} \quad \frac{PB}{OP} = \sin \beta$$

Therefore:

$$\cos(\alpha + \beta) = \cos \alpha \cos \beta - \sin \alpha \sin \beta$$

Difference of Two Angles

We can obtain a formula for the sine of the difference of two angles merely by substituting $-\beta$ for β in the equation previously derived for $\sin(\alpha + \beta)$.

$$\sin[\alpha + (-\beta)] = \sin \alpha \cos(-\beta) + \cos \alpha \sin(-\beta)$$

But for β in the first quadrant, $(-\beta)$ is in the fourth, so

$$\cos(-\beta) = \cos \beta \quad \text{and} \quad \sin(-\beta) = -\sin \beta$$

Therefore,

$$\sin(\alpha - \beta) = \sin \alpha \cos \beta + \cos \alpha(-\sin \beta)$$

which is rewritten as follows:

$$\sin(\alpha - \beta) = \sin \alpha \cos \beta - \cos \alpha \sin \beta$$

We see that the result is identical to the formula for $\sin(\alpha + \beta)$ except for a change in sign. This enables us to write the two identities in a single expression using the $\pm$ sign.

When the double signs are used, it is understood that the upper signs correspond and the lower signs correspond.

| Sine of Sum or Difference of Two Angles | $\sin(\alpha \pm \beta) = \sin \alpha \cos \beta \pm \cos \alpha \sin \beta$ | 167 |

Similarly, finding $\cos(\alpha - \beta)$, we have

$$\cos[\alpha + (-\beta)] = \cos \alpha \cos(-\beta) - \sin \alpha \sin(-\beta)$$

Thus:

$$\cos(\alpha - \beta) = \cos \alpha \cos \beta + \sin \alpha \sin \beta$$

or:

| Cosine of Sum or Difference of Two Angles | $\cos(\alpha \pm \beta) = \cos \alpha \cos \beta \mp \sin \alpha \sin \beta$ | 168 |

◆◆◆ **Example 12:** Expand the expression $\sin(x + 3y)$.

Solution: By Eq. 167,

$$\sin(x + 3y) = \sin x \cos 3y + \cos x \sin 3y \qquad ◆◆◆$$

◆◆◆ **Example 13:** Simplify

$$\cos 5x \cos 3x - \sin 5x \sin 3x$$

Solution: We see that this has a similar form to Eq. 168, with $\alpha = 5x$ and $\beta = 3x$, so

$$\cos 5x \cos 3x - \sin 5x \sin 3x = \cos(5x + 3x)$$
$$= \cos 8x \qquad ◆◆◆$$

◆◆◆ **Example 14:** Prove that

$$\cos(180° - \theta) = -\cos \theta$$

Solution: Expanding the left side by means of Eq. 168, we get

$$\cos(180° - \theta) = \cos 180° \cos \theta + \sin 180° \sin \theta$$

But $\cos 180° = -1$ and $\sin 180° = 0$, so

$$\cos(180° - \theta) = (-1) \cos \theta + (0) \sin \theta = -\cos \theta \qquad ◆◆◆$$

◆◆◆ **Example 15:** Prove that

$$\frac{\tan x - \tan y}{\tan x + \tan y} = \frac{\sin(x - y)}{\sin(x + y)}$$

Solution: The right side contains functions of the sum of two angles, but the left side contains functions of single angles. We thus start by expanding the right side by using Eq. 167.

$$\frac{\sin(x - y)}{\sin(x + y)} = \frac{\sin x \cos y - \cos x \sin y}{\sin x \cos y + \cos x \sin y}$$

Our expression now contains $\sin x$, $\sin y$, $\cos x$, and $\cos y$, but we want an expression containing only $\tan x$ and $\tan y$.

To have $\tan x$ instead of $\sin x$, we can divide numerator and denominator by $\cos x$. Similarly, to obtain $\tan y$ instead of $\sin y$, we can divide by $\cos y$. We thus divide numerator and denominator by $\cos x \cos y$.

$$\frac{\sin x \cos y - \cos x \sin y}{\sin x \cos y + \cos x \sin y} = \frac{\dfrac{\sin x \cancel{\cos y}}{\cos x \cancel{\cos y}} - \dfrac{\cancel{\cos x} \sin y}{\cancel{\cos x} \cos y}}{\dfrac{\sin x \cancel{\cos y}}{\cos x \cancel{\cos y}} + \dfrac{\cancel{\cos x} \sin y}{\cancel{\cos x} \cos y}}$$

Then, by Eq. 162,

$$= \frac{\tan x - \tan y}{\tan x + \tan y}$$ ◆◆◆

Tangent of the Sum or Difference of Two Angles

Since, by Eq. 162,

$$\tan \theta = \frac{\sin \theta}{\cos \theta}$$

we simply divide Eq. 167 by Eq. 168.

$$\tan(\alpha + \beta) = \frac{\sin(\alpha + \beta)}{\cos(\alpha + \beta)} = \frac{\sin \alpha \cos \beta + \cos \alpha \sin \beta}{\cos \alpha \cos \beta - \sin \alpha \sin \beta}$$

Dividing numerator and denominator by $\cos \alpha \cos \beta$ yields

$$\tan(\alpha + \beta) = \frac{\dfrac{\sin \alpha}{\cos \alpha} + \dfrac{\sin \beta}{\cos \beta}}{1 - \dfrac{\sin \alpha}{\cos \alpha} \cdot \dfrac{\sin \beta}{\cos \beta}}$$

Applying Eq. 162 again, we get

$$\tan(\alpha + \beta) = \frac{\tan \alpha + \tan \beta}{1 - \tan \alpha \tan \beta}$$

A similar derivation (which we will not do) will show that $\tan(\alpha - \beta)$ is identical to the expression just derived, except, as we might expect, for a reversal of signs. We combine the two expressions using double signs as follows:

| Tangent of Sum or Difference of Two Angles | $\tan(\alpha \pm \beta) = \dfrac{\tan \alpha \pm \tan \beta}{1 \mp \tan \alpha \tan \beta}$ | **169** |

◆◆◆ **Example 16:** Simplify

$$\frac{\tan 3x + \tan 2x}{\tan 2x \tan 3x - 1}$$

Solution: This can be put into the form of Eq. 169 if we factor (-1) from the denominator.

$$\frac{\tan 3x + \tan 2x}{\tan 2x \tan 3x - 1} = \frac{\tan 3x + \tan 2x}{-(-\tan 2x \tan 3x + 1)}$$

$$= -\frac{\tan 3x + \tan 2x}{1 - \tan 3x \tan 2x}$$

$$= -\tan(3x + 2x)$$

$$= -\tan 5x \qquad \text{◆◆◆}$$

◆◆◆ **Example 17:** Prove that

$$\tan(45° + x) = \frac{1 + \tan x}{1 - \tan x}$$

Solution: Expanding the left side by Eq. 169, we get

$$\tan(45° + x) = \frac{\tan 45° + \tan x}{1 - \tan 45° \tan x}$$

But $\tan 45° = 1$, so

$$\tan(45° + x) = \frac{1 + \tan x}{1 - \tan x} \qquad \text{◆◆◆}$$

◆◆◆ **Example 18:** Prove that

$$\frac{\cot y - \cot x}{\cot x \cot y + 1} = \tan(x - y)$$

Solution: Using Eq. 152c on the left side yields

$$\frac{\dfrac{1}{\tan y} - \dfrac{1}{\tan x}}{\dfrac{1}{\tan x} \cdot \dfrac{1}{\tan y} + 1} = \tan(x - y)$$

Multiply numerator and denominator by $\tan x \tan y$.

$$\frac{\tan x - \tan y}{1 + \tan x \tan y} = \tan(x - y)$$

Then, by Eq. 169:

$$\tan(x - y) = \tan(x - y) \qquad \text{◆◆◆}$$

Adding a Sine Wave and a Cosine Wave of the Same Frequency

In Sec. 17–3, we used vectors to show that the sum of a sine wave and a cosine wave of the same frequency could be written as a single sine wave, at the original frequency, but with some phase angle. The resulting equation is useful in electrical applications. Here we use the formula for the sum or difference of two angles to derive that equation.

Let $A \sin \omega t$ be a sine wave of amplitude A, and $B \cos \omega t$ be a cosine wave of amplitude B, each of frequency $\omega/2\pi$. If we draw a right triangle (Fig. 18–6) with sides A and B and hypotenuse R, then

$$A = R \cos \phi \quad \text{and} \quad B = R \sin \phi$$

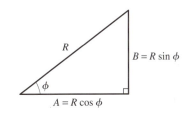

FIGURE 18–6

The sum of the sine wave and the cosine wave is then

$$A \sin \omega t + B \cos \omega t = R \sin \omega t \cos \phi + R \cos \omega t \sin \phi$$
$$= R(\sin \omega t \cos \phi + \cos \omega t \sin \phi)$$

the expression on the right will be familiar. It is the relationship of the sine of the *sum* of two quantities, ωt and ϕ. Continuing, we have

$$A \sin \omega t + B \cos \omega t = R(\sin \omega t \cos \phi + \cos \omega t \sin \phi)$$
$$= R \sin(\omega t + \phi)$$

by Eq. 167. Thus we have the following equation:

This is sometimes referred to as *magnitude and phase* form. As we mentioned in Sec. 17–3, this calculation can be done using the rectangular-to-polar conversion keys on the calculator. Enter the rectangular coordinates A and B, and read the polar coordinates R and θ.

Addition of a Sine Wave and a Cosine Wave	$A \sin \omega t + B \cos \omega t = R \sin(\omega t + \phi)$ where $R = \sqrt{A^2 + B^2}$ and $\phi = \arctan \dfrac{B}{A}$	**214**

••• **Example 19:** Express the following as a single sine function:

$$y = 3.46 \sin \omega t + 2.28 \cos \omega t$$

Solution: The magnitude of the resultant is

$$R = \sqrt{(3.46)^2 + (2.28)^2} = 4.14$$

and the phase angle is

$$\phi = \arctan \frac{2.28}{3.46} = \arctan 0.659 = 33.4°$$

So

$$y = 3.46 \sin \omega t + 2.28 \cos \omega t$$
$$= 4.14 \sin(\omega t + 33.4°)$$ •••

Exercise 2 • Sum or Difference of Two Angles

Expand by means of the addition and subtraction formulas, and simplify.

1. $\sin(\theta + 30°)$
2. $\cos(45° - x)$
3. $\sin(x + 60°)$
4. $\tan(\pi + \theta)$
5. $\cos\left(x + \dfrac{\pi}{2}\right)$
6. $\tan(2x + y)$
7. $\sin(\theta + 2\phi)$
8. $\cos[(\alpha + \beta) + \gamma]$
9. $\tan(2\theta - 3\alpha)$

Simplify.

10. $\cos 2x \cos 9x + \sin 2x \sin 9x$
11. $\cos(\pi + \theta) + \sin(\pi + \theta)$
12. $\sin 3\theta \cos 2\theta - \cos 3\theta \sin 2\theta$
13. $\sin\left(\dfrac{\pi}{3} - x\right) - \cos\left(\dfrac{\pi}{6} - x\right)$

Prove each identity.

14. $\cos x = \sin(x + 90°)$

15. $\sin(30° - x) + \cos(60° - x) = \cos x$

16. $2 \sin \alpha \cos \beta = \sin(\alpha + \beta) + \sin(\alpha - \beta)$

17. $\sin(\theta + 60°) + \cos(\theta + 30°) = \sqrt{3} \cos \theta$

18. $\cos(2\pi - x) = \cos x$

19. $\sin \alpha = \cos(30° - \alpha) - \sin(60° - \alpha)$

20. $\cos(x + y) + \cos(x - y) = 2 \cos x \cos y$

21. $\cos x = \sin\left(x + \dfrac{\pi}{6}\right) - \sin\left(x - \dfrac{\pi}{6}\right)$

22. $\tan(360° - \beta) = -\tan \beta$

23. $\cos(60° + \alpha) + \sin(330° + \alpha) = 0$

24. $\sin 5x \sec x \csc x = \dfrac{\sin 4x}{\sin x} + \dfrac{\cos 4x}{\cos x}$

25. $\dfrac{\sin(x - y)}{\sin(x + y)} = \dfrac{\tan x - \tan y}{\tan x + \tan y}$

26. $\cos(x + 60°) + \cos(60° - x) = \cos x$

27. $\dfrac{\cos(x - y)}{\sin x \cos y} = \tan y + \cot x$

28. $\cot\left(x + \dfrac{\pi}{4}\right) + \tan\left(x - \dfrac{\pi}{4}\right) = 0$

29. $\tan(\alpha + 45°) = \dfrac{\cos \alpha + \sin \alpha}{\cos \alpha - \sin \alpha}$

30. $\dfrac{\cot \alpha \cot \beta - 1}{\cot \alpha + \cot \beta} = \cot(\alpha + \beta)$

31. $\dfrac{1 + \tan x}{1 - \tan x} = \tan\left(\dfrac{\pi}{4} + x\right)$

Express as a single sine function.

32. $y = 47.2 \sin \omega t + 64.9 \cos \omega t$

33. $y = 8470 \sin \omega t + 7360 \cos \omega t$

34. $y = 1.83 \sin \omega t + 2.74 \cos \omega t$

35. $y = 84.2 \sin \omega t + 74.2 \cos \omega t$

Computer and Graphics Calculator

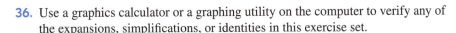

36. Use a graphics calculator or a graphing utility on the computer to verify any of the expansions, simplifications, or identities in this exercise set.

18–3 Functions of Double Angles

Sine of 2α

An equation for the sine of 2α is easily derived by setting $\beta = \alpha$ in Eq. 167.

$$\sin(\alpha + \alpha) = \sin \alpha \cos \alpha + \cos \alpha \sin \alpha$$

which is rewritten in the following form:

| Sine of Twice an Angle | $\sin 2\alpha = 2 \sin \alpha \cos \alpha$ | 170 |

Cosine of 2α

Similarly, setting $\beta = \alpha$ in Eq. 168, we have

$$\cos(\alpha + \alpha) = \cos \alpha \cos \alpha - \sin \alpha \sin \alpha$$

which is also given as follows:

Cosine of Twice an Angle	$\cos 2\alpha = \cos^2 \alpha - \sin^2 \alpha$	171a

There are two alternative forms for Eq. 171a. From Eq. 164,

$$\cos^2 \alpha = 1 - \sin^2 \alpha$$

Substituting into Eq. 171a yields

$$\cos 2\alpha = 1 - \sin^2 \alpha - \sin^2 \alpha$$

Cosine of Twice an Angle	$\cos 2\alpha = 1 - 2 \sin^2 \alpha$	171b

We can similarly use Eq. 164 to eliminate the $\sin^2 \alpha$ term from Eq. 171a.

$$\cos 2\alpha = \cos^2 \alpha - (1 - \cos^2 \alpha)$$

Thus:

Cosine of Twice an Angle	$\cos 2\alpha = 2 \cos^2 \alpha - 1$	171c

◆◆◆ **Example 20:** Express $\sin 3x$ in terms of the single angle x.

Solution: We can consider $3x$ to be $x + 2x$, and, using Eq. 167, we have

$$\sin 3x = \sin(x + 2x) = \sin x \cos 2x + \cos x \sin 2x$$

We now replace $\cos 2x$ by $\cos^2 x - \sin^2 x$, and $\sin 2x$ by $2 \sin x \cos x$.

$$
\begin{aligned}
\sin 3x &= \sin x(\cos^2 x - \sin^2 x) + \cos x(2 \sin x \cos x) \\
&= \sin x \cos^2 x - \sin^3 x + 2 \sin x \cos^2 x \\
&= 3 \sin x \cos^2 x - \sin^3 x
\end{aligned}
$$

◆◆◆

◆◆◆ **Example 21:** Prove that

$$\cos 2A + \sin(A - B) = 0$$

where A and B are the two acute angles of a right triangle.

Solution: By Eqs. 171a and 167,

$$\cos 2A + \sin(A - B) = \cos^2 A - \sin^2 A + \sin A \cos B - \cos A \sin B$$

But, by the cofunctions of Eq. 154,

$$\cos B = \sin A \quad \text{and} \quad \sin B = \cos A$$

So

$$
\begin{aligned}
\cos 2A + \sin(A - B) &= \cos^2 A - \sin^2 A + \sin A \sin A - \cos A \cos A \\
&= \cos^2 A - \sin^2 A + \sin^2 A - \cos^2 A = 0
\end{aligned}
$$

◆◆◆

◆◆◆ **Example 22:** Simplify the expression

$$\frac{\sin 2x}{1 + \cos 2x}$$

Solution: By Eqs. 170 and 171c,

$$\frac{\sin 2x}{1 + \cos 2x} = \frac{2 \sin x \cos x}{1 + 2 \cos^2 x - 1}$$

$$= \frac{2 \sin x \cos x}{2 \cos^2 x}$$

$$= \frac{\sin x}{\cos x} = \tan x$$

◆◆◆

Tangent of 2α

Setting $\beta = \alpha$ in Eq. 169 gives

$$\tan(\alpha + \alpha) = \frac{\tan \alpha + \tan \alpha}{1 - \tan \alpha \tan \alpha}$$

Therefore:

Tangent of Twice an Angle	$\tan 2\alpha = \dfrac{2 \tan \alpha}{1 - \tan^2 \alpha}$	172

◆◆◆ **Example 23:** Prove that

$$\frac{2 \cot x}{\csc^2 x - 2} = \tan 2x$$

Solution: By Eq. 172,

$$\tan 2x = \frac{2 \tan x}{1 - \tan^2 x}$$

and, by Eq. 152c,

$$\tan 2x = \frac{\dfrac{2}{\cot x}}{1 - \dfrac{1}{\cot^2 x}}$$

Multiply numerator and denominator by $\cot^2 x$.

$$\tan 2x = \frac{2 \cot x}{\cot^2 x - 1}$$

Then, by Eq. 166,

$$\tan 2x = \frac{2 \cot x}{(\csc^2 x - 1) - 1} = \frac{2 \cot x}{\csc^2 x - 2}$$

◆◆◆

Common Error	The sine of twice an angle is not twice the sine of the angle. Nor is the cosine (or tangent) of twice an angle equal to twice the cosine (or tangent) of that angle. Remember to use the formulas from this section for all of the trig functions of double angles.

Exercise 3 ◆ Functions of Double Angles

Simplify.

1. $2 \sin^2 x + \cos 2x$

2. $2 \sin 2\theta \cos 2\theta$

3. $\dfrac{2 \tan x}{1 + \tan^2 x}$

4. $\dfrac{2 - \sec^2 x}{\sec^2 x}$

5. $\dfrac{\sin 6x}{\sin 2x} - \dfrac{\cos 6x}{\cos 2x}$

Prove.

6. $\dfrac{2 \tan \theta}{1 - \tan^2 \theta} = \tan 2\theta$

7. $\dfrac{1 - \tan^2 x}{1 + \tan^2 x} = \cos 2x$

8. $2 \csc 2\theta = \tan \theta + \cot \theta$

9. $2 \cot 2x = \cot x - \tan x$

10. $\sec 2x = \dfrac{1 + \cot^2 x}{\cot^2 x - 1}$

11. $\dfrac{1 - \tan x}{\tan x + 1} = \dfrac{\cos 2x}{\sin 2x + 1}$

12. $\dfrac{\sin 2\theta + \sin \theta}{1 + \cos \theta + \cos 2\theta} = \tan \theta$

13. $1 + \cot x = \dfrac{2 \cos 2x}{\sin 2x - 2 \sin^2 x}$

14. $4 \cos^3 x - 3 \cos x = \cos 3x$

15. $\sin \alpha + \cos \alpha = \dfrac{\sin 2\alpha + 1}{\cos \alpha + \sin \alpha}$

16. $\dfrac{\cot^2 x - 1}{2 \cot x} = \cot 2x$

If A and B are the two acute angles in a right triangle, show that:

17. $\sin 2A = \sin 2B$.

18. $\tan(A - B) = -\cot 2A$.

19. $\sin 2A = \cos(A - B)$.

Computer and Graphics Calculator

20. Use a graphics calculator or a graphing utility on the computer to verify any of the simplifications or identities in this exercise set.

18–4   Functions of Half-Angles

Sine of α/2

The double-angle formulas derived in Sec. 18–3 can also be regarded as half-angle formulas, because if one angle is double another, the second angle must be half the first.

Starting with Eq. 171b, we obtain

$$\cos 2\theta = 1 - 2 \sin^2 \theta$$

We solve for $\sin \theta$.

$$2 \sin^2 \theta = 1 - \cos 2\theta$$

$$\sin \theta = \pm \sqrt{\dfrac{1 - \cos 2\theta}{2}}$$

For emphasis, we replace θ by $\alpha/2$.

Sine of Half an Angle	$\sin \dfrac{\alpha}{2} = \pm \sqrt{\dfrac{1 - \cos \alpha}{2}}$	173

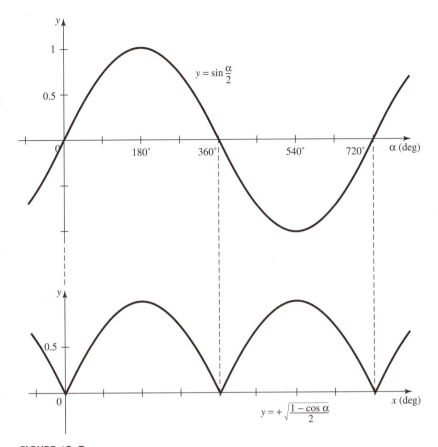

FIGURE 18–7

The $\pm$ sign in Eq. 173 (and in Eqs. 174 and 175c as well) is to be read as plus *or* minus, but *not both*. This sign is different from the $\pm$ sign in the quadratic formula, for example, where we took *both* the positive and the negative values.

The reason for this difference is clear from Fig. 18–7, which shows a graph of $\sin \alpha/2$ and a graph of $+\sqrt{(1 - \cos \alpha)/2}$. Note that the two curves are the same only when $\sin \alpha/2$ is positive. When $\sin \alpha/2$ is negative, it is necessary to use the negative of $\sqrt{(1 - \cos \alpha)/2}$. This occurs when $\alpha/2$ is in the third or fourth quadrant. Thus we choose the plus or the minus according to the quadrant in which $\alpha/2$ is located.

◆◆◆ **Example 24:** Given that $\cos 200° = -0.9397$, find $\sin 100°$.

Solution: From Eq. 173,

$$\sin \frac{200°}{2} = \pm \sqrt{\frac{1 - \cos 200°}{2}} = \pm \sqrt{\frac{1 - (-0.9397)}{2}} = \pm 0.9848$$

Since $100°$ is in the second quadrant, the sine is positive, so we drop the minus sign and get

$$\sin 100° = 0.9848 \qquad \text{◆◆◆}$$

Cosine of $\alpha/2$

Similarly, starting with Eq. 171c, we have

$$\cos 2\theta = 2 \cos^2 \theta - 1$$

Then we solve for $\cos \theta$.

$$2 \cos^2 \theta = 1 + \cos 2\theta$$

$$\cos \theta = \pm \sqrt{\frac{1 + \cos 2\theta}{2}}$$

We next replace θ by $\alpha/2$ and obtain the following:

Here also, we choose the plus *or* the minus according to the quadrant in which $\alpha/2$ is located.

Cosine of Half an Angle	$\cos \dfrac{\alpha}{2} = \pm \sqrt{\dfrac{1 + \cos \alpha}{2}}$	174

Tangent of α/2

There are three formulas for the tangent of a half-angle; we will show the derivation of each in turn. First, from Eq. 162,

$$\tan \frac{\alpha}{2} = \frac{\sin \dfrac{\alpha}{2}}{\cos \dfrac{\alpha}{2}}$$

Multiply numerator and denominator by $2 \sin(\alpha/2)$.

$$\tan \frac{\alpha}{2} = \frac{2 \sin^2 \dfrac{\alpha}{2}}{2 \sin \dfrac{\alpha}{2} \cos \dfrac{\alpha}{2}}$$

Then, by Eqs. 171b and 170, we have the following:

Tangent of Half an Angle	$\tan \dfrac{\alpha}{2} = \dfrac{1 - \cos \alpha}{\sin \alpha}$	175a

Another form of this identity is obtained by multiplying numerator and denominator by $1 + \cos \alpha$.

$$\tan \frac{\alpha}{2} = \frac{1 - \cos \alpha}{\sin \alpha} \cdot \frac{1 + \cos \alpha}{1 + \cos \alpha}$$

$$= \frac{1 - \cos^2 \alpha}{\sin \alpha (1 + \cos \alpha)} = \frac{\sin^2 \alpha}{\sin \alpha (1 + \cos \alpha)}$$

also written in the following form:

Tangent of Half an Angle	$\tan \dfrac{\alpha}{2} = \dfrac{\sin \alpha}{1 + \cos \alpha}$	175b

We can obtain a third formula for the tangent by dividing Eq. 173 by Eq. 174.

$$\tan \frac{\alpha}{2} = \frac{\sin \dfrac{\alpha}{2}}{\cos \dfrac{\alpha}{2}} = \frac{\pm \sqrt{\dfrac{1 - \cos \alpha}{2}}}{\pm \sqrt{\dfrac{1 + \cos \alpha}{2}}}$$

| Tangent of Half an Angle | $\tan \dfrac{\alpha}{2} = \pm \sqrt{\dfrac{1 - \cos \alpha}{1 + \cos \alpha}}$ | 175c |

◆◆◆ **Example 25:** Prove that

$$\frac{1 + \sin^2 \dfrac{\theta}{2}}{1 + \cos^2 \dfrac{\theta}{2}} = \frac{3 - \cos \theta}{3 + \cos \theta}$$

Solution: By Eqs. 173 and 174,

$$\frac{1 + \sin^2 \dfrac{\theta}{2}}{1 + \cos^2 \dfrac{\theta}{2}} = \frac{1 + \dfrac{1 - \cos \theta}{2}}{1 + \dfrac{1 + \cos \theta}{2}}$$

Multiply numerator and denominator by 2.

$$= \frac{2 + 1 - \cos \theta}{2 + 1 + \cos \theta}$$

$$= \frac{3 - \cos \theta}{3 + \cos \theta}$$

◆◆◆

Exercise 4 ◆ Functions of Half-Angles

Prove each identity.

1. $2 \sin^2 \dfrac{\theta}{2} + \cos \theta = 1$

2. $4 \cos^2 \dfrac{x}{2} \sin^2 \dfrac{x}{2} = 1 - \cos^2 x$

3. $2 \cos^2 \dfrac{x}{2} = \dfrac{\sec x + 1}{\sec x}$

4. $\cot \dfrac{\theta}{2} = \csc \theta + \cot \theta$

5. $\dfrac{\sin \alpha}{\cos \alpha + 1} = \sin \dfrac{\alpha}{2} \sec \dfrac{\alpha}{2}$

6. $\dfrac{\cos^2 \dfrac{\theta}{2} - \cos \theta}{\sin^2 \dfrac{\theta}{2}} = 1$

7. $\sec \theta = \tan \dfrac{\theta}{2} \tan \theta + 1$

8. $\sin x + 1 = \left(\cos \dfrac{x}{2} + \sin \dfrac{x}{2} \right)^2$

9. $2 \sin \alpha + \sin 2\alpha = 4 \sin \alpha \cos^2 \dfrac{\alpha}{2}$

10. $\dfrac{1 - \cos x}{\sin x} = \tan \dfrac{x}{2}$

11. $\dfrac{1 - \tan^2 \dfrac{\theta}{2}}{1 + \tan^2 \dfrac{\theta}{2}} = \cos \theta$

In right triangle *ABC*, show the following:

12. $\tan \dfrac{A}{2} = \dfrac{a}{b + c}$

13. $\sin \dfrac{A}{2} = \sqrt{\dfrac{c - b}{2c}}$

Computer and Graphics Calculator

14. Use a graphics calculator or a graphing utility on the computer to verify any of the identities in this exercise set.

18–5   Trigonometric Equations

Solving Trigonometric Equations

One use for the trigonometric identities we have just studied is in the solution of trigonometric equations.

A trigonometric equation will usually have an infinite number of roots. However, it is customary to list only *nonnegative values less than 360°* that satisfy the equation.

♦♦♦ **Example 26:** Solve the equation $\sin \theta = \frac{1}{2}$.

Solution: Even though there are infinitely many values of θ,

$$\theta = 30°, \ 150°, \ 390°, \ 510°, \ \ldots$$

whose sine is $\frac{1}{2}$, we will follow the common practice of listing only the nonnegative values less than 360°. Thus our solution is

$$\theta = 30°, \ 150°$$

The solution could also be expressed in radian measure: $\theta = \pi/6, \ 5\pi/6$ radians.

♦♦♦

Graphical Solution of Trigonometric Equations

In Sec. 5–6, we used a graphics calculator or a graphing utility on the computer to get an approximate solution to an algebraic equation. Here we use exactly the same procedure to solve a trigonometric equation.

We put the given equation into the form $f(\theta) = 0$ and then graph $y = f(\theta)$. We then use the $\boxed{\text{TRACE}}$ and $\boxed{\text{ZOOM}}$ features to locate the zeros as accurately as we wish.

♦♦♦ **Example 27:** Find, to the nearest tenth of a degree, any nonnegative values less than 360° of the zeros of the equation in Example 26.

$$\sin \theta = \frac{1}{2}$$

Solution: We rewrite the given equation in the form

$$\sin \theta - \frac{1}{2} = 0$$

and graph the function $y = \sin \theta - \frac{1}{2}$ as shown in Fig. 18–8. We see zeros at about

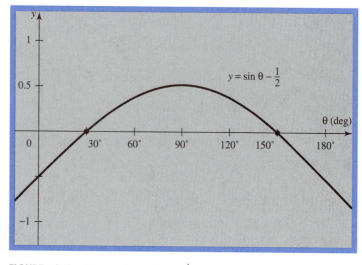

FIGURE 18–8   Graph of $y = \sin \theta - \frac{1}{2}$.

30° and 150°. Using $\boxed{\text{TRACE}}$ and $\boxed{\text{ZOOM}}$ at each zero, we get

$$\theta = 30.0°, \ 150.0°$$

to the nearest tenth of a degree.   ◆◆◆

It is difficult to give general rules for solving the great variety of possible trigonometric equations, but a study of the following examples should be helpful.

Equations Containing a Single Trigonometric Function and a Single Angle

We start with the simplest type. These equations contain, after simplification, only one trigonometric function (only sine, or only cosine, for example) and have only one angle, as in Example 26. If an equation does not *appear* to be of this type, simplify it first. Then isolate the trigonometric function on one side of the equation, and find the angles.

◆◆◆ **Example 28:** Solve the equation

$$\frac{2.73 \sec \theta}{\csc \theta} + \frac{1.57}{\cos \theta} = 0$$

Solution: As with identities, it is helpful to rewrite the given expression with only sines and cosines. Thus

$$\frac{2.73 \sin \theta}{\cos \theta} + \frac{1.57}{\cos \theta} = 0$$

Multiplying by $\cos \theta$ and rearranging gives an expression with only one function, sine.

$$\sin \theta = -\frac{1.57}{2.73} = -0.575$$

Our reference angle is arcsin(0.575) or 35.1°. The sine is negative in the third and fourth quadrants, so we give the third and fourth quadrant values for θ.

$$\theta = 180° + 35.1° = 215.1°$$

and

$$\theta = 360° - 35.1° = 324.9°$$

Figure 18–9 shows a graph of the function and its zeros.

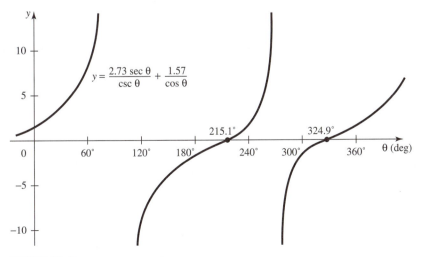

FIGURE 18–9   ◆◆◆

Our next example contains a double angle.

◆◆◆ **Example 29:** Solve the equation $2 \cos 2\theta - 1 = 0$.

Solution: Rearranging and dividing, we have

$$\cos 2\theta = \tfrac{1}{2}$$
$$2\theta = 60°, 300°, 420°, 660°, \ldots$$
$$\theta = 30°, 150°, 210°, \text{ and } 330°$$

Note that although this problem contained a double angle, *we did not need the double-angle formula.* It would have been needed, however, if the same problem contained both a double angle *and* a single angle.

if we limit our solution to angles less than 360°. Figure 18–10 shows a graph of the function $y = 2 \cos 2\theta - 1$ and its zeros.

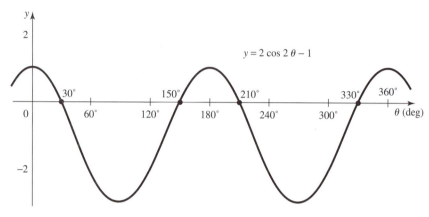

FIGURE 18–10

◆◆◆

> **Common Error**
>
> It is easy to forget to find *all of the angles* less than 360°, especially when the given equation contains a double angle, such as in Example 29.

If one of the functions is *squared,* we may have an equation in *quadratic form,* which can be solved by the methods of Chapter 14.

◆◆◆ **Example 30:** Solve the equation $\sec^2 \theta = 4$.

Solution: Taking the square root of both sides gives us

$$\sec \theta = \pm\sqrt{4} = \pm 2$$

We thus have two solutions. By the reciprocal relations, we get

$$\frac{1}{\cos \theta} = 2 \qquad \text{and} \qquad \frac{1}{\cos \theta} = -2$$

$$\cos \theta = \tfrac{1}{2} \qquad \text{and} \quad \cos \theta = -\tfrac{1}{2}$$
$$\theta = 60°, 300° \quad \text{and} \qquad \theta = 120°, 240°$$

Our solution is then

$$\theta = 60°, 120°, 240°, 300°$$

Figure 18–11 shows a graph of the function $y = \sec^2 \theta - 4$ and its zeros.

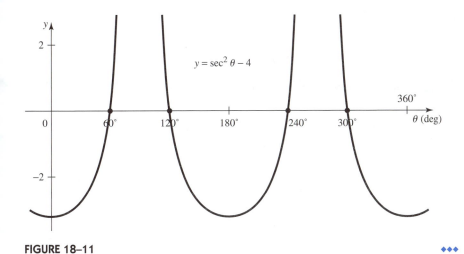

FIGURE 18–11 ◆◆◆

◆◆◆ **Example 31:** Solve the equation $2 \sin^2 \theta - \sin \theta = 0$.

Solution: This is an *incomplete quadratic* in $\sin \theta$. Factoring, we obtain

$$\sin \theta (2 \sin \theta - 1) = 0$$

Setting each factor equal to zero yields

$\sin \theta = 0$	$2 \sin \theta - 1 = 0$
$\theta = 0°, 180°$	$\sin \theta = \frac{1}{2}$
	Since sine is positive in the first and second quadrants,
	$\theta = 30°, 150°$

Figure 18–12 shows a graph of the function $y = 2 \sin^2 \theta - \sin \theta$ and its zeros.

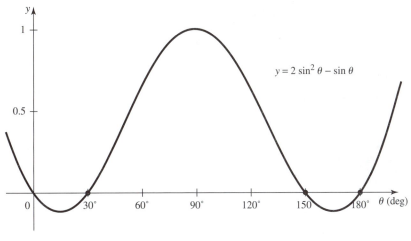

FIGURE 18–12 ◆◆◆

If an equation is in the form of a quadratic that cannot be factored, use the quadratic formula.

◆◆◆ **Example 32:** Solve the equation $\cos^2 \theta = 3 + 5 \cos \theta$.

Solution: Rearranging into standard quadratic form, we have

$$\cos^2 \theta - 5 \cos \theta - 3 = 0$$

This cannot be factored, so we use the quadratic formula.

$$\cos \theta = \frac{5 \pm \sqrt{25 - 4(-3)}}{2}$$

There are two values for $\cos \theta$.

$\cos \theta = 5.54$	$\cos \theta = -0.541$
(not possible)	Since cosine is negative in the second and third quadrants, $\theta = 123°, 237°$

Figure 18–13 shows a graph of the function $y = \cos^2 \theta - 3 - 5 \cos \theta$ and its zeros.

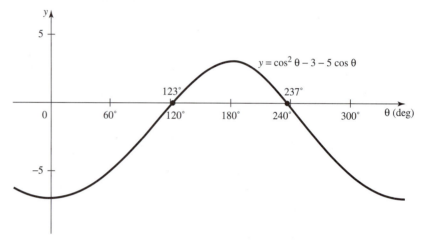

FIGURE 18–13   ◆◆◆

Equations with One Angle But More Than One Function

If an equation contains two or more trigonometric functions of the same angle, first transpose all of the terms to one side and try to factor that side into factors, *each containing only a single function,* and proceed as above.

◆◆◆ **Example 33:** Solve $\sin \theta \sec \theta - 2 \sin \theta = 0$.

Solution: Factoring, we have

$$\sin \theta(\sec \theta - 2) = 0$$

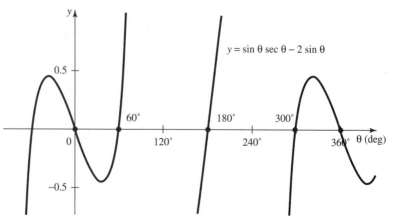

FIGURE 18–14

$$\begin{array}{c|l} \sin \theta = 0 & \sec \theta = 2 \\ & \cos \theta = \frac{1}{2} \\ & \text{Since cosine is positive in the first and} \\ & \text{fourth quadrants,} \\ \theta = 0°, 180° & \theta = 60°, 300° \end{array}$$

Figure 18–14 shows a graph of the function $y = \sin \theta \sec \theta - 2 \sin \theta$ and its zeros. ♦♦♦

If an expression is *not factorable* at first, use the fundamental identities to *express everything in terms of a single trigonometric function,* and proceed as above.

♦♦♦ **Example 34:** Solve $\sin^2 \theta + \cos \theta = 1$.

Solution: By Eq. 164, $\sin^2 \theta = 1 - \cos^2 \theta$. Substituting gives

$$1 - \cos^2 \theta + \cos \theta = 1$$
$$\cos \theta - \cos^2 \theta = 0$$

Factoring, we obtain

$$\cos \theta (1 - \cos \theta) = 0$$

$$\begin{array}{c|c} \cos \theta = 0 & \cos \theta = 1 \\ \theta = 90°, 270° & \theta = 0° \end{array}$$

Figure 18–15 shows a graph of the function $y = \sin^2 \theta + \cos^2 \theta - 1$ and its zeros.

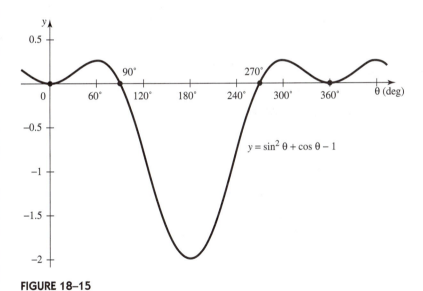

FIGURE 18–15 ♦♦♦

In order to simplify an equation using the Pythagorean relations, it is sometimes necessary to *square both sides.*

♦♦♦ **Example 35:** Solve $\sec \theta + \tan \theta = 1$.

Solution: We have no identity that enables us to write $\sec \theta$ in terms of $\tan \theta$, but we do have an identity for $\sec^2 \theta$. We rearrange and square both sides, getting

$$\sec \theta = 1 - \tan \theta$$

$$\sec^2 \theta = 1 - 2 \tan \theta + \tan^2 \theta$$

Replacing $\sec^2 \theta$ by $1 + \tan^2 \theta$ yields

$$1 + \tan^2 \theta = 1 - 2 \tan \theta + \tan^2 \theta$$

$$\tan \theta = 0$$

$$\theta = 0, 180°$$

Since squaring can introduce extraneous roots that do not satisfy the original equation, we substitute back to check our answers. We find that the only angle that satisfies the given equation is $\theta = 0°$. Figure 18–16 shows a graph of the function and the zeros.

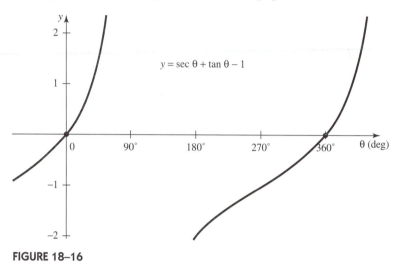

FIGURE 18–16   ◆◆◆

For our final example, let us find the roots of a trigonometric equation that would be difficult or impossible to solve exactly. We will do an approximate graphical solution.

◆◆◆ **Example 36:** Find, to the nearest tenth of a degree, any nonnegative values less than 360° of the zeros of

$$2.43 \sin^3(1.89 \, \theta) = 1.74 \cos(1.12 \, \theta) - 1.15$$

Solution: The graph of the function

$$f(\theta) = 2.43 \sin^3(1.89 \, \theta) - 1.74 \cos(1.12\theta) + 1.15$$

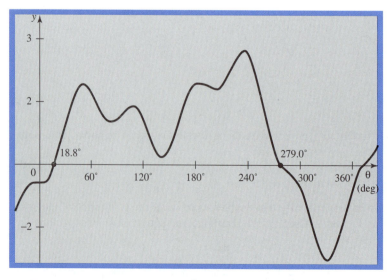

FIGURE 18–17   Graph of $y = 2.43 \sin^3(1.89\theta) - 1.74 \cos(1.12\theta) + 1.15$

is shown in Fig. 18–17. We see zeros at about 20° and 280°. Using $\boxed{\text{TRACE}}$ and $\boxed{\text{ZOOM}}$ at each zero, we get

$$\theta = 18.8°, \ 279.0°$$

to the nearest tenth of a degree. 　　　　　　　　　　　　◆◆◆

Exercise 5 ◆ Trigonometric Equations

Solve each equation for all nonnegative values of θ less than 360°.

1. $\sin \theta = \frac{1}{2}$
2. $2 \cos \theta - \sqrt{3} = 0$
3. $1 - \tan \theta = 0$
4. $\sin \theta = \sqrt{3} \cos \theta$
5. $4 \sin^2 \theta = 3$
6. $2 \sin 3\theta = \frac{1}{2}$
7. $3 \sin \theta - 1 = 2 \sin \theta$
8. $\csc^2 \theta = 4$
9. $2 \cos^2 \theta = 1 + 2 \sin^2 \theta$
10. $2 \sec \theta = -3 - \cos \theta$
11. $4 \sin^4 \theta = 1$
12. $2 \csc \theta - \cot \theta = \tan \theta$
13. $1 + \tan \theta = \sec^2 \theta$
14. $1 + \cot^2 \theta = \sec^2 \theta$
15. $3 \cot \theta = \tan \theta$
16. $\sin^2 \theta = 1 - 6 \sin \theta$
17. $3 \sin(\theta/2) - 1 = 2 \sin^2(\theta/2)$
18. $\sin \theta = \cos \theta$
19. $4 \cos^2 \theta + 4 \cos \theta = -1$
20. $1 + \sin \theta = \sin \theta \cos \theta + \cos \theta$
21. $3 \tan \theta = 4 \sin^2 \theta \tan \theta$
22. $3 \sin \theta \tan \theta + 2 \tan \theta = 0$
23. $\sec \theta = -\csc \theta$
24. $\sin \theta = 2 \cos(\theta/2)$
25. $\sin \theta = 2 \sin \theta \cos \theta$
26. $\cos \theta \sin 2\theta = 0$

Computer and Graphics Calculator

27. Use your program for solving equations by the midpoint method (Sec. 5–6) to find the roots of any of the trigonometric equations in this exercise set.

28. Use a graphics calculator or a graphing utility on the computer to graphically locate the roots of any of the above equations.

29. Use a CAS to solve any of the above equations.

18–6　Inverse Trigonometric Functions

Inverse of the Sine Function: The Arc Sine

Recall from Sec. 4–3 that to find the inverse of a function $y = f(x)$, we perform the following steps:

1. Solve the given equation for x.
2. Interchange x and y.

Glance back at what we have already said about inverse trigonometric functions in Chapters 7 and 15.

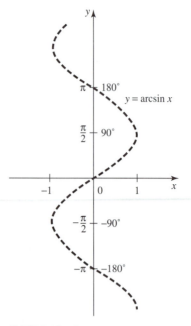

FIGURE 18–18 Graph of arcsin x.

Thus, starting with the sine function.

$$y = \sin x$$

we solve for x and get

$$x = \arcsin y$$

Now we interchange x and y so that y is, as usual, the dependent variable. The inverse of $y = \sin x$ is thus

$$y = \arcsin x$$

The graph of $y = \arcsin x$ is shown in Fig. 18–18. Note that it is identical to the graph of the sine function but with the x and y axes interchanged.

Recalling the distinction between a relation and a function from Sec. 4–1, we see that $y = \arcsin x$ is a relation, but not a function, because for a single x there may be more than one y. For this relation to also be a function, we must limit the range so that repeated values of y are not encountered.

The arc sine, for example, is customarily limited to the interval between $-\pi/2$ and $\pi/2$ (or $-90°$ to $90°$), as shown in Fig. 18–19. The numbers within this interval are called the *principal values* of the arc sine. It is also, of course, the *range* of arcsin x.

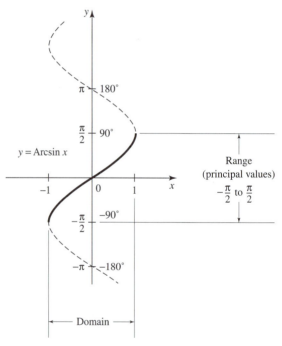

FIGURE 18–19 Principal values.

Given these limits, our relation $y = \arcsin x$ now becomes a *function*. The first letter of the function is sometimes capitalized. Thus, $y = \arcsin x$ would be the relation, and $y = \text{Arcsin } x$ would be the function.

Arc Cosine and Arc Tangent

In a similar way, the inverse of the cosine function and the inverse of the tangent function are limited to principal values, as shown by the solid lines in Figs. 18–20 and 18–21.

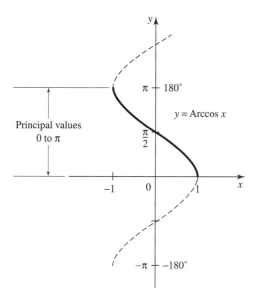

FIGURE 18–20

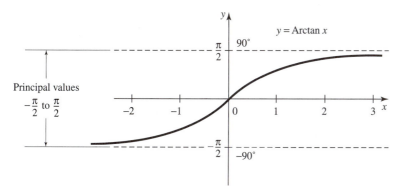

FIGURE 18–21

The numerical values of any of the arc functions are found by calculator, as was shown in Sec. 7–3 and 15–1.

To summarize, if *x is positive,* then Arcsin *x* is an angle between 0 and 90° and hence lies in quadrant I. The same is true of Arccos *x* and Arctan *x.* If *x is negative,* then Arcsin *x* is an angle between −90° and 0° and hence is in quadrant IV. The same is true of Arctan *x.* However, Arccos *x* is in quadrant II when *x* is negative. These facts are summarized in the following table:

x	**Arcsin *x***	**Arccos *x***	**Arctan *x***
Positive	Quadrant I	Quadrant I	Quadrant I
Negative	Quadrant IV	Quadrant II	Quadrant IV

◆◆◆ **Example 37:** Find (a) Arcsin 0.638, (b) Tan⁻¹(−1.63), and (c) Arcsec(−2.34). Give your answer in degrees to one decimal place.

Solution:

(a) By calculator,

$$\text{Arcsin } 0.638 = 39.6°$$

Most calculators will give the principal value directly, for first ≠ and second ≠ quadrant angles, and as a negative angle for fourth ≠ quadrant angles. Try your calculator to see what it does.

(b) By calculator,

$$\text{Tan}^{-1}(-1.63) = -58.5°$$

(c) Using the reciprocal relations, we get

$$\text{Arcsec}(-2.34) = \text{Arccos}\ \frac{1}{-2.34} = \text{Arccos}(-0.427)$$

$$= 115.3°$$

◆◆◆

Exercise 6 ◆ Inverse Trigonometric Functions

Evaluate each expression. Give your answer in degrees to one decimal place.

1. Arcsin 0.374   **2.** Arccos 0.826   **3.** Arctan(−4.92)
4. Cos^{-1} 0.449   **5.** Tan^{-1} 6.92   **6.** $\text{Sin}^{-1}(-0.822)$
7. Arcsec 3.96   **8.** Cot^{-1} 4.97   **9.** Csc^{-1} 1.824
10. Sec^{-1} 2.89

Computer and Graphics Calculator

Use a graphics calculator or a graphing utility on a computer to graph the following functions.

11. $y = \text{Arcsin}\ x$ **12.** $y = \text{Arccos}\ x$ **13.** $y = \text{Arctan}\ x$
14. $y = 2\ \text{Arcsin}\ x$ **15.** $y = \text{Arccos}\ 2x$ **16.** $y = 3\ \text{Arctan}\ 2x$

◆◆◆ CHAPTER 18 REVIEW PROBLEMS ◆◆◆◆◆◆◆◆◆◆◆◆◆◆◆◆◆◆◆◆◆◆◆◆◆◆◆◆◆◆◆◆◆◆

Prove.

1. $\sec^2 \theta + \tan^2 \theta = \sec^4 \theta - \tan^4 \theta$

2. $\dfrac{1 + \csc \theta}{\cot \theta} = \dfrac{\cot \theta}{\csc \theta - 1}$

3. $\tan^2 \theta \sin^2 \theta = \tan^2 \theta - \sin^2 \theta$

4. $\cot \theta + \csc \theta = \dfrac{1}{\csc \theta - \cot \theta}$

5. $\sin(45° + \theta) - \sin(45° - \theta) = \sqrt{2} \sin \theta$

6. $\dfrac{\cot \alpha \cot \beta + 1}{\cot \beta - \cot \alpha} = \cot(\alpha - \beta)$

7. $\dfrac{\sin^3 \theta + \cos^3 \theta}{\sin \theta + \cos \theta} = 1 - \dfrac{\sin 2\theta}{2}$

8. $\dfrac{1 + \tan \theta}{1 - \tan \theta} = \sec 2\theta + \tan 2\theta$

9. $\sin \theta \cot \dfrac{\theta}{2} = \cos \theta + 1$

10. $\sin^4 \theta + 2 \sin^2 \theta \cos^2 \theta + \cos^4 \theta = 1$

11. $\cot \alpha \cot \beta = \dfrac{\csc \beta + \cot \alpha}{\tan \alpha \sec \beta + \tan \beta}$

12. $\cos \theta \cot \theta = \dfrac{\cos \theta + \cot \theta}{\tan \theta + \sec \theta}$

Simplify.

13. $\dfrac{\sin \theta \sec \theta}{\tan \theta}$

14. $\sec^2 \theta - \sin^2 \theta \sec^2 \theta$

15. $\dfrac{\sec \theta}{\sec \theta - 1} + \dfrac{\sec \theta}{\sec \theta + 1}$

16. $\dfrac{\sin^3 \theta + \cos^3 \theta}{\sin \theta + \cos \theta}$

17. $\cos \theta \tan \theta \csc \theta$

18. $(\sec \theta - 1)(\sec \theta + 1)$

19. $\dfrac{\cot \theta}{1 + \cot^2 \theta} \cdot \dfrac{\tan^2 \theta + 1}{\tan \theta}$

20. $\dfrac{\cos \theta}{1 - \tan \theta} + \dfrac{\sin \theta}{1 - \cot \theta}$

21. $(1 - \sin^2 \theta) \sec^2 \theta$

22. $\tan^2 \theta (1 + \cot^2 \theta)$

23. $\dfrac{\sin \theta - 2 \sin \theta \cos \theta}{2 - 2 \sin^2 \theta - \cos \theta}$

Solve for all positive values of θ less than $360°$.

24. $1 + 2 \sin^2 \theta = 3 \sin \theta$
25. $3 + 5 \cos \theta = 2 \cos^2 \theta$
26. $\cos \theta - 2 \cos^3 \theta = 0$
27. $\tan \theta = 2 - \cot \theta$
28. $\sin 3\theta - \sin 6\theta = 0$

29. $\cot 2\theta - 2 \cos 2\theta = 0$
30. $\sin \theta + \cos 2\theta = 1$
31. $\sin \theta = 1 - 3 \cos \theta$
32. $5 \tan \theta = 6 + \tan^2 \theta$
33. $\sin^2 \theta = 1 + 6 \sin \theta$
34. $\tan^2 2\theta = 1 + \sec 2\theta$

35. $16 \cos^4 \dfrac{\theta}{2} = 9$

36. $\tan^2 \theta + 1 = \sec \theta$

Evaluate.

37. Arcsin 0.825
38. $\text{Cos}^{-1} 0.232$
39. Arctan 3.55

Express as a single sine function.

40. $y = 1.84 \sin \omega t + 2.18 \cos \omega t$
41. $y = 47.6 \sin \omega t + 62.1 \cos \omega t$

Writing

42. Leaf through this chapter and locate any points that are not clear. Write a letter to the authors in which you give specific examples of what you think can be explained better. Point out where you would like more examples, and be specific about the type. Instead of putting this letter into your journal, you are urged to send it to the authors at Vermont Technical College, Randolph Center, VT 05061. We promise an answer. (You may, of course, do the same for anything in this text.)

Team Project

43. Several sections of rail were welded together to form a continuous straight rail 255 ft long. It was installed when the temperature was 0°F, and the crew neglected to allow a gap for thermal expansion. The ends of the rail are fixed so that they cannot move outward, so the rail took the shape shown in Fig. 18–22 when the temperature rose to 120°F. Assuming the curve to be circular, find the height x at the midpoint of the track.

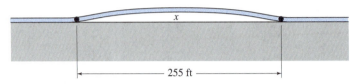

FIGURE 18–22

Ratio, Proportion, and Variation

◆◆◆ **OBJECTIVES** ◆◆

When you have completed this chapter, you should be able to:

- Set up and solve a proportion for a missing quantity.
- Solve applied problems using proportions.
- Set up and solve problems involving direct variation, inverse variation, joint variation, and combined variation.
- Solve applied problems involving variation.
- Set up and solve power function problems.
- Find dimensions, areas, and volumes of similar geometric figures.

In Chapter 9 we introduced the idea of a *ratio* as the quotient of two quantities. Here we expand upon that idea, and we also do problems where two ratios are set equal to each other, that is, problems involving *proportions.* The computations will be no different, but we will learn another way of looking at problems involving fractions.

When two quantities, say, x and y, are connected by some functional relation, $y = f(x)$, some variation in the independent variable x will imply a variation in the dependent variable y. In our study of *functional variation,* we study the changes in y brought about by changes in x, for several simple functions. We often want to know the amount by which y will change when we make a certain change in x. The quantities x and y will not be abstract quantities but quantities we care about, such as: By how much will the deflection y of a beam increase when we decrease the beam thickness x by 50%?

We will see that the methods in this chapter also give us another powerful tool for making *estimates.* In particular, we will be able to estimate a new value for one variable when we change the other by a certain amount. Again, we will be interested in such real quantities as the velocity of an object or the resistance of a circuit.

We have, in fact, already studied some aspects of functional variation—first when we substituted numerical values for x into a function and computed corresponding values of y and later when we graphed functions in Chapter 5. We do some graphing here, too, and also learn some better techniques for solving numerical problems.

19–1   Ratio and Proportion

Ratio

In our work so far, we have often dealt with expressions of the form

$$\frac{a}{b}$$

There are several different ways of looking at such an expression. We can say that a/b is

- A *fraction*, with a numerator a and a denominator b
- A *quotient*, where a is *divided* by b
- The *ratio* of a to b

Thus a ratio can be thought of as a fraction or as the quotient of two quantities. For a ratio of two physical quantities, it is usual to express the numerator and denominator *in the same units,* so that they cancel and leave the ratio *dimensionless.* The quantities a and b are called the *terms* of the ratio.

◆◆◆ **Example 1:** A corridor is 8 ft wide and 12 yd long. Find the ratio of length to width.

Solution: We first express the length and width in the same units, say, feet.

$$12 \text{ yd} = 36 \text{ ft}$$

So the ratio of length to width is

$$\frac{36}{8} = \frac{9}{2} \qquad\qquad ◆◆◆$$

Another way of writing a ratio is with the colon (:). The ratio of a to b is

$$\frac{a}{b} = a:b$$

In Example 1 we could write that the ratio of length to width is $9:2$.

Dimensionless ratios are handy because you do not have to worry about units, and they are often used in technology. Some examples are as follows:

Poisson's ratio	fuel-air ratio	turn ratio
endurance ratio	load ratio	gear ratio
radian measure of angles	pi	trigonometric ratio

The word *specific* is often used to denote a ratio when there is a standard unit to which a given quantity is being compared under standard conditions. The following are a few such ratios:

specific heat	specific volume	specific conductivity
specific weight	specific gravity	specific speed

For example, the specific gravity of a solid or liquid is the ratio of the density of the substance to the density of water at a standard temperature.

◆◆◆ **Example 2:** Aluminum has a density of 165 lb/ft³, and water has a density of 62.4 lb/ft³. Find the specific gravity (SG) of aluminum.

Solution: Dividing gives

$$SG = \frac{165 \text{ lb/ft}^3}{62.4 \text{ lb/ft}^3} = 2.64$$

Note that specific gravity is a *dimensionless ratio.* ◆◆◆

Proportion

A *proportion* is an equation obtained when two ratios are set equal to each other. If the ratio $a:b$ is equal to the ratio $c:d$, we have the proportion

$$a:b = c:d$$

which reads "the ratio of a to b equals the ratio of c to d" or "a is to b as c is to d." We will often write such a proportion in the form

$$\frac{a}{b} = \frac{c}{d}$$

> This ratio can also be written $a:b::c:d$.

The two inside terms of a proportion are called the *means,* and the two outside terms are the *extremes*.

$$
\begin{array}{c}
\overbrace{}^{\text{extremes}} \\
a:b \quad = \quad c:d \\
\underbrace{}_{\text{means}}
\end{array}
$$

Finding a Missing Term

Solve a proportion just as you would any other fractional equation.

◆◆◆ **Example 3:** Find x if $\dfrac{3}{x} = \dfrac{7}{9}$.

Solution: Multiplying both sides by the LCD, $9x$, we obtain

$$27 = 7x$$
$$x = \frac{27}{7}$$

◆◆◆

◆◆◆ **Example 4:** Find x if $\dfrac{x+2}{3} = \dfrac{x-1}{5}$.

Solution: Multiplying through by 15 gives

$$5(x + 2) = 3(x - 1)$$
$$5x + 10 = 3x - 3$$
$$2x = -13$$
$$x = -\frac{13}{2}$$

◆◆◆

◆◆◆ **Example 5:** Insert the missing quantity in the following equation:

$$\frac{x-3}{2x} = \frac{?}{4x^2}$$

Solution: Replacing the question mark with a variable, say, z, and solving for z gives us

$$z = \frac{4x^2(x-3)}{2x}$$
$$= 2x(x - 3)$$

◆◆◆

Mean Proportional

When the means (the two inside terms) of a proportion are equal, as in

$$a:b = b:c$$

The mean proportional b is also called the *geometric mean* between a and c, because a, b, and c form a *geometric progression* (a series of numbers in which each term is obtained by multiplying the previous term by the same quantity).

Some books define b as positive if both a and c are positive. Here, we'll show both values.

the term b is called the *mean proportional* between a and c. Solving for b, we get

$$b^2 = ac$$

Mean Proportional	$b = \pm\sqrt{ac}$	59

◆◆◆ **Example 6:** Find the mean proportional between 3 and 12.

Solution: From Eq. 59,

$$b = \pm\sqrt{3(12)} = \pm 6 \qquad \text{◆◆◆}$$

Applications

Use the same guidelines for setting up these word problems as were given in Sec. 3–2 for any word problem. Just be careful not to reverse the terms in the proportion. Match the first item in the verbal statement with the first number in the ratio, and the second with the second.

◆◆◆ **Example 7:** A man is 25 years old and his brother is 15. In how many years will their ages be in the ratio 5:4?

Solution: Let x = required number of years. After x years, the man's age will be

$$25 + x$$

and the brother's age will be

$$15 + x$$

Forming a proportion, we have

$$\frac{25 + x}{15 + x} = \frac{5}{4}$$

Multiplying by $4(15 + x)$, we obtain

$$4(25 + x) = 5(15 + x)$$
$$100 + 4x = 75 + 5x$$
$$x = 25 \text{ years} \qquad \text{◆◆◆}$$

Exercise 1 ◆ Ratio and Proportion

Find the value of x.

1. $3:x = 4:6$
2. $x:5 = 3:10$
3. $4:6 = x:4$
4. $3:x = x:12$
5. $x:(14 - x) = 4:3$
6. $x:12 = (x - 12):3$
7. $x:6 = (x + 6):10\frac{1}{2}$
8. $(x - 7):(x + 7) = 2:9$

Insert the missing quantity.

9. $\dfrac{x}{3} = \dfrac{?}{9}$
10. $\dfrac{?}{4x} = \dfrac{7}{16x}$

11. $\dfrac{5a}{7b} = \dfrac{?}{-7b}$
12. $\dfrac{a - b}{c - d} = \dfrac{b - a}{?}$

13. $\dfrac{x + 2}{5x} = \dfrac{?}{5}$

Find the mean proportional between the following.

14. 2 and 50
15. 3 and 48
16. 6 and 150
17. 5 and 45
18. 4 and 36

Applications

19. Find two numbers that are to each other as $5:7$ and whose sum is 72.

20. Separate $150 into two parts so that the smaller may be to the greater as 7 is to 8.

21. Into what two parts may the number 56 be separated so that one may be to the other as 3 is to 4?

22. The difference between two numbers is 12, and the larger is to the smaller as 11 is to 7. What are the numbers?

23. An estate of $75,000 was divided between two heirs so that the elder's share was to the younger's as 8 is to 7. What was the share of each?

24. Two partners jointly bought some stock. The first put in $400 more than the second, and the stock of the first was to that of the second as 5 is to 4. How much money did each invest?

25. The sum of two numbers is 20, and their sum is to their difference as 10 is to 1. What are the numbers?

26. The sum of two numbers is a, and their sum is to their difference as m is to n. What are the numbers?

27. For the transformer shown in Fig. 19–1, the ratio of the number of turns in the secondary winding to the number of turns in the primary winding is 15. The secondary winding has 4500 turns. Find the number of turns in the primary.

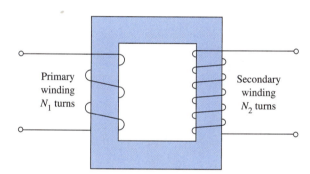

Primary winding
N_1 turns

Secondary winding
N_2 turns

FIGURE 19–1 A transformer.

28. A line, as shown in Fig. 19–2, is subdivided into two segments a and b such that the ratio of the smaller segment a to the larger segment b equals the ratio of the larger b to the whole $(a + b)$. The ratio a/b is called the *golden ratio* or *golden section*. Set up a proportion, based on the above definition, and compute the numerical value of this ratio.

FIGURE 19–2 A line subdivided by the golden ratio.

19–2 Direct Variation

Constant of Proportionality

If two variables are related by an equation of the following form:

Direct Variation	$y = kx$ or $y \propto x$	60

where k is a constant, we say that *y varies directly* as *x*, or that *y* is *directly proportional* to *x*. The constant *k* is called the *constant* of *proportionality*. As shown above, direct variation may also be written using the special symbol $\propto$:

$$y \propto x$$

Glance back at Sec. 5–4 where we plotted the equation (Eq. 289) of straight line

$$y = mx + b$$

We see that Equation 60 is the equation of a straight line with a slope of k and a y intercept of zero.

It is read "y varies directly as x" or "y is directly proportional to x."

Solving Variation Problems

Variation problems can be solved with or without evaluating the constant of proportionality. We first show a solution in which the constant *is* found by substituting the given values into Eq. 60.

◆◆◆ **Example 8:** If y is directly proportional to x, and y is 27 when x is 3, find y when x is 6.

Solution: Since y varies directly as x, we use Eq. 60.

$$y = kx$$

To find the constant of proportionality, we substitute the given values for x and y, 3 and 27.

$$27 = k(3)$$

So $k = 9$. Our equation is then

$$y = 9x$$

When $x = 6$,

$$y = 9(6) = 54 \qquad\qquad ◆◆◆$$

We now show how to solve such a problem without finding the constant of proportionality.

◆◆◆ **Example 9:** Solve Example 8 *without* finding the constant of proportionality.

Solution: When quantities vary directly with each other, we can set up a proportion and solve it. Let us represent the initial values of x and y by x_1 and y_1, and the second set of values by x_2 and y_2. Substituting each set of values into Eq. 60 gives us

$$y_1 = kx_1$$

and

$$y_2 = kx_2$$

We divide the second equation by the first, and k cancels.

$$\frac{y_2}{y_1} = \frac{x_2}{x_1}$$

The proportion says, The new y is to the old y as the new x is to the old x. We now substitute the old x and y (3 and 27), as well as the new x (6).

$$\frac{y_2}{27} = \frac{6}{3}$$

Solving yields

$$y_2 = 27\left(\frac{6}{3}\right) = 54 \quad \text{as before} \qquad ◆◆◆$$

Of course, the same proportion can also be written in the form $y_1/x_1 = y_2/x_2$. Two other forms are also possible:

$$\frac{x_1}{y_1} = \frac{x_2}{y_2}$$

and

$$\frac{x_1}{x_2} = \frac{y_1}{y_2}$$

◆◆◆ **Example 10:** If y varies directly as x, fill in the missing numbers in the table of values.

x	1	2			5	
y				16	20	28

Solution: We find the constant of proportionality from the given pair of values (5, 20). Starting with Eq. 60, we have

$$y = kx$$

and substituting gives

$$20 = k(5)$$

$$k = 4$$

So

$$y = 4x$$

With this equation we find the missing values.

When $x = 1$: $y = 4$
When $x = 2$: $y = 8$

When $y = 16$: $x = \dfrac{16}{4} = 4$

When $y = 28$: $x = \dfrac{28}{4} = 7$

So the completed table is

x	1	2	4	5	7
y	4	8	16	20	28

◆◆◆

Applications

Many practical problems can be solved using the idea of direct variation. Once you know that two quantities are directly proportional, you may assume an equation of the form of Eq. 60. Substitute the two given values to obtain the constant of proportionality, which you then put back into Eq. 60 to obtain the complete equation. From it you may find any other corresponding values.

Alternatively, you may decide not to find k, but to form a proportion in which three values will be known, enabling you to find the fourth.

◆◆◆ **Example 11:** The force F needed to stretch a spring (Fig. 19–3) is directly proportional to the distance x stretched. If it takes 15 N to stretch a certain spring 28 cm, how much force is needed to stretch it 34 cm?

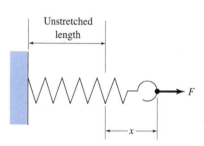

FIGURE 19–3

Estimate: We see that 15 N will stretch the spring 28 cm, or about 2 cm per newton. Thus a stretch of 34 cm should take about $34 \div 2$, or 17 N.

Solution: Assuming an equation of the form

$$F = kx$$

and substituting the first set of values, we have

$$15 = k(28)$$

$$k = \dfrac{15}{28}$$

So the equation is $F = \dfrac{15}{28}x$. When $x = 34$,

$$F = \left(\dfrac{15}{28}\right) 34 = 18 \text{ N} \quad \text{(rounded)}$$

which is close to our estimated value of 17 N.

◆◆◆

Exercise 2 ◆ Direct Variation

1. If y varies directly as x, and y is 56 when x is 21, find y when x is 74.
2. If w is directly proportional to z, and w has a value of 136 when z is 10.8, find w when z is 37.3.
3. If p varies directly as q, and p is 846 when q is 135, find q when p is 448.
4. If y is directly proportional to x, and y has a value of 88.4 when x is 23.8:
   (a) Find the constant of proportionality.
   (b) Write the equation $y = f(x)$.
   (c) Find y when $x = 68.3$.
   (d) Find x when $y = 164$.

Assuming that y varies directly as x, fill in the missing values in each table of ordered pairs.

5.

x	9	11	
y	45		75

6.

x	3.40	7.20		12.3
y		50.4	68.6	

7.

x	115	125		154
y			167	187

8. Graph the linear function $y = 2x$ for values of x from -5 to 5.

Applications

9. The distance between two cities is 828 km, and they are 29.5 cm apart on a map. Find the distance between two points 15.6 cm apart on the same map.
10. If the weight of 2500 steel balls is 3.65 kg, find the number of balls in 10.0 kg.
11. If 80 transformer laminations make a stack 1.75 cm thick, how many laminations are contained in a stack 3.00 cm thick?
12. If your car now gets 21.0 mi/gal of gas, and if you can go 250 mi on a tank of gasoline, how far could you drive with the same amount of gasoline with a car that gets 35.0 mi/gal?
13. A certain automobile engine delivers 53 hp and has a displacement (the total volume swept out by the pistons) of 3.0 liters. If the power is directly proportional to the displacement, what horsepower would you expect from a similar engine that has a displacement of 3.8 liters?
14. The resistance of a conductor is directly proportional to its length. If the resistance of 2.60 mi of a certain transmission line is 155 Ω, find the resistance of 75.0 mi of that line.
15. The resistance of a certain spool of wire is 1120 Ω. A piece 10.0 m long is found to have a resistance of 12.3 Ω. Find the length of wire on the spool.
16. If a certain machine can make 1850 parts in 55 min, how many parts can it make in 7.5 h? Work to the nearest part.
17. In Fig. 19–4, the constant force on the plunger keeps the pressure of the gas in the cylinder constant. The piston rises when the gas is heated and falls when the gas is cooled. If the volume of the gas is 1520 cm^3 when the temperature is 302 K, find the volume when the temperature is 358 K.
18. The power generated by a hydroelectric plant is directly proportional to the flow rate through the turbines, and a flow rate of 5625 gallons of water per minute produces 41.2 MW. How much power would you expect when a drought reduces the flow to 5000 gal/min?

Weight

Plunger

Compressed gas

FIGURE 19–4

For problem 17, use Charles' law: *The volume of a gas at constant pressure is directly proportional to its absolute temperature.* K is the abbreviation for kelvin, the SI absolute temperature scale. Add 273.15 to Celsius temperatures to obtain temperatures on the kelvin scale.

Computer and Graphics Calculator

19. Use a graphics calculator or a graphing utility on a computer to graph, on the same axes, the *family of curves*

$$y_1 = x \qquad y_2 = 2x$$
$$y_3 = 3x \qquad y_4 = \tfrac{1}{2}x$$

for $x = -1$ to 5. In what way do the four curves differ? What quantity in the equations causes the difference?

19–3   The Power Function

Definition

In Sec. 19–2, we saw that we could represent the statement "y varies directly as x" by Eq. 60,

$$y = kx$$

Similarly, if y varies directly as the *square* of x, we have

$$y = kx^2$$

or if y varies directly as the *square root* of x,

$$y = k\sqrt{x} = kx^{1/2}$$

These are all examples of the *power function*.

Power Function	$y = kx^n$	196

The constants k and n can be any positive or negative number. This simple function gives us a great variety of relations that are useful in technology and whose forms depend on the value of the exponent, n. A few of these are shown in the following table:

> The constants can, of course, be represented by any letter. Appendix A shows a instead of k.

> For exponents of 4 and 5, we have the *quartic* and *quintic* functions, respectively.

When:	We Get the:		Whose Graph Is a:
$n = 1$	Linear function	$y = kx$ (direct variation)	Straight line
$n = 2$	Quadratic function	$y = kx^2$	Parabola
$n = 3$	Cubic function	$y = kx^3$	Cubical parabola
$n = -1$		$y = \dfrac{k}{x}$ (inverse variation)	Hyperbola

The first three of these are special cases of the *general polynomial function of degree n.*

$y = a_0 x^n + a_1 x^{n-1} + \cdots + a_{n-3}x^3 + a_{n-2}x^2 + a_{n-1}x + a_n$	102

where n is a positive integer and the a's are constants. For example, the quadratic function $y = kx^2$ is a polynomial function of second degree, where $a_{n-2} = k$ and all of the other *a*'s are zeros. It is called *incomplete* because it is lacking an *x* term and a constant term.

We also consider power functions that are not polynomials. Those with noninteger positive powers will be covered in this section, and those with negative exponents in the following section on inverse variation.

Graph of the Power Function

The graph of a power function varies greatly, depending on the exponent. We first show some power functions with positive, integral exponents.

◆◆◆ **Example 12:** Graph the power functions $y = x^2$ and $y = x^3$ for a range of x from -3 to $+3$.

Solution: Make a table of values.

x	-3	-2	-1	0	1	2	3
$y = x^2$	9	4	1	0	1	4	9
$y = x^3$	-27	-8	-1	0	1	8	27

> You can, of course, use a graphics calculator or a computer to do this graph and the later ones, instead of doing a point-by-point plot.

> The graph of $y = kx^3$ is sometimes called a *cubical parabola*.

The curves $y = x^2$ and $y = x^3$ (shown dashed) are plotted in Fig. 19–5. Notice that the plot of $y = x^2$ has no negative values of y. This is typical of power functions that have *even* positive integers for exponents. The plot of $y = x^3$ does have negative y's for negative values of x. The shape of this curve is typical of a power function that has an *odd* positive integer as exponent.

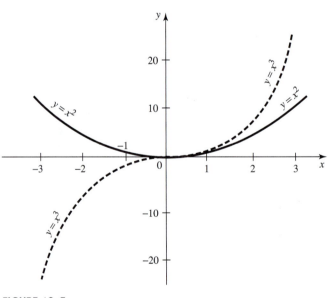

FIGURE 19–5

◆◆◆

When we are graphing a power function with a *fractional* exponent, the curve may not exist for negative values of *x*.

◆◆◆ **Example 13:** Graph the function $y = x^{1/2}$ for $x = -9$ to 9.

Solution: We see that for negative values of x, there are no real number values of y. For the nonnegative values, we get

x	0	1	2	3	4	5	6	7	8	9
y	0	1	1.41	1.73	2.00	2.24	2.45	2.65	2.83	3.00

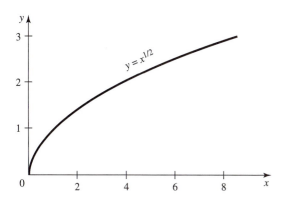

FIGURE 19–6

which are plotted in Fig. 19–6.   ◆◆◆

Solving Power Function Problems

As with direct variation (and with inverse variation and combined variation, treated later), we can solve problems involving the power function with or without finding the constant of proportionality. The following example will illustrate both methods.

◆◆◆ **Example 14:** If y varies directly as the $\frac{3}{2}$ power of x, and y is 54 when x is 9, find y when x is 25.

Solution by Solving for the Constant of Proportionality: Here we set up an equation with k and solve for the constant of proportionality. We let the exponent n in Eq. 196 be equal to $\frac{3}{2}$.

$$y = kx^{3/2}$$

As before, we evaluate the constant of proportionality by substituting the known values into the equation. Using $x = 9$ and $y = 54$, we obtain

$$54 = k(9)^{3/2} = k(27)$$

so $k = 2$. Our power function is then

$$y = 2x^{3/2}$$

This equation can then be used to find other pairs of corresponding values. For example, when $x = 25$,

$$y = 2(25)^{3/2}$$
$$= 2(\sqrt{25})^3$$
$$= 2(5)^3 = 250$$

Solution by Setting Up a Proportion: Here we set up a proportion. Three values are known, and we then solve for the fourth. If

$$y_1 = kx_1^{3/2}$$

and

$$y_2 = kx_2^{3/2}$$

then y_2 is to y_1 as $kx_2^{3/2}$ is to $kx_1^{3/2}$.

$$\frac{y_2}{y_1} = \left(\frac{x_2}{x_1}\right)^{3/2}$$

Substituting $y_1 = 54$, $x_1 = 9$, and $x_2 = 25$ yields

$$\frac{y_2}{54} = \left(\frac{25}{9}\right)^{3/2}$$

from which

$$y_2 = 54\left(\frac{25}{9}\right)^{3/2} = 250$$

◆◆◆

◆◆◆ **Example 15:** The horizontal distance S traveled by a projectile is directly proportional to the square of its initial velocity V. If the distance traveled is 1250 m when the initial velocity is 190 m/s, find the distance traveled when the initial velocity is 250 m/s.

Solution: The problem statement implies a power function of the form $S = kV^2$. Substituting the initial set of values gives us

$$1250 = k(190)^2$$

$$k = \frac{1250}{(190)^2} = 0.03463$$

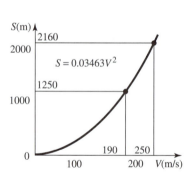

So our relationship is

$$S = 0.03463V^2$$

When $V = 250$ m/s,

$$S = 0.03463(250)^2 = 2160 \text{ m}$$

Check: We use our graphics calculator to plot $S = 0.03463V^2$, as in Fig. 19–7. Using the $\boxed{\text{TRACE}}$ feature, we check that $S = 2160$ when $V = 250$, that $S = 1250$ when $V = 190$, and that $S = 0$ when $V = 0$.   ◆◆◆

FIGURE 19–7

Some problems may contain no numerical values at all, as in the following example.

◆◆◆ **Example 16:** If y varies directly as the cube of x, by what factor will y change if x is increased by 25%?

Solution: The relationship between x and y is

$$y = kx^3$$

If we give subscripts 1 to the initial values and subscripts 2 to the final values, we may write the proportion

$$\frac{y_2}{y_1} = \frac{kx_2^3}{kx_1^3} = \left(\frac{x_2}{x_1}\right)^3$$

If the new x is 25% greater than the old x, then $x_2 = x_1 + 0.25x_1$ or $1.25x_1$. We thus substitute

$$x_2 = 1.25x_1$$

so

$$\frac{y_2}{y_1} = \left(\frac{1.25x_1}{x_1}\right)^3 = \left(\frac{1.25}{1}\right)^3 = 1.95$$

or

$$y_2 = 1.95y_1$$

So y has increased by a factor of 1.95. ◆◆◆

Similar Figures

We considered similar triangles in Chapter 6 and said that corresponding sides were in proportion. We now expand the idea to cover similar plane figures of *any* shape, and also similar *solids*. Similar figures (plane or solid) are those in which the distance between any two points on one of the figures is a *constant multiple* of the distance between two corresponding points on the other figure. In other words, if two corresponding dimensions are in the ratio of, say, 2:1, all other corresponding dimensions must be in the ratio of 2:1. In other words:

> We are using the term *dimension* to refer only to linear dimensions, such as lengths of sides or perimeters. It does not refer to angles, areas, or volumes.

Dimensions of Similar Figures	Corresponding dimensions of plane or solid similar figures are in proportion.	**134**

◆◆◆ **Example 17:** Two similar solids are shown in Fig. 19–8. Find the hole diameter D.

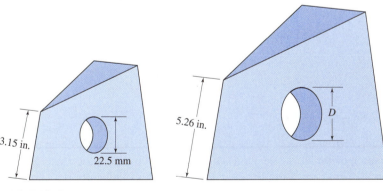

FIGURE 19–8 Similar solids.

Solution: By Statement 134,

$$\frac{D}{22.5} = \frac{5.26}{3.15}$$

$$D = \frac{22.5(5.26)}{3.15} = 37.6 \text{ mm}$$

◆◆◆

Note that since we are dealing with *ratios* of corresponding dimensions, it was not necessary to convert all dimensions to the same units.

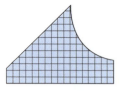

FIGURE 19–9 We can, in our minds, make the squares so small that they completely fill any irregular area. We use a similar idea in calculus when finding areas by integration.

Areas of Similar Figures

The area of a square of side s is equal to the square of one side, that is, $s \times s$ or s^2. Thus if a side is multiplied by a factor k, the area of the larger square is $(ks) \times (ks)$ or $k^2 s^2$. We see that the area has increased by a factor of k^2.

An area more complicated than a square can be thought of as made up of many small squares, as shown in Fig. 19–9. Then, if a dimension of that area is multiplied by k, the area of each small square increases by a factor of k^2, and hence the entire area of the figure increases by a factor of k^2. Thus:

Areas of Similar Figures	Areas of similar plane or solid figures are proportional to the *squares* of any two corresponding dimensions.	135

Thus if a figure has its sides doubled, the new area would be *four* times the original area. This relationship is valid not only for plane areas and for surface areas of solids, but also for cross-sectional areas of solids.

The units of measure for area are square inches (in.^{2}), square centimeters (cm^2), and so on, depending on the units used for the sides of the figure.

♦♦♦ **Example 18:** The triangular top surface of the smaller solid shown in Fig. 19–8 has an area of 4.84 in.^{2}. Find the area A of the corresponding surface on the larger solid.

Estimate: To bracket our answer, note that if the dimensions of the larger solid were double those of the smaller solid, the area of the top would be greater by a factor of 4, or about 19 in.^{2}. But the larger dimensions are less than double, so the area will be less than 19 in.^{2}. Thus we expect the area to be between 5 and 19 in.^{2}.

Solution: By Statement 135, the area A is to 4.84 as the *square* of the ratio of 5.26 to 3.15.

$$\frac{A}{4.84} = \left(\frac{5.26}{3.15}\right)^2$$

$$A = 4.84\left(\frac{5.26}{3.15}\right)^2 = 13.5 \text{ in.}^2$$

♦♦♦

Volumes of Similar Solids

A cube of side s has volume s^3. We can think of any solid as being made up of many tiny cubes, each of which has a volume equal to the cube of its side. Thus if the dimensions of the solid are multiplied by a factor of k, the volume of each cube (and hence the entire solid) will increase by a factor of k^3.

Volumes of Similar Figures	Volumes of similar solid figures are proportional to the *cubes* of any two corresponding dimensions.	136

Note that we know very little about the size and shape of these solids, yet we are able to compute the volume of the larger solid. The methods of this chapter give us another powerful tool for making *estimates*.

The units of measure for volume are cubic inches (in.^{3}), cubic centimeters (cm^3), and so on, depending on the units used for the sides of the solid figure.

♦♦♦ **Example 19:** If the volume of the smaller solid in Fig. 19–8 is 15.6 in.^{3}, find the volume V of the larger solid.

Solution: By Statement 136, the volume V is to 15.6 as the *cube* of the ratio of 5.26 to 3.15.

$$\frac{V}{15.6} = \left(\frac{5.26}{3.15}\right)^3$$

$$V = 15.6\left(\frac{5.26}{3.15}\right)^3 = 72.6 \text{ in.}^3$$

•••

Common Error	Students often forget to *square* corresponding dimensions when finding areas and to *cube* corresponding dimensions when finding volumes.

Scale Drawings

An important application of similar figures is in the use of *scale drawings,* such as maps, engineering drawings, surveying layouts, and so on. The ratio of distances on the drawing to corresponding distances on the actual object is called the *scale* of the drawing. Use Statement 135 to convert between the areas on the drawing and areas on the actual object.

••• **Example 20:** A certain map has a scale of 1:5000. How many acres on the land are represented by 168 in.^{2} on the map?

Solution: If $A =$ the area on the land, then, by Statement 135,

$$\frac{A}{168} = \left(\frac{5000}{1}\right)^2 = 25,000,000$$

$$A = 168(25,000,000) = 42.0 \times 10^8 \text{ in.}^2$$

Converting to acres, we have

$$A = 42.0 \times 10^8 \text{ in.}^2\left(\frac{1 \text{ ft}^2}{144 \text{ in.}^2}\right)\left(\frac{1 \text{ acre}}{43,560 \text{ ft}^2}\right) = 670 \text{ acres}$$

•••

Exercise 3   The Power Function

1. If y varies directly as the square of x, and y is 726 when x is 163, find y when x is 274.
2. If y is directly proportional to the square of x, and y is 5570 when x is 172, find y when x is 382.
3. If y varies directly as the cube of x, and y is 4.83 when x is 1.33, find y when x is 3.38.
4. If y is directly proportional to the cube of x, and y is 27.2 when x is 11.4, find y when x is 24.9
5. If y varies directly as the square of x, and y is 285.0 when x is 112.0, find y when x is 351.0.
6. If y varies directly as the square root of x, and y is 11.8 when x is 342, find y when x is 288.
7. If y is directly proportional to the cube of x, and y is 638 when x is 145, find y when x is 68.3
8. If y is directly proportional to the five-halves power of x, and y has the value 55.3 when x is 17.3:
   (a) Find the constant of proportionality.
   (b) Write the equation $y = f(x)$.
   (c) Find y when $x = 27.4$.
   (d) Find x when $y = 83.6$.

9. If y varies directly as the fourth power of x, fill in the missing values in the following table of ordered pairs:

x	18.2	75.6	
y	29.7		154

10. If y is directly proportional to the $\frac{3}{2}$ power of x, fill in the missing values in the following table of ordered pairs:

x	1.054	1.135	
y		4.872	6.774

11. If y varies directly as the cube root of x, fill in the missing values in the following table of ordered pairs:

x	315		782
y		148	275

12. Graph the power function $y = 1.04x^2$ for $x = -5$ to 5.
13. Graph the power function $y = 0.553x^5$ for $x = -3$ to 3.
14. Graph the power function $y = 1.25x^{3/2}$ for $x = 0$ to 5.

Freely Falling Body

From Eq. A18, we see that the distance fallen by a body (from rest) varies directly as the square of the elapsed time. Assume an initial velocity of zero for each of these problems.

15. If a body falls 176 m in 6.00 s, how far will it fall in 9.00 s?
16. If a body falls 4.90 m during the first second, how far will it fall during the third second?
17. If a body falls 129 ft in 2.00 s, how many seconds will it take to fall 525 ft?
18. If a body falls 738 units in 3.00 s, how far will it fall in 6.00 s?

Power in a Resistor

The power dissipated in a resistor varies directly as the square of the current in the resistor (Eq. A67).

19. If the power dissipated in a resistor is 486 W when the current is 2.75 A, find the power when the current is 3.45 A.
20. If the current through a resistor is increased by 28%, by what percent will the power increase?
21. By what factor must the current in an electric heating coil be increased to triple the power consumed by the heater?

Photographic Exposures

The exposure time for a photograph is directly proportional to the square of the f stop. (The f stop of a lens is its focal length divided by its diameter.)

Photographer's rule of thumb: *Double the exposure time for each increase in f stop.*

22. A certain photograph will be correctly exposed at a shutter speed of 1/100 s with a lens opening of f5.6. What shutter speed is required if the lens opening is changed to f8?
23. For the photograph in problem 22, what lens opening is needed for a shutter speed of 1/50 s?
24. Most cameras have the following f stops: f2.8, f4, f5.6, f8, f11, and f16. To keep the same correct exposure, by what factor must the shutter speed be increased when the lens is opened one stop?
25. A certain enlargement requires an exposure time of 24 s when the enlarger's lens is set at f22. What lens opening is needed to reduce the exposure time to 8 s?

Similar Figures

26. A container for storing propane gas is 0.755 m high and contains 20.0 liters. How high must a container of similar shape be to have a volume of 40.0 liters?

27. A certain wood stove has a firebox volume of 4.25 ft^3. What firebox volume would be expected if all dimensions of the stove were increased by a factor of 1.25?

28. If the stove in problem 27 weighed 327 lb, how much would the larger stove be expected to weigh?

29. A certain solar house stores heat in 155 metric tons of stones which are in a chamber beneath the house. Another solar house is to have a chamber of similar shape but with all dimensions increased by 15%. How many metric tons of stone will it hold?

30. Each side of a square is increased by 15.0 mm, and the area is seen to increase by 2450 mm². What were the dimensions of the original square?

31. The floor plan of a certain building has a scale of $\frac{1}{4}$ in. = 1 ft and shows a room having an area of 40 in.². What is the actual room area in square feet?

32. The area of a window of a car is 18.2 in.² on a drawing having a scale of 1:4. Find the actual window area in square feet.

33. A pipe 3.00 inches in diameter discharges 500 gal of water in a certain time. What must be the diameter of a pipe that will discharge 750 gal in the same time?

> Assume that the amount of flow through a pipe is proportional to its cross-sectional area.

34. If it cost $756 to put a fence around a circular pond, what will it cost to enclose another pond having $\frac{1}{5}$ the area?

35. A triangular field whose base is 215 m contains 12,400 m². Find the area of a field of similar form whose base is 328 m.

Computer and Graphics Calculator

36. The power function $y = 11.9x^{2.32}$ has been proposed as an equation to fit the following data:

x	1	2	3	4	5	6	7	8	9	10
y	11.9	59.4	152	292	496	754	1097	1503	1901	2433

Write a program or use a spreadsheet that will compute and print the following for each given x:
(a) The value of y from the formula.
(b) The difference between the value of y from the table and that obtained from the formula. This is called a *residual.*

37. Use your graphics calculator or computer to graph the family of curves

$$y = x \qquad y = x^2$$
$$y = x^3 \qquad y = x^4$$

in the same viewing window.

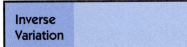 Inverse Variation

Definition

When we say that "y varies inversely as x" or that "y is inversely proportional to x," we mean that x and y are related by the following equation, where, as before, k is a constant of proportionality:

Inverse Variation	$y = \dfrac{k}{x} \quad \text{or} \quad y \propto \dfrac{1}{x}$	**61**

> We could also say that we have inverse variation when the product of x and y is a constant, that is, when $xy = k$.

The Hyperbola

The graph of the function $y = k/x$ is called a *hyperbola.* It is one of the *conic sections,* which you will study in more detail when you study analytic geometry (Chapter 22).

◆◆◆ **Example 21:** Graph the function $y = 2/x$ from $x = 0$ to $x = 5$.

Solution: Make a table of point pairs.

x	0	$\frac{1}{2}$	1	2	3	4	5
y	Not defined	4	2	1	$\frac{2}{3}$	$\frac{1}{2}$	$\frac{2}{5}$

These points are graphed in Fig. 19–10. Note that as x takes on smaller and smaller values, y gets larger and larger; and as x gets very large, y gets smaller and smaller but never reaches zero. Thus the curve gets ever closer to the x and y axes but never touches them. The x and y axes are *asymptotes* for this curve.

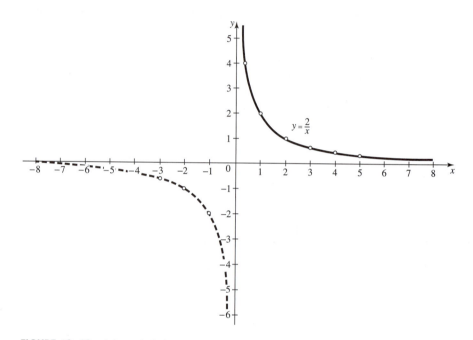

FIGURE 19–10   A hyperbola.

We did not plot the curve for negative values of x, but, if we had, we would have gotten the second branch of the hyperbola, shown dashed in the figure.

Power Function with Negative Exponent

Suppose, for example, that y varies inversely as the cube of x. This would be written

$$y = \frac{k}{x^3}$$

which is the same as

$$y = kx^{-3}$$

Note that this equation is a power function with a negative exponent. Such power functions are sometimes referred to as *hyperbolic*.

◆◆◆ **Example 22:** Rewrite $yx^2 = 10$.

Solution: Dividing by x^2, we see that

$$y = \frac{10}{x^2}$$

which can be rewritten as

$$y = 10x^{-2}$$

Therefore y is inversely proportional to the square of x. ◆◆◆

Solving Inverse Variation Problems

Inverse variation problems are solved by the same methods as for any other power function. As before, we can work these problems with or without finding the constant of proportionality.

◆◆◆ **Example 23:** If y is inversely proportional to x, and y is 8 when x is 3, find y when x is 12.

Solution: x and y are related by $y = k/x$. We find the constant of proportionality by substituting the given values of x and y (3 and 8).

$$8 = \frac{k}{3}$$
$$k = 3(8) = 24$$

So $y = 24/x$. When $x = 12$,

$$y = \frac{24}{12} = 2$$

◆◆◆

◆◆◆ **Example 24:** If y is inversely proportional to the square of x, by what factor will y change if x is tripled?

Solution: From the problem statement we know that $y = k/x^2$ or $k = x^2y$. If we use the subscript 1 for the original values and 2 for the new values, we can write

$$k = x_1^2 y_1 = x_2^2 y_2$$

Since the new x is triple the old, we substitute $3x_1$ for x_2.

$$x_1^2 y_1 = (3x_1)^2 y_2$$
$$= 9x_1^2 y_2$$
$$y_1 = 9y_2$$

or

$$y_2 = \frac{y_1}{9}$$

So the new y is one-ninth of its initial value. ◆◆◆

◆◆◆ **Example 25:** If y varies inversely as x, fill in the missing values in the table:

x	1			9	12
y			12	4	

Solution: Substitute the ordered pair (9, 4) into Eq. 61.

$$4 = \frac{k}{9}$$
$$k = 36$$

So

$$y = \frac{36}{x}$$

Filling in the table, we have

x	1	3	9	12
y	36	12	4	3

◆◆◆ **Example 26:** The number of oscillations N that a pendulum (Fig. 19–11) makes per unit time is inversely proportional to the square root of the length L of the pendulum, and $N = 2.00$ oscillations per second when $L = 85.0$ cm.

(a) Write the equation $N = f(L)$.
(b) Find N when $L = 115$ cm.

Solution:

(a) The equation will be of the form

$$N = \frac{k}{\sqrt{L}}$$

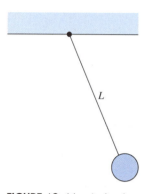

FIGURE 19–11   A simple pendulum.

Substituting $N = 2.00$ and $L = 85.0$ gives us

$$2.00 = \frac{k}{\sqrt{85.0}}$$

$$k = 2.00\sqrt{85.0} = 18.4$$

So the equation is

$$N = \frac{18.4}{\sqrt{L}}$$

(b) When $L = 115$,

$$N = \frac{18.4}{\sqrt{115}} = 1.72 \text{ oscillations per second}$$

Check: Does the answer seem reasonable? We see that N has decreased, which we expect with a longer pendulum, but it has not decreased by much. As a check, we note that L has increased by a factor of $115 \div 85$, or 1.35, so we expect N to decrease by a factor of $\sqrt{1.35}$, or 1.16. Dividing, we find that $2.00 \div 1.72 = 1.16$, so the answer checks.   ◆◆◆

Exercise 4 ◆ Inverse Variation

1. If y varies inversely as x, and y is 385 when x is 832, find y when x is 226.
2. If y is inversely proportional to the square of x, and y has the value 1.55 when x is 7.38, find y when x is 44.2.
3. If y is inversely proportional to x, how does y change when x is doubled?
4. If y varies inversely as x, and y has the value 104 when x is 532:
   (a) Find the constant of proportionality.
   (b) Write the equation $y = f(x)$.
   (c) Find y when x is 668.
   (d) Find x when y is 226.
5. If y is inversely proportional to x, fill in the missing values.

x	306	622	
y	125		418

6. Fill in the missing values, assuming that y is inversely proportional to the square of x.

x	2.69		7.83
y		1.16	1.52

7. If y varies inversely as the square root of x, fill in the missing values.

x		3567	5725
y	1136	1828	

8. If y is inversely proportional to the cube root of x, by what factor will y change when x is tripled?

9. If y is inversely proportional to the square root of x, by what percentage will y change when x is decreased by 50.0%?

10. Plot the curve $y = 5/x^2$ for the domain 0 to 5.

11. Plot the function $y = 1/x$ for the domain 0 to 10.

Boyle's Law

Boyle's law states that for a confined gas at a constant temperature, the product of the pressure and the volume is a constant. Another way of stating this law is that the pressure is inversely proportional to the volume or that the volume is inversely proportional to the pressure. Assume a constant temperature in the following problems.

12. A certain quantity of gas, when compressed to a volume of 2.50 m^3, has a pressure of 184 Pa. Find the pressure resulting when that gas is further compressed to 1.60 m^3.

13. The air in the cylinder in Fig. 19–4 is at a pressure of 14.7 lb/in.^{2} and occupies a volume of 175 in.^{3}. Find the pressure when it is compressed to 25.0 in.^{3}.

14. A balloon contains 320 m^3 of gas at a pressure of 140,000 Pa. What would the volume be if the same quantity of gas were at a pressure of 250,000 Pa?

> The *pascal* (Pa) is the SI unit of pressure. It equals 1 newton per square meter.

Gravitational Attraction

15. The force of attraction between two certain steel spheres is 3.75×10^{-5} dyne when the spheres are placed 18.0 cm apart. Find the force of attraction when they are 52.0 cm apart.

16. How far apart must the spheres in problem 15 be placed to cause the force of attraction between them to be 3.00×10^{-5} dyne?

17. The force of attraction between the earth and some object is called the *weight* of that object. The law of gravitation states, then, that the weight of an object is inversely proportional to the square of its distance from the center of the earth. If a person weighs 150 lb on the surface of the earth (assume this to be 3960 mi from the center), how much will he weigh 1500 mi above the surface of the earth?

18. How much will a satellite, whose weight on earth is 675 N, weigh at an altitude of 250 km above the surface of the earth?

> Newton's law of gravitation states that any two bodies attract each other with a force that is inversely proportional to the square of the distance between them.

Illumination

19. A certain light source produces an illumination of 800 lux (a lux is 1 lumen per square meter) on a surface. Find the illumination on that surface if the distance to the light source is doubled.

20. A light source located 2.75 m from a surface produces an illumination of 528 lux on that surface. Find the illumination if the distance is changed to 1.55 m.

21. A light source located 7.50 m from a surface produces an illumination of 426 lux on that surface. At what distance must that light source be placed to give an illumination of 850 lux?

22. When a document is photographed on a certain copy stand, an exposure time of $\frac{1}{25}$ s is needed, with the light source 0.750 m from the document. At what distance must the light be located to reduce the exposure time to $\frac{1}{100}$ s?

> The *inverse square law* states that for a surface illuminated by a light source, the intensity of illumination on the surface is inversely proportional to the *square* of the distance between the source and the surface.

Electrical

23. The current in a resistor is inversely proportional to the resistance. By what factor will the current change if a resistor increases 10.0% due to heating?

24. The resistance of a wire is inversely proportional to the square of its diameter. If an AWG size 12 conductor (0.0808-in. diameter) has a resistance of 14.8 Ω, what will be the resistance of an AWG size 10 conductor (0.1019-in. diameter) of the same length and material?

25. The capacitive reactance X_C of a circuit varies inversely as the capacitance C of the circuit. If the capacitance of a certain circuit is decreased by 25.0%, by what percentage will X_C change?

Computer and Graphics Calculator

26. Use a graphics calculator or computer to graph

$$y = 1/x \quad \text{and} \quad y = 1/x^2$$

in the same viewing window from $x = -4$ to 4. How do you explain the different behavior of the two functions for negative values of x?

19–5   Functions of More Than One Variable

Joint Variation

So far in this chapter, we have considered only cases where y was a function of a *single* variable x. In functional notation, this is represented by $y = f(x)$. In this section we cover functions of two or more variables, such as

$$y = f(x, w)$$
$$y = f(x, w, z)$$

and so forth. When y varies directly as x and w, we say that y varies *jointly* as x and w. The three variables are related by the following equation, where, as before, k is a constant of proportionality:

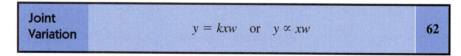

Joint Variation	$y = kxw \quad \text{or} \quad y \propto xw$	62

◆◆◆ **Example 27:** If y varies jointly as x and w, how will y change when x is doubled and w is one-fourth of its original value?

Solution: Let y' be the new value of y obtained when x is replaced by $2x$ and w is replaced by $w/4$, while the constant of proportionality k, of course, does not change. Substituting in Eq. 62, we obtain

$$y' = k(2x)\left(\frac{w}{4}\right)$$
$$= \frac{kxw}{2}$$

But, since $kxw = y$, then

$$y' = \frac{y}{2}$$

So the new y is half as large as its former value.   ◆◆◆

Combined Variation

When one variable varies with two or more variables in ways that are more complex than in Eq. 62, it is referred to as *combined* variation. This term is applied to many different kinds of relationships, so a formula cannot be given that will cover all types. You must carefully read the problem statement in order to write the equation. Once you have the equation, the solution of combined variation problems is no different than for other types of variation.

◆◆◆ **Example 28:** If y varies directly as the cube of x and inversely as the square of z, we write

$$y = \frac{kx^3}{z^2}$$

◆◆◆

◆◆◆ **Example 29:** If y is directly proportional to x and the square root of w, and inversely proportional to the square of z, we write

$$y = \frac{kx\sqrt{w}}{z^2}$$

◆◆◆

◆◆◆ **Example 30:** If y varies directly as the square root of x and inversely as the square of w, by what factor will y change when x is made four times larger and w is tripled?

Solution: First write the equation linking y to x and w, including a constant of proportionality k.

$$y = \frac{k\sqrt{x}}{w^2}$$

We get a new value for y (let's call it y') when x is replaced by $4x$ and w is replaced by $3w$. Thus

$$y' = \frac{k\sqrt{4x}}{(3w)^2} = k\frac{2\sqrt{x}}{9w^2}$$

$$= \frac{2}{9}\left(\frac{k\sqrt{x}}{w^2}\right) = \frac{2}{9}y$$

We see that y' is $\frac{2}{9}$ as large as the original y.

◆◆◆

◆◆◆ **Example 31:** If y varies directly as the square of w and inversely as the cube root of x, and $y = 225$ when $w = 103$ and $x = 157$, find y when $w = 126$ and $x = 212$.

Solution: From the problem statement,

$$y = k\frac{w^2}{\sqrt[3]{x}}$$

Solving for k, we have

$$225 = k\frac{(103)^2}{\sqrt[3]{157}} = 1967k$$

or $k = 0.1144$. So

$$y = \frac{0.1144w^2}{\sqrt[3]{x}}$$

When $w = 349$ and $x = 212$,

$$y = \frac{0.1144(126)^2}{\sqrt[3]{212}} = 305$$

◆◆◆

◆◆◆ **Example 32:** y is directly proportional to the square of x and inversely proportional to the cube of w. By what factor will y change if x is increased by 15% and w is decreased by 20%?

Solution: The relationship between x, y, and w is

$$y = k\frac{x^2}{w^3}$$

so we can write the proportion

$$\frac{y_2}{y_1} = \frac{k\dfrac{x_2^2}{w_2^3}}{k\dfrac{x_1^2}{w_1^3}} = \frac{x_2^2}{w_2^3} \cdot \frac{w_1^3}{x_1^2} = \left(\frac{x_2}{x_1}\right)^2\left(\frac{w_1}{w_2}\right)^3$$

We now replace x_2 with $1.15x_1$ and replace w_2 with $0.8w_1$.

$$\frac{y_2}{y_1} = \left(\frac{1.15x_1}{x_1}\right)^2\left(\frac{w_1}{0.8w_1}\right)^3 = \frac{(1.15)^2}{(0.8)^3} = 2.58$$

So y will increase by a factor of 2.58.   ◆◆◆

Exercise 5 ◆ Functions of More Than One Variable

Joint Variation

1. If y varies jointly as w and x, and y is 483 when x is 742 and w is 383, find y when x is 274 and w is 756.

2. If y varies jointly as x and w, by what factor will y change if x is tripled and w is halved?

3. If y varies jointly as w and x, by what percent will y change if w is increased by 12% and x is decreased by 7.0%?

4. If y varies jointly as w and x, and y is 3.85 when w is 8.36 and x is 11.6, evaluate the constant of proportionality, and write the complete expression for y in terms of w and x.

5. If y varies jointly as w and x, fill in the missing values.

w	x	y
46.2	18.3	127
19.5	41.2	
	8.86	155
12.2		79.8

Combined Variation

6. If y is directly proportional to the square of x and inversely proportional to the cube of w, and y is 11.6 when x is 84.2 and w is 28.4, find y when x is 5.38 and w is 2.28.

7. If y varies directly as the square root of w and inversely as the cube of x, by what factor will y change if w is tripled and x is halved?

8. If y is directly proportional to the cube root of x and to the square root of w, by what percent will y change if x and w are both increased by 7.0%?

9. If y is directly proportional to the $\frac{3}{2}$ power of x and inversely proportional to w, and y is 284 when x is 858 and w is 361, evaluate the constant of proportionality, and write the complete equation for y in terms of x and w.

10. If y varies directly as the cube of x and inversely as the square root of w, fill in the missing values in the table.

w	x	y
1.27		3.05
	5.66	1.93
4.66	2.75	3.87
7.07	1.56	

Geometry

11. The area of a triangle varies jointly as its base and altitude. By what percent will the area change if the base is increased by 15% and the altitude decreased by 25%?

12. If the base and the altitude of a triangle are both halved, by what factor will the area change?

Electrical Applications

13. When an electric current flows through a wire, the resistance to the flow varies directly as the length and inversely as the cross-sectional area of the wire. If the length and the diameter are both tripled, by what factor will the resistance change?

14. If 750 m of 3.00-mm-diameter wire has a resistance of 27.6 Ω, what length of similar wire 5.00 mm in diameter will have the same resistance?

Gravitation

15. Newton's law of gravitation states that every body in the universe attracts every other body with a force that varies directly as the product of their masses and inversely as the square of the distance between them. By what factor will the force change when the distance is doubled and each mass is tripled?

16. If both masses are increased by 60% and the distance between them is halved, by what percent will the force of attraction increase?

Illumination

17. The intensity of illumination at a given point is directly proportional to the intensity of the light source and inversely proportional to the square of the distance from the light source. If a desk is properly illuminated by a 75.0-W lamp 8.00 ft from the desk, what size lamp will be needed to provide the same lighting at a distance of 12.0 ft?

18. How far from a 150-candela light source would a picture have to be placed so as to receive the same illumination as when it is placed 12 m from an 85-candela source?

Gas Laws

19. The volume of a given weight of gas varies directly as the absolute temperature t and inversely as the pressure p. If the volume is 4.45 m^3 when $p = $ 225 kilopascals (kPa) and $t = $ 305 K, find the volume when $p = $ 325 kPa and $t = $ 354 K.

20. If the volume of a gas is 125 ft^3, find its volume when the absolute temperature is increased 10% and the pressure is doubled.

Work

21. The amount paid to a work crew varies jointly as the number of persons working and the length of time worked. If 5 workers earn $5123.73 in 3.0 weeks, in how many weeks will 6 workers earn a total of $6148.48?

22. If 5 bricklayers take 6.0 days to finish a certain job, how long would it take 7 bricklayers to finish a similar job requiring 4 times the number of bricks?

Strength of Materials

23. The maximum safe load of a rectangular beam (Fig. 19–12) varies jointly as the width and the square of the depth and inversely as the length of the beam. If a beam 8.00 in. wide, 11.5 in. deep, and 16.0 ft long can safely support 15,000 lb, find the safe load for a beam 6.50 in. wide, 13.4 in. deep, and 21.0 ft long made of the same material.

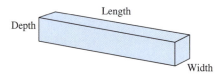

FIGURE 19–12   A rectangular beam.

24. If the width of a rectangular beam is increased by 11%, the depth decreased by 8%, and the length increased 6%, by what percent will the safe load change?

Mechanics

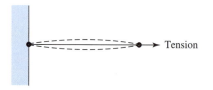

FIGURE 19–13   A stretched wire.

25. The number of vibrations per second made when a stretched wire (Fig. 19–13) is plucked varies directly as the square root of the tension in the wire and inversely as the length. If a 1.00-m-long wire will vibrate 325 times a second when the tension is 115 N, find the frequency of vibration if the wire is shortened to 0.750 m and the tension is decreased to 95.0 N.

26. The kinetic energy of a moving body is directly proportional to its mass and the square of its speed. If the mass of a bullet is halved, by what factor must its speed be increased to have the same kinetic energy as before?

Fluid Flow

27. The time needed to empty a vertical cylindrical tank (Fig. 19–14) varies directly as the square root of the height of the tank and the square of the radius. By what factor will the emptying time change if the height is doubled and the radius increased by 25%?

28. The power available in a jet of liquid is directly proportional to the cross-sectional area of the jet and to the cube of the velocity. By what factor will the power increase if the area and the velocity are both increased 50%?

FIGURE 19–14   A vertical cylindrical tank.

✦✦✦ CHAPTER 19 REVIEW PROBLEMS ✦✦✦✦✦✦✦✦✦✦✦✦✦✦✦✦✦✦✦✦✦✦✦✦✦✦✦✦✦

1. If y varies inversely as x, and y is 736 when x is 822, find y when x is 583.

2. If y is directly proportional to the $\frac{5}{2}$ power of x, by what factor will y change when x is tripled?

3. If y varies jointly as x and z, by what percent will y change when x is increased by 15% and z is decreased by 4%?

4. The braking distance of an automobile varies directly as the square of the speed. If the braking distance of a certain automobile is 34.0 ft at 25.0 mi/h, find the braking distance at 55.0 mi/h.

5. The rate of flow of liquid from a hole in the bottom of a tank is directly proportional to the square root of the liquid depth. If the flow rate is 225 L/min when the depth is 3.46 m, find the flow rate when the depth is 1.00 m.

6. The power needed to drive a ship varies directly as the cube of the speed of the ship, and a 77.4-hp engine will drive a certain ship at 11.2 knots. Find the horsepower needed to propel that ship at 18.0 knots.

7. If the tensile strength of a cylindrical steel bar varies as the square of its diameter, by what factor must the diameter be increased to triple the strength of the bar?

8. The life of an incandescent lamp varies inversely as the 12th power of the applied voltage, and the light output varies directly as the 3.5th power of the applied voltage. By what factor will the life increase if the voltage is lowered by an amount that will decrease the light output by 10%?

9. One of Kepler's laws states that the time for a planet to orbit the sun varies directly as the $\frac{3}{2}$ power of its distance from the sun. How many years will it take for Saturn, which is about $9\frac{1}{2}$ times as far from the sun as is the earth, to orbit the sun?

10. The volume of a cone varies directly as the square of the base radius and as the altitude. By what factor will the volume change if the altitude is doubled and the base radius is halved?

11. The number of oscillations made by a pendulum in a given time is inversely proportional to the length of the pendulum. A certain clock with a 30.00-in.-long pendulum is losing 15.00 min/day. Should the pendulum be lengthened or shortened, and by how much?

12. A trucker usually makes a trip in 18.0 h at an average speed of 60.0 mi/h. Find the traveling time if the speed were reduced to 50.0 mi/h.

13. The force on the vane of a wind generator varies directly as the area of the vane and the square of the wind velocity. By what factor must the area of a vane be increased so that the wind force on it will be the same in a 12-mi/h wind as it was in a 35-mi/h wind?

14. The maximum deflection of a rectangular beam varies inversely as the product of the width and the cube of the depth. If the deflection of a beam having a width of 15 cm and a depth of 35 cm is 7.5 mm, find the deflection if the width is made 20 cm and the depth 45 cm.

15. When an object moves at a constant speed, the travel time is inversely proportional to the speed. If a satellite now circles the earth in 26 h 18 min, how long will it take if booster rockets increase the speed of the satellite by 15%?

16. The life of an incandescent lamp is inversely proportional to the 12th power of the applied voltage. If a lamp has a life of 2500 h when run at 115 V, what will its life expectancy be when it is run at 110 V?

17. How far from the earth will a spacecraft be equally attracted by the earth and the moon? The distance from the earth to the moon is approximately 239,000 mi, and the mass of the earth is about 82.0 times that of the moon.

18. Ohm's law states that the electric current flowing in a circuit varies directly as the applied voltage and inversely as the resistance. By what percent will the current change when the voltage is increased by 25% and the resistance increased by 15%?

19. The allowable strength F of a column varies directly as its length L and inversely as its radius of gyration r. If $F = 35$ kg when L is 12 m and r is 3.6 cm, find F when L is 18 m and r is 4.5 cm.

20. The time it takes a pendulum to go through one complete oscillation (the *period*) is directly proportional to the square root of its length. If the period of a 1-m pendulum is 1.25 s, how long must the pendulum be to have a period of 2.5 s?

21. If the maximum safe current in a wire is directly proportional to the $\frac{3}{2}$ power of the wire diameter, by what factor will the safe current increase when the wire diameter is doubled?

22. Four workers take ten 6-h days to finish a job. How many workers are needed to finish a similar job which is 3 times as large, in five 8-h days?

23. When a jet of water strikes the vane of a water turbine and is deflected through an angle θ, the force on the vane varies directly as the square of the jet velocity and the sine of $\theta/2$. If θ is decreased from 55° to 40°, and if the jet velocity increases by 40%, by what percentage will the force change?

24. By what factor will the kinetic energy change if the speed of a projectile is doubled and its mass is halved? Use the fact that the kinetic energy of a moving body is directly proportional to the mass and the square of the velocity of the body.

25. The fuel consumption for a ship is directly proportional to the cube of the ship's speed. If a certain tanker uses 1584 gal of diesel fuel on a certain run at 15.0 knots, how much fuel would it use for the same run when the speed is reduced to 10.0 knots? A *knot* is equal to 1 nautical mile per hour. (1 nautical mile = 1.151 statute miles.)

26. The diagonal of a certain square is doubled. By what factor does the area change?

27. A certain parachute is made from 52.0 m² of fabric. If all of its dimensions are increased by a factor of 1.40, how many square meters of fabric will be needed?

28. The rudder of a certain airplane has an area of 7.50 ft². By what factor must the dimensions be scaled up to triple the area?

29. A 1.5-in.-diameter pipe fills a cistern in 5.0 h. Find the diameter of a pipe that will fill the same cistern in 9.0 h.

30. If 315 m of fence will enclose a circular field containing 2.0 acres, what length will enclose 8.0 acres?

31. The surface area of a one-quarter model of an automobile measures 1.10 m². Find the surface area of the full-sized car.

32. A certain water tank is 25.0 ft high and holds $50\overline{0}0$ gal. How much would a 35.0-ft-high tank of similar shape hold?

33. A certain tractor weighs 5.80 tons. If all of its dimensions were scaled up by a factor of 1.30, what would the larger tractor be expected to weigh?

34. If a trough 5.0 ft long holds 12 pailfuls of water, how many pailfuls will a similar trough hold that is 8.0 ft long?

35. A ball 4.50 inches in diameter weighs 18.0 oz. What is the weight of another ball of the same density that is 9.00 inches in diameter?

36. A ball 4.00 inches in diameter weighs 9.00 lb. What is the weight of a ball 25.0 cm in diameter made of the same material?

37. A worker's contract has a cost-of-living clause that requires that his salary be proportional to the cost-of-living index. If he earned $450 per week when the index was 9.40, how much should he earn when the index is 12.7?

38. A woodsman pays a landowner $8.00 for each cord of wood he cuts on the landowner's property, and he sells the wood for $75 per cord. When the price of firewood rose to $95, the landowner asked for a proportional share of the price. How much should he get per cord?

39. The series of numbers 3.5, 5.25, 7.875, . . . form a *geometric progression,* where the ratio of any term to the one preceding it is a constant. This is called the *common ratio.* Find this ratio.

40. The *specific gravity* (SG) of a solid or liquid is the ratio of the density of the substance to the density of water at a standard temperature (Eq. A45). Taking the density of water as 62.4 lb/ft³, find the density of copper having a specific gravity of 8.89.

Specific Gravity	$SG = \dfrac{\text{density of substance}}{\text{density of water}}$	**A45**

41. An iron casting having a surface area of 746 cm^2 is heated so that all its dimensions increase by 1.00%. Find the new area.
42. A certain boiler pipe will permit a flow of 35.5 gal/min. Buildup of scale inside the pipe eventually reduces its diameter to three-fourths of its previous value. Assuming that the flow rate is proportional to the cross-sectional area, find the new flow rate for the pipe.

Writing

43. Suppose that your company makes plastic trays and is planning new ones with dimensions double those now being made. Your company president is convinced that they will need only twice as much plastic as the older version. "Twice the size, twice the plastic," he proclaims, and no one is willing to challenge him. Your job is to make a presentation to the president where you tactfully point out that he is wrong and where you explain that the new trays will require eight times as much plastic. Write your presentation.

Team Project

44. We saw that the volumes (and hence the weights) of solids are proportional to the *cube* of corresponding dimensions. Applying that idea, suppose that a sporting goods company has designed a new line of equipment (ski packages, wind surfers, diving gear, jet skis, clothing, etc.) based on the following statement:

> The weights of people of similar build are proportional to the cube of their heights.

The goal of this project is to prove or disprove the given statement. Your team will use data gathered from students on your campus in reaching a conclusion.

20

Exponential and Logarithmic Functions

◆◆◆ OBJECTIVES ◆◆◆

When you have completed this chapter, you should be able to:

- Graph the exponential function.
- Solve exponential growth and decay problems by formula or by the universal growth and decay curves.
- Convert expressions between exponential and logarithmic form.
- Evaluate common and natural logarithms and antilogarithms.
- Evaluate, manipulate, and simplify logarithmic expressions.
- Solve exponential and logarithmic equations.
- Solve applied problems involving exponential or logarithmic equations.
- Make graphs on logarithmic or semilogarithmic paper.
- Graph empirical data and determine an approximate equation describing the data.

◆◆◆

You may have heard the story about the inventor of chess, who asked that the reward for his invention be a grain of wheat on the first square of the chessboard, two grains on the second square, four on the third, and so on, each square getting double the amount as on the previous square. It turns out that he would receive more wheat than has been grown by man during all history—an illustration of the remarkable properties of the *exponential function,* which we study in this chapter.

We also study the *logarithmic function* and show how it is related to the exponential function, and we do a few logarithmic computations. We then solve exponential and logarithmic equations and use them in some interesting applications. Finally we do some graphing on logarithmic and semilogarithmic paper.

20–1 The Exponential Function

Definition

An *exponential function* is one in which the independent variable appears in the exponent. The quantity that is raised to the power is called the *base*.

◆◆◆ **Example 1:** The following are exponential functions if a, b, and e represent positive constants:

$$y = 5^x \qquad y = b^x \qquad y = e^x \qquad y = 10^x$$
$$y = 3a^x \qquad y = 7^{x-3} \qquad y = 5e^{-2x}$$

◆◆◆

Realize that these are different from the *power function, $y = ax^n$.* Here the unknown is *in the exponent.*

We shall see that e, like π, has a specific value which we will discuss later in this chapter.

Graph of the Exponential Function

We may make a graph of an exponential function in the same way as we graphed other functions in Chapter 5. Choose values of the independent variable x throughout the region of interest, and for each x compute the corresponding value of the dependent variable y. Plot the resulting table of ordered pairs, and connect the points with a smooth curve.

◆◆◆ **Example 2:** Graph the exponential function $y = 2^x$ from $x = -4$ to $x = +3$. For comparison, plot the power function $y = x^2$ on the same axes.

Solution: We make our graphs, shown in Fig. 20–1, by using a graphics calculator or a computer or by computing and plotting point pairs.

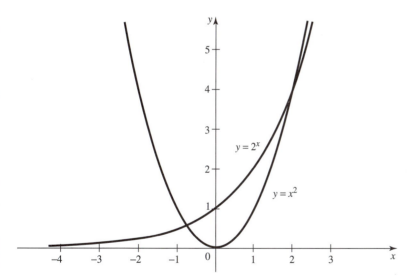

FIGURE 20–1

Note that the graph of the exponential function is completely different from the graph of the power function whose graph is a parabola. For the exponential function, note that as x increases, y increases even more greatly, and that the curve gets steeper and steeper. This is a characteristic of *exponential growth,* which we cover more fully below. Also note that the y intercept is 1 and that there is no x intercept, the x axis being an asymptote. Finally, note that each successive y value is two times the previous one. We say that y *increases geometrically.* ◆◆◆

Sometimes the exponent in an exponential function is negative. We will see in the following example what effect a negative exponent has on the graph of the function.

◆◆◆ **Example 3:** Graph the exponential functions $y = 2^x$ and $y = 2^{-x}$, for $x = -4$ to 4, on the same axes.

Solution: We have already graphed $y = 2^x$ in the preceding example, and now we add the graph of $y = 2^{-x}$, shown in Fig. 20–2.

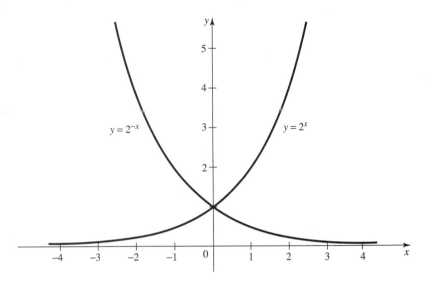

FIGURE 20–2

Note that when the exponent is *positive,* the graph *increases exponentially,* and that when the exponent is *negative,* the graph *decreases exponentially.* Also notice that neither curve reaches the x axis. Thus for the exponential functions $y = 2^x$ and $y = 2^{-x}$, y cannot equal zero.   ◆◆◆

Exponential functions often have the irrational number e as the base. We will see why in the following section. For now, we will treat e like any other number, using $e \approx 2.718$ for its approximate value.

◆◆◆ **Example 4:** Graph the exponential function $y = 1.85e^{-x/2}$, for $x = -4$ to 4, if e equals 2.718.

Solution: Scientific calculators have a key, $\boxed{e^x}$, for raising e to any power. On some calculators you press the $\boxed{e^x}$ key and then enter the exponent; on other calculators you enter the exponent and then press the $\boxed{e^x}$ key. With the aid of that key, we make our graph, shown in Fig. 20–3, using a graphing utility, or by making a table of point pairs.

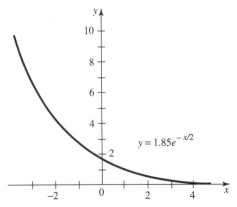

FIGURE 20–3

Note that the negative sign in the exponent has made the graph *decrease* exponentially as x increases, and that the curve approaches the x axis as an asymptote. The shape of the curve, however, is that of an exponential function. ◆◆◆

Compound Interest

In this and the next few sections, we will see some problems that require exponential equations for their solution.

When an amount of money a (called the *principal*) is invested at an interest rate n (expressed as a decimal), it will earn an amount of interest (an) at the end of the first interest period. If this interest is then added to the principal, and if both continue to earn interest, we have what is called *compound interest,* the type of interest commonly given on bank accounts.

Suppose that the interest rate is 5% (0.05). As the balance increases, the interest received for the current period (5% of a or $0.05a$) is added to the previous balance (100% of a or $1.00a$) to give a new total of $1.05a$. So we could say we multiply a by $(1 + 0.05)$ or 1.05.

Similarly, for an interest rate of n, the balance y_t at the end of any period t is obtained by multiplying the preceding balance y_{t-1} by $(1 + n)$.

Recursion Relation for Compound Interest	$y_t = y_{t-1}(1 + n)$	A10b

This equation is called a *recursion relation,* from which each term in a series of numbers may be found from the term immediately preceding it. We'll do more with recursion relations in Chapter 25, but now we will use Eq. A10b to make a table showing the balance at the end of a number of interest periods.

Period	Balance y at End of Period
1	$a(1 + n)$
2	$a(1 + n)^2$
3	$a(1 + n)^3$
4	$a(1 + n)^4$
.	.
.	.
.	.
t	$a(1 + n)^t$

We thus get the following exponential function, with a equal to the principal, n equal to the rate of interest, and t the number of interest periods:

Compound Interest Computed Once per Period	$y = a(1 + n)^t$	A10a

◆◆◆ **Example 5:** To what amount (to the nearest cent) will an initial deposit of $500 accumulate if invested at a compound interest rate of $6\frac{1}{2}\%$ per year for 8 years?

Solution: We are given

Principal $a = \$500$

Interest rate $n = 0.065$, expressed in decimal form

Number of interest periods $t = 8$

Substituting into Eq. A10a gives

$$y = 500(1 + 0.065)^8$$
$$= \$827.50$$

◆◆◆

Compound Interest Computed *m* Times per Period

In Sec. 20–2, we will use this equation to derive an equation for exponential (continuous) growth.

We can use Eq. A10a to compute, say, interest on a savings account for which the interest is computed once a year. But what if the interest is computed more often, as most banks do, for example, monthly, weekly, or even daily?

Let us modify the compound interest formula. Instead of computing interest *once* per period, let us compute it *m times* per period. The exponent in Eq. A10 thus becomes *mt*. The interest rate *n*, however, has to be reduced to (1/*m*)th of its old value, or to *n/m*.

Compound Interest Computed *m* Times per Period	$y = a\left(1 + \dfrac{n}{m}\right)^{mt}$	**A11**

◆◆◆ **Example 6:** Repeat the computation of Example 5 with the interest computed monthly rather than annually.

Solution: As in Example 5, $a = \$500$, annual interest rate $n = 0.065$, and $t = 8$. But *n* is an annual rate, so we must use Eq. A11, with $m = 12$.

$$y = 500\left(1 + \frac{0.065}{12}\right)^{12(8)} = 500(1.00542)^{96}$$
$$= \$839.83$$

or \$12.33 more than in Example 5.

◆◆◆

Exercise 1 ◆ The Exponential Function

Exponential Function

Graph each exponential function over the given domains of *x*. Take $e = 2.718$.

1. $y = 0.2(3.2)^x$ $(x = -4 \text{ to } +4)$
2. $y = 3(1.5)^{-2x}$ $(x = -1 \text{ to } 5)$
3. $y = 5(1 - e^{-x})$ $(x = 0 \text{ to } 10)$
4. $y = 4e^{x/2}$ $(x = 0 \text{ to } 4)$
5. A rope or chain between two points hangs in the shape of a curve called a *catenary* (Fig. 20–4). Its equation is

$$y = \frac{a(e^{x/a} + e^{-x/a})}{2}$$

Using values of $a = 2.5$ and $e = 2.718$, plot *y* for $x = -5$ to 5.

FIGURE 20–4 A rope or chain hangs in the shape of a catenary. Mount your finished graph for problem 5 vertically, and hang a fine chain from the endpoints. Does its shape (after you have adjusted its length) match that of your graph?

Compound Interest

6. Find the amount to which \$500 will accumulate in 6 years at a compound interest rate of 6% per year compounded annually.

7. If our compound interest formula is solved for a, we get

$$a = \frac{y}{(1 + n)^t}$$

where a is the amount that must be deposited now at interest rate n to produce an amount y in t years. How much money would have to be invested at $7\frac{1}{2}\%$ per year to accumulate to \$10,000 in 10 years?

8. What annual compound interest rate (compounded annually) is needed to enable an investment of \$5000 to accumulate to \$10,000 in 12 years?

9. Using Eq. A11, find the amount to which \$1 will accumulate in 20 years at a compound interest rate of 10% per year **(a)** compounded annually, **(b)** compounded monthly, and **(c)** compounded daily.

10. If an amount of money R is deposited *every year* for t years at a compound interest rate n, it will accumulate to an amount y, where

$$y = R\left[\frac{(1 + n)^t - 1}{n}\right]$$

If a worker pays \$2000 per year into a retirement plan yielding 6% annual interest, what will the value of this annuity be after 25 years?

11. If the equation for an annuity in problem 10 is solved for R, we get

$$R = \frac{ny}{(1 + n)^t - 1}$$

where R is the annual deposit required to produce a total amount y at the end of t years. How much would the worker in problem 10 have to deposit each year (at 6% per year) to have a total of \$100,000 after 25 years?

12. The yearly payment R that can be obtained for t years from a present investment a is

$$R = a\left[\frac{n}{(1 + n)^t - 1} + n\right]$$

If a person has \$85,000 in a retirement fund, how much can be withdrawn each year so that the fund will be exhausted in 20 years if the amount remaining in the fund is earning $6\frac{1}{4}\%$ per year?

> The amount a is called the *present worth* of the future amount y. This formula, and the three that follow, are standard formulas used in business and finance.

> A series of annual payments such as these is called an *annuity*.

> This is called a *sinking fund*.

> This is called *capital recovery*.

20–2 Exponential Growth and Decay

Exponential Growth

Equation A11, $y = a(1 + n/m)^{mt}$, allows us to compute the amount obtained, with interest compounded any number of times per period that we choose. We could use this formula to compute the interest if it were compounded every week, or every day, or every second. But what about *continuous* compounding, or continuous growth? What would happen to Eq. A11 if m got very, very large, in fact, infinite? Let us make m increase to large values; but first, to simplify the work, we make the substitution. Let

$$k = \frac{m}{n}$$

Equation A11 then becomes

$$y = a\left(1 + \frac{1}{k}\right)^{knt}$$

which can be written

$$y = a\left[\left(1 + \frac{1}{k}\right)^k\right]^{nt}$$

Then as m grows large, so will k. Let us try some values on the calculator and evaluate only the quantity in the square brackets, to four decimal places.

k	$\left(1 + \dfrac{1}{k}\right)^k$
1	2
10	2.5937 . . .
1000	2.7048 . . .
1,000	2.7169 . . .
10,000	2.7181 . . .
100,000	2.7183 . . .
1,000,000	2.7183 . . .

We get the surprising result that as m (and k) continue to grow infinitely large, the value of $(1 + 1/k)^k$ does *not* grow without limit, but approaches a specific, well-known value, the value 2.7183. . . . This important number is given the special symbol e. We can express the same idea in *limit notation* by writing the following equation:

Definition of *e*	$$\lim_{k \to \infty}\left(1 + \frac{1}{k}\right)^k = e$$	185

As k grows without bound, the value of $(1 + 1/k)^k$ approaches e (≈ 2.7183 . . .).

Thus when m gets infinitely large, the quantity $(1 + 1/k)^k$ approaches e, and the formula for continuous growth becomes the following:

Exponential Growth	$y = ae^{nt}$	199

The number e is named after a Swiss mathematician, Leonhard Euler (1707–83). Its value has been computed to thousands of decimal places. In problem 19 of Exercise 2, we show how to find e using series, and in problem 21 we show how to find the value of e graphically.

◆◆◆ **Example 7:** Repeat the computation of Example 5 if the interest were compounded *continuously* rather than monthly.

Solution: From Eq. 199,

$$y = 500e^{0.065(8)}$$
$$= 500e^{0.52}$$
$$= \$841.01$$

or \$1.18 more than when compounded monthly.   ◆◆◆

Banks that offer "continuous compounding" of your account do not have a clerk constantly computing your interest. They use Eq. 199.

Common Error	When you are using Eq. 199, the time units of n and t *must agree*.

◆◆◆ **Example 8:** A quantity grows exponentially at the rate of 2% per minute. By how many times will it have increased after 4 h?

Solution: The units of n and t do *not* agree (yet).

$$n = 2\% \text{ per } minute \quad \text{and} \quad t = 4 \ hours$$

So we convert one of them to agree with the other.

$$t = 4 \text{ h} = 240 \text{ min}$$

Using Eq. 199,

$$y = ae^{nt} = ae^{0.02(240)}$$

We can find the ratio of y to a by dividing both sides by a.

$$\frac{y}{a} = e^{0.02(240)} = e^{4.8} = 122$$

So the final amount is 122 times as great as the initial amount. ◆◆◆

Exponential Decay

When a quantity *decreases* so that the amount of decrease is proportional to the present amount, we have what is called *exponential decay*. Take, for example, a piece of steel just removed from a furnace. The amount of heat leaving the steel depends upon its temperature: The hotter it is, the faster heat will leave, and the faster it will cool. So as the steel cools, its lower temperature causes slower heat loss. Slower heat loss causes slower rate of temperature drop, and so on. The result is the typical *exponential decay curve* shown in Fig. 20–5. Finally, the temperature is said to *asymptotically approach* the outside (room) temperature.

The equation for exponential decay is the same as for exponential growth, except that the exponent is negative.

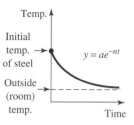

FIGURE 20–5 Exponential decrease in temperature. This is called *Newton's law of cooling.*

| Exponential Decay | $y = ae^{-nt}$ | 201 |

◆◆◆ **Example 9:** A room initially 80°F above the outside temperature cools exponentially at the rate of 25% per hour. Find the temperature of the room (above the outside temperature) at the end of 135 min.

Solution: We first make the units consistent: 135 min = 2.25 h. Then, by Eq. 201,

$$y = 80e^{-0.25(2.25)} = 45.6°F$$ ◆◆◆

Exponential Growth to an Upper Limit

Suppose that the bucket in Fig. 20–6 initially contains a gallons of water. Then assume that y_1, the amount remaining in the bucket, decreases exponentially. By Eq. 201,

$$y_1 = ae^{-nt}$$

The amount of water in bucket 2 thus increases from zero to a final amount a. The amount y in bucket 2 at any instant is equal to the amount that has left bucket 1, or

$$y = a - y_1$$
$$= a - ae^{-nt}$$

which is defined formally in the following equation:

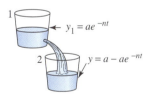

FIGURE 20–6 The amount in bucket 2 grows exponentially to an upper limit a.

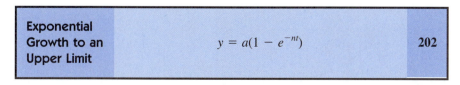

| Exponential Growth to an Upper Limit | $y = a(1 - e^{-nt})$ | 202 |

So the equation for this type of growth to an upper limit is exponential, where a equals the upper limit quantity, n is equal to the rate of growth, and t represents time (see Fig. 20–7).

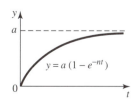

FIGURE 20–7 Exponential growth to an upper limit.

♦♦♦ Example 10: The voltage across a certain capacitor grows exponentially from 0 V to a maximum of 75 V, at a rate of 7% per second. Write the equation for the voltage at any instant, and find the voltage at 2.0 s.

Solution: From Eq. 202, with $a = 75$ and $n = 0.07$,

$$v = 75(1 - e^{-0.07t})$$

where v represents the instantaneous voltage. When $t = 2.0$ s,

$$v = 75(1 - e^{-0.14})$$
$$= 9.80 \text{ V}$$

♦♦♦

Time Constant

Our equations for exponential growth and decay all contained a rate of growth n. Often it is more convenient to work with the *reciprocal* of n, rather than n itself. This reciprocal is called the *time constant, T.*

Time Constant	$T = \dfrac{1}{n}$	203

♦♦♦ Example 11: If a quantity decays exponentially at a rate of 25% per second, what is the time constant?

Solution:

$$T = \frac{1}{0.25} = 4 \text{ s}$$

Notice that the time constant has the units of *time.*

♦♦♦

Universal Growth and Decay Curves

If we replace n by $1/T$ in Eq. 201, we get

$$y = ae^{-t/T}$$

Then we divide both sides by a.

$$\frac{y}{a} = e^{-t/T}$$

We plot this curve with the dimensionless ratio t/T for our horizontal axis and the dimensionless ratio y/a for the vertical axis (Fig. 20–8). Thus the horizontal

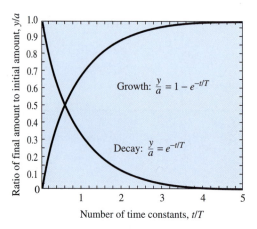

FIGURE 20–8 Universal growth and decay curves.

axis is *the number of time constants* that have elapsed, and the vertical axis is *the ratio of the final amount to the initial amount*. It is called the *universal decay curve*.

We can use the universal decay curve to obtain *graphical solutions* to exponential decay problems.

◆◆◆ **Example 12:** A current decays from an initial value of 300 mA at an exponential rate of 20% per second. Using the universal decay curve, graphically find the current after 7 s.

Solution: We have $n = 0.20$, so the time constant is

$$T = \frac{1}{n} = \frac{1}{0.20} = 5 \text{ s}$$

Our given time $t = 7$ s is then equivalent to

$$\frac{t}{T} = \frac{7}{5} = 1.4 \text{ time constants}$$

From the universal decay curve, at $t/T = 1.4$, we read

$$\frac{y}{a} = 0.25$$

Since $a = 300$ mA,

$$y = 300(0.25) = 75 \text{ mA} \qquad\qquad ◆◆◆$$

Figure 20–8 also shows the *universal growth curve*

$$\frac{y}{a} = 1 - e^{-t/T}$$

It is used in the same way as the universal decay curve.

A Recursion Relation for Exponential Growth

We have found that an amount a will grow to an amount y in t periods of time, where

$$y = ae^{nt} \qquad\qquad (199)$$

where n is the rate of growth. Computing y for different values of t gives the following table:

t	y
0	$y_0 = ae^0 = a$
1	$y_1 = ae^n = y_0 e^n$
2	$y_2 = ae^{2n} = (e^n)(ae^n) = e^n y_1$
3	$y_3 = ae^{3n} = (e^n)(ae^{2n}) = e^n y_2$
.	.
.	.
.	.
t	$y_t = e^n y_{t-1}$

Thus each value y_1 may be obtained by multiplying the preceding value y_{t-1} by the constant e^n. If we replace e^n by a constant B (sometimes called the *birthrate*), we have the recursion relation, given in the following equation:

Recursion Relation for Exponential Growth	$y_t = By_{t-1}$	**204**

◆◆◆ **Example 13:** Compute and graph values of y versus t using a starting value of 1 and a birthrate of 2, for five time periods.

Solution: We get each new y by multiplying the preceding y by 2.

t	0	1	2	3	4	5
y	1	2	4	8	16	32

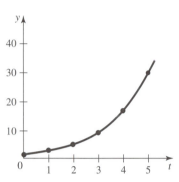

FIGURE 20–9

The graph of y versus t, shown in Fig. 20–9, is a typical exponential growth curve.

Nonlinear Growth Equation

Equations 199 and 204 represent a population that can grow without being limited by shortage of food or space, by predators, and so forth. This is rarely the case, so biologists have modified the equation to take those limiting factors into account. One model of the growth equation that takes these limits into account is obtained by multiplying the right side of Eq. 204 by the factor $(1 - y_{t-1})$.

Nonlinear Growth Equation	$y_t = By_{t-1}(1 - y_{t-1})$	**205**

This equation is one form of what is called the *Logistic Function,* or the *Verhulst Population Model*. As y_{t-1} gets larger, the factor $(1 - y_{t-1})$ gets smaller. This equation will cause the predicted population y to stabilize at some value, for a range of values of the birthrate B from 1 to 3, regardless of the starting value for y.

The population of 0.8 is a *normalized* population, considered to be 80% of some reference value.

◆◆◆ **Example 14:** Compute the population for 20 time periods using the nonlinear growth equation. Use a starting population of 0.8 and a birthrate of 2.

Solution: Make a table of values.

t	Population y
0	0.8
1	$2(0.8)(1 - 0.8)\quad = 0.32$
2	$2(0.32)(1 - 0.32) = 0.44$
3	$2(0.44)(1 - 0.44) = 0.49$
4	$2(0.49)(1 - 0.49) = 0.50$
5	$2(0.50)(1 - 0.50) = 0.50$
.	.
.	.
.	.
20	$2(0.50)(1 - 0.50) = 0.50$

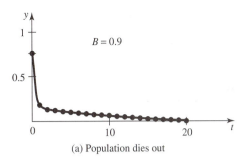

(a) Population dies out

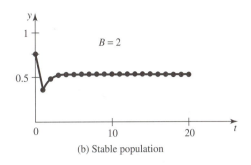

(b) Stable population

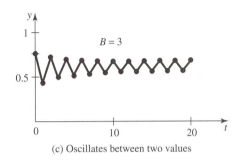

(c) Oscillates between two values

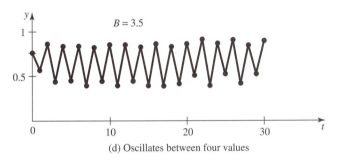

(d) Oscillates between four values

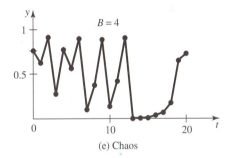

(e) Chaos

FIGURE 20–10

As shown in Fig. 20–10(b), the population y stabilizes at a value of 0.50.   ◆◆◆

For a birthrate less than 1, the nonlinear growth equation predicts that the population eventually dies out [Fig. 20–10(a)]. For a birthrate between 1 and 3, the population y will stabilize at some value between 0 and 0.667. But for a birthrate of 3 or greater, strange things start to happen.

◆◆◆ **Example 15:** Repeat the computation of Example 14 with a birthrate of 3.

Solution: Again using a starting value of 0.8 and $B = 3$, we get the following table:

t	y	t	y
0	0.8	.	.
1	0.48	.	.
2	0.75	.	.
3	0.56	17	0.62
4	0.74	18	0.71
5	0.58	19	0.62
		20	0.71

We see that the value *y oscillates* between two different values [Fig. 20–10(c)]. We have what is called *period doubling* or *bifurcation.*   ◆◆◆

Chaos

For *B* between 3.4495 and 3.56, the population has been found to oscillate between *four* different values. Figure 20–10(d), for *B* = 3.5, shows oscillation between values of 0.38, 0.5, 0.83, and 0.87. At higher values of *B*, the population oscillates between 8 values, then 16. Above *B* = 3.57, the population jumps around in a way that is not predictable at all.

◆◆◆ **Example 16:** Figure 20 10(e) shows the first 20 values of *y* using *B* = 4 and an initial *y* of 0.8. Note that the values do not repeat. The same simple equation that gave us predictable populations at some values of *B* now gives us *chaos* at other values of *B*, just slightly different from before.   ◆◆◆

Exercise 2 ◆ Exponential Growth and Decay

Exponential Growth

Use either the formulas or the universal growth and decay curves, as directed by your instructor.

1. A quantity grows exponentially at the rate of 5.00% per year for 7 years. Find the final amount if the initial amount is 200 units.

2. A yeast manufacturer finds that the yeast will grow exponentially at a rate of 15.0% per hour. How many pounds of yeast must the manufacturer start with to obtain 500 lb at the end of 8 h?

3. If the population of a country was 11.4 million in 1995 and grows at an annual rate of 1.63%, find the population by the year 2000.

4. If the rate of inflation is 12.0% per year, how much could a camera that now costs $225 be expected to cost 5 years from now?

5. The oil consumption of a certain country was 12.0 million barrels in 1995, and it grows at an annual rate of 8.30%. Find the oil consumption by the year 2000.

6. A pharmaceutical company makes a vaccine-producing bacterium that grows at a rate of 2.5% per hour. How many units of this organism must the company have initially to have 1000 units after 5 days?

Exponential Growth to an Upper Limit

7. A steel casting initially at 0°C is placed in a furnace at $800°C$. If it increases in temperature at the rate of 5.00% per minute, find its temperature after 20.0 min.

8. Equation 202 approximately describes the hardening of concrete, where *y* is the compressive strength (lb/in.^{2}) *t* days after pouring. Using values of *a* = 4000 and *n* = 0.0696 in the equation, find the compressive strength after 14 days.

9. When the switch in Fig. 20–11 is closed, the current *i* will grow exponentially according to the following equation:

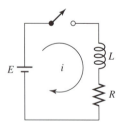

FIGURE 20–11

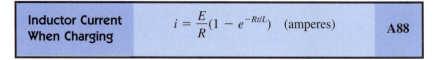

Inductor Current When Charging	$i = \dfrac{E}{R}(1 - e^{-Rt/L})$ (amperes)	A88

where *L* is the inductance in henries and *R* is the resistance in ohms. Find the current at 0.0750 s if *R* = 6.25 Ω, *L* = 186 H, and *E* = 250 V.

Exponential Decay

10. A flywheel is rotating at a speed of 1800 rev/min. When the power is disconnected, the speed decreases exponentially at the rate of 32.0% per minute. Find the speed after 10.0 min.

11. A steel forging is 1500°F above room temperature. If it cools exponentially at the rate of 2.00% per minute, how much will its temperature drop in 1 h?

12. As light passes through glass or water, its intensity decreases exponentially according to the equation

$$I = I_0 e^{-kx}$$

where I is the intensity at a depth x and I_0 is the intensity before entering the glass or water. If, for a certain filter glass, $k = 0.500$/cm (which means that each centimeter of filter thickness removes half the light reaching it), find the fraction of the original intensity that will pass through a filter glass 2.00 cm thick.

13. The atmospheric pressure p decreases exponentially with the height h (in miles) above the earth according to the function $p = 29.92e^{-h/5}$ in. of mercury. Find the pressure at a height of 30,000 ft.

14. A certain radioactive material loses its radioactivity at the rate of $2\frac{1}{2}$% per year. What fraction of its initial radioactivity will remain after 10.0 years?

15. When a capacitor C, charged to a voltage E, is discharged through a resistor R (Fig. 20–12), the current i will decay exponentially according to the following equation:

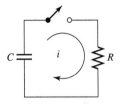

FIGURE 20–12

Capacitor Current When Charging or Discharging	$i = \dfrac{E}{R}e^{-t/RC}$ (amperes)	A83

Find the current after 45 ms (0.045 s) in a circuit where $E = 220$ V, $C = 130$ μF (130×10^{-6} F), and $R = 2700$ Ω.

16. When a fully discharged capacitor C (Fig. 20–13) is connected across a battery, the current i flowing into the capacitor will decay exponentially according to Equation A83. If $E = 115$ V, $R = 350$ Ω, and $C = 0.000750$ F, find the current after 75.0 ms.

17. In a certain fabric mill, cloth is removed from a dye bath and is then observed to dry exponentially at the rate of 24% per hour. What percent of the original moisture will still be present after 5 h?

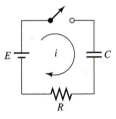

FIGURE 20–13

Computer and Graphics Calculator

18. Program the computer to print a two-column table, the first column giving the value of k and the second column the value of $(1 + 1/k)^k$. Take a starting value for k equal to 1 and increase it each time by a factor of 10. Stop the run when $1/k$ is less than 0.00001. Does the value of $(1 + 1/k)^k$ seem to be converging on some value? If not, why?

19. We can get an approximate value for e from the following infinite series:

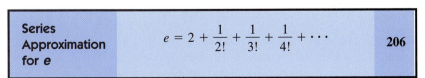

Series Approximation for e	$e = 2 + \dfrac{1}{2!} + \dfrac{1}{3!} + \dfrac{1}{4!} + \cdots$	206

where 4! (read "4 factorial") is $4(3)(2)(1) = 24$. Write a program to compute e using the first five terms of the series.

20. The infinite series often used to calculate e^x is as follows:

Series Approximation for e^x	$e^x = 1 + x + \dfrac{x^2}{2!} + \dfrac{x^3}{3!} + \dfrac{x^4}{4!} + \cdots$	**207**

Write a program to compute e^x by using the first 15 terms of this series, and use it to find e^5.

21. Use a graphics calculator or graphing utility on a computer to plot $y = (1 + 1/x)^x$. What happens to the value of y as x increases? Use ⎡TRACE⎤ to find the coordinates of points on the curve as x increases from zero to, say, 100.

20–3   Logarithms

In Sec. 20–1, we studied the exponential function

$$y = b^x$$

Given b and x, we know how to find y. Thus if $b = 2.48$ and $x = 3.50$, then, by calculator,

$$y = 2.48^{3.50} = 24.0$$

Given y and x, we are also able to find b. For example, if y were 24.0 and x were 3.50,

$$24.0 = b^{3.50}$$
$$b = 24.0^{1/3.50} = 2.48$$

We will finish this problem in Sec. 20–5.

Suppose, now, that we have y and b and want to find x. Say that $y = 24.0$ when $b = 2.48$. Then

$$24.0 = 2.48^x$$

Try it. You will find that none of the math we have learned so far will enable us to find x. To solve this and similar problems, we need *logarithms,* which are explained in this section.

We have the same relationship between the same quantities as we had before but a different way of expressing it. We have a similar situation in ordinary language. Take "Mary is John's *sister.*" If we had only the word *sister* (and not *brother*) to describe the relationship between children, we would have to say something like "John is the person whom Mary is the sister of." That is as awkward as "the exponent to which the base must be raised to obtain the number." So we introduce new words, such as *brother* and *logarithm.*

Definition of a Logarithm

The logarithm of some positive number y *is the exponent* to which a base b must be raised to obtain y.

◆◆◆ **Example 17:** Since $10^2 = 100$, we say that "2 is the logarithm of 100, to the base 10" because 2 is the exponent to which the base 10 must be raised to obtain 100. The logarithm is written

$$\log_{10} 100 = 2$$

Notice that the base is written in small numerals below the word *log.* The statement is read "The log of 100 (to the base 10) is 2." This means that 2 is an exponent because *a logarithm is an exponent;* 100 is the result of raising the base to that power. Therefore these two expressions are equivalent:

$$10^2 = 100$$
$$\log_{10}100 = 2$$

◆◆◆

Given the exponential function $y = b^x$, we say that "x is the logarithm of y, to the base b" because x is the exponent to which the base b must be raised to obtain y.

So, in general, the following equation applies:

Definition of a Logarithm	If $\quad\quad y = b^x \quad\quad (y > 0, \quad b > 0, \quad b \neq 1)$ Then $\quad x = \log_b y$ (x is the logarithm of y to the base b.)	**186**

In practice, the base b is taken as either 10 for common logarithms or e for natural logarithms. If "log" is written without a base, the base is assumed to be 10.

◆◆◆ **Example 18:**

$$\log 5 \equiv \log_{10} 5$$

◆◆◆

Converting between Logarithmic and Exponential Forms

In working with logarithmic and exponential expressions, we often have to switch between exponential and logarithmic forms. When converting from one form to the other, remember that (1) the base in exponential form is also the base in logarithmic form and (2) "the *log* is the *exponent.*"

◆◆◆ **Example 19:** Change the exponential expression $3^2 = 9$ to logarithmic form.

Solution: The base (3) in the exponential expression remains the base in the logarithmic expression.

$$3^2 \quad = 9$$
$$\downarrow$$
$$\log_3 (\ \ ) = (\ \ )$$

The log is the exponent.

$$3^2 \quad\searrow \quad = 9$$
$$\log_3 (\ \ ) = 2$$

So our expression is

$$\log_3 9 \quad = 2$$

This is read "the log *of* 9 (to the base 3) *is* 2."

◆◆◆

We could also write the same expression in *radical form* as $\sqrt{9} = 3$.

◆◆◆ **Example 20:** Change $\log_e x = 3$ to exponential form.

Solution:

$$\log_e \ x = 3$$

$$e^3 \ = x$$

◆◆◆

◆◆◆ **Example 21:** Change $10^3 = 1000$ to logarithmic form.

Solution:

$$\log_{10} 1000 = 3$$

◆◆◆

Common logarithms are also called *Briggs' logarithms*, after Henry Briggs (1561–1630).

Remember that when a logarithm is written without a base (such as log x), we assume the base to be 10.

A logarithm such as 1.8740 can be expressed as the sum of an integer 1, called the *characteristic*, and a positive decimal 0.8740, called the *mantissa*. These terms were useful when tables were used to find logarithms.

Common Logarithms

Although we may write logarithms to any base, only two bases are in regular use: the base 10 and the base e. Logarithms to the base e, called *natural* logarithms, are discussed later. In this section we study logarithms to the base 10, called *common* logarithms.

Common Logarithms by Calculator

We use the calculator key marked $\boxed{\text{LOG}}$ to find common logarithms. On some calculators we press $\boxed{\text{LOG}}$, enter the number we wish to take the log of (called the *argument*), and then press $\boxed{\text{ENTER}}$. On other calculators we enter the argument first and then press $\boxed{\text{LOG}}$.

◆◆◆ **Example 22:** Verify the following common logarithms, found to five significant digits:

(a) log 74.82 = 1.8740
(b) log 0.03755 = −1.4254   ◆◆◆

◆◆◆ **Example 23:** Try to take the common logarithm of −2. You should get an error indication. We cannot take the logarithm of a negative number.   ◆◆◆

Antilogarithms by Calculator

Suppose that we have the logarithm of a number and want to find the number itself, say, log x = 1.852. This process is known as *finding the antilogarithm (antilog)* of 1.852. First we switch to exponential form.

$$x = 10^{1.852}$$

Some calculators use $\boxed{\text{INV}}$ $\boxed{\text{LOG}}$ to find antilogarithms.

This quantity can be found using the $\boxed{10^x}$ key on the calculator. On some calculators, press $\boxed{10^x}$ and then enter the number. On other calculators, enter the number first and then press $\boxed{10^x}$.

◆◆◆ **Example 24:** Verify the following antilogarithms to four significant digits:

(a) The antilog of 1.852 is 71.12.
(b) The antilog of −1.738 is 0.01828.   ◆◆◆

◆◆◆ **Example 25:** Find x to four significant digits if log x = 2.74.

Solution: By calculator,

$$x = 10^{2.74} = 549.5$$ ◆◆◆

The Logarithmic Function

We wish to graph the logarithmic function $x = \log_b y$, but first let us follow the usual convention of calling the dependent variable y and the independent variable x. Our function then becomes the following:

Logarithmic Function	$y = \log_b x$ $(x > 0, \quad b > 0, \quad b \neq 1)$	**208**

◆◆◆ **Example 26:** Graph the logarithmic function $y = \log x$ for $x = -1$ to 14.

Solution: We use a graphing utility or make a table of point pairs, getting the values of y by using the common log key on the calculator.

x	−1	0	1	2	3	4	5	6	7	8	9	10	11	12	13	14
y			0	0.30	0.48	0.60	0.70	0.78	0.85	0.90	0.95	1.00	1.04	1.08	1.11	1.15

First we notice that when x is negative or zero, our calculator gives an error indication. In other words, the domain of x is the positive numbers only. *We cannot take the common logarithm of a negative number or zero.* We'll see that this is true for logarithms to other bases as well.

If we try a value of x between 0 and 1, say, $x = 0.5$, we get a negative y $(-0.3$ in this case). Plotting our point pairs gives the curve in Fig. 20–14.

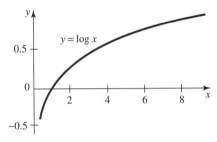

FIGURE 20–14

◆◆◆

Natural Logarithms

You may want to refer to Sec. 20–2 for our earlier discussion of the number e (2.718 approximately). After we have defined e, we can use it as a base; logarithms using e as a base are called *natural* logarithms. They are written $\ln x$. Thus $\ln x$ is understood to mean $\log_e x$.

> Natural logarithms are also called *Napierian logarithms,* after John Napier (1550–1617), who first wrote logarithms.

Natural Logarithms by Calculator

To find the natural logarithm of a (positive) number by calculator, we use the $\boxed{\text{LN}}$ key. As with common logarithms, you must check your calculator to see if the number whose logarithm you want is to be entered before or after pressing the $\boxed{\text{LN}}$ key.

◆◆◆ **Example 27:** Find $\ln 19.3$ to four significant digits.

Solution: Enter 19.3, press $\boxed{\text{LN}}$, and read 2.960 (rounded). Or press $\boxed{\text{LN}}$, enter 19.3, and press $\boxed{\text{ENTER}}$. ◆◆◆

◆◆◆ **Example 28:** Verify the following to four significant digits:

(a) $\ln 25.4 = 3.235$
(b) $\ln 0.826 = -0.1912$
(c) $\ln(-3)$ gives an error indication. ◆◆◆

Antilogarithms by Calculator

If the natural logarithm of a number is known, we can find the number itself by using the $\boxed{e^x}$ key. Again, some calculators require entering the number before pressing the $\boxed{e^x}$ key, and some after.

Some calculators do not have an $\boxed{e^x}$ key. Instead, use $\boxed{\text{INV}}$ $\boxed{\text{LN}}$ or $\boxed{\text{2nd}}$ $\boxed{\text{LN}}$.

◆◆◆ **Example 29:** Find x to four digits if $\ln x = 3.825$.

Solution: Rewriting the expression $\ln x = 3.825$ in exponential form, we have

$$x = e^{3.825} = 45.83$$

by calculator.   ◆◆◆

Exercise 3 ◆ Logarithms

Converting between Logarithmic and Exponential Forms

Convert to logarithmic form.

1. $3^4 = 81$	**2.** $5^3 = 125$
3. $4^6 = 4096$	**4.** $7^3 = 343$
5. $x^5 = 995$	**6.** $a^3 = 6.83$

Convert to exponential form.

7. $\log_{10} 100 = 2$	**8.** $\log_2 16 = 4$
9. $\log_5 125 = 3$	**10.** $\log_4 1024 = 5$
11. $\log_3 x = 57$	**12.** $\log_x 54 = 285$
13. $\log_5 x = y$	**14.** $\log_w 3 = z$

Logarithmic Function

Graph each logarithmic function from $x = \frac{1}{2}$ to 4.

15. $y = \log_e x$	**16.** $y = \log_{10} x$

Common Logarithms

Find the common logarithm of each number to four decimal places.

17. 27.6	**18.** 4.83	**19.** 5.93
20. 9.26	**21.** 48.3	**22.** 385
23. 836	**24.** 1.03	**25.** 27.4
26. 0.0573	**27.** 0.00737	**28.** 9738
29. 34,970	**30.** 5.28×10^6	

Find x to three significant digits if the common logarithm of x is equal to the given value.

31. 1.584	**32.** 2.957	**33.** 5.273
34. 0.886	**35.** -2.857	**36.** 0.366
37. -2.227	**38.** 4.973	**39.** 3.979
40. 1.295	**41.** -3.972	**42.** -0.8835

Natural Logarithms

Find the natural logarithm of each number to four decimal places.

43. 48.3	**44.** 846	**45.** 2365
46. 1.005	**47.** 0.845	**48.** 4.97
49. 0.00836	**50.** 45,900	**51.** 0.0000462
52. 82,900	**53.** 3.84×10^4	**54.** 8.24×10^{-3}

Find to four significant digits the number whose natural logarithm is the given value.

55. 2.846	**56.** 4.263	**57.** 0.879
58. -2.846	**59.** -0.365	**60.** 5.937
61. 0.936	**62.** -4.97	**63.** 15.84
64. -9.47	**65.** -18.36	**66.** 21.83

Computer

67. The series approximation for the natural logarithm of a positive number x is as follows:

| **Series Approximation for 1n x** | $$\ln x = 2a + \frac{2a^3}{3} + \frac{2a^5}{5} + \frac{2a^7}{7} + \cdots$$ where $$a = \frac{x-1}{x+1}$$ | **209** |

Write a program to compute $\ln x$ using the first 10 terms of the series.

68. The series expansion for b^x is as follows:

| **Series Approximation for b^x** | $$b^x = 1 + x \ln b + \frac{(x \ln b)^2}{2!} + \frac{(x \ln b)^3}{3!} + \cdots$$ $(b > 0)$ | **198** |

Write a program to compute b^x using the first x terms of the series.

69. To enter a logarithm when using a CAS, use the following commands:

Program	Natural Log	Log to Base b
Derive	LN(x)	LOG(x, b)
Maple	ln(x)	ln(x)/ln(b)
Mathematica	Log[x]	Log[b, x]

To find the antilogarithm of y, simply raise the base, e or b, to the power y. Use the command exp(y) if y is a natural logarithm, and $b \wedge y$ if y is a logarithm to the base b. Try your CAS on any of the problems in this exercise set.

20–4   Properties of Logarithms

In this section we use the properties of logarithms along with our ability to convert back and forth between exponential and logarithmic forms. These together will enable us to later solve equations that we could not previously handle, such as exponential equations.

Products

We wish to write an expression for the log of the product of two positive numbers, say, M and N.

$$\log_b MN = ?$$

Let us substitute: $M = b^c$ and $N = b^d$. Then

$$MN = b^c b^d = b^{c+d}$$

Here, $b > 0$ and $b \neq 1$.

by the laws of exponents. Writing this expression in logarithmic form, we have

$$\log_b MN = c + d$$

But, since $b^c = M$, $c = \log_b M$. Similarly, $d = \log_b N$. Substituting, we have the following equation:

Log of a Product	$\log_b MN = \log_b M + \log_b N$	187

The log of a product equals the sum of the logs of the factors.

◆◆◆ **Example 30:**

(a) $\log 7x = \log 7 + \log x$
(b) $\log 3 + \log x + \log y = \log 3xy$ ◆◆◆

Common Error	The log of a sum is *not* equal to the sum of the logs.
	$\log_b(M + N) \neq \log_b M + \log_b N$
	The product of two logs is *not* equal to the sum of the logs.
	$(\log_b M)(\log_b N) \neq \log_b M + \log_b N$

◆◆◆ **Example 31:** Multiply $(2.947)(9.362)$ by logarithms.

Solution: We take the log of the product

$$\log(2.947 \times 9.362) = \log 2.947 + \log 9.362$$
$$= 0.46938 + 0.97137 = 1.44075$$

This gives us the logarithm of the product. To get the product itself, we take the antilog.

$$(2.947)(9.362) = 10^{1.44075} = 27.59 \qquad ◆◆◆$$

We don't pretend that this is a practical problem, now that we have calculators. But we recommend doing a few logarithmic computations just to get practice in using the properties of logarithms, which we'll need later to solve logarithmic and exponential equations.

Quotients

Let us now consider the quotient M divided by N. Using the same substitution as above, we obtain

$$\frac{M}{N} = \frac{b^c}{b^d} = b^{c-d}$$

Going to logarithmic form yields

$$\log_b \frac{M}{N} = c - d$$

Finally, substituting back gives us the following equation:

Log of a Quotient	$\log_b \dfrac{M}{N} = \log_b M - \log_b N$	188

The log of a quotient equals the log of the dividend minus the log of the divisor.

◆◆◆ **Example 32:**

$$\log \frac{3}{4} = \log 3 - \log 4 \qquad ◆◆◆$$

◆◆◆ **Example 33:** Express the following as the sum or difference of two or more logarithms:

$$\log \frac{ab}{xy}$$

Solution: By Eq. 188,

$$\log \frac{ab}{xy} = \log ab - \log xy$$

and by Eq. 187,

$$\log ab - \log xy = \log a + \log b - (\log x + \log y)$$
$$= \log a + \log b - \log x - \log y \qquad ◆◆◆$$

◆◆◆ **Example 34:** The expression $\log 10x^2 - \log 5x$ can be expressed as a single logarithm as follows:

$$\log \frac{10x^2}{5x} \quad \text{or} \quad \log 2x \qquad ◆◆◆$$

Common Errors	Similar errors are made with quotients as with products. $$\log_b(M - N) \neq \log_b M - \log_b N$$ $$\frac{\log_b M}{\log_b N} \neq \log_b M - \log_b N$$

◆◆◆ **Example 35:** Divide using logarithms:

$$28.47 \div 1.883$$

Solution: By Eq. 188,

$$\log \frac{28.47}{1.883} = \log 28.47 - \log 1.883$$

$$= 1.4544 - 0.2749 = 1.1795$$

This gives us the logarithm of the quotient. To get the quotient itself, we take the antilog.

$$28.47 \div 1.883 = 10^{1.1795} = 15.12 \qquad ◆◆◆$$

Powers

Consider now the quantity M raised to a power p. Substituting b^c for M, as before, we have

$$M^p = (b^c)^p = b^{cp}$$

In logarithmic form,

$$\log_b M^p = cp$$

Substituting back gives us the following equation:

Log of a Power	$\log_b M^p = p \log_b M$	**189**

The log of a number raised to a power equals the power times the log of the number.

◆◆◆ **Example 36:**

$$\log 3.85^{1.4} = 1.4 \log 3.85$$

◆◆◆

◆◆◆ **Example 37:** The expression $7 \log x$ can be expressed as a single logarithm with a coefficient of 1 as follows:

$$\log x^7$$

◆◆◆

◆◆◆ **Example 38:** Express $2 \log 5 + 3 \log 2 - 4 \log 1$ as a single logarithm.

Solution:

$$2 \log 5 + 3 \log 2 - 4 \log 1 = \log 5^2 + \log 2^3 - \log 1^4$$

$$= \log \frac{25(8)}{1} = \log 200$$

◆◆◆

◆◆◆ **Example 39:** Express as a single logarithm with a coefficient of 1:

$$2 \log \frac{2}{3} + 4 \log \frac{a}{b} - 3 \log \frac{x}{y}$$

Solution:

$$2 \log \frac{2}{3} + 4 \log \frac{a}{b} - 3 \log \frac{x}{y} = \log \frac{2^2}{3^2} + \log \frac{a^4}{b^4} - \log \frac{x^3}{y^3}$$

$$= \log \frac{4a^4 y^3}{9b^4 x^3}$$

◆◆◆

◆◆◆ **Example 40:** Rewrite without logarithms the equation

$$\log(a^2 - b^2) + 2 \log a = \log(a - b) + 3 \log b$$

Solution: We first try to combine all of the logarithms into a single logarithm. Rearranging gives

$$\log(a^2 - b^2) - \log(a - b) + 2 \log a - 3 \log b = 0$$

By the properties of logarithms,

$$\log \frac{a^2 - b^2}{a - b} + \log \frac{a^2}{b^3} = 0$$

or

$$\log \frac{a^2(a - b)(a + b)}{b^3(a - b)} = 0$$

Simplifying yields

$$\log \frac{a^2(a + b)}{b^3} = 0$$

We then rewrite without logarithms by going from logarithmic to exponential form.

$$\frac{a^2(a + b)}{b^3} = 10^0 = 1$$

or

$$a^2(a + b) = b^3$$

◆◆◆

♦♦♦ **Example 41:** Find $35.82^{1.4}$ by logarithms to four significant digits.

Solution: By Eq. 189,

$$\log 35.82^{1.4} = 1.4 \log 35.82$$
$$= 1.4(1.5541) = 2.1758$$

Taking the antilog yields

$$35.82^{1.4} = 10^{2.1758} = 149.9 \qquad \blacklozenge\blacklozenge\blacklozenge$$

Roots

We can write a given radical expression in exponential form and then use the rule for powers. Thus, by Eq. 36,

$$\sqrt[q]{M} = M^{1/q}$$

Taking the logarithm of both sides, we obtain

$$\log_b \sqrt[q]{M} = \log_b M^{1/q}$$

Then, by Eq. 189, the following equation applies:

> We may take the logarithm of both sides of an equation, just as we took the square root of both sides of an equation or took the sine of both sides of an equation.

Log of a Root	$\log_b \sqrt[q]{M} = \dfrac{1}{q} \log_b M$	190

The log of the root of a number equals the log of the number divided by the index of the root.

♦♦♦ **Example 42:**

(a) $\log \sqrt[5]{8} = \dfrac{1}{5} \log 8 \simeq \dfrac{1}{5}(0.9031) \simeq 0.1806$

(b) $\dfrac{\log x}{2} + \dfrac{\log z}{4} = \log \sqrt{x} + \log \sqrt[4]{z}$ $\qquad \blacklozenge\blacklozenge\blacklozenge$

♦♦♦ **Example 43:** Find $\sqrt[5]{3824}$ by logarithms.

Solution: By Eq. 190,

$$\log \sqrt[5]{3824} = \frac{1}{5} \log 3824 = \frac{3.5825}{5}$$
$$= 0.7165$$

Taking the antilog gives us

$$\sqrt[5]{3824} = 10^{0.7165} = 5.206 \qquad \blacklozenge\blacklozenge\blacklozenge$$

Log of 1

Let us take the log to the base b of 1 and call it x. Thus

$$x = \log_b 1$$

Switching to exponential form, we have

$$b^x = 1$$

This equation is satisfied, for any positive b, by $x = 0$. Therefore:

Log of 1	$\log_b 1 = 0$	191

The log of 1 is zero.

Log of the Base

We now take the $\log_b$ of its own base b. Let us call this quantity x.

$$\log_b b = x$$

Going to exponential form gives us

$$b^x = b$$

So x must equal 1.

Log of the Base	$\log_b b = 1$	192

The log (to the base b) of b is equal to 1.

◆◆◆ Example 44:

(a) $\log 1 = 0$ (b) $\ln 1 = 0$
(c) $\log_5 5 = 1$ (d) $\log 10 = 1$
(e) $\ln e = 1$ ◆◆◆

Log of the Base Raised to a Power

Consider the expression $\log_b b^n$. *We set this expression equal to x and, as before, go to exponential form.*

$$\log_b b^n = x$$
$$b^x = b^n$$
$$x = n$$

Therefore:

Log of the Base Raised to a Power	$\log_b b^n = n$	193

The log (to the base b) of b raised to a power is equal to the power.

◆◆◆ Example 45:

(a) $\log_2 2^{4.83} = 4.83$
(b) $\log_e e^{2x} = 2x$
(c) $\log_{10} 0.0001 = \log_{10} 10^{-4} = -4$ ◆◆◆

Base Raised to a Logarithm of the Same Base

We want to evaluate an expression of the form

$$b^{\log_b x}$$

Setting this expression equal to y and taking the logarithms of both sides, gives us

$$y = b^{\log_b x}$$

$$\log_b y = \log_b b^{\log_b x} = \log_b x \log_b b$$

Since $\log_b b = 1$, we have $\log_b y = \log_b x$. Taking the antilog of both sides, we have

$$x = y = b^{\log_b x}$$

Thus:

Base Raised to a Logarithm of the Same Base	$b^{\log_b x} = x$	194

◆◆◆ **Example 46:**

(a) $10^{\log w} = w$ (b) $e^{\log_e 3x} = 3x$ ◆◆◆

Change of Base

We can convert between natural logarithms and common logarithms (or between logarithms to any base) by the following procedure. Suppose that $\log N$ is the common logarithm of some number N. We set it equal to x.

$$\log N = x$$

Going to exponential form, we obtain

$$10^x = N$$

Now we take the natural log of both sides.

$$\ln 10^x = \ln N$$

By Eq. 189,

$$x \ln 10 = \ln N$$

$$x = \frac{\ln N}{\ln 10}$$

But $x = \log N$, so we have the following equation:

Change of Base	$\log N = \dfrac{\ln N}{\ln 10}$ where $\ln 10 \simeq 2.3026$	195

Some computer languages enable us to find the natural log but not the common log. This equation lets us convert from one to the other.

The common logarithm of a number is equal to the natural log of that number divided by the natural log of 10.

◆◆◆ **Example 47:** Find $\log N$ if $\ln N = 5.824$.

Solution: By Eq. 195,

$$\log N = \frac{5.824}{2.3026} = 2.529$$

◆◆◆

◆◆◆ **Example 48:** What is $\ln N$ if $\log N = 3.825$?

Solution: By Eq. 195,

$$\ln N = \ln 10(\log N)$$
$$= 2.3026(3.825) = 8.807$$ ◆◆◆

Exercise 4 ◆ Properties of Logarithms

Write as the sum or difference of two or more logarithms.

1. $\log \frac{2}{3}$

2. $\log 4x$

3. $\log ab$

4. $\log \frac{x}{2}$

5. $\log xyz$

6. $\log 2ax$

7. $\log \frac{3x}{4}$

8. $\log \frac{5}{xy}$

9. $\log \frac{1}{2x}$

10. $\log \frac{2x}{3y}$

11. $\log \frac{abc}{d}$

12. $\log \frac{x}{2ab}$

Express as a single logarithm with a coefficient of 1. Assume that the logarithms in each problem have the same base.

13. $\log 3 + \log 4$

14. $\log 7 - \log 5$

15. $\log 2 + \log 3 - \log 4$

16. $3 \log 2$

17. $4 \log 2 + 3 \log 3 - 2 \log 4$

18. $\log x + \log y + \log z$

19. $3 \log a - 2 \log b + 4 \log c$

20. $\frac{\log x}{3} + \frac{\log y}{2}$

21. $\log \frac{x}{a} + 2 \log \frac{y}{b} + 3 \log \frac{z}{c}$

22. $2 \log x + \log(a + b) - \frac{1}{2} \log(a + bx)$

23. $\frac{1}{2} \log(x + 2) + \log(x - 2)$

Rewrite each equation so that it contains no logarithms.

24. $\log x + 3 \log y = 0$

25. $\log_2 x + 2 \log_2 y = x$

26. $2 \log(x - 1) = 5 \log(y + 2)$

27. $\log_2(a^2 + b^2) + 1 = 2 \log_2 a$

28. $\log(x + y) - \log 2 = \log x + \log(x^2 - y^2)$

29. $\log(p^2 - q^2) - \log(p + q) = 2$

30. $\log_2 2$

31. $\log_e e$

32. $\log_{10} 10$

33. $\log_3 3^2$

34. $\log_e e^x$

35. $\log_{10} 10^x$

36. $e^{\log_e x}$

37. $2^{\log_2 3y}$

38. $10^{\log x^2}$

Logarithmic Computation

Evaluate using logarithms.

39. 5.937×92.47

40. 6923×0.003846

41. $3.97 \times 8.25 \times 9.82$

42. $88.25 \div 42.94$

This may not seem like hot stuff in the age of calculators and computers, but the use of logarithms revolutionized computation in its time. The French mathematician Laplace wrote: "The method of logarithms, by reducing to a few days the labor of many months, doubles, as it were, the life of the astronomer, besides freeing him from the errors and disgust inseparable from long calculation."

43. $\sqrt[9]{8563}$

44. $4.836^{3.970}$

45. $83.62^{0.5720}$

46. $\sqrt[3]{587}$

47. $\sqrt[7]{8364}$

48. $\sqrt[4]{62.4}$

Change of Base

Find the common logarithm of the number whose natural logarithm is the given value.

49. 8.36

50. −3.846

51. 3.775

52. 15.36

53. 5.26

54. −0.638

Find the natural logarithm of the number whose common logarithm is the given value.

55. 84.9

56. 2.476

57. −3.82

58. 73.9

59. 2.37

60. −2.63

Graphics Calculator

61. Use a graphing utility to separately graph each each side of the equation

$$\log 2x = \log 2 + \log x$$

from $x = 0$ to 10. Then graph both sides of

$$\log(2 + x) = \log 2 + \log x$$

and

$$(\log 2)(\log x) = \log 2 + \log x$$

What do your graphs tell you about these three expressions?

Computer

62. A CAS can simplify or manipulate logarithmic expressions. Enter the logarithms as was shown in Exercise 3, and use the *Simplify* and *Expand* commands to change the form of the expression. Use your CAS to try to simplify any of the expressions in this exercise set. You may find some expressions that the computer cannot simplify.

20–5 Exponential Equations

Solving Exponential Equations

We return now to the problem we started in Sec. 20–3 but could not finish, that of finding the *exponent* when the other quantities in an exponential equation were known. We tried to solve the equation

$$24.0 = 2.48^x$$

The key to solving exponential equations is to *take the logarithm of both sides*. This enables us to use Eq. 189 to extract the unknown from the exponent. Taking the logarithm of both sides gives

$$\log 24.0 = \log 2.48^x$$

By Eq. 189,

$$\log 24.0 = x \log 2.48$$

$$x = \frac{\log 24.0}{\log 2.48} = \frac{1.380}{0.3945} = 3.50$$

In this example we took *common* logarithms, but *natural* logarithms would have worked just as well.

◆◆◆ **Example 49:** Solve for x to three significant digits:

$$3.25^{x+2} = 1.44^{3x-1}$$

Solution: We take the log of both sides. This time choosing to use natural logs, we get

$$(x + 2) \ln 3.25 = (3x - 1) \ln 1.44$$

$$\frac{\ln 3.25}{\ln 1.44}(x + 2) = 3x - 1$$

By calculator,

$$3.232(x + 2) = 3x - 1$$

Removing parentheses and collecting terms yields

$$0.232x = -7.464$$

$$x = -32.2$$ ◆◆◆

Solving Exponential Equations with Base e

If an exponential equation contains the base e, taking *natural* logs rather than common logarithms will simplify the work.

◆◆◆ **Example 50:** Solve for x:

$$157 = 112e^{3x+2}$$

Solution: Dividing by 112, we have

$$1.402 = e^{3x+2}$$

Taking natural logs gives us

$$\ln 1.402 = \ln e^{3x+2}$$

By Eq. 189,

$$\ln 1.402 = (3x + 2) \ln e$$

But $\ln e = 1$, so

$$\ln 1.402 = 3x + 2$$

$$x = \frac{\ln 1.402 - 2}{3} = -0.554$$ ◆◆◆

◆◆◆ **Example 51:** Solve for x to three significant digits:

$$3e^x + 2e^{-x} = 4(e^x - e^{-x})$$

Solution: Removing parentheses gives

$$3e^x + 2e^{-x} = 4e^x - 4e^{-x}$$

Combining like terms yields

$$e^x - 6e^{-x} = 0$$

Factoring gives

$$e^{-x}(e^{2x} - 6) = 0$$

This equation will be satisfied if either of the two factors is equal to zero. So we set each factor equal to zero (just as when solving a quadratic by factoring) and get

$$e^{-x} = 0 \quad \text{and} \quad e^{2x} = 6$$

There is no value of x that will make e^{-x} equal to zero, so we get no root from that equation. We solve the other equation by taking the natural log of both sides.

$$2x \ln e = \ln 6 = 1.792$$

$$x = 0.896 \qquad \text{◆◆◆}$$

Our exponential equation might be of *quadratic type,* in which case it can be solved by substitution as we did in Chapter 14.

◆◆◆ Example 52: Solve for x to three significant digits:

$$3e^{2x} - 2e^x - 5 = 0$$

Solution: If we substitute

$$w = e^x$$

our equation becomes the quadratic

$$3w^2 - 2w - 5 = 0$$

By the quadratic formula,

$$w = \frac{2 \pm \sqrt{4 - 4(3)(-5)}}{6} = -1 \quad \text{and} \quad \frac{5}{3}$$

Substituting back, we have

$$e^x = -1 \quad \text{and} \quad e^x = \frac{5}{3}$$

So $x = \ln(-1)$, which we discard, and

$$x = \ln \frac{5}{3} = 0.511 \qquad \text{◆◆◆}$$

Solving Exponential Equations with Base 10

If an equation contains 10 as a base, the work will be simplified by taking *common* logarithms.

◆◆◆ Example 53: Solve $10^{5x} = 2(10^{2x})$ to three significant digits.

Solution: Taking common logs of both sides gives us

$$\log 10^{5x} = \log[2(10^{2x})]$$

By Eqs. 187 and 189,

$$5x \log 10 = \log 2 + 2x \log 10$$

By Eq. 192, $\log 10 = 1$, so

$$5x = \log 2 + 2x$$
$$3x = \log 2$$
$$x = \frac{\log 2}{3} = 0.100 \qquad \text{◆◆◆}$$

Doubling Time

Being able to solve an exponential equation for the exponent allows us to return to the formulas for exponential growth and decay (Sec. 20–2) and to derive two interesting quantities: doubling time and half-life.

If a quantity grows exponentially according to the following function:

| Exponential Growth | $y = ae^{nt}$ | 199 |

it will eventually double (y will be twice a). Setting $y = 2a$, we get

$$2a - ae^{nt}$$

or $2 = e^{nt}$. Taking natural logs, we have

$$\ln 2 = \ln e^{nt} = nt \ln e = nt$$

also given in the following form:

| Doubling Time | $t = \dfrac{\ln 2}{n}$ | 200 |

Since $\ln 2 \cong 0.693$, if we let P be the rate of growth expressed as a percent (*not* as a decimal, $P = 100n$), we get a convenient *rule of thumb*:

$$\text{doubling time} \simeq \frac{70}{P}$$

◆◆◆ **Example 54:** In how many years will a quantity double if it grows at the rate of 5.0% per year?

Solution:

$$\text{doubling time} = \frac{\ln 2}{0.050}$$

$$= \frac{0.693}{0.050}$$

$$= 14 \text{ years} \qquad ◆◆◆$$

Half-Life

When a material decays exponentially according to the following function:

| Exponential Decay | $y = ae^{-nt}$ | 201 |

the time it takes for the material to be half gone is called the *half-life*. If we let $y = a/2$,

$$\frac{1}{2} = e^{-nt} = \frac{1}{e^{nt}}$$

$$2 = e^{nt}$$

Taking natural logarithms gives us

$$\ln 2 = \ln e^{nt} = nt$$

Notice that the equation for half-life is the same as for doubling time. The rule of thumb is, of course, also the same. The rule of thumb gives us another valuable tool for *estimation*.

| Half-Life | $t = \dfrac{\ln 2}{n}$ | 200 |

◆◆◆ **Example 55:** Find the half-life of a radioactive material that decays exponentially at the rate of 2.0% per year.

Solution:

$$\text{half-life} = \frac{\ln 2}{0.020} = 35 \text{ years}$$

◆◆◆

Change of Base

If we have an exponential expression, such as 5^x, we can convert it to another exponential expression with base e or any other base.

◆◆◆ **Example 56:** Convert 5^x into an exponential expression with base e.

Solution: We write 5 as e raised to some power, say, n.

$$5 = e^n$$

Taking the natural logarithm, we obtain

$$\ln 5 = \ln e^n = n$$

So $n = \ln 5 \approx 1.61$, and $5 = e^{1.61}$. Then

$$5^x = (e^{1.61})^x = e^{1.61x}$$

◆◆◆

◆◆◆ **Example 57:** Convert 2^{4x} to an exponential expression with base e.

Solution: We first express 2 as e raised to some power.

$$2 = e^n$$
$$\ln 2 = \ln e^n = n$$
$$n = \ln 2 = 0.693$$

Then

$$2^{4x} = (e^{0.693})^{4x} = e^{2.773x}$$

◆◆◆

◆◆◆ **Example 58:** A bacteria sample contains 2850 bacteria and grows at a rate of 5.85% per hour. Assuming that the bacteria grow exponentially:

(a) Write the equation for the number N of bacteria as a function of time.
(b) Find the time for the sample to grow to 10,000 bacteria.

Estimate: Note that the doubling time here is $70 \div 5.85$ or about 12 hours. To grow from 2850 to 10,000 will require nearly two doublings, or nearly 24 hours.

Solution:

(a) We substitute into the equation for exponential growth, $y = ae^{nt}$, with $y = N$, $a = 2850$, and $n = 0.0585$, getting

$$N = 2850e^{0.0585t}$$

(b) Setting $N = 10,000$ and solving for t, we have

$$10,000 = 2850e^{0.0585t}$$
$$e^{0.0585t} = 3.51$$

We take the natural log of both sides and set $\ln e = 1$.

$$0.0585t \ln e = \ln 3.51$$

$$t = \frac{\ln 3.51}{0.0585} = 21.5 \text{ h}$$

which agrees with our estimate.   ◆◆◆

Exercise 5 ◆ Exponential Equations

Solve for x to three significant digits.

1. $2^x = 1$

2. $(7.26)^x = 86.8$

3. $(1.15)^{x+2} = 12.5$

4. $(2.75)^x = (0.725)^{x^2}$

5. $(15.4)^{\sqrt{x}} = 72.8$

6. $e^{5x} = 125$

7. $5.62e^{3x} = 188$

8. $1.05e^{4x+1} = 5.96$

9. $e^{2x-1} = 3e^{x+3}$

10. $14.8e^{3x^2} = 144$

11. $5^{2x} = 7^{3x-2}$

12. $3^{x^2} = 175^{x-1}$

13. $10^{3x} = 3(10^x)$

14. $e^x + e^{-x} = 2(e^x - e^{-x})$

15. $2^{3x+1} = 3^{2x+1}$

16. $5^{2x} = 3^{3x+1}$

17. $7e^{1.5x} = 2e^{2.4x}$

18. $e^{4x} - 2e^{2x} - 3 = 0$

19. $e^x + e^{-x} = 2$

20. $e^{6x} - e^{3x} - 2 = 0$

Hint: Problems 18 through 20 are *equations of quadratic type.*

Convert to an exponential expression with base e.

21. $3(4^{3x})$

22. $4(2.2^{4x})$

Applications

23. The current i in a certain circuit is given by

$$i = 6.25e^{-125t} \quad \text{(amperes)}$$

where t is the time in seconds. At what time will the current be 1.00 A?

24. The current through a charging capacitor is given by

$$i = \frac{E}{R}e^{-t/RC} \tag{A83}$$

If $E = 325$ V, $R = 1.35$ Ω, and $C = 3210$ μF, find the time at which the current through the capacitor is 0.0165 A.

25. The voltage across a charging capacitor is given by

$$v = E(1 - e^{-t/RC}) \tag{A84}$$

If $E = 20.3$ V and $R = 4510$ Ω, find the time when the voltage across a 545-μF capacitor is equal to 10.1 V.

26. The temperature above its surroundings of an iron casting initially at $2\overline{0}00°$F will be

$$T = 2\overline{0}00e^{-0.620t}$$

after t seconds. Find the time for the casting to be at a temperature of $5\overline{0}0°$F above its surroundings.

27. A certain long pendulum, released from a height y_0 above its rest position, will be at a height

$$y = y_0e^{-0.75t}$$

at t seconds. If the pendulum is released at a height of 15 cm, at what time will the height be 5.0 cm?

28. A population growing at a rate of 2.0% per year from an initial population of 9000 will grow in t years to an amount

$$P = 9000e^{0.02t}$$

How many years will it take the population to triple?

29. The barometric pressure in inches of mercury at a height of h feet above sea level is

$$p = 30e^{-kh}$$

where $k = 3.83 \times 10^{-5}$. At what height will the pressure be 10 in. of mercury?

30. The approximate density of seawater at a depth of h mi is

$$d = 64.0e^{0.00676h} \quad (\text{lb/ft}^3)$$

Find the depth at which the density will be 64.5 lb/ft^3.

31. A rope passing over a rough cylindrical beam (Fig. 20–15) supports a weight W. The force F needed to hold the weight is

$$F = We^{-\mu\theta}$$

where μ is the coefficient of friction and θ is the angle of wrap in radians. If $\mu = 0.150$, what angle of wrap is needed for a force of 100 lb to hold a weight of 200 lb?

32. Using the formula for compound interest, Eq. A10, calculate the number of years it will take a sum of money to triple when invested at a rate of 12% per year.

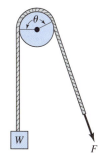

FIGURE 20–15

33. Using the formula for present worth, given in problem 7 of Exercise 1, in how many years will $50,000 accumulate to $70,000 at 15% interest?

34. Using the annuity formula, given in problem 10 of Exercise 1, find the number of years it will take a worker to accumulate $100,000 with an annual payment of $3000 if the interest rate is 8.0%.

35. Using the capital recovery formula from problem 12 of Exercise 1, calculate the number of years a person can withdraw $10,000/yr from a retirement fund containing $60,000 if the rate of interest is $6\frac{3}{4}$%.

36. Find the half-life of a material that decays exponentially at the rate of 3.50% per year.

37. How long will it take the U.S. annual oil consumption to double if it is increasing exponentially at a rate of 7.0% per year?

38. How long will it take the world population to double at an exponential growth rate of 1.64% per year?

39. What is the maximum annual growth in energy consumption permissible if the consumption is not to double in the next 20 years?

Computer

40. An iron ball with a mass of 1 kg is cut into two equal pieces. It is cut again, the second cut producing four equal pieces, the third cut making eight pieces, and so on. Write a program to compute the number of cuts needed to make each piece smaller than one atom (mass $= 9.4 \times 10^{-26}$ kg). Assume that no material is removed during cutting.

41. Assuming that the present annual world oil consumption is 17×10^9 barrels/yr, that this rate of consumption is increasing at a rate of 5% per year, and that the total world oil reserves are 1700×10^9 barrels, compute and print the following table:

Year	Annual Consumption	Oil Remaining
0	17	1700
1	17.85	1682.15
.	.	.
.	.	.
.	.	.

Have the computation stop when the oil reserves are gone.

42. Use your program for finding roots of equations by simple iteration or the mid point method to solve any of problems 1 through 20.

43. Use a CAS to solve any of the equations in problems 1 through 20. Simply type in the equation and give the *Solve* command. Enter e^x by typing Exp(x), and b^x by typing b $\wedge$ x. The computer may also give you some complex solutions, containing the imaginary number i, which we will not consider here.

Graphics Calculator

44. Use a graphics utility to get an approximate solution to any of the equations in this exercise set. Recall that we first put the equation into the form $f(x) = 0$. Next we graph $y = f(x)$ and note where the curve crosses the x axis. We then use TRACE and ZOOM to find the roots to whatever accuracy we wish.

20–6 Logarithmic Equations

Often a logarithmic expression can be evaluated, or an equation containing logarithms can be solved, simply by rewriting it in exponential form.

◆◆◆ **Example 59:** Evaluate $x = \log_5 25$.

Solution: Changing to exponential form, we have

$$5^x = 25 = 5^2$$
$$x = 2$$

Since $5^2 = 25$, the answer checks.   ◆◆◆

◆◆◆ **Example 60:** Solve for x:

$$\log_x 4 = \frac{1}{2}$$

Solution: Going to exponential form gives us

$$x^{1/2} = 4$$

Squaring both sides yields

$$x = 16$$

Since $16^{1/2} = 4$, the answer checks.   ◆◆◆

◆◆◆ **Example 61:** Solve for x:

$$\log_{25} x = -\frac{3}{2}$$

Solution: Writing the equation in exponential form, we obtain

$$25^{-3/2} = x$$

$$x = \frac{1}{25^{3/2}} = \frac{1}{5^3} = \frac{1}{125}$$

Since $25^{-3/2} = \frac{1}{125}$, the answer checks.   ◆◆◆

◆◆◆ **Example 62:** Evaluate $x = \log_{81} 27$.

Solution: Changing to exponential form yields

$$81^x = 27$$

But 81 and 27 are both powers of 3.

$$(3^4)^x = 3^3$$
$$3^{4x} = 3^3$$
$$4x = 3$$
$$x = 3/4$$

To check, we verify that $81^{3/4} = 3^3 = 27$.   ◆◆◆

If only one term in our equation contains a logarithm, we isolate that term on one side of the equation and then go to exponential form.

◆◆◆ **Example 63:** Solve for x to three significant digits:

$$3 \log(x^2 + 2) - 6 = 0$$

Solution: Rearranging and dividing by 3 gives

$$\log(x^2 + 2) = 2$$

Going to exponential form, we obtain

$$x^2 + 2 = 10^2 = 100$$
$$x^2 = 98$$
$$x = \pm 9.90$$ ◆◆◆

If every term contains "log," we use the properties of logarithms to combine those terms into a single logarithm on each side of the equation and then take the antilog of both sides.

◆◆◆ **Example 64:** Solve for x to three significant digits:

$$3 \log x - 2 \log 2x = 2 \log 5$$

Solution: Using the properties of logarithms gives

$$\log \frac{x^3}{(2x)^2} = \log 5^2$$

$$\log \frac{x}{4} = \log 25$$

Taking the antilog of both sides, we have

$$\frac{x}{4} = 25$$

$$x = 100$$ ◆◆◆

If one or more terms do not contain a log, combine all of the terms that do contain a log on one side of the equation. Then go to exponential form.

◆◆◆ **Example 65:** Solve for x:

$$\log(5x + 2) - 1 = \log(2x - 1)$$

Solution: Rearranging yields

$$\log(5x + 2) - \log(2x - 1) = 1$$

By Eq. 188,

$$\log \frac{5x + 2}{2x - 1} = 1$$

Expressing in exponential form gives

$$\frac{5x + 2}{2x - 1} = 10^1 = 10$$

Solving for x, we have

$$5x + 2 = 20x - 10$$
$$12 = 15x$$
$$x = \frac{4}{5}$$

◆◆◆

◆◆◆ **Example 66:** *The Richter Scale.* The Richter magnitude R, used to rate the intensity of an earthquake, is given by

$$R = \log \frac{a}{a_0}$$

where a and a_0 are the vertical amplitudes of the ground movement of the measured earthquake and of an earthquake taken as reference. If two earthquakes measure 4 and 5 on the Richter scale, by what factor is the amplitude of the stronger quake greater than that of the weaker?

Solution: Let us rewrite the equation for the Richter magnitude using the laws of logarithms.

$$R = \log a - \log a_0$$

Then, for the first earthquake,

$$4 = \log a_1 - \log a_0$$

and for the second,

$$5 = \log a_2 - \log a_0$$

Subtracting the first equation from the second gives

$$1 = \log a_2 - \log a_1 = (\log a_2/a_1)$$

Going to exponential form, we have

$$\frac{a_2}{a_1} = 10^1 = 10$$
$$a_2 = 10 \, a_1$$

So the second earthquake has 10 times the amplitude of the first.

Check: Does this seem reasonable? From our study of logarithms, we know that as a number increases tenfold, its common logarithm increases only by 1. For example, $\log 45 = 1.65$ and $\log 450 = 2.65$. Since the Richter magnitude is proportional to the common log of the amplitude, it seems reasonable that a tenfold increase in amplitude increases the Richter magnitude by just 1 unit.   ◆◆◆

Exercise 6 ◆ Logarithmic Equations

Find the value of x in each expression.

1. $x = \log_3 9$ **2.** $x = \log_2 8$ **3.** $x = \log_8 2$

4. $x = \log_9 27$ **5.** $x = \log_{27} 9$ **6.** $x = \log_4 8$

7. $x = \log_8 4$ **8.** $x = \log_{27} 81$ **9.** $\log_x 8 = 3$

10. $\log_3 x = 4$ **11.** $\log_x 27 = 3$ **12.** $\log_x 16 = 4$

13. $\log_5 x = 2$ **14.** $\log_x 2 = \frac{1}{2}$ **15.** $\log_{36} x = \frac{1}{2}$

16. $\log_2 x = 3$ **17.** $x = \log_{25} 125$ **18.** $x = \log_5 125$

Solve for x. Give any approximate results to three significant digits. Check your answers.

19. $\log(2x + 5) = 2$ **20.** $2 \log(x + 1) = 3$

21. $\log(2x + x^2) = 2$ **22.** $\ln x - 2 \ln x = \ln 64$

23. $\ln 6 + \ln(x - 2) = \ln(3x - 2)$ **24.** $\ln(x + 2) - \ln 36 = \ln x$

25. $\ln(5x + 2) - \ln(x + 6) = \ln 4$ **26.** $\log x + \log 4x = 2$

27. $\ln x + \ln(x + 2) = 1$ **28.** $\log 8x^2 - \log 4x = 2.54$

29. $2 \log x - \log(1 - x) = 1$ **30.** $3 \log x - 1 = 3 \log(x - 1)$

31. $\log(x^2 - 4) - 1 = \log(x + 2)$ **32.** $2 \log x - 1 = \log(20 - 2x)$

33. $\log(x^2 - 1) - 2 = \log(x + 1)$ **34.** $\ln 2x - \ln 4 + \ln(x - 2) = 1$

35. $\log(4x - 3) + \log 5 = 6$

Applications

36. An amount of money a invested at a compound interest rate of n per year will take t years to accumulate to an amount y, where t is

$$t = \frac{\log y - \log a}{\log(1 + n)} \quad \text{(years)}$$

How many years will it take an investment to triple in value when deposited at 8.00% per year?

37. If an amount R is deposited once every year at a compound interest rate n, the number of years it will take to accumulate to an amount y is

$$t = \frac{\log\left(\dfrac{ny}{R} + 1\right)}{\log(1 + n)} \quad \text{(years)}$$

How many years will it take an annual payment of $1500 to accumulate to $13,800 at 9.0% per year?

38. If an amount a is invested at a compound interest rate n, it will be possible to withdraw a sum R at the end of every year for t years until the deposit is exhausted. The number of years is given by

$$t = \frac{\log\left(\dfrac{an}{R - an} + 1\right)}{\log(1 + n)} \quad \text{(years)}$$

If $200,000 is invested at 12% interest, for how many years can an annual withdrawal of $30,000 be made before the money is used up?

39. The *magnitude M* of a star of intensity I is

$$M = 2.5 \log\frac{I_1}{I} + 1$$

where I_1 is the intensity of a first-magnitude star. What is the magnitude of a star whose intensity is one-tenth that of a first-magnitude star?

The equation in problem 36 is derived from the compound interest formula. The equations in problems 37 and 38 are obtained from the equations for an annuity and for capital recovery from Exercise 1. Can you see how they were derived?

40. The difference in elevation h (ft) between two locations having barometer readings of B_1 and B_2 inches of mercury is given by the logarithmic equation $h = 60,470 \log(B_2/B_1)$, where B_1 is the pressure at the *upper* station. Find the difference in elevation between two stations having barometer readings of 29.14 in. at the lower station and 26.22 in. at the upper.

41. What will the barometer reading be 815.0 ft above a station having a reading of 28.22 in.?

Use the following information for problems 42 through 45. If the power input to a network or device is P_1 and the power output is P_2, the number of decibels gained or lost in the device is given by the following logarithmic equation:

Decibels Gained or Lost	$G = 10 \log_{10} \dfrac{P_2}{P_1}$ (dB)	A104

42. A certain amplifier gives a power output of $\overline{10}00$ W for an input of $\overline{50}$ W. Find the dB gain.

43. A transmission line has a loss of 3.25 dB. Find the power transmitted for an input of 2750 kW.

44. What power input is needed to produce a 250-W output with an amplifier having a 50-dB gain?

45. The output of a certain device is half the input. How many decibels does this loss represent?

46. The heat loss q per foot of cylindrical pipe insulation (Fig. 20–16) having an inside radius r_1 and outside radius r_2 is given by the logarithmic equation

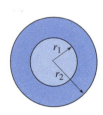

$$q = \frac{2\pi k(t_1 - t_2)}{\ln(r_2/r_1)} \quad \text{(Btu/h)}$$

FIGURE 20–16 An insulated pipe.

where t_1 and t_2 are the inside and outside temperatures (°F) and k is the conductivity of the insulation. Find q for a 4.0-in.-thick insulation having a conductivity of 0.036 and wrapped around a 9.0-in.-diameter pipe at 550°F if the surroundings are at 90°F.

47. If a resource is being used up at a rate that increases exponentially, the time it takes to exhaust the resource (called the *exponential expiration time,* EET) is

$$\text{EET} = \frac{1}{n} \ln\left(\frac{nR}{r} + 1\right)$$

where n is the rate of increase in consumption, R the total amount of the resource, and r the initial rate of consumption. If we assume that the United States has oil reserves of 207×10^9 barrels and that our present rate of consumption is 6.00×10^9 barrels/yr, how long will it take to exhaust these reserves if our consumption increases by 7.00% per year?

Use the following information for problems 48 through 50. The pH value of a solution having a concentration C of hydrogen ions is given by the following equation:

pH	$\text{pH} = -\log_{10} C$	A49

48. Find the concentration of hydrogen ions in a solution having a pH of 4.65.

49. A pH of 7 is considered *neutral,* while a lower pH is *acidic* and a higher pH is *alkaline.* What is the hydrogen ion concentration at a pH of 7.0?

50. The acid rain during a certain storm had a pH of 3.5. Find the hydrogen ion concentration. How does it compare with that for a pH of 7.0? Show that:
(a) when the pH doubles, the hydrogen ion concentration is *squared.*
(b) when the pH increases by a factor of *n,* the hydrogen ion concentration is raised *to the nth power.*

51. Two earthquakes have Richter magnitudes of 4.6 and 6.2. By what factor is the amplitude of the stronger quake greater than that of the weaker? See Example 66.

Computer

52. Use your program for solving equations by simple iteration or by the midpoint method to solve any of the problems 19 through 35.

53. Use a CAS to solve any of the equations in problems 19 through 35. Simply type in the equation and give the *Solve* command.

Graphics Calculator

54. Use a graphics utility to get an approximate solution to any of the equations in this exercise set. Recall that we first put the equation into the form $f(x) = 0$. Next we graph $y = f(x)$ and note where the curve crosses the x axis. We then use $\boxed{\text{TRACE}}$ and $\boxed{\text{ZOOM}}$ to find the roots to whatever accuracy we wish.

20–7 Graphs on Logarithmic and Semilogarithmic Paper

Logarithmic and Semilogarithmic Paper

Our graphing so far has all been done on ordinary graph paper, on which the lines are equally spaced. For some purposes, though, it is better to use *logarithmic* paper (Fig. 20–17), also called *log-log* paper, or *semilogarithmic* paper (Fig. 20–18), also called *semilog* paper. Looking at the logarithmic scales of these graphs, we note the following:

1. The lines are not equally spaced. The distance in inches from, say, 1 to 2, is equal to the distance from 2 to 4, which, in turn, is equal to the distance from 4 to 8.
2. Each tenfold increase in the scale, say, from 1 to 10 or from 10 to 100, is called a *cycle.* Each cycle requires the same distance in inches along the scale.
3. The log scales do not include zero.

Looking at Fig. 20–18, notice that although the numbers on the vertical scale are in equal increments (1, 2, 3, . . . , 10), the spacing on that scale is proportional to the logarithms of those numbers. So the numeral "2" is placed at a position corresponding to log 2 (which is 0.3010, or about one-third of the distance along the vertical); "4" is placed at log 4 (about 0.6 of the way); and "10" is at log 10 (which equals 1, at the top of the scale).

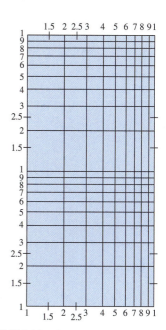

FIGURE 20–17 Logarithmic graph paper.

When to Use Logarithmic or Semilog Paper

We use these special papers when:

1. The range of the variables is too large for ordinary paper.

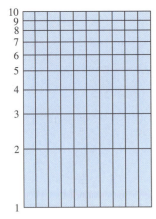

FIGURE 20–18 Semilogarithmic graph paper.

2. We want to graph a power function or an exponential function. Each of these will plot as a straight line on the appropriate paper, as shown in Fig. 20–19.

3. We want to find an equation that will approximately represent a set of empirical data.

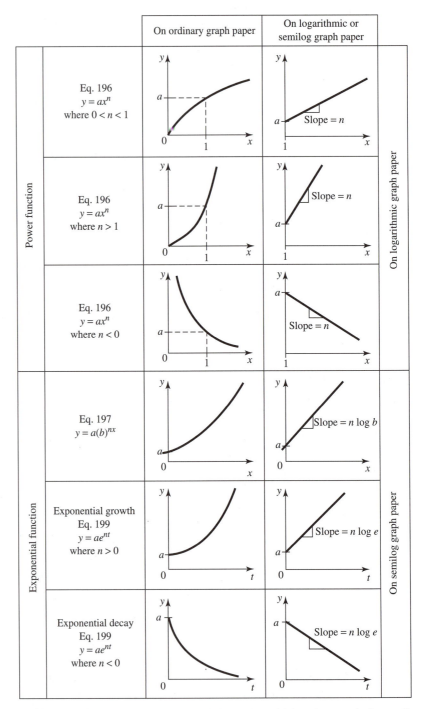

FIGURE 20–19 The power function and the exponential function, graphed on ordinary paper and log-log or semilog paper.

Graphing the Power Function

A *power function* is one whose equation is of the following form:

Power Function	$y = ax^n$	196

where a and n are nonzero constants. This equation is nonlinear (except when $n = 1$), and the shape of its graph depends upon whether n is positive or negative and whether n is greater than or less than 1. Figure 20–19 shows the shapes that this curve can have for various ranges of n. If we take the logarithm of both sides of Eq. 196, we get

$$\log y = \log(ax^n)$$
$$= \log a + n \log x$$

If we now make the substitution $X = \log x$ and $Y = \log y$, our equation becomes

$$Y = nX + \log a$$

This equation *is* linear and, on rectangular graph paper, graphs as a straight line with a slope of n and a y intercept of $\log a$ (Fig. 20–20). However, we do not have to make the substitutions shown above *if we use logarithmic paper,* where the scales are proportional to the logarithms of the variables x and y. We simply have to plot the original equation on log-log paper, and it will be a straight line which has slope n and which has a value of $y = a$ when $x = 1$ (see Fig. 20–19). This will be illustrated in the following example.

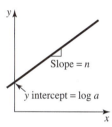

FIGURE 20–20   Graph of $y = ax^n$ on rectangular graph paper, where $X = \log x$ and $Y = \log y$.

◆◆◆ **Example 67:** Plot the equation $y = 2.5x^{1.4}$ for values of x from 1 to 10. Choose graph paper so that the equation plots as a straight line.

Solution: We make a table of point pairs. Since the graph will be a straight line, we need only two points, with a third as a check. Here, we will plot four points to show that all points do lie on a straight line. We choose values of x and for each compute the value of y.

x	0	1	4	7	10
y	0	2.5	17.4	38.1	62.8

We choose log-log rather than semilog paper because we are graphing a power function, which plots as a straight line on this paper (see Fig. 20–19). We choose the number of cycles for each scale by looking at the range of values for x and y. Note that the logarithmic scales do not contain zero, so we cannot plot the point $(0, 0)$. Thus on the x axis we need one cycle. On the y axis we must go from 2.5 to 62.8. With two cycles we can span a range of 1 to 100. Thus we need log-log paper, one cycle by two cycles.

We mark the scales, plot the points as shown in Fig. 20–21, and get a straight line as expected. We note that the value of y at $x = 1$ is equal to 2.5 and is the same as the coefficient of $x^{1.4}$ in the given equation.

We can get the slope of the straight line by measuring the rise and run with a scale and dividing rise by run. Or we can use the values from the graph. But since the spacing on the graph really tells the logarithm value of the position of the

Recall that we studied the power function in Sec. 19–3.

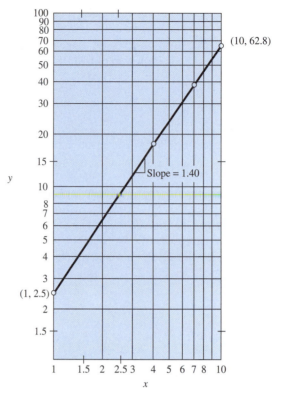

FIGURE 20–21   Graph of $y = 2.5x^{1.4}$.

pictured points, we must remember to take the logarithm of those values. (Either common or natural will give the same result.) Thus

$$\text{slope} = \frac{\text{rise}}{\text{run}} = \frac{\ln 62.8 - \ln 2.5}{\ln 10 - \ln 1} = \frac{3.22}{2.30} = 1.40$$

The slope of the line is thus equal to the power of x, as expected from Fig. 20–20. We will use these ideas later when we try to write an equation to fit a set of data.   ◆◆◆

Common Error	Be sure to take the *logs* of the values on the x and y axes when computing the slope.

Graphing the Exponential Function

Consider the exponential function given by the following equation:

Exponential Function	$y = a(b)^{nx}$	**197**

If we take the logarithm of both sides, we get

$$\log y = \log b^{nx} + \log a$$
$$= (n \log b)x + \log a$$

If we replace $\log y$ with Y, we get the linear equation

$$Y = (n \log b)\, x + \log a$$

If we graph the given equation on semilog paper with the logarithmic scale along the y axis, we get a straight line with a slope of $n \log b$ which cuts the y axis at a (Fig. 20–19). Also shown are the special cases where the base b is equal to e (2.718 . . .). Here, the independent variable is shown as t, because exponential growth and decay are usually functions of time.

◆◆◆ **Example 68:** Plot the exponential function

$$y = 100e^{-0.2x}$$

for values of x from 0 to 10.

Solution: We make a table of point pairs.

x	0	2	4	6	8	10
y	100	67.0	44.9	30.1	20.2	13.5

We choose semilog paper for graphing the exponential function and use the linear scale for x. The range of y is from 13.5 to 100; thus we need *one cycle* of the logarithmic scale. The graph is shown in Fig. 20–22. Note that the line obtained has a

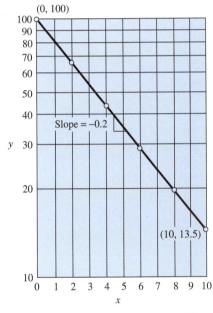

FIGURE 20–22 Graph of $y = 100e^{-0.2x}$.

y intercept of 100. Also, the slope is equal to $n \log e$ or, if we use natural logs, is equal to n.

$$\text{slope} = n = \frac{\ln 13.5 - \ln 100}{10} = -0.2$$

This is the coefficient of x in our given equation. ◆◆◆

When computing the slope on semilog paper, we take the logarithms of the values on the log scale, but not on the linear scale. Computing the slope using common logs, we get

$$\text{slope} = n \log e$$

$$= \frac{\log 13.5 - \log 100}{10}$$

$$= -0.0870$$

$$n = \frac{-0.0870}{\log e} = -0.2$$

as we got using natural logs.

Empirical Functions

We choose logarithmic or semilog paper to plot a set of empirical data when:

1. The range of values is too large for ordinary paper.
2. We suspect that the relation between our variables may be a power function or an exponential function, and we want to find that function.

We show the second case by means of an example.

The process of fitting an approximate equation to fit a set of data points is called *curve fitting*. In statistics it is referred to as *regression*. Here we will do only some very simple cases.

◆◆◆ **Example 69:** A test of a certain electronic device shows it to have an output current i versus input voltage v as shown in the following table:

v (V)	1	2	3	4	5
i (A)	5.61	15.4	27.7	42.1	58.1

Plot the given empirical data, and try to find an approximate formula for y in terms of x.

Solution: We first make a graph on linear graph paper (Fig. 20–23) and get a curve that is concave upward. Comparing its shape with the curves in Fig. 20–19, we suspect that the equation of the curve (if we can find one at all) may be either a power function or an exponential function.

Next, we make a plot on semilog paper (Fig. 20–24) and do *not* get a straight line. However, a plot on log-log paper (Fig. 20–25) *is* linear. We thus assume that our equation has the form

$$i = av^n \quad \text{or} \quad \ln i = n \ln v + \ln a$$

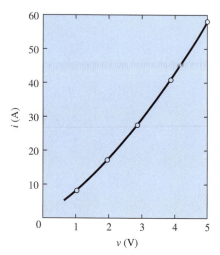

FIGURE 20–23 Plot of table of points on linear graph paper.

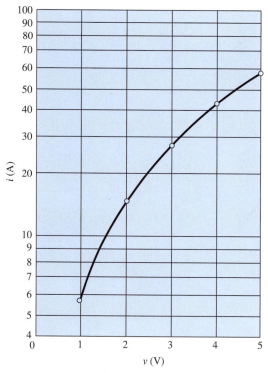

FIGURE 20–24 Plot of table of points on semilog paper.

In Example 61 we showed how to compute the slope of the line to get the exponent n, and we also saw that the coefficient a was the value of the function at $x = 1$. Now we show a different method for finding a and n which can be used even if we do not have the y value at $x = 1$. We choose two points on the curve, say, $(5, 58.1)$ and $(2, 15.4)$, and substitute each into

$$\ln i = n \ln v + \ln a$$

getting

$$\ln 58.1 = n \ln 5 + \ln a$$

and

$$\ln 15.4 = n \ln 2 + \ln a$$

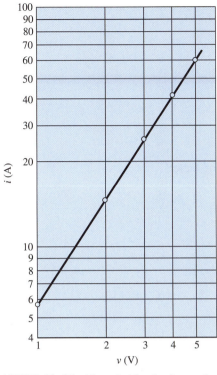

FIGURE 20–25 Plot of table of points on log-log paper.

Here our data plotted as a nice straight line on logarithmic paper. But with real data we are often unable to draw a straight line that passes through every point. The *method of least squares*, not shown here, is often used to draw a line that is considered the "best fit" for a scattering of data points.

A simultaneous solution, not shown, for n and a yields

$$n = 1.45 \quad \text{and} \quad a = 5.64$$

Our equation is then

$$i = 5.64v^{1.45}$$

Here, again, we could have used common logarithms and gotten the same result.

Finally, we test this formula by computing values of i and comparing them with the original data, as shown in the following table:

v	1	2	3	4	5
Original i	5.61	15.4	27.7	42.1	58.1
Calculated i	5.64	15.4	27.7	42.1	58.2

The fitting of a power function to a set of data is called *power regression* in statistics. You may be able to do this on your graphics calculator. See problem 27 of Exercise 7.

We get values very close to the original. ◆◆◆

Exercise 7 ◆ Graphs on Logarithmic and Semilogarithmic Paper

Graphing the Power Function

Graph each power function on log-log paper for $x = 1$ to 10.

1. $y = 2x^3$
2. $y = 3x^4$
3. $y = x^5$
4. $y = x^{3/2}$
5. $y = 5\sqrt{x}$
6. $y = 2\sqrt[3]{x}$
7. $y = 1/x$
8. $y = 3/x^2$
9. $y = 2x^{2/3}$

Graph each set of data on log-log paper, determine the coefficients graphically, and write an approximate equation to fit the given data.

10.

x	2	4	6	8	10
y	3.48	12.1	25.2	42.2	63.1

11.

x	2	4	6	8	10
y	240	31.3	9.26	3.91	2.00

Graphing the Exponential Function

Graph each exponential function on semilog paper.

12. $y = 3^x$
13. $y = 5^x$
14. $y = e^x$
15. $y = e^{-x}$
16. $y = 4^{x/2}$
17. $y = 3^{x/4}$
18. $y = 3e^{2x/3}$
19. $y = 5e^{-x}$
20. $y = e^{x/2}$

Graphing Empirical Functions

You may be able to do this on your graphics calculator. See Problem 27.

Graph each set of data on log-log or semilog paper, determine the coefficients graphically, and write an approximate equation to fit the given data.

21.

x	0	1	2	3	4	5
y	1.00	2.50	6.25	15.6	39.0	97.7

22.

x	1	2	3	4
y	48.5	29.4	17.6	10.8

23. Current in a tungsten lamp, i, for various voltages, v:

v (V)	2	8	25	50	100	150	200
i (mA)	24.6	56.9	113	172	261	330	387

24. Difference in temperature, T, between a cooling body and its surroundings at various times, t:

t (s)	0	3.51	10.9	19.0	29.0	39.0	54.0	71.0
T (°F)	20.0	19.0	17.0	15.0	13.0	11.0	8.50	6.90

25. Pressure, p, of 1 lb of saturated steam at various volumes, v:

v (ft^3)	26.4	19.1	14.0	10.5	8.00
p (lb/in.^{2})	14.7	20.8	28.8	39.3	52.5

26. Maximum height y reached by a long pendulum t seconds after being set in motion:

t (s)	0	1	2	3	4	5	6
y (in.)	10	4.97	2.47	1.22	0.61	0.30	0.14

Graphics Calculator

27. Some graphics calculators can fit a straight line, a power function, an exponential function, or a logarithmic function to a given set of data points, and can give the two constants in the function. Such fitting goes by the statistics name of *regression.* The TI-81, for example, can do linear, logarithmic, exponential, and power regression. You must enter the data points and then choose the type of function that you think will fit. The calculator will give the two constants. It will also give the *correlation coefficient,* which is a measure of goodness of fit. If this coefficient is close to 1 or -1, the fit is good; if it is close to 0, the fit is bad.

Study your calculator manual to learn how to do regression. Then use it for any of the problems 21 through 26 in this exercise set.

••• CHAPTER 20 REVIEW PROBLEMS ••••••••••••••••••••••••••••••••

Convert to logarithmic form.

1. $x^{5.2} = 352$
2. $4.8^x = 58$
3. $24^{1.4} = x$

Convert to exponential form.

4. $\log_3 56 = x$
5. $\log_x 5.2 = 124$
6. $\log_2 x = 48.2$

Solve for x.

7. $\log_{81} 27 = x$
8. $\log_3 x = 4$
9. $\log_x 32 = -\frac{5}{7}$

Write as the sum or difference of two or more logarithms.

10. $\log xy$

11. $\log \dfrac{3x}{z}$

12. $\log \dfrac{ab}{cd}$

Express as a single logarithm.

13. $\log 5 + \log\ 2$
14. $2 \log x - 3 \log y$
15. $\frac{1}{2} \log p - \frac{1}{4} \log q$

Find the common logarithm of each number to four decimal places.

16. 6.83
17. 364
18. 0.00638

Find x if $\log x$ is equal to the given value.

19. 2.846
20. 1.884
21. -0.473

Evaluate using logarithms

22. 5.836×88.24

23. $\dfrac{8375}{2846}$

24. $(4.823)^{1.5}$

Find the natural logarithm of each number to four decimal places.

25. 84.72
26. 2846
27. 0.00873

Find the number whose natural logarithm is the given value.

28. 5.273
29. 1.473
30. −4.837

Find log x if:

31. ln x = 4.837
32. ln x = 8.352

Find ln x if:

33. log x = 5.837
34. log x = 7.264

Solve for x to three significant digits.

35. $(4.88)^x = 152$
36. $e^{2x+1} = 72.4$
37. $3 \log x - 3 = 2 \log x^2$
38. $\log(3x - 6) = 3$
39. A sum of $1500 is deposited at a compound interest rate of $6\frac{1}{2}\%$ compounded quarterly. How much will it accumulate to in 5 years?
40. A quantity grows exponentially at a rate of 2.00% per day for 10.0 weeks. Find the final amount if the initial amount is 500 units.
41. A flywheel decreases in speed exponentially at the rate of 5.00% per minute. Find the speed after 20.0 min if the initial speed is 2250 rev/min.
42. Find the half-life of a radioactive substance that decays exponentially at the rate of 3.50% per year.
43. Find the doubling time for a population growing at the rate of 3.0% per year.
44. Graph the power function $y = 3x^{2/3}$ on log-log paper for x = 1 to 6.
45. Graph the exponential function $y = e^{x/3}$ on semilog paper for x = 0 to 6.

Writing

46. In this chapter we have introduced two more kinds of equations: the *exponential* equation and the *logarithmic* equation. Earlier we have had five other types. List the seven types, give an example of each, and write one sentence for each telling how it differs from the others.

Team Project

47. The temperatures of a kettle of water, measured every minute during cooling from 212°F at t = 0 to a room temperature of 70°F, are as follows:

212.0	206.5	198.4	185.6
171.2	155.3	136.4	121.4
110.0	100.6	90.1	86.3
79.1	74.6	73.1	72.5
71.8	70.0	70.9	70.1

Write an equation for the kettle temperature as a function of time, with all constants evaluated.

21

Complex Numbers

OBJECTIVES

When you have completed this chapter, you should be able to:

- Simplify radicals having negative radicands.
- Write complex numbers in rectangular, polar, trigonometric, and exponential forms.
- Evaluate powers of j.
- Find the sums, differences, products, quotients, powers, and roots of complex numbers.
- Solve quadratic equations having complex roots.
- Factor polynomials that have complex factors.
- Add, subtract, multiply, and divide vectors using complex numbers.
- Solve alternating current problems using complex numbers.

So far we have dealt entirely with *real numbers,* which are all of the numbers that correspond to points on the number line, that is, the rational and irrational numbers. The real numbers include integers, fractions, roots, and the negatives of these numbers. In this chapter we consider *imaginary* and *complex numbers.* We learn what they are, how to express them in several different forms, and how to perform the basic operations with them.

Why bother with a new type of number when, as we have seen, the real numbers can do so much? The *complex number system* is a natural extension of the real number system. We use complex numbers mainly to manipulate *vectors,* which are so important in many branches of technology. Complex numbers (despite their unfortunate name) actually simplify computations with vectors, as we see especially with the alternating current calculations at the end of this chapter.

21–1   Complex Numbers in Rectangular Form

The Imaginary Unit

Recall that in the real number system, the equation

$$x^2 = -1$$

had no solution because there was no real number such that its square was -1. Now we extend our number system to allow the quantity $\sqrt{-1}$ to have a meaning. We define the *imaginary unit* as the square root of -1 and represent it by the symbol j.

$$\text{Imaginary unit:} \quad j \equiv \sqrt{-1}$$

Complex Numbers

A *complex number* is any number, real or imaginary, that can be written in the form

$$a + jb$$

where a and b are real numbers, and $j = \sqrt{-1}$ is the *imaginary unit*.

The letter i is often used for the imaginary unit. In technical work, however, we save i for electric current. Further, j (or i) is sometimes written before the b, and sometimes after. So the number $j5$ may also be written $5i$, $5j$, or $i5$.

◆◆◆ **Example 1:** The following numbers are complex numbers:

$$4 + j2 \qquad -7 + j8 \qquad 5.92 - j2.93 \qquad 83 \qquad j27 \qquad\qquad ◆◆◆$$

Real and Imaginary Numbers

When $b = 0$ in a complex number $a + jb$, we have a *real number.* When $a = 0$, the number is called a *pure imaginary number.*

◆◆◆ **Example 2:**
(a) The complex number 48 is also a real number.
(b) The complex number $j9$ is a pure imaginary number.   ◆◆◆

The two parts of a complex number are called the *real part* and the *imaginary part.*

This is called the *rectangular form* of a complex number.

Complex Number	$a + jb$ real part ⎤ ⎡ imaginary part	218

Addition and Subtraction of Complex Numbers

To combine complex numbers, separately combine the real parts, then combine the imaginary parts, and express the result in the form $a + jb$.

Addition of Complex Numbers	$(a + jb) + (c + jd) = (a + c) + j(b + d)$	219
Subtraction of Complex Numbers	$(a + jb) - (c + jd) = (a - c) + j(b - d)$	220

◆◆◆ **Example 3:** Add or subtract, as indicated.

(a) $j3 + j5 = j8$
(b) $j2 + (6 - j5) = 6 - j3$

(c) $(2 - j5) + (-4 + j3) = (2 - 4) + j(-5 + 3) = -2 - j2$
(d) $(-6 + j2) - (4 - j) = (-6 - 4) + j[2 - (-1)] = -10 + j3$ ◆◆◆

Powers of j

We often have to evaluate powers of the imaginary unit, especially the square of j. Since

$$j = \sqrt{-1}$$

then

$$j^2 = \sqrt{-1}\,\sqrt{-1} = -1$$

Higher powers are easily found.

$$j^3 = j^2 j \quad = (-1)j = -j$$
$$j^4 = (j^2)^2 = (-1)^2 = \quad 1$$
$$j^5 = j^4 j \quad = (1)j \quad = \quad j$$
$$j^6 = j^4 j^2 \quad = (1)j^2 \quad = -1$$

We see that the values are starting to repeat: $j^5 = j$, $j^6 = j^2$, and so on. The first four values keep repeating.

Note that when the exponent n is a multiple of 4, then $j^n = 1$.

Powers of j	$j = \sqrt{-1}$ $\quad j^2 = -1$ $\quad j^3 = -j$ $\quad$ $j^4 = 1$ $\quad\quad j^5 = j\dots$	**217**

◆◆◆ **Example 4:** Evaluate j^{17}.

Solution: Using the laws of exponents, we express j^{17} in terms of one of the first four powers of j.

$$j^{17} = j^{16} j = (j^4)^4 j = (1)^4 j = j$$ ◆◆◆

Multiplication of Imaginary Numbers

Multiply as with ordinary numbers, but use Eq. 217 to simplify any powers of j.

◆◆◆ **Example 5:**

(a) $5 \times j3 = j15$
(b) $j2 \times j4 = j^2 8 = (-1)8 = -8$
(c) $3 \times j4 \times j5 \times j = j^3 60 = (-j)60 = -j60$
(d) $(j3)^2 = j^2 3^2 = (-1)9 = -9$ ◆◆◆

Multiplication of Complex Numbers

Multiply complex numbers as you would any algebraic expressions: Replace j^2 by -1, and put the expression into the form $a + jb$.

Multiplication of Complex Numbers	$(a + jb)(c + jd) = (ac - bd) + j(ad + bc)$	**221**

✦✦✦ Example 6:

(a) $3(5 + j2) = 15 + j6$
(b) $(j3)(2 - j4) = j6 - j^2 12 = j6 - (-1)12 = 12 + j6$
(c) $(3 - j2)(-4 + j5) = 3(-4) + 3(j5) + (-j2)(-4) + (-j2)(j5)$
$$= -12 + j15 + j8 - j^2 10$$
$$= -12 + j15 + j8 - (-1)10$$
$$= -2 + j23$$
(d) $(3 - j5)^2 = (3 - j5)(3 - j5)$
$$= 9 - j15 - j15 + j^2 25 = 9 - j30 - 25$$
$$= -16 - j30$$
                                                                                        ◆◆◆

To multiply radicals that contain negative quantities in the radicand, first express all quantities in terms of j; then proceed to multiply. Always be sure to convert radicals to imaginary numbers *before* performing other operations, or contradictions may result.

✦✦✦ Example 7: Multiply $\sqrt{-4}$ by $\sqrt{-4}$.

Solution: Converting to imaginary numbers, we obtain
$$\sqrt{-4} \, \sqrt{-4} = (j2)(j2) = j^2 4$$

> We will see in the next section that multiplication and division are easier in *polar* form.

Since $j^2 = -1$,
$$j^2 4 = -4$$
                                                                                        ◆◆◆

Common Error	It is *incorrect* to write $$\sqrt{-4} \, \sqrt{-4} = \sqrt{(-4)(-4)} = \sqrt{16} = +4$$ Our previous rule of $\sqrt{a} \cdot \sqrt{b} = \sqrt{ab}$ applied only to positive a and b.

The Conjugate of a Complex Number

The *conjugate* of a complex number is obtained by changing the sign of its imaginary part.

✦✦✦ Example 8:

> A computer algebra system can find the conjugate of a complex number. See problem 68 in Exercise 1.

(a) The conjugate of $2 + j3$ is $2 - j3$.
(b) The conjugate of $-5 - j4$ is $-5 + j4$.
(c) The conjugate of $a + jb$ is $a - jb$.
                                                                                        ◆◆◆

Multiplying any complex number by its conjugate will eliminate the j term.

✦✦✦ Example 9:

> This equation has the same form as the difference of two squares.

$$(2 + j3)(2 - j3) = 4 - j6 + j6 - j^2 9 = 4 - (-1)(9) = 13$$
                                                                                        ◆◆◆

Division of Complex Numbers

Division involving single terms, real or imaginary, is shown by examples.

••• Example 10:

(a) $j8 \div 2 = j4$
(b) $j6 \div j3 = 2$
(c) $(4 - j6) \div 2 = 2 - j3$ ◆◆◆

••• Example 11: Divide 6 by $j3$.

Solution:

$$\frac{6}{j3} = \frac{6}{j3} \times \frac{j3}{j3} = \frac{j18}{j^2 9} = \frac{j18}{-9} = -j2$$ ◆◆◆

To divide by a complex number, multiply dividend and divisor by the conjugate of the divisor. This will make the divisor a real number, as in the following example.

••• Example 12: Divide $3 - j4$ by $2 + j$.

Solution: We multiply numerator and denominator by the conjugate $(2 - j)$ of the denominator.

$$\frac{3 - j4}{2 + j} = \frac{3 - j4}{2 + j} \cdot \frac{2 - j}{2 - j}$$

$$= \frac{6 - j3 - j8 + j^2 4}{4 + j2 - j2 - j^2}$$

$$= \frac{2 - j11}{5}$$

$$= \frac{2}{5} - j\frac{11}{5}$$ ◆◆◆

This is very similar to *rationalizing the denominator* of a fraction containing radicals (Sec. 13–2).

In general, the following equation applies:

Division of Complex Numbers	$\dfrac{a + jb}{c + jd} = \dfrac{ac + bd}{c^2 + d^2} + j\dfrac{bc - ad}{c^2 + d^2}$	222

A CAS will give complex roots of a quadratic equation. See problem 69 in Exercise 1.

Quadratics with Complex Roots

When we solved quadratic equations in Chapter 14, the roots were real numbers. Some quadratics, however, will yield complex roots, as in the following example.

••• Example 13: Solve for x to three significant digits:

$$2x^2 - 5x + 9 = 0$$

Solution: By the quadratic formula (Eq. 100),

$$x = \frac{5 \pm \sqrt{25 - 4(2)(9)}}{4} = \frac{5 \pm \sqrt{-47}}{4}$$

$$= \frac{5 \pm j6.86}{4} = 1.25 \pm j1.72$$

A graph of the function $y = 2x^2 - 5x + 9$ (Fig. 21–1) shows that the curve does not cross the x axis and hence has no real zeros. ◆◆◆

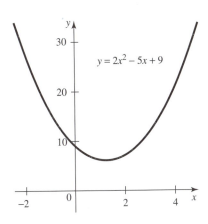

FIGURE 21–1

Complex Factors

It is sometimes necessary to factor an expression into *complex factors,* as shown in the following example.

◆◆◆ **Example 14:** Factor the expression $a^2 + 4$ into complex factors.

Solution:

$$a^2 + 4 = a^2 - j^2 4 = a^2 - (j2)^2$$
$$= (a - j2)(a + j2) \qquad\qquad ◆◆◆$$

Exercise 1 ◆ Complex Numbers in Rectangular Form

Write as imaginary numbers.

1. $\sqrt{-9}$
2. $\sqrt{-81}$
3. $\sqrt{-\dfrac{4}{16}}$
4. $\sqrt{-\dfrac{1}{5}}$

A CAS can find complex factors. See problem 70 in Exercise 1.

Write as a complex number in rectangular form.

5. $4 + \sqrt{-4}$
6. $\sqrt{-25} + 3$
7. $\sqrt{-\dfrac{9}{4}} - 5$
8. $\sqrt{-\dfrac{1}{3}} + 2$

Combine and simplify.

9. $\sqrt{-9} + \sqrt{-4}$
10. $\sqrt{-4a^2} - a\sqrt{-25}$
11. $(3 - j2) + (-4 + j3)$
12. $(-1 - j2) - (j + 6)$
13. $(a - j3) + (a + j5)$
14. $(p + jq) + (q + jp)$
15. $\left(\dfrac{1}{2} + \dfrac{j}{3}\right) + \left(\dfrac{1}{4} - \dfrac{j}{6}\right)$
16. $(-84 + j91) - (28 + j72)$
17. $(2.28 - j1.46) + (1.75 + j2.66)$

Evaluate each power of j.

18. j^{11}
19. j^5
20. j^{10}
21. j^{21}
22. j^{14}

Multiply and simplify.

23. $7 \times j2$
24. $9 \times j3$
25. $j3 \times j5$
26. $j \times j4$
27. $4 \times j2 \times j3 \times j4$
28. $j \times 5 \times j4 \times j3 \times 5$
29. $(j5)^2$
30. $(j3)^2$
31. $2(3 - j4)$
32. $-3(7 + j5)$
33. $j4(5 - j2)$
34. $j5(2 + j3)$
35. $(3 - j5)(2 + j6)$
36. $(5 + j4)(4 + j2)$
37. $(6 + j3)(3 - j8)$
38. $(5 - j3)(8 + j2)$
39. $(5 - j2)^2$
40. $(3 + j6)^2$

Write the conjugate of each complex number.

41. $2 - j3$
42. $-5 - j7$
43. $p + jq$
44. $-j5 + 6$
45. $-jm + n$
46. $5 - j8$

Divide and simplify.

47. $j8 \div 4$

48. $9 \div j3$

49. $j12 \div j6$

50. $j44 \div j2$

51. $(4 + j2) \div 2$

52. $8 \div (4 - j)$

53. $(-2 + j3) \div (1 - j)$

54. $(5 - j6) \div (-3 + j2)$

55. $(j7 + 2) \div (j3 - 5)$

56. $(-9 + j3) \div (2 - j4)$

Quadratics with Complex Roots

Solve for x to three significant digits.

57. $3x^2 - 5x + 7 = 0$

58. $2x^2 + 3x + 5 = 0$

59. $x^2 - 2x + 6 = 0$

60. $4x^2 + x + 8 = 0$

Complex Factors

Factor each expression into complex factors.

61. $x^2 + 9$

62. $b^2 + 25$

63. $4y^2 + z^2$

64. $25a^2 + 9b^2$

Computer

65. Modify your program for solving quadratic equations (Chapter 14) so that the run will not end when the discriminant is negative, but will instead compute and print complex roots in the form $a + jb$.

66. To convert a number from radical to complex form in a CAS, simply type it in and give the *Simplify* command. Thus

$$\text{SQRT}(-4) \quad \text{simplifies to} \quad j2$$

Different systems have different ways to display a complex number. For example, j is displayed as î in *Derive,* I in *Maple,* and I in *Mathematica.* Use this process to do any of problems 1 through 10 in this exercise set.

67. Complex expressions can be entered in a CAS by typing the imaginary unit j. In *Derive,* type #i or Ctrl I, or select î from the menu of symbols. In each case, it will display as î. In *Mathematica,* type capital I, which will display as capital I. The command *Simplify* will then return the simplest form of the expression. Use these steps to do any of problems 11 through 40 and 47 through 56.

68. A complex conjugate is obtained by the command CONJ( ) in *Derive* and Conjugate [ ] in *Mathematica.* Use this command to do any of problems 41 through 46.

69. To obtain complex roots of a quadratic in a CAS, simply enter the equation and give the *Solve* command. Use this process to solve any of the quadratics in problems 57 through 60.

70. To get complex factors of an expression in a CAS, first set the factoring level to COMPLEX. Then enter the expression and give the *Factor* command. Use these steps to factor any of the expressions in problems 61 through 64.

21–2 Graphing Complex Numbers

The Complex Plane

A complex number in rectangular form is made up of two parts: a real part a and an imaginary part jb. To graph a complex number, we can use a rectangular coordinate system (Fig. 21–2) in which the horizontal axis is the *real* axis and the vertical axis is the *imaginary* axis. Such a coordinate system defines what is called the

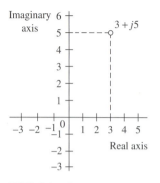

FIGURE 21–2 The complex plane.

Such a plot is called an *Argand diagram*, named for Jean Robert Argand (1768–1822).

complex plane. Real numbers are graphed as points on the horizontal axis; pure imaginary numbers are graphed as points on the vertical axis. Complex numbers, such as $3 + j5$ (Fig. 21–2), are graphed elsewhere within the plane.

Graphing a Complex Number

To plot a complex number $a + jb$ in the complex plane, simply locate a point with a horizontal coordinate of a and a vertical coordinate of b.

♦♦♦ **Example 15:** Plot the complex numbers $(2 + j3)$, $(-1 + j2)$, $(-3 - j2)$, and $(1 - j3)$.

Solution: The points are plotted in Fig. 21–3.   ♦♦♦

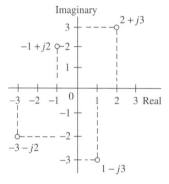

FIGURE 21–3

Exercise 2 ♦ Graphing Complex Numbers

Graph each complex number.

1. $2 + j5$ **2.** $-1 - j3$
3. $3 - j2$ **4.** $2 - j$
5. $j5$ **6.** $2.25 - j3.62$
7. $-2.7 - j3.4$ **8.** $5.02 + j$
9. $1.46 - j2.45$

21–3   Complex Numbers in Trigonometric and Polar Forms

In some books, the terms *polar form* and *trigonometric form* are used interchangeably. Here, we distinguish between them.

Polar Form

In Chapter 17 we saw that a point in a plane could be located by *polar coordinates,* as well as by rectangular coordinates, and we learned how to convert from one set of coordinates to the other. We will do the same thing now with complex numbers, converting between rectangular and polar form.

In Fig. 21–4, we plot a complex number $a + jb$. We connect that point to the origin with a line of length r at an angle of θ with the horizontal axis. Now, in addition to expressing the complex number in terms of a and b, we can express it as follows in terms of r and θ:

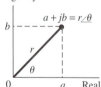

FIGURE 21–4 Polar form of a complex number shown on a complex plane.

Polar Form	$a + jb = r\underline{/\theta}$	228

Since $\cos \theta = a/r$ and $\sin \theta = b/r$, we have the following equations:

$$a = r \cos \theta \qquad\qquad 224$$

$$b = r \sin \theta \qquad\qquad 225$$

The radius r is called the *absolute value*. It can be found from the Pythagorean theorem.

$$r = \sqrt{a^2 + b^2} \qquad\qquad 226$$

The angle θ is called the *argument* of the complex number. Since $\tan \theta = b/a$:

$$\theta = \arctan \frac{b}{a} \qquad\qquad 227$$

◆◆◆ **Example 16:** Write the complex number $2 + j3$ in polar form.

Solution: The absolute value is

$$r = \sqrt{2^2 + 3^2} \cong 3.61$$

The argument is found from

$$\tan \theta = \frac{3}{2}$$

$$\theta \simeq 56.3°$$

So

$$2 + j3 = 3.61 \;\underline{/56.3°} \qquad\qquad ◆◆◆$$

If your calculator has keys for converting between rectangular and polar coordinates, you can use these keys to convert complex numbers between rectangular and polar forms.

Trigonometric Form

If we substitute the values of a and b from Eqs. 224 and 225 into $a + jb$, we get

$$a + jb = (r \cos \theta) + j(r \sin \theta)$$

Factoring gives us the following equation:

| **Trigonometric Form** | $a + jb = r(\cos \theta + j \sin \theta)$
 rectangular   trigonometric <br> form   form | 223 |

The angle θ is sometimes written $\theta = \arg z$, which means "θ is the argument of the complex number z." Also, the expression in parentheses in Eq. 223 is sometimes abbreviated as cis θ, so cis $\theta = \cos \theta + j \sin \theta$.

◆◆◆ **Example 17:** Write the complex number $2 + j3$ in polar and trigonometric forms.

Solution: We already have r and θ from Example 16.

$$r = 3.61 \quad \text{and} \quad \theta = 56.3°$$

That is,

$$2 + j3 = 3.61 \;\underline{/56.3°}$$

So, by Eq. 224,

$$2 + j3 = 3.61(\cos 56.3° + j \sin 56.3°)$$ ◆◆◆

◆◆◆ **Example 18:** Write the complex number $6(\cos 30° + j \sin 30°)$ in polar and rectangular forms.

Solution: By inspection,

$$r = 6 \quad \text{and} \quad \theta = 30°$$

So

$$a = r \cos \theta = 6 \cos 30° = 5.20$$

and

$$b = r \sin \theta = 6 \sin 30° = 3.00$$

Thus our complex number in rectangular form is $5.20 + j3.00$, and in polar form, $6\underline{/30°}$. ◆◆◆

Arithmetic Operations in Polar Form

It is common practice to switch back and forth between rectangular and polar forms during a computation, using that form in which a particular operation is easier. We saw that addition and subtraction are fast and easy in rectangular form, and we now show that multiplication, division, and raising to a power are best done in polar form.

Products

We will use trigonometric form to work out the formula for multiplication and then express the result in the simpler polar form.

Let us multiply $r(\cos \theta + j \sin \theta)$ by $r'(\cos \theta' + j \sin \theta')$.

$$[r(\cos \theta + j \sin \theta)][r'(\cos \theta' + j \sin \theta')]$$
$$= rr'(\cos \theta \cos \theta' + j \cos \theta \sin \theta' + j \sin \theta \cos \theta' + j^2 \sin \theta \sin \theta')$$
$$= rr'[(\cos \theta \cos \theta' - \sin \theta \sin \theta') + j(\sin \theta \cos \theta' + \cos \theta \sin \theta')]$$

But, by Eq. 168,

$$\cos \theta \cos \theta' - \sin \theta \sin \theta' = \cos(\theta + \theta')$$

and by Eq. 167,

$$\sin \theta \cos \theta' + \cos \theta \sin \theta' = \sin(\theta + \theta')$$

So

$$r(\cos \theta + j \sin \theta) \cdot r'(\cos \theta' + j \sin \theta') = rr'[\cos(\theta + \theta') + j \sin (\theta + \theta')]$$

Switching now to polar form, we get the following equation:

Products	$r\underline{/\theta} \cdot r'\underline{/\theta'} = rr'\underline{/\theta + \theta'}$	229

The absolute value of the product of two complex numbers is the product of their absolute values, and the argument is the sum of the individual arguments.

••• **Example 19:** Multiply $5\underline{/30°}$ by $3\underline{/20°}$.

Solution: The absolute value of the product will be $5(3) = 15$, and the argument of the product will be $30° + 20° = 50°$. So

$$5\underline{/30°} \cdot 3\underline{/20°} = 15\underline{/50°} \qquad \text{•••}$$

The angle θ is not usually written greater than 360°. Subtract multiples of 360° if necessary.

••• **Example 20:** Multiply $6.27\underline{/300°}$ by $2.75\underline{/125°}$.

Solution:

$$6.27\underline{/300°} \times 2.75\underline{/125°} = (6.27)(2.75)\underline{/300° + 125°}$$
$$= 17.2\underline{/425°}$$
$$= 17.2\underline{/65°} \qquad \text{•••}$$

Quotients

The rule for division of complex numbers in trigonometric form is similar to that for multiplication.

Quotients	$$\dfrac{r\underline{/\theta}}{r'\underline{/\theta'}} = \dfrac{r}{r'}\underline{/\theta - \theta'}$$	230

You might try deriving this yourself.

The absolute value of the quotient of two complex numbers is the quotient of their absolute values, and the argument is the difference (numerator minus denominator) of their arguments.

••• **Example 21:**

$$\frac{6\underline{/70°}}{2\underline{/50°}} = 3\underline{/20°} \qquad \text{•••}$$

Powers

To raise a complex number to a power, we merely have to multiply it by itself the proper number of times, using Eq. 229.

$$(r\underline{/\theta})^2 = r\underline{/\theta} \cdot r\underline{/\theta} = r \cdot r\underline{/\theta + \theta} = r^2\underline{/2\theta}$$
$$(r\underline{/\theta})^3 = r\underline{/\theta} \cdot r^2\underline{/2\theta} = r \cdot r^2\underline{/\theta + 2\theta} = r^3\underline{/3\theta}$$

Do you see a pattern developing? In general, we get the following equation:

DeMoivre's Theorem	$$(r\underline{/\theta})^n = r^n\underline{/n\theta}$$	231

DeMoivre's theorem is named after Abraham DeMoivre (1667–1754).

When a complex number is raised to the nth power, the new absolute value is equal to the original absolute value raised to the nth power, and the new argument is n times the original argument.

••• **Example 22:**

$$(2\underline{/10°})^5 = 2^5\underline{/5(10°)} = 32\underline{/50°} \qquad \text{•••}$$

Roots

We know that $1^4 = 1$, so, conversely, the fourth root of 1 is 1.

$$\sqrt[4]{1} = 1$$

But we have seen that $j^4 = 1$, so shouldn't it also be true that

$$\sqrt[4]{1} = j \quad ?$$

In fact, both 1 and j are fourth roots of 1, and there are two other roots as well. In this section we learn how to use DeMoivre's theorem to find all roots of a number.

First we note that a complex number $r\underline{/\theta}$ in polar form is unchanged if we add multiples of 360° to the angle θ (in degrees). Thus

$$r\underline{/\theta} = r\underline{/\theta + 360°} = r\underline{/\theta + 720°} = r\underline{/\theta + 360k}$$

where k is an integer. Rewriting DeMoivre's theorem with an exponent $1/p$ (where p is an integer), we get

$$(r\underline{/\theta})^{1/p} = (r\underline{/\theta + 360k})^{1/p} = r^{1/p}\underline{/(\theta + 360k)/p}$$

so

$$(r\underline{/\theta})^{1/p} = r^{1/p}\underline{/(\theta + 360k)/p} \tag{1}$$

We use this equation to find all of the roots of a complex number by letting k take on the values $k = 0, 1, 2, \ldots, (p - 1)$. We'll see in the following example that values of k greater than $(p - 1)$ will give duplicate roots.

◆◆◆ **Example 23:** Find the fourth roots of 1.

Solution: We write the given number in polar form; thus $1 = 1\underline{/0°}$. Then, using Equation (1) with $p = 4$,

$$1^{1/4} = (1\underline{/0})^{1/4} = 1^{1/4}\underline{/(0 + 360k)/4} = 1\underline{/90k}$$

We now let $k = 0, 1, 2, \ldots$.

		Root
k	**Polar Form**	**Rectangular Form**
0	$1\underline{/0°}$	1
1	$1\underline{/90°}$	j
2	$1\underline{/180°}$	-1
3	$1\underline{/270°}$	$-j$
4	$1\underline{/360°}$	$1\underline{/0°} = 1$
5	$1\underline{/450°}$	$1\underline{/90°} = j$

Notice that the roots repeat for $k = 4$ and higher, so we look no further. Thus the fourth roots of 1 are $1, j, -1,$ and $-j$. We check them by noting that each raised to the fourth power equals 1.   ◆◆◆

Thus the number 1 has four fourth roots. In general, any number has two square roots, three cube roots, and so on.

> There are p pth roots of a complex number.

If a complex number is in rectangular form, we convert to polar form before taking the root.

◆◆◆ Example 24: Find $\sqrt[3]{256 + j192}$.

Solution: We convert the given complex number to polar form.

$$r = \sqrt{(256)^2 + (192)^2} = 320 \qquad \theta = \arctan(192/256) = 36.9°$$

Then, using Equation (1), we have

$$\sqrt[3]{256 + j192} = (320\underline{/36.9°})^{1/3} = (320)^{1/3}\underline{/(36.9° + 360k)/3}$$
$$= 6.84\underline{/12.3° + 120k}$$

We let $k = 0, 1, 2$, to obtain the three roots.

	Root	
k	**Polar Form**	**Rectangular Form**
0	$6.84\underline{/12.3° + 0°}\ \ = 6.84\underline{/12.3°}$	$6.68 + j1.46$
1	$6.84\underline{/12.3° + 120°} = 6.84\underline{/132.3°}$	$-4.60 + j5.06$
2	$6.84\underline{/12.3° + 240°} = 6.84\underline{/252.3°}$	$-2.08 - j6.52$

◆◆◆

Exercise 3 ◆ Complex Numbers in Trigonometric and Polar Forms

Write each complex number in polar and trigonometric forms.

1. $5 + j4$
2. $-3 - j7$
3. $4 - j3$
4. $8 + j4$
5. $-5 - j2$
6. $-4 + j7$
7. $-9 - j5$
8. $7 - j3$
9. $-4 - j7$

Round to three significant digits, where necessary, in this exercise.

Write in rectangular and in polar forms.

10. $4(\cos 25° + j \sin 25°)$
11. $3(\cos 57° + j \sin 57°)$
12. $2(\cos 110° + j \sin 110°)$
13. $9(\cos 150° + j \sin 150°)$
14. $7(\cos 12° + j \sin 12°)$
15. $5.46(\cos 47.3° + j \sin 47.3°)$

Write in rectangular and trigonometric forms.

16. $5\underline{/28°}$
17. $9\underline{/59°}$
18. $4\underline{/63°}$
19. $7\underline{/-53°}$
20. $6\underline{/-125°}$

Multiplication and Division

Multiply.

21. $3(\cos 12° + j \sin 12°)$ by $5(\cos 28° + j \sin 28°)$
22. $7(\cos 48° + j \sin 48°)$ by $4(\cos 72° + j \sin 72°)$

23. $5.82(\cos 44.8° + j \sin 44.8°)$ by $2.77(\cos 10.1° + j \sin 10.1°)$

24. $5\underline{/30°}$ by $2\underline{/10°}$

25. $8\underline{/45°}$ by $7\underline{/15°}$

26. $2.86\underline{/38.2°}$ by $1.55\underline{/21.1°}$

Divide.

27. $8(\cos 46° + j \sin 46°)$ by $4(\cos 21° + j \sin 21°)$

28. $49(\cos 27° + j \sin 27°)$ by $7(\cos 15° + j \sin 15°)$

29. $58.3(\cos 77.4° + j \sin 77.4°)$ by $12.4(\cos 27.2° + j \sin 27.2°)$

30. $24\underline{/50°}$ by $12\underline{/30°}$

31. $50\underline{/72°}$ by $5\underline{/12°}$

32. $71.4\underline{/56.4°}$ by $27.7\underline{/15.2°}$

Powers and Roots

Evaluate.

33. $[2(\cos 15° + j \sin 15°)]^3$

34. $[9(\cos 10° + j \sin 10°)]^2$

35. $(7\underline{/20°})^2$

36. $(1.55\underline{/15°})^3$

37. $\sqrt{57\underline{/52°}}$

38. $\sqrt{22\underline{/12°}}$

39. $\sqrt[3]{38\underline{/73°}}$

40. $\sqrt[3]{15\underline{/89°}}$

41. $\sqrt{135 + j204}$

42. $\sqrt[3]{56.3 + j28.5}$

Calculator

43. Learn how to convert between rectangular and polar forms on your calculator. Use these keys to aid you in doing problems 1 through 20 in this exercise set.

Computer

44. In *Derive,* the command ABS(z) will give the absolute value r of a complex number z, and the command PHASE(z) will give the argument θ of that complex number, in radians. Thus

$$ABS(3 + î4) \quad \text{gives the value} \quad 5$$

and

$$PHASE(3 + î4) \quad \text{gives the value} \quad 0.927$$

the radian equivalent of 53.1°. The equivalent commands in *Mathematica* are Abs[ ] and Arg[ ]. Thus these two commands can be used to convert a complex number from rectangular form to polar and trigonometric forms. Use them to do any of problems 1 through 9 in this exercise set.

45. In *Derive* or *Mathematica,* the command RE(z) will give the real part of a complex number z, and the command IM(z) will give the imaginary part. Thus the expression

$$RE(z) + î \, IM(z)$$

will convert a complex number z from trigonometric to rectangular form. Use this command to convert the complex numbers in problems 10 through 15.

46. An easier way to convert a complex number from trigonometric to rectangular form is to highlight the expression and select

SIMPLIFY/APPROXIMATE

Use this command to convert the complex numbers in problems 10 through 15.

47. Algebraic operations on complex expressions in rectangular or trigonometric form are done by entering the expressions and giving the *Simplify* command. The result will be given in rectangular form. If polar or trigonometric forms are wanted, convert as shown in problem 44 above. Use this process to simplify any of the expressions in problems 21 through 23, 27 through 29, 33, and 34.

21–4 Complex Numbers in Exponential Form

Euler's Formula

We have already expressed a complex number in *rectangular* form

$$a + jb$$

in *polar* form

$$r\underline{/\theta}$$

and in *trigonometric* form

$$r(\cos \theta + j \sin \theta)$$

Our next (and final) form for a complex number is *exponential* form, given by Euler's formula.

Euler's Formula	$re^{j\theta} = r(\cos \theta + j \sin \theta)$	232

This formula is named after Leonard Euler (1707–83). We give it without proof here.

Here e is the base of natural logarithms (approximately 2.718), discussed in Chapter 20, and θ is the argument expressed in radians.

◆◆◆ **Example 25:** Write the complex number $5(\cos 180° + j \sin 180°)$ in exponential form.

Solution: We first convert the angle to radians: $180° = \pi$ rad. So

$$5(\cos \pi + j \sin \pi) = 5e^{j\pi} \qquad \text{◆◆◆}$$

Common Error	Make sure that θ is in *radians* when using Eq. 232.

Setting $\theta = \pi$ and $r = 1$ in Euler's formula gives

$$e^{j\pi} = \cos \pi + j \sin \pi$$
$$= -1 + j0$$
$$= -1$$

The constant e arises out of natural growth, j is the square root of -1, and π is the ratio of the circumference of a circle to its diameter. Thus we get the astounding result that the irrational number e raised to the product of the imaginary unit j and the irrational number π give the integer -1.

◆◆◆ **Example 26:** Write the complex number $3e^{j2}$ in rectangular, trigonometric, and polar forms.

Solution: Changing θ to degrees gives us

$$\theta = 2 \text{ rad} = 115°$$

and from Eq. 232,

$$r = 3$$

From Eq. 224,

$$a = r \cos \theta = 3 \cos 115° = -1.27$$

and from Eq. 225,

$$b = r \sin \theta = 3 \sin 115° = 2.72$$

So, in rectangular form,

$$3e^{j2} = -1.27 + j2.72$$

In polar form,

$$3e^{j2} = 3\underline{/115°}$$

In trigonometric form,

$$3e^{j2} = 3(\cos 115° + j \sin 115°) \qquad \blacklozenge\blacklozenge\blacklozenge$$

Products

We now find products, quotients, powers, and roots of complex numbers in exponential form. These operations are quite simple in this form, as we merely have to use the laws of exponents. Thus by Eq. 29:

Products	$r_1 e^{j\theta_1} \cdot r_2 e^{j\theta_2} = r_1 r_2 e^{j(\theta_1 + \theta_2)}$	233

◆◆◆ Example 27:

(a) $2e^{j3} \cdot 5e^{j4} = 10e^{j7}$ \qquad (b) $2.83e^{j2} \cdot 3.15e^{j4} = 8.91e^{j6}$

(c) $-13.5e^{-j3} \cdot 2.75e^{j5} = -37.1e^{j2}$ \qquad $\blacklozenge\blacklozenge\blacklozenge$

Quotients

By Eq. 30:

Quotients	$\dfrac{r_1 e^{j\theta_1}}{r_2 e^{j\theta_2}} = \dfrac{r_1}{r_2} e^{j(\theta_1 - \theta_2)}$	234

◆◆◆ Example 28:

(a) $\dfrac{8e^{j5}}{4e^{j2}} = 2e^{j3}$ \qquad (b) $\dfrac{63.8e^{j2}}{13.7e^{j5}} = 4.66e^{-j3}$

(c) $\dfrac{5.82e^{-j4}}{9.83e^{-j7}} = 0.592e^{j3}$ \qquad $\blacklozenge\blacklozenge\blacklozenge$

Powers and Roots

By Eq. 31:

Powers and Roots	$(re^{j\theta})^n = r^n e^{jn\theta}$	235

◆◆◆ Example 29:

(a) $(2e^{j3})^4 = 16e^{j12}$ \qquad (b) $(-3.85e^{-j2})^3 = -57.1e^{-j6}$

(c) $(0.233e^{-j3})^{-2} = \dfrac{e^{j6}}{(0.223)^2} = \dfrac{e^{j6}}{0.0497} = 20.1e^{j6}$ \qquad $\blacklozenge\blacklozenge\blacklozenge$

Exercise 4 ◆ Complex Numbers in Exponential Form

Express each complex number in exponential form.

1. $2 + j3$

2. $-1 + j2$

3. $3(\cos 50° + j \sin 50°)$

4. $12\underline{/14°}$

5. $2.5\underline{/\pi/6}$

6. $7\left(\cos \dfrac{\pi}{3} + j \sin \dfrac{\pi}{3}\right)$

7. $5.4\underline{/\pi/12}$

8. $5 + j4$

Express in rectangular, polar, and trigonometric forms.

9. $5e^{j3}$

10. $7e^{j5}$

11. $2.2e^{j1.5}$

12. $4e^{j2}$

Operations in Exponential Form

Multiply.

13. $9e^{j2} \cdot 2e^{j4}$

14. $8e^{j} \cdot 6e^{j3}$

15. $7e^{j} \cdot 3e^{j3}$

16. $6.2e^{j1.1} \cdot 5.8e^{j2.7}$

17. $1.7e^{j5} \cdot 2.1e^{j2}$

18. $4e^{j7} \cdot 3e^{j5}$

Divide.

19. $18e^{j6}$ by $6e^{j3}$

20. $45e^{j4}$ by $9e^{j2}$

21. $55e^{j9}$ by $5e^{j6}$

22. $123e^{j6}$ by $105e^{j2}$

23. $21e^{j2}$ by $7e^{j}$

24. $7.7e^{j4}$ by $2.3e^{j2}$

Evaluate.

25. $(3e^{j5})^2$

26. $(4e^{j2})^3$

27. $(2e^{j})^3$

Computer

28. To enter a complex number such as $3e^{j2}$ in exponential form in *Derive,* type

$$3ê^{\wedge}(î2)$$

Another way is to use EXP( ). Thus 3 EXP(î2) is the same as $3e^{j2}$. Learn how to enter such complex numbers on your CAS, and try entering the numbers given in problems 9 through 12.

29. In *Derive,* the command ABS(z) will give the absolute value *r* of a complex number *z*, and the command PHASE(z) will give the argument θ of that complex number in radians. Thus

$$\text{ABS}(3 + î4) \quad \text{gives the value} \quad 5$$

and

$$\text{PHASE}(3 + î4) \quad \text{gives the value} \quad 0.927$$

the radian equivalent of 53.1°. Therefore these two commands can be used to convert a complex number from rectangular form to exponential form, as well as to polar and trigonometric forms. Where applicable, use ABS and PHASE to do any of problems 1 through 8 in this exercise set.

30. To convert a complex number from trigonometric to exponential form in *Derive,* enter the number and give the *Simplify* command. The *Approximate* command will further convert the number to rectangular form. Where appropriate, use these commands to convert the numbers in problems 1 through 8.

31. To convert a number from exponential to rectangular form in *Derive,* use the *Simplify/Approximate* command. Use this command to do problems 9 through 12.

32. To do arithmetic operations on complex numbers in exponential form in *Derive,* enter the expressions and give the *Simplify* command. The result will be given in exponential form. The *Approximate* command will convert the result to rectangular form. Use these commands to do any of problems 13 through 27.

21–5   Vector Operations Using Complex Numbers

Vectors Represented by Complex Numbers

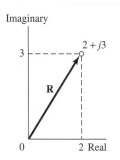

FIGURE 21–5   A vector represented by a complex number.

One of the major uses of complex numbers is that they can represent vectors and, as we will soon see, can enable us to manipulate vectors in ways that are easier than we learned when studying oblique triangles.

Take the complex number $2 + j3$, for example, which is plotted in Fig. 21–5. If we connect that point with a line to the origin, we can think of the complex number $2 + j3$ *as representing a vector* **R** *having a horizontal component of 2 units and a vertical component of 3 units.* The complex number used to represent a vector can, of course, be expressed in any of the forms of a complex number.

Vector Addition and Subtraction

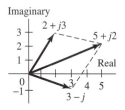

FIGURE 21–6   Vector addition with complex numbers.

Let us place a second vector on our diagram, represented by the complex number $3 - j$ (Fig. 21–6). We can add them graphically by the parallelogram method, and we get a resultant $5 + j2$. But let us add the two original complex numbers, $2 + j3$ and $3 - j$.

$$(2 + j3) + (3 - j) = 5 + j2$$

This result is the same as what we obtained graphically. In other words, *the resultant of two vectors is equal to the sum of the complex numbers representing those vectors.*

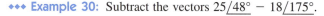

◆◆◆ **Example 30:** Subtract the vectors $25\underline{/48°} - 18\underline{/175°}$.

Solution: Vector addition and subtraction are best done in rectangular form. Converting the first vector, we have

$$a_1 = 25 \cos 48° = 16.7$$
$$b_1 = 25 \sin 48° = 18.6$$

so

$$25\underline{/48°} = 16.7 + j18.6$$

Similarly, for the second vector,

$$a_2 = 18 \cos 175° = -17.9$$
$$b_2 = 18 \sin 175° = 1.57$$

so

$$18\underline{/175°} = -17.9 + j1.57$$

Combining, we have

$$(16.7 + j18.6) - (-17.9 + j1.57) = (16.7 + 17.9) + j(18.6 - 1.57)$$
$$= 34.6 + j17.0$$

These vectors are shown in Fig. 21–7.

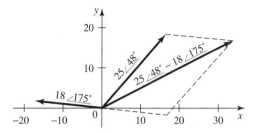

FIGURE 21–7 ◆◆◆

Multiplication and Division of Vectors

Multiplication or division is best done if the vector is expressed as a complex number in polar form.

◆◆◆ **Example 31:** Multiply the vectors $2\underline{/25°}$ and $3\underline{/15°}$.

Solution: By Eq. 229, the product vector will have a magnitude of $3(2) = 6$ and an angle of $25° + 15° = 40°$. The product is thus

$$6\underline{/40°}$$ ◆◆◆

◆◆◆ **Example 32:** Divide the vector $8\underline{/44°}$ by $4\underline{/12°}$.

Solution: By Eq. 230,

$$(8\underline{/44°}) \div (4\underline{/12°}) = \frac{8}{4}\underline{/44° - 12°} = 2\underline{/32°}$$ ◆◆◆

◆◆◆ **Example 33:** A certain current **I** is represented by the complex number $1.15\underline{/23.5°}$ amperes, and a complex impedance **Z** is represented by $24.6\underline{/14.8°}$ ohms. Multiply **I** by **Z** to obtain the voltage **V**.

Solution:

$$\mathbf{V} = \mathbf{IZ} = (1.15\underline{/23.5°})(24.6\underline{/14.8°}) = 28.3\underline{/38.3°} \quad (\text{V})$$ ◆◆◆

The *j* Operator

Let us multiply a complex number $r\underline{/\theta}$ by j. First we must express j in polar form. By Eq. 224,

$$j = 0 + j1 = 1(\cos 90° + j \sin 90°) = 1\underline{/90°}$$

After we multiply by j, the new absolute value will be

$$r \cdot 1 = r$$

and the argument will be

$$\theta + 90°$$

So

$$r\underline{/\theta} \cdot j = r\underline{/\theta + 90°}$$

Thus the only effect of multiplying our original complex number by j was to increase the angle by 90°. If our complex number represents a vector, *we may think of* j *as an operator that causes the vector to rotate through one-quarter revolution.*

Exercise 5 ◆ Vector Operations Using Complex Numbers

Express each vector in rectangular and polar forms.

	Magnitude	Angle
1.	7.00	49.0°
2.	28.0	136°
3.	193	73.5°
4.	34.2	1.10 rad
5.	39.0	2.50 rad
6.	59.4	58.0°

Combine the vectors.

7. $(3 + j2) + (5 - j4)$
8. $(7 - j3) - (4 + j)$
9. $58\underline{/72°} + 21\underline{/14°}$
10. $8.60\underline{/58°} - 4.20\underline{/160°}$
11. $9(\cos 42° + j \sin 42°) + 2(\cos 8° + j \sin 8°)$
12. $8(\cos 15° + j \sin 15°) - 5(\cos 9° - j \sin 9°)$

Round to three significant digits, where needed, in this exercise set.

Multiply the vectors.

13. $(7 + j3)(2 - j5)$
14. $(2.50\underline{/18°})(3.70\underline{/48°})$
15. $2(\cos 25° - j \sin 25°) \cdot 6(\cos 7° - j \sin 7°)$

Divide the vectors.

16. $(25 - j2) \div (3 + j4)$
17. $(7.70\underline{/47°}) \div (2.50\underline{/15°})$
18. $5(\cos 72° + j \sin 72°) \div 3(\cos 31° + j \sin 31°)$

21–6 Alternating Current Applications

Rotating Vectors in the Complex Plane

Glance back at Sec. 17–3 where we first introduced alternating current.

We have already shown that a vector can be represented by a complex number, $r\underline{/\theta}$. For example, the complex number $5.00\underline{/28°}$ represents a vector of magnitude 5.00 at an angle of 28° with the real axis.

A phasor (a rotating vector) may also be represented by a complex number $R\underline{/\omega t}$ by replacing the angle θ by ωt, where ω is the angular velocity and t is time.

◆◆◆ **Example 34:** The complex number

$$11\underline{/5t}$$

represents a phasor of magnitude 11 rotating with an angular velocity of 5 rad/s. ◆◆◆

In Sec. 17–3 we showed that a phasor of magnitude R has a projection on the y axis of $R \sin \omega t$. Similarly, a phasor $R\underline{/\omega t}$ in the complex plane (Fig. 21–8) will have a projection on the imaginary axis of $R \sin \omega t$. Thus

$$R \sin \omega t = \text{imaginary part of } R\omega t$$

Similarly,

$$R \cos \omega t = \text{real part of } R\underline{/\omega t}$$

Thus either a sine or a cosine wave can be represented by a complex number $R\underline{/\omega t}$, depending upon whether we project onto the imaginary or the real axis. It

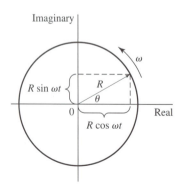

FIGURE 21–8

usually does not matter which we choose, because the sine and cosine functions are identical except for a phase difference of 90°. What does matter is that we are consistent. Here we will follow the convention of projecting the phasor onto the imaginary axis, and we will drop the phrase "imaginary part of." Thus we say that

$$R \sin \omega t = R\underline{/\omega t}$$

Similarly, if there is a phase angle ϕ,

$$R \sin(\omega t + \phi) = R\underline{/\omega t + \phi}$$

One final simplification: It is customary to draw phasors at $t = 0$, so that ωt vanishes from the expression. Thus we write

$$R \sin(\omega t + \phi) = R\underline{/\phi}$$

Effective or Root Mean Square (rms) Values

Before we write currents and voltages in complex form, we must define the *effective value* of current and voltage. The power P delivered to a resistor by a direct current I is $P = I^2R$. We now define an effective current I_{eff} that delivers the same power from an alternating current as that delivered by a direct current of the same number of amperes.

$$P = I_{eff}^2R$$

Figure 21–9 shows an alternating current $i = I_m \sin \omega t$ and i^2, the square of that current. The mean value of this alternating quantity is found to deliver the same power in a resistor as a steady quantity of the same value.

We get the effective value by taking the square root of the mean value squared. Hence it is also called a *root mean squared value*. Thus I_{eff} is often written I_{rms} instead.

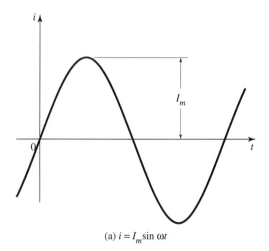

(a) $i = I_m \sin \omega t$

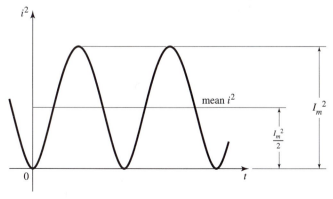

(b) $i^2 = I_m^2\sin^2 \omega t$

FIGURE 21–9

$$P = I_{eff}^2 R = (\text{mean } i^2)R$$

But the mean value of i^2 is equal to $I_m^2/2$, half the peak value, so

$$I_{eff} = \sqrt{I_m^2/2} = I_m/\sqrt{2}$$

Thus *the effective current* I_{eff} *is equal to the peak current* I_m *divided by* $\sqrt{2}$. Similarly, *the effective voltage* V_{eff} *is equal to the peak voltage* V_m *divided by* $\sqrt{2}$.

◆◆◆ **Example 35:** An alternating current has a peak value of 2.84 A. What is the effective current?

Solution:

$$I_{eff} = \frac{I_m}{\sqrt{2}} = \frac{2.84}{\sqrt{2}} = 2.01 \text{ A}$$

◆◆◆

Alternating Current and Voltage in Complex Form

Next we will write a sinusoidal current or voltage expression in the form of a complex number, because computations are easier in that form. At the same time, we will express the magnitude of the current or voltage in terms of effective value. The effective current I_{eff} is the quantity that is always implied when a current is given, rather than the peak current I_m. It is the quantity that is read when using a meter. The same is true of effective voltage.

To write an alternating current in complex form, we give the effective value of the current and the phase angle. The current is written in boldface type: **I**. Thus we have the following equations:

Complex Voltage and Current	The current is represented by $$i = I_m \sin(\omega t + \phi_2)$$ $$\mathbf{I} = I_{eff}\underline{/\phi_2} = \frac{I_m}{\sqrt{2}}\underline{/\phi_2}$$	**A76**
	Similarly, the voltage is represented by $$v = V_m \sin(\omega t + \phi_1)$$ $$\mathbf{V} = V_{eff}\underline{/\phi_1} = \frac{V_m}{\sqrt{2}}\underline{/\phi_1}$$	**A75**

Note that the complex expressions for voltage and current do not contain t. We say that the voltage and current have been converted from the *time domain* to the *phasor domain*.

◆◆◆ **Example 36:** The complex expression for the alternating current

$$i = 2.84 \sin(\omega t + 33°) \text{ A}$$

is

Note that the "complex" expressions are the simpler. This happy situation is used in the ac computations to follow, which are much more cumbersome without complex numbers.

$$\mathbf{I} = I_{eff}\underline{/33°} = \frac{2.84}{\sqrt{2}}\underline{/33°} \text{ A}$$
$$= 2.01\underline{/33°} \text{ A}$$

◆◆◆

◆◆◆ **Example 37:** The sinusoidal expression for the voltage

$$\mathbf{V} = 84.2\underline{/-49°} \text{ V}$$
$$v = 84.2\sqrt{2} \sin(\omega t - 49°) \text{ V}$$
$$= 119 \sin(\omega t - 49°) \text{ V}$$

◆◆◆

Complex Impedance

In Sec. 7–7 we drew a vector impedance diagram in which the impedance was the resultant of two perpendicular vectors: the resistance R along the horizontal axis and the reactance X along the vertical axis. If we now use the complex plane and draw the resistance along the *real* axis and the reactance along the *imaginary* axis, we can represent impedance by a complex number.

Complex Impedance	$\mathbf{Z} = R + jX = Z\underline{/\phi} = Ze^{j\phi}$	A101

where

$$R = \text{circuit resistance}$$
$$X = \text{circuit reactance} = X_L - X_C \qquad \text{(A98)}$$
$$Z = \text{magnitude of impedance} = \sqrt{R^2 + X^2} \qquad \text{(A99)}$$
$$\phi = \text{phase angle} = \arctan \frac{X}{R} \qquad \text{(A100)}$$

◆◆◆ **Example 38:** A circuit has a resistance of 5 Ω in series with a reactance of 7 Ω, as shown in Fig. 21–10(a). Represent the impedance by a complex number.

Solution: We draw the vector impedance diagram as shown in Fig. 21–10(b). The impedance in rectangular form is

$$\mathbf{Z} = 5 + j7$$

The magnitude of the impedance is

$$\sqrt{5^2 + 7^2} = 8.60$$

The phase angle is

$$\arctan \frac{7}{5} = 54.5°$$

So we can write the impedance as a complex number in polar form as follows:

$$\mathbf{Z} = 8.60\underline{/54.5°} \qquad \text{◆◆◆}$$

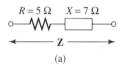

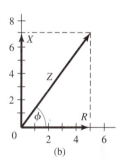

FIGURE 21–10

Ohm's Law for Alternating Current (ac)

We stated at the beginning of this section that the use of complex numbers would make calculations with alternating current almost as easy as for direct current. We do this by means of the following relationship:

Ohm's Law for ac	$\mathbf{V} = \mathbf{ZI}$	A103

Note the similarity between this equation and Ohm's law, Eq. A62. It is used in the same way.

◆◆◆ **Example 39:** A voltage of 142 sin 200t is applied to the circuit of Fig. 21–10(a). Write a sinusoidal expression for the current i.

Solution: We first write the voltage in complex form. By Eq. A75,

$$\mathbf{V} = \frac{142}{\sqrt{2}}\underline{/0°} = 100\underline{/0°}$$

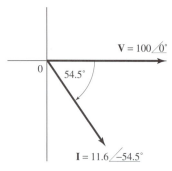

FIGURE 21–11

Next we find the complex impedance **Z**. From Example 38,

$$\mathbf{Z} = 8.60\underline{/54.5°}$$

The complex current **I**, by Ohm's law for ac, is

$$\mathbf{I} = \frac{\mathbf{V}}{\mathbf{Z}} = \frac{100\underline{/0°}}{8.60\underline{/54.5°}} = 11.6\underline{/-54.5°}$$

Then the current in sinusoidal form is (Eq. A76)

$$i = 11.6\ \sqrt{2}\ \sin(200t - 54.5°)$$
$$= 16.4\ \sin(200t - 54.5°)$$

The current and voltage phasors are plotted in Fig. 21–11, and the instantaneous current and voltage are plotted in Fig. 21–12. Note the phase difference between the voltage and current waves. This phase difference, 54.5°, converted to time, is

$$54.5° = 0.951\ \text{rad} = 200t$$
$$t = 0.951/200 = 0.00475\ \text{s} = 4.75\ \text{ms}$$

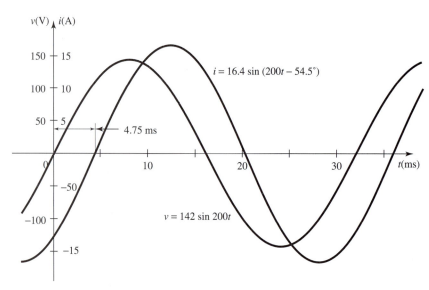

FIGURE 21–12

We say that the current lags the voltage by 54.5° or 4.75 ms.   ◆◆◆

Exercise 6 ◆ Alternating Current Applications

Express each current or voltage in complex form.

1. $i = 250\ \sin(\omega t + 25°)$ **2.** $v = 1.5\ \sin(\omega t - 30°)$
3. $v = 57\ \sin(\omega t - 90°)$ **4.** $i = 2.7\ \sin \omega t$
5. $v = 144\ \sin \omega t$ **6.** $i = 2.7\ \sin(\omega t - 15°)$

Express each current or voltage in sinusoidal form.

7. $\mathbf{V} = 150\underline{/0°}$ **8.** $\mathbf{V} = 1.75\underline{/70°}$
9. $\mathbf{V} = 300\underline{/-90°}$ **10.** $\mathbf{I} = 25\underline{/30°}$

11. $I = 7.5\underline{/0°}$　　　　　　　　**12.** $I = 15\underline{/-130°}$

Express the impedance of each circuit as a complex number in rectangular form and in polar form.

13. Fig. 21–13(a)　　　　　　　**14.** Fig. 21–13(b)
15. Fig. 21–13(c)　　　　　　　**16.** Fig. 21–13(d)
17. Fig. 21–13(e)　　　　　　　**18.** Fig. 21–13(f)
19. Fig. 21–13(g)

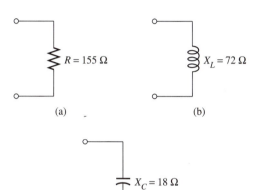

(a)　　　　　　　　　　　(b)

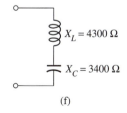

(c)

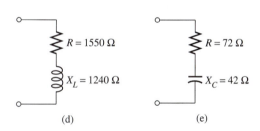

(d)　　　　　　　　(e)

(f)

(g)

FIGURE 21–13

20. Write a sinusoidal expression for the current i in each part of Fig. 21–14.

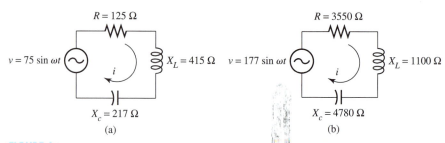

FIGURE 21–14

21. Write a sinusoidal expression for the voltage v in each part of Fig. 21–15.

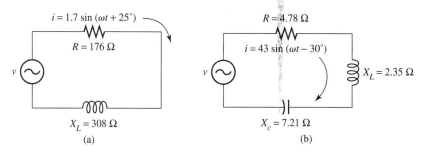

FIGURE 21–15

22. Find the complex impedance **Z** in each part of Fig. 21–16.

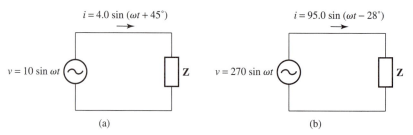

FIGURE 21–16

◆◆◆ CHAPTER 21 REVIEW PROBLEMS ◆◆◆◆◆◆◆◆◆◆◆◆◆◆◆◆◆◆◆◆◆◆◆◆◆◆◆◆◆

Express the following as complex numbers in rectangular, polar, trigonometric, and exponential forms.

1. $3 + \sqrt{-4}$
2. $-\sqrt{-9} - 5$
3. $-2 + \sqrt{-49}$
4. $3.25 + \sqrt{-11.6}$

Evaluate.

5. j^{17}
6. j^{25}

Combine the complex numbers. Leave your answers in rectangular form.

7. $(7 - j3) + (2 + j5)$

8. $4.8\underline{/28°} - 2.4\underline{/72°}$

9. $52(\cos 50° + j \sin 50°) + 28(\cos 12° + j \sin 12°)$

10. $2.7e^{j7} - 4.3e^{j5}$

Multiply. Leave your answer in the same form as the complex numbers.

11. $(2 - j)(3 + j5)$

12. $(7.3\underline{/21°})(2.1\underline{/156°})$

13. $2(\cos 20° + j \sin 20°) \cdot 6(\cos 18° + j \sin 18°)$

14. $(93e^{j2})(5e^{j7})$

Divide and leave your answer in the same form as the complex numbers.

15. $(9 - j3) \div (4 + j)$

16. $(18\underline{/72°}) \div (6\underline{/22°})$

17. $16(\cos 85° + j \sin 85°) \div 8(\cos 40° + j \sin 40°)$

18. $127e^{j8} \div 4.75e^{j5}$

Graph each complex number.

19. $7 + j4$

20. $2.75\underline{/44°}$

21. $6(\cos 135° + j \sin 135°)$

22. $4.75e^{j2.2}$

Evaluate each power.

23. $(-4 + j3)^2$

24. $(5\underline{/12°})^3$

25. $[5(\cos 10° + j \sin 10°)]^3$

26. $[2e^{j3}]^5$

27. Express in complex form: $i = 45 \sin(\omega t + 32°)$

28. Express in sinusoidal form: $\mathbf{V} = 283\underline{/-22°}$

29. Write a complex expression for the current i in Fig. 21–17.

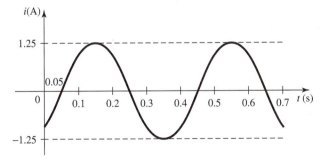

FIGURE 21–17

Writing

30. Suppose that you are completing a job application to an electronics company which asks you to, "Explain, in writing, how you would use complex numbers in an electrical calculation, and illustrate your explanation with an example." How would you respond?

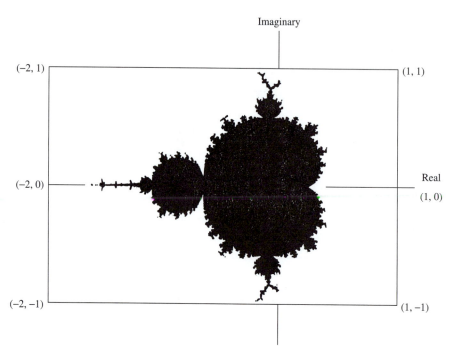

FIGURE 21–18 Plot of the Mandelbrot set, named for Benoit Mandelbrot, who did much pioneering work in the new fields of chaos and fractals.

Team Project: The Mandelbrot Set

31. Figure 21–18 shows a plot of the Mandelbrot set in the complex plane. One way to define this set of numbers is as follows: If we take a complex number, square it, add the original number, square the result, add the original number, square the result, and so forth, for an infinite number of iterations, and if the total does not grow without bound, then the original number is in the Mandelbrot set.

This project is to write a computer program to plot the Mandelbrot set. Instead of an infinite number of iterations, we will do 50. Our test will be to see if the absolute value of $p^2 + q^2$ of each succeeding iteration $p + jq$ does not exceed 2.

Your program should do the following:

(a) Choose a point (a, b) representing a complex number $a + jb$. Start with the point $(-2, 0)$ in the window (Fig. 21–18), and proceed in steps of about 0.002 in both a and b until the opposite corner $(1, 1)$ is reached.

(b) Square the complex number.

(c) Add this square to the original number $a + jb$, getting a new complex number with real part p and imaginary part q.

(d) Test $p + jq$. If, at any point in the iteration, $p^2 + q^2$ gets larger than 4, return to Step (a) and proceed to the next point.

(e) Repeat Steps (b), (c), and (d) 50 times. If in those iterations the test in Step (d) has not been failed, then the original number $a + jb$ is in the Mandelbrot set. Plot the point (a, b). Also plot its mirror image $(a, -b)$.

(f) Go to Step (a) and proceed to the next point.

Run your program on the fastest computer you have available. It may take hours to run.

Analytic Geometry

22

OBJECTIVES

When you have completed this chapter, you should be able to:

- Calculate the distance between two points.
- Determine the slope of a line given two points on the line.
- Find the slope of a line given its angle of inclination, and vice versa.
- Determine the slope of a line perpendicular to a given line.
- Calculate the angle between two lines.
- Write the equation of a line using the slope-intercept form, the point-slope form, or the two-point form.
- Solve applied problems involving the straight line.
- Write the equation of a circle, ellipse, parabola, or hyperbola from given information.
- Write an equation in standard form given the equation of any of the conic sections.
- Determine all of the features of interest from the standard equation of any conic section, and make a graph.
- Make a graph of any of the conic sections.
- Tell, by inspection, whether a given second-degree equation represents a circle, ellipse, parabola, or hyperbola.
- Write a new equation for a curve with the axes shifted when given the equation of that curve.
- Solve applied problems involving any of the conic sections.

This chapter begins our study of *analytic geometry*. The main idea in analytic geometry is to place geometric figures (points, lines, circles, and so forth) on coordinate axes, where they may be studied using the methods of algebra.

We start with the *straight line,* building on what we have already covered in Chapter 5. We study the idea of *slope* of a line and of a *curve,* an idea that will be

You might want to try making these shapes by cutting a cone from a form made of modeling clay or damp sand. Also, what position of the plane would you use to make a straight line? A point?

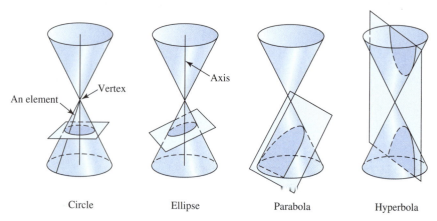

FIGURE 22–1   Conic sections.

crucial to our understanding of the *derivative.* Next come the *conic sections:* the circle, parabola, ellipse, and hyperbola. They are called *conic sections,* or just *conics,* because each can be obtained by passing a plane through a right circular cone, as shown in Fig. 22–1.

When the plane is perpendicular to the cone's axis, it intercepts a *circle.* When the plane is tilted a bit, but not so much as to be parallel to an element (a line on the cone that passes through the vertex) of the cone, we get an *ellipse.* When the plane is parallel to an element of the cone, we get a *parabola,* and when the plane is steep enough to cut the upper nappe of the cone, we get the two branched *hyperbolas.*

In this chapter we learn to write the equations of the straight line and of the conics. These equations enable us to understand and to use these geometric figures in ways that are not possible without the equations. We also give some interesting applications for each curve.

22–1   The Straight Line

Length of a Line Segment Parallel to a Coordinate Axis

When we speak about the length of a line segment or about the distance between two points, we usually mean the *magnitude* of that length or distance.

To find the magnitude of the length of a line segment lying on or parallel to the *x* axis, simply subtract the abscissa (the *x* value) of either endpoint from the abscissa of the other endpoint, and take the absolute value of this result.

The procedure is similar for a line segment lying on or parallel to the *y* axis: Subtract the ordinate (the *y* value) of either endpoint from the ordinate of the other endpoint, and take the absolute value of this result.

◆◆◆ **Example 1:** The magnitudes of the lengths of the lines in Fig. 22–2 are as follows:

(a) $AB = |5 - 2| = 3$
(b) $PQ = |5 - (-2)| = 7$
(c) $RS = |-1 - (-5)| = 4$

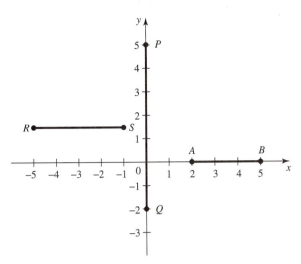

FIGURE 22–2 ◆◆◆

Note that it doesn't matter if we reverse the order of the endpoints. We get the same result.

◆◆◆ **Example 2:** Repeating Example 1(a) with the endpoints reversed gives

$$AB = |2 - 5| = |-3| = 3$$

as before. ◆◆◆

Directed Distance

Sometimes when speaking about the length of a line segment or the distance between two points, it is necessary to specify *direction* as well as magnitude. When we specify the *directed distance AB*, for example, we mean the distance *from A to B*, sometimes written $\overline{AB}$.

◆◆◆ **Example 3:** For the two points $P(5, 2)$ and $Q(1, 2)$, the directed distance PQ is $PQ = 1 - 5 = -4$, and the directed distance QP is $QP = 5 - 1 = 4$. ◆◆◆

Increments

Let us say that a particle is moving along a curve from point P to point Q, as shown in Fig. 22–3. As it moves, its abscissa changes from x_1 to x_2. We call this change an *increment* in x and label it Δx (read "delta x"). Similarly, the ordinate changes from y_1 to y_2 and is labeled Δy. The increments are found simply by subtracting the coordinates at P from those at Q.

Note that directed distance applies only to a line parallel to a coordinate axis.

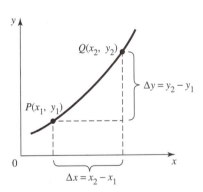

FIGURE 22–3 Increments.

Increments	$\Delta x = x_2 - x_1$ and $\Delta y = y_2 - y_1$	337

◆◆◆ **Example 4:** A particle moves from $P_1(2, 5)$ to $P_2(7, 3)$. The increments in its coordinates are

$$\Delta x = x_2 - x_1 = 7 - 2 = 5$$

and

$$\Delta y = y_2 - y_1 = 3 - 5 = -2$$

◆◆◆

Distance Formula

We now find the length of a line inclined at any angle, such as PQ in Fig. 22–4. We first draw a horizontal line through P and drop a perpendicular from Q, forming a right triangle PQR. The sides of the triangle are d, Δx, and Δy. By the Pythagorean theorem,

$$d^2 = (\Delta x)^2 + (\Delta y)^2 = (x_2 - x_1)^2 + (y_2 - y_1)^2$$

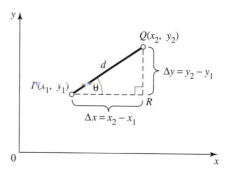

FIGURE 22–4   Length of a line segment.

Since we want the magnitude of the distance, we take only the positive root and get the following:

Distance Formula	$d = \sqrt{(\Delta x)^2 + (\Delta y)^2}$ $= \sqrt{(x_2 - x_1)^2 + (y_2 - y_1)^2}$	283

◆◆◆ **Example 5:** Find the length of the line segment with the endpoints

$$(3, -5) \text{ and } (-1, 6).$$

Solution: Let us give the first point $(3, -5)$ the subscripts 1, and the other point the subscripts 2 (it does not matter which we label 1). So

$$x_1 = 3 \qquad y_1 = -5 \qquad x_2 = -1 \qquad y_2 = 6$$

Substituting into the distance formula, we obtain

$$d = \sqrt{(-1 - 3)^2 + [6 - (-5)]^2} = \sqrt{(-4)^2 + (11)^2} = 11.7 \quad \text{(rounded)} \quad ◆◆◆$$

Common Error	Do not take the square root of each term separately. $d \neq \sqrt{(x_2 - x_1)^2} + \sqrt{(y_2 - y_1)^2}$

Slope and Angle of Inclination of a Straight Line

Recall from Sec. 15–3 that the slope of a straight line for which

$$\text{rise} = \Delta y = y_2 - y_1$$
$$\text{run} = \Delta x = x_2 - x_1$$

is given by the following equation:

Slope	$m = \dfrac{\text{rise}}{\text{run}} = \dfrac{\Delta y}{\Delta x} = \dfrac{y_2 - y_1}{x_2 - x_1}$	284

The slope is equal to the rise divided by the run.

We now relate the slope to the *angle of inclination*. The smallest positive angle that a line makes with the positive x direction is called the *angle of inclination θ* of the line (Fig. 22–5). A line parallel to the x axis has an angle of inclination of zero. Thus the angle of inclination can have values from 0° up to 180°. From Fig. 22–4,

$$\tan \theta = \frac{\text{opposite side}}{\text{adjacent side}} = \frac{y_2 - y_1}{x_2 - x_1} = \frac{\text{rise}}{\text{run}} = \frac{\Delta y}{\Delta x}$$

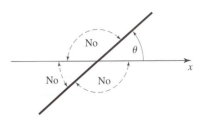

FIGURE 22–5   Angle of inclination, θ.

But this is the slope m of the line, so we have the following equations:

$$m = \tan \theta$$
$$0° \leq \theta < 180°$$

285

The slope of a line is equal to the tangent of the angle of inclination θ (except when $\theta = 90°$).

◆◆◆ **Example 6:** The slope of a line having an angle of inclination of 50° (Fig. 22–6) is, by Eq. 285,

$$m = \tan 50° = 1.192 \quad \text{(rounded)} \qquad \text{◆◆◆}$$

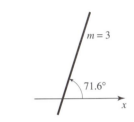

FIGURE 22–6

◆◆◆ **Example 7:** Find the angle of inclination of a line having a slope of 3 (Fig. 22–7).

Solution: By Eq. 285, $\tan \theta = 3$, so

$$\theta = \arctan 3 = 71.6° \quad \text{(rounded)} \qquad \text{◆◆◆}$$

When the slope is negative, our calculator will give us a negative angle, which we then use to obtain a positive angle of inclination less than 180°.

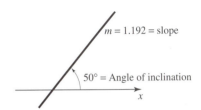

FIGURE 22–7

◆◆◆ **Example 8:** For a line having a slope of -2 (Fig. 22–8), $\arctan (-2) = -63.4°$ (rounded), so

$$\theta = 180° - 63.4° = 116.6° \qquad \text{◆◆◆}$$

◆◆◆ **Example 9:** Find the angle of inclination of the line passing through $(-2.47, 1.74)$ and $(3.63, -4.26)$ (Fig. 22–9).

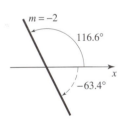

FIGURE 22–8

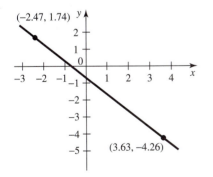

FIGURE 22–9

Solution: The slope, from Eq. 285, is

$$m = \frac{-4.26 - 1.74}{3.63 - (-2.47)} = -0.984$$

from which $\theta = 135.5°$ ◆◆◆

When *different scales* are used for the *x* and *y* axes, the angle of inclination *will appear distorted.*

◆◆◆ **Example 10:** The angle of inclination of the line in Fig. 22–10 is $\theta = \arctan 10 = 84.3°$. Notice that the angle in the graph appears much smaller than 84.3° because of the different scales on each axis.

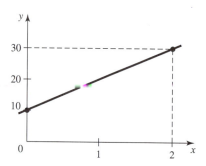

FIGURE 22–10

◆◆◆

Slopes of Parallel and Perpendicular Lines

Parallel lines have, of course, *equal* slopes.

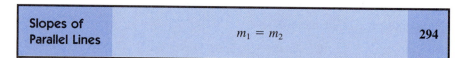

| Slopes of Parallel Lines | $m_1 = m_2$ | 294 |

Parallel lines have equal slopes.

Now consider a line L_1, with slope m_1 and angle of inclination θ_1 (Fig. 22–11), and a *perpendicular* line L_2, with slope m_2 and angle of inclination $\theta_2 = \theta_1 + 90°$. Note that $\angle QOS = \theta_2 - 90° = \theta_1 + 90° - 90° = \theta_1$, so right triangles *OPR* and *OQS* are congruent. Thus the magnitudes of their corresponding sides *a* and *b* are equal. The slope m_1 of L_1 is a/b, and the slope m_2 of L_2 is $-b/a$, so we have the following equation:

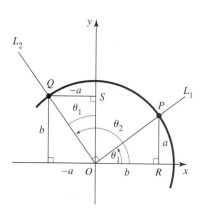

FIGURE 22–11 Slopes of perpendicular lines.

| Slopes of Perpendicular Lines | $m_1 = -\dfrac{1}{m_2}$ | 295 |

The slope of a line is the negative reciprocal of the slope of a perpendicular to that line.

◆◆◆ **Example 11:** Any line perpendicular to a line whose slope is 5 has a slope of $-\frac{1}{5}$.

◆◆◆

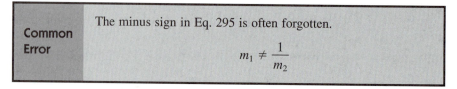

| Common Error | The minus sign in Eq. 295 is often forgotten. $m_1 \neq \dfrac{1}{m_2}$ |

◆◆◆ **Example 12:** Is angle *ACB* of Fig. 22–12 a right angle?

Solution: The slope of *AC* is

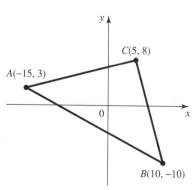

FIGURE 22–12

$$\frac{8 - 3}{5 - (-15)} = \frac{5}{20} = \frac{1}{4}$$

To be perpendicular, BC must have a slope of -4. Its actual slope is

$$\frac{-10 - 8}{10 - 5} = \frac{-18}{5} = -3.6$$

Although lines AC and BC *appear* to be perpendicular, we have shown that they are not. Thus $\angle ACB$ is not a right angle. ◆◆◆

Angle of Intersection between Two Lines

Figure 22–13 shows two lines L_1 and L_2 intersecting at an angle ϕ, measured counterclockwise from line 1 to line 2. We want a formula for ϕ in terms of the slopes m_1 and m_2 of the lines. By Eq. 142, θ_2 equals the sum of the angle of inclination θ_1 and the angle of intersection ϕ. So $\phi = \theta_2 - \theta_1$. Taking the tangent of both sides gives $\tan\phi = \tan(\theta_2 - \theta_1)$, or

$$\tan\phi = \frac{\tan\theta_2 - \tan\theta_1}{1 + \tan\theta_1\tan\theta_2}$$

by the trigonometric identity for the tangent of the difference of two angles (Eq. 169). But the tangent of an angle of inclination is the slope, so we have the following equation:

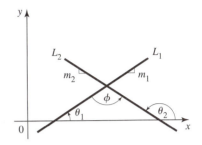

FIGURE 22–13 Angle of intersection between two lines.

Angle between Two Lines	$\tan\phi = \dfrac{m_2 - m_1}{1 + m_1 m_2}$	296

◆◆◆ **Example 13:** The tangent of the angle of intersection between line L_1, having a slope of 4, and line L_2, having a slope of -1, is

$$\tan\phi = \frac{-1 - 4}{1 + 4(-1)} = \frac{-5}{-3} = \frac{5}{3}$$

from which $\phi = \arctan\frac{5}{3} = 59.0°$ (rounded). This angle is measured counterclockwise from line 1 to line 2, as shown in Fig. 22–14. ◆◆◆

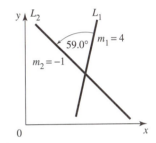

FIGURE 22–14

Tangents and Normals to Curves

We know what the tangent to a circle is, and we probably have an intuitive idea of what the *tangent to a curve* is. We'll speak a lot about tangents, so let us try to get a clear idea of what one is.

To determine the tangent line at P in Fig. 22–15, we start by selecting a second point, Q, on the curve. As shown, Q can be on either side of P. The line PQ is called

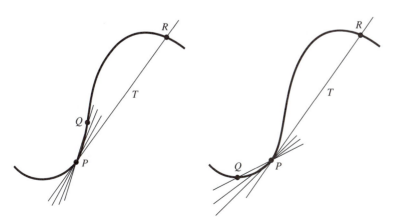

FIGURE 22–15 A tangent line T at point P, here defined as the limiting position of a secant line PQ, as Q approaches P.

a *secant line*. We then let Q approach P (from either side). If the secant lines PQ approach a single line T as Q approaches P, then line T is called the *tangent line at P*.

Note that the tangent line can intersect the curve at more than one point, such as point R. However, at points "near" P, the tangent line intersects the curve just once while a secant line intersects the curve twice. The *slope of a curve* at some point P is defined as the *slope of the tangent* to the curve drawn through that point.

In calculus you will see that the slope gives the *rate of change* of the function, an extremely useful quantity to find. You will see that it is found by taking the *derivative* of the function.

◆◆◆ **Example 14:** Plot the curve $y = x^2$ for integer values of x from 0 to 4. By eye, draw the tangent to the curve at $x = 2$, and find the approximate value of the slope.

Solution: We make a table of point pairs.

x	0	1	2	3	4
y	0	1	4	9	16

The plot and the tangent line are shown in Fig. 22–16. Using the scales on the x and y axes, we measure a rise of 12 units for the tangent line in a run of 3 units. The slope of the tangent line is then

$$m_t \simeq \frac{12}{3} = 4$$

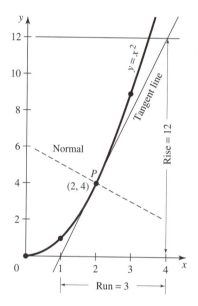

FIGURE 22–16   Slope of a curve.   ◆◆◆

The notion of a tangent to a curve can be clearly demonstrated by using a graphics utility to zoom in on a portion of the curve.

◆◆◆ **Example 15:** Zoom in on the curve of Example 14 at the point (2, 4), and estimate the slope of the tangent.

Solution: As we zoom in at (2, 4), the curve appears straighter and straighter. With a viewing window of $x = 1.9$ to 2.1 and $y = 3.9$ to 4.1, the curve appears as in Fig. 22–17. Using TRACE, we get the coordinates of two points on the curve, (1.9873, 3.9494) and (2.0143, 4.0573). Calculating the slope gives

$$m = \frac{4.0573 - 3.9494}{2.0143 - 1.9873} = 3.996$$

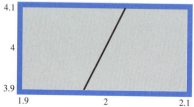

FIGURE 22–17

which agrees with the value found in Example 14.   ◆◆◆

The *normal* to a curve at a given point is the line perpendicular to the tangent at that point, as in Fig. 22–16. Its slope is the negative reciprocal of the slope of the tangent. For Example 14, $m_n = -\frac{1}{4}$.

Equation of a Straight Line

In Sec. 5–4 we introduced the *slope-intercept form* for the equation of a straight line. The equation for a line with a slope m and y intercept b is as follows:

This is a good time to glance back at Sec. 5–4 and review some of the exercises there.

Slope-Intercept Form	$y = mx + b$	289

Using this equation, we can quickly write the equation of a straight line if we know the slope and y intercept. However, we may have *other* information about the line, such as the coordinates of two points on that line, and want to write the equation using that information. To enable us to do this, we'll now derive several other forms for the straight-line equation.

General Form

The slope-intercept form is the equation of a straight line in *explicit* form, $y = f(x)$. We can also write the equation of a straight line in *implicit* form, $f(x, y) = 0$, simply by transposing all terms to one side of the equals sign and simplifying. We usually write the x term first, then the y term, and finally the constant term. This form is referred to as the *general form* of the equation of a straight line.

General Form	$Ax + By + C = 0$	286

Here A, B, and C are constants.

You will see as we go along that there are several forms for the equation of a straight line, and we use the one that is most convenient in a particular problem. But to make it easy to compare answers, we usually rearrange the equation into general form.

◆◆◆ **Example 16:** Change the equation $y = \frac{6}{7}x - 5$ from slope-intercept form to general form.

Solution: Subtracting y from both sides, we have

$$0 = \frac{6}{7}x - 5 - y$$

Multiplying by 7 and rearranging gives $6x - 7y - 35 = 0$.   ◆◆◆

Point-Slope Form

Let us write an equation for a line that has slope m, but that passes through a given point (x_1, y_1) which is *not,* in general, on either axis, as in Fig. 22–18. Again using the definition of slope (Eq. 284), with a general point (x, y), we get the following form:

Point-Slope Form	$m = \dfrac{y - y_1}{x - x_1}$ or $y - y_1 = m(x - x_1)$	291

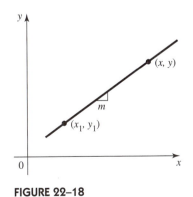

FIGURE 22–18

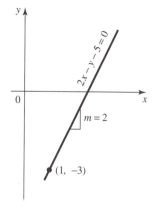

FIGURE 22–19

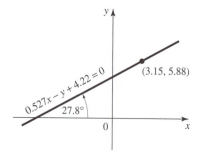

FIGURE 22–20

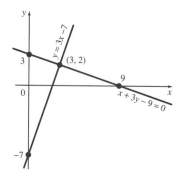

FIGURE 22–21

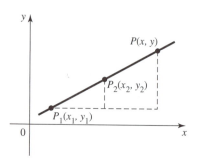

FIGURE 22–22

This form of the equation of a straight line is most useful when we know the slope of a line and one point through which the line passes.

◆◆◆ **Example 17:** Write the equation in general form of the line having a slope of 2 and passing through the point $(1, -3)$ (Fig. 22–19).

Solution: Substituting $m = 2$, $x_1 = 1$, $y_1 = -3$ into Eq. 291 gives us

$$2 = \frac{y - (-3)}{x - 1}$$

Multiplying by $x - 1$ yields

$$2x \quad 2 = y + 3$$

or, in general form, $2x - y - 5 = 0$.　　　　　　　　◆◆◆

◆◆◆ **Example 18:** Write the equation in general form of the line passing through the point $(3, 2)$ and perpendicular to the line $y = 3x - 7$ (Fig. 22–20).

Solution: The slope of the given line is 3, so the slope of our perpendicular line is $-\frac{1}{3}$. Using the point-slope form, we obtain

$$-\frac{1}{3} = \frac{y - 2}{x - 3}$$

Going to the general form, $x - 3 = -3y + 6$, or

$$x + 3y - 9 = 0$$　　　　　　　　◆◆◆

◆◆◆ **Example 19:** Write the equation in general form of the line passing through the point $(3.15, 5.88)$ and having an angle of inclination of $27.8°$ (Fig. 22–21).

Solution: From Eq. 285, $m = \tan 27.8° = 0.527$. From Eq. 291,

$$0.527 = \frac{y - 5.88}{x - 3.15}$$

Changing to general form, $0.527x - 1.66 = y - 5.88$, or

$$0.527x - y + 4.22 = 0$$　　　　　　　　◆◆◆

Two-Point Form

If two points on a line are known, the equation of the line is easily written using the two-point form, which we now derive. If we call the points P_1 and P_2 in Fig. 22–22, the slope of the line is

$$m = \frac{y_2 - y_1}{x_2 - x_1}$$

The slope of the line segment connecting P_1 with any other point P on the same line is

$$m = \frac{y - y_1}{x - x_1}$$

Since these slopes must be equal, we get the following equation:

Two-Point Form	$\dfrac{y - y_1}{x - x_1} = \dfrac{y_2 - y_1}{x_2 - x_1}$	290

◆◆◆ **Example 20:** Write the equation in general form of the line passing through the points $(1, -3)$ and $(-2, 5)$ (Fig. 22–23).

Solution: Calling the first given point P_1 and the second P_2 and substituting into Eq. 290, we have

$$\frac{y - (-3)}{x - 1} = \frac{5 - (-3)}{-2 - 1} = \frac{8}{-3}$$

$$\frac{y + 3}{x - 1} = -\frac{8}{3}$$

Putting the equation into general form, we have $3y + 9 = -8x + 8$, or

$$8x + 3y + 1 = 0$$

◆◆◆

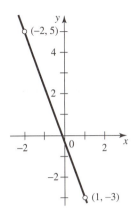

FIGURE 22–23

Lines Parallel to the Coordinate Axes

Line 1 in Fig. 22–24 is parallel to the x axis. Its slope is 0 and it cuts the y axis at $(0, b)$. From the point-slope form, $y - y_1 = m(x - x_1)$, we get $y - b = 0(x - 0)$, or the following:

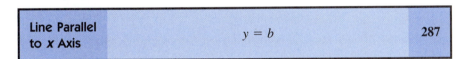

| Line Parallel to *x* Axis | $y = b$ | 287 |

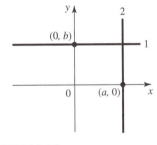

FIGURE 22–24

Line 2 has an undefined slope, but we get its equation by noting that $x = a$ at every point on the line, *regardless of the value of y.* Thus:

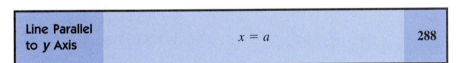

| Line Parallel to *y* Axis | $x = a$ | 288 |

◆◆◆ **Example 21:**

(a) A line that passes through the point $(5, -2)$ and is parallel to the x axis has the equation $y = -2$.
(b) A line that passes through the point $(5, -2)$ and is parallel to the y axis has the equation $x = 5$.

◆◆◆

Exercise 1 ◆ The Straight Line

Length of a Line Segment

Find the distance between the given points.

1. $(5, 0)$ and $(2, 0)$
2. $(0, 3)$ and $(0, -5)$
3. $(-2, 0)$ and $(7, 0)$
4. $(-4, 0)$ and $(-6, 0)$
5. $(0, -2.74)$ and $(0, 3.86)$
6. $(55.34, 0)$ and $(25.38, 0)$
7. $(5.59, 3.25)$ and $(8.93, 3.25)$
8. $(-2.06, -5.83)$ and $(-2.06, -8.34)$
9. $(8.38, -3.95)$ and $(2.25, -4.99)$

Find the directed distance *AB*.

10. *A*(3,0); *B*(5, 0)

11. *A*(3.95, −2.07); *B*(−3.95, −2.07)

12. *B*(−8, −2); *A*(−8, −5)

13. *A*(−9, −2); *B*(17, −2)

14. *B*(−6, −6); *A*(−6, −7)

15. *B*(11.5, 3.68); *A*(11.5, −5.38)

Slope and Angle of Inclination

Find the slope of the line having the given angle of inclination.

16. 38.2°

17. 77.9°

18. 1.83 rad

19. 58°14′

20. 156.3°

21. 132.8°

Find the angle of inclination, in decimal degrees to three significant digits, of a line having the given slope.

22. $m = 3$

23. $m = 1.84$

24. $m = -4$

25. $m = -2.75$

26. $m = 0$

27. $m = -15$

Find the slope of a line perpendicular to a line having the given slope.

28. $m = 5$

29. $m = 2$

30. $m = 4.8$

31. $m = -1.85$

32. $m = -2.85$

33. $m = -5.372$

Find the slope of a line perpendicular to a line having the given angle of inclination.

34. 58.2°

35. 1.84 rad

36. 136°44′

Find the angle of inclination, in decimal degrees to three significant digits, of a line passing through the given points.

37. (5, 2) and (−3, 4)

38. (−2.5, −3.1) and (5.8, 4.2)

39. (6, 3) and (−1, 5)

40. (*x*, 3) and (*x* + 5, 8)

Angle between Two Lines

41. Find the angle of intersection between line L_1 having a slope of 1 and line L_2 having a slope of 6.

42. Find the angle of intersection between line L_1 having a slope of 3 and line L_2 having a slope of −2.

43. Find the angle of intersection between line L_1 having an angle of inclination of 35° and line L_2 having an angle of inclination of 160°.

44. Find the angle of intersection between line L_1 having an angle of inclination of 22° and line L_2 having an angle of inclination of 86°.

Tangents to Curves

Plot the functions from *x* = 0 to *x* = 4. Graphically find the slope of the curve at *x* = 2.

45. $y = x^3$

46. $y = -x^2 + 3$

47. $y = \dfrac{x^2}{4}$

Equation of a Straight Line

Write the equation, in general form, of each line.

48. slope = 4; *y* intercept = −3

49. slope = 2.25; *y* intercept = −1.48

50. passes through points (3, 5) and (−1, 2)

51. passes through points (4.24, −1.25) and (3.85, 4.27)

52. slope = −4; passes through (−2, 5)

53. slope = −2; passes through (−2, −3)

54. *x* intercept = 5; *y* intercept = −3

55. x intercept $= -2$; y intercept $= 6$

56. y intercept $= 3$; parallel to $y = 5x - 2$

57. y intercept $= -2.3$; parallel to $2x - 3y + 1 = 0$

58. y intercept $= -5$; perpendicular to $y = 3x - 4$

59. y intercept $= 2$; perpendicular to $4x - 3y = 7$

60. passes through $(-2, 5)$; parallel to $y = 5x - 1$

61. passes through $(4, -1)$; parallel to $4x - y = -3$

62. passes through $(-4, 2)$; perpendicular to $y = 5x - 3$

63. passes through $(6, 1)$; perpendicular to $6y - 2x = 3$

64. y intercept $= 4.0$; angle of inclination $= 48°$

65. y intercept $= -3.52$; angle of inclination $= 154°44'$

66. passes through $(4, -1)$; angle of inclination $= 22.8°$

67. passes through $(-2.24, 5.17)$; angle of inclination $= 68°14'$

68. passes through $(5, 2)$; is parallel to the x axis.

69. passes through $(-3, 6)$; is parallel to the y axis.

70. line A in Fig. 22–25

71. line B in Fig. 22–25

72. line C in Fig. 22–25

73. line D in Fig. 22–25

74. Find the length of girder AB in Fig. 22–26.

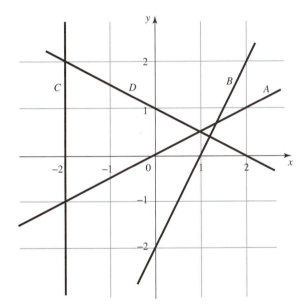

FIGURE 22–25

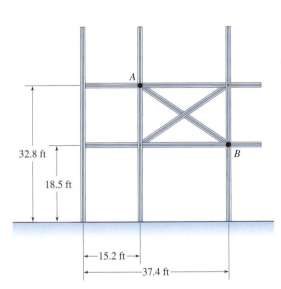

FIGURE 22–26

75. Find the distance between the centers of the holes in Fig. 22–27.

76. A triangle has vertices at $(3, 5)$, $(-2, 4)$, and $(4, -3)$. Find the length of each side. Then compute the area using Hero's formula (Eq. 138). Work to three significant digits.

77. The distance between two stakes on a slope is taped at 2055 ft, and the angle of the slope with the horizontal is $12.3°$. Find the horizontal distance between the stakes.

78. What is the angle of inclination with the horizontal of a roadbed that rises 15.0 ft in each 250 ft, measured horizontally?

79. How far apart must two stakes on a $7°$ slope be placed so that the horizontal distance between them is 1250 m?

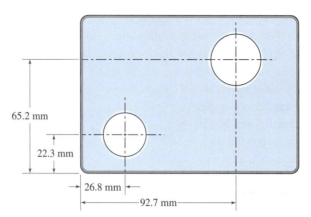

FIGURE 22–27

80. On a 5% road grade, at what angle is the road inclined to the horizontal? How far does one rise in traveling upward 500 ft, measured along the road?

81. A straight tunnel under a river is 755 ft long and descends 12 ft in this distance. What angle does the tunnel make with the horizontal?

82. A straight driveway slopes downward from a house to a road and is 28.0 m in length. If the angle of inclination from the road to the house is 3.6°, find the height of the house above the road.

83. An escalator is built so as to rise 2 m for each 3 m of horizontal travel. Find its angle of inclination.

84. A straight highway makes an angle of 4.5° with the horizontal. How much does the highway rise in a distance of 2500 ft, measured along the road?

In some fields, such as highway work, the word *grade* is used instead of slope. It is usually expressed as a percent:

$$\text{percent grade} = 100 \times \text{slope}$$

Thus a 5% grade rises 5 units for every 100 units of run.

Spring Constant

85. A spring whose length is L_0 with no force applied (Fig. 22–28) stretches an amount x with an applied force of F, where $F = kx$. The constant k is called the *spring constant*. Write an equation, in slope-intercept form, for F in terms of k, L, and L_0.

86. What force would be needed to stretch a spring ($k = 14.5$ lb/in.) from an unstretched length of 8.50 in. to a length of 12.50 in.?

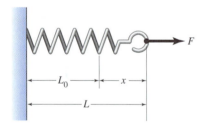

FIGURE 22–28

Velocity of Uniformly Accelerated Body

87. When a body moves with constant acceleration a (such as in free fall), its velocity v at any time t is given by $v = v_0 + at$, where v_0 is the initial velocity. Note that this is the equation of a straight line. If a body has a constant acceleration of 2.15 m/s² and has a velocity of 21.8 m/s at 5.25 s, find (a) the initial velocity and (b) the velocity at 25.0 s.

Resistance Change with Temperature

88. The resistance of metals is a linear function of the temperature (Fig. 22–29) for certain ranges of temperature. The slope of the line is $R_1\alpha$, where α is the temperature coefficient of resistance at temperature t_1. If R is the resistance of any temperature t, write an equation for R as a function of t.

89. Using a value for α of $\alpha = 1/(234.5t_1)$ (for copper), find the resistance of a copper conductor at 75.0°C if its resistance at 20.0°C is 148.4 Ω.

90. If the resistance of the copper conductor in problem 89 is 1255 Ω at 20.0°C, what temperature will the resistance be 1265 Ω?

Thermal Expansion

91. When a bar is heated, its length will increase from an initial length L_0 at temperature t_0 to a new length L at temperature t. The plot of L versus t is a straight

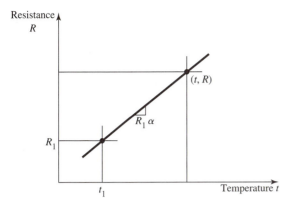

FIGURE 22–29 Resistance change with temperature.

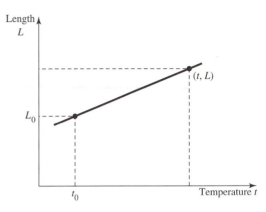

FIGURE 22–30 Thermal expansion.

line (Fig. 22–30) with a slope of $L_0\alpha$, where α is the coefficient of thermal expansion. Derive the equation $L = L_0(1 + \alpha\,\Delta t)$, where Δt is the change in temperature, $t - t_0$.

92. A steel pipe is 21.50 m long at 0°C. Find its length at 75.0°C if α for steel is 12.0×10^{-6} per Celsius degree.

Fluid Pressure

93. The pressure at a point located at a depth x ft from the surface of the liquid varies directly as the depth. If the pressure at the surface is 20.6 lb/in.^{2} and increases by 0.432 lb/in.^{2} for every foot of depth, write an equation for P as a function of the depth x (in feet). At what depth will the pressure be 30.0 lb/in.^{2}?

94. A straight pipe slopes downward from a reservoir to a water turbine (Fig. 22–31). The pressure head at any point in the pipe, expressed in feet, is equal

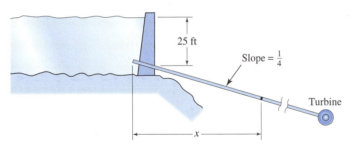

FIGURE 22–31

to the vertical distance between the point and the surface of the reservoir. If the reservoir surface is 25 ft above the upper end of the pipe, write an expression for the head H as a function of the horizontal distance x. At what distance x will the head be 35 ft?

Temperature Gradient

95. Figure 22–32 shows a uniform wall whose inside face is at temperature t_i and whose outside face is at t_o. The temperatures within the wall plot as a straight line connecting t_i and t_o. Write the equation $t = f(x)$ of that line, taking $x = 0$ at the inside face, if $t_i = 25.0°C$ and $t_o = -5.0°C$. At what x will the temperature be 0°C? What is the slope of the line?

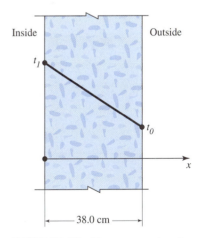

FIGURE 22–32 Temperature drop in a wall. The slope of the line (in °C/cm) is called the *temperature gradient*. The amount of heat flowing through the wall is proportional to the temperature gradient.

Hooke's Law

96. The increase in length of a wire in tension is directly proportional to the applied load P. Write an equation for the length L of a wire that has an initial length of 3.00 m and that stretches 1.00 mm for each 12.5 N. Find the length of the wire with a load of 750 N.

Straight-Line Depreciation

97. The straight-line method is often used to depreciate a piece of equipment for tax purposes. Starting with the purchase price P, the item is assumed to drop in value the same amount each year (the annual depreciation) until the salvage value S is reached (Fig. 22–33). Write an expression for the book value y (the value at any time t) as a function of the number of years t. For a lathe that cost $15,428 and has a salvage value of $2264, find the book value after 15 years if it is depreciated over a period of 20 years.

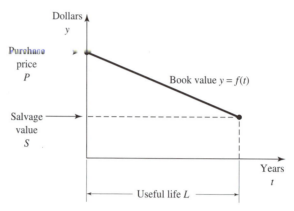

FIGURE 22–33   Straight-line depreciation.

Computer and Graphics Calculator

98. Write a program that will accept as input the coordinates of the vertices of a triangle, and then compute and print the length of each side and the area of the triangle.

99. Use a graphics calculator or a graphics utility on the computer to graph any of the straight lines in this exercise set. The equation of the line must first be written in explicit form, that is, the slope-intercept form.

22–2   The Circle

Definition

Anyone who has drawn a circle using a compass will not be surprised by the following definition of the circle:

Definition of a Circle	A *circle* is a plane curve all points of which are at a fixed distance (the *radius*) from a fixed point (the *center*).	303

Circle with Center at the Origin

We will now derive an equation for a circle of radius r, starting with the simplest case of a circle whose center is at the origin (Fig. 22–34). Let x and y be the coordinates of any point P on the circle. The equation we develop will give a relationship between x and y that will have two meanings: geometrically, (x, y) will represent a point on the circle; and algebraically, the numbers corresponding to those points will satisfy that equation.

We observe that, by the definition of a circle, the distance OP must be constant and equal to r. But, by the distance formula (Eq. 283),

$$OP = \sqrt{x^2 + y^2} = r$$

FIGURE 22–34   Circle with center at origin.

Squaring, we get a *standard equation of a circle* (also called *standard form* of the equation).

| Standard Equation, Circle of Radius r: Center at Origin (O) | $x^2 + y^2 = r^2$ | 304 |

Note that both x^2 and y^2 have the same coefficient. Otherwise, the graph is not a circle.

◆◆◆ **Example 22:** Write, in standard form, the equation of a circle of radius 3, whose center is at the origin.

Solution:

$$x^2 + y^2 = 3^2 = 9$$

◆◆◆

Circle with Center Not at the Origin

Figure 22–35 shows a circle whose center has the coordinates (h, k). We can think of the difference between this circle and the one in Fig. 22–34 as having its center moved or *translated* h units to the right and k units upward. Our derivation is similar to the preceding one.

$$CP = r = \sqrt{(x - h)^2 + (y - k)^2}$$

Squaring, we get the following equation:

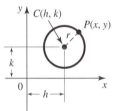

FIGURE 22–35 Circle with center at (h, k).

| Standard Equation, Circle of Radius r: Center at (h, k) | $(x - h)^2 + (y - k)^2 = r^2$ | 305 |

◆◆◆ **Example 23:** Write, in standard form, the equation of a circle of radius 5 whose center is at $(3, -2)$.

Solution: We substitute into Eq. 305 with $r = 5$, $h = 3$, and $k = -2$.

$$(x - 3)^2 + [y - (-2)]^2 = 5^2$$
$$(x - 3)^2 + (y + 2)^2 = 25$$

◆◆◆

◆◆◆ **Example 24:** Find the radius and the coordinates of the center of the circle $(x + 5)^2 + (y - 3)^2 = 16$.

Solution: We see that $r^2 = 16$, so the radius r is 4. Also, since

$$x - h = x + 5$$

then

$$h = -5$$

and since

$$y - k = y - 3$$

then

$$k = 3$$

So the center is at $(-5, 3)$ as shown in Fig. 22–36.

◆◆◆

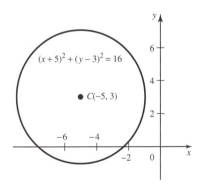

FIGURE 22–36

Common Error	It is easy to get the signs of h and k wrong. In Example 24, do *not* take $$h = +5 \quad \text{and} \quad k = -3$$

Translation of Axes

We see that Equation 305 for a circle with its center at (h, k) is almost identical to Eq. 304 for a circle with center at the origin, *except that x has been replaced by* $(x - h)$ *and y has been replaced by* $(y - k)$. This same substitution will, of course, work for curves other than the circle, and we will use it to translate or *shift* the axes for the other conic sections.

Translation of Axes	To translate or shift the axes of a curve to the left by a distance h and downward by a distance k, replace x by $(x - h)$ and y by $(y - k)$ in the equation of the curve.	298

General Equation of a Circle

A second-degree equation in x and y which has all possible terms would have an x^2 term, a y^2 term, an xy term, and all terms of lesser degree as well. The general second-degree equation is usually written with terms in the following order, where A, B, C, D, E, and F are constants:

There are six constants in this equation, but only five are independent. We can divide through by any constant and thus make it equal to 1.

General Second-Degree Equation	$Ax^2 + Bxy + Cy^2 + Dx + Ey + F = 0$	297

If we expand Eq. 305, we get

$$(x - h)^2 + (y - k)^2 = r^2$$
$$x^2 - 2hx + h^2 + y^2 - 2ky + k^2 = r^2$$

Rearranging gives

$$x^2 + y^2 - 2hx - 2ky + (h^2 + k^2 - r^2) = 0$$

or an equation with D, E, and F as constants.

General Equation of a Circle	$x^2 + y^2 + Dx + Ey + F = 0$	306

Comparing this with the general second-degree equation, we see that the general second-degree equation

$$Ax^2 + Bxy + Cy^2 + Dx + Ey + F = 0$$

represents a circle if $B = 0$ and $A = C$.

◆◆◆ **Example 25:** Write the equation of Example 24 in general form.

Solution: The equation, in standard form, was

$$(x + 5)^2 + (y - 3)^2 = 16$$

Expanding, we obtain

$$x^2 + 10x + 25 + y^2 - 6y + 9 = 16$$

or

$$x^2 + y^2 + 10x - 6y + 18 = 0 \qquad \blacklozenge\blacklozenge\blacklozenge$$

Changing from General to Standard Form

When we want to go from general form to standard form, we must *complete the square,* both for x and for y.

◆◆◆ **Example 26:** Write the equation $2x^2 + 2y^2 - 18x + 16y + 60 = 0$ in standard form. Find the radius and center, and plot the curve.

Solution: We first divide by 2:

$$x^2 + y^2 - 9x + 8y + 30 = 0$$

and separate the x and y terms:

$$(x^2 - 9x \quad) + (y^2 + 8y \quad) = -30$$

We complete the square for the x terms with $\frac{81}{4}$ because we take $\frac{1}{2}$ of (-9) and square it. Similarly, for completing the square for the y terms, $(\frac{1}{2} \times 8)^2 = 16$. Completing the square on the left and compensating on the right gives

$$\left(x^2 - 9x + \frac{81}{4}\right) + (y^2 + 8y + 16) = -30 + \frac{81}{4} + 16$$

Factoring each group of terms yields

$$\left(x - \frac{9}{2}\right)^2 + (y + 4)^2 = \left(\frac{5}{2}\right)^2$$

So

$$r = \frac{5}{2} \qquad h = \frac{9}{2} \qquad k = -4$$

The circle is shown in Fig. 22–37. ◆◆◆

We learned how to complete the square in Sec. 14–2 of Chapter 14, "Quadratic Equations." Glance back there if you need to refresh your memory. Some computer algebra systems have commands for completing the square. See problem 35 in Exercise 2.

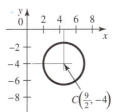

FIGURE 22–37

Graphics Utility

To use a graphics calculator or a graphics utility on the computer to graph a circle, we must first solve the equation of the circle for y.

$$y = k \pm \sqrt{r^2 - (x - h)^2}$$

We then plot the two functions

$$y_1 = k + \sqrt{r^2 - (x - h)^2}$$

and

$$y_2 = k - \sqrt{r^2 - (x - h)^2}$$

Note that the graph may not *appear* circular on a particular grapher, but some have a provision for equalizing the vertical and horizontal scales so that circles do have the proper shape.

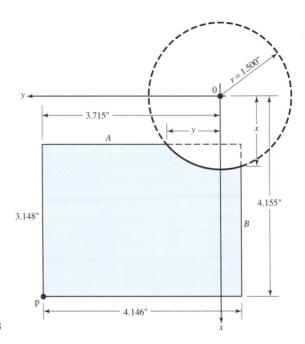

FIGURE 22–38

Applications

◆◆◆ **Example 27:** Write an equation for the circle shown in Fig. 22–38, and find the dimensions A and B produced by the circular cutting tool.

Estimate: The easiest way to get an estimate here is to make a sketch and measure the distances. We get

$$A \approx 2.6 \text{ in.} \quad \text{and} \quad B \approx 2.7 \text{ in.}$$

Solution: In order to write an equation, we must have coordinate axes. Otherwise, the quantities x and y in the equation will have no meaning. Since no axes are given, we are free to draw them anywhere we please. Our first impulse might be to place the origin at the corner p, since most of the dimensions are referenced from that point, but our equation will be simpler if we place the origin at the center of the circle. Let us draw axes as shown, so that not all our numbers will be negative. The equation of the circle is then

$$x^2 + y^2 = (1.500)^2$$
$$= 2.2500$$

The distance from the y axis to the top edge of the block is $4.155 - 3.148 = 1.007$ in. Substituting this value for x into the equation of the circle will give the corresponding value for y. When $x = 1.007$,

$$y^2 = 2.250 - (1.007)^2 = 1.236$$
$$y = 1.112 \text{ in.}$$

Similarly, the distance from the x axis to the right edge of the block is $4.146 - 3.715 = 0.431$ in. When $y = -0.431$,

$$x^2 = 2.250 - (-0.431)^2 = 2.064$$
$$x = 1.437$$

Finally,

$$A = 3.715 - 1.112 = 2.603 \text{ in.}$$

and

$$B = 4.155 - 1.437 = 2.718 \text{ in.} \qquad \text{◆◆◆}$$

Exercise 2 ◆ The Circle

Equation of a Circle

Write the equation of each circle in standard form.

1. center at $(0, 0)$; radius $= 7$
2. center at $(0, 0)$; radius $= 4.82$
3. center at $(2, 3)$; radius $= 5$
4. center at $(5, 2)$; radius $= 10$
5. center at $(5, -3)$; radius $= 4$
6. center at $(-3, -2)$; radius $= 11$

Find the center and radius of each circle.

7. $x^2 + y^2 = 49$
8. $x^2 + y^2 = 64.8$
9. $(x - 2)^2 + (y + 4)^2 = 16$
10. $(x + 5)^2 + (y - 2)^2 = 49$
11. $(y + 5)^2 + (x - 3)^2 = 36$
12. $(x - 2.22)^2 + (y + 7.16)^2 = 5.93$
13. $x^2 + y^2 - 8x = 0$
14. $x^2 + y^2 - 2x - 4y = 0$
15. $x^2 + y^2 - 10x + 12y + 25 = 0$
16. $x^2 + y^2 - 4x + 2y = 36$
17. $x^2 + y^2 + 6x - 2y = 15$
18. $x^2 + y^2 - 2x - 6y = 39$

Write the equation of each circle in general form.

19. center at $(1, 1)$; passes through $(4, -3)$
20. diameter joins the points $(-2, 5)$ and $(6, -1)$

Tangent to a Circle

Write the equation of the tangent to each circle at the given point. (*Hint:* The slope of the tangent is the negative reciprocal of the slope of the radius to the given point.)

21. $x^2 + y^2 = 25$ at $(4, 3)$
22. $(x - 5)^2 + (y - 6)^2 = 100$ at $(11, 14)$
23. $x^2 + y^2 + 10y = 0$ at $(-4, -2)$

Intercepts

Find the x and y intercepts for each circle. (*Hint:* Set x and y, in turn, equal to zero to find the intercepts.)

24. $x^2 + y^2 - 6x + 4y + 4 = 0$
25. $x^2 + y^2 - 5x - 7y + 6 = 0$

Intersecting Circles

Find the point(s) of intersection. (*Hint:* Solve each pair of equations simultaneously as we did in Exercise 8 in Chapter 14.)

26. $x^2 + y^2 - 10x = 0$ and $x^2 + y^2 + 2x - 6y = 0$
27. $x^2 + y^2 - 3y - 4 = 0$ and $x^2 + y^2 + 2x - 5y - 2 = 0$
28. $x^2 + y^2 + 2x - 6y + 2 = 0$ and $x^2 + y^2 - 4x + 2 = 0$

Even though you may be able to solve
some of these with only the Pythagorean theorem, we suggest that you use
analytic geometry for the practice.

Applications

29. Prove that any angle inscribed in a semicircle is a right angle.

30. Write the equation of the circle in Fig. 22–39, taking the axes as shown. Use your equation to find *A* and *B*.

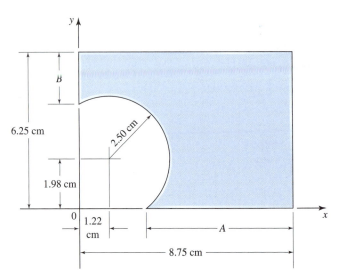

FIGURE 22–39

31. Write the equations for each of the circular arches in Fig. 22–40, taking the axes as shown. Solve simultaneously to get the point of intersection *P*, and compute the height *h* of the column.

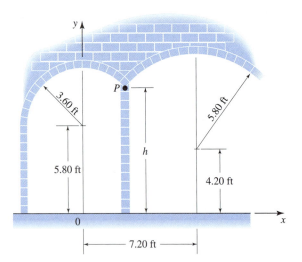

FIGURE 22–40   Circular arches.

32. Write the equation of the centerline of the circular street shown in Fig. 22–41, taking the origin at the intersection *O*. Use your equation to find the distance *y*.

33. Each side of a Gothic arch is a portion of a circle. Write the equation of one side of the Gothic arch shown in Fig. 22–42. Use the equation to find the width *w* of the arch at a height of 3.00 ft.

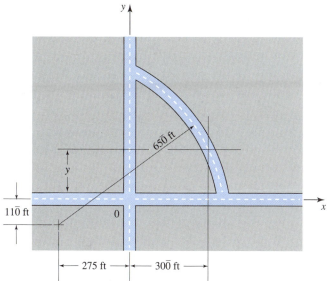

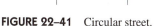

FIGURE 22–41 Circular street.

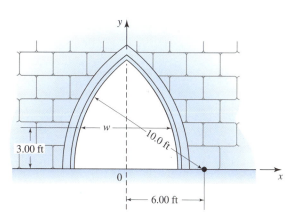

FIGURE 22–42 Gothic arch.

Computer and Graphics Calculator

34. Use a graphics calculator or a graphics utility on the computer to graph any of the circles in this exercise set.

35. Some computer algebra systems can complete the square. *Maple,* for example, uses the *completesquare* command. If the expression has two variables, however, it is necessary to complete the square in two separate operations, once for *x* and once for *y.* If you have such a system, use it to go from general to standard form whenever needed in this exercise set.

22–3   The Parabola

Definition

A parabola has the following definition:

Definition of a Parabola	A *parabola* is the set of points in a plane, each of which is equidistant from a fixed point, the *focus,* and a fixed line, the *directrix.*	**307**

◆◆◆ **Example 28:** Use Definition 307 above to construct a parabola for which the distance from focus to directrix is 2.0 in.

Solution: We draw a line to represent the directrix as shown in Fig. 22–43, and we indicate a focus 2.0 in. from that line. Then we draw a line *L* parallel to the directrix at some arbitrary distance, say, 3.0 in. With that same (3.0-in.) distance as radius and *F* as center, we use a compass to draw arcs intersecting *L* at P_1 and P_2. Each of these points is now at the same distance (3.0 in.) from *F* and from the directrix, and is hence a point on the parabola. Repeat the construction with distances other than 3.0 in. to get more points on the parabola.   ◆◆◆

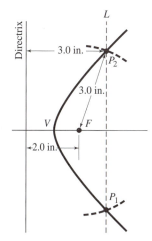

FIGURE 22–43 Construction of a parabola.

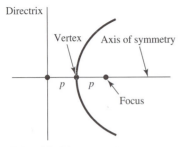

FIGURE 22–44 Parabola.

Figure 22–44 shows the typical shape of a parabola. The parabola has an *axis of symmetry* which intersects it at the *vertex*. The distance p from directrix to vertex is equal to the directed distance from the vertex to the focus.

Standard Equation of a Parabola with Vertex at the Origin

Let us place the parabola on coordinate axes with the vertex at the origin and with the axis of symmetry along the x axis, as shown in Fig. 22–45. Choose any point P

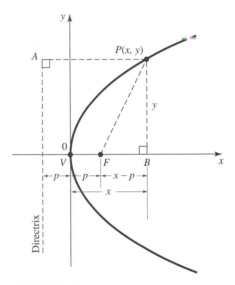

FIGURE 22–45

on the parabola. Then, by the definition of a parabola, $FP = AP$. But in right triangle FBP,

$$FP = \sqrt{(x - p)^2 + y^2}$$

and

$$AP = p + x$$

But, since $FP = AP$,

$$\sqrt{(x - p)^2 + y^2} = p + x$$

Squaring both sides yields

$$(x - p)^2 + y^2 = p^2 + 2px + x^2$$
$$x^2 - 2px + p^2 + y^2 = p^2 + 2px + x^2$$

Collecting terms, we get the standard equation of a parabola with vertex at the origin.

Standard Equation of a Parabola; Vertex at Origin, Axis Horizontal		$y^2 = 4px$	308

We have defined p as the directed distance VF from the vertex V to the focus F. Thus if p is positive, the focus must lie to the right of the vertex, and hence the parabola opens to the right. Conversely, if p is negative, the parabola opens to the left.

◆◆◆ Example 29: Find the coordinates of the focus of the parabola $2y^2 + 7x = 0$.

Solution: Subtracting $7x$ from both sides and dividing by 2 gives us

$$y^2 = -3.5x$$

Thus $4p = -3.5$ and $p = -0.875$. Since p is negative, the parabola opens to the left. The focus is thus 0.875 unit to the left of the vertex (Fig. 22–46) and has the coordinates $(-0.875, 0)$.

◆◆◆

◆◆◆ Example 30: Find the coordinates of the focus and write the equation of a parabola that has its vertex at the origin, that has a horizontal axis of symmetry, and that passes through the point $(5, -6)$.

Solution: Our given point must satisfy the equation $y^2 = 4px$. Substituting 5 for x and -6 for y gives

$$(-6)^2 = 4p(5)$$

So $4p = \frac{36}{5}$ and $p = \frac{9}{5}$. The equation of the parabola is then $y^2 = 36x/5$, or

$$5y^2 = 36x$$

Since the axis is horizontal and p is positive, the focus must be on the x axis and is a distance p $\left(\frac{9}{5}\right)$ to the right of the origin. The coordinates of F are thus $\left(\frac{9}{5}, 0\right)$, as shown in Fig. 22–47.

◆◆◆

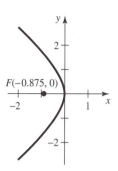

FIGURE 22–46

We have graphed the parabola before. Glance back at Sec. 5–2.

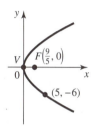

FIGURE 22–47

Standard Equation of a Parabola with Vertical Axis

The standard equation for a parabola having a vertical axis is obtained by switching the positions of x and y in Eq. 308.

Standard Equation of a Parabola: Vertex at Origin, Axis Vertical		$x^2 = 4py$	309

Note that only one variable is squared in any parabola equation. This gives us the best way to recognize such equations.

When p is positive, the parabola opens upward; when p is negative, it opens downward.

◆◆◆ Example 31: A parabola has its vertex at the origin and passes through the points $(3, 2)$ and $(-3, 2)$. Write its equation and find the focus.

Solution: The sketch of Fig. 22–48 shows that the axis must be vertical, so our equation is of the form $x^2 = 4py$. Substituting 3 for x and 2 for y gives

$$3^2 = 4p(2)$$

from which $4p = \frac{9}{2}$ and $p = \frac{9}{8}$. Our equation is then $x^2 = 9y/2$ or $2x^2 = 9y$. The focus is on the y axis at a distance $p = \frac{9}{8}$ from the origin, so its coordinates are $\left(0, \frac{9}{8}\right)$.

◆◆◆

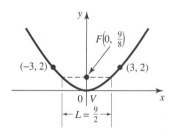

FIGURE 22–48

Focal Width of a Parabola

The *latus rectum* of a parabola is a line through the focus which is perpendicular to the axis of symmetry, such as line AB in Fig. 22–49. The length of the latus rectum is also called the *focal width*. We will find the focal width, or length L of the latus rectum, by substituting the coordinates (p, h) of point A into Eq. 308.

$$h^2 = 4p(p) = 4p^2$$

$$h = \pm 2p$$

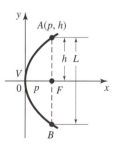

FIGURE 22–49

The focal width is twice h, so we have the following equation:

Focal Width of a Parabola	$L = |4p|$	**313**

The focal width (length of the latus rectum) of a parabola is four times the distance from vertex to focus.

◆◆◆ **Example 32:** The focal width for the parabola in Fig. 22–48 is $\frac{9}{2}$, or 4.5 units.

◆◆◆

Shift of Axes

As with the circle, when the vertex of the parabola is not at the origin but at (h, k), our equations will be similar to Eqs. 308 and 309, except that x is replaced with $x - h$ and y is replaced with $y - k$.

Standard Equations of a Parabola: Vertex at (*h, k*)	Axis Horizontal		$(y - k)^2 = 4p(x - h)$	**310**
	Axis Vertical		$(x - h)^2 = 4p(y - k)$	**311**

◆◆◆ **Example 33:** Find the vertex, focus, focal width, and equation of the axis for the parabola $(y - 3)^2 = 8(x + 2)$.

Solution: The given equation is of the same form as Eq. 310, so the axis is horizontal. Also, $h = -2$, $k = 3$, and $4p = 8$. So the vertex is at $V(-2, 3)$ (Fig. 22–50). Since $p = \frac{8}{4} = 2$, the focus is 2 units to the right of the vertex, at $F(0, 3)$. The focal width is $4p$, so $L = 8$. The axis is horizontal and 3 units from the x axis, so its equation is $y = 3$.

◆◆◆

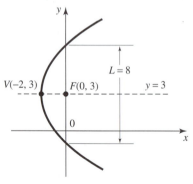

FIGURE 22–50

◆◆◆ **Example 34:** Write the equation of a parabola that opens upward, with vertex at $(-1, 2)$, and that passes through the point $(1, 3)$ (Fig. 22–51). Find the focus and the focal width.

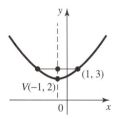

FIGURE 22–51

Solution: We substitute $h = -1$ and $k = 2$ into Eq. 311.

$$(x + 1)^2 = 4p(y - 2)$$

Now, since $(1, 3)$ is on the parabola, these coordinates must satisfy our equation. Substituting yields

$$(1 + 1)^2 = 4p(3 - 2)$$

Solving for p, we obtain $2^2 = 4p$, or $p = 1$. So the equation is

$$(x + 1)^2 = 4(y - 2)$$

The focus is p units above the vertex, at $(-1, 3)$. The focal width is, by Eq. 313,

$$L = |4p| = 4(1) = 4 \text{ units}$$

◆◆◆

General Equation of a Parabola

We get the general equation of the parabola by expanding the standard equation (Eq. 310) as follows:

$$(y - k)^2 = 4p(x - h)$$
$$y^2 - 2ky + k^2 = 4px - 4ph$$

or

$$y^2 - 4px - 2ky + (k^2 + 4ph) = 0$$

which is of the following general form (C, D, E, and F are constants):

General Equation of a Parabola with Horizontal Axis	$Cy^2 + Dx + Ey + F = 0$	312

Compare this with the general second-degree equation.

We see that the equation of a parabola having a horizontal axis of symmetry has a y^2 term but no x^2 term. Conversely, the equation for a parabola with vertical axis has an x^2 term but not a y^2 term. The parabola is the only conic for which there is only one variable squared.

If the coefficient B of the xy term in the general second-degree equation (Eq. 297) were not zero, it would indicate that the axis of symmetry was *rotated* by some amount and was no longer parallel to a coordinate axis. The presence of an xy term indicates rotation of the ellipse and hyperbola as well.

Completing the Square

As with the circle, we go from general to standard form by completing the square.

◆◆◆ **Example 35:** Find the vertex, focus, and focal width for the parabola

$$x^2 + 6x + 8y + 1 = 0$$

Solution: Separating the x and y terms, we have

$$x^2 + 6x = -8y - 1$$

Completing the square by adding 9 to both sides, we obtain

$$x^2 + 6x + 9 = -8y - 1 + 9$$

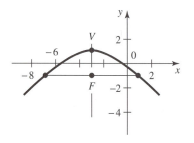

FIGURE 22–52

Factoring yields

$$(x + 3)^2 = -8(y - 1)$$

which is the form of Eq. 311, with $h = -3$, $k = 1$, and $p = -2$.

The vertex is $(-3, 1)$. Since the parabola opens downward (Fig. 22–52), the focus is 2 units below the vertex, at $(-3, -1)$. The focal width is $4|p|$, or 8 units.   ◆◆◆

Exercise 3 ◆ The Parabola

Parabola with Vertex at Origin

Find the focus and focal width of each parabola.

1. $y^2 = 8x$ **2.** $x^2 = 16y$
3. $7x^2 + 12y = 0$ **4.** $3y^2 + 5x = 0$

Write the equation of each parabola.

5. focus at $(0, -2)$
6. passes through $(25, 20)$; axis horizontal
7. passes through $(3, 2)$ and $(3, -2)$
8. passes through $(3, 4)$ and $(-3, 4)$

Parabola with Vertex Not at Origin

Find the vertex, focus, focal width, and equation of the axis for each parabola.

9. $(y - 5)^2 = 12(x - 3)$ **10.** $(x + 2)^2 = 16(y - 6)$
11. $(x - 3)^2 = 24(y + 1)$ **12.** $(y + 3)^2 = 4(x + 5)$
13. $3x + 2y^2 + 4y - 4 = 0$ **14.** $y^2 + 8y + 4 - 6x = 0$
15. $y - 3x + x^2 + 1 = 0$ **16.** $x^2 + 4x - y - 6 = 0$

Write the equation, in general form, for each parabola.

17. vertex at $(1, 2)$; $L = 8$; axis is $y = 2$; opens to the right
18. axis is $y = 3$; passes through $(6, -1)$ and $(3, 1)$
19. passes through $(-3, 3)$, $(-6, 5)$, and $(-11, 7)$; axis horizontal
20. vertex $(0, 2)$; axis is $x = 0$; passes through $(-4, -2)$
21. axis is $y = -1$; passes through $(-4, -2)$ and $(2, 1)$

Trajectories

22. A ball thrown into the air will, neglecting air resistance, follow a parabolic path, as shown in Fig. 22–53. Write the equation of the path, taking axes as shown. Use your equation to find the height of the ball when it is at a horizontal distance of 95.0 ft from O.

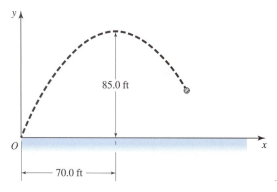

FIGURE 22–53   Ball thrown into the air.

23. Some comets follow a parabolic orbit with the sun at the focal point (Fig. 22–54). Taking axes as shown, write the equation of the path if the distance p is 75 million kilometers.

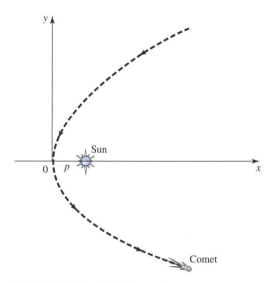

FIGURE 22–54 Path of a comet.

24. An object dropped from a moving aircraft (Fig. 22–55) will follow a parabolic path if air resistance is negligible. A weather instrument released at a height of 3520 m is observed to strike the water at a distance of 2150 m from the point of release. Write the equation of the path, taking axes as shown. Find the height of the instrument when x is 1000 m.

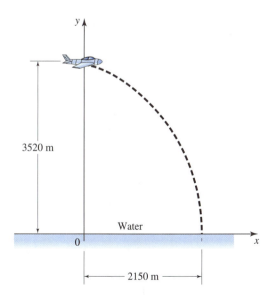

FIGURE 22–55 Object dropped from an aircraft.

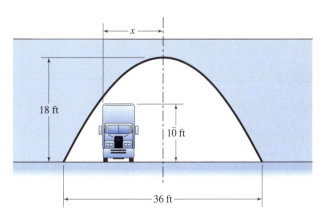

FIGURE 22–56 Parabolic arch.

Parabolic Arch

25. A 10-ft-high truck passes under a parabolic arch, as shown in Fig. 22–56. Find the maximum distance x that the side of the truck can be from the center of the road.

26. Assuming the bridge cable *AB* of Fig. 22–57 to be a parabola, write its equation, taking axes as shown.

A cable will hang in the shape of a parabola if the vertical load per horizontal foot is constant.

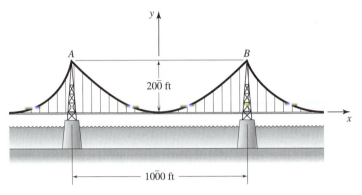

FIGURE 22–57   Parabolic bridge cable.

27. A parabolic arch supports a roadway as shown in Fig. 22–58. Write the equation of the arch, taking axes as shown. Use your equation to find the vertical distance from the roadway to the arch at a horizontal distance of 50.0 m from the center.

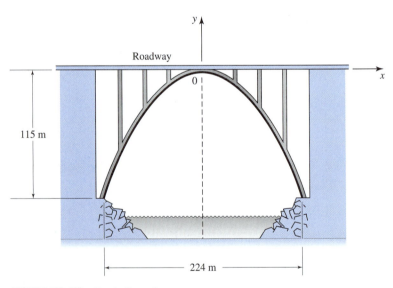

FIGURE 22–58   Parabolic arch.

Parabolic Reflector

28. A certain solar collector consists of a long panel of polished steel bent into a parabolic shape (Fig. 22–59), which focuses sunlight onto a pipe *P* at the focal point of the parabola. At what distance *x* should the pipe be placed?

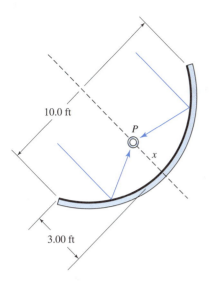

FIGURE 22–59 Parabolic solar collector.

29. A parabolic collector for receiving television signals from a satellite is shown in Fig. 22–60. The receiver R is at the focus, 1.00 m from the vertex. Find the depth d of the collector.

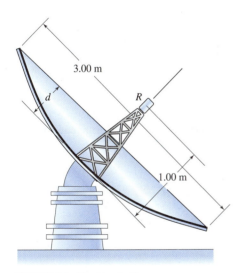

FIGURE 22–60 Parabolic antenna.

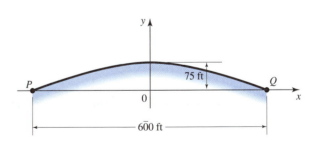

FIGURE 22–61 Road over a hill.

Vertical Highway Curves

30. A parabolic curve is to be used at a dip in a highway. The road dips 32.0 m in a horizontal distance of 125 m and then rises to its previous height in another 125 m. Write the equation of the curve of the roadway, taking the origin at the bottom of the dip and the y axis vertical.

31. Write the equation of the vertical highway curve in Fig. 22–61, taking axes as shown.

Beams

32. A simply supported beam with a concentrated load at its midspan will deflect in the shape of a parabola, as shown in Fig. 22–62. If the deflection at the midspan is 1.00 in., write the equation of the parabola (called the *elastic curve*), taking axes as shown.

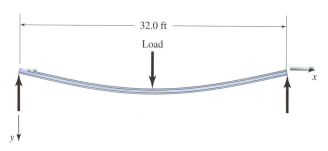

FIGURE 22–62   Deflection of a beam.

33. Using the equation found in problem 32, find the deflection of the beam in Fig. 22–62 at a distance of 10.0 ft from the left end.

Computer and Graphics Calculator

34. Use a graphics calculator or a graphics utility on the computer to graph any of the parabolas in this exercise set. As with the circle, you must first solve the equation for *y*. For a parabola whose axis is vertical, simply plot that equation. For a parabola whose axis is horizontal, you must plot two functions: one for the upper portion and another for the lower portion of the curve.

35. Some computer algebra systems have commands, such as the *complete-square* command in *Maple,* to complete the square. If you have such a system, use it to go from general to standard form wherever needed in this exercise set.

22–4   The Ellipse

Definition

The ellipse is defined in the following way:

Definition of an Ellipse	An *ellipse* is the set of all points in a plane such that the sum of the distances from each point on the ellipse to two fixed points (called the *foci*) is constant.	315

One way to draw an ellipse that is true to its definition is to pass a loop of string around two tacks, pull the string taut with a pencil, and trace the curve (Fig. 22–63).

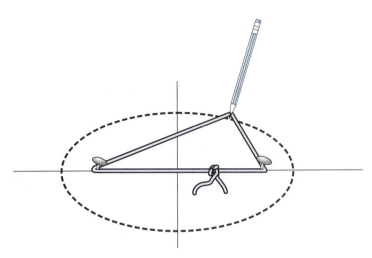

FIGURE 22–63 Construction of an ellipse.

Figure 22–64 shows the typical shape of the ellipse. An ellipse has two axes of symmetry: the *major axis* and the *minor axis,* which intersect at the *center* of the ellipse. A *vertex* is a point where the ellipse crosses the major axis.

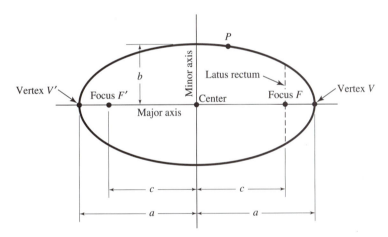

FIGURE 22–64 Ellipse.

It is often convenient to speak of *half* the lengths of the major and minor axes, and these are called the *semimajor* and *semiminor* axes, whose lengths we label a and b, respectively. The distance from either focus to the center is labeled c.

Distance to Focus

Before deriving an equation for the ellipse, let us first write an expression for the distance c from the center to a focus in terms of the semimajor axis a and the semiminor axis b.

From our definition of an ellipse, if P is any point on the ellipse, then

$$PF + PF' = k \qquad (1)$$

where k is a constant. If P is taken at a vertex V, then Equation (1) becomes

$$VF + VF' = k = 2a \qquad (2)$$

because $VF + VF'$ is equal to the length of the major axis. Substituting back into (1) gives us

$$PF + PF' = 2a \qquad (3)$$

Figure 22–65 shows our point P moved to the intersection of the ellipse and the minor axis. Here PF and PF' are equal. But since their sum is $2a$, PF and PF' must each equal a. By the Pythagorean theorem, $c^2 + b^2 = a^2$, or

$$c^2 = a^2 - b^2$$

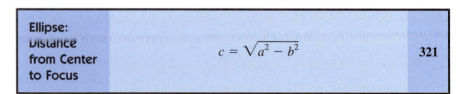

| Ellipse: Distance from Center to Focus | $c = \sqrt{a^2 - b^2}$ | 321 |

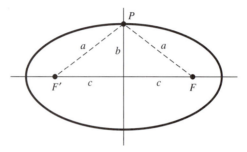

FIGURE 22–65

Ellipse with Center at the Origin, Major Axis Horizontal

Let us place an ellipse on coordinate axes with its center at the origin and major axis along the x axis, as shown in Fig. 22–66. If $P(x, y)$ is any point on the ellipse, then by the definition of the ellipse,

$$PF + PF' = 2a$$

A line from a focus to a point on the ellipse is called a *focal radius*.

FIGURE 22–66   Ellipse with center at origin.

To get PF and PF' in terms of x and y, we first drop a perpendicular from P to the x axis. Then in triangle PQF,

$$PF = \sqrt{(c - x)^2 + y^2}$$

and in triangle PQF',

$$PF' = \sqrt{(c + x)^2 + y^2}$$

Substituting yields

$$PF + PF' = \sqrt{(c - x)^2 + y^2} + \sqrt{(c + x)^2 + y^2} = 2a$$

Rearranging, we obtain

$$\sqrt{(c + x)^2 + y^2} = 2a - \sqrt{(c - x)^2 + y^2}$$

Squaring, and then expanding the binomials, we have

$$x^2 + 2cx + c^2 + y^2 = 4a^2 - 4a\sqrt{(c - x)^2 + y^2} + c^2 - 2cx + x^2 + y^2$$

Collecting terms, we get

$$cx = a^2 - a\sqrt{(c - x)^2 + y^2}$$

Dividing by a and rearranging yields

$$a - \frac{cx}{a} = \sqrt{(c - x)^2 + y^2}$$

Squaring both sides again yields

$$a^2 - 2cx + \frac{c^2x^2}{a^2} = c^2 - 2cx + x^2 + y^2$$

Collecting terms, we have

$$a^2 + \frac{c^2x^2}{a^2} = c^2 + x^2 + y^2$$

But $c = \sqrt{a^2 - b^2}$. Substituting, we have

$$a^2 + \frac{x^2}{a^2}(a^2 - b^2) = a^2 - b^2 + x^2 + y^2$$

or

$$a^2 + x^2 - \frac{b^2x^2}{a^2} = a^2 - b^2 + x^2 + y^2$$

Collecting terms and rearranging gives us

$$\frac{b^2x^2}{a^2} + y^2 = b^2$$

Finally, dividing through by b^2, we get the standard form of the equation of an ellipse with center at origin.

| Standard Equation of an Ellipse: Center at Origin, Major Axis Horizontal | | $$\frac{x^2}{a^2} + \frac{y^2}{b^2} = 1$$ $a > b$ | 316 |

Note that if $a = b$, this equation reduces to the equation of a circle.

Here a is half the length of the major axis, and b is half the length of the minor axis.

◆◆◆ **Example 36:** Find the vertices and foci for the ellipse

$$16x^2 + 36y^2 = 576$$

Solution: To be in standard form, our equation must have 1 (unity) on the right side. Dividing by 576 and simplifying gives us

$$\frac{x^2}{36} + \frac{y^2}{16} = 1$$

from which $a = 6$ and $b = 4$. The vertices are then $V(6, 0)$ and $V'(-6, 0)$, as shown in Fig. 22–67. The distance c from the center to a focus is

$$c = \sqrt{6^2 - 4^2} = \sqrt{20} \approx 4.47$$

So the foci are $F(4.47, 0)$ and $F'(-4.47, 0)$.

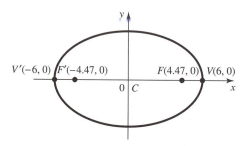

FIGURE 22–67

◆◆◆ **Example 37:** An ellipse whose center is at the origin and whose major axis is on the x axis has a minor axis of 10 units and passes through the point $(6, 4)$, as shown in Fig. 22–68. Write the equation of the ellipse in standard form.

Solution: Our equation will have the form of Eq. 316. Substituting, with $b = 5$, we have

$$\frac{x^2}{a^2} + \frac{y^2}{25} = 1$$

Since the ellipse passes through $(6, 4)$, these coordinates must satisfy our equation. Substituting yields

$$\frac{36}{a^2} + \frac{16}{25} = 1$$

Solving for a^2, we multiply by the LCD, $25a^2$.

$$36(25) + 16a^2 = 25a^2$$
$$9a^2 = 36(25)$$
$$a^2 = 100$$

So our final equation is

$$\frac{x^2}{100} + \frac{y^2}{25} = 1$$

◆◆◆

Ellipse with Center at Origin, Major Axis Vertical

When the major axis is vertical rather than horizontal, the only effect on the standard equation is to interchange the positions of x and y.

Standard Equation of an Ellipse: Center at Origin, Major Axis Vertical		$\dfrac{y^2}{a^2} + \dfrac{x^2}{b^2} = 1$ $a > b$	**317**

Figure 22-68 (margin):

FIGURE 22–68

Notice that the quantities a and b are dimensions of the semimajor and semiminor axes and remain so as the ellipse is turned or shifted. Therefore for an ellipse in any position, the distance c from center to focus is found the same way as before.

◆◆◆ **Example 38:** Find the lengths of the major and minor axes and the distance from center to focus for the ellipse

$$\frac{x^2}{16} + \frac{y^2}{49} = 1$$

Solution: How can we tell which denominator is a^2 and which is b^2? It is easy: a is always greater than b. So $a = 7$ and $b = 4$. Thus the major and minor axes are 14 units and 8 units long (Fig. 22–69). From Eq. 321,

$$c = \sqrt{a^2 - b^2} = \sqrt{49 - 16} = \sqrt{33}$$ ◆◆◆

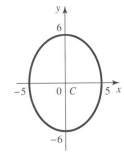

FIGURE 22–69

◆◆◆ **Example 39:** Write the equation of an ellipse with center at the origin, whose major axis is 12 units on the y axis and whose minor axis is 10 units (Fig. 22–70).

Solution: Substituting into Eq. 317, with $a = 6$ and $b = 5$, we obtain

$$\frac{y^2}{36} + \frac{x^2}{25} = 1$$ ◆◆◆

FIGURE 22–70

Common Error	Do not confuse a and b in the ellipse equations. The *larger* denominator is always a^2. Also, the variable (x or y) in the same term with a^2 tells the direction of the major axis.

Shift of Axes

Now consider an ellipse whose center is not at the origin, but at (h, k). The equation for such an ellipse will be the same as before, except that x is replaced by $x - h$ and y is replaced by $y - k$.

Standard Equations of an Ellipse: Center at (h, k)	Major Axis Horizontal		$\dfrac{(x - h)^2}{a^2} + \dfrac{(y - k)^2}{b^2} = 1$ $a > b$	**318**
	Major Axis Vertical		$\dfrac{(y - k)^2}{a^2} + \dfrac{(x - h)^2}{b^2} = 1$ $a > b$	**319**

◆◆◆ **Example 40:** Find the center, vertices, and foci for the ellipse

$$\frac{(x - 5)^2}{9} + \frac{(y + 3)^2}{16} = 1$$

Solution: From the given equation, $h = 5$ and $k = -3$, so the center is $C(5, -3)$, as shown in Fig. 22–71. Also, $a = 4$ and $b = 3$, and the major axis is vertical because a is with the y term. By setting $x = 5$ in the equation of the ellipse, we find that the vertices are $V(5, 1)$ and $V'(5, -7)$. The distance c to the foci is

$$c = \sqrt{4^2 - 3^2} = \sqrt{7} \cong 2.66$$

so the foci are $F(5, -0.34)$ and $F'(5, -5.66)$. ◆◆◆

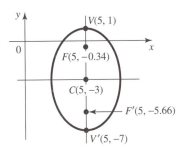

FIGURE 22–71

◆◆◆ **Example 41:** Write the equation in standard form of an ellipse with vertical major axis 10 units long, center at (3, 5), and whose distance between focal points is 8 units (Fig. 22–72).

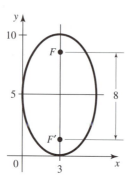

FIGURE 22–72

Yes, Equation 321 still applies here.

Solution: From the information given, $h = 3$, $k = 5$, $a = 5$, and $c = 4$. By Eq. 321,

$$b = \sqrt{a^2 - c^2} = \sqrt{25 - 16} = 3 \text{ units}$$

Substituting into Eq. 319, we get

$$\frac{(y - 5)^2}{25} + \frac{(x - 3)^2}{9} = 1$$

◆◆◆

General Equation of an Ellipse

As we did with the circle, we now expand the standard equation for the ellipse to get the general equation. Starting with Eq. 318,

$$\frac{(x - h)^2}{a^2} + \frac{(y - k)^2}{b^2} = 1$$

we multiply through by a^2b^2 and expand the binomials:

$$b^2(x^2 - 2hx + h^2) + a^2(y^2 - 2ky + k^2) = a^2b^2$$
$$b^2x^2 - 2b^2hx + b^2h^2 + a^2y^2 - 2a^2ky + a^2k^2 - a^2b^2 = 0$$

or

$$b^2x^2 + a^2y^2 - 2b^2hx - 2a^2ky + (b^2h^2 + a^2k^2 - a^2b^2) = 0$$

which is of the following general form:

General Equation of an Ellipse	$Ax^2 + Cy^2 + Dx + Ey + F = 0$	320

Comparing this with the general second-degree equation, we see that $B = 0$. As for the parabola (and the hyperbola, as we will see), this indicates that the axes of the curve are parallel to the coordinate axes. In other words, the curve is not rotated.

Also note that neither A nor C is zero. That tells us that the curve is not a parabola. Further, A and C have different values, telling us that the curve cannot be a circle. We'll see that A and C will have the same sign for the ellipse and opposite signs for the hyperbola.

Completing the Square

As before, we go from general to standard form by completing the square.

◆◆◆ **Example 42:** Find the center, foci, vertices, and major and minor axes for the ellipse $9x^2 + 25y^2 + 18x - 50y - 191 = 0$.

Solution: Grouping the x terms and the y terms gives us

$$(9x^2 + 18x) + (25y^2 - 50y) = 191$$

Factoring yields

$$9(x^2 + 2x \quad) + 25(y^2 - 2y \quad) = 191$$

Completing the square, we obtain

$$9(x^2 + 2x + 1) + 25(y^2 - 2y + 1) = 191 + 9 + 25$$

Factoring gives us

$$9(x + 1)^2 + 25(y - 1)^2 = 225$$

Finally, dividing by 225, we have

$$\frac{(x + 1)^2}{25} + \frac{(y - 1)^2}{9} = 1$$

We see that $h = -1$ and $k = 1$, so the center is at $(-1, 1)$. Also, $a = 5$, so the major axis is 10 units and is horizontal, and $b = 3$, so the minor axis is 6 units. A vertex is located 5 units to the right of the center, at $(4, 1)$, and 5 units to the left, at $(-6, 1)$. From Eq. 321,

$$c = \sqrt{a^2 - b^2}$$
$$= \sqrt{25 - 9}$$
$$= 4$$

So the foci are at $(3, 1)$ and $(-5, 1)$. This ellipse is shown in Fig. 22–73.

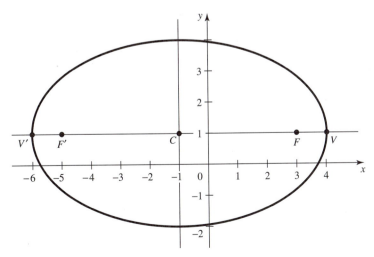

FIGURE 22-73

◆◆◆

> **Common Error**
>
> In Example 42 we needed 1 to complete the square on x, but we added 9 to the right side because the expression containing the 1 was multiplied by a factor of 9. Similarly, for y, we needed 1 on the left but added 25 to the right. It is very easy to forget to multiply by those factors.

Focal Width of an Ellipse

The focal width L, or length of the latus rectum, is the width of the ellipse through the focus (Fig. 22–74). The sum of the focal radii PF and PF' from a point P at one end of the latus rectum must equal $2a$, by the definition of an ellipse. So $PF' = 2a - PF$, or, since $L/2 = PF$, $PF' = 2a - L/2$. Squaring both sides, we obtain

$$(PF')^2 = 4a^2 - 2aL + \frac{L^2}{4} \tag{1}$$

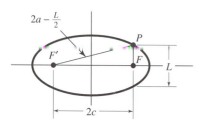

FIGURE 22–74   Derivation of focal width.

But in right triangle PFF',

$$(PF')^2 = \left(\frac{L}{2}\right)^2 + (2c)^2 = \frac{L^2}{4} + 4c^2 = \frac{L^2}{4} + 4a^2 - 4b^2 \tag{2}$$

since $c^2 = a^2 - b^2$. Equating (1) and (2) and collecting terms gives $2aL = 4b^2$, or the following:

As with the parabola, the main use for L is for quick sketching of the ellipse.

Focal Width of an Ellipse	$L = \dfrac{2b^2}{a}$	322

The focal width of an ellipse is twice the square of the semiminor axis, divided by the semimajor axis.

◆◆◆ **Example 43:** The focal width of an ellipse that is 25 m long and 10 m wide is

$$L = \frac{2(5^2)}{12.5} = \frac{50}{12.5} = 4 \text{ m}$$

◆◆◆

Exercise 4 ◆ The Ellipse

Ellipse with Center at Origin

Find the coordinates of the vertices and foci for each ellipse.

1. $\dfrac{x^2}{25} + \dfrac{y^2}{16} = 1$
2. $\dfrac{x^2}{49} + \dfrac{y^2}{36} = 1$
3. $3x^2 + 4y^2 = 12$
4. $64x^2 + 15y^2 = 960$
5. $4x^2 + 3y^2 = 48$
6. $8x^2 + 25y^2 = 200$

Write the equation of each ellipse in standard form.

7. vertices at (±5, 0); foci at (±4, 0)

8. vertical major axis 8 units long; a focus at (0, 2)

9. horizontal major axis 12 units long; passes through (3, $\sqrt{3}$)

10. passes through (4, 6) and (2, $3\sqrt{5}$)

11. horizontal major axis 26 units long; distance between foci = 24

12. vertical minor axis 10 units long; distance from focus to vertex = 1

13. passes through (1, 4) and (−6, 1)

14. distance between foci = 18; sum of axes = 54; horizontal major axis

Ellipse with Center Not at Origin

Find the coordinates of the center, vertices, and foci for each ellipse. Round to three significant digits where needed.

15. $\dfrac{(x-2)^2}{16} + \dfrac{(y+2)^2}{9} = 1$

16. $\dfrac{(x+5)^2}{25} + \dfrac{(y-3)^2}{49} = 1$

17. $5x^2 + 20x + 9y^2 - 54y + 56 = 0$

18. $16x^2 - 128x + 7y^2 + 42y = 129$

19. $7x^2 - 14x + 16y^2 + 32y = 89$

20. $3x^2 - 6x + 4y^2 + 32y + 55 = 0$

21. $25x^2 + 150x + 9y^2 - 36y + 36 = 0$

Write the equation of each ellipse.

22. minor axis = 10; foci at (13, 2) and (−11, 2)

23. center at (0, 3); vertical major axis = 12; length of minor axis = 6

24. center at (2, −1); a vertex at (2, 5); length of minor axis = 3

25. center at (−2, −3); a vertex at (−2, 1); a focus halfway between vertex and center

Intersections of Curves

Solve simultaneously to find the points of intersection of each ellipse with the given curve.

26. $3x^2 + 6y^2 = 11$ with $y = x + 1$

27. $2x^2 + 3y^2 = 14$ with $y^2 = 4x$

28. $x^2 + 7y^2 = 16$ with $x^2 + y^2 = 10$

Construction of an Ellipse

29. Draw an ellipse using tacks and string, as shown in Fig. 22–63, with major axis of 84.0 cm and minor axis of 58.0 cm. How far apart must the tacks be placed?

Applications

30. A certain bridge arch is in the shape of half an ellipse $\overline{120}$ ft wide and 30.0 ft high. At what horizontal distance from the center of the arch is the height equal to 15.0 ft?

For problems 30 through 33, first write the equation for each ellipse.

31. A curved mirror in the shape of an ellipse will reflect all rays of light coming from one focus onto the other focus (Fig. 22–75). A certain spot heater is to be made with a heating element at *A* and the part to be heated at *B*, contained within an ellipsoid (a solid obtained by rotating an ellipse about one axis). Find the width *x* of the chamber if its length is 25 cm and the distance from *A* to *B* is 15 cm.

32. The paths of the planets and certain comets are ellipses, with the sun at one focal point. The path of Halley's comet is an ellipse with a major axis of 36.18 AU and a minor axis of 9.12 AU. What is the greatest distance that Halley's comet gets from the sun?

An *astronomical unit*, AU, is the distance between the earth and the sun, about 92.6 million miles.

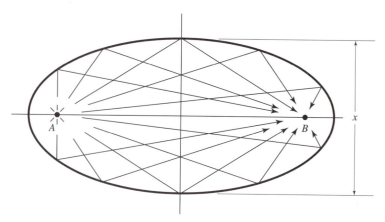

FIGURE 22–75   Focusing property of the ellipse.

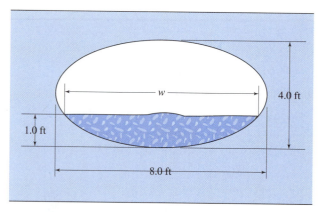

FIGURE 22–76

33. An elliptical culvert (Fig. 22–76) is filled with water to a depth of 1.0 ft. Find the width w of the stream.

Computer and Graphics Calculator

34. Use a graphics calculator or a graphics utility on the computer to graph any of the ellipses in this exercise set. As before, you must first solve the equation for y and then plot two functions, one for the upper portion and another for the lower portion of the curve.

35. Find a graphical solution to problems 26 through 28 by graphing each pair of curves and then using TRACE and ZOOM to find the coordinates of the points of intersection.

36. Use a CAS to complete the square in problems 17 through 21 to put those equations into standard form.

22–5   The Hyperbola

The hyperbola is defined as follows:

| Definition of a Hyperbola | A *hyperbola* is the set of all points in a plane such that the distances from each point to two fixed points, the *foci*, have a constant difference. | 325 |

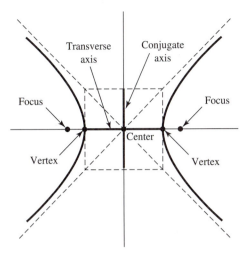

FIGURE 22–77 Hyperbola.

Figure 22–77 shows the typical shape of the hyperbola. The point midway between the foci is called the *center* of the hyperbola. The line passing through the foci and the center is one *axis* of the hyperbola. The hyperbola crosses that axis at points called the *vertices*. The line segment connecting the vertices is called the *transverse axis*. A second axis of the hyperbola passes through the center and is perpendicular to the transverse axis. The segment of this axis, shown in Fig. 22–77, is called the *conjugate axis*. Half the lengths of the transverse and conjugate axes are the *semitransverse* and *semiconjugate* axes, respectively. They are also referred to as *semiaxes*.

Standard Equations of a Hyperbola with Center at Origin

We place the hyperbola on coordinate axes, with its center at the origin and its transverse axis on the x axis (Fig. 22–78). Let a be half the transverse axis, and let c be half the distance between foci. Now take any point P on the hyperbola

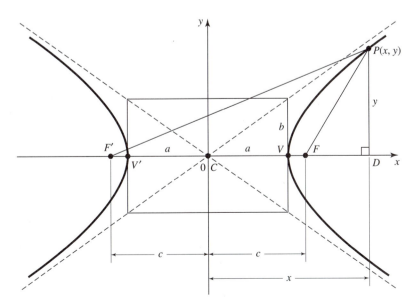

FIGURE 22–78 Hyperbola with center at origin.

and draw the focal radii PF and PF'. Then, by the definition of the hyperbola, $|PF' - PF| = $ constant. The constant can be found by moving P to the vertex V, where

$$|PF' - PF| = VF' - VF$$
$$= 2a + V'F' - VF$$
$$= 2a$$

since $V'F'$ and VF are equal. So $|PF' - PF| = 2a$. But in right triangle $PF'D$,

$$PF' = \sqrt{(x + c)^2 + y^2}$$

and in right triangle PFD,

$$PF = \sqrt{(x - c)^2 + y^2}$$

so

$$\sqrt{(x + c)^2 + y^2} - \sqrt{(x - c)^2 + y^2} = 2a$$

Notice that this equation is almost identical to the one we had when deriving the equation for the ellipse in Sec. 22–4, and the derivation is almost identical as well. However, to eliminate c from our equation, we define a new quantity b, such that $b^2 = c^2 - a^2$. We'll soon give a geometric meaning to the quantity b.

The remainder of the derivation is left as an exercise. The resulting equations are very similar to those for the ellipse.

Try to derive these equations. Follow the same steps we used for the ellipse.

Standard Equations of a Hyperbola: Center at Origin	Transverse Axis Horizontal		$\dfrac{x^2}{a^2} - \dfrac{y^2}{b^2} = 1$	**326**
	Transverse Axis Vertical		$\dfrac{y^2}{a^2} - \dfrac{x^2}{b^2} = 1$	**327**

Asymptotes of a Hyperbola

In Fig. 22–78, the dashed diagonal lines are asymptotes of the hyperbola. (The two branches of the hyperbola approach, but do not intersect, the asymptotes.)

We find the slope of these asymptotes from the equation of the hyperbola. Solving Eq. 326 for y gives $y^2 = b^2(x^2/a^2 - 1)$, or

$$y = \pm b\sqrt{\frac{x^2}{a^2} - 1}$$

As x gets large, the 1 under the radical sign becomes insignificant in comparison with x^2/a^2, and the equation for y becomes

$$y = \pm\frac{b}{a}x$$

This is the equation of a straight line having a slope of b/a. Thus as x increases, the branches of the hyperbola more closely approach straight lines of slopes $\pm b/a$.

In general, if the distance from a point P on a curve to some line L approaches zero as the distance from P to the origin increases without bound, then L is an *asymptote*. The hyperbola has two such asymptotes, of slopes $\pm b/a$.

For a hyperbola whose transverse axis is *vertical,* the slopes of the asymptotes can be shown to be $\pm a/b$.

Slope of the Asymptotes of a Hyperbola	Transverse Axis Horizontal	slope $= \pm\dfrac{b}{a}$	332
	Transverse Axis Vertical	slope $= \pm\dfrac{a}{b}$	333

Common Error	The asymptotes are not (usually) perpendicular to each other. The slope of one is **not** the negative reciprocal of the other.

Looking back at Fig. 22–78, we can now give meaning to the quantity b. If an asymptote has a slope b/a, it must have a rise of b in a run equal to a. Thus b is the distance, perpendicular to the transverse axis, from vertex to asymptote. It is half the length of the conjugate axis.

Hyperbola: Distance from Center to Focus	$c = \sqrt{a^2 + b^2}$	331

◆◆◆ **Example 44:** Find a, b, and c, the coordinates of the center, the vertices, and the foci, and the slope of the asymptotes for the hyperbola

$$\frac{x^2}{25} - \frac{y^2}{36} = 1$$

Solution: This equation is of the same form as Eq. 326, so we know that the center is at the origin and that the transverse axis is on the x axis. Also, $a^2 = 25$ and $b^2 = 36$, so

$$a = 5 \quad \text{and} \quad b = 6$$

The vertices then have the coordinates

$$V(5, 0) \quad \text{and} \quad V'(-5, 0)$$

Then, from Eq. 331,

$$c^2 = a^2 + b^2 = 25 + 36 = 61$$

so $c = \sqrt{61} \approx 7.81$. The coordinates of the foci are then

$$F(7.81, 0) \quad \text{and} \quad F'(-7.81, 0)$$

The slopes of the asymptotes are

$$\pm\frac{b}{a} = \pm\frac{6}{5}$$

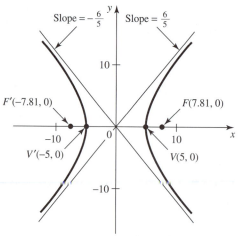

FIGURE 22–79   Graph of $\dfrac{x^2}{25} - \dfrac{y^2}{36} = 1$.

This hyperbola is shown in Fig. 22–79. In the following section we show how to make such a graph.   ◆◆◆

Graphing a Hyperbola

A good way to start a sketch of the hyperbola is to draw a rectangle whose dimension along the transverse axis is $2a$ and whose dimension along the conjugate axis is $2b$. The asymptotes are then drawn along the diagonals of this rectangle. Half the diagonal of the rectangle has a length $\sqrt{a^2 + b^2}$, which is equal to c, the distance to the foci. Thus an arc of radius c will cut extensions of the transverse axis at the focal points.

As with the parabola and ellipse, a perpendicular through a focus connecting two points on the hyperbola is called a *latus rectum*. Its length, called the *focal width*, is $2b^2/a$, the same as for the ellipse.

◆◆◆ **Example 45:** Find the vertices, foci, semiaxes, slope of the asymptotes, and focal width, and graph the hyperbola $64x^2 - 49y^2 = 3136$.

Solution: We write the equation in standard form by dividing by 3136.

$$\frac{x^2}{49} - \frac{y^2}{64} = 1$$

This is the form of Eq. 326, so the transverse axis is horizontal, with $a = 7$ and $b = 8$. We draw a rectangle of width 14 and height 16 (shown shaded in Fig. 22–80),

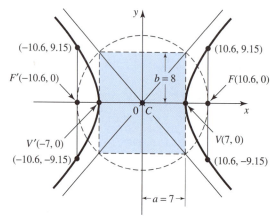

FIGURE 22–80

thus locating the vertices at $(\pm 7, 0)$. Diagonals through the rectangle give us the asymptotes of slopes $\pm \frac{8}{7}$. We locate the foci by swinging an arc of radius c, equal to half the diagonal of the rectangle. Thus

$$c = \sqrt{7^2 + 8^2} = \sqrt{113} \approx 10.6$$

The foci then are at $(\pm 10.6, 0)$. We obtain a few more points by computing the focal width.

$$L = \frac{2b^2}{a} = \frac{2(64)}{7} \approx 18.3$$

This gives us the additional points $(10.6, 9.15)$, $(10.6, -9.15)$, $(-10.6, 9.15)$, and $(-10.6, -9.15)$. ♦♦♦

♦♦♦ **Example 46:** A hyperbola whose center is at the origin has a focus at $(0, -5)$ and a transverse axis 8 units long, as shown in Fig. 22–81. Write the standard equation of the hyperbola, and find the slope of the asymptotes.

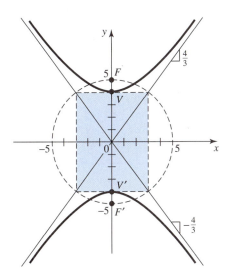

FIGURE 22–81

Solution: The transverse axis is 8, so $a = 4$. A focus is 5 units below the origin, so the transverse axis must be vertical, and $c = 5$. From Eq. 331,

$$b = \sqrt{c^2 - a^2} = \sqrt{25 - 16} = 3$$

Substituting into Eq. 327 gives us

$$\frac{y^2}{16} - \frac{x^2}{9} = 1$$

and from Eq. 333,

$$\text{slope of asymptotes} = \pm \frac{a}{b} = \pm \frac{4}{3} \qquad ♦♦♦$$

Common Error	Be sure that you know the direction of the transverse axis before computing the slopes of the asymptotes, which are $\pm b/a$ when the axis is horizontal, but $\pm a/b$ when the axis is vertical.

Shift of Axes

As with the other conics, we shift the axes by replacing x by $(x - h)$ and y by $(y - k)$ in Eqs. 326 and 327.

Standard Equations of a Hyperbola: Center at (h, k)	Transverse Axis Horizontal		$\dfrac{(x - h)^2}{a^2} - \dfrac{(y - k)^2}{b^2} = 1$	**328**
	Transverse Axis Vertical		$\dfrac{(y - k)^2}{a^2} - \dfrac{(x - h)^2}{b^2} = 1$	**329**

The equations for c and for the slope of the asymptotes are still valid for these cases.

Common Error

Do not confuse these equations with those for the ellipse. Here the terms have **opposite signs**. Also, a^2 is always the denominator of the **positive** term, even though it may be smaller than b^2. As with the ellipse, the variable in the same term as a^2 tells the direction of the transverse axis.

◆◆◆ **Example 47:** A certain hyperbola has a focus at $(1, 0)$, passes through the origin, has its transverse axis on the x axis, and has a distance of 10 between its focal points. Write its standard equation.

Solution: In Fig. 22–82 we plot the given focus $F(1, 0)$. Since the hyperbola passes through the origin, and the transverse axis also passes through the origin, a vertex

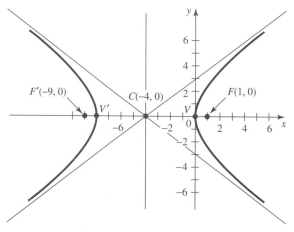

FIGURE 22–82

must be at the origin as well. Then, since $c = 5$, we go 5 units to the left of F and plot the center, $(-4, 0)$. Thus $a = 4$ units, from center to vertex. Then, by Eq. 331,

$$b = \sqrt{c^2 - a^2}$$
$$= \sqrt{25 - 16} = 3$$

Substituting into Eq. 328 with $h = -4$, $k = 0$, $a = 4$, and $b = 3$, we get

$$\frac{(x + 4)^2}{16} - \frac{y^2}{9} = 1$$

as the required equation.   ◆◆◆

General Equation of a Hyperbola

Since the standard equations for the ellipse and hyperbola are identical except for the minus sign, it is not surprising that their general equations should be identical except for a minus sign. The general second-degree equation for ellipse or hyperbola is as follows:

It's easy to feel overwhelmed by the large number of formulas in this chapter. Study the summary of formulas in Appendix A, where you'll find them side by side and can see where they differ and where they are alike.

| General Equation of a Hyperbola | $Ax^2 + Cy^2 + Dx + Ey + F = 0$ | 330 |

Comparing this with the general second-degree equation, we see that $B = 0$, indicating that the axes of the curve are parallel to the coordinate axes.

Also note that neither A nor C is zero, which tells us that the curve is not a parabola. Further, A and C have different values, telling us that the curve cannot be a circle. But here we will see that A and C have opposite signs, while for the ellipse, A and C have the same sign.

As before, we complete the square to go from general form to standard form.

◆◆◆ **Example 48:** Write the equation $x^2 - 4y^2 - 6x + 8y - 11 = 0$ in standard form. Find the center, vertices, foci, and asymptotes.

Solution: Rearranging gives us

$$(x^2 - 6x \quad) - 4(y^2 - 2y \quad) = 11$$

Completing the square and factoring yields

$$(x^2 - 6x + 9) - 4(y^2 - 2y + 1) = 11 + 9 - 4$$
$$(x - 3)^2 - 4(y - 1)^2 = 16$$

(Take care, when completing the square, to add -4 to the right side to compensate for 1 times -4 added on the left.) Dividing by 16 gives us

$$\frac{(x - 3)^2}{16} - \frac{(y - 1)^2}{4} = 1$$

from which $a = 4$, $b = 2$, $h = 3$, and $k = 1$.
From Eq. 331, $c = \sqrt{a^2 + b^2} = \sqrt{16 + 4} \approx 4.47$, and from Eq. 332,

$$\text{slope of asymptotes} = \pm\frac{b}{a} = \pm\frac{1}{2}$$

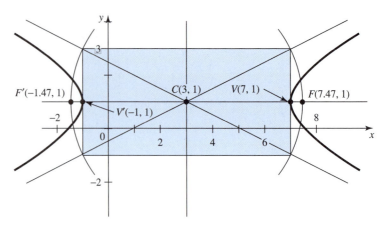

FIGURE 22–83

This hyperbola is plotted in Fig. 22–83.   ◆◆◆

Hyperbola Whose Asymptotes Are the Coordinate Axes

The graph of an equation of the form

$$xy = k$$

where k is a constant, is a hyperbola similar to those we have studied in this section, but rotated so that the coordinate axes are now the asymptotes, and the transverse and conjugate axes are at 45°.

Hyperbola: Axes Rotated 45°	$xy = k$	335

◆◆◆ **Example 49:** Plot the equation $xy = -4$.

Solution: We select values for x and evaluate $y = -4/x$.

x	-4	-3	-2	-1	0	1	2	3	4
y	1	$\frac{4}{3}$	2	4	—	-4	-2	$-\frac{4}{3}$	-1

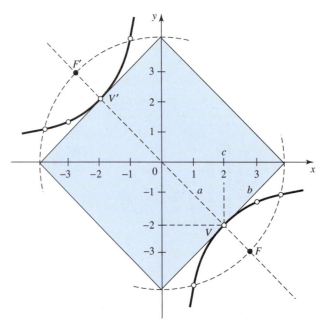

FIGURE 22–84   Hyperbola whose asymptotes are the coordinate axis.

The graph of Fig. 22–84 shows vertices at $(-2, 2)$ and $(2, -2)$. We see from the graph that a and b are equal and that each is the hypotenuse of a right triangle of side 2. Therefore

$$a = b = \sqrt{2^2 + 2^2} = \sqrt{8}$$

and from Eq. 331,

$$c = \sqrt{a^2 + b^2} = \sqrt{16} = 4 \qquad \text{◆◆◆}$$

When the constant k in Eq. 335 is negative, the hyperbola lies in the second and fourth quadrants, as in Example 49. When k is positive, the hyperbola lies in the first and third quadrants.

Exercise 5 ◆ The Hyperbola

Hyperbola with Center at Origin

Find the vertices, the foci, the lengths a and b of the semiaxes, and the slope of the asymptotes for each hyperbola.

1. $\dfrac{x^2}{16} - \dfrac{y^2}{25} = 1$

2. $\dfrac{y^2}{25} - \dfrac{x^2}{49} = 1$

3. $16x^2 - 9y^2 = 144$
4. $9x^2 - 16y^2 = 144$
5. $x^2 - 4y^2 = 16$
6. $3x^2 - y^2 + 3 = 0$

Write the equation of each hyperbola in standard form.

7. vertices at $(\pm 5, 0)$; foci at $(\pm 13, 0)$

8. vertices at $(0, \pm 7)$; foci at $(0, \pm 10)$

9. distance between foci $= 8$; transverse axis $= 6$ and is horizontal

10. conjugate axis $= 4$; foci at $(\pm 2.5, 0)$

11. transverse and conjugate axes equal; passes through $(3, 5)$; transverse axis vertical

12. transverse axis $= 16$ and is horizontal; conjugate axis $= 14$
13. transverse axis $= 10$ and is vertical; curve passes through $(8, 10)$
14. transverse axis $= 8$ and is horizontal; curve passes through $(5, 6)$

Hyperbola with Center Not at Origin

Find the center, lengths of the semiaxes, foci, slope of the asymptotes, and vertices for each hyperbola.

15. $\dfrac{(x - 2)^2}{25} - \dfrac{(y + 1)^2}{16} = 1$

16. $\dfrac{(y + 3)^2}{25} - \dfrac{(x - 2)^2}{49} = 1$

17. $16x^2 - 64x - 9y^2 - 54y = 161$

18. $16x^2 + 32x - 9y^2 + 592 = 0$

19. $4x^2 + 8x - 5y^2 - 10y + 19 = 0$

20. $y^2 - 6y - 3x^2 - 12x = 12$

Write the equation of each hyperbola in standard form.

21. center at (3, 2); length of the transverse axis = 8 and is vertical; length of the conjugate axis = 4

22. foci at $(-1, -1)$ and $(-5, -1)$; $a = b$

23. foci at $(5, -2)$ and $(-3, -2)$; vertex halfway between center and focus

24. length of the conjugate axis = 8 and is horizontal; center at $(-1, -1)$; length of the transverse axis = 16

Hyperbola Whose Asymptotes Are Coordinate Axes

25. Write the equation of a hyperbola with center at the origin and with asymptotes on the coordinate axes, whose vertex is at (6, 6).

26. Write the equation of a hyperbola with center at the origin and with asymptotes on the coordinate axes, which passes through the point $(-9, 2)$.

Applications

27. A hyperbolic mirror (Fig. 22–85) has the property that a ray of light directed at one focus will be reflected so as to pass through the other focus. Write the equation of the mirror shown, taking the axes as indicated.

28. A ship (Fig. 22–86) receives simultaneous radio signals from stations P and Q, on the shore. The signal from station P is found to arrive 375 microseconds (μs) later than that from Q. Assuming that the signals travel at a rate of 0.186 mi/μs, find the distance from the ship to each station (*Hint:* The ship will be on one branch of a hyperbola H whose foci are at P and Q. Write the equation of this hyperbola, taking axes as shown, and then substitute 75.0 mi for y to obtain the distance x.)

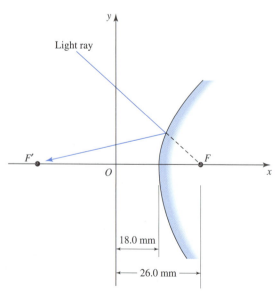

FIGURE 22–85   Hyperbolic mirror. This type of mirror is used in the *Cassegrain* form of reflecting telescope.

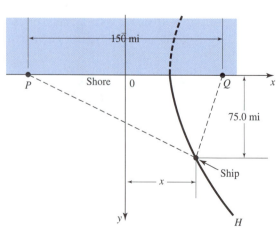

FIGURE 22–86

29. Boyle's law states that under certain conditions, the product of the pressure and volume of a gas is constant, or $pv = c$. This equation has the same form as the hyperbola (Eq. 335). If a certain gas has a pressure of 25.0 lb/in.^{2} at a volume of 1000 in.^{3}, write Boyle's law for this gas, and make a graph of pressure versus volume.

Computer and Graphics Calculator

30. Write a program that will accept as input the constants A, B, and C in the general second-degree equation (Eq. 297), and then identify and print the type of conic represented.

31. Some computer algebra systems have commands, such as the *complete-square* command in *Maple,* to complete the square. If you have such a system, use it to go from general to standard form wherever needed in this exercise.

32. Use a graphics calculator or a graphics utility on the computer to graph any of the hyperbolas in this exercise set. As for the circle, you must first solve the equation for y, and you must plot two functions: one for the upper portion and another for the lower portion of the curve.

◆◆◆ CHAPTER 22 REVIEW PROBLEMS ◆◆◆◆◆◆◆◆◆◆◆◆◆◆◆◆◆◆◆◆◆◆◆◆◆◆◆◆◆◆

1. Find the distance between the points $(3, 0)$ and $(7, 0)$.

2. Find the distance between the points $(4, -4)$ and $(1, -7)$.

3. Find the slope of the line perpendicular to a line that has an angle of inclination of $34.8°$.

4. Find the angle of inclination in degrees of a line with a slope of -3.

5. Find the angle of inclination of a line perpendicular to a line having a slope of 1.55.

6. Find the angle of inclination of a line passing through $(-3, 5)$ and $(5, -6)$.

7. Find the slope of a line perpendicular to a line having a slope of $a/2b$.

8. Write the equation of a line having a slope of -2 and a y intercept of 5.

9. Find the slope and y intercept of the line $2y - 5 = 3(x - 4)$.

10. Write the equation of a line having a slope of $2p$ and a y intercept of $p - 3q$.

11. Write the equation of the line passing through $(-5, -1)$ and $(-2, 6)$.

12. Write the equation of the line passing through $(-r, s)$ and $(2r, -s)$.

13. Write the equation of the line having a slope of 5 and passing through the point $(-4, 7)$.

14. Write the equation of the line having a slope of $3c$ and passing through the point $(2c, c - 1)$.

15. Write the equation of the line having an x intercept of -3 and a y intercept of 7.

16. Find the acute angle between two lines if one line has a slope of 1.50 and the other has a slope of 3.40.

17. Write the equation of the line that passes through $(2, 5)$ and is parallel to the x axis.

18. Find the angle of intersection between line L_1 having a slope of 2 and line L_2 having a slope of 7.

19. Find the directed distance AB between the points $A(-2, 0)$ and $B(-5, 0)$.

20. Find the angle of intersection between line L_1 having an angle of inclination of $18°$ and line L_2 having an angle of inclination of $75°$.

21. Write the equation of the line that passes through $(-3, 6)$ and is parallel to the y axis.

22. Find the increments in the coordinates of a particle that moves along a curve from $(3, 4)$ to $(5, 5)$.

23. Find the area of a triangle with vertices at $(6, 4)$, $(5, -2)$, and $(-3, -4)$.

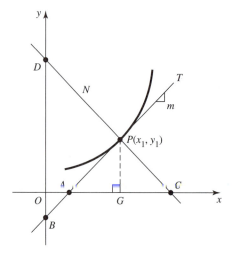

FIGURE 22–87   Tangent and normal to a curve.

In problems 24 through 33, a tangent T, of slope m, and a normal N are drawn to a curve at the point $P(x_1, y_1)$, as shown in Fig. 22–87. Show the following.

24. The equation of the tangent is

$$y - y_1 = m(x - x_1)$$

25. The equation of the normal is

$$x - x_1 + m(y - y_1) = 0$$

26. The x intercept A of the tangent is

$$x_1 - y_1/m$$

27. The y intercept B of the tangent is

$$y_1 - mx_1$$

28. The length of the tangent from P to the x axis is

$$PA = \frac{y_1}{m}\sqrt{1 + m^2}$$

29. The length of the tangent from P to the y axis is

$$PB = x_1\sqrt{1 + m^2}$$

30. The x intercept C of the normal is

$$x_1 + my_1$$

31. The y intercept D of the normal is

$$y_1 + x_1/m$$

32. The length of the normal from P to the x axis is

$$PC = y_1\sqrt{1 + m^2}$$

33. The length of the normal from P to the y axis is

$$PD = \frac{x_1}{m}\sqrt{1 + m^2}$$

Identify the curve represented by each equation. Find, where applicable, the center, vertices, foci, radius, semiaxes, and so on.

34. $x^2 - 2x - 4y^2 + 16y = 19$

35. $x^2 + 6x + 4y = 3$

36. $x^2 + y^2 = 8y$

37. $25x^2 - 200x + 9y^2 - 90y = 275$

38. $16x^2 - 9y^2 = 144$

39. $x^2 + y^2 = 9$

40. Write the equation for an ellipse whose center is at the origin, whose major axis ($2a$) is 20 and is horizontal, and whose minor axis ($2b$) equals the distance ($2c$) between the foci.

41. Write an equation for the circle passing through $(0, 0)$, $(8, 0)$, and $(0, -6)$.

42. Write the equation for a parabola whose vertex is at the origin and whose focus is $(-4.25, 0)$.

43. Write an equation for a hyperbola whose transverse axis is horizontal, with center at $(1, 1)$ passing through $(6, 2)$ and $(-3, 1)$.

44. Write the equation of a circle whose center is $(-5, 0)$ and whose radius is 5.

45. Write the equation of a hyperbola whose center is at the origin, whose transverse axis = 8 and is horizontal, and passing through $(10, 25)$.

46. Write the equation of an ellipse whose foci are $(2, 1)$ and $(-6, 1)$, and the sum of the focal radii is 10.

47. Write the equation of a parabola whose axis is the line $y = -7$, whose vertex is 3 units to the right of the y axis, and passing through $(4, -5)$.

48. Find the intercepts of the curve $y^2 + 4x - 6y = 16$.

49. Find the points of intersection of $x^2 + y^2 + 2x + 2y = 2$ and $3x^2 + 3y^2 + 5x + 5y = 10$.

50. A stone bridge arch is in the shape of half an ellipse and is 15.0 m wide and 5.00 m high. Find the height of the arch at a distance of 6.00 m from its center.

51. Write the equation of a hyperbola centered at the origin, where the conjugate axis = 12 and is vertical, and the distance between foci is 13.

52. A parabolic arch is 5.00 m high and 6.00 m wide at the base. Find the width of the arch at a height of 2.00 m above the base.

53. A stone thrown in the air follows a parabolic path and reaches a maximum height of 56.0 ft in a horizontal distance of 48.0 ft. At what horizontal distances from the launch point will the height be 25.0 ft?

Writing

54. Write a short paragraph explaining, in your own words, what is meant by the *slope of a curve*.

55. Suppose that the day before your visit to your former high school math class, the teacher unexpectedly asks you to explain how that day's topic, the conic sections, got that name. Write a paragraph on what you will tell the class about how the circle, ellipse, parabola, and hyperbola (and the point and straight line, too) can be formed by intersecting a plane and a cone. You may plan to illustrate your talk with a clay model or a piece of cardboard rolled into a cone. You will probably be asked what the conic sections are good for, so write out at least one use for each curve.

Team Projects

56. There are an infinite number of number pairs p and q (such as 3 and 1.5) whose product equals their sum. Write an expression for q as a function of p, and make a graph of p versus q for all of the real numbers, not just integer values, from -5 to 5. With a suitable shift of axes, show that the graph is a hyperbola of the form $y = 1/x$.

57. We have already shown how to graph each of the conic sections. Now graph, in the same viewing window, the ellipse

$$\frac{x^2}{a^2} + \frac{y^2}{b^2} = 1$$

the hyperbola

$$\frac{x^2}{a^2} - \frac{y^2}{b^2} = 1$$

and the hyperbola

$$\frac{y^2}{a^2} - \frac{x^2}{b^2} = 1$$

using the same values of a and b for each. Also graph the asymptotes of the hyperbolas.

58. If the area of an ellipse is twice the area of the inscribed circle, find the length of the semimajor axis of the ellipse.

59. If a projectile is launched with a horizontal velocity v_x and vertical velocity v_y, its position after t seconds is given by

$$x = v_x t$$
$$y = v_y t - (g/2)t^2$$

where g is the acceleration due to gravity. Eliminate t from this pair of equations to get $y = f(x)$, and show that this equation represents a parabola.

23

Binary, Hexadecimal, Octal, and BCD Numbers

••• OBJECTIVES •••

When you have completed this chapter, you should be able to:

- Convert between binary and decimal numbers.
- Convert between decimal and binary fractions.
- Convert between binary and hexadecimal numbers.
- Convert between decimal and hexadecimal numbers.
- Convert between binary and octal numbers.
- Convert between binary and 8421 BCD numbers.

••

The decimal number system of Chapter 1 is fine for calculations done by humans, but it is not the easiest system for a computer to use. A digital computer contains elements that can be in either of two states: on or off, magnetized or not magnetized, and so on. For such devices, calculations are most conveniently done using *binary numbers*. In this chapter we learn what binary numbers are and how to convert between binary and decimal numbers.

Binary numbers are useful in a computer, where each binary digit *(bit)* can be represented by one state of a "binary switch" that is either on or off. However, binary numbers are hard to read, partly because of their great length. To represent a nine-digit Social Security number, for example, requires a binary number *29 bits long.* So, in addition to binary numbers, we also study other ways in which numbers can be represented. *Hexadecimal, octal,* and *binary-coded decimal* systems allow us to express binary numbers more compactly, and they make the transfer of data between computers and people much easier. Here we learn how to convert numbers between each of these systems, and between decimal and binary as well.

23–1   The Binary Number System

Binary Numbers

A *binary number* is a sequence of the digits 0 and 1, such as

$$1101001$$

The number shown has no fractional part and so is called a *binary integer.* A binary number having a fractional part contains a *binary point* (also called a *radix point*), as in the number

$$1001.01$$

Base or Radix

The *base* of a number system (also called the *radix*) is equal to the number of digits used in the system.

◆◆◆ Example 1:
(a) The decimal system uses the ten digits

$$0 \quad 1 \quad 2 \quad 3 \quad 4 \quad 5 \quad 6 \quad 7 \quad 8 \quad 9$$

and has a base of 10.
(b) The binary system uses two digits

$$0 \quad 1$$

and has a base of 2.   ◆◆◆

Bits, Bytes, and Words

Each of the digits is called a *bit,* from *binary digit.* A *byte* is a group of 8 bits, and a *word* is the largest string of bits that a computer can handle in one operation. The number of bits in a word is called the *word length.* Different computers have different word lengths, with 8, 16, or 32 bits being common for desktop or personal computers. The longer words are often broken down into bytes for easier handling. Half a byte (4 bits) is called a *nibble.*

A *kilobyte* (Kbyte or KB) is 1024 (2^{10}) bytes, and a *megabyte* (Mbyte or MB) is 1,048,575 (2^{20}) bytes.

Note that this is different from the usual meaning of these prefixes, where *kilo* means 1000 and *mega* means 1,000,000.

Writing Binary Numbers

A binary number is sometimes written with a subscript 2 when there is a chance that the binary number would otherwise be mistaken for a decimal number.

◆◆◆ **Example 2:** The binary number 110 could easily be mistaken for the decimal number 110, unless we write it

$$110_2$$

Similarly, a decimal number that may be mistaken for binary is often written with a subscript 10, as in

$$101_{10}$$ ◆◆◆

Long binary numbers are sometimes written with their bits in groups of four for easier reading.

◆◆◆ **Example 3:** The number 100100010100.001001 is easier to read when written as

$$1001 \quad 0001 \quad 0100.0010 \quad 01 \qquad ◆◆◆$$

The leftmost bit in a binary number is called the *high-order* or *most significant bit (MSB)*. The bit at the extreme right of the number is the *low-order* or *least significant bit (LSB)*.

Place Value

As noted in Chapter 1, a positional number system is one in which the position of a digit determines its value, and each position in a number has a *place value* equal to the base of the number system raised to the position number.

The place values in the binary number system are

$$\ldots 2^4 \quad 2^3 \quad 2^2 \quad 2^1 \quad 2^0 \quad . \quad 2^{-1} \quad 2^{-2} \ldots$$

or

$$\ldots 16 \quad 8 \quad 4 \quad 2 \quad 1 \quad . \quad \tfrac{1}{2} \quad \tfrac{1}{4} \quad \ldots$$

binary point $\longrightarrow$

A more complete list of place values for a binary number is given in Table 23–1.

Expanded Notation

Thus the value of any digit in a number is the product of that digit and the place value. The value of the entire number is then the sum of these products.

◆◆◆ **Example 4:**
(a) The decimal number 526 can be expressed as

$$5 \times 10^2 + 2 \times 10^1 + 6 \times 10^0$$

or

$$5 \times 100 + 2 \times 10 + 6 \times 1$$

or

$$500 \quad + \quad 20 \quad + \quad 6$$

(b) The binary number 1011 can be expressed as

$$1 \times 2^3 + 0 \times 2^2 + 1 \times 2^1 + 1 \times 2^0$$

or

$$1 \times 8 + 0 \times 4 + 1 \times 2 + 1 \times 1$$

or

$$8 \quad + \quad 0 \quad + \quad 2 \quad + \quad 1$$

Numbers written in this way are said to be in *expanded notation*. ◆◆◆

Converting Binary Numbers to Decimal

To convert a binary number to decimal, simply write the binary number in expanded notation (omitting those where the bit is 0), and add the resulting values.

◆◆◆ **Example 5:** Convert the binary number

$$1001.011$$

to decimal.

TABLE 23–1

Position	Place Value
10	$2^{10} = 1024$
9	$2^9 = 512$
8	$2^8 = 256$
7	$2^7 = 128$
6	$2^6 = 64$
5	$2^5 = 32$
4	$2^4 = 16$
3	$2^3 = 8$
2	$2^2 = 4$
1	$2^1 = 2$
0	$2^0 = 1$
-1	$2^{-1} = 0.5$
-2	$2^{-2} = 0.25$
-3	$2^{-3} = 0.125$
-4	$2^{-4} = 0.0625$
-5	$2^{-5} = 0.03125$
-6	$2^{-6} = 0.015625$

Compare these place values to those in the decimal system shown in Fig. 1–1.

Solution: In expanded notation,

$$1001.011 = 1 \times 8 + 1 \times 1 + 1 \times \frac{1}{4} + 1 \times \frac{1}{8}$$

$$= 8 + 1 + \frac{1}{4} + \frac{1}{8}$$

$$= 9\frac{3}{8}$$

$$= 9.375 \qquad \qquad \text{◆◆◆}$$

Largest Decimal Number Obtainable with n Bits

The largest possible 3-bit binary number is

$$1\ 1\ 1 = 7$$

or

$$2^3 - 1$$

The largest possible 4-bit number is

$$1\ 1\ 1 = 15$$

or

$$2^4 - 1$$

Similarly, the largest n-bit binary number is given by the following:

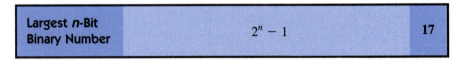

Largest n-Bit Binary Number	$2^n - 1$	17

◆◆◆ **Example 6:** If a computer stores numbers with 15 bits, what is the largest decimal number that can be represented?

Solution: The largest decimal number is

$$2^{15} - 1 = 32{,}767 \qquad \qquad \text{◆◆◆}$$

Significant Digits

In Example 6 we saw that a 15-bit binary number could represent a decimal number no greater than 32,767. Thus it took 15 binary digits to represent 5 decimal digits. As a rule of thumb, *it takes about 3 bits for each decimal digit.*

Stated another way, if we have a computer that stores numbers with 15 bits, we should assume that the decimal numbers that it prints do not contain more than 5 significant digits.

◆◆◆ **Example 7:** If we want a computer to print decimal numbers containing 7 significant digits, how many bits must it use to store those numbers?

Solution: Using our rule of thumb, we need

$$3 \times 7 = 21 \text{ bits} \qquad \qquad \text{◆◆◆}$$

Converting Decimal Integers to Binary

To convert a decimal *integer* to binary, we first divide it by 2, obtaining a quotient and a remainder. We write down the remainder and divide the quotient by 2, getting a new quotient and remainder. We then repeat this process until the quotient is zero.

◆◆◆ **Example 8:** Convert the decimal integer 59 to binary.

Solution: We divide 59 by 2, getting a quotient of 29 and a remainder of 1. Then dividing 29 by 2 gives a quotient of 14 and a remainder of 1. These calculations, and those that follow, can be arranged in a table, as follows:

Remainder

$$
\begin{array}{rcl}
2\overline{)59} & & \\
2\overline{)29} & 1 & \text{LSB} \\
2\overline{)14} & 1 & \\
2\overline{)7} & 0 & \text{Read} \\
2\overline{)3} & 1 & \text{up} \\
2\overline{)1} & 1 & \\
2\overline{)0} & 1 & \text{MSB}
\end{array}
$$

Our binary number then consists of the digits in the remainders, with those at the *top* of the column appearing to the *right* in the binary number. Thus

$$59_{10} = 111011_2$$

◆◆◆

The conversion can, of course, be checked by converting back to decimal.

To convert a decimal *fraction* to binary, we first multiply it by 2, remove the integer part of the product, and multiply by 2 again. We then repeat the procedure.

◆◆◆ **Example 9:** Convert the decimal fraction 0.546875 to binary.

Solution: We multiply the given number by 2, getting 1.09375. We remove the integer part, 1, leaving 0.09375, which we again multiply by 2, getting 0.1875. We repeat the computation until we get a product that has a fractional part of zero, as in the following table:

Integer Part

$$
\begin{array}{rcl}
0.546875 & & \\
\times \quad 2 & & \\
\hline
\mathbf{1}.09375 & 1 & \text{MSB} \\
\times \quad 2 & & \\
\hline
\mathbf{0}.1875 & 0 & \\
\times \quad 2 & & \\
\hline
\mathbf{0}.375 & 0 & \\
\times \quad 2 & & \text{Read} \\
\hline
\mathbf{0}.75 & 0 & \text{down} \\
\times \quad 2 & & \\
\hline
\mathbf{1}.50 & 1 & \\
\times \quad 2 & & \\
\hline
\mathbf{1}.00 & 1 & \text{LSB}
\end{array}
$$

Notice that we remove the integral part, shown in boldface, before multiplying by 2.

We stop now that the fractional part of the product (1.00) is zero. The column containing the integer parts is now our binary number, with the digits at the top appearing at the *left* of the binary number. So

$$0.546875_{10} = 0.1000\ 11_2$$

◆◆◆

Note that this is the reverse of what we did when converting a decimal integer to binary.

In Example 9 we were able to find an exact binary equivalent of a decimal fraction. This is not always possible, as shown in the following example.

◆◆◆ **Example 10:** Convert the decimal fraction 0.743 to binary.

Solution: We follow the same procedure as before and get the following values.

$$
\begin{array}{rc}
\textbf{Integral Part} & \\
0.743 & \\
\times\quad 2 & \\
\hline
\mathbf{1}.486 & 1 \\
\times\quad 2 & \\
\hline
\mathbf{0}.972 & 0 \\
\times\quad 2 & \\
\hline
\mathbf{1}.944 & 1 \\
\times\quad 2 & \\
\hline
\mathbf{1}.888 & 1 \\
\times\quad 2 & \\
\hline
\mathbf{1}.776 & 1 \\
\times\quad 2 & \\
\hline
\mathbf{1}.552 & 1 \\
\times\quad 2 & \\
\hline
\mathbf{1}.104 & 1 \\
\times\quad 2 & \\
\hline
\mathbf{0}.208 & 0 \\
\times\quad 2 & \\
\hline
\mathbf{0}.416 & 0 \\
\end{array}
$$

It is becoming clear that this computation can continue indefinitely, demonstrating that *not all decimal fractions can be exactly converted to binary*. This inability to make an exact conversion is an unavoidable source of inaccuracy in some computations. To decide how far to carry out computation, we use the rule of thumb that each decimal digit requires about 3 binary bits to give the same accuracy. Thus our original 3-digit number requires about 9 bits, which we already have. The result of our conversion is then

$$0.743_{10} \approx 0.1011\ 1110\ 0_2 \qquad\qquad ◆◆◆$$

To convert a decimal number having both an integer part and a fractional part to binary, convert each part separately as shown above, and combine.

◆◆◆ **Example 11:** Convert the number 59.546875 to binary.

Solution: From the preceding examples,

$$59 = 11\ 1011$$

and

$$0.546875 = 0.1000\ 11$$

So

$$59.546875 = 11\ 1011.1000\ 11 \qquad\qquad ◆◆◆$$

Converting Binary Fractions to Decimal

To convert a binary fraction to decimal, we use Table 23–1 to find the decimal equivalent of each binary bit located to the right of the binary point, and add.

◆◆◆ **Example 12:** Convert the binary fraction 0.101 to decimal.

Solution: From Table 23–1,

$$0.1 \quad = 2^{-1} = 0.5$$
$$0.001 = 2^{-3} = \underline{0.125}$$
$$\text{Add:} \quad 0.101 \qquad = 0.625 \qquad \text{◆◆◆}$$

The procedure is no different when converting a binary number that has both a whole and a fractional part.

◆◆◆ **Example 13:** Convert the binary number 10.01 to decimal.

Solution: From Table 23–1,

$$10 \quad = 2$$
$$\underline{0.01 = 0.25}$$
$$\text{Add:} \quad 10.01 = 2.25 \qquad \text{◆◆◆}$$

Exercise 1 ◆ The Binary Number System

Conversion between Binary and Decimal

Convert each binary number to decimal.

1. 10	**2.** 01	**3.** 11
4. 1001	**5.** 0110	**6.** 1111
7. 1101	**8.** 0111	**9.** 1100
10. 1011	**11.** 0101	**12.** 1101 0010
13. 0110 0111	**14.** 1001 0011	**15.** 0111 0111
16. 1001 1010 1000 0010		

Convert each decimal number to binary.

17. 5	**18.** 9
19. 2	**20.** 7
21. 72	**22.** 28
23. 93	**24.** 17
25. 274	**26.** 937
27. 118	**28.** 267
29. 8375	**30.** 2885
31. 82740	**32.** 72649

Convert each decimal fraction to binary. Retain 8 bits to the right of the binary point when the conversion is not exact.

33. 0.5	**34.** 0.25	**35.** 0.75
36. 0.375	**37.** 0.3	**38.** 0.8
39. 0.55	**40.** 0.35	**41.** 0.875
42. 0.4375	**43.** 0.3872	**44.** 0.8462

Convert each binary fraction to decimal.

45. 0.1	**46.** 0.11	**47.** 0.01
48. 0.011	**49.** 0.1001	**50.** 0.0111

Convert each decimal number to binary. Keep 8 bits to the right of the binary point when the conversion is not exact.

51. 5.5	**52.** 2.75
53. 4.375	**54.** 29.381
55. 948.472	**56.** 2847.22853

Convert each binary number to decimal.

57. 1.1   **58.** 10.11
59. 10.01   **60.** 101.011
61. 1 1001.0110 1   **62.** 10 1001.1001 101

Calculator

63. Some calculators can convert directly between binary and decimal numbers. Some will do this only for integers. Check your calculator manual to see if you can perform these operations, and if you have this capability, use it to solve any of the problems in this exercise set.

23–2   The Hexadecimal Number System

Hexadecimal Numbers

Hexadecimal numbers (or *hex* for short) are obtained by grouping the bits in a binary number into sets of four and representing each such set by a single number or letter. A hex number one-fourth the length of the binary number is thus obtained.

Base 16

Since a 4-bit group of binary digits can have a value between 0 and 15, we need 16 symbols to represent all of these values. The base of hexadecimal numbers is thus 16. We use the digits from 0 to 9 and the capital letters A to F, as shown in Table 23–2.

TABLE 23–2   Hexadecimal and octal number systems.

Decimal	Binary	Hexadecimal	Octal
0	0000	0	0
1	0001	1	1
2	0010	2	2
3	0011	3	3
4	0100	4	4
5	0101	5	5
6	0110	6	6
7	0111	7	7
8	1000	8	10
9	1001	9	11
10	1010	A	12
11	1011	B	13
12	1100	C	14
13	1101	D	15
14	1110	E	16
15	1111	F	17

Table 23–2 also shows octal numbers, which we will use later.

Converting Binary to Hexadecimal

To convert binary to hexadecimal, group the bits into sets of four starting at the binary point, adding zeros as needed to fill out the groups. Then assign to each group the appropriate letter or number from Table 23–2.

◆◆◆ **Example 14:** Convert 10110100111001 to hexadecimal.

Solution: Grouping, we get

$$10\ 1101\ 0011\ 1001$$

or

$$0010 \ 1101 \ 0011 \ 1001$$

From Table 23–2, we obtain

$$2 \quad D \quad 3 \quad 9$$

So the hexadecimal equivalent is 2D39. Hexadecimal numbers are sometimes written with the subscript 16, as in

$$2D39_{16} \qquad \qquad \text{◆◆◆}$$

The procedure is no different for binary fractions.

◆◆◆ **Example 15:** Convert 101111.0011111 to hexadecimal.

Solution: Grouping from the binary point gives us

$$10 \ 1111 \ . \ 0011 \ 111$$

or

$$0010 \ 1111 \ . \ 0011 \ 1110$$

From Table 23–2, we have

$$2 \quad F \quad . \quad 3 \quad E$$

or 2F.3E. ◆◆◆

Converting Hexadecimal to Binary

To convert hexadecimal to binary, we simply reverse the procedure.

◆◆◆ **Example 16:** Convert 3B25.E to binary.

Solution: We write the group of 4 bits corresponding to each hexadecimal symbol.

$$3 \quad B \quad 2 \quad 5 \quad . \quad E$$
$$0011 \ 1011 \ 0010 \ 0101 \ . \ 1110$$

or 11 1011 0010 0101.111. ◆◆◆

Converting Hexadecimal to Decimal

As with decimal and binary numbers, each hex digit has a *place value,* equal to the base, 16, raised to the position number. Thus the place value of the first position to the left of the decimal point is

$$16^0 = 1$$

and the next is

$$16^1 = 16$$

and so on.

To convert from hex to decimal, first replace each letter in the hex number by its decimal equivalent. Write the number in expanded notation, multiplying each hex digit by its place value. Add the resulting numbers.

◆◆◆ **Example 17:** Convert the hex number 3B.F to decimal.

Solution: We replace the hex B with 11, and the hex F with 15, and write the number in expanded form.

$$3 \qquad B \qquad . \qquad F$$
$$3 \qquad 11 \qquad . \qquad 15$$
$$(3 \times 16^1) + (11 \times 16^0) + (15 \times 16^{-1})$$
$$= \quad 48 \quad + \quad 11 \quad + \quad 0.9375$$
$$= 59.9375 \qquad \qquad \text{◆◆◆}$$

Instead of converting directly between decimal and hex, many find it easier to first convert to binary.

Converting Decimal to Hexadecimal

To convert decimal to hexadecimal, we repeatedly divide the given decimal number by 16 and convert each remainder to hex. The remainders form our hex number, the last remainder obtained being the most significant digit.

◆◆◆ **Example 18:** Convert the decimal number 83759 to hex.

Solution:

$$
\begin{array}{rll}
83759 \div 16 = 5234 & \text{with remainder of} & 15 \\
5234 \div 16 = 327 & \text{with remainder of} & 2 \\
327 \div 16 = 20 & \text{with remainder of} & 7 \\
20 \div 16 = 1 & \text{with remainder of} & 4 \\
1 \div 16 = 0 & \text{with remainder of} & 1
\end{array}
$$

Read up

Changing the number 15 to hex F and reading the remainders from bottom up, we find that our hex equivalent is 1472F.   ◆◆◆

Exercise 2 ◆ The Hexadecimal Number System

Binary-Hex Conversions

Convert each binary number to hexadecimal.

1. 1101	**2.** 1010
3. 1001	**4.** 1111
5. 1001 0011	**6.** 0110 0111
7. 1101 1000	**8.** 0101 1100
9. 1001 0010 1010 0110	**10.** 0101 1101 0111 0001
11. 1001.0011	**12.** 10.0011
13. 1.0011 1	**14.** 101.101

Convert each hex number to binary.

15. 6F	**16.** B2
17. 4A	**18.** CC
19. 2F35	**20.** D213
21. 47A2	**22.** ABCD
23. 5.F	**24.** A4.E
25. 9.AA	**26.** 6D.7C

Decimal-Hex Conversions

Convert each hex number to decimal.

27. F2	**28.** 5C
29. 33	**30.** DF
31. 37A4	**32.** A3F6
33. F274	**34.** C721
35. 3.F	**36.** 22.D
37. ABC.DE	**38.** C.284

Convert each decimal number to hex.

39. 39	**40.** 13
41. 921	**42.** 554
43. 2741	**44.** 9945
45. 1736	**46.** 2267

Calculator

47. Some calculators can convert directly between hexadecimal, decimal, and binary numbers. Some will do this only for integers. Check your calculator manual to see if you can perform these operations, and if you have this capability, use it to solve any of the problems in this exercise set.

23–3   The Octal Number System

The Octal System

The *octal* number system uses eight digits, 0 to 7, and hence has a base of eight. A comparison of the decimal, binary, hex, and octal digits is given in Table 23–2.

Binary-Octal Conversions

To convert from binary to octal, write the bits of the binary number in groups of three, starting at the binary point. Then write the octal equivalent for each group.

◆◆◆ **Example 19:** Convert the binary number 1 1010 0011 0110 to octal.

Solution: Grouping the bits in sets of three from the binary point, we obtain

$$001 \ 101 \ 000 \ 110 \ 110$$

and the octal equivalents,

$$1 \quad 5 \quad 0 \quad 6 \quad 6$$

so the octal equivalent of 1 1010 0011 0110 is 15066.   ◆◆◆

To convert from octal to binary, simply reverse the procedure shown in Example 19.

◆◆◆ **Example 20:** Convert the octal number 7364 to binary.

Solution: We write the group of 3 bits corresponding to each octal digit.

$$7 \quad 3 \quad 6 \quad 4$$
$$111 \ 011 \ 110 \ 100$$

or 1110 1111 0100.   ◆◆◆

Exercise 3 ◆ The Octal Number System

Convert each binary number to octal.

1. 110	**2.** 010	**3.** 111
4. 101	**5.** 11 0011	**6.** 01 1010
7. 0110 1101	**8.** 1 1011 0100	**9.** 100 1001
10. 1100 1001		

Convert each octal number to binary.

11. 26	**12.** 35	**13.** 623
14. 621	**15.** 5243	**16.** 1153
17. 63150	**18.** 2346	

Calculator

19. Some calculators can convert directly between octal, binary, and decimal numbers. Some will do this only for integers. Check your calculator manual to see

if you can perform these operations, and if you have this capability, use it to solve any of the problems in this exercise set.

23–4   BCD Codes

In Sec. 23–1, we converted decimal numbers into binary. Recall that each bit had a place value equal to 2 raised to the position number. Thus the binary number 10001 is equal to

$$2^4 + 2^0 = 16 + 1 = 17$$

We shall now refer to numbers such as 10001 as *straight binary*.

With a BCD or *binary-coded decimal* code, a decimal number is not converted *as a whole* to binary, but rather *digit by digit*. For example, the 1 in the decimal number 17 is 0001 in 4-bit binary, and the 7 is equal to 0111. Thus

$$17 = 10001 \quad \text{in straight binary}$$

and

$$17 = 0001\ 0111 \quad \text{in BCD}$$

When we converted the digits in the decimal number 17, we used the same 4-bit binary equivalents as in Table 23–2. The bits have the place values 8, 4, 2, and 1, and BCD numbers written in this manner are said to be in *8421 code*. The 8421 code and its decimal equivalents are listed in Table 23–3.

TABLE 23–3 Binary-coded-decimal numbers.

Decimal	8421	2421	5211
0	0000	0000	0000
1	0001	0001	0001
2	0010	0010	0011
3	0011	0011	0101
4	0100	0100	0111
5	0101	1011	1000
6	0110	1100	1010
7	0111	1101	1100
8	1000	1110	1110
9	1001	1111	1111

Converting Decimal Numbers to BDC

To convert decimal numbers to BCD, simply convert each decimal digit to its equivalent 4-bit code.

◆◆◆ **Example 21:** Convert the decimal number 25.3 to 8421 BCD code.

Solution: We find the BCD equivalent of each decimal digit from Table 23–3.

$$2 \quad\quad 5 \quad . \quad 3$$
$$0010\ 0101\ .\ 0011$$

or 0010 0101.0011.   ◆◆◆

Converting from BCD to Decimal

To convert from BCD to decimal, separate the BCD number into 4-bit groups, and write the decimal equivalent of each group.

◆◆◆ **Example 22:** Convert the 8421 BCD number 1 0011.0101 to decimal.

Solution: We have

$$0001\ 0011\ .\ 0101$$

From Table 23–3, we have

$$1 \quad\quad 3 \quad . \quad 5$$

or 13.5.   ◆◆◆

Other BCD Codes

Other BCD codes are in use which represent each decimal digit by a 6-bit binary number or an 8-bit binary number, and other 4-bit codes in which the place values are other than 8421. Some of these are shown in Table 23–3. Each is used in the same way as shown for the 8421 code.

••• **Example 23:** Convert the decimal number 825 to 2421 BCD code.

Solution: We convert each decimal digit separately,

$$8 \quad 2 \quad 5$$
$$1110 \ 0010 \ 1011 \qquad\qquad ﹙•••﹚$$

Other Computer Codes

In addition to the binary, hexadecimal, octal, and BCD codes we have discussed, many other codes are used in the computer industry. Numbers can be represented by the *excess-3 code* or the *Gray code.* There are error-detecting and parity-checking codes. Other codes are used to represent letters of the alphabet and special symbols, as well as numbers. Some of these are *Morse code,* the American Standard Code for Information Interchange or *ASCII* (pronounced "as-key") *code,* the Extended Binary-Coded-Decimal Interchange Code or *EBCDIC* (pronounced "eb-si-dik") *code,* and the *Hollerith code* used for punched cards.

Space does not permit discussion of each code. However, if you have understood the manipulation of the codes we have described, you should have no trouble when faced with a new one.

Exercise 4 ◆ BCD Codes

Convert each decimal number to 8421 BCD.

1. 62	**2.** 25	**3.** 274
4. 284	**5.** 42.91	**6.** 5.014

Convert each 8421 BCD number to decimal.

7. 1001	**8.** 101	**9.** 110 0001
10. 100 0111	**11.** 11 0110.1000	**12.** 11 1000.1000 1

••• CHAPTER 23 REVIEW PROBLEMS •••••••••••••••••••••••••••••••••••

1. Convert the binary number 110.001 to decimal.
2. Convert the binary number 1100.0111 to hex.
3. Convert the hex number 2D to BCD.
4. Convert the binary number 1100 1010 to decimal.
5. Convert the hex number 2B4 to octal.
6. Convert the binary number 110 0111.011 to octal.
7. Convert the decimal number 26.875 to binary.
8. Convert the hexadecimal number 5E.A3 to binary.
9. Convert the octal number 534 to decimal.
10. Convert the BCD number 1001 to hex.
11. Convert the octal number 276 to hex.
12. Convert the 8421 BCD number 1001 0100 0011 to decimal.
13. Convert the decimal number 8362 to hex.
14. Convert the octal number 2453 to BCD.
15. Convert the BCD number 1001 0010 to octal.
16. Convert the decimal number 482 to octal.
17. Convert the octal number 47135 to binary.
18. Convert the hexadecimal number 3D2.A4F to decimal.

19. Convert the BCD number 0111 0100 to straight binary.

20. Convert the decimal number 382.4 to 8421 BCD.

Writing

21. A legislator, trying to trim the college budget, has suggested dropping the study of binary numbers. "Why do we need other numbers?" he asks. "Plain numbers were good enough when I went to school."

Write a letter to this legislator, explaining why, in this computer age, "plain" numbers are not enough. Include specific examples of the use of binary numbers. Be polite but forceful, and limit your letter to one page.

Inequalities and Linear Programming

When you have completed this chapter, you should be able to:

- Graph linear inequalities on the number line.
- Graph linear inequalities on the coordinate axes.
- Graph a system of linear inequalities.
- Solve conditional inequalities graphically and algebraically.
- Use inequalities to determine the range of a function or relation.
- Solve absolute value inequalities.
- Solve linear programming problems.

So far we have dealt only with equations, where one expression is equal to another, and our aim was to solve them—that is, to find any value of the variables that makes the equation true. In this chapter we work with *inequalities,* in which one expression is not necessarily equal to another, but may be greater than or less than the other expression. We will graph inequalities and also solve them. Here we seek a whole range of values of the variable to satisfy the inequality.

We then introduce *linear programming* (a term that should not be confused with *computer programming*), which plays an important part in certain business decisions. It helps us decide how to allocate resources in order to reduce costs and maximize profit—a concern of technical people as well as management. We then write a computer program to do linear programming.

24–1 Definitions

Inequalities

An *inequality* is a statement that one quantity is greater than or less than another quantity.

◆◆◆ **Example 1:** The statement $a > b$ means "a is greater than b."   ◆◆◆

Parts of an Inequality

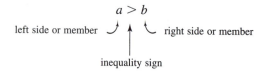

Inequality Symbols

In Sec. 1–1, we introduced the inequality symbols $>$ and $<$. These symbols can be combined with the equals sign to produce the compound symbols $\geq$ and $\leq$, which mean "greater than or equal to" and "less than or equal to." Further, a slash through any of these symbols indicates the negation of the inequality.

◆◆◆ **Example 2:** The statement $a \not> b$ means "a is not greater than b."   ◆◆◆

Sense of an Inequality

The *sense* of an inequality refers to the *direction* in which the inequality sign points. We use this term mostly when operating on an inequality, where certain operations will *change the sense* but others will not.

Conditional and Unconditional Inequalities

Just as we have conditional equations and identities, we also have conditional and unconditional inequalities. As with equations, a *conditional inequality* is one that is satisfied only by certain values of the unknown, whereas *unconditional equalities* are true for any values of the unknown.

◆◆◆ **Example 3:** The statement $x - 2 > 5$ is a conditional inequality because it is true only for values of x greater than 7.   ◆◆◆

◆◆◆ **Example 4:** The statement $x^2 + 5 > 0$ is an unconditional inequality (also called an *absolute* inequality) because for any value of x, positive, negative, or zero, x^2 cannot be negative. Thus $x^2 + 5$ is always greater than zero.   ◆◆◆

In this chapter we limit ourselves to the *real* numbers.

Linear and Nonlinear Inequalities

A *linear* inequality is defined in the same way as a linear equation, where all variables are of *first degree* or less. Other inequalities are called *nonlinear*.

◆◆◆ **Example 5:** The inequalities

$$x + 2 > 3x - 5$$

and

$$2x - 3y \geq 4$$

are linear, while the inequalities

$$x^2 - 3 < 2x$$

and

$$y \le x^3 + 2x - 3$$

are nonlinear. ◆◆◆

Inequalities with Three Members

The inequality $3 < x$ has two members, left and right. However, many inequalities have *three* members.

◆◆◆ **Example 6:** The statement $3 < x < 9$ means "*x* is greater than 3 and less than 9," or "*x* is a number between 3 and 9." The given inequality is really a combination of the two inequalities

$$x > 3$$

and

$$x < 9$$

An inequality such as the one in Example 6 is sometimes called a *double inequality*. ◆◆◆

Common Error

It is incorrect to write the two inequalities

$$x < 3 \quad \text{and} \quad x > 9$$

as

$$3 > x > 9$$

because there is no number that is less than 3 and also greater than 9.

◆◆◆ **Example 7:** Combine the two inequalities

$$x > -6 \quad \text{and} \quad x \le 22$$

into a double inequality.

Solution: We write

$$-6 < x \le 22$$ ◆◆◆

Intervals

The *interval* of an inequality is the set of all numbers between the left member and the right member, called the *endpoints* of the interval. The interval may include one or both endpoints. An interval is called *open* or *closed* depending on whether or not the endpoint is included.

An interval is often denoted by enclosing the endpoints in parentheses or brackets. We use parentheses for an open interval and brackets for a closed interval, or a combination of the two.

◆◆◆ **Example 8:** Some inequalities and their notations are as follows:

Inequality	Notation	Open or Closed
(a) $2 \leq x \leq 3$	[2, 3]	Closed
(b) $4 < x < 5$	(4, 5)	Open
(c) $6 < x \leq 7$	(6, 7]	Open on left, closed on right

◆◆◆

Graphing Intervals

An inequality can be represented on the number line by graphing the interval. An open endpoint is usually shown as an open circle, while a closed endpoint is shown solid.

◆◆◆ **Example 9:** The three inequalities of Example 8 are graphed in Fig. 24–1.

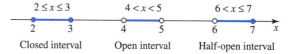

FIGURE 24–1   ◆◆◆

An interval is called *unbounded* if it has only one endpoint and extends indefinitely in one direction.

◆◆◆ **Example 10:**

(a) The inequality $x < 1$ is shown graphically in Fig. 24–2. It is said to be *open on the right and unbounded on the left*. The interval notation for the inequality $x < 1$ is $(-\infty, 1)$.

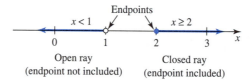

FIGURE 24–2

(b) The inequality $x \geq 2$ is also shown in Fig. 24–2. It is said to be *closed on the left and unbounded on the right*. The interval notation for this inequality is $[2, \infty)$.

◆◆◆

Graphing Inequalities Having Two Variables

An inequality having *two variables* can be represented by a *region* in the *xy* plane. To graph an inequality, we first graph the corresponding equality with a dashed curve. Then, if the equality is included in the graph, we replace the dashed curve by a solid one. If the points above (or below) the curve satisfy the inequality, we shade the region above (or below) the curve.

◆◆◆ **Example 11:** Graph the inequality $y > x + 1$.

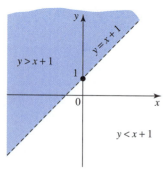

FIGURE 24–3 Points within the shaded region satisfy the inequality $y > x + 1$.

Solution: We graph the line $y = x + 1$ as shown in Fig. 24–3 using a solid line if points on the line satisfy the given inequality, or a dashed line if they do not. Here, we need a dashed line. All points in the shaded region above the dashed line satisfy the given equality. For example, if we substitute the values for the point (1, 3) into our inequality, we get

$$3 > 1 + 1$$

which is true. So (1, 3) will be in the region that satisfies the inequality $y > x + 1$. This trial point helps us choose the region to shade (here, the region above the line) and serves as a check for our work. ◆◆◆

◆◆◆ **Example 12:** A certain automobile company can make a sedan in 3 h and a hatchback in 2 h. How many of each type of car can be made if the working time is not to exceed 12 h?

Solution: We let

$$x = \text{number of sedans made in 12 h}$$

and

$$y = \text{number of hatchbacks made in 12 h}$$

We look at the time it takes to make each vehicle. Since each sedan takes 3 h to make, it will take $3x$ hours to make x sedans. Similarly, it takes $2y$ hours for the hatchbacks. So we get the inequality which states that the sum of the times is less than or equal to 12 h.

$$3x + 2y \leq 12$$

We plot the equation $3x + 2y = 12$ in Fig. 24–4. Any point on the line or in the shaded region satisfies the inequality. For example, the company could make two sedans and three hatchbacks in the 12-h period, or it could make no cars at all, or it could make four sedans and no hatchbacks, and so on. It could not, however, make three sedans and three hatchbacks in the allotted 12 h. ◆◆◆

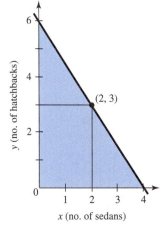

FIGURE 24–4

Sometimes we want to graph two inequalities on the same axes. This is a way to solve two inequalities simultaneously.

◆◆◆ **Example 13:** Figure 24–5 shows a graph of the two inequalities

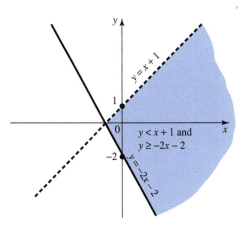

FIGURE 24–5

$$y < x + 1 \quad \text{and} \quad y \geq -2x - 2$$

Points within the shaded region simultaneously satisfy both inequalities. ◆◆◆

Exercise 1 ◆ Definitions

Kinds of Inequalities

Label each inequality as conditional, unconditional, linear, or nonlinear.

1. $x^2 > -4$

2. $x + 2 < 3 - 4x$

3. $3x^2 - 2x \leq 6$

4. $3x^4 + 5 > 0$

5. $3x + 3 \geq x - 7$

6. $x^2 - 3x + 2 > 7$

Inequalities with Three Members

Rewrite each double inequality as two separate inequalities.

7. $5 < x < 9$

8. $-2 \leq x < 7$

9. $-11 < x \leq 1$

10. $22 < x \leq 53$

Rewrite each pair of inequalities as a double inequality.

11. $x > 5, x < 8$

12. $x > -4, x < 1$

13. $x \geq -1, x < 24$

14. $x > 12, x \leq 13$

Intervals

Graph each inequality on the number line.

15. $x > 5$

16. $2 < x < 5$

17. $x < 3$

18. $-1 \leq x \leq 3$

19. $x \geq -2$

20. $4 \leq x < 7$

21. $x \leq 1$

22. $-4 < x \leq 0$

Write the inequality corresponding to each given interval, and graph.

23. $(-\infty, 8)$

24. $(-\infty, 5]$

25. $(-3, \infty)$

26. $[10, \infty)$

27. $(3, 12)$

28. $[-2, 7]$

29. $(4, 22]$

30. $[-11, 5)$

Graphing Inequalities Having Two Variables

Graph each inequality.

31. $y < 3x - 2$

32. $x + y \geq 1$

33. $2x - y < 3$

34. $y > -x^2 + 4$

Graph the region in which both inequalities are satisfied.

35. $y > -x + 3$
$y < 2x - 2$

36. $y < x^2 - 2$
$y \geq x$

37. $y > x^2 - 4$
$y < 4 - x^2$

38. The supply voltage V to a certain device is required to be over 100 V, but not over 150 V. Express this as an inequality.

39. A manufacturer makes skis and snowshoes. A pair of skis takes 2.5 h to make, and a pair of snowshoes takes 3.1 h. Write and graph an inequality to represent the number of pairs of skis and snowshoes that can be made in 8 h.

Calculator

40. Some graphics calculators can graph inequalities and can even shade the region in which the inequalities are satisfied. Check your manual, and if you have this capability, use it to graph any of the inequalities in this exercise set.

24–2   Solving Inequalities

Unlike the solution to a conditional equation, which is one or more discrete values of the variable, the solution to a conditional inequality will usually be an infinite set of values. We first show graphical methods for finding these values, and then we give an algebraic method. For each method we will first solve linear inequalities and then nonlinear inequalities.

Graphical Solution of Inequalities

We will show two graphical methods. The first method is to graph both sides of the inequality and then identify the intervals for which the inequality is true. We can make our graphs by a point-to-point plot or with a graphics calculator or graphing utility on a computer.

Our first example will be the solution of a linear inequality.

♦♦♦ **Example 14:** Graphically solve the linear inequality

$$3x - 4 > x + 2$$

Solution: On the same axes, we graph

$$y = 3x - 4 \quad \text{and} \quad y = x + 2$$

as shown in Fig. 24–6. The interval for which $3x - 4$ is greater than $x + 2$ is clearly

$$x > 3$$

or

$$(3, \infty)$$

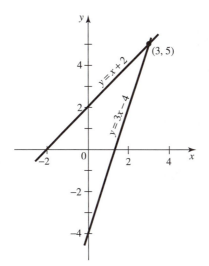

FIGURE 24–6   ♦♦♦

Our second graphical method is to move all terms to one side of the inequality and to graph the resulting expression. We then identify the intervals for which this inequality is true. We will illustrate this method by solving a nonlinear inequality.

◆◆◆ **Example 15:** Solve the nonlinear inequality

$$x^2 - 3 > 2x$$

Solution: Rearranging gives

$$x^2 - 2x - 3 > 0$$

We now plot the function

$$y = x^2 - 2x - 3$$

as shown in Fig. 24–7.

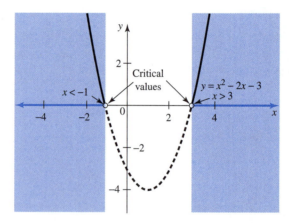

FIGURE 24–7   The points $x = -1$ and $x = 3$ where the curve crosses the x axis are called *critical values.* $x < -1$ and $x > 3$ are solutions because those x values satisfy the inequality.

Note that our solutions to the inequality are *all values of x* that make the inequality true, that is, those values of x that here make y positive. The curve (a parabola) is above the x axis when $x < -1$ and $x > 3$. These values of x are the solutions to our inequality, because, for these values, $x^2 - 2x - 3 > 0$. The intervals containing the solutions are thus

$$(-\infty, -1) \quad \text{and} \quad (3, \infty) \qquad ◆◆◆$$

Checking a Solution

We can gain confidence in our solution by substituting values of x into the inequality. Those values within the range of our solution should check, while all others should not.

◆◆◆ **Example 16:** For the inequality of Example 15, the values of, say, $x = 4$ and $x = -2$ should check, while $x = 1$ should not. Substituting $x = 4$ into our original inequality gives

$$4^2 - 3 > 2(4)$$

or

$$16 - 3 > 8$$

which is true. Similarly, the value $x = -2$ gives

$$(-2)^2 - 3 > 2(-2)$$

or

$$4 - 3 > -4$$

which is also true. On the other hand, a value between -1 and 3, such as $x = 1$, gives

$$1^2 - 3 > 2(1)$$

or

$$-2 \ngtr 2$$

which is not true, as expected. ◆◆◆

Algebraic Solution

The object of an *algebraic solution* to a conditional inequality is to locate the *critical values* (the points such as in Fig. 24–7 where the curve crosses the x axis). To do this, it is usually a good idea to transpose all terms to one side of the inequality and simplify the expression as much as possible. We use the following principles for operations with inequalities:

1. *Addition or subtraction.* The sense of an inequality is not changed if the same quantity is added to or subtracted from both sides.
2. *Multiplication or division by a positive quantity.* The sense of an inequality is not changed if both sides are multiplied or divided by the same *positive* quantity.
3. *Multiplication or division by a negative quantity.* The sense of an inequality *is reversed* if both sides are multiplied or divided by the same *negative* quantity.

As we did with the graphical solution, we will first solve a linear inequality and then some nonlinear inequalities.

◆◆◆ **Example 17:** Solve the linear inequality $21x - 23 < 2x + 15$.

Solution: Rearranging gives us

$$19x < 38$$

Dividing by 19, we get

$$x < 2$$

or $(-\infty, 2)$, as shown in Fig. 24–8. ◆◆◆

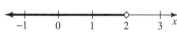

FIGURE 24–8

When solving a nonlinear inequality, we usually get more than one critical value, which thus divide the x axis into more than two intervals. To decide which of these intervals contain the correct values, we test each for algebraic sign, as shown in the following example.

◆◆◆ **Example 18:** Solve the nonlinear inequality $2x - \dfrac{x^2}{2} > -\dfrac{21}{2}$.

Solution: Multiplying by 2, we have

$$4x - x^2 > -21$$

Rearranging yields

$$4x - x^2 + 21 > 0$$

or

$$-x^2 + 4x + 21 > 0$$

Multiplying by -1, which reverses the sense of the inequality, we obtain

$$x^2 - 4x - 21 < 0$$

Factoring gives us

$$(x - 7)(x + 3) < 0$$

The critical values are those that make the left side equal zero, or $x = 7$ and $x = -3$. These divide the x axis into three intervals.

$$x < -3 \qquad -3 < x < 7 \qquad x > 7$$

Which of these is correct? We can decide by computing the sign of $(x - 7)(x + 3)$ in each interval. Since this product is supposed to be less than zero, we choose the interval(s) in which the product is negative. Thus the signs are as follows:

Interval	$(x - 7)$	$(x + 3)$	$(x - 7)(x + 3)$
$x < -3$	$-$	$-$	$+$
$-3 < x < 7$	$-$	$+$	$-$
$x > 7$	$+$	$+$	$+$

The *sign pattern* for $(x - 7)(x + 3)$ is shown in Fig. 24–9. The signs are correct in only one interval, so our solution is

$$x > -3 \quad \text{and} \quad x < 7$$

FIGURE 24–9

or $(-3, 7)$ in interval notation. Note that this is the interval for which the graph of $y = x^2 - 4x - 21$ is *below* the x axis (Fig. 24–10).   ◆◆◆

Inequalities containing the $\geq$ or $\leq$ signs are solved the same way as in the preceding examples, except that now the endpoint of an interval may be included in the solution.

◆◆◆ **Example 19:** Solve the nonlinear inequality $x^3 - x^2 \geq 2x$.

Solution: Rearranging and factoring gives

$$x^3 - x^2 - 2x \geq 0$$

$$x(x^2 - x - 2) \geq 0$$

$$x(x + 1)(x - 2) \geq 0$$

The critical values are then

$$x = -1 \qquad x = 0 \qquad x = 2$$

We now analyze the signs, as in Example 18, looking for those intervals in which the product $x(x + 1)(x - 2)$ is positive.

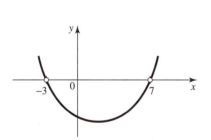

FIGURE 24–10

Interval	x	$(x + 1)$	$(x - 2)$	$x(x + 1)(x - 2)$
$x < -1$	$-$	$-$	$-$	$-$
$-1 < x < 0$	$-$	$+$	$-$	$+$
$0 < x < 2$	$+$	$+$	$-$	$-$
$x > 2$	$+$	$+$	$+$	$+$

Our solution is then

$$-1 \leq x \leq 0 \quad \text{and} \quad x \geq 2$$

Note that the endpoints $x = -1$, 0, and 2 are included in the solution. In interval notation, the solutions are $[-1, 0]$ and $[2, \infty)$.

The inequality is graphed in Fig. 24–11. Note that we selected the two intervals that correspond to values on the curve that are on or above the x axis. ◆◆◆

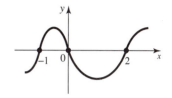

FIGURE 24–11

Our next example shows the algebraic and graphical solutions of a double inequality.

◆◆◆ **Example 20:** Solve the inequality

$$-5 < 3x + 4 \leq 9$$

Algebraic Solution: We can manipulate both inequalities at the same time.

$$-9 < 3x \leq 5$$

$$-3 < x \leq 5/3$$

or $(-3, 5/3]$ in interval notation.

Graphical Solution: We graph the three functions

$$y = -5 \qquad y = 3x + 4 \qquad y = 9$$

on the same axes (Fig. 24–12). The interval for which $3x + 4$ lies between -5 and 9 is $(-3, 5/3]$. ◆◆◆

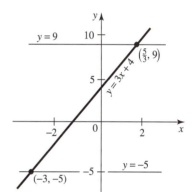

FIGURE 24–12

Our final example shows a graphical solution to a difficult nonlinear inequality.

◆◆◆ **Example 21:** Graphically solve the following inequality. Work to three significant digits.

$$3.72x^3 - 6.26x^2 > 3.27x - 4.57$$

Solution: On the same axes we graph

$$y = 3.72x^3 - 6.26x^2 \quad \text{and} \quad y = 3.27x - 4.57$$

(Fig. 24–13). Using $\boxed{\text{TRACE}}$ and $\boxed{\text{ZOOM}}$, we find the coordinates of the three critical points to be $(-0.884, -7.46)$, $(0.776, -2.03)$, and $(1.79, 1.29)$. The intervals for which the inequality is true are then

$$(-0.884, 0.776) \quad \text{and} \quad (1.79, \infty)$$

◆◆◆

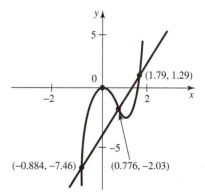

FIGURE 24–13

Inequalities Containing Absolute Values

Recall from Sec. 1–1 that the *absolute value* of a quantity is its *magnitude* and is always positive.

◆◆◆ **Example 22:**

(a) $|5| = 5$

(b) $|-5| = 5$ ◆◆◆

It would thus be correct to say that in Example 22, the quantity inside the absolute value sign could be 5 or −5. So if we see the expression $|x| > 2$, we know that either x or $-x$ will be greater than 2.

$$|x| > 2 \quad \text{is equivalent to} \quad x > 2 \quad \text{and} \quad -x > 2$$

expression with
absolute value signs

two inequalities without
absolute value signs

The two inequalities can be written $x > 2$ and $x < -2$ because multiplying the right-hand inequality by -1 reverses the sense of the inequality.

Similarly, $|x| < 2$ is equivalent to $x < 2$ and $-x < 2$, or (after multiplying on the right by -1) $x < 2$ and $x > -2$. This rule is stated as follows:

> $|x| < a$ is equivalent to $-a < x < a$
> $|x| > a$ is equivalent to $x < -a$ or $x > a$

These rules are, of course, equally true for the ≤ or ≥ relationships.

To solve algebraically an inequality containing an absolute value sign, replace the inequality by a double inequality or by two inequalities having no absolute value signs. Solve each inequality as shown before.

To graphically solve an inequality containing an absolute value sign, simply graph both sides of the inequality, and note the intervals in which the inequality is true. If you wish, you may move all terms to one side of the inequality before graphing, as in the second graphical method.

We will show both methods in the following examples.

◆◆◆ **Example 23:** Solve the inequality $|3x - 2| \leq 5$.

Algebraic Solution: We replace the given inequality with the double inequality

$$-5 \leq 3x - 2 \leq 5$$

Solving, we get

$$-3 \leq 3x \leq 7$$

$$-1 \leq x \leq 7/3$$

In interval notation, our solution is $[-1, 7/3]$.

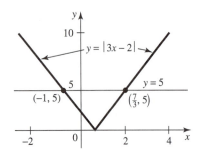

FIGURE 24–14

Graphical Solution: We graph

$$y = |3x - 2| \quad \text{and} \quad y = 5$$

on the same axes (Fig. 24–14). The interval for which $|3x - 2|$ is less than or equal to 5 is the same as found algebraically.   ◆◆◆

On a graphics calculator, use the $\boxed{\text{ABS}}$ **key to obtain an absolute value. Thus we would enter**

$$Y1 = ABS(3X - 2)$$

◆◆◆ **Example 24:** Solve the inequality $|7 - 3x| > 4$.

Algebraic Solution: We replace this inequality with

$$(7 - 3x) > 4 \quad \text{or} \quad (7 - 3x) < -4$$

Subtracting 7, we obtain

$$-3x > -3 \quad \text{or} \quad -3x < -11$$

Dividing by -3 and reversing the sense yields

$$x < 1 \quad \text{or} \quad x > \frac{11}{3}$$

So x may have values less than 1 or greater than $3\frac{2}{3}$.
In interval notation, the solution is

$$(-\infty, 1) \quad \text{or} \quad (11/3, \infty)$$

Graphical Solution: We graph

$$y = |7 - 3x| \quad \text{and} \quad y = 4$$

on the same axes (Fig. 24–15). The interval for which $|7 - 3x|$ is greater than 4 is the same as found algebraically. ◆◆◆

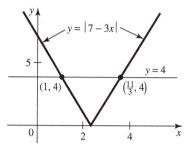

FIGURE 24–15

Finding the Domain of a Function

We can use inequalities to find the *domain* of a function or relation, those values of x for which the function or relation has real values. The domain is useful in graphing to locate, in advance, those regions in which the function exists.

◆◆◆ **Example 25:** Find the values of x for which the relation $x^2 - y^2 - 4x + 3 = 0$ is real.

Solution: We solve for y.

$$y^2 = x^2 - 4x + 3$$
$$y = \pm\sqrt{x^2 - 4x + 3}$$

For y to be real, the quantity under the radical sign must be nonnegative, so we seek the values of x that satisfy the inequality

$$x^2 - 4x + 3 \geq 0$$

Factoring gives

$$(x - 1)(x - 3) \geq 0$$

The critical values are then $x = 1$ and $x = 3$. Analyzing the signs, we obtain the following table:

Interval	$(x - 1)$	$(x - 3)$	$(x - 1)(x - 3)$
$x < 1$	$-$	$-$	$+$
$1 < x < 3$	$+$	$-$	$-$
$x > 3$	$+$	$+$	$+$

We conclude that the function exists for the ranges $x \leq 1$ and $x \geq 3$, but not for $1 < x < 3$. This information helps when graphing the curve, for we now know where the curve does and does not exist. (The curve is actually a hyperbola, as shown in Fig. 24–16, which is studied in analytic geometry.) ◆◆◆

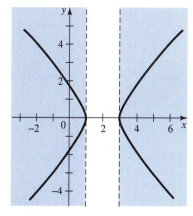

FIGURE 24–16

Exercise 2 ◆ Solving Inequalities

Solve each inequality.

1. $2x - 5 > x + 4$
2. $x + 2 < 3 - 2x$
3. $x^2 + 4 \geq 2x^2 - 5$
4. $2x^2 + 15x \leq -25$

5. $2x^2 - 5x + 3 > 0$

6. $6x^2 < 19x - 10$

7. $\dfrac{x}{5} + \dfrac{x}{2} \geq 4$

8. $\dfrac{x+1}{2} - \dfrac{x-2}{3} \leq 4$

9. $1 < x + 3 < 8$

10. $12 > 2x - 1 > -3$

11. $-2 \leq 2x + 4 \leq 6$

12. $10 > 8 + x \geq 2$

Inequalities Containing Absolute Values

Solve each inequality.

13. $|3x| > 9$

14. $2|x + 3| < 8$

15. $5|x| > 8$

16. $|3x + 2| \leq 11$

17. $|2 - 3x| \geq 8$

18. $3|x + 2| > 21$

19. $|x + 5| > 2x$

20. $|x + 7| > 3x$

21. $|3 - 5x| \leq 4 - 3x$

22. $|x + 1| \geq 2x - 1$

Finding the Domain of a Function

For what values of x is each expression real?

23. $\sqrt{x^2 + 4x - 5}$

24. $\sqrt{x^3 - 4x^2 - 12x}$

25. $y^2 = x^2 - x - 6$

26. $x^2 + y^2 + x - 2 = 0$

Applications

27. A certain machine is rented for $155 per day, and it costs $25 per hour to operate. The total cost per day for this machine is not to exceed $350. Write an inequality for this situation, and find the maximum permissible hours of operation.

28. A computer salesperson earns an annual salary of $25,500 and also gets a 37.5% commission on sales. How much must she sell in order to make an annual income of at least $50,000?

29. A company can make a camera lens for $12.75 and sell it for $39.99. It also has overhead costs of $2884 per week. Find the minimum number of lenses that must be sold for the company to make a profit.

Computer and Graphics Calculator

30. Use a graphics calculator or a graphing utility on a computer to solve graphically any of the inequalities in this exercise set. Also use it to solve the following inequalities. Work to three significant digits.

31. $2.81x^2 - 4.73 < 1.75$

32. $11.2 - 4.72x^2 \geq 2.23x + 3.05$

33. $4.77x^2 - 7.14 < 4.72 - 2.33x$

34. $4.73 \leq 5.75x - 1.17x^2$

24–3 Linear Programming

Linear programming is a method for finding the maximum value of some function (usually representing *profit*) when the variables that it contains are themselves restricted to within certain limits. The function that is maximized is called the *objective function,* and the limits on the variables are called *constraints.* The constraints will be in the form of inequalities.

We show the method by means of examples.

We will do problems with only two variables here. To see how to *solve* problems with more than two variables, look up the *simplex method* in a text on linear programming.

◆◆◆ **Example 26:** Find the values of x and y that will make z a maximum, where the objective function is

$$z = 5x + 10y$$

and x and y are positive and subject to the constraints

$$x + y \leq 5$$

and

$$2y - x \leq 4$$

In applications, x and y usually represent amounts of a certain resource, or the number of items produced, and thus will be restricted to *positive* values.

Solution: We plot the two inequalities as shown in Fig. 24–17. The only permissible values for x and y are the coordinates of points on the edges of or within the shaded region. These are called *feasible solutions.*

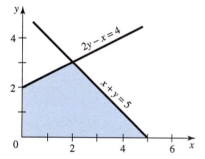

FIGURE 24–17 Feasible solutions.

But which (x, y) pair selected from all those in the shaded region will give the greatest value of z? In other words, which of the infinite number of feasible solutions is the *optimum* solution?

To help us answer this, let us plot the equation $z = 5x + 10y$ for different values of z. For example:

When $z = 0$,

$$5x + 10y = 0$$

$$y = -\frac{x}{2}$$

When $z = 10$,

$$5x + 10y = 10$$

$$y = -\frac{x}{2} + 1$$

and so on.

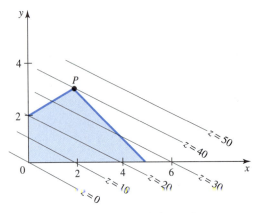

FIGURE 24–18   Optimum solution at P.

We get a *family* of parallel straight lines, each with a slope of $-\frac{1}{2}$ (Fig. 24–18). The point P at which the line having the greatest z value intersects the region gives us the x and y values we seek. Such a point will always occur at a *corner* or *vertex* of the region. Thus instead of plotting the family of lines, it is necessary only to compute z at the vertices of the region. The coordinates of the vertices are called the *basic feasible solutions*. Of these, the one that gives the greatest z is the *optimum solution*.

> If we had used a graphics utility to make our graph, we would now use ⟨TRACE⟩ and ⟨ZOOM⟩ to find the points of intersection. Otherwise, we find the vertices of the region by simultaneously solving pairs of equations, as shown here.

In our problem, the vertices are $(0, 0)$, $(0, 2)$, $(5, 0)$, and P, whose coordinates we get by simultaneous solution of the equations of the boundary lines.

$$
\begin{array}{rcr}
x + y &=& 5 \\
x - 2y &=& -4 \\
\hline
\text{Subtract:} \qquad 3y &=& 9 \\
y &=& 3
\end{array}
$$

Substituting back, we have

$$x = 2$$

So P has the coordinates $(2, 3)$. Now compute z at each vertex.

$$
\begin{array}{lll}
\text{At } (0, 0) & z = 5(0) + 10(0) = & 0 \\
\text{At } (0, 2) & z = 5(0) + 10(2) = & 20 \\
\text{At } (5, 0) & z = 5(5) + 10(0) = & 25 \\
\text{At } (2, 3) & z = 5(2) + 10(3) = & 40
\end{array}
$$

So, $x = 2$, $y = 3$, which gives the largest z, is our optimum solution.   ◆◆◆

◆◆◆ **Example 27:** Find the nonnegative values of x and y that will maximize the function $z = 2x + y$ with the following *four* constraints:

$$
\begin{array}{rcr}
y - x &\le& 6 \\
x + 4y &\le& 40 \\
x + y &\le& 16 \\
2x - y &\le& 20
\end{array}
$$

Solution: We graph the inequalities as shown in Fig. 24–19 and compute the points of intersection. This gives the six basic feasible solutions.

$$(0, 0) \quad (0, 6) \quad (3.2, 9.2) \quad (8, 8) \quad (12, 4) \quad (10, 0)$$

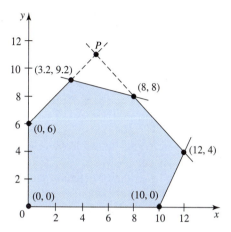

We have not shown the work in finding the points of intersection. They are found, as before, by using TRACE and ZOOM on a graphics utility, or by simultaneously solving the equations of the intersecting lines.

FIGURE 24–19 The coordinates of the corners of the shaded region are the basic feasible solutions. One of these is the optimum solution.

We now compute z at each point of intersection.

Point	$z = 2x + y$
(0, 0)	$z = 2(0) + 0 \quad = 0$
(0, 6)	$z = 2(0) + 6 \quad = 6$
(3.2, 9.2)	$z = 2(3.2) + 9.2 = 15.6$
(8, 8)	$z = 2(8) + 8 \quad = 24$
(12, 4)	$z = 2(12) + 4 \quad = 28$
(10, 0)	$z = 2(10) + 0 \quad = 20$

The greatest z (28) occurs at $x = 12$, $y = 4$, which is thus our optimum solution.

◆◆◆

Applications

Our applications will usually be business problems, where our objective function z (the one we maximize) represents profit, and our variables x and y represent the quantities of two products to be made. A typical problem is to determine how much of each product should be manufactured so that the profit is a maximum.

◆◆◆ **Example 28:** Each month a company makes x stereos and y television sets. The manufacturing hours, material costs, testing time, and profit for each are as follows:

	Mfg. Time	Material Costs	Test Time	Profit
Stereo	4.00 h	$95.00	1.00 h	$45.00
Television	5.00 h	$48.00	2.00 h	$38.00
Available	180 h	$3600	60.0 h	

Also shown in the table are the maximum amount of manufacturing and testing time and the cost of material available per month. How many of each item should be made for maximum profit?

Solution: First, why should we make any TV sets at all if we get more profit on stereos? The answer is that we are limited by the money available. In 180 h we could make

$$180 \div 4.00 = 45.0 \text{ stereos}$$

But 45 stereos, at $95 each, would require $4275, but there is only $3600 available. Then why not make

$$\$3600 \div 95.00 = 37.0 \text{ stereos}$$

using all our money on the most profitable item? However, this would leave us with manufacturing time unused but no money left to make any TV sets.

Thus we wonder if we could do better by making fewer than 37 stereos and using the extra time and money to make some TV sets. We seek the *optimum* solution. We start by writing the equations and inequalities to describe our situation.

Our total profit z will be the profit per item times the number of items made.

$$z = 45.00x + 38.00y$$

The manufacturing time must not exceed 180 h, so

$$4.00x + 5.00y \le 180$$

The materials cost must not exceed $3600, so

$$95.00x + 48.00y \le 3600$$

The test time must not exceed 60 h, so

$$x + 2.00y \le 60.0$$

This gives us our objective function and three constraints. We plot the constraints in Fig. 24–20.

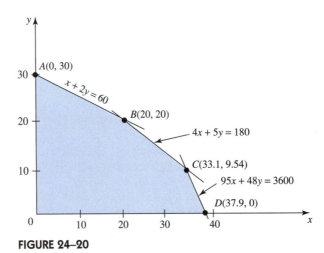

FIGURE 24–20

If we had used a graphics utility to make our graph, we would now use TRACE and ZOOM to find the points of intersection to three significant digits.

Otherwise, we find the vertices of the region by simultaneously solving pairs of equations. To find point A, we set $x = 0$ in the equation

$$x + 2.00y = 60.0$$

and get $y = 30.0$, or

$$A(0, 30.0)$$

To find point B, we solve simultaneously

$$x + 2.00y = 60.0$$
$$4.00x + 5.00y = 180$$

and get

$$B(20.0, 20.0)$$

Again, the actual work is omitted here. To find point C, we solve the equations

$$4.00x + 5.00y = 180$$
$$95.00x + 48.00y = 3600$$

and get $x = 33.1$, $y = 9.54$, or

$$C(33.1, 9.54)$$

To find point D, we use the equation

$$95.00x + 48.00y = 3600$$

Letting $y = 0$ gives $x = 37.9$, or

$$D(37.9, 0)$$

We then evaluate the profit z at each vertex (omitting the point 0, 0 at which $z = 0$).

At $A(0, 30.0)$ $z = 45.00(0)$ $+ 38.00(30.0) = \$1140$
At $B(20.0, 20.0)$ $z = 45.00(20.0) + 38.00(20.0) = \$1660$
At $C(33.1, 9.54)$ $z = 45.00(33.1) + 38.00(9.54) = \$1852$
At $D(37.9, 0)$ $z = 45.00(37.9) + 38.00(0)$ $= \$1706$

Thus our maximum profit, \$1852, is obtained when we make 33.1 stereos and 9.54 television sets. ◆◆◆

Linear Programming on the Computer

The computer can reduce the work of linear programming, especially when the number of constraints gets large. The following algorithm is for a problem with two variables and any number of constraints:

1. Using Eq. 63, find the point of intersection of any two of the given boundary lines.
2. If the point of intersection is within the region of permissible solutions, go to Step 3. If not, go to Step 5. (This will eliminate points such as P in Fig. 24–19).
3. Compute z at the point of intersection.
4. If z is the largest so far, save it and the coordinates of the point. If not, discard it.
5. If there are combinations of equations yet unused, return to Step 1.
6. Print the largest z and the coordinates of the point at which it occurs.

A flowchart for this computation is shown in Fig. 24–21.

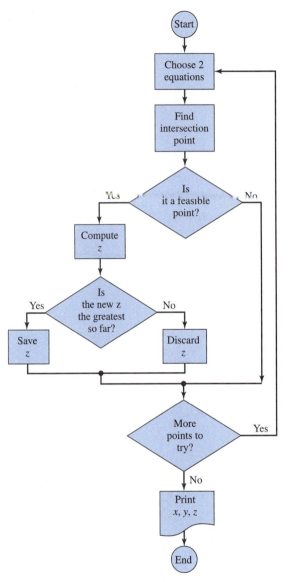

FIGURE 24–21   Flowchart for linear programming by computer.

Exercise 3 ◆ Linear Programming

As in the examples, we seek only non-negative values of x and y.

1. Maximize the function $z = x + 2y$ under the constraints

$$x + y \leq 5 \quad \text{and} \quad x - y \geq -2$$

2. Maximize the function $z = 4x + 3y$ under the constraints

$$x - 2y \geq -2 \quad \text{and} \quad x + 2y \leq 4$$

3. Maximize the function $z = 5x + 4y$ under the constraints

$$7x + 4y \leq 35 \qquad 2x - 3y \leq 7 \quad \text{and} \quad x - 2y \geq -5$$

4. Maximize the function $z = x + 2y$ under the constraints

$$x - y \geq -5 \qquad x + 5y \leq 45 \quad \text{and} \quad x + y \leq 15$$

For problems 5, 6, and 7, round your answer to the nearest integer.

5. A company makes pulleys and sprockets. A pulley takes 3 min to make, requires $1.25 in materials, and gives a profit of $2.10. A sprocket takes 4 min to make, requires $1.30 in materials, and gives a profit of $2.35. The time to make x pulleys and y sprockets is not to exceed 8 h, and the cost for materials must not exceed $175. How many of each item should be made for maximum profit?

6. A company makes two computers. Model A requires 225 h of labor and $223 in material and yields a profit of $1975. Model B requires 67 h of labor and $118 in material and yields a profit of $824. There are 172,000 h and $240,000 available per month to make the computers. How many of each computer must be made each month for maximum profit?

7. Each week a company makes x disk drives and y monitors. The manufacturing hours, material costs, inspection time, and profit for each are as follows:

	Manufacturing Time (h)	Material Costs	Inspection Time (h)	Profit
Disk drive	5	$ 21	4.5	$34
Monitor	3	28	4	22
Available	50	320	50	

Also shown in the table are the maximum amount of manufacturing and inspection time available per week and the cost of materials. How many of each item should be made per week for maximum profit?

Computer

8. Using the flowchart of Fig. 24–21 as a guide, write a computer program to do linear programming. Try it on any of the problems in this exercise set.

9. Some computer algebra systems, such as *Mathematica,* can solve linear programming problems. Consult your manual for the proper instruction, and use it to solve any of the problems in this exercise set.

••• CHAPTER 24 REVIEW PROBLEMS •••••••••••••••••••••••••••••••••••

1. Rewrite as two separate inequalities:

$$-11 < x \le 1$$

2. Rewrite as an inequality:

$$x \ge -1 \qquad x < 24$$

Graph on the number line, and give the interval notation for each.

3. $x < 8$ 4. $2 \le x < 6$

Write the inequality corresponding to each given interval, and graph.

5. $(8, \infty)$ 6. $[-4, 9)$

Solve each inequality algebraically or graphically.

7. $2x - 5 > x + 3$

8. $|2x| < 8$

9. $|x - 4| > 7$

10. $x^2 > x + 6$

11. Maximize the function $z = 7x + y$ under the constraints

$$2x + 3y \leq 5 \quad \text{and} \quad x - 4y \geq -4$$

12. A company makes four-cylinder engine blocks and six-cylinder engine blocks. A "four" requires 4 h of machine time and 5 h of labor and gives a profit of $172. A "six" needs 5 h on the machine and 3 h of labor and yields $195 in profit. If 130 h of labor and 155 h of machine time are available, how many (to the nearest integer) of each engine block should be made for maximum profit?

Find the values of x for which each relation is real.

13. $\sqrt{x^2 + 8x + 15}$

14. $\sqrt{x^2 - 10x + 16}$

15. $y^2 = x^2 + 5x - 14$

16. $x^2 - y^2 - 10x + 9 = 0$

Solve each double inequality.

17. $-4.73 < x - 3 \leq 5.82$

18. $2.84 \leq x + 5 < 7.91$

19. Solve graphically to three significant digits:

$$1.27 - 2.24x + 1.84x^2 < 3.87x - 2.44x^2$$

20. A gear factory can make a bevel gear for $1.32 and sell it for $4.58. It also has overhead costs of $1845 per week for the bevel gear department. Find the minimum number that must be sold for the company to make a profit on bevel gears.

Writing

21. A job applicant says on her résumé that she knows linear programming, and the personnel director of your company insists that she fits your request for a "programmer." Write a memo to the personnel director explaining the difference between linear programming and computer programming.

Team Project

22. Search newspapers, magazines, and advertising flyers for inequalities in verbal form. Clip as many as you can find. For each, write an algebraic expression with inequality symbols, as we've done in this chapter, and make a graph.

Sequences, Series, and the Binomial Theorem

◆◆◆ **OBJECTIVES** ◆◆◆

When you have completed this chapter, you should be able to:

• Identify various types of sequences and series.

• Write the general term or a recursion relation for many series.

• Compute any term or the sum of any number of terms of an arithmetic progression or a geometric progression.

• Compute any term of a harmonic progression.

• Insert any number of arithmetic means, harmonic means, or geometric means between two given numbers.

• Compute the sum of an infinite geometric progression.

• Solve applications problems using series.

• Raise a binomial to a power using the binomial theorem.

• Find any term in a binomial expansion.

◆◆

We start this chapter with a general introduction to *sequences* and *series*. Sequences and series are of great interest because a computer or calculator uses series internally to calculate many functions, such as the sine of an angle, the logarithm of a number, and so on. We then cover some specific sequences, such as the *arithmetic progression* and the *geometric progression*. *Progressions* are used to describe such things as the sequence of heights reached by a swinging pendulum on subsequent swings or the monthly balance on a savings account that is growing with compound interest.

We finish this chapter with an introduction to the *binomial theorem*. The binomial theorem enables us to expand a binomial, such as $(3x^2 - 2y)^5$, without actually having to multiply the terms. It is also useful in deriving formulas in calculus, statistics, and probability.

Sequences

A *sequence* is a set of quantities, called *terms*, arranged in the same order as the positive integers.

$$u_1, u_2, u_3, \ldots, u_n$$

◆◆◆ Example 1:

(a) The sequence $1, \frac{1}{2}, \frac{1}{3}, \frac{1}{4}, \frac{1}{5}, \ldots, 1/n$ is called a *finite* sequence because it has a finite number of terms, n,

(b) The sequence $1, \frac{1}{2}, \frac{1}{3}, \frac{1}{4}, \frac{1}{5}, \ldots, 1/n, \ldots$ is called an *infinite* sequence. The three dots at the end indicate that the sequence continues indefinitely.

(c) The sequence $3, 7, 11, 15, \ldots$ is called an *arithmetic sequence,* or *arithmetic progression* (AP), because each term after the first is equal to the sum of the preceding term and a constant. That constant (4, in this example) is called the *common difference.*

(d) The sequence $2, 6, 18, 54, \ldots$ is called a *geometric sequence,* or *geometric progression* (GP), because each term after the first is equal to the *product* of the preceding term and a constant. That constant (3, in this example) is called the *common ratio.*

(e) The sequence $1, 1, 2, 3, 5, 8, \ldots$ is called a *Fibonacci* sequence. Each term after the first is the sum of the two preceding terms.   ◆◆◆

Series

A *series* is the indicated sum of a sequence.

$u_1 + u_2 + u_3 + \cdots + u_n + \cdots$	**441**

◆◆◆ Example 2:

(a) The series $1 + \frac{1}{2} + \frac{1}{3} + \frac{1}{4} + \frac{1}{5}$ is called a *finite series*. It is also called a *positive* series because all of its terms are positive.

(b) The series $x - x^2 + x^3 - \cdots + x^n \cdots$ is an *infinite series*. It is also called an *alternating* series because the signs of the terms alternate in sign.

(c) The series $6 + 9 + 12 + 15 + \cdots$ is an infinite arithmetic series. The terms of this series form an AP.

(d) $1 - 2 + 4 - 8 + 16 - \cdots$ is an infinite, alternating, geometric series. The terms of this series form a GP.   ◆◆◆

General Term

The *general term* u_n in a sequence or series is an expression involving n (where $n = 1, 2, 3, \ldots$) by which we can obtain any term. The general term is also called the *n*th term. If we have an expression for the general term, we can then find any specific term of the sequence or series.

◆◆◆ Example 3: The general term of a certain series is

$$u_n = \frac{n}{2n + 1}$$

Write the first three terms of the series.

Solution: Substituting $n = 1$, $n = 2$, and $n = 3$, in turn, we get

$$u_1 = \frac{1}{2(1) + 1} = \frac{1}{3} \qquad u_2 = \frac{2}{2(2) + 1} = \frac{2}{5} \qquad u_3 = \frac{3}{2(3) + 1} = \frac{3}{7}$$

so our series is

$$\frac{1}{3} + \frac{2}{5} + \frac{3}{7} + \cdots + \frac{n}{2n + 1} + \cdots$$

◆◆◆

Graphing a Sequence

We can graph a sequence if, as in Example 3, the general term is known.

◆◆◆ **Example 4:** Graph the sequence given in Example 3.

Solution: *Graphing by hand:* We let $n = 1, 2, 3, \ldots$ and compute the value of as many terms as we wish. We then plot the points,

$$(1, \tfrac{1}{3}), (2, \tfrac{2}{5}), (3, \tfrac{3}{7}), \ldots$$

as in Fig. 25–1.

Using a graphing utility: We graph the function

$$u = \frac{n}{2n + 1}$$

as in Fig. 25–2 and place dots at $x = 1, 2, 3, \ldots$ to represent the terms of the series.

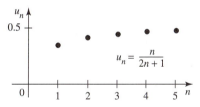

FIGURE 25–1

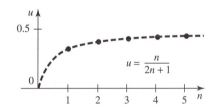

FIGURE 25–2

◆◆◆

Recursion Relations

We have seen that we can find the terms of a series, given an expression for the nth term. Sometimes we can find each term from one or more immediately preceding terms. The relationship between a term and those preceding it is called a *recursion relation* or *recursion formula*.

◆◆◆ **Example 5:** Each term (after the first) in the series $1 + 4 + 13 + 40 + \cdots$ is found by multiplying the preceding term by 3 and adding 1. The recursion relation is then

$$u_n = 3u_{n-1} + 1$$

◆◆◆

◆◆◆ **Example 6:** In a *Fibonacci sequence*

$$1, 1, 2, 3, 5, 8, 13, 21, 34, \ldots$$

each term after the first is the sum of the two preceding terms. Its recursion relation is

$$u_n = u_{n-1} + u_{n-2}$$

◆◆◆

The words *recurrence* and *recursive* are sometimes used instead of *recursion*. We have used recursion relations before, for compound interest in Sec. 20–1 and for exponential growth in Sec. 20–2.

The Fibonacci sequence is named for Leonardo Fibonacci (ca. 1170–ca. 1250), an Italian number theorist and algebraist who studied this sequence. He is also known as *Leonardo of Pisa*.

Series by Computer

Computers now make series easier to study than before. If we have either an expression for the general term of a series or the recursion formula, we can use a computer to generate large numbers of terms. By inspecting them, we can get a good idea about how the series behaves.

♦♦♦ **Example 7:** Table 25–1 shows a portion of a computer printout of 80 terms of the series

$$\frac{1}{e} + \frac{2}{e^2} + \frac{3}{e^3} + \cdots + \frac{n}{e^n} + \cdots$$

TABLE 25–1

n	Term u_n	Partial Sum S_n	Ratio
1	0.367879	0.367879	
2	0.270671	0.638550	0.735759
3	0.149361	0.787911	0.551819
4	0.073263	0.861174	0.490506
5	0.033690	0.894864	0.459849
6	0.014873	0.909736	0.441455
7	0.006383	0.916119	0.429193
8	0.002684	0.918803	0.420434
9	0.001111	0.919914	0.413864
10	0.000454	0.920368	0.408755
⋮	⋮	⋮	⋮
75	0.000000	0.920674	0.372851
76	0.000000	0.920674	0.372785
77	0.000000	0.920674	0.372720
78	0.000000	0.920674	0.372657
79	0.000000	0.920674	0.372596
80	0.000000	0.920674	0.372536

Also shown for each term u_n is the sum of that term and all preceding terms, called the *partial sum,* and the ratio of that term to the one preceding. We notice several things about this series.

1. The terms get smaller and smaller and approach a value of zero (Fig. 25–3).

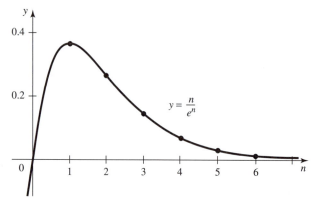

FIGURE 25–3

2. The partial sums approach a limit (0.920674).
3. The ratio of two successive terms approaches a limit that is less than 1.

What do those facts tell us about this series? We'll see in the following section.

◆◆◆

Convergence or Divergence of an Infinite Series

When we use a series to do computations, we cannot, of course, work with an infinite number of terms. In fact, for practical computation, we prefer as few terms as possible, as long as we get the needed accuracy. We want a series in which the terms decrease rapidly and approach zero, and for which the sum of the first several terms is not too different from the sum of all of the terms of the series. Such a series is said to *converge* on some limit. A series that does not converge is said to *diverge*.

There are many tests for convergence. A particular test may tell us with certainty if the series converges or diverges. Sometimes, though, a test will fail and we must try another test. Here we consider only three tests.

1. *Magnitude of the terms.* If the terms of an infinite series do not approach zero, the series diverges. However, if the terms do approach zero, we cannot say with certainty if the series converges. Thus having the terms approach zero is a necessary but not a sufficient condition for convergence.
2. *Sum of the terms.* If the sum of the first *n* terms of a series reaches a limiting value as we take more and more terms, the series converges. Otherwise, the series diverges.
3. *Ratio test.* We compute the ratio of each term to the one preceding it and see how that ratio changes as we go further out in the series. If that ratio approaches a number that is

    (a) less than 1, the series converges
    (b) greater than 1, the series diverges.
    (c) equal to 1, the test is not conclusive.

◆◆◆ **Example 8:** Does the series of Example 7 converge or diverge?

Solution: We apply the three tests by inspecting the computed values.

1. The terms approach zero, so this test is not conclusive.
2. The sum of the terms appears to approach a limiting value (0.920674), indicating convergence.
3. The ratio of the terms appears to approach a value of less than 1, indicating convergence.

Thus the series appears to converge.

◆◆◆

Note that we say *appears to* converge. The computer run, while convincing, is still not a *proof* of convergence.

◆◆◆ **Example 9:** Table 25–2 shows part of a computer run for the series

$$1 + \frac{1}{2} + \frac{1}{3} + \cdots + \frac{1}{n} + \cdots$$

Does the series converge or diverge?

Solution: Again we apply the three tests.

1. We see that the terms are getting smaller, so this test is not conclusive.
2. The ratio of the terms appears to approach 1, so this test also is not conclusive.
3. The partial sum S_n does not seem to approach a limit, but appears to keep growing even after 495 terms.

It thus appears that this series diverges.

TABLE 25–2

n	Term u_n	Partial Sum S_n	Ratio
1	1.000000	1.000000	
2	0.500000	1.500000	0.500000
3	0.333333	1.833333	0.666667
4	0.250000	2.083334	0.750000
5	0.200000	2.283334	0.800000
6	0.166667	2.450000	0.833333
7	0.142857	2.592857	0.857143
8	0.125000	2.717857	0.875000
9	0.111111	2.828969	0.888889
10	0.100000	2.928969	0.900000
⋮	⋮	⋮	⋮
75	0.013333	4.901356	0.986667
76	0.013158	4.914514	0.986842
77	0.012987	4.927501	0.987013
78	0.012821	4.940322	0.987180
79	0.012658	4.952980	0.987342
80	0.012500	4.965480	0.987500
⋮	⋮	⋮	⋮
495	0.002020	6.782784	0.997980
496	0.002016	6.784800	0.997984
497	0.002012	6.786813	0.997988
498	0.002008	6.788821	0.997992
499	0.002004	6.790825	0.997996
500	0.002000	6.792825	0.998000

◆◆◆

Exercise 1 ◆ Sequences and Series

Write the first five terms of each series, given the general term.

1. $u_n = 3n$

2. $u_n = 2n + 3$

3. $u_n = \dfrac{n + 1}{n^2}$

4. $u_n = \dfrac{2^n}{n}$

Deduce the general term of each series. Use it to predict the next two terms.

5. $2 + 4 + 6 + \cdots$

6. $1 + 8 + 27 + \cdots$

7. $\dfrac{2}{4} + \dfrac{4}{5} + \dfrac{8}{6} + \dfrac{16}{7} + \cdots$

8. $2 + 5 + 10 + \cdots$

Deduce a recursion relation for each series. Use it to predict the next two terms.

9. $1 + 5 + 9 + \cdots$

10. $5 + 15 + 45 + \cdots$

11. $3 + 9 + 81 + \cdots$

12. $5 + 8 + 14 + 26 + \cdots$

Computer

Write a program or use a spreadsheet to generate the terms of a series, given the general term or a recursion relation. Have the program compute and print each term, the partial sum, and the ratio of that term to the preceding one. Use the program to determine if each of the following series converges or diverges.

13. $1 + \dfrac{1}{2} + \dfrac{1}{3} + \dfrac{1}{4} + \cdots + \dfrac{1}{n} + \cdots$

14. $\dfrac{3}{2} + \dfrac{9}{8} + \dfrac{27}{24} + \dfrac{81}{64} + \cdots + \dfrac{3^n}{n \cdot 2^n} + \cdots$

15. $1 + \dfrac{4}{7} + \dfrac{9}{49} + \dfrac{16}{343} + \cdots + \dfrac{n^2}{7^{n-1}} + \cdots$

Graphics Calculator

16. Some graphics calculators will generate the terms of a sequence if you enter the expression for the nth term, the first and last values of n, and the increment for n (usually 1). Check your manual for the proper instructions, and use your calculator to do problems 1 through 4 of this exercise set.

25–2 Arithmetic Progressions

Recursion Formula

We stated earlier that an *arithmetic progression* (or AP) is a sequence of terms in which each term after the first equals the sum of the preceding term and a constant, called the *common difference, d.* If a_n is any term of an AP, the recursion formula for an AP is as follows:

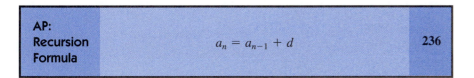

| AP: Recursion Formula | $a_n = a_{n-1} + d$ | 236 |

Each term of an AP after the first equals the sum of the preceding term and the common difference.

◆◆◆ **Example 10:** The following sequences are arithmetic progressions. The common difference for each is given in parentheses:

(a) $1, 5, 9, 13, \ldots$ ($d = 4$)
(b) $20, 30, 40, 50, \ldots$ ($d = 10$)
(c) $75, 70, 65, 60, \ldots$ ($d = -5$)

We see that each series is increasing when d is positive and decreasing when d is negative.   ◆◆◆

General Term

For an AP whose first term is a and whose common difference is d, the terms are

$$a, a + d, a + 2d, a + 3d, a + 4d, \ldots$$

We see that each term is the sum of the first term and a multiple of d, where the coefficient of d is one less than the number n of the term. So the nth term a_n is given by the following equation:

The general term is sometimes called the *last* term, but this is not an accurate name since the AP continues indefinitely.

| AP: General Term | $a_n = a + (n - 1)d$ | 237 |

The nth term of an AP is found by adding the first term and $(n - 1)$ times the common difference.

◆◆◆ **Example 11:** Find the twentieth term of an AP that has a first term of 5 and common difference of 4.

Solution: By Eq. 237, with $a = 5$, $n = 20$, and $d = 4$,

$$a_{20} = 5 + 19(4) = 81$$ ◆◆◆

Of course, Equation 237 can be used to find any of the four quantities (a, n, d, or a_n) given the other three.

◆◆◆ **Example 12:** Write the AP whose eighth term is 19 and whose fifteenth term is 33.

Solution: Applying Eq. 237 twice gives

$$33 = a + 14d$$
$$19 = a + 7d$$

We now have two equations in two unknowns, which we solve simultaneously. Subtracting the second from the first gives $14 = 7d$, so $d = 2$. Also, $a = 33 - 14d = 5$, so the general term of our AP is

$$a_n = 5 + (n - 1)2$$

and the AP is then

$$5, 7, 9, \ldots$$ ◆◆◆

AP: Sum of n Terms

Let us derive a formula for the sum s_n of the first n terms of an AP. Adding term by term gives

$$s_n = a + (a + d) + (a + 2d) + \cdots + (a_n - d) + a_n \qquad (1)$$

or, written in reverse order,

$$s_n = a_n + (a_n - d) + (a_n - 2d) + \cdots + (a + d) + a \qquad (2)$$

Adding Equations (1) and (2) term by term gives

$$2s_n = (a + a_n) + (a + a_n) + (a + a_n) + \cdots$$
$$= n(a + a_n)$$

Dividing both sides by 2 gives the following formula:

| AP: Sum of n Terms | $s_n = \dfrac{n}{2}(a + a_n)$ | 238 |

The sum of n terms of an AP is half the product of n and the sum of the first and nth terms.

◆◆◆ **Example 13:** Find the sum of 10 terms of the AP

$$2, 5, 8, 11, \ldots$$

Solution: From Eq. 237, with $a = 2$, $d = 3$, and $n = 10$,

$$a_{10} = 2 + 9(3) = 29$$

Then, from Eq. 238,

$$s_{10} = \frac{10(2 + 29)}{2} = 155$$

◆◆◆

We get another form of Eq. 238 by substituting the expression for a_n from Eq. 237, as follows:

AP: Sum of n Terms	$s_n = \dfrac{n}{2}[2a + (n - 1)d]$	239

This form is useful for finding the sum without first computing the nth term.

◆◆◆ **Example 14:** We repeat Example 13 without first having to find the tenth term. From Eq. 239,

$$s_{10} = \frac{10[2(2) + 9(3)]}{2} = 155$$

◆◆◆

Sometimes the sum may be given and we must find one of the other quantities in the AP.

◆◆◆ **Example 15:** How many terms of the AP 5, 9, 13, . . . give a sum of 275?

Solution: We seek n so that $s_n = 275$. From Eq. 239, with $a = 5$ and $d = 4$,

$$275 = \frac{n}{2}[2(5) + (n - 1)4]$$

$$= 5n + 2n(n - 1)$$

Removing parentheses and collecting terms gives the quadratic equation

$$2n^2 + 3n - 275 = 0$$

From the quadratic formula,

$$n = \frac{-3 \pm \sqrt{9 - 4(2)(-275)}}{2(2)} = 11 \quad \text{or} \quad -12.5$$

We discard the negative root and get 11 terms as our answer.   ◆◆◆

Arithmetic Means

The first term a and the last (nth) term a_n of an AP are sometimes called the *extremes,* while the intermediate terms, $a_2, a_3, \ldots, a_{n-1}$, are called *arithmetic means.* We now show, by example, how to insert any number of arithmetic means between two extremes.

◆◆◆ **Example 16:** Insert five arithmetic means between 3 and -9.

Solution: Our AP will have seven terms, with a first term of 3 and a seventh term of -9. From Eq. 237,

$$-9 = 3 + 6d$$

from which $d = -2$. The progression is then

$$3, 1, -1, -3, -5, -7, -9$$

and the five arithmetic means are $1, -1, -3, -5,$ and -7.   ◆◆◆

Average Value

Let us insert a single arithmetic mean between two numbers.

◆◆◆ **Example 17:** Find a single arithmetic mean m between the extremes a and b.

Solution: The sequence is a, m, b. The common difference d is $m - a = b - m$. Solving for m gives

$$2m = b + a$$

Dividing by 2 gives

$$m = \frac{a + b}{2}$$

which agrees with the common idea of an *average* of two numbers as the sum of those numbers divided by 2.   ◆◆◆

Harmonic Progressions

A sequence is called a *harmonic progression* if the reciprocals of its terms form an arithmetic progression.

◆◆◆ **Example 18:** The sequence

$$1, \frac{1}{3}, \frac{1}{5}, \frac{1}{7}, \frac{1}{9}, \frac{1}{11}, \cdot \cdot \cdot$$

is a harmonic progression because the reciprocals of the terms, $1, 3, 5, 7, 9, 11, \ldots,$ form an AP.   ◆◆◆

It is not possible to derive an equation for the nth term or for the sum of a harmonic progression. However, we can solve problems involving harmonic progressions by taking the reciprocals of the terms and using the formulas for the AP.

◆◆◆ **Example 19:** Find the tenth term of the harmonic progression

$$2, \frac{2}{3}, \frac{2}{5}, \cdot \cdot \cdot$$

Solution: We write the reciprocals of the terms,

$$\frac{1}{2}, \frac{3}{2}, \frac{5}{2}, \cdot \cdot \cdot$$

and note that they form an AP with $a = \frac{1}{2}$ and $d = 1$. The tenth term of the AP is then

$$a_n = a + (n - 1)d$$
$$a_{10} = \frac{1}{2} + (10 - 1)(1)$$
$$= \frac{19}{2}$$

The tenth term of the harmonic progression is the reciprocal of $\frac{19}{2}$, or $\frac{2}{19}$.   ◆◆◆

Harmonic Means

To insert *harmonic means* between two terms of a harmonic progression, we simply take the reciprocals of the given terms, insert arithmetic means between those terms, and take reciprocals again.

◆◆◆ **Example 20:** Insert three harmonic means between $\frac{2}{9}$ and 2.

Solution: Taking reciprocals, our AP is

$$\frac{9}{2}, \underline{\quad}, \underline{\quad}, \underline{\quad}, \frac{1}{2}$$

In this AP, $a = \frac{9}{2}$, $n = 5$, and $a_5 = \frac{1}{2}$. We can find the common difference d from the equation

$$a_n = a + (n - 1)d$$
$$\frac{1}{2} = \frac{9}{2} + 4d$$
$$-4 = 4d$$
$$d = -1$$

Having the common difference, we can fill in the missing terms of the AP,

$$\frac{9}{2}, \frac{7}{2}, \frac{5}{2}, \frac{3}{2}, \frac{1}{2}$$

Taking reciprocals again, our harmonic progression is

$$\frac{2}{9}, \frac{2}{7}, \frac{2}{5}, \frac{2}{3}, 2$$

◆◆◆

Exercise 2 ◆ Arithmetic Progressions

General Term

1. Find the fifteenth term of an AP with first term 4 and common difference 3.
2. Find the tenth term of an AP with first term 8 and common difference 2.
3. Find the twelfth term of an AP with first term -1 and common difference 4.
4. Find the ninth term of an AP with first term -5 and common difference -2.
5. Find the eleventh term of the AP
$$9, 13, 17, \ldots$$
6. Find the eighth term of the AP
$$-5, -8, -11, \ldots$$
7. Find the ninth term of the AP
$$x, x + 3y, x + 6y, \ldots$$
8. Find the fourteenth term of the AP
$$1, \frac{6}{7}, \frac{5}{7}, \ldots$$

For problems 9 through 12, write the first five terms of each AP.

9. First term is 3 and thirteenth term is 55.
10. First term is 5 and tenth term is 32.
11. Seventh term is 41 and fifteenth term is 89.
12. Fifth term is 7 and twelfth term is 42.
13. Find the first term of an AP whose common difference is 3 and whose seventh term is 11.
14. Find the first term of an AP whose common difference is 6 and whose tenth term is 77.

Sum of an Arithmetic Progression

15. Find the sum of the first 12 terms of the AP

$$3, 6, 9, 12, \ldots$$

16. Find the sum of the first five terms of the AP

$$1, 5, 9, 13, \ldots$$

17. Find the sum of the first nine terms of the AP

$$5, 10, 15, 20, \ldots$$

18. Find the sum of the first 20 terms of the AP

$$1, 3, 5, 7, \ldots$$

19. How many terms of the AP 4, 7, 10, . . . will give a sum of 375?

20. How many terms of the AP 2, 9 , 16, . . . will give a sum of 270?

Arithmetic Means

21. Insert two arithmetic means between 5 and 20.

22. Insert five arithmetic means between 7 and 25.

23. Insert four arithmetic means between -6 and -9.

24. Insert three arithmetic means between 20 and 56.

Harmonic Progressions

25. Find the fourth term of the harmonic progression

$$\frac{3}{5}, \frac{3}{8}, \frac{3}{11}, \ldots$$

26. Find the fifth term of the harmonic progression

$$\frac{4}{19}, \frac{4}{15}, \frac{4}{11}, \ldots$$

Harmonic Means

27. Insert two harmonic means between $\frac{7}{9}$ and $\frac{7}{15}$.

28. Insert three harmonic means between $\frac{6}{21}$ and $\frac{6}{5}$.

Applications

29. *Loan Repayment:* A person agrees to repay a loan of $10,000 with an annual payment of $1000 plus 8% of the unpaid balance.
(a) Show that the interest payments alone form the AP: $800, $720, $640, ...
(b) Find the total amount of interest paid.

30. *Simple Interest:* A person deposits $50 in a bank on the first day of each month, at the same time withdrawing all interest earned on the money already in the account.
(a) If the rate is 1% per month, computed monthly, write an AP whose terms are the amounts withdrawn each month.
(b) How much interest will have been earned in the 36 months following the first deposit?

> To find the amount of depreciation for each year, divide the total depreciation (initial value – scrap value) by the number of years of depreciation.

31. *Straight-Line Depreciation:* A certain milling machine has an initial value of $150,000 and a scrap value of $10,000 twenty years later. Assuming that the machine depreciates the same amount each year, find its value after 8 years.

32. *Salary or Price Increase:* A person is hired at a salary of $40,000 and receives a raise of $2500 at the end of each year. Find the total amount earned during 10 years.

33. *Freely Falling Body:* A freely falling body falls $g/2$ feet during the first second, $3g/2$ feet during the next second, $5g/2$ feet during the third second, and so on, where $g \approx 32.2$ ft/s². Find the total distance the body falls during the first 10 s.

34. Using the information of problem 33, show that the total distance s fallen in t seconds is $s = \frac{1}{2}gt^2$.

Graphics Calculator

35. Some graphics calculators will compute the sum of a progression. You first indicate the progression by entering the expression for the nth term, the first and last values of n, and the increment for n (usually 1). Check your manual for the proper instructions, and use your calculator to solve problems 15 through 20 of this exercise set.

Computer

36. Some computer algebra systems, such as *Maple*, can compute the sum of a progression once the progression is entered. If your CAS has this capability, use it to solve problems 15 through 20 of this exercise set.

25–3 Geometric Progressions

Recursion Formula

A geometric sequence or *geometric progression* (GP) is one in which each term after the first is formed by multiplying the preceding term by a factor r, called the *common ratio*. Thus if a_n is any term of a GP, the recursion relation is as follows:

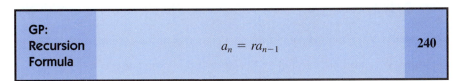

GP: Recursion Formula	$a_n = ra_{n-1}$	240

Each term of a GP after the first equals the product of the preceding term and the common ratio.

◆◆◆ **Example 21:** Some geometric progressions, with their common ratios given, are as follows:

(a) 2, 4, 8, 16, . . . $(r = 2)$
(b) 27, 9, 3, 1, $\frac{1}{3}$, . . . $(r = \frac{1}{3})$
(c) $-1, 3, -9, 27, . . .$ $(r = -3)$ ◆◆◆

General Term

For a GP whose first term is a and whose common ratio is r, the terms are

$$a, ar, ar^2, ar^3, ar^4, . . .$$

We see that each term after the first is the product of the first term and a power of r, where the power of r is one less than the number n of the term. So the nth term a_n is given by the following equation:

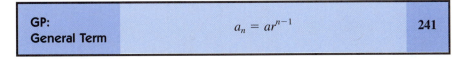

GP: General Term	$a_n = ar^{n-1}$	241

The nth term of a GP is found by multiplying the first term by the $n - 1$ power of the common ratio.

♦♦♦ **Example 22:** Find the sixth term of a GP with first term 5 and common ratio 4.

Solution: We substitute into Eq. 241 for the general term of a GP, with $a = 5$, $n = 6$, and $r = 4$.

$$a_6 = 5(4^5) = 5(1024) = 5120$$

♦♦♦

GP: Sum of n Terms

We find a formula for the sum s_n of the first n terms of a GP (also called the sum of n terms of a geometric series) by adding the terms of the GP.

$$s_n = a + ar + ar^2 + ar^3 + \cdots + ar^{n-2} + ar^{n-1} \qquad (1)$$

Multiplying each term in Equation (1) by r gives

$$rs_n = ar + ar^2 + ar^3 + \cdots + ar^{n-1} + ar^n \qquad (2)$$

Subtracting (1) from (2) term by term, we get

$$(1 - r)s_n = a - ar^n$$

Dividing both sides by $(1 - r)$ gives us the following formula:

GP: Sum of n Terms	$s_n = \dfrac{a(1 - r^n)}{1 - r}$	242

♦♦♦ **Example 23:** Find the sum of the first six terms of the GP in Example 22.

Solution: We substitute into Eq. 242 using $a = 5$, $n = 6$, and $r = 4$.

$$s_n = \frac{a(1 - r^n)}{1 - r} = \frac{5(1 - 4^6)}{1 - 4} = \frac{5(-4095)}{-3} = 6825$$

♦♦♦

We can get another equation for the sum of a GP in terms of the nth term a_n. We substitute into Eq. 242, using $ar^n = r(ar^{n-1}) = ra_n$, as follows:

GP: Sum of n Terms	$s_n = \dfrac{a - ra_n}{1 - r}$	243

♦♦♦ **Example 24:** Repeat Example 23, given that the sixth term (found in Example 22) is $a_6 = 5120$.

Solution: Substitution, with $a = 5$, $r = 4$, and $a_6 = 5120$, yields

$$s_n = \frac{a - ra_n}{1 - r} = \frac{5 - 4(5120)}{1 - 4} = 6825$$

as before.

♦♦♦

Geometric Means

As with the AP, the intermediate terms between any two terms are called *means*. A single number inserted between two numbers is called the *geometric mean* between those numbers.

◆◆◆ **Example 25:** Insert a geometric mean b between two numbers a and c.

Solution: Our GP is a, b, c. The common ratio r is then

$$r = \frac{b}{a} = \frac{c}{b}$$

from which $b^2 = ac$, which is rewritten as follows:

Geometric Mean	$b = \pm\sqrt{ac}$	59

This is not new. We studied the mean proportional in Chapter 19.

The geometric mean, or mean proportional, between two numbers is equal to the square root of their product. ◆◆◆

◆◆◆ **Example 26:** Find the geometric mean between 3 and 48.

Solution: Letting $a = 3$ and $c = 48$ gives us

$$b = \pm\sqrt{3(48)} = \pm 12$$

Our GP is then

$$3, \mathbf{12}, 48$$

or

$$3, \mathbf{-12}, 48$$

Note that we get *two* solutions. ◆◆◆

To insert *several* geometric means between two numbers, we first find the common ratio.

◆◆◆ **Example 27:** Insert four geometric means between 2 and $15\frac{3}{16}$.

Solution: Here $a = 2$, $a_6 = 15\frac{3}{16}$, and $n = 6$. Then

$$a_6 = ar^5$$
$$15\frac{3}{16} = 2r^5$$
$$r^5 = \frac{243}{32}$$
$$r = \frac{3}{2}$$

Having r, we can write the terms of the GP. They are

$$2, \mathbf{3}, \mathbf{4\frac{1}{2}}, \mathbf{6\frac{3}{4}}, \mathbf{10\frac{1}{8}}, 15\frac{3}{16}$$

◆◆◆

Exercise 3 ◆ Geometric Progressions

1. Find the fifth term of a GP with first term 5 and common ratio 2.
2. Find the fourth term of a GP with first term 7 and common ratio -4.
3. Find the sixth term of a GP with first term -3 and common ratio 5.
4. Find the fifth term of a GP with first term -4 and common ratio -2.
5. Find the sum of the first ten terms of the GP in problem 1.
6. Find the sum of the first nine terms of the GP in problem 2.

7. Find the sum of the first eight terms of the GP in problem 3.

8. Find the sum of the first five terms of the GP in problem 4.

Geometric Means

9. Insert a geometric mean between 5 and 45.

10. Insert a geometric mean between 7 and 112.

11. Insert a geometric mean between -10 and -90.

12. Insert a geometric mean between -21 and -84.

13. Insert two geometric means between 8 and 216.

14. Insert two geometric means between 9 and -243.

15. Insert three geometric means between 5 and 1280.

16. Insert three geometric means between 144 and 9.

Applications

17. *Exponential Growth:* Using the equation for exponential growth:

$$y = ae^{nt} \qquad\qquad \textbf{199}$$

with $a = 1$ and $n = 0.5$, compute values of y for $t = 0, 1, 2, \ldots, 10$. Show that while the values of t form an AP, the values of y form a GP. Find the common ratio.

18. *Exponential Decay:* Repeat problem 17 with the formula for exponential decay:

$$y = ae^{-nt} \qquad\qquad \textbf{201}$$

19. *Cooling:* A certain iron casting is at 1800°F and cools so that its temperature at each minute is 10% less than its temperature the preceding minute. Find its temperature after 1 h.

20. *Light through an Absorbing Medium:* Sunlight passes through a glass filter. Each millimeter of glass absorbs 20% of the light passing through it. What percentage of the original sunlight will remain after passing through 5.0 mm of the glass?

21. *Radioactive Decay:* A certain radioactive material decays so that after each year the radioactivity is 8% less than at the start of that year. How many years will it take for its radioactivity to be 50% of its original value?

22. *Pendulum:* Each swing of a certain pendulum is 85.0% as long as the one before. If the first swing is 12.0 in., find the entire distance traveled in eight swings.

23. *Bouncing Ball:* A ball dropped from a height of 10.0 ft rebounds to half its height on each bounce. Find the total distance traveled when it hits the ground for the fifth time.

24. *Population Growth:* Each day the size of a certain colony of bacteria is 25% larger than on the preceding day. If the original size of the colony was 10,000 bacteria, find its size after 5 days.

25. A person has two parents, and each parent has two parents, and so on. We can write a GP for the number of ancestors as 2, 4, 8, ... Find the total number of ancestors in five generations, starting with the parents' generation.

26. *Musical Scale:* The frequency of the "A" note above middle C is, by international agreement, equal to 440 Hz. A note one *octave* higher is at twice that frequency, or 880 Hz. The octave is subdivided into 12 *half-tone* intervals, where each half-tone is higher than the one preceding by a factor equal to the twelfth root of 2. Write a GP showing the frequency of each half-tone, from 440 to 880 Hz. Work to two decimal places.

One of the most famous and controversial references to arithmetic and geometric progressions was made by Thomas Malthus in 1798. He wrote: *"Population, when unchecked, increases in a geometrical ratio, and subsistence for man in an arithmetical ratio."*

This is called the *equally tempered scale* and is usually attributed to Johann Sebastian Bach (1685–1750).

27. *Chemical Reactions:* Increased temperature usually causes chemicals to react faster. If a certain reaction proceeds 15% faster for each 10°C increase in temperature, by what factor is the reaction speed increased when the temperature rises by 50°C?

28. *Mixtures:* A radiator contains 30% antifreeze and 70% water. One-fourth of the mixture is removed and replaced by pure water. If this procedure is repeated three more times, find the percent antifreeze in the final mixture.

29. *Energy Consumption:* If the U.S. energy consumption is 7.00% higher each year, by what factor will the energy consumption have increased after 10.0 years?

30. *Atmospheric Pressure:* The pressures measured at 1-mi intervals above sea level form a GP, with each value smaller than the preceding by a factor of 0.819. If the pressure at sea level is 29.92 in. Hg, find the pressure at an altitude of 5 mi.

31. *Compound Interest:* A person deposits $10,000 in a bank giving 6% interest, compounded annually. Find to the nearest dollar the value of the deposit after 50 years.

32. *Inflation:* The price of a certain house, now $126,000, is expected to increase by 5% each year. Write a GP whose terms are the value of the house at the end of each year, and find the value of the house after 5 years.

33. *Depreciation:* When calculating depreciation by the *declining-balance method,* a taxpayer claims as a deduction a fixed percentage of the book value of an asset each year. The new book value is then the last book value less the amount of the depreciation. Thus for a machine having an initial book value of $100,000 and a depreciation rate of 40%, the first year's depreciation is 40% of $100,000, or $40,000, and the new book value is $100,000 − $40,000 = $60,000. Thus the book values for each year form the following GP:

$$\$100{,}000, \$60{,}000, \$36{,}000, \ldots$$

Find the book value after 5 years.

Graphics Calculator

34. Some graphics calculators will compute the sum of a progression. You first indicate the progression by entering the expression for the nth term, the first and last values of n, and the increment for n (usually 1). Check your manual for the proper instructions, and use your calculator to solve problems 5 through 8 of this exercise set.

Computer

35. Some computer algebra systems, such as *Maple,* can compute the sum of a progression once the progression is entered. If your CAS has this capability, use it to solve problems 5 through 8 of this exercise set.

25–4   Infinite Geometric Progressions

Sum of an Infinite Geometric Progression

Before we derive a formula for the sum of an infinite geometric progression, let us explore the idea graphically and numerically.

We have already determined that the sum of n terms of any geometric progression with a first term a and a common ratio r is

$$s_n = \frac{a(1 - r^n)}{1 - r} \tag{242}$$

Thus a graph of s_n versus n should tell us about the sum changes as n increases.

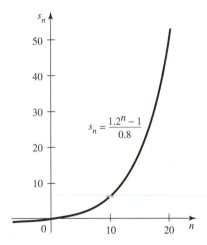

FIGURE 25–4

◆◆◆ **Example 28:** Graph the sum s_n versus n for the GP

$$1, 1.2, 1.2^2, 1.2^3, \ldots$$

for $n = 0$ to 20.

Solution: Here, $a = 1$ and $r = 1.2$, so

$$s_n = \frac{1 - 1.2^n}{1 - 1.2} = \frac{1.2^n - 1}{0.8}$$

The sum s_n is graphed in Fig. 25–4. Note that the sum continues to increase. We say that this progression *diverges*. ◆◆◆

Decreasing GP

If the common ratio r in a GP is less than 1, each term in the progression will be less than those preceding it. Such a progression is called a *decreasing GP*.

◆◆◆ **Example 29:** Graph the sum s_n versus n for the GP

$$1, 0.8, 0.8^2, 0.8^3, \ldots$$

for $n = 0$ to 20.

Solution: Here, $a = 1$ and $r = 0.8$, so

$$s_n = \frac{1 - 0.8^n}{1 - 0.8} = \frac{1 - 0.8^n}{0.2}$$

The sum is graphed in Fig. 25–5. Note that the sum appears to reach a limiting value, so we say that this progression *converges*. The sum appears to be approaching a value of 5.

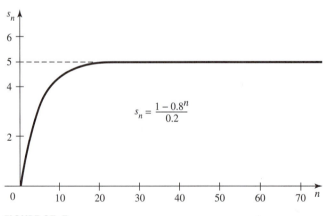

FIGURE 25–5 ◆◆◆

TABLE 25–3

	Term	Sum
1	9.0000	9.0000
2	3.0000	12.0000
3	1.0000	13.0000
4	0.3333	13.3333
5	0.1111	13.4444
6	0.0370	13.4815
7	0.0123	13.4938
8	0.0041	13.4979
9	9.0014	13.4993
10	0.0005	13.4998
11	0.0002	13.4999
12	0.0001	13.5000
13	0.0000	13.5000
14	0.0000	13.5000
15	0.0000	13.5000
16	0.0000	13.5000
17	0.0000	13.5000
18	0.0000	13.5000

Our graphs have thus indicated that the sum of an infinite number of terms of a decreasing GP appears to approach a limit. Let us now verify that fact numerically.

◆◆◆ **Example 30:** Find the sum of infinitely many terms of the GP

$$9, 3, 1, \tfrac{1}{3}, \ldots$$

Solution: Knowing that $a = 9$ and $r = \frac{1}{3}$, we can use a computer to compute each term and keep a running sum as shown in Table 25–3. Notice that the terms get smaller and smaller, and the sum appears to approach a value of around 13.5. ◆◆◆

We will confirm the value found in Example 30 after we have derived a formula for the sum of an infinite geometric progression.

Limit Notation

In order to derive a formula for the sum of an infinite geometric progression, we need to introduce limit notation. Such notation will also be of great importance in the study of calculus.

First, we use an arrow ($\rightarrow$) to indicate that a quantity *approaches* a given value.

We used limit notation briefly in Sec. 20–2, and here we show a bit more. Limits are usually covered more completely in a calculus course.

◆◆◆ **Example 31:**

(a) $x \rightarrow 5$ means that x gets closer and closer to the value 5.
(b) $n \rightarrow \infty$ means that n grows larger and larger, without bound.
(c) $y \rightarrow 0$ means that y continuously gets closer and closer to zero. ◆◆◆

If some function $f(x)$ approaches some value L as x approaches a, we could write this as

$$f(x) \rightarrow L \quad \text{as} \quad x \rightarrow a$$

which is written in more compact notation as follows:

Limit Notation	$\lim\limits_{x \to a} f(x) = L$	336

Limit notation lets us compactly write the value to which a function is tending, as in the following example.

◆◆◆ **Example 32:**

(a) $\lim\limits_{b \to 0} b = 0$

(b) $\lim\limits_{b \to 0}(a + b) = a$

(c) $\lim\limits_{b \to 0}\dfrac{a + b}{c + d} = \dfrac{a}{c + d}$

(d) $\lim\limits_{x \to 0}\dfrac{x + 1}{x + 2} = \dfrac{1}{2}$ ◆◆◆

A Formula for the Sum of an Infinite, Decreasing Geometric Progression

As the number of terms n of an infinite, decreasing geometric progression increases without bound, the term a_n will become smaller and will approach zero. Using our new notation, we may write

$$\lim_{n \to \infty} a_n = 0 \quad \text{when} \quad |r| < 1$$

Thus in the equation for the sum of n terms,

$$s_n = \frac{a - ra_n}{1 - r} \tag{243}$$

the expression ra_n will approach zero as n approaches infinity, so the sum s_∞ of infinitely many terms (called the *sum to infinity*) is

$$s_\infty = \lim_{n \to \infty} \frac{a - ra_n}{1 - r} = \frac{a}{1 - r}$$

Thus letting $S = s_\infty$, the sum to infinity is given by the following equation:

GP: Sum to Infinity	$S = \dfrac{a}{1-r}$ when $|r| < 1$	244

♦♦♦ **Example 33:** Find the sum of infinitely many terms of the GP

$$9, 3, 1, \frac{1}{3}, \ldots$$

Solution: Here $a = 9$ and $r = \frac{1}{3}$, so

$$S = \frac{a}{1-r} = \frac{9}{1 - 1/3} = \frac{9}{2/3} = 13\frac{1}{2}$$

This agrees with the value we found numerically in Example 30.   ♦♦♦

Repeating Decimals

A *finite decimal* (or terminating decimal) is one that has a finite number of digits to the right of the decimal point, followed by zeros. An *infinite decimal* (or nonterminating decimal) has an infinite string of digits to the right of the decimal point. A *repeating decimal* (or periodic decimal) is one that has a block of one or more digits that repeats indefinitely. It can be proved that when one integer is divided by another, we get either a finite decimal or a repeating decimal. Irrational numbers, on the other hand, yield *nonrepeating decimals.*

♦♦♦ **Example 34:**

(a) $\frac{3}{4} = 0.75000000000 \ldots$ (finite decimal)

(b) $\frac{3}{7} = 0.428571\ 428571\ 428571 \ldots$ (infinite repeating decimal)

(c) $\sqrt{2} = 1.414213562 \ldots$ (infinite nonrepeating decimal)   ♦♦♦

We can write a repeating decimal as a rational fraction by writing the repeating part as an infinite geometric series, as shown in the following example.

♦♦♦ **Example 35:** Write the number $4.85151 \ldots$ as a rational fraction.

Solution: We rewrite the given number as

$$4.8 + 0.051 + 0.00051 + \cdots$$

The terms of this series, except for the first, form an infinite GP where $a = 0.051$ and $r = 0.01$. The sum S to infinity is then

$$S = \frac{a}{1-r} = \frac{0.051}{1 - 0.01} = \frac{0.051}{0.99} = \frac{51}{990} = \frac{17}{330}$$

Our given number is then equal to $4.8 + S$, so

$$4.85151 = 4 + \frac{8}{10} + \frac{17}{330} = 4\frac{281}{330}$$ ♦♦♦

Exercise 4 ♦ Infinite Geometric Progressions

Evaluate each limit.

1. $\displaystyle\lim_{b \to 0}(b - c + 5)$

2. $\displaystyle\lim_{b \to 0}(a + b^2)$

3. $\displaystyle\lim_{b \to 0}\frac{3 + b}{c + 4}$

4. $\displaystyle\lim_{b \to 0}\frac{a + b + c}{b + c - 5}$

Find the sum of the infinitely many terms of each GP.

5. 144, 72, 36, 18, . . .   **6.** 8, 4, 1, $\frac{1}{4}$, . . .

7. 10, 2, 0.4, 0.08, . . .   **8.** 1, $\frac{1}{4}$, $\frac{1}{16}$, . . .

Write each decimal number as a rational fraction.

9. 0.57$\overline{57}$ . . .   **10.** 0.69$\overline{69}$ . . .

11. 7.68$\overline{18}$1 . . .   **12.** 5.86111 . . .

13. Each swing of a certain pendulum is 78% as long as the one before. If the first swing is 10 in., find the entire distance traveled by the pendulum before it comes to rest.

14. A ball dropped from a height of 20 ft rebounds to three-fourths of its height on each bounce. Find the total distance traveled by the ball before it comes to rest.

25–5   The Binomial Theorem

Powers of a Binomial

Recall that a *binomial,* such as $(a + b)$, is a polynomial with two terms. By actual multiplication we can show that

$$(a + b)^1 = a + b$$
$$(a + b)^2 = a^2 + 2ab + b^2$$
$$(a + b)^3 = a^3 + 3a^2b + 3ab^2 + b^3$$
$$(a + b)^4 = a^4 + 4a^3b + 6a^2b^2 + 4ab^3 + b^4$$

We now want a formula for $(a + b)^n$ with which to expand a binomial without actually carrying out the multiplication. In the expansion of $(a + b)^n$, where n is a positive integer, we note the following patterns:

1. There are $n + 1$ terms.
2. The power of a is n in the first term, decreases by 1 in each later term, and reaches 0 in the last term.
3. The power of b is 0 in the first term, increases by 1 in each later term, and reaches n in the last term.
4. Each term has a total degree of n. (That is, the sum of the degrees of the variables is n.)
5. The first coefficient is 1. Further, the product of the coefficient of any term and its power of a, divided by the number of the term, gives the coefficient of the next term. (This property gives a *recursion formula* for the coefficients of the binomial expansion.)

The formula is expressed as follows:

$$(a + b)^n = a^n + na^{n-1}b + \frac{n(n - 1)}{2}a^{n-2}b^2 + \frac{n(n - 1)(n - 2)}{2(3)}a^{n-3}b^3$$
$$+ \cdots + b^n$$

◆◆◆ **Example 36:** Use the binomial theorem to expand $(a + b)^5$.

Solution: From the binomial theorem,

$$(a + b)^5 = a^5 + 5a^4b + \frac{5(4)}{2}a^3b^2 + \frac{5(4)(3)}{2(3)}a^2b^3 + \frac{5(4)(3)(2)}{2(3)(4)}ab^4 + b^5$$
$$= a^5 + 5a^4b + 10a^3b^2 + 10a^2b^3 + 5ab^4 + b^5$$

which can be verified by actual multiplication.   ◆◆◆

If the expression contains only one variable, we can verify the expansion graphically.

••• **Example 37:** Use the result from Example 36 to expand $(x + 1)^5$, and verify graphically.

Solution: Substituting $a = x$ and $b = 1$ into the preceding result gives

$$(x + 1)^5 = x^5 + 5x^4 + 10x^3 + 10x^2 + 5x + 1$$

To verify graphically, we plot the following on the same axes:

$$y_1 = (x + 1)^5$$

and

$$y_2 = x^5 + 5x^4 + 10x^3 + 10x^2 + 5x + 1$$

The two graphs, shown in Fig. 25–6, are identical. This is not a *proof* that our expansion is correct, but it certainly gives us more confidence in our result.

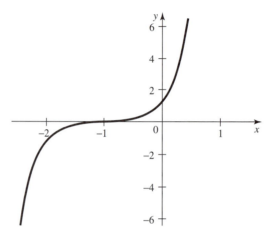

FIGURE 25–6   Graphs of $y = (x + 1)^5$ and $y = x^5 + 5x^4 + 10x^3 + 10x^2 + 5x + 1$.   •••

Binomial Expansion on the Computer

With a CAS, simply type in the expression to be expanded, and give the *Expand* command.

Factorial Notation

We now introduce *factorial notation.* For a positive integer n, factorial n, written $n!$, is the product of all of the positive integers less than or equal to n.

$$n! = 1 \cdot 2 \cdot 3 \cdots n$$

Of course, it does not matter in which order we multiply the numbers. We may also write

$$n! = n(n - 1) \cdots 3 \cdot 2 \cdot 1$$

••• **Example 38:**

(a) $3! = 1 \cdot 2 \cdot 3 = 6$
(b) $5! = 1 \cdot 2 \cdot 3 \cdot 4 \cdot 5 = 120$
(c) $\dfrac{6!\,3!}{7!} = \dfrac{6!\,(2)(3)}{7(6!)} = \dfrac{6}{7}$
(d) $0! = 1$ by definition   •••

We will use the binomial theorem to obtain the values in a binomial probability distribution in the following chapter on statistics.

Many calculators have a key for evaluating factorials.

Binomial Theorem Written with Factorial Notation

The use of factorial notation allows us to write the binomial theorem more compactly. Thus:

| Binomial Theorem | $(a + b)^n = a^n + na^{n-1}b + \dfrac{n(n-1)}{2!}a^{n-2}b^2 + \dfrac{n(n-1)(n-2)}{3!}a^{n-3}b^3 + \cdots + b^n$ | 245 |

◆◆◆ **Example 39:** Using the binomial formula, expand $(a + b)^6$.

Solution: There will be $(n + 1)$ or seven terms. Let us first find the seven coefficients. The first coefficient is always 1, and the second coefficient is always n, or 6. We calculate the remaining coefficients in Table 25–4.

TABLE 25–4

Term	Coefficient
1	1
2	$n = 6$
3	$\dfrac{n(n-1)}{2!} = \dfrac{6(5)}{1(2)} = 15$
4	$\dfrac{n(n-1)(n-2)}{3!} = \dfrac{6(5)(4)}{1(2)(3)} = 20$
5	$\dfrac{n(n-1)(n-2)(n-3)}{4!} = \dfrac{6(5)(4)(3)}{1(2)(3)(4)} = 15$
6	$\dfrac{n(n-1)(n-2)(n-3)(n-4)}{5!} = \dfrac{6(5)(4)(3)(2)}{1(2)(3)(4)(5)} = 6$
7	$\dfrac{n(n-1)(n-2)(n-3)(n-4)(n-5)}{6!} = \dfrac{6(5)(4)(3)(2)(1)}{1(2)(3)(4)(5)(6)} = 1$

The seven terms thus have the coefficients

$$1 \quad 6 \quad 15 \quad 20 \quad 15 \quad 6 \quad 1$$

and our expansion is

$$(a + b)^6 = a^6 + 6a^5b + 15a^4b^2 + 20a^3b^3 + 15a^2b^4 + 6ab^5 + b^6 \quad ◆◆◆$$

Pascal's Triangle

We now have expansions for $(a + b)^n$, for values of n from 1 to 6, to which we add $(a + b)^0 = 1$. Let us now write the coefficients only of the terms of each expansion.

```
Exponent n

0                        1
1                     1     1
2                  1     2     1
3               1     3     3     1
4            1     4     6     4     1
5         1     5    10    10     5     1
6      1     6    15    20    15     6     1
```

Pascal's triangle is named for Blaise Pascal (1623–62), the French geometer, probabilist, combinatorist, physicist, and philosopher.

This array is called *Pascal's triangle.* We may use it to predict the coefficients of expansions with powers higher than 6 by noting that each number in the triangle is equal to the sum of the two numbers above it (thus the 15 in the lowest row is the sum of the 5 and the 10 above it). When expanding a binomial, we may take the coefficients directly from Pascal's triangle, as shown in the following example.

◆◆◆ **Example 40:** Expand $(1 - x)^6$, obtaining the binomial coefficients from Pascal's triangle.

Solution: From Pascal's triangle, for $n = 6$, we read the coefficients

$$1 \quad 6 \quad 15 \quad 20 \quad 15 \quad 6 \quad 1$$

We now substitute into the binomial formula with $n = 6$, $a = 1$, and $b = -x$. When one of the terms of the binomial is negative, as here, we must be careful to include the minus sign whenever that term appears in the expansion. Thus

$$(1 - x)^6 = [1 + (-x)]^6$$
$$= 1^6 + 6(1^5)(-x) + 15(1^4)(-x)^2 + 20(1^3)(-x)^3$$
$$+ 15(1^2)(-x)^4 + 6(1)(-x)^5 + (-x)^6$$
$$= 1 - 6x + 15x^2 - 20x^3 + 15x^4 - 6x^5 + x^6$$

◆◆◆

If either term in the binomial is itself a power, a product, or a fraction, it is a good idea to enclose that entire term in parentheses before substituting.

◆◆◆ **Example 41:** Expand $(3/x^2 + 2y^3)^4$.

Solution: The binomial coefficients are

$$1 \quad 4 \quad 6 \quad 4 \quad 1$$

We substitute $3/x^2$ for a and $2y^3$ for b.

$$\left(\frac{3}{x^2} + 2y^3\right)^4 = \left[\left(\frac{3}{x^2}\right) + (2y^3)\right]^4$$
$$= \left(\frac{3}{x^2}\right)^4 + 4\left(\frac{3}{x^2}\right)^3(2y^3) + 6\left(\frac{3}{x^2}\right)^2(2y^3)^2 + 4\left(\frac{3}{x^2}\right)(2y^3)^3 + (2y^3)^4$$
$$= \frac{81}{x^8} + \frac{216y^3}{x^6} + \frac{216y^6}{x^4} + \frac{96y^9}{x^2} + 16y^{12}$$

◆◆◆

Sometimes we may not need the entire expansion but only the first several terms.

◆◆◆ **Example 42:** Find the first four terms of $(x^2 - 2y^3)^{11}$.

Solution: From the binomial formula, with $n = 11$, $a = x^2$, and $b = (-2y^3)$,

$$(x^2 - 2y^3)^{11} = [(x^2) + (-2y^3)]^{11}$$
$$= (x^2)^{11} + 11(x^2)^{10}(-2y^3) + \frac{11(10)}{2}(x^2)^9(-2y^3)^2$$
$$+ \frac{11(10)(9)}{2(3)}(x^2)^8(-2y^3)^3 + \cdots$$
$$= x^{22} - 22x^{20}y^3 + 220x^{18}y^6 - 1320x^{16}y^9 + \cdots$$

◆◆◆

A *trinomial* may be expanded by the binomial formula by grouping the terms, as in the following example.

◆◆◆ **Example 43:** Expand $(1 + 2x - x^2)^3$ by the binomial formula.

Solution: Let $z = 2x - x^2$. Then

$$(1 + 2x - x^2)^3 = (1 + z)^3 = 1 + 3z + 3z^2 + z^3$$

Substituting back, we have

$$(1 + 2x - x^2)^3 = 1 + 3(2x - x^2) + 3(2x - x^2)^2 + (2x - x^2)^3$$
$$= 1 + 6x + 9x^2 - 4x^3 - 9x^4 + 6x^5 - x^6 \qquad ◆◆◆$$

General Term

We can write the general or *r*th term of a binomial expansion $(a + b)^n$ by noting the following pattern:

1. The power of b is 1 less than r. Thus a fifth term would contain b^4, and the rth term would contain b^{r-1}.
2. The power of a is n minus the power of b. Thus a fifth term would contain a^{n-5+1}, and the rth term would contain a^{n-r+1}.
3. The coefficient of the rth term is

$$\frac{n!}{(r - 1)!(n - r + 1)!}$$

The formula is expressed as follows:

General Term	In the expansion for $(a + b)^n$, $$r\text{th term} = \frac{n!}{(r - 1)!(n - r + 1)!}a^{n-r+1}b^{r-1}$$	246

◆◆◆ **Example 44:** Find the eighth term of $(3a + b^5)^{11}$.

Solution: Here, $n = 11$ and $r = 8$. So $n - r + 1 = 4$. Substitution yields

$$\frac{n!}{(r - 1)!(n - r + 1)!} = \frac{11!}{7! \, 4!} = 330$$

Then

$$\text{the eighth term} = 330(3a)^{11-8+1}(b^5)^{8-1}$$
$$= 330(81a^4)(b^{35}) = 26{,}730a^4b^{35} \qquad ◆◆◆$$

Proof of the Binomial Theorem

We have shown that the binomial theorem is true for $n = 1, 2, 3, 4,$ or 5. But what about some other exponent, say, $n = 50$? To show that the theorem works for any positive integral exponent, we give the following proof.

We first show that if the binomial theorem is true for some exponent $n = k$, it is also true for $n = k + 1$. We assume that

$$(a + b)^k = a^k + ka^{k-1}b + \cdots + Pa^{k-r}b^r + Qa^{k-r-1}b^{r+1}$$
$$+ Ra^{k-r-2}b^{r+2} + \cdots + b^k \qquad (1)$$

where we show three intermediate terms with coefficients P, Q, and R. Multiplying both sides of Equation (1) by $(a + b)$ gives

$$(a + b)^{k+1} = a^{k+1} + ka^k b + \cdots + Pa^{k-r+1}b^r + Qa^{k-r}b^{r+1}$$
$$+ Ra^{k-r-1}b^{r+2} + \cdots + ab^k + \cdots + a^k b + \cdots$$
$$+ Pa^{k-r}b^{r+1} + Qa^{k-r-1}b^{r+2} + Ra^{k-r-2}b^{r+3} + \cdots + b^{k+1}$$
$$(a + b)^{k+1} = a^{k+1} + (k + 1)a^k b + \cdots + (P + Q)a^{k-r}b^{r+1}$$
$$+ (Q + R)a^{k-r-1}b^{r+2} + \cdots + ab^k + b^{k+1} \tag{2}$$

We have not written the b^r term or the b^{r+3} term because we do not have expressions for their coefficients, and they are not needed for the proof.

after combining like terms.

Does Equation (2) obey the binomial theorem? First, the power of b is 0 in the first term and increases by 1 in each later term, reaching $k + 1$ in the last term, so property 3 is met. Thus there are $k + 1$ terms containing b, plus the first term that does not, making a total of $k + 2$ terms, so property 1 is met.

Next, the power of a is $k + 1$ in the first term and decreases by 1 in each succeeding term until it equals 0 in the last term, making the total degree of each term equal to $k + 1$, so properties 2 and 4 are met. Further, the coefficient of the first term is 1, and the coefficient of the second term is $k + 1$, which agrees with property 5.

Finally, in a binomial expansion, the product of the coefficient of any term and its power of a, divided by the number of the term, must give the coefficient of the next term. Here, the first of our two intermediate terms has a coefficient $(P + Q)$. The coefficient of its a term is $k - r$, the number of that term is $r + 2$, and the coefficient of the next term is $(Q + R)$. Thus the identity

$$(P + Q)\frac{k - r}{r + 2} = Q + R$$

must be true. But since

$$Q = \frac{P(k - r)}{r + 1} \quad \text{and} \quad R = \frac{Q(k - r - 1)}{r + 2}$$

we must show that

$$\left[P + \frac{P(k - r)}{r + 1} \right] \frac{k - r}{r + 2} = \frac{P(k - r)}{r + 1} + \frac{P(k - r)(k - r - 1)}{(r + 1)(r + 2)}$$

Working each side separately, we combine each over a common denominator.

$$\frac{P(k - r)(r + 1)}{(r + 1)(r + 2)} + \frac{P(k - r)(k - r)}{(r + 1)(r + 2)} = \frac{P(k - r)(r + 2)}{(r + 1)(r + 2)} + \frac{P(k - r)(k - r - 1)}{(r + 1)(r + 2)}$$

Since the denominators are the same on both sides, we simply have to show that the numerators are identical. Factoring gives us

$$P(k - r)(r + 1 + k - r) = P(k - r)(r + 2 + k - r - 1)$$
$$P(k - r)(k + 1) = P(k - r)(k + 1)$$

This type of proof is called *mathematical induction*.

This completes the proof that if the binomial theorem is true for some exponent $n = k$, it is also true for the exponent $n = k + 1$. But since we have shown, by direct multiplication, that the binomial theorem *is true* for $n = 5$, it must be true for $n = 6$. And since it is true for $n = 6$, it must be true for $n = 7$, and so on, for all of the positive integers.

Fractional and Negative Exponents

We have shown that the binomial formula is valid when the exponent n is a positive integer. We shall not prove it here, but the binomial formula also holds when the exponent is negative or fractional. However, we now obtain an infinite series, called a *binomial series,* with no last term. Further, the binomial series is equal to $(a + b)^n$ only if the series converges.

◆◆◆ **Example 45:** Write the first four terms of the infinite binomial expansion for $\sqrt{5 + x}$.

Solution: We replace the radical with a fractional exponent and substitute into the binomial formula in the usual way.

$$(5 + x)^{1/2} = 5^{1/2} + \left(\frac{1}{2}\right)5^{-1/2}x + \frac{(1/2)(-1/2)}{2}5^{-3/2}x^2$$

$$+ \frac{(1/2)(-1/2)(-3/2)}{6}5^{-5/2}x^3 + \cdots$$

$$= \sqrt{5} + \frac{x}{2\sqrt{5}} - \frac{x^2}{8(5)^{3/2}} + \frac{x^3}{16(5)^{5/2}} - \cdots$$

Switching to decimal form and working to five decimal places, we get

$$(5 + x)^{1/2} \approx 2.23607 + 0.22361x - 0.01118x^2 + 0.00112x^3 - \cdots \quad ◆◆◆$$

We now ask if this series is valid for any x. Let's try a few values. For $x = 0$,

$$(5 + 0)^{1/2} \approx 2.23607$$

which checks within the number of digits retained. For $x = 1$,

$$(5 + 1)^{1/2} = \sqrt{6} \approx 2.23607 + 0.22361 - 0.01118 + 0.00112 - \cdots$$

$$\approx 2.44962$$

which compares with a value for $\sqrt{6}$ of 2.44949. Not bad, considering that we are using only four terms of an infinite series.

To determine the effect of larger values of x, we have used a computer to generate eight terms of the binomial series as listed in Table 25–5.

TABLE 25–5

x	$\sqrt{5 + x}$ by Series	$\sqrt{5 + x}$ by Calculator	Error
0	2.236068	2.236068	0
2	2.645766	2.645751	0.000015
4	3.002957	3	0.002957
6	3.379908	3.316625	0.063283
8	4.148339	3.605551	0.543788
11	9.723964	4	5.723964

Into it we have substituted various x values, from 0 to 11, and compared the value obtained by series with that from a calculator. The values are plotted in Fig. 25–7.

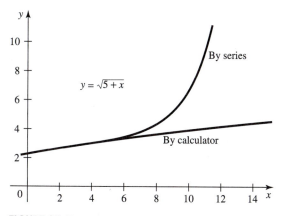

FIGURE 25–7

Notice that the accuracy gets worse as x gets larger, with useless values being obtained when x is 6 or greater. This illustrates the following:

> A binomial series is equal to $(a + b)^n$ only if the series converges, and that occurs when $|a| > |b|$.

Our series in Example 45 thus converges when $x < 5$.

The binomial series, then, is given by the following formula:

| Binomial Series | $$(a + b)^n = a^n + na^{n-1}b + \frac{n(n-1)}{2!}a^{n-2}b^2 + \frac{n(n-1)(n-2)}{3!}a^{n-3}b^3 + \cdots$$ $$\text{where } |a| > |b|$$ | 247 |

The binomial series is often written with $a = 1$ and $b = x$, so we give that form here also.

| Binomial Series | $$(1 + x)^n = 1 + nx + \frac{n(n-1)}{2!}x^2 + \frac{n(n-1)(n-2)}{3!}x^3 + \cdots$$ $$\text{where } |x| < 1$$ | 248 |

◆◆◆ **Example 46:** Expand to four terms:

$$\frac{1}{\sqrt[3]{1/a - 3b^{1/3}}}$$

Solution: We rewrite the given expression without radicals or fractions, and then expand.

$$\frac{1}{\sqrt[3]{1/a - 3b^{1/3}}} = [(a^{-1}) + (-3b^{1/3})]^{-1/3}$$

$$= (a^{-1})^{-1/3} - \frac{1}{3}(a^{-1})^{-4/3}(-3b^{1/3}) + \frac{2}{9}(a^{-1})^{-7/3}(-3b^{1/3})^2$$

$$- \frac{14}{81}(a^{-1})^{-10/3}(-3b^{1/3})^3 + \cdots$$

$$= a^{1/3} + a^{4/3}b^{1/3} + 2a^{7/3}b^{2/3} + \frac{14}{3}a^{10/3}b + \cdots$$

◆◆◆

Exercise 5 ◆ The Binomial Theorem

Factorial Notation

Evaluate each factorial.

1. $6!$
2. $8!$
3. $\dfrac{7!}{5!}$
4. $\dfrac{11!}{9!\,2!}$
5. $\dfrac{7!}{3!\,4!}$
6. $\dfrac{8!}{3!\,5!}$

Binomials Raised to an Integral Power

Verify each expansion. Obtain the binomial coefficients by formula or from Pascal's triangle as directed by your instructor.

7. $(x + y)^7 = x^7 + 7x^6y + 21x^5y^2 + 35x^4y^3 + 35x^3y^4 + 21x^2y^5 + 7xy^6 + y^7$

8. $(4 + 3b)^4 = 256 + 768b + 864b^2 + 432b^3 + 81b^4$

9. $(3a - 2b)^4 = 81a^4 - 216a^3b + 216a^2b^2 - 96ab^3 + 16b^4$
10. $(x^3 + y)^7 = x^{21} + 7x^{18}y + 21x^{15}y^2 + 35x^{12}y^3 + 35x^9y^4 + 21x^6y^5$
    $+ 7x^3y^6 + y^7$
11. $(x^{1/2} + y^{2/3})^5 = x^{5/2} + 5x^2y^{2/3} + 10x^{3/2}y^{4/3} + 10xy^2 + 5x^{1/2}y^{8/3} + y^{10/3}$
12. $(1/x + 1/y^2)^3 = 1/x^3 + 3/x^2y^2 + 3/xy^4 + 1/y^6$
13. $(a/b - b/a)^6 = (a/b)^6 - 6(a/b)^4 + 15(a/b)^2 - 20 + 15(b/a)^2 - 6(b/a)^4$
    $+ (b/a)^6$
14. $(1/\sqrt{x} + y^2)^4 = x^{-2} + 4x^{-3/2}y^2 + 6x^{-1}y^4 + 4x^{-1/2}y^6 + y^8$
15. $(2a^2 + \sqrt{b})^5 = 32a^{10} + 80a^8b^{1/2} + 80a^6b + 40a^4b^{3/2} + 10a^2b^2 + b^{5/2}$
16. $(\sqrt{x} - c^{2/3})^4 = x^2 - 4x^{3/2}c^{2/3} + 6xc^{4/3} - 4x^{1/2}c^2 + c^{8/3}$

Verify the first four terms of each binomial expansion.

17. $(x^2 + y^3)^8 = x^{16} + 8x^{14}y^3 + 28x^{12}y^6 + 56x^{10}y^9 + \cdots$
18. $(x^2 - 2y^3)^{11} = x^{22} - 22x^{20}y^3 + 220x^{18}y^6 - 1320x^{16}y^9 + \cdots$
19. $(a - b^4)^9 = a^9 - 9a^8b^4 + 36a^7b^8 - 84a^6b^{12} + \cdots$
20. $(a^3 + 2b)^{12} = a^{36} + 24a^{33}b + 264a^{30}b^2 + 1760a^{27}b^3 + \cdots$

Trinomials

Verify each expansion.

21. $(2a^2 + a + 3)^3 = 8a^6 + 12a^5 + 42a^4 + 37a^3 + 63a^2 + 27a + 27$
22. $(4x^2 - 3xy - y^2)^3 = 64x^6 - 144x^5y + 60x^4y^2 + 45x^3y^3 - 15x^2y^4$
    $- 9xy^5 - y^6$
23. $(2a^2 + a - 4)^4 = 16a^8 + 32a^7 - 104a^6 - 184a^5 + 289a^4 + 368a^3$
    $- 416a^2 - 256a + 256$

General Term

Write the requested term of each binomial expansion, and simplify.

24. Seventh term of $(a^2 - 2b^3)^{12}$
25. Eleventh term of $(2 - x)^{16}$
26. Fourth term of $(2a - 3b)^7$
27. Eighth term of $(x + a)^{11}$
28. Fifth term of $(x - 2\sqrt{y})^{25}$
29. Ninth term of $(x^2 + 1)^{15}$

Binomials Raised to a Fractional or Negative Power

Verify the first four terms of each infinite binomial series.

30. $(1 - a)^{2/3} = 1 - 2a/3 - a^2/9 - 4a^3/81 \ldots$
31. $\sqrt{1 + a} = 1 + a/2 - a^2/8 + a^3/16 \ldots$
32. $(1 + 5a)^{-5} = 1 - 25a + 375a^2 - 4375a^3 \ldots$
33. $(1 + a)^{-3} = 1 - 3a + 6a^2 - 10a^3 \ldots$
34. $1/\sqrt[6]{1 - a} = 1 + a/6 + 7a^2/72 + 91a^3/1296 \ldots$

Computer and Graphics Calculator

35. Use a graphics calculator or a graphing utility on a computer to graph each binomial and four terms of the expansion for problems 30 through 34, and compare the results. Choose a value of a with an absolute value less than 1.
36. Use a CAS to expand any of the binomials in this exercise set.

❖❖❖ **CHAPTER 25 REVIEW PROBLEMS** ❖❖❖❖❖❖❖❖❖❖❖❖❖❖❖❖❖❖❖❖❖❖❖❖❖❖❖❖❖❖❖❖❖

1. Find the sum of seven terms of the AP: $-4, -1, 2, \ldots$
2. Find the tenth term of $(1 - y)^{14}$.
3. Insert four arithmetic means between 3 and 18.

4. How many terms of the AP -5, -2, 1, . . . will give a sum of 63?

5. Insert four harmonic means between 2 and 12.

6. Find the twelfth term of an AP with first term 7 and common difference 5.

7. Find the sum of the first 10 terms of the AP 1, $2\frac{2}{3}$, $4\frac{1}{3}$, ...

Expand each binomial.

8. $(2x^2 - \sqrt{x}/2)^5$;

9. $(a - 2)^5$

10. $(x - y^3)^7$

11. Find the fifth term of a GP with first term 3 and common ratio 2.

12. Find the fourth term of a GP with first term 8 and common ratio -4.

Evaluate each limit.

13. $\lim\limits_{b \to 0}(2b - x + 9)$

14. $\lim\limits_{b \to 0}(a^2 + b^2)$

Find the sum of infinitely many terms of each GP.

15. 4, 2, 1, . . .

16. $\frac{1}{4}$, $-\frac{1}{16}$, $\frac{1}{64}$, . . .

17. 1, $-\frac{2}{5}$, $\frac{4}{25}$, . . .

Write each decimal number as a rational fraction.

18. $0.8\overline{86}$ . . .

19. $0.5\overline{44}$ . . .

20. Find the seventh term of the harmonic progression 3, $3\frac{3}{7}$, 4, ...

21. Find the sixth term of a GP with first term 5 and common ratio 3.

22. Find the sum of the first seven terms of the GP in problem 21.

23. Find the fifth term of a GP with first term -4 and common ratio 4.

24. Find the sum of the first seven terms of the GP in problem 23.

25. Insert two geometric means between 8 and 125.

26. Insert three geometric means between 14 and 224.

Evaluate each factorial.

27. $\dfrac{4!\,5!}{2!}$

28. $\dfrac{8!}{4!\,4!}$

29. $\dfrac{7!\,3!}{5!}$

30. $\dfrac{7!}{2!\,5!}$

Find the first four terms of each binomial expansion.

31. $(\dfrac{2x}{y^2} - y\sqrt{x})^7$

32. $(1 + a)^{-2}$

33. $\dfrac{1}{\sqrt{1 + a}}$

34. Expand using the binomial theorem:

$$(1 - 3a + 2a^2)^4$$

35. Find the eighth term of $(3a - b)^{11}$.

36. Find the tenth term of an AP with first term 12 if the sum of the first ten terms is 10.

Writing

37. State in your own words the difference between an arithmetic progression and a geometric progression. Give a real-world example of each.

Team Project

38. *Zeno's Paradox:* "A runner can never reach a finish line 1 mile away because first he would have to run half a mile, and then must run half of the remaining distance, or $\frac{1}{4}$ mile, and then half of that, and so on. Since there are an infinite number of distances that must be run, it will take an infinite length of time, and so the runner will never reach the finish line." Show that the distances to be run form an infinite series,

$$\frac{1}{2}, \frac{1}{4}, \frac{1}{8}, \cdots$$

Then disprove the paradox by actually finding the sum of that series and showing that the sum is *not* infinite.

Zeno's Paradox is named for Zeno of Elea (ca. 490–ca. 435 B.C.), a Greek philosopher and mathematician. He wrote several famous paradoxes, including this one.

26

Introduction to Statistics and Probability

••• **OBJECTIVES** ••

When you have completed this chapter, you should be able to:

- Identify data as continuous, discrete, or categorical.
- Represent data as an *x-y* graph, scatter plot, bar graph, pie chart, stem and leaf plot, or boxplot.
- Organize data into frequency distributions, frequency histograms, frequency polygons, and cumulative frequency distributions.
- Calculate the mean, the weighted mean, the median, and the mode.
- Find the range, the quartiles, the deciles, the percentiles, the variance, and the standard deviation.
- Calculate the probabilities for frequency distributions, including the binomial and normal distributions.
- Estimate population means, standard deviations, and proportions within a given confidence interval.
- Make control charts for statistical process control.
- Fit a straight line to a set of data using the method of least squares.

••

Why study statistics? There are several reasons. First, for your work you may have to interpret statistical data or even need to collect such data and make inferences from it. For example, you may be asked to determine if replacing a certain machine on a production line actually reduced the number of defective parts made. Further, a knowledge of statistics may help you to evaluate the claims made in news reports, polls, and advertisements. A toothpaste manufacturer may claim that their product "significantly reduces" tooth decay. A pollster may claim that 52% of the electorate is going to vote for candidate Jones. A newspaper may show a chart that pictures oil consumption going down. It is easy to use statistics to distort the truth, either on purpose or unintentionally. Some knowledge of the subject may help you to separate fact from distortion.

In this chapter we introduce many new terms. We then show how to display statistical data in graphical and numerical forms. How reliable are those statistical data? To answer that question, we include a brief introduction to *probability,* which enables us to describe the *confidence* we can place on those statistics.

26–1 Definitions and Terminology

Populations and Samples

In statistics, an entire group of people or things is called a *population* or *universe.* A population can be *infinite* or *finite.* A small part of the entire group chosen for study is called a *sample.*

◆◆◆ **Example 1:** In a study of the heights of students at Tech College, the entire student body is our *population.* This is a *finite* population. Instead of studying every student in the population, we may choose to work with a *sample* of only 100 students. ◆◆◆

◆◆◆ **Example 2:** Suppose that we want to determine what percentage of tosses of a coin will be heads. Then all possible tosses of the coin are an example of an *infinite* population. Theoretically, there is no limit to the number of tosses that can be made. ◆◆◆

Parameters and Statistics

We often use just a few numbers to describe an entire population. Such numbers are called *parameters.* The branch of statistics in which we use parameters to describe a population is referred to as *descriptive statistics.*

◆◆◆ **Example 3:** For the population of all students at Tech College, a computer read the height of each student and computed the average height of all students. It found this *parameter,* average height, to be 68.2 in. ◆◆◆

Similarly, we use just a few numbers to describe a sample. Such numbers are called *statistics.* When we then use these sample statistics to *infer* things about the entire population, we are engaging in what is called *inductive statistics.*

◆◆◆ **Example 4:** From the population of all students in Tech College, a sample of 100 students was selected. The average of their heights was found to be 67.9 in. This is a sample *statistic.* From it we infer that for the entire population, the average height is 67.9 in. plus or minus some *standard error.* We show how to compute standard error in Sec. 26–6. ◆◆◆

Thus *parameters* describe *populations,* while *statistics* describe *samples.* The first letters of the words help us to keep track of which goes with which.

Population **P**arameter

versus

Sample **S**tatistic

Variables

In statistics we distinguish between data that are *continuous* (obtained by measurement), *discrete* (obtained by counting), or *categorical* (belonging to one of several categories).

◆◆◆ **Example 5:** A survey at Tech College asked each student for (a) height, *continuous data;* (b) number of courses taken, *discrete data;* and (c) whether they were in favor (yes or no) of eliminating final exams, *categorical data.* ◆◆◆

A *variable,* as in algebra, is a quantity that can take on different values within a problem; *constants* do not change value. We call variables *continuous, discrete,* or *categorical,* depending on the type of data they represent.

TABLE 26–1

Month	Pairs Made
Jan.	13,946
Feb.	7,364
Mar.	5,342
Apr.	3,627
May	1,823
June	847
July	354
Aug.	425
Sept.	3,745
Oct.	5,827
Nov.	11,837
Dec.	18,475
Total	73,612

◆◆◆ **Example 6:** If, in the survey of Example 5, we let

$$H = \text{student's height}$$
$$N = \text{number of courses taken}$$
$$F = \text{opinion on eliminating finals}$$

then H is a continuous variable, N is a discrete variable, and F is a categorical variable.   ◆◆◆

Ways to Display Data

Statistical data are usually presented as a table of values or are displayed in a graph or a chart. Common types of graphs are the x-y graph, scatter plot, bar graph, or pie chart. Here we give examples of the display of discrete and categorical data. We cover display of continuous data in Sec. 26–2.

◆◆◆ **Example 7:** Table 26–1 shows the number of skis made per month by the Ace Ski Company. We show these discrete data as an x-y graph, a scatter plot, a bar chart, and a pie chart in Fig. 26–1.   ◆◆◆

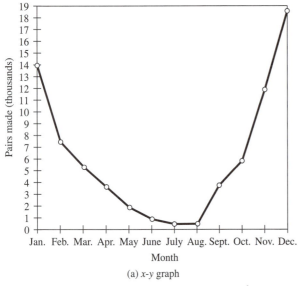

(a) x-y graph

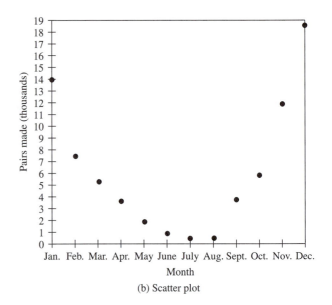

(b) Scatter plot

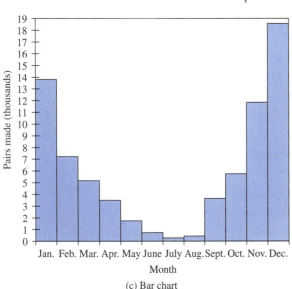

(c) Bar chart

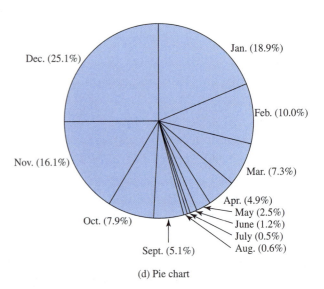

(d) Pie chart

FIGURE 26–1   ◆◆◆

◆◆◆ **Example 8:** Table 26–2 shows the number of skis of various types made in a certain year by the Ace Ski Company. We show these discrete data as an *x-y* graph, a bar chart, and a pie chart (Fig. 26–2).

TABLE 26–2

Type	Pairs Made
Downhill	35,725
Racing	7,735
Jumping	2,889
Cross-country	27,263
Total	73,612

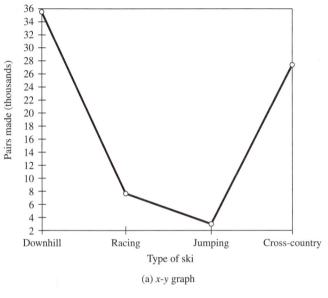

(a) *x-y* graph

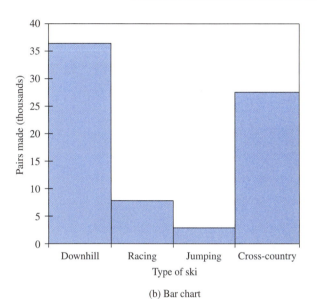

(b) Bar chart

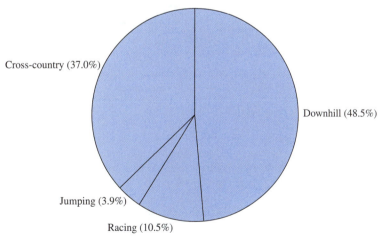

(c) Pie chart

FIGURE 26–2   ◆◆◆

Exercise 1 ◆ Definitions and Terminology

Types of Data

Label each type of data as continuous, discrete, or categorical.

1. The number of people in each county who voted for Jones.
2. The life of certain 100-W light bulbs.
3. The blood types of patients at a certain hospital.

4. The number of Ford cars sold each day.

5. The models of Ford cars sold each day.

6. The weights of steers in a given herd.

Graphical Representation of Data

7. The following table shows the population of a certain town for the years 1920–30:

Year	Population
1920	5364
1921	5739
1922	6254
1923	7958
1924	7193
1925	6837
1926	7245
1927	7734
1928	8148
1929	8545
1930	8623

Make a scatter plot and an *x-y* graph of the population versus the year.

8. Make a bar graph for the data of problem 7.

9. In a certain election, the tally was as listed in the following table:

Candidate	Number of Votes
Smith	746,000
Jones	623,927
Doe	536,023
Not voting	163,745

Make a bar chart showing the election results.

10. Make a pie chart showing the election results of problem 9. To make the pie chart, simply draw a circle and subdivide the area of the circle into four sectors. Make the central angle (and hence the area) of each sector proportional to the votes obtained by each candidate as a percentage of the total voting population. Label the percent in each sector.

Graphics Calculator

11. Many graphics calculators can draw *x-y* graphs, scatter plots, and bar graphs. The data to be entered, however, must be in numerical form; thus the months January, February, . . . would be entered as 1, 2, ... Use your graphics calculator to draw any of the graphs in this exercise set.

Computer

12. Many spreadsheet programs can draw *x-y* graphs, scatter plots, and bar graphs. Unlike with the graphics calculator, the data on the *x* axis can, for certain graphs, be words. Use a spreadsheet to draw any of the graphs in this exercise set.

13. There are a number of computer software packages, such as *Minitab,* designed especially for statistics. These, of course, can draw all of the common statistical graphs. Use one of these programs to draw any of the graphs in this exercise set.

26–2 Frequency Distributions

Raw Data

Data that have not been organized in any way are called *raw data.* Data that have been sorted into ascending or descending order are called an *array.* The *range* of the data is the difference between the largest and the smallest number in that array.

◆◆◆ **Example 9:** Five students were chosen at random and their heights (in inches) were measured. The *raw data* obtained were

$$56.2 \quad 72.8 \quad 58.3 \quad 56.9 \quad 67.5$$

If we sort the numbers into ascending order, we get the *array*

$$56.2 \quad 56.9 \quad 58.3 \quad 67.5 \quad 72.8$$

The *range* of the data is

$$\text{range} = 72.8 - 56.2 = 16.6 \text{ in.} \qquad \qquad ◆◆◆$$

Grouped Data

For discrete and categorical data, we have seen that *x-y* plots and bar charts are effective. These graphs can also be used for *continuous* data, but first we must collect those data into groups. Such groups are usually called *classes.* We create classes by dividing the range into a convenient number of *intervals.* Each interval has an upper and a lower *limit* or *class boundary.* The *class width* is equal to the upper class boundary minus the lower class limit. The intervals are chosen so that a given data point cannot fall directly on a class boundary. We call the center of each interval the *class midpoint.*

Frequency Distribution

Once the classes have been defined, we may tally the number of measurements falling within each class. Such a tally is called a *frequency distribution.* The number of entries in a given class is called the *absolute frequency.* That number, divided by the total number of entries, is called the *relative frequency.* The relative frequency is often expressed as a percent.

To Make a Frequency Distribution:

1. Subtract the smallest number in the data from the largest to find the range of the data.
2. Divide the range into class intervals, usually between five and twenty. Avoid class limits that coincide with actual data.
3. Get the class frequencies by tallying the number of observations that fall within each class.
4. Get the relative frequency of each class by dividing its class frequency by the total number of observations.

◆◆◆ **Example 10:** The grades of 30 students are as follows:

85.0	62.0	94.6	76.3	77.4	82.3	58.4	86.4	84.4	69.8
91.3	75.0	69.6	86.4	84.1	74.7	65.7	86.4	90.6	69.7
86.7	70.5	78.4	86.4	81.5	60.7	79.8	97.2	85.7	78.5

Group these raw data into convenient classes. Choose the class width, the class midpoints, and the class limits. Find the absolute frequency and the relative frequency of each class.

Solution:

1. The highest value in the data is 97.2 and the lowest is 58.4, so

$$\text{range} = 97.2 - 58.4 = 38.8$$

2. If we divide the range into five classes, we get a class width of 38.8/5 or 7.76. If we divide the range into twenty classes, we get a class width of 1.94. Picking a convenient width between those two values, we let

$$\text{class width} = 6$$

Then we choose class limits of

$$56, 62, 68, \ldots, 98$$

Sometimes a given value, such as 62.0 in the above grade list, will fall right on a class limit. Although it does not matter much into which of the two adjoining classes we put it, we can easily avoid such ambiguity. One way is to make our class limits the following:

$$56\text{–}61.9, \quad 62\text{–}67.9, \quad 68\text{–}73.9, \quad \ldots, \quad 92\text{–}97.9$$

Thus 62.0 would fall into the second class. Another way to say the same thing is to use the interval notation from Sec. 24–1.

$$[56, 62), [62, 68), [68, 74), \ldots, [92, 98)$$

Recall that the interval [56, 62) is *closed* on the left, meaning that it includes 56, and *open* on the right, meaning that it does *not* include 62. Again, a data value of 62.0 would go into the second class.

   Our class midpoints are then

$$59, 65, 71, \ldots, 95$$

3. Our next step is to tally the number of grades falling within each class and get a frequency distribution, shown in Table 26–3.

TABLE 26–3   Frequency distribution of student grades.

Class Limits	Class Midpoint	Tally	Absolute Frequency	Relative Frequency
56–61.9	59	//	2	2/30 = 6.7%
62–67.9	65	//	2	2/30 = 6.7%
68–73.9	71	////	4	4/30 = 13.3%
74–79.9	77	ⅧⅡ //	7	7/30 = 23.3%
80–85.9	83	ⅧⅡ //	7	7/30 = 23.3%
86–91.9	89	//// /	6	6/30 = 20.0%
92–97.9	95	//	2	2/30 = 6.7%
		Total	30	30/30 = 100 %

4. We then divide each frequency by 30 to get the relative frequency.   ◆◆◆

Frequency Histogram

A *frequency histogram* is a bar chart showing the frequency of occurrence of each class. The width of each bar is equal to the class width, and the height of each bar is equal to the frequency. The horizontal axis can show the class limits, the class midpoints, or both. The vertical axis can show the absolute frequency, the relative frequency, or both.

◆◆◆ **Example 11:** We show a frequency histogram for the data of Example 10 in Fig. 26–3. Note that the horizontal scale shows both the class limits and the class midpoints, and the vertical scales show both absolute frequency and relative frequency.

Since all of the bars in a frequency histogram have the same width, the *areas* of the bars are proportional to the heights of the bars. The areas are thus proportional to the class frequencies. So if one bar in a relative frequency histogram has a height of 20%, that bar will contain 20% of the area of the whole graph. Thus there is a 20% chance that one grade chosen at random will fall within that class. We will make use of this connection between areas and probabilities in Sec. 26–4.

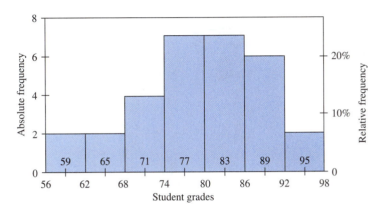

FIGURE 26–3　◆◆◆

Frequency Polygon

A *frequency polygon* is simply an *x-y* graph in which frequency is plotted against class midpoint. As with the histogram, it can show either absolute or relative frequency, or both.

◆◆◆ **Example 12:** We show a frequency polygon for the data of Example 10 in Fig. 26–4.

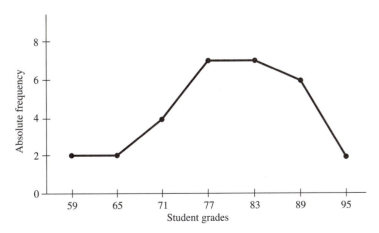

FIGURE 26–4　A frequency polygon for the data of Example 10. Note that this polygon could have been obtained by connecting the midpoints of the tops of the bars in Fig. 26–3.

◆◆◆

Cumulative Frequency Distribution

We obtain a *cumulative* frequency distribution by summing all values *less than* a given class limit. The graph of a cumulative frequency distribution is called a *cumulative frequency polygon,* or *ogive*.

◆◆◆ **Example 13:** Make and graph a cumulative frequency distribution for the data of Example 10.

Solution: We compute the cumulative frequencies by adding the values below the given limits as shown in Table 26–4. Then we plot them in Fig. 26–5.

TABLE 26–4 Cumulative frequency distribution for student grades.

Grade	Cumulative Absolute Frequency	Cumulative Relative Frequency
Under 62.45	2	2/30 (6.7%)
Under 67.45	4	4/30 (13.3%)
Under 72.45	8	8/30 (26.7%)
Under 77.45	12	12/30 (40%)
Under 82.45	17	17/30 (56.7%)
Under 87.45	23	23/30 (76.7%)
Under 92.45	27	27/30 (90%)
Under 97.45	30	30/30 (100%)

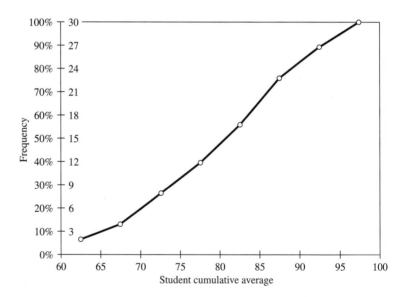

FIGURE 26–5   ◆◆◆

Stem and Leaf Plot

A fast and easy way to get an idea of the distribution of a new set of data is to make a *stem and leaf* plot. We will show how by means of an example.

◆◆◆ **Example 14:** Make a stem and leaf plot for the data in Example 10.

Solution: Let us choose as the *stem* of our plot the first digit of each grade. These range from 5 to 9, representing grades from 50 to 90. Then opposite each number on the stem we write all of the values in the table starting with that number, as shown in Table 26–5.

TABLE 26–5

Stem	Leaves
5	8.4
6	2.5 9.8 9.6 5.7 9.7 0.7
7	6.3 7.4 8.9 5.0 9.5 4.7 0.5 8.4 9.8 8.5
8	5.0 6.4 8.4 6.4 8.6 6.7 6.4 5.7
9	4.6 1.3 0.6 4.6 7.2

We sometimes go on to arrange the leaves in numerical order (not shown here). You may also choose to drop the decimal points.

To see the shape of the distribution, simply turn the page on its side. Notice the similarity in shape to that of Fig. 26–3.

Also note that the stem and leaf plot *preserves the original data.* Thus we could exactly reconstruct the data list of Example 10 from this plot, whereas such a reconstruction is not possible from the grouped data of Table 26–3.

Exercise 2 ◆ Frequency Distributions

Frequency Distributions

1. The weights (in pounds) of 40 students at Tech College are

127	136	114	147	158	149	155	162
154	139	144	114	163	147	155	165
172	168	146	154	111	149	117	152
166	172	158	149	116	127	162	153
118	141	128	153	166	125	117	161

   (a) Determine the range of the data.
   (b) Make an absolute frequency distribution using class widths of 5 lb.
   (c) Make a relative frequency distribution.

2. The times (in minutes) for 30 racers to complete a cross-country ski race are

61.4	72.5	88.2	71.2	48.5	48.9	54.8	71.4	99.2	74.5
84.6	73.6	69.3	49.6	59.3	71.4	89.4	92.4	48.4	66.3
85.7	59.3	74.9	59.3	72.7	49.4	83.8	50.3	72.8	69.3

   (a) Determine the range of the data.
   (b) Make an absolute frequency distribution using class widths of 5 min.
   (c) Make a relative frequency distribution.

3. The prices (in dollars) of various computer printers in a distributer's catalog are

850	625	946	763	774	789	584	864	884	698
913	750	696	864	795	747	657	886	906	697
867	705	784	864	946	607	798	972	857	785

   (a) Determine the range of the data.
   (b) Make an absolute frequency distribution using class widths of $50.
   (c) Make a relative frequency distribution.

Frequency Histograms and Frequency Polygons

Draw a histogram, showing both absolute and relative frequency, for the data of:

4. problem 1.

5. problem 2.

6. problem 3.

Draw a frequency polygon, showing both absolute and relative frequency, for the data of:

7. problem 1.

8. problem 2.

9. problem 3.

Cumulative Frequency Distributions

Make a cumulative frequency distribution showing

(a) absolute frequency and
(b) relative frequency, for the data of:

10. problem 1.

11. problem 2.

12. problem 3.

Draw a cumulative frequency polygon, showing both absolute and relative frequency, for the data of:

13. problem 10.

14. problem 11.

15. problem 12.

Stem and Leaf Plots

Make a stem and leaf plot for the data of:

16. problem 1.

17. problem 2.

18. problem 3.

Graphics Calculator

19. Use your graphics calculator to produce any of the histograms or frequency polygons in this exercise set.

Computer

20. Some spreadsheets, such as *Lotus 123,* have commands for automatically making a frequency distribution, and they can also draw frequency histograms and polygons. If you have such a spreadsheet available, use it to solve any of the problems in this exercise set.

21. Use any available statistical software for the computer, such as *Minitab,* to produce any of the distributions or plots in this exercise set.

26–3 Numerical Description of Data

We saw in Sec. 26–2 how we can describe a set of data by frequency distribution, a frequency histogram, a frequency polygon, or a cumulative frequency distribution. We can also describe a set of data with just a few numbers, and this more compact description is more convenient for some purposes. For example, if, in a report, you wanted to describe the heights of a group of students and did not want to give the entire frequency distribution, you might simply say:

The mean height is 58 inches, with a standard deviation of 3.5 inches.

The *mean* is a number that shows the *center* of the data; the *standard deviation* is a measure of the *spread* of the data. As mentioned earlier, the mean and the standard deviation may be found either for an entire population (and are thus population parameters) or for a sample drawn from that population (and are thus sample statistics).

Thus to describe a population or a sample, we need numbers that give both the center of the data and the spread. Further, for sample statistics, we need to give the uncertainty of each figure. This will enable us to make inferences about the larger population.

◆◆◆ **Example 15:** A student recorded the running times for a sample of participants in a race. From that sample she inferred (by methods we'll learn later) that for the entire population of racers

$$\text{mean time} = 23.65 \pm 0.84 \text{ minutes}$$
$$\text{standard deviation} = 5.83 \pm 0.34 \text{ minutes} \qquad ◆◆◆$$

The mean time is called a *measure of central tendency.* We show how to calculate the mean, and other measures of central tendency, later in this section. The standard deviation is called a *measure of dispersion.* We also show how to compute it, and other measures of dispersion, in this section. The "plus-or-minus" values show a degree of uncertainty called the *standard error.* The uncertainty ± 0.84 min is called the *standard error of the mean,* and the uncertainty ± 0.34 min is called the *standard error of the standard deviation.* We show how to calculate standard errors in Sec. 26–6.

Measures of Central Tendency: The Mean

Some common measures of the center of a distribution are the mean, the median, and the mode. The arithmetic mean, or simply the *mean,* of a set of measurements is equal to the sum of the measurements divided by the number of measurements. It is what we commonly call the *average.* It is the most commonly used measure of central tendency. If our n measurements are $x_1, x_2, \ldots, x_n$, then

$$\Sigma x = x_1 + x_2 + \cdots + x_n$$

where we use the Greek symbol Σ (sigma) to represent "the sum of." The mean, which we call $\bar{x}$ (read "x bar"), is then given by the following:

Arithmetic Mean	$\bar{x} = \dfrac{\Sigma x}{n}$	249

There is a story, probably untrue, about a statistician who drowned in a lake that had an average depth of 1 ft.

The arithmetic mean of n measurements is the sum of those measurements divided by n.

We can calculate the mean for a sample or for the entire population. However, we use a *different symbol* in each case.

$$\bar{x} \text{ is the sample mean}$$

$$\mu \text{ (mu) is the population mean}$$

◆◆◆ **Example 16:** Find the mean of the following sample:

$$746 \quad 574 \quad 645 \quad 894 \quad 736 \quad 695 \quad 635 \quad 794$$

Solution: Adding the values gives

$$\Sigma x = 746 + 574 + 645 + 894 + 736 + 695 + 635 + 794 = 5719$$

Then, with $n = 8$,

$$\bar{x} = \frac{5719}{8} = 715$$

rounded to three digits. ◆◆◆

Weighted Mean

When not all of the data are of equal importance, we may compute a *weighted mean,* where the values are weighted according to their importance.

◆◆◆ **Example 17:** A student has grades of 83, 59, and 94 on 3 one-hour exams, a grade of 82 on the final exam which is equal in weight to 2 one-hour exams, and a grade of 78 on a laboratory report, which is worth 1.5 one-hour exams. Compute the weighted mean.

Solution: If a one-hour exam is assigned a weight of 1, then we have a total of the weights of

$$\Sigma w = 1 + 1 + 1 + 2 + 1.5 = 6.5$$

To get a weighted mean, we add the products of each grade and its weight, and divide by the total weight.

$$\text{weighted mean} = \frac{83(1) + 59(1) + 94(1) + 82(2) + 78(1.5)}{6.5} = 79.5$$ ◆◆◆

In general, the weighted mean is given by the following:

$$\text{weighted mean} = \frac{\Sigma(wx)}{\Sigma w}$$

Midrange

We have already noted that the range of a set of data is the difference between the highest and the lowest numbers in the set. The *midrange* is simply the value midway between the two extreme values.

$$\text{midrange} = \frac{\text{highest value} + \text{lowest value}}{2}$$

◆◆◆ **Example 18:** The midrange for the values

$$3, 5, 6, 6, 7, 11, 11, 15$$

is

$$\text{midrange} = \frac{3 + 15}{2} = 9$$

◆◆◆

Mode

Our next measure of central tendency is the *mode.*

Mode	The mode of a set of numbers is the measurement(s) that occur most often in the set.	251

A set of numbers may have no mode, one mode, or more than one mode.

◆◆◆ **Example 19:**

(a) The set   1   3   3   5   6   7   7   7   9   has the mode 7.
(b) The set   1   1   3   3   5   5   7   7   has no mode.
(c) The set   1   3   3   3   5   6   7   7   7   9   has two modes, 3 and 7. It is called *bimodal.* ◆◆◆

Median

To find the *median,* we simply arrange the data in order of magnitude and pick the *middle value.* For an even number of measurements, we take the mean of the two middle values.

Median	The median of a set of numbers arranged in order of magnitude is the middle value of an odd number of measurements, or the mean of the two middle values of an even number of measurements.	250

◆◆◆ **Example 20:** Find the median of the data in Example 16.

Solution: We rewrite the data in order of magnitude.

$$574 \quad 635 \quad 645 \quad 695 \quad 736 \quad 746 \quad 794 \quad 894$$

The two middle values are 695 and 736. Taking their mean gives

$$\text{median} = \frac{695 + 736}{2} = 715.5$$

◆◆◆

Five-Number Summary

The median of that half of a set of data from the minimum value up to and including the median is called the *lower hinge.* Similarly, the median of the upper half of the data is called the *upper hinge.*

     The lowest value in a set of data, together with the highest value, the median, and the two hinges, is called a *five-number summary* of the data.

◆◆◆ **Example 21:** For the data

$$12 \quad 17 \quad 18 \quad 20 \quad 22 \quad 28 \quad 32 \quad 34 \quad 49 \quad 52 \quad 59 \quad 66$$

the median is $(28 + 32)/2 = 30$, so

12 17 18 **20** 22 28 **30 30** 32 34 **49** 52 59 **66**

| minimum value | lower hinge | median | upper hinge | maximum value |

The five-number summary for this set of data may be written

$$[12, 20, 30, 49, 66] \qquad \text{◆◆◆}$$

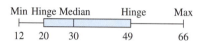

FIGURE 26–6   A boxplot. It is also called a *box and whisker* diagram.

Boxplots

A graph of the values in the five-number summary is called a *boxplot*.

◆◆◆ **Example 22:** A boxplot for the data of Example 21 is given in Fig. 26–6. ◆◆◆

Measures of Dispersion

We will usually use the mean to describe a set of numbers, but used alone it can be misleading.

◆◆◆ **Example 23:** The sets of numbers

$$1 \quad 1 \quad 1 \quad 1 \quad 1 \quad 1 \quad 1 \quad 1 \quad 1 \quad 91$$

and the set

$$10 \quad 10 \quad 10 \quad 10 \quad 10 \quad 10 \quad 10 \quad 10 \quad 10 \quad 10$$

each have a mean value of 10 but are otherwise quite different. Each set has a sum of 100. In the first set most of this sum is concentrated in a single value, but in the second set the sum is *dispersed* among all the values. ◆◆◆

Thus we need some *measure of dispersion* of a set of numbers. Four common ones are the range, the percentile range, the variance, and the standard deviation. We will cover each of these.

Range

We have already introduced the range in Sec. 26–2, and we define it here.

Range	The range of a set of numbers is the difference between the largest and the smallest number in the set.	**252**

◆◆◆ **Example 24:** For the set of numbers

$$6 \quad 3 \quad 9 \quad 12 \quad 44 \quad 2 \quad 53 \quad 1 \quad 8$$

the largest value is 53 and the smallest is 1, so the range is

$$\text{range} = 53 - 1 = 52 \qquad \text{◆◆◆}$$

We sometimes give the range by stating the end values themselves. Thus in Example 24 we might say that the range is from 1 to 53.

Quartiles, Deciles, and Percentiles

We have seen that for a set of data arranged in order of magnitude, the median divides the data into two equal parts. There are as many numbers below the median as there are above it.

Similarly, we can determine two more values that divide each half of the data in half again. We call these values *quartiles*. Thus one-fourth of the values will fall into each quartile. The quartiles are labeled Q_1, Q_2, and Q_3. Thus Q_2 is the median.

Those values that divide the data into 10 equal parts we call *deciles* and label D_1, D_2, ... Those values that divide the data into 100 equal parts are *percentiles*, labeled P_1, P_2, ... Thus the 25th percentile is the same as the first quartile ($P_{25} = Q_1$). The median is the 50th percentile, the fifth decile, and the second quartile ($P_{50} = D_5 = Q_2$). The 70th percentile is the seventh decile ($P_{70} = D_7$).

One measure of dispersion sometimes used is to give the range of values occupied by some given percentiles. Thus we have a *quartile range, decile range,* and *percentile range.* For example, the quartile range is the range from the first to the third quartile.

♦♦♦ **Example 25:** Find the quartile range for the set of data

$$2 \quad 4 \quad 5 \quad 7 \quad 8 \quad 11 \quad 15 \quad 18 \quad 19 \quad 21 \quad 24 \quad 25$$

Solution: The quartiles are

$$Q_1 = \frac{5 + 7}{2} = 6$$

$$Q_2 = \text{the median} = \frac{11 + 15}{2} = 13$$

$$Q_3 = \frac{19 + 21}{2} = 20$$

Then

$$\text{quartile range} = Q_3 - Q_1 = 20 - 6 = 14 \qquad \blacklozenge\blacklozenge\blacklozenge$$

> Note that the quartiles are not in the same locations as the upper and lower hinges. The hinges here would be at 7 and 19.

We see that half the values in Example 25 fall within the quartile range. We may similarly compute the range of any percentiles or deciles. The range from the 10th to the 90th percentile is the one commonly used.

Variance

We now give another measure of dispersion called the *variance,* but we must first mention deviation. We define the *deviation* of any number x in a set of data as the difference between that number and the mean $\bar{x}$ of that set of data.

♦♦♦ **Example 26:** A certain set of measurements has a mean of 48.3. What are the deviations of the values 24.2 and 69.3 in that set?

Solution: The deviation of 24.2 is

$$24.2 - 48.3 = -24.1$$

and the deviation of 69.3 is

$$69.3 - 48.3 = 21.0 \qquad \blacklozenge\blacklozenge\blacklozenge$$

To get the *variance* of a population of n numbers, we add up the squares of the deviations of each number in the set and divide by n.

Population Variance	$\sigma^2 = \dfrac{\Sigma(x - \bar{x})^2}{n}$	253

To find the variance of a *sample*, it is more accurate to divide by $n - 1$ rather than n. As with the mean, we use one symbol for the sample variance and a different symbol for the population variance.

s^2 is the sample variance

σ^2 is the population variance

Sample Variance	$s^2 = \dfrac{\Sigma(x - \bar{x})^2}{n - 1}$	254

◆◆◆ **Example 27:** Compute the variance for the population

$$1.74 \quad 2.47 \quad 3.66 \quad 4.73 \quad 5.14 \quad 6.23 \quad 7.29 \quad 8.93 \quad 9.56$$

Solution: We first compute the mean, $\bar{x}$.

$$\bar{x} = \frac{1.74 + 2.47 + 3.66 + 4.73 + 5.14 + 6.23 + 7.29 + 8.93 + 9.56}{9}$$

$$= 5.53$$

We then subtract the mean from each of the nine values to obtain deviations. The deviations are then squared and added, as shown in Table 26–6.

Computers are ideal for long, tedious statistical computations.

TABLE 26–6

Measurement x	Deviation $x - \bar{x}$	Deviation Squared $(x - \bar{x})^2$
1.74	−3.79	14.35
2.47	−3.06	9.35
3.66	−1.87	3.49
4.73	−0.80	0.64
5.14	−0.39	0.15
6.23	0.70	0.49
7.29	1.76	3.11
8.93	3.40	11.58
9.56	4.03	16.26
$\Sigma x = 49.75$	$\Sigma(x - \bar{x}) = 0$	$\Sigma(x - \bar{x})^2 = 59.41$

The variance σ^2 is then

$$\sigma^2 = \frac{59.41}{9} = 6.60$$

◆◆◆

Standard Deviation

Once we have the variance, it is a simple matter to get the *standard deviation*. It is the most common measure of dispersion.

Standard Deviation	The standard deviation of a set of numbers is the positive square root of the variance.	255

As before, we use s for the sample standard deviation and σ for the population standard deviation.

◆◆◆ **Example 28:** Find the standard deviation for the data of Example 27.

Solution: We have already found the variance in Example 27.

$$\sigma^2 = 6.60$$

Taking the square root gives the standard deviation.

$$\text{standard deviation} = \sqrt{6.60} = 2.57 \qquad ◆◆◆$$

To get an intuitive feel for the standard deviation, we have computed it in Fig. 26–7 for several data sets consisting of 12 numbers which can range from 1 to 12.

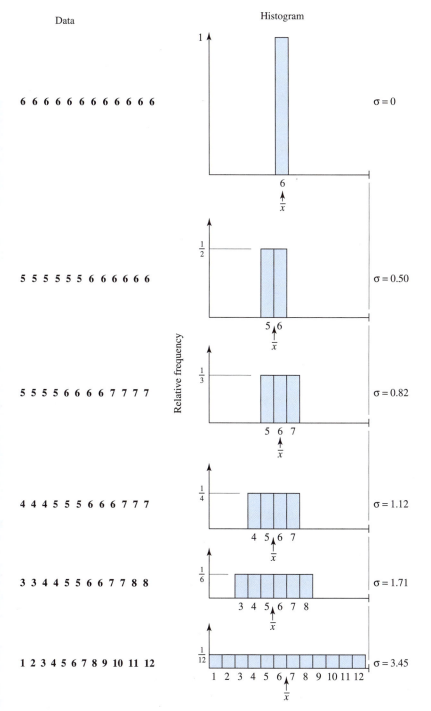

FIGURE 26–7

In the first set, all of the numbers have the same value, 6, and in the last set every number is different. The data sets in between have differing amounts of repetition. To the right of each data set are a relative frequency histogram and the population standard deviation (computation not shown).

Note that the most compact distribution has the lowest standard deviation, and as the distribution spreads, the standard deviation increases.

For a final demonstration, let us again take 12 numbers in two groups of six equal values, as shown in Fig. 26–8. Now let us separate the two groups, first by one interval and then by two intervals. Again notice that the standard deviation increases as the data move further from the mean.

The mean shifts slightly from one distribution to the next, but this does not affect our conclusions.

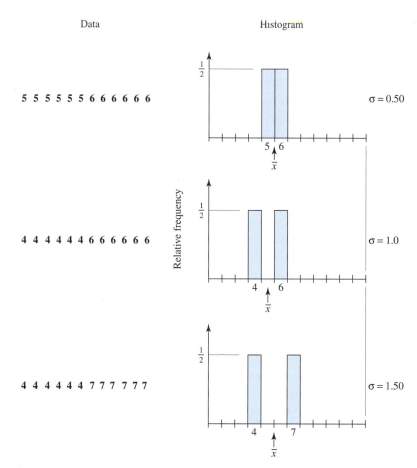

FIGURE 26–8

From these two demonstrations we may conclude that *the standard deviation increases whenever data move away from the mean.*

Exercise 3 ◆ Numerical Description of Data

Mean

1. Find the mean of the following set of grades:

   85 74 69 59 60 96 84 48 89 76 96 68 98 79 76

2. Find the mean of the following set of weights:

   173 127 142 164 163 153 116 199

3. Find the mean of the weights in problem 1 of Exercise 2.
4. Find the mean of the times in problem 2 of Exercise 2
5. Find the mean of the prices in problem 3 of Exercise 2.

Weighted Mean

6. A student's grades and the weight of each grade are given in the following table. Find their weighted mean.

	Grade	Weight
Hour exam	83	5
Hour exam	74	5
Quiz	93	1
Final exam	79	10
Report	88	7

7. A student receives hour-test grades of 86, 92, 68, and 75, a final exam grade of 82, and a project grade of 88. Find the weighted mean if each hour-test counts for 15% of his grade, the final exam counts for 30%, and the project counts for 10%.

Midrange

8. Find the midrange of the grades in problem 1.
9. Find the midrange of the weights in problem 2.

Mode

10. Find the mode of the grades in problem 1.
11. Find the mode of the weights in problem 2.
12. Find the mode of the weights in problem 1 of Exercise 2.
13. Find the mode of the times in problem 2 of Exercise 2.
14. Find the mode of the prices in problem 3 of Exercise 2.

Median

15. Find the median of the grades in problem 1.
16. Find the median of the weights in problem 2.
17. Find the median of the weights in problem 1 of Exercise 2.
18. Find the median of the times in problem 2 of Exercise 2.
19. Find the median of the prices in problem 3 of Exercise 2.

Five-Number Summary

20. Give the five-number summary for the grades in problem 1.
21. Give the five-number summary for the weights in problem 2.

Boxplot

22. Make a boxplot using the results of problem 20.
23. Make a boxplot using the results of problem 21.

Range

24. Find the range of the grades in problem 1.
25. Find the range of the weights in problem 2.

Percentiles

26. Find the quartiles and give the quartile range of the following data:

28   39   46   53   69   71   83   94   102   117   126

27. Find the quartiles and give the quartile range of the following data:

1.33   2.28   3.59   4.96   5.23   6.89

7.91   8.13   9.44   10.6   11.2   12.3

Variance and Standard Deviation

28. Find the variance and standard deviation of the grades in problem 1. Assume that these grades are a sample drawn from a larger population.

29. Find the variance and standard deviation of the weights in problem 2. Assume that these weights are a sample drawn from a larger population.

30. Find the population variance and standard deviation of the weights in problem 1 of Exercise 2.

31. Find the population variance and standard deviation of the times in problem 2 of Exercise 2.

32. Find the population variance and standard deviation of the prices in problem 3 of Exercise 2.

Calculator

33. Many calculators, both graphic and nongraphic, have built-in functions for finding the mean and the sample or population variance and standard deviation. If you have such a calculator, use it to solve any of the problems in this exercise set.

Computer

34. Most spreadsheet programs can find the mean, the sample or population variance and standard deviation, counts, and the range. If you have such a program, use it to solve any of the problems in this exercise set.

35. A statistics program for a computer can, of course, do any of the above calculations, and create boxplots as well. Use such a program to solve any of the problems in this exercise set.

36. Many computer algebra systems can compute the mean, standard deviation, variance, and other statistical measures. Check your manual for the proper instructions, and use your CAS to solve any of the problems in this exercise set.

26–4   Introduction to Probability

Why do we need probability to learn statistics? In Sec. 26–3 we learned how to compute certain sample statistics, such as the mean and the standard deviation. Knowing, for example, that the mean height $\bar{x}$ of a sample of students is 69.5 in., we might *infer* that the mean height μ of the entire population is 69.5 in. But how reliable is that number? Is μ equal to 69.5 in. *exactly?* We'll see later that we give population parameters such as μ as a *range* of values, say, 69.5 ± 0.6 in., and we state the *probability* that the true mean lies within that range. We might say, for example, that there is a 68% chance that the true value lies within the stated range.

Further, we may report that the sample standard deviation is, say, 2.6 in. What does that mean? Of 1000 students, for example, how many can be expected to have a height falling within, say, 1 standard deviation of the mean height at that college? We use probability to help us answer questions such as that.

Further, statistics are often used to "prove" many things, and it takes a knowledge of probability to help decide whether the claims are valid or are merely a result of chance.

◆◆◆ **Example 29:** Suppose that a nurse measures the heights of 14-year-old students in a town located near a chemical plant. Of 100 students measured, she finds 75 students to be shorter than 59 in. Parents then claim that this is caused by chemicals in the air, but the company claims that such heights can occur by chance. We cannot resolve this question without a knowledge of probability. ◆◆◆

We distinguish between *statistical* probability and *mathematical* probability.

◆◆◆ **Example 30:** We may toss a die 6000 times and count the number of times that "one" comes up, say, 1006 times. We conclude that the chance of tossing a "one" is 1006 out of 6000, or about 1 in 6. This is an example of what is called *statistical* probability.

On the other hand, we may reason, without even tossing the die, that each face has an equal chance of coming up. Since there are six faces, the chance of tossing a "one" is 1 in 6. This is an example of *mathematical* probability. ◆◆◆

We first give a brief introduction to mathematical probability and then apply it to statistics.

Probability of a Single Event Occurring

Suppose that an event A can happen in m ways out of a total of n equally likely ways. The probability $P(A)$ of A occurring is defined as follows:

Probability of a Single Event Occurring	$P(A) = \dfrac{m}{n}$	256

Note that the probability $P(A)$ is a number between 0 and 1. $P(A) = 0$ means that there is no chance that A will occur. $P(A) = 1$ means that it is certain that A will occur. Also note that if the probability of A occurring is $P(A)$, the probability of A *not* occurring is $1 - P(A)$.

◆◆◆ **Example 31:** A bag contains 15 balls, 6 of which are red. Assuming that any ball is as likely to be drawn as any other, the probability of drawing a red ball from the bag is $\frac{6}{15}$ or $\frac{2}{5}$. The probability of drawing a ball that is not red is $\frac{9}{15}$ or $\frac{3}{5}$. ◆◆◆

◆◆◆ **Example 32:** A nickel and a dime are each tossed once. Determine (a) the probability $P(A)$ that both coins will show heads and (b) the probability $P(B)$ that not both coins will show heads.

Solution: The possible outcomes of a single toss of each coin are as follows:

Nickel	Dime
Heads	Heads
Heads	Tails
Tails	Heads
Tails	Tails

Thus there are four equally likely outcomes. For event A (both heads), there is only one favorable outcome (heads, heads). Thus

$$P(A) = \frac{1}{4} = 0.25$$

Event B (that not both coins show heads) can happen in three ways (HT, TH, and TT). Thus $P(B) = \frac{3}{4} = 0.75$. Or, alternatively,

$$P(B) = 1 - P(A) = 0.75$$

◆◆◆

Probability of Two Events Both Occurring

If $P(A)$ is the probability that A will occur, and $P(B)$ is the probability that B will occur, the probability $P(A, B)$ that *both* A and B will occur is as follows:

Probability of Two Events Both Occurring	$P(A, B) = P(A)P(B)$	257

Independent events are those in which the outcome of the first in no way affects the outcome of the second.

The probability that two independent events will both occur is the product of their individual probabilities.

The rule above can be extended as follows:

Probability of Several Events All Occurring	$P(A, B, C, \ldots) = P(A)P(B)P(C) \cdots$	258

The probability that several independent events, A, B, C, . . . , will all occur is the product of the probabilities for each.

◆◆◆ **Example 33:** What is the probability that in three tosses of a coin, all will be heads?

Solution: Each toss is independent of the others, and the probability of getting a head on a single toss is $\frac{1}{2}$. Thus the probability of three heads being tossed is

$$\left(\frac{1}{2}\right)\left(\frac{1}{2}\right)\left(\frac{1}{2}\right) = \frac{1}{8}$$

or 1 in 8. If we were to make, say, 800 sets of tosses, we would expect to toss three heads about 100 times.

◆◆◆

Probability of Either of Two Events Occurring

If $P(A)$ and $P(B)$ are the probabilities that events A and B will occur, the probability $P(A + B)$ that *either A or B* or both will occur is as follows:

Probability of Either or Both of Two Events Occurring	$P(A + B) = P(A) + P(B) - P(A, B)$	259

Why subtract $P(A,B)$ in this formula? Since $P(A)$ includes all occurances of A, it must include $P(A,B)$. Since $P(B)$ includes all occurances of B, it also includes $P(A,B)$. Thus if we add $P(A)$ and $P(B)$ we will mistakenly be counting $P(A,B)$ twice.

For example, if we count all aces and all spades in a deck of cards, we get 16 cards (13 spades plus 3 additional aces). But if we had added the number of aces (4) to the number of spades (13), we would have gotten the incorrect answer of 17 by counting the ace of spades twice.

The probability that either A or B or both will occur is the sum of the individual probabilities for A and B, less the probability that both will occur.

◆◆◆ **Example 34:** In a certain cup production line, 6% are chipped, 8% are cracked, and 2% are both chipped and cracked. What is the probability that a cup picked at random will be either chipped or cracked, or both chipped and cracked?

Solution: If

$$P(C) = \text{probability of being chipped} = 0.06$$

and

$$P(S) = \text{probability of being cracked} = 0.08$$

and

$$P(C, S) = 0.02$$

then

$$P(C + S) = 0.06 + 0.08 - 0.02 = 0.12$$

We would thus predict that 12 cups out of 100 chosen would be chipped or cracked, or both chipped and cracked.   ◆◆◆

Probability of Two Mutually Exclusive Events Occurring

Mutually exclusive events A and B are those that cannot both occur at the same time. Thus the probability $P(A, B)$ of both A and B occurring is zero.

◆◆◆ **Example 35:** A tossed coin lands either heads or tails. Thus the probability $P(H, T)$ of a single toss being both heads and tails is

$$P(H, T) = 0$$ ◆◆◆

Setting $P(A, B) = 0$ in Eq. 259 gives the following equation:

Probability of Two Mutually Exclusive Events Occurring	$P(A + B) = P(A) + P(B)$	260

The probability that either of two mutually exclusive events will occur in a certain trial is the sum of the two individual probabilities.

We can generalize this to n mutually exclusive events. The probability that one of n mutually exclusive events will occur in a certain trial is the sum of the n individual probabilities.

◆◆◆ **Example 36:** In a certain school, 10% of the students have red hair, and 20% have black hair. If we select one student at random, what is the probability that this student will have either red hair or black hair?

Solution: The probability $P(A)$ of having red hair is 0.1, and the probability $P(B)$ of having black hair is 0.2. The two events (having red hair or having black hair) are mutually exclusive. Thus the probability $P(A + B)$ of a student having either red or black hair is

$$P(A + B) = 0.1 + 0.2 = 0.3$$ ◆◆◆

Random Variables

In Sec. 26–1 we defined continuous, discrete, and categorical variables. Now we define a *random* variable.

Suppose that there is a process or an experiment whose outcome is a real number X. If the process is repeated, then X can assume specific values, each determined by chance and each with a certain probability of occurrence. X is then called a *random variable*.

◆◆◆ **Example 37:** The process of rolling a pair of dice once results in a single numerical value X, whose value is determined by chance but whose value has a definite probability of occurrence. Repeated rollings of the dice may result in other values of X, which can take on the values

$$2, 3, 4, 5, 6, 7, 8, 9, 10, 11, 12$$

Here X is a random variable.   ◆◆◆

Discrete and Continuous Random Variables

A *discrete* random variable can take on a finite number of variables. Thus the value X obtained by rolling a pair of dice is a discrete random variable.

A *continuous* random variable may have infinitely many values on a continuous scale, with no gaps.

◆◆◆ **Example 38:** An experiment measuring the temperatures T of the major U.S. cities at any given hour gives values that are determined by chance, depending on the whims of nature, but each value has a definite probability of occurrence. The random variable T is continuous because temperatures are read on a continuous scale.   ◆◆◆

Probability Distributions

A probability distribution is sometimes called a *probability density function*, a *relative frequency function*, or a *frequency function*.

If a process or an experiment has X different outcomes, where X is a discrete random variable, we say that the probability of a particular outcome is $P(X)$. If we calculate $P(X)$ for each X, we have what is called a *probability distribution*. Such a distribution may be presented as a table, a graph, or a formula.

◆◆◆ **Example 39:** Make a probability distribution for a single roll of a single die.

Solution: Let X equal the number on the face that turns up, and $P(X)$ be the probability of that face turning up. That is, $P(5)$ is the probability of rolling a five. Since each number has an equal chance of being rolled, we have the following:

X	$P(X)$
1	$\frac{1}{6}$
2	$\frac{1}{6}$
3	$\frac{1}{6}$
4	$\frac{1}{6}$
5	$\frac{1}{6}$
6	$\frac{1}{6}$

We graph this probability distribution in Fig. 26–9.

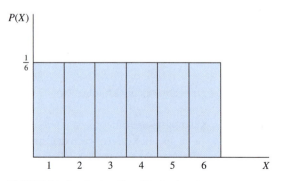

FIGURE 26–9   Probability distribution for the toss of a single die.   ◆◆◆

◆◆◆ **Example 40:** Make a probability distribution for the two-coin toss experiment of Example 32.

Solution: Let X be the number of heads tossed, and $P(X)$ be the probability of X heads being tossed. The experiment has four possible outcomes (HH, HT, TH, TT). The outcome $X = 0$ (no heads) can occur in only one way (TT), so the probability $P(0)$ of not tossing a head is one in four, or $\frac{1}{4}$. The probabilities of the other outcomes are found in a similar way. They are as follows:

X	$P(X)$
0	$\frac{1}{4}$
1	$\frac{1}{2}$
2	$\frac{1}{4}$

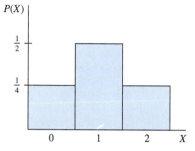

FIGURE 26–10 Probability distribution.

They are shown graphically in Fig. 26–10. ◆◆◆

Probabilities as Areas on a Probability Distribution

The probability of an event occurring can be associated with a geometric *area* of the probability distribution or relative frequency distribution.

◆◆◆ **Example 41:** For the distribution in Fig. 26–10, each bar has a width of 1 unit, so the total shaded area is

$$1(\tfrac{1}{4}) + 1(\tfrac{1}{2}) + 1(\tfrac{1}{4}) = 1 \text{ square unit}$$

The bar representing the probability of tossing a "zero" has an area of $\frac{1}{4}$, which is equal to the probability of that event occurring. The areas of the other two bars also give the value of their probabilities. ◆◆◆

Probabilities from a Relative Frequency Histogram

Suppose we were to repeat the two-coin toss experiment of Example 40 a very large number of times and draw a relative frequency histogram of our results. That relative frequency histogram *would be identical to the probability distribution of Fig. 26–10.* The relative frequency of tossing a "zero" would be $\frac{1}{4}$, of a "one" would be $\frac{1}{2}$, and of a "two" would be $\frac{1}{4}$.

We thus may think of a probability distribution as being the limiting value of a relative frequency distribution as the number of observations increases—in fact, the distribution for a *population.*

Thus we can obtain probabilities from a relative frequency histogram for a population or from one obtained using a large number of observations.

◆◆◆ **Example 42:** For the relative frequency histogram for the population of students' grades (Fig. 26–3), find the probability that a student chosen at random will have received a grade between 92 and 98.

Solution: The given grades are represented by the bar having a class midpoint of 95 and a relative frequency of 6.7%. Since the relative frequency distribution for a population probability is no different from the probability distribution, the grade of a student chosen at random has a 6.7% chance of falling within that class. ◆◆◆

Probability of a Variable Falling between Two Limits

Further, the probability of a variable falling *between two limits* is given by the area of the probability distribution or relative frequency histogram for that variable, between those limits.

◆◆◆ **Example 43:** Using the relative frequency histogram shown in Fig. 26–3, find the chance of a student chosen at random having a grade that falls between 62 and 80.

Solution: This range spans three classes in Fig. 26–3, having relative frequencies of 6.7%, 13.3%, and 23.3%. The chance of falling within one of the three is

$$6.7 + 13.3 + 23.3 = 43.3\%$$

So the probability is 0.433.   ◆◆◆

Thus we have shown that *the area between two limits on probability distribution or relative frequency histogram for a population is a measure of the probability of a measurement falling within those limits.* This idea will be used again when we compute probabilities from a *continuous* probability distribution such as the normal curve.

Binomial Experiments

A *binomial experiment* is one that can have only two possible outcomes.

◆◆◆ **Example 44:** The following are examples of binomial experiments:

(a) A tossed coin shows either heads or tails.
(b) A manufactured part is either accepted or rejected.
(c) A voter either chooses Smith for Congress or does not choose Smith for Congress.   ◆◆◆

Binomial Probability Formula

We call the occurrence of an event a *success* and the nonoccurrence of that event a *failure.* Let us denote the probability of success of a single trial of a binomial experiment by p. Then the probability of failure, q, must be $(1 - p)$.

◆◆◆ **Example 45:** If the probability is 80% ($p = 0.8$) that a single part drawn at random from a production line is accepted, then the probability is 20% ($q = 1 - 0.8$) that a part is rejected.   ◆◆◆

If we repeat a binomial experiment n times, we will have x successes, where

$$x = 0, 1, 2, 3, \ldots, n$$

The probability $P(x)$ of having x successes in n trials is given by the following:

Binomial Probability Formula	$P(x) = \dfrac{n!}{(n - x)!\, x!} p^x q^{n-x}$	262

◆◆◆ **Example 46:** For the production line of Example 45, find the probability that of 10 parts taken at random from the line, exactly 6 will be acceptable.

Solution: Here, $n = 10$, $x = 6$, and $p = 0.8$, $q = 0.2$, from before. Substituting into the formula gives

$$P(6) = \frac{10!}{(10 - 6)!\, 6!} (0.8)^6 (0.2)^{10-6}$$

$$= \frac{10!}{4!\, 6!} (0.8)^6 (0.2)^4 = 0.0881$$

or about 9%.   ◆◆◆

Think of a sheet of paper on which the distribution is drawn lying face up on the ground during a snowfall. The prob-ability of a snowflake falling within any region is proportional to the area of that region.

There are tables given in statistics books that can be used instead of this formula.

Does this formula look a little familiar? We will soon connect the binomial dis-tribution with what we have already learned about the binomial theorem.

Binomial Distribution

In Example 46, we computed a single probability, that 6 out of 10 parts drawn at random will be acceptable. We can also compute the probabilities of having 0, 1, 2, . . . , 10 acceptable parts in a group of 10. If we compute all 10 possible probabilities, we will then have a *probability distribution*.

◆◆◆ **Example 47:** For the production line of Example 45, find the probability that of 10 parts taken at random from the line, 0, 1, 2, . . . , 10 will be acceptable.

Solution: Computing the probabilities in the same way as in the preceding example (work not shown), we get the following binomial distribution:

x	$P(x)$
0	0
1	0
2	0.0001
3	0.0008
4	0.0055
5	0.0264
6	0.0881
7	0.2013
8	0.3020
9	0.2684
10	0.1074
Total	1.0000

This binomial distribution is shown graphically in Fig. 26–11.

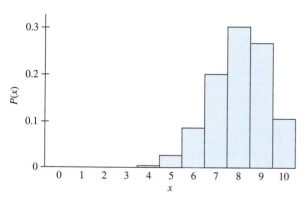

FIGURE 26–11

◆◆◆

Binomial Distribution Obtained by Binomial Theorem

The n values in a binomial distribution for n trials may be obtained by using the binomial theorem to expand the binomial

$$(q + p)^n$$

where p is the probability of success of a single trial and $q = (1 - p)$ is the probability of failure of a single trial, as before.

Refer to Sec. 25–6 for a refresher on the binomial theorem.

◆◆◆ **Example 48:** Use the binomial theorem, Eq. 245, to obtain the binomial distribution of Example 47.

Solution: We expand $(q + p)^n$ using $q = 0.2$, $p = 0.8$, and $n = 10$.

$(0.2 + 0.8)^{10}$

$$= (0.2)^{10}(0.8)^0 + 10(0.2)^9(0.8)^1 + \frac{10(9)}{2!}(0.2)^8(0.8)^2$$

$$+ \frac{10(9)(8)}{3!}(0.2)^7(0.8)^3 + \frac{10(9)(8)(7)}{4!}(0.2)^6(0.8)^4$$

$$+ \frac{10(9)(8)(7)(6)}{5!}(0.2)^5(0.8)^5 + \frac{10(9)(8)(7)(6)(5)}{6!}(0.2)^4(0.8)^6$$

$$+ \frac{10(9)(8)(7)(6)(5)(4)}{7!}(0.2)^3(0.8)^7 + \frac{10(9)(8)(7)(6)(5)(4)(3)}{8!}(0.2)^2(0.8)^8$$

$$+ \frac{10(9)(8)(7)(6)(5)(4)(3)(2)}{9!}(0.2)^1(0.8)^9 + \frac{10!}{10!}(0.2)^0(0.8)^{10}$$

$$= 0 + 0 + 0.0001 + 0.0008 + 0.0055 + 0.0264 + 0.0881$$

$$+ 0.2013 + 0.3020 + 0.2684 + 0.1074$$

As expected, we get the same values as given by binomial probability formula. ◆◆◆

Exercise 4 ◆ Introduction to Probability

Probability of a Single Event

1. We draw a ball from a bag that contains 8 green balls and 7 blue balls. What is the probability that a ball drawn at random will be green?
2. A card is drawn from a deck containing 13 hearts, 13 diamonds, 13 clubs, and 13 spades. What is the chance that a card drawn at random will be a heart?
3. If we toss four coins, what is the probability of getting two heads and two tails? [*Hint:* This experiment has 16 possible outcomes (HHHH, HHHT, . . . , TTTT).]
4. If we toss four coins, what is the probability of getting one head and three tails?
5. If we throw two dice, what is the probability that their sum is 9? [*Hint:* List all possible outcomes, (1, 1), (1, 2), and so on, and count those that have a sum of 9.]
6. If two dice are thrown, what is the probability that their sum is 7?

Probability That Several Events Will All Occur

7. A die is rolled twice. What is the probability that both rolls will give a six?
8. We draw four cards from a deck, replacing each before the next is drawn. What chance is there that all four draws will be a red card?

Probability That Either of Two Events Will Occur

9. At a certain school, 55% of the students have brown hair, 15% have blue eyes, and 7% have both brown hair and blue eyes. What is the probability that a student chosen at random will have either brown hair or blue eyes, or both brown hair and blue eyes?
10. Find the probability that a card drawn from a deck will be either a "picture" card (jack, queen, or king) or a spade picture card.

Probability of Mutually Exclusive Events

11. What chance is there of tossing a sum of 7 or 9 when two dice are tossed? You may use your results from problems 5 and 6 here.
12. On a certain production line, 5% of the parts are underweight and 8% are overweight. What is the probability that one part selected at random is either underweight or overweight?

Probability Distributions

13. For the relative frequency histogram for the population of students' grades (Fig. 26–3), find the probability that a student chosen at random will have gotten a grade between 68 and 74.

14. Using the relative frequency histogram of Fig. 26–3, find the chance of a student chosen at random having a grade between 74 and 86.

Binomial Distribution

A binomial experiment is repeated n times, with a probability p of success on one trial. Find the probability $P(x)$ of x successes, if:

15. $n = 5$, $p = 0.4$, and $x = 3$.
16. $n = 7$, $p = 0.8$, and $x = 4$.
17. $n = 7$, $p = 0.8$, and $x = 6$.
18. $n = 9$, $p = 0.3$, and $x = 6$.

For problems 19 and 20, find all of the terms in each probability distribution, and graph.

19. $n = 5$, $p = 0.8$
20. $n = 6$, $p = 0.3$
21. What is the probability of tossing 7 heads in 10 tosses of a fair coin?
22. If male and female births are equally likely, what is the probability of five births being all girls?
23. If the probability of producing a defective floppy disk is 0.15, what is the probability of getting 10 bad disks in a sample of 15?
24. A certain multiple-choice test has 20 questions, each of which has four choices, only one of which is correct. If a student were to guess every answer, what is the probability of getting 10 correct?

Computer

25. Any mathematical software that can evaluate factorials and exponents can be used to give values from the binomial probability formula. Some software, however, has the formula built in, and one just has to enter the required information. *Derive,* for example, takes the command

BINOMIAL_DENSITY(k,n,p)

and returns the kth value in the binomial distribution having n terms and a probability of success of p. Some statistical software will give the *entire* distribution. In *Minitab,* for example, the commands

MTB>PDF;
SUBC>BINOMIAL n = 10 p = 0.8

would return the probability distribution for our production line in Example 45. If such software is available, use your computer to find the distributions in problems 19 and 20.

26–5 The Normal Curve

In the preceding section we showed some discrete probability distributions and studied the binomial distribution, one of the most useful of the discrete distributions. Now we introduce *continuous* probabilities and go on to study the most useful of these, the *normal curve.*

Continuous Probability Distributions

Suppose now that we collect data from a "large" population. For example, suppose that we record the weight of every college student in a certain state. Since the population is large, we can use very narrow class widths and still have many observations that fall within each class. If we do so, our relative frequency histogram, Fig. 26–12(a), will have very narrow bars, and our relative frequency polygon, Fig. 26–12(b), will be practically a *smooth curve.*

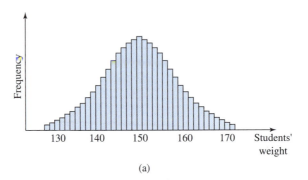

(a)

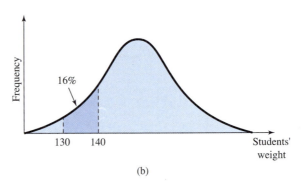

(b)

FIGURE 26–12

We get probabilities from the nearly smooth curve just as we did from the relative frequency histogram. *The probability of an observation falling between two limits a and b is proportional to the area bounded by the curve and the* x *axis, between those limits.* Let us designate the total area under the curve as 100% (or $P = 1$). Then the probability of an observation falling between two limits is equal to the area between those two limits.

◆◆◆ **Example 49:** Suppose that 16% of the entire area under the curve in Fig. 26–12(b) lies between the limits 130 and 140. That means there is a 16% chance ($P = 0.16$) that one student chosen at random will weigh between 130 and 140 lb.

Normal Distribution

The frequency distribution for many real-life observations, such as a person's height, is found to have a typical "bell" shape. We have seen that the chance of one observation falling between two limits is equal to the area under the relative frequency histogram [such as Fig. 26–12(a)] between those limits. But how are we to get those areas for a curve such as in Fig. 26–12(b)?

One way is to make a *mathematical model* of the probability distribution and from it get the areas of any section of the distribution. One such mathematical model is called the *normal distribution, normal curve,* or *Gaussian distribution.*

The Gaussian distribution is named for Carl Friedrich Gauss (1777–1855), whose *Gauss-Seidel method* we studied in Chapter 10. Some consider him to be one of the greatest mathematicians of all time. Many theorems and proofs are named after him.

The normal distribution is obtained by plotting the following equation:

Normal Distribution	$$y = \frac{1}{\sigma\sqrt{2\pi}} e^{-(x-\mu)^2/(2\sigma^2)}$$	261

Don't be frightened by this equation. We will not be calculating with it but give it only to show where the normal curve and Table 26–7 come from.

The normal curve, shown in Fig. 26–13, is bell-shaped and is symmetrical about its mean, μ. Here, σ is the standard deviation of the population. This curve closely matches many frequency distributions in the real world.

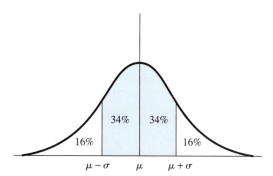

FIGURE 26–13   The normal curve.

To use the normal curve to determine probabilities, we must know the areas between any given limits. Since the equation of the curve is known, it is possible, by methods too advanced to show here, to compute these areas. They are given in Table 26–7. The areas are given "within z standard deviations of the mean." This means that each area given is that which lies between the mean and z standard deviations *on one side* of the mean, as shown in Fig. 26–14.

TABLE 26–7   Areas under the normal curve within z standard deviations of the mean.

z	Area	z	Area
0.1	0.0398	2.1	0.4821
0.2	0.0793	2.2	0.4861
0.3	0.1179	2.3	0.4893
0.4	0.1554	2.4	0.4918
0.5	0.1915	2.5	0.4938
0.6	0.2258	2.6	0.4952
0.7	0.2580	2.7	0.4965
0.8	0.2881	2.8	0.4974
0.9	0.3159	2.9	0.4981
1.0	0.3413	3.0	0.4987
1.1	0.3643	3.1	0.4990
1.2	0.3849	3.2	0.4993
1.3	0.4032	3.3	0.4995
1.4	0.4192	3.4	0.4997
1.5	0.4332	3.5	0.4998
1.6	0.4452	3.6	0.4998
1.7	0.4554	3.7	0.4999
1.8	0.4641	3.8	0.4999
1.9	0.4713	3.9	0.5000
2.0	0.4772	4.0	0.5000

More extensive tables can be found in most statistics books.

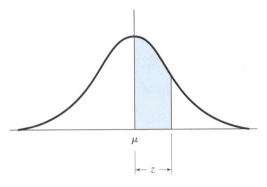

FIGURE 26–14

The number z is called the *standardized variable,* or *normalized variable.* When used for educational testing, it is often called the *standard score,* or *z score.* Since the normal curve is symmetrical about the mean, the given values apply to either the left or the right side.   ◆◆◆

◆◆◆ **Example 50:** Find the area under the normal curve between the following limits: from one standard deviation to the left of the mean, to one standard deviation to the right of the mean.

Solution: From Table 26–7, the area between the mean and one standard deviation is 0.3413. By symmetry, the same area lies to the other side of the mean. The total is thus 2(0.3413), or 0.6826. Thus *about 68% of the area under the normal curve lies within one standard deviation of the mean* (Fig. 26–13). Therefore $100 - 68$ or 32% of the area lies in the "tails," or 16% in each tail.   ◆◆◆

◆◆◆ **Example 51:** Find the area under the normal curve between 0.8 standard deviation to the left of the mean, to 1.1 standard deviations to the right of the mean (Fig. 26–15).

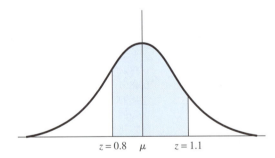

FIGURE 26–15

Solution: From Table 26–7, the area between the mean and 0.8 standard deviation ($z = 0.8$) is 0.2881. Also, the area between the mean and 1.1 standard deviations ($z = 1.1$) is 0.3643. The combined area is thus

$$0.2881 + 0.3643 = 0.6524$$

or 65.24% of the total area.   ◆◆◆

◆◆◆ **Example 52:** Find the area under the tail of the normal curve to the right of $z = 1.4$ (Fig. 26–16).

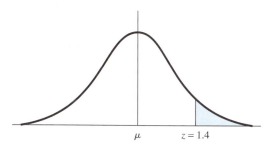

FIGURE 26–16

Solution: From Table 26–7, the area between the mean and $z = 1.4$ is 0.4192. But since the total area to the right of the mean is $\frac{1}{2}$, the area in the tail is

$$0.5000 - 0.4192 = 0.0808$$

or 8.08% of the total area.   ◆◆◆

Since, for a normal distribution, the probability of a measurement falling within a given interval is equal to the area under the normal curve within that interval, we can use the areas to assign probabilities. We must first convert the measurement x to the standard variable z, which tells the number of standard deviations between x and the mean $\bar{x}$. To get the value of z, we simply find the difference between x and $\bar{x}$ and divide that difference by the standard deviation.

$$z = \frac{x - \bar{x}}{s}$$

◆◆◆ **Example 53:** Suppose that the heights of 2000 students have a mean of 69.3 in. with a standard deviation of 3.2 in. Assuming the heights to have a normal distribution, predict the number of students (a) shorter than 65 in., (b) taller than 70 in., and (c) between 65 in. and 70 in.

Solution:

(a) Let us compute z for $x = 65$ in. The mean $\bar{x}$ is 69.3 in. and the standard deviation s is 3.2 in., so

$$z = \frac{65 - 69.3}{3.2} = -1.3$$

From Table 26–7, with $z = 1.3$, we read an area of 0.4032. The area in the tail to the left of $z = -1.3$ is thus

$$0.5000 - 0.4032 = 0.0968$$

Thus there is a 9.68% chance that a single student will be shorter than 65 in. Since there are 2000 students, we predict that

$$9.68\% \text{ of 2000 students} = 194 \text{ students}$$

will be within this range.

(b) For a height of 70 in., the value of z is

$$z = \frac{70 - 69.3}{3.2} = 0.2$$

The area between the mean and $z = 0.2$ is 0.0793, so the area to the right of $z = 0.2$ is

$$0.5000 - 0.0793 = 0.4207$$

Since the normal curve is symmetrical about the mean, either a positive or a negative value of z will give the same area. Thus for compactness, Table 26–7 shows only positive values of z.

Since there is a 42.07% chance that a student will be taller than 70 in., we predict that 0.4207(2000) or 841 students will be within that range.

(c) Since 194 + 841 or 1035 students are either shorter than 65 in. or taller than 70 in., we predict that 965 students will be between these two heights.   ◆◆◆

Exercise 5 ◆ The Normal Curve

Use Table 26–7 for the following problems.

1. Find the area under the normal curve between the mean and 1.5 standard deviations in Fig. 26–17.

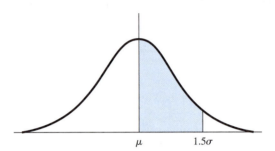

FIGURE 26–17

2. Find the area under the normal curve between 1.6 standard deviations on both sides of the mean in Fig. 26–18.

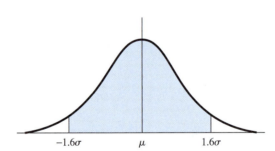

FIGURE 26–18

3. Find the area in the tail of the normal curve to the left of $z = -0.8$ in Fig. 26–19.

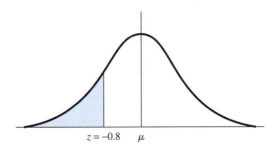

FIGURE 26–19

4. Find the area in the tail of the normal curve to the right of $z = 1.3$ in Fig. 26–20.

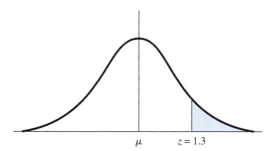

FIGURE 26–20

5. The distribution of the weights of 1000 students at Tech College has a mean of 163 lb with a standard deviation of 18 lb. Assuming that the weights are normally distributed, predict the number of students who have weights between 130 and 170 lb.
6. For the data of problem 5, predict the number of students who will weigh less than 130 lb.
7. For the data of problem 5, predict the number of students who will weigh more than 195 lb.
8. On a test given to 500 students, the mean grade was 82.6 with a standard deviation of 7.4. Assuming a normal distribution, predict the number of A grades (a grade of 90 or over).
9. For the test of problem 8, predict the number of failing grades (a grade of 60 or less).
10. Computation of the gas consumption of a new car model during a series of trials gives a set of data that is normally distributed with a mean of 32.2 MPG and a standard deviation of 3.2 MPG. Find the probability that the same car will get at least 35 MPG on the next trial.

Graphics Calculator

11. Some graphics calculators will draw the normal curve. Check your manual to see how to do this, if possible. If you can draw the curve, you can then use `TRACE` to find values on the curve. Use this feature to solve any of the problems in this exercise set.

Computer

12. Some mathematics software has the normal curve built in. In *Derive,* for example, the command

$$\text{NORMAL}(z, -, 1)$$

will return the value of z from $-\infty$ to $+z$, for a normal curve with a mean of 0 and a standard deviation of 1. To get the value of z as measured *from the mean,* as in Table 26–7, we must subtract $\frac{1}{2}$ from the value returned by this command. Use this feature to solve any of the problems in this exercise set.

26–6 Standard Errors

When we draw a random sample from a population, it is usually to *infer* something about that population. Typically, from our sample we compute a *statistic* (such as the sample mean $\bar{x}$) and use it to infer a population *parameter* (such as the population mean μ).

◆◆◆ **Example 54:** A researcher measured the heights of a randomly drawn sample of students at Tech College and calculated the mean height of that sample. It was

$$\text{mean height } \bar{x} = 67.50 \text{ in.}$$

From this he inferred that the mean height of the entire population of students at Tech College was

$$\text{mean height } \mu = 67.50 \text{ in.} \qquad\qquad\text{◆◆◆}$$

In general,

$$\text{population parameter} \approx \text{sample statistic}$$

How accurate is the population parameter that we find in this way? We know that all measurements are approximate, and we usually give some indication of the accuracy of a measurement. In fact, a population parameter such as the mean height is often given in the form

$$\text{mean height } \mu = 67.50 \pm 0.24 \text{ in.}$$

where ± 0.24 is called the *standard error* of the mean. In this section we show how to compute the standard error of the mean and the standard error of the standard deviation.

Further, when we give a range of values for a statistic, we know that it is possible that another sample can have a mean that lies outside the given range. In fact, the true mean itself can lie outside the given range. That is why a statistician will *give the probability* that the true mean falls within the given range.

◆◆◆ **Example 55:** Using the same example of heights at Tech College, we might say that

$$\text{mean height } \mu = 67.50 \pm 0.24 \text{ in.}$$

with a 68% chance that the mean student height μ falls within the range

$$67.50 - 0.24 \quad \text{to} \quad 67.50 + 0.24$$

or from 67.26 to 67.74 in.   ◆◆◆

We call the interval 67.50 ± 0.24 in Example 55 a *68% confidence interval.* This means that there is a 68% chance that the true population mean falls within the given range. So the complete way to report a population parameter is

$$\text{population parameter} = \text{sample statistic} \pm \text{standard error}$$

within a given confidence interval. We will soon show how to compute this confidence interval, and others as well, but first we must lay some groundwork.

Frequency Distribution of a Statistic

Suppose that you draw a sample of, say, 40 heights from the population of students in your school, and find the mean $\bar{x}$ of those 40 heights, say, 65.6 in. Now measure another 40 heights from the same population and get another $\bar{x}$, say, 69.3 in. Then repeat, drawing one sample of 40 after another, and for each compute the statistic $\bar{x}$. This collection of $\bar{x}$'s now has a frequency distribution of its own (called the $\bar{x}$ distribution) with its own mean and standard deviation.

We now ask, *How is the $\bar{x}$ distribution related to the original distribution?* To answer this, we need a statistical theorem called the *central limit theorem,* which we give here without proof.

Standard Error of the Mean

Suppose that we were to draw *all possible* samples of size *n* from a given population, and for each sample compute the mean $\bar{x}$, and then make a frequency distribution of the $\bar{x}$'s. The central limit theorem states the following:

1. The mean of the $\bar{x}$ distribution is the same as the mean μ of the population from which the samples were drawn.
2. The standard deviation of the $\bar{x}$ distribution is equal to the standard deviation σ of the population divided by the square root of the sample size *n*. The standard error or a sampling distribution of a statistic such as $\bar{x}$ is often called its *standard error*. We will denote the standard error of $\bar{x}$ by $\text{SE}_{\bar{x}}$. So

$$\text{SE}_{\bar{x}} = \frac{\sigma}{\sqrt{n}}$$

3. For a "large" sample (over 30 or so), the $\bar{x}$ distribution is approximately a normal distribution, even if the population from which the samples were drawn is not a normal distribution.

We will illustrate parts of the central limit theorem with an example.

◆◆◆ **Example 56:** Consider the population of 256 integers from 100 to 355 given in Table 26–8. The frequency distribution of these integers [Fig 26–21(a)] shows that each class has a frequency of 32. We computed the population mean μ by adding all of the integers and dividing by 256 and got 227. We also got a population standard deviation σ of 74.

TABLE 26–8

174	230	237	173	214	242	113	177	238	175	303	172	114	179	307	176
190	246	253	189	194	258	129	193	254	191	319	188	130	111	108	103
206	262	269	205	210	274	145	209	270	207	335	204	146	127	124	119
222	278	285	221	226	290	161	225	286	223	351	220	162	143	140	135
300	229	302	228	296	233	298	232	301	239	236	231	304	241	306	240
316	245	318	244	312	249	314	248	317	255	252	247	320	257	322	256
332	261	334	260	328	265	330	264	333	271	268	263	336	273	338	272
348	277	350	276	344	281	346	280	349	287	284	279	352	289	354	288
166	102	109	165	170	234	105	169	110	167	295	164	106	171	299	168
182	118	125	181	186	250	121	185	126	183	311	180	122	187	315	184
198	134	141	197	202	266	137	201	142	199	327	196	138	203	331	200
178	150	157	213	218	282	153	217	158	215	343	212	154	219	347	216
292	101	294	100	297	107	104	235	293	195	323	192	305	115	112	243
308	117	310	116	313	123	120	251	309	211	339	208	321	131	128	259
324	133	326	132	329	139	136	267	325	227	355	224	337	147	144	275
340	149	342	148	345	155	152	283	341	159	156	151	353	163	160	291

Then 32 samples of 16 integers each were drawn from the population. For each sample we got the sample mean $\bar{x}$. The 32 values we got for $\bar{x}$ are as follows:

234 229 236 235 227 232 229 224 266 230 218 212 234 226 237 216
170 222 220 218 232 247 262 277 265 252 238 225 191 204 216 223

We now make a frequency distribution for the sample means, which is called a *sampling distribution of the mean* [Fig. 26–21(b)]. We further compute the mean of the sample means and get 229, nearly the same value as for the population mean μ.

The calculations in this example were done by computer, but you will not need a computer to make use of the results of this demonstration.

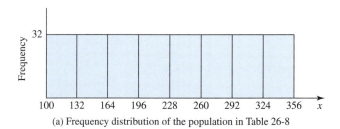

(a) Frequency distribution of the population in Table 26-8

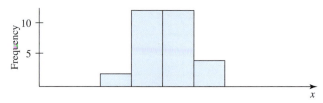

(b) Sampling distribution of the mean (frequency distribution
of the means of 32 samples drawn from the population in
Table 26-8)

FIGURE 26–21

The standard deviation of the sample means is computed to be 20. Thus we have the following results:

1. The mean of the $\bar{x}$ distribution (229) is approximately equal to the mean μ of the population (227).
2. By the central limit theorem, the standard deviation $\text{SE}_{\bar{x}}$ of the $\bar{x}$ distribution is found by dividing σ by the square root of the sample size.

$$\text{SE}_{\bar{x}} = \frac{\sigma}{\sqrt{n}} = \frac{74}{\sqrt{16}} = 18.5$$

This compares well with the actual standard deviation of the $\bar{x}$ distribution, which was 20.   ◆◆◆

We now take the final step and show how to predict the population mean μ from the mean $\bar{x}$ and standard deviation s *of a single sample*. We also give the probability that our prediction is correct.

Predicting μ from a Single Sample

If the standard deviation of the $\bar{x}$ distribution, $\text{SE}_{\bar{x}}$, is small, most of the $\bar{x}$'s will be near the center μ of the population. Thus a particular $\bar{x}$ has a good chance of being close to μ and will hence be a good estimator of μ. Conversely, a large $\text{SE}_{\bar{x}}$ means that a given $\bar{x}$ will be a poor estimator of μ.

However, since the frequency distribution of the $\bar{x}$'s is normal, the chance of a single $\bar{x}$ lying within one standard deviation of the mean μ is 68%. Conversely, the mean μ has a 68% chance of lying within one standard deviation of *a single randomly chosen $\bar{x}$*. Thus there is a 68% chance that the true population mean μ falls within the interval

$$\bar{x} \pm \text{SE}_{\bar{x}}$$

In this way we can estimate the population mean μ from a *single sample*. Not only that, but we are able to give a range of values within which μ must lie and to give *the probability* that μ will fall within that interval.

There is just one difficulty: To compute the standard error $\text{SE}_{\bar{x}}$ by the central limit theorem, we must divide the population standard deviation σ by the square root of the sample size. But σ *is not usually known.* Thus it is common practice to use the sample standard deviation s instead. Thus the standard error of the mean is approximated by the following formula (valid if the population is large):

Standard Error of the Mean $\bar{x}$	$SE_{\bar{x}} = \dfrac{\sigma}{\sqrt{n}} \approx \dfrac{s}{\sqrt{n}}$	263

◆◆◆ **Example 57:** For the population of 256 integers (Table 26–8), the central limit theorem says that 68% of the sample means should fall within one standard error of μ, or within the interval 227 ± 18.5. Is this statement correct?

Solution: We count the number of sample means that actually do fall within this interval and get 23, which is 72% of the total number of samples.   ◆◆◆

◆◆◆ **Example 58:** One of the 32 samples in Example 56 has a mean of 234 and a standard deviation s of 67. Use these figures to estimate the mean μ of the population.

Solution: Computing the standard error from Eq. 263, with $s = 67$ and $n = 16$, gives

$$SE_{\bar{x}} \approx \frac{67}{\sqrt{16}} = 17$$

We may then predict that there is a 68% chance that μ will fall within the range

$$\mu = 234 \pm 17$$

We see that the true population mean 227 does fall within that range.   ◆◆◆

◆◆◆ **Example 59:** The heights of 64 randomly chosen students at Tech College were measured. The mean $\bar{x}$ was found to be 68.25 in., and the standard deviation s was 2.21 in. Estimate the mean μ of the entire population of students at Tech College.

Solution: From Eq. 263, with $s = 2.21$ and $n = 64$,

$$SE_{\bar{x}} \approx \frac{s}{\sqrt{n}} = \frac{2.21}{\sqrt{64}} = 0.28$$

We may then state that there is a 68% chance that the mean height μ of all students at Tech College is

$$\mu = 68.25 \pm 0.28 \text{ in.}$$

In other words, the true mean has a 68% chance of falling within the interval

$$67.97 \text{ to } 68.53 \text{ in.}$$ ◆◆◆

Confidence Intervals

The interval 67.97 to 68.53 in. found in Example 59 is called the *68% confidence interval.* Similarly, there is a 95% chance that the population mean falls within *two* standard errors of the mean.

$$\mu = \bar{x} \pm 2SE_{\bar{x}}$$

So we call it the *95% confidence interval.*

◆◆◆ Example 60: For the data of Example 59, find the 95% confidence interval.

Solution: We had found that $SE_{\bar{x}}$ was 0.28 in. Thus we may state that μ will lie within the interval

$$68.25 \pm 2(0.28)$$

or

$$68.25 \pm 0.56 \text{ in.}$$

This is the 95% confidence interval.   ◆◆◆

We find other confidence intervals in a similar way.

Standard Error of the Standard Deviation

We may compute standard errors for sample statistics other than the mean. The most common is for the standard deviation. The standard error of the standard deviation is called SE_s and is given by the following equation:

Standard Error of the Standard Deviation	$SE_s = \dfrac{\sigma}{\sqrt{2n}}$	264

where again, we use the sample standard deviation s if we don't know σ.

◆◆◆ Example 61: A single sample of size 32 drawn from a population is found to have a standard deviation s of 21.55. Give the population standard deviation σ with a confidence level of 68%.

Solution: From Eq. 264, with $\sigma \approx s = 21.55$ and $n = 32$,

$$SE_s = \frac{21.55}{\sqrt{2(32)}} = 2.69$$

Thus there is a 68% chance that σ falls within the interval

$$21.55 \pm 2.69$$ ◆◆◆

Standard Error of a Proportion

Consider a binomial experiment in which the probability of occurrence (called a *success*) of an event is p and the probability of nonoccurrence (a *failure*) of that event is q (or $1 - p$). For example, for the toss of a die, the probability of success in rolling a three is $p = \frac{1}{6}$, and the probability of failure in rolling a three is $q = \frac{5}{6}$.

If we threw the die n times and recorded the proportion of successes, it would be close to, but probably not equal to, p. Suppose that we then threw the die another n times and got another proportion, and then another n times, and so on. If we did this enough times, the proportion of successes of our samples would form a normal distribution with a mean equal to p and a standard deviation given by the following:

Standard Error of a Proportion	$SE_p = \sqrt{\dfrac{p(1 - p)}{n}}$	265

◆◆◆ Example 62: Find the 68% confidence interval for rolling a three for a die tossed 150 times.

Solution: Using Eq. 265, with $p = \frac{1}{6}$ and $n = 150$,

$$SE_p = \sqrt{\frac{\left(\frac{1}{6}\right)\left(\frac{5}{6}\right)}{150}} = 0.030$$

Thus there is a 68% probability that in 150 rolls of a die, the proportion of threes will lie between $\frac{1}{6} + 0.03$ and $\frac{1}{6} - 0.03$.   ◆◆◆

We can use the proportion of success in a single sample to estimate the proportion of success of an entire population. Since we will not usually know p for the population, we use the proportion of successes found from the sample when computing SE_p.

◆◆◆ **Example 63:** In a poll of 152 students at Tech College, 87 said that they would vote for Jones for president of the student union. Estimate the support for Jones among the entire student body.

Solution: In the sample, $\frac{87}{152} = 0.572$ support Jones. So we estimate that her support among the entire student body is 57.2%. To compute the confidence interval, we need the standard error. With $n = 152$, and using 0.572 as an approximation to p, we get

$$SE_p \approx \sqrt{\frac{0.572(1 - 0.572)}{152}} = 0.040$$

We thus expect that there is a 68% chance that the support for Jones is

$$0.572 \pm 0.040$$

or that she will capture between 53.2% and 61.2% of the vote of the entire student body.   ◆◆◆

Exercise 6 ◆ Standard Errors

1. The heights of 49 randomly chosen students at Tech College were measured. Their mean $\bar{x}$ was found to be 69.47 in., and their standard deviation s was 2.35 in. Estimate the mean μ of the entire population of students at Tech College with a confidence level of 68%.

2. For the data of problem 1, estimate the population mean with a 95% confidence interval.

3. For the data of problem 1, estimate the population standard deviation with a 68% confidence interval.

4. For the data of problem 1, estimate the population standard deviation with a 95% confidence interval.

5. A single sample of size 32 drawn from a population is found to have a mean of 164.0 and a standard deviation s of 16.31. Give the population mean with a confidence level of 68%.

6. For the data of problem 5, estimate the population mean with a 95% confidence interval.

7. For the data of problem 5, estimate the population standard deviation with a 68% confidence interval.

8. For the data of problem 5, estimate the population standard deviation with a 95% confidence interval.

9. Find the 68% confidence interval for drawing a heart from a deck of cards for 200 draws from the deck, replacing the card each time before the next draw.

10. In a survey of 950 viewers, 274 said that they watched a certain program. Estimate the proportion of viewers in the entire population that watched that program, given the 68% confidence interval.

26–7   Process Control

Statistical Process Control

Statistical process control, or SPC, is perhaps the most important application of statistics for technology students. Process control involves continuous testing of samples from a production line. Any manufactured item will have chance variations in weight, dimensions, color, and so forth. As long as the variations are due only to chance, we say that the process is *in control.*

◆◆◆ **Example 64:** The diameters of steel balls on a certain production line have harmless chance variations between 1.995 mm and 2.005 mm. The production process is said to be in control.   ◆◆◆

However, as soon as there is variation due to some cause other than chance, we say that the process is no longer in control.

◆◆◆ **Example 65:** During a certain week, some of the steel balls from the production line of Example 64 were found to have diameters over 2.005. The process was then *out of control.*   ◆◆◆

One problem of process control is to detect when a process is out of control so that those other causes may be eliminated.

◆◆◆ **Example 66:** The period during which the process was out of control in Example 65 was found to occur when the factory air conditioning was out of operation during a heat wave. Now the operators are instructed to stop the production line during similar occurrences.   ◆◆◆

Control Charts

The main tool of SPC is the *control chart.* The idea behind a control chart is simple. We pull samples off a production line at regular time intervals, measure them in some way, and graph those measurements over time. We then establish upper and lower limits between which the sample is acceptable, but outside of which it is not acceptable.

◆◆◆ **Example 67:** Figure 26–22 shows a control chart for the diameters of samples of steel balls of the preceding examples. One horizontal line shows the mean value of the diameter, while the other two give the upper and lower *control limits.* Note

To cover the subject of statistics usually requires an entire textbook, and SPC is often an entire chapter in such a book or may take an entire textbook by itself. Obviously we can only scratch the surface here.

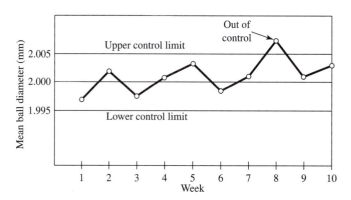

FIGURE 26–22   A control chart.

that the diameters fluctuate randomly between the control limits until the start of the heat wave, and then go out of control. ◆◆◆

To draw a control chart, we must decide what variables we want to measure and what statistical quantities we want to compute. Then we must calculate the control limits. A sample can be tested for a categorical variable (such as pass/fail) or for a continuous variable (such as the ball diameter). When testing a categorical variable, we usually compute the *proportion* of those items that pass. When testing a continuous variable, we usually compute the *mean, standard deviation, or range.*

We will draw three control charts: one for the proportion of a categorical variable, called a *p chart;* a second for the mean of a continuous variable, called an $\overline{X}$ *chart;* and a third for the range of a continuous variable, called an *R chart.* The formulas for computing the central line and the control limits are summarized in Fig. 26–23. We will illustrate the construction of each chart with an example.

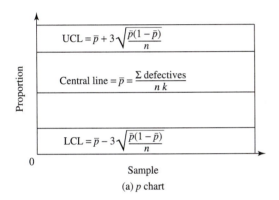

(a) *p* chart

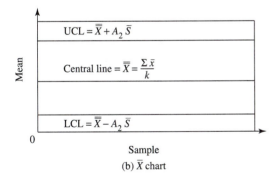

(b) $\overline{X}$ chart

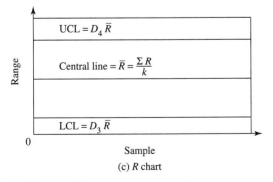

(c) *R* chart

FIGURE 26–23

The *p* Chart

◆◆◆ **Example 68:** Make a control chart for the proportion of defective light bulbs coming off a certain production line.

Solution:

1. *Choose a sample size n and the testing frequency.*

Let us choose to test a sample of 1000 bulbs every day for 21 days.

2. *Collect the data. Count the number d of defectives in each sample of 100. Obtain the proportion defective p for each sample by dividing the number of defectives d by the sample size n.*

The collected data and the proportion defective are given in Table 26–9.

TABLE 26–9   Daily samples of 1000 light bulbs per day for 21 days.

Day	Number Defective	Proportion Defective
1	63	0.063
2	47	0.047
3	59	0.059
4	82	0.082
5	120	0.120
6	73	0.073
7	58	0.058
8	55	0.055
9	39	0.039
10	99	0.099
11	85	0.085
12	47	0.047
13	46	0.046
14	87	0.087
15	90	0.090
16	67	0.067
17	85	0.085
18	24	0.024
19	77	0.077
20	58	0.058
21	103	0.103
Total	1464	

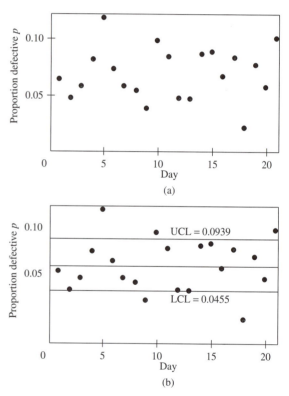

FIGURE 26–24

3. *Start the control chart by plotting p versus time.*

The daily proportion defective *p* is plotted in Fig. 26–24(a)

4. *Compute the average proportion defective $\bar{p}$ by dividing the total number of defectives for all samples by the total number of items measured.*

The total number of defectives is 1464, and the total number of bulbs tested is $1000(21) = 21{,}000$. So

$$\bar{p} = \frac{1464}{21{,}000} = 0.0697$$

We are assuming here that our sampling distribution is approximately normal.

If the calculation of the lower limit gives a negative number, then we have no lower control limit.

5. *Compute the standard error SE_p. From Sec. 26–6, the standard error for a proportion is in which the probability p of success of a single event is given by*

$$SE_p = \sqrt{\frac{p(1-p)}{n}} \qquad (265)$$

Since we do not know q, we use $\bar{p}$ as an estimator for p.

$$SE_p = \sqrt{\frac{\bar{p}(1-\bar{p})}{n}} = \sqrt{\frac{0.0697(1-0.0697)}{1000}} = 0.00805$$

6. *Choose a confidence interval. Recall from Sec. 26–6 that one standard error will give a 68% confidence interval, two standard errors a 95% confidence interval, three a 99.7% level, and so forth (Fig. 26–25). Set the control limits at these values.*

We will use three-sigma limits, which are the most commonly used. Thus

$$\text{upper control limit} = \bar{p} + 3SE_p = 0.0697 + 3(0.00805) = 0.0939$$
$$\text{lower control limit} = \bar{p} - 3SE_p = 0.0697 - 3(0.00805) = 0.0455$$

The next step would be to *analyze* the finished control chart. We would look for points that are out of control and try to determine their cause. We do not have space here for this discussion, which can be found in any SPC book.

We would also *adjust the control limits* if in our calculation of the limits we used values that we now see are outside those limits. Since those values are not the result of chance but are caused by some production problem, they should not be used in calculating the permissible variations due to chance. Thus we would normally recompute the control limits without using those values. We omit that step here.

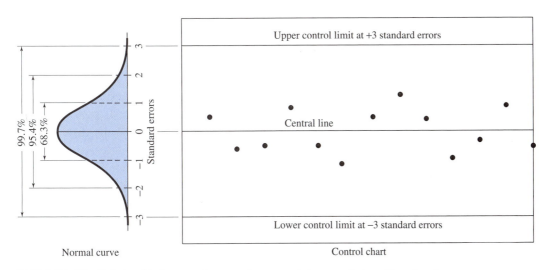

Normal curve   Control chart

FIGURE 26–25 Relationship between the normal curve and the control chart. Note that the upper and lower limits are placed at three standard errors from the mean, so that there is a 99.7% chance that a sample will fall between those limits.

7. *Draw a horizontal at $\bar{p}$, and upper and lower control limits at $\bar{p} \pm 3SE_p$.*

These are shown in the control chart in Fig. 26–24(b).   ◆◆◆

Control Chart for a Continuous Variable

We construct a control chart for a continuous variable in much the same way as for a categorical variable. The variables commonly charted are the mean and either the range or the standard deviation. The range is usually preferred over the standard deviation because it is simple to calculate and easier to understand by factory personnel who may not be familiar with statistics. We will now make control charts for the mean and range, commonly called the $\overline{X}$ and *R charts*.

The $\overline{X}$ and *R* Charts

We again illustrate the method with an example. As is usually done, we will construct both charts at the same time.

◆◆◆ **Example 69:** Make $\overline{X}$ and R control charts for a sampling of the wattages of the lamps on a production line.

Solution:

1. *Choose a sample size n and the testing frequency.*

Since the measurement of wattage is more time-consuming than just simple counting of defectives, let us choose a smaller sample size than before, say, 5 bulbs every day for 21 days.

2. *Collect the data.*

We measure the wattage in each sample of 5 bulbs. For each sample we compute the sample mean $\overline{X}$ and the sample range R. These are given in Table 26–10.

TABLE 26–10 Daily samples of wattages of 5 lamps per day for a period of 21 days.

Day	Wattage of 5 Samples					Mean X	Range R
1	105.6	92.8	92.6	101.5	102.5	99.0	13.0
2	106.2	100.4	106.6	108.3	109.2	106.1	8.8
3	108.6	103.3	101.8	96.0	98.8	101.7	12.6
4	95.5	106.3	106.1	97.8	101.0	101.3	10.8
5	98.4	91.5	106.3	91.6	93.2	96.2	14.8
6	101.9	102.5	94.5	107.7	108.6	103.0	14.1
7	96.7	94.8	106.9	103.4	96.2	99.6	12.2
8	99.3	107.5	94.8	102.1	108.2	102.4	13.3
9	98.0	108.3	94.8	98.6	102.8	100.5	13.5
10	109.3	98.6	92.2	106.7	96.9	100.8	17.1
11	98.8	95.0	99.4	104.8	96.2	98.9	9.7
12	91.2	103.3	95.3	103.2	108.9	100.4	17.7
13	98.5	102.8	106.4	108.6	94.2	102.1	14.4
14	100.8	108.1	104.7	108.6	97.0	103.8	11.6
15	109.0	102.6	90.5	91.3	90.5	96.8	18.6
16	100.0	104.0	99.5	92.9	93.4	98.0	11.1
17	106.3	106.2	107.6	90.4	102.9	102.7	17.2
18	103.4	91.8	105.7	106.9	106.3	102.8	15.2
19	99.7	106.8	99.6	106.6	90.1	100.6	16.7
20	98.4	104.5	94.5	101.7	96.2	99.0	10.0
21	102.2	103.0	90.3	94.0	103.5	98.6	13.2
					Sums	2114	285.3
					Averages	100.7	13.6

3. *Start the control chart by plotting $\overline{X}$ and R versus time.*

The daily mean and range are plotted as shown in Fig. 26–26.

4. *Compute the average mean $\overline{\overline{X}}$ and the average range $\overline{R}$.*

We take the average of the 21 means.

$$\overline{\overline{X}} = \frac{\Sigma \overline{X}}{21} = \frac{2114}{21} = 100.7$$

The average of the ranges is

$$\overline{R} = \frac{\Sigma R}{21} = \frac{285.3}{21} = 13.6$$

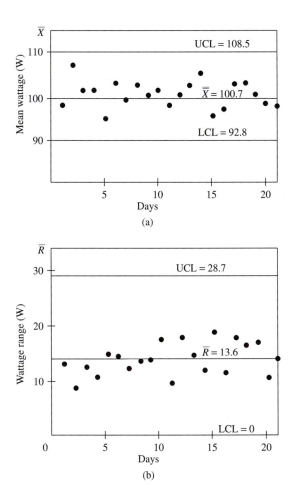

FIGURE 26–26

5. *Compute the control limits.*

When making an $\overline{X}$ and R chart, we normally do not have the standard deviation to use in computing the control limits. The average range $\overline{R}$, multiplied by a suitable constant, is usually used to give the control limits for both the mean and the range charts. For the mean, the three-sigma control limits are

$$\overline{\overline{X}} \pm A_2\overline{R}$$

where A_2 is a constant depending on sample size, given in Table 26–11. The control limits for the range are

$$D_3\overline{R} \quad \text{and} \quad D_4\overline{R}$$

where D_3 and D_4 are also from the table.

TABLE 26–11   Factors for computing control chart lines.

n	2	3	4	5	6	7	8	9	10
A_2	1.880	1.023	0.729	0.577	0.483	0.419	0.373	0.337	0.308
D_3	0	0	0	0	0	0.076	0.136	0.184	0.223
D_4	3.268	2.574	2.282	2.114	2.004	1.924	1.864	1.816	1.777

Excerpted from the *ASTM Manual on Presentation of Data and Control Chart Analysis* (1976). For larger sample sizes, consult a book on SPC.

For a sample size of 5, $A_2 = 0.577$, so the control limits for the mean are

$$UCL = 100.7 + 0.577(13.58) = 108.5$$

and

$$LCL = 100.7 - 0.577(13.58) = 92.8$$

Also from the table, $D_3 = 0$ and $D_4 = 2.114$, so the control limits for the range are

$$UCL = 2.114(13.58) = 28.7$$

and

$$LCL = 0(13.58) = 0$$

6. *Draw the central line and the upper and lower control limits.*

These are shown in the control chart [Fig. 26–26(b)].   ◆◆◆

Exercise 7 ◆ Process Control

1. The proportion of defectives for samples of 1000 tennis balls per day for 20 days is as follows:

Day	Number Defective	Proportion Defective
1	31	0.031
2	28	0.028
3	26	0.026
4	27	0.027
5	30	0.030
6	26	0.026
7	26	0.026
8	27	0.027
9	29	0.029
10	27	0.027
11	27	0.027
12	26	0.026
13	31	0.031
14	30	0.030
15	27	0.027
16	27	0.027
17	25	0.025
18	26	0.026
19	30	0.030
20	28	0.028
Total	554	

Find the values for the central line, and determine the upper and lower control limits.

2. Draw a *p* chart for the data of problem 1.

3. The proportion of defectives for samples of 500 calculators per day for 20 days is as follows:

Day	Number Defective	Proportion Defective
1	153	0.306
2	174	0.348
3	139	0.278
4	143	0.286
5	156	0.312
6	135	0.270
7	141	0.282
8	157	0.314
9	125	0.250
10	126	0.252
11	155	0.310
12	174	0.348
13	138	0.276
14	165	0.330
15	144	0.288
16	166	0.332
17	145	0.290
18	153	0.306
19	132	0.264
20	169	0.338
Total	2990	

Find the values for the central line, and determine the upper and lower control limits.

4. Draw a p chart for the data of problem 3.

5. Five pieces of pipe are taken from a production line each day for 20 days, and their lengths are measured as shown in the following table:

Day	Measurements of 5 Samples (in.)				
1	201.7	184.0	201.3	183.9	192.6
2	184.4	184.2	207.2	194.4	193.1
3	217.2	212.9	198.7	185.1	196.9
4	201.4	201.8	182.5	195.0	215.4
5	214.3	197.2	219.8	219.9	194.5
6	190.9	180.6	181.9	203.6	185.1
7	189.5	209.5	207.6	215.2	200.0
8	182.2	216.4	190.7	189.0	204.2
9	184.9	217.3	185.7	183.0	189.7
10	211.2	194.8	208.6	209.4	183.3
11	197.1	189.5	200.1	197.7	187.9
12	207.1	203.1	218.4	199.3	219.1
13	199.1	207.1	205.8	197.4	189.4
14	198.9	196.1	210.3	210.6	217.0
15	215.2	200.8	205.0	186.6	181.6
16	180.2	216.8	182.5	206.3	191.7
17	219.3	195.1	182.2	207.4	191.9
18	204.0	209.9	196.4	202.3	206.9
19	202.2	216.3	217.8	200.7	215.1
20	185.0	194.1	191.2	186.3	201.6
21	181.6	187.8	188.2	191.7	200.1

Find the values for the central line, and determine the upper and lower control limits for the mean.

6. Find the values for the central line for the data of problem 5, and determine the upper and lower control limits for the range.

7. Draw an $\overline{X}$ chart for the data of problem 5.

8. Draw an R chart for the data of problem 5.

9. Five circuit boards are taken from a production line each day for 20 days and are weighed (in grams) as shown in the following table:

Day	Measurements of 5 Samples (g)				
1	16.7	19.5	11.1	18.8	11.8
2	15.0	14.5	14.5	18.4	11.5
3	14.3	16.5	13.3	14.0	14.4
4	19.3	14.0	14.6	15.0	18.0
5	16.2	13.6	16.8	12.4	16.6
6	14.3	16.7	18.4	18.3	16.5
7	13.5	15.5	13.7	19.8	11.6
8	15.6	12.3	16.8	18.3	11.8
9	13.3	15.4	16.1	17.0	12.2
10	17.6	18.6	12.4	19.9	19.3
11	19.0	12.8	14.7	19.6	15.8
12	12.3	12.2	14.9	15.5	12.9
13	16.3	12.3	18.6	14.0	11.3
14	11.4	15.8	13.9	19.1	10.5
15	12.5	11.4	14.1	12.4	14.9
16	18.1	15.1	19.2	15.2	17.6
17	18.7	13.7	11.0	17.0	12.7
18	11.3	11.5	13.4	13.1	13.9
19	16.6	18.7	12.1	14.7	11.2
20	11.2	10.5	16.2	11.0	16.2
21	12.8	18.7	10.4	16.8	17.1

Find the values for the central line, and determine the upper and lower control limits for the mean.

10. Find the values for the central line for the data of problem 9, and determine the upper and lower control limits for the range.

11. Draw an $\overline{X}$ chart for the data of problem 9.

12. Draw an R chart for the data of problem 9.

Computer

13. Some statistics software for the computer, such as *Minitab,* can be used to draw control charts. One simply enters the raw data and specifies which kind of chart is wanted, and the program automatically sets the central line and the control limits and draws the chart. Some software will even detect out-of-control or nonrandom behavior in the data. If you have such software, use it to draw any of the control charts in this exercise set.

26–8   Regression

Curve Fitting

The word *regression* was first used in the nineteenth century when these methods were used to determine the extent by which the heights of children of tall or short parents *regressed* or got closer to the mean height of the population.

In statistics, *curve fitting,* the fitting of a curve to a set of data points, is called *regression.* The fitting of a straight line to data points is called *linear regression,* while the fitting of some other curve is called *nonlinear regression.* Fitting a curve to *two* sets of data is called *multiple regression.* We will cover linear regression in this section.

Scatter Plot

We mentioned *scatter plots* in Sec. 26–1. A scatter plot is simply a plot of all of the data points. It can be made by hand, by graphics calculator, or by computer.

••• **Example 70:** Scatter plots for three sets of data are shown in Fig. 26–27. In Fig. 26–27(b), the point (6, 7) is called an *outlier*. Such points are usually suspected as being the result of an error and are sometimes discarded.

Glance back at Sec. 20–7 where we did some curve fitting using logarithmic and semilogarithmic graph paper.

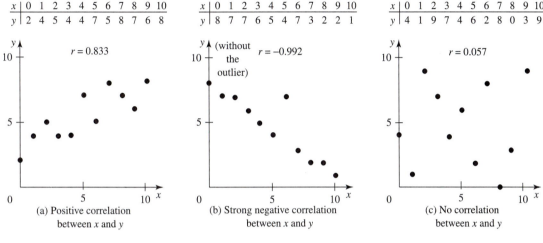

x	0	1	2	3	4	5	6	7	8	9	10
y	2	4	5	4	4	7	5	8	7	6	8

x	0	1	2	3	4	5	6	7	8	9	10
y	8	7	7	6	5	4	7	3	2	2	1

x	0	1	2	3	4	5	6	7	8	9	10
y	4	1	9	7	4	6	2	8	0	3	9

(a) Positive correlation between x and y — r = 0.833

(b) Strong negative correlation between x and y — r = −0.992 (without the outlier)

(c) No correlation between x and y — r = 0.057

FIGURE 26–27

◆◆◆

Correlation Coefficient

In Fig. 26–27 we see that some data are more scattered than others. The *correlation coefficient r* gives a numerical measure of this scattering. For a set of n xy pairs, the correlation coefficient is given by the following equation:

Correlation Coefficient	$$r = \frac{n \sum xy - \sum x \sum y}{\sqrt{n \sum x^2 - (\sum x)^2}\,\sqrt{n \sum y^2 - (\sum y)^2}}$$	266

The formula looks complicated but is actually easy to use when doing the computation in table form.

••• **Example 71:** Find the correlation coefficient for the data in Fig. 26–27(a).

Solution: We make a table giving x and y, the product of x and y, and the squares of x and y. We then sum each column as shown in Table 26–12.
Then, from Eq. 266, with $n = 11$,

$$r = \frac{n \sum xy - \sum x \sum y}{\sqrt{n \sum x^2 - (\sum x)^2}\,\sqrt{n \sum y^2 - (\sum y)^2}}$$

$$= \frac{11(353) - 55(60)}{\sqrt{11(385) - (55)^2}\,\sqrt{11(364) - (60)^2}} = 0.833$$

The correlation coefficient can vary from + 1 (perfect positive correlation) through 0 (no correlation) to −1 (perfect negative correlation). For comparison with the data in Fig. 26–27(a), the correlation coefficients for the data in (b) and (c) are

TABLE 26–12

x	y	xy	x^2	y^2
0	2	0	0	4
1	4	4	1	16
2	5	10	4	25
3	4	12	9	16
4	4	16	16	16
5	7	35	25	49
6	5	30	36	25
7	8	56	49	64
8	7	56	64	49
9	6	54	81	36
10	8	80	100	64
Sums 55	60	353	385	364
Σx	Σy	Σxy	Σx^2	Σy^2

found to be $r = -0.992$ (after discarding the outlier) and $r = 0.057$, respectively (work not shown).   ♦♦♦

Linear Regression

In *linear regression,* the object is to find the constants m and b for the equation of a straight line

$$y = mx + b \tag{289}$$

that best fits a given set of data points. The simplest way to do this is to fit a straight line by eye on a scatter plot and to measure the slope and the y intercept of that line.

♦♦♦ **Example 72:** Find, by eye, the equation of the straight line that best fits the following data:

x	1	2	3	4	5	6	7	8
y	3.249	3.522	4.026	4.332	4.900	5.121	5.601	5.898

Solution: We draw a scatter diagram as shown in Fig. 26–28. Then using a straight-edge, we draw a line through the data, trying to balance those points above

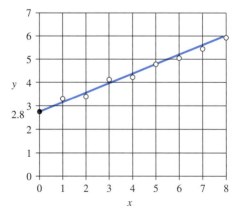

FIGURE 26–28

the line with those below. We then read the y intercept and the rise of the line in a chosen run.

$$y \text{ intercept } b = 2.8$$

Our line appears to pass through $(8, 6)$, so in a run of 8 units, the line rises 3.2 units. Its slope is then

$$\text{slope } m = \frac{\text{rise}}{\text{run}} = \frac{3.2}{8} = 0.40$$

The equation of the fitting line is then

$$y = 0.40x + 2.8$$ ◆◆◆

Method of Least Squares

Sometimes the points may be too scattered to enable drawing a line with a good fit, or perhaps we desire more accuracy than can be obtained when fitting by eye. Then we may want a method that does not require any manual steps, so that it can be computerized. One such method is the *method of least squares.*

Let us define a *residual* as the vertical distance between a data point and the approximating curve (Fig. 26–29). The method of least squares is a method to fit a straight line to a data set so that *the sum of the squares of the residuals is a minimum*—hence the name *least squares.*

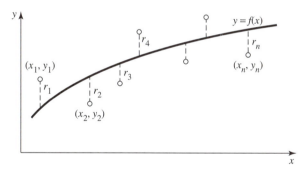

FIGURE 26–29   Definition of a residual.

With this method, the slope and the y intercept of the least squares line are given by the following equations:

$$\text{slope } m = \frac{n \sum xy - \sum x \sum y}{n \sum x^2 - (\sum x)^2}$$

|267|

$$y \text{ intercept } b = \frac{\sum x^2 \sum y - \sum x \sum xy}{n \sum x^2 - (\sum x)^2}$$

We give these equations without proof here. They are derived using calculus.

These equations are easier to use than they look, as you will see in the following example.

◆◆◆ **Example 73:** Repeat Example 72 using the method of least squares.

Solution: We tabulate the given values in Table 26–13.

TABLE 26–13

x	y	x^2	xy
1	3.249	1	3.249
2	3.522	4	7.044
3	4.026	9	12.08
4	4.332	16	17.33
5	4.900	25	24.50
6	5.121	36	30.73
7	5.601	49	39.21
8	5.898	64	47.18
$\Sigma x = 36$	$\Sigma y = 36.649$	$\Sigma x^2 = 204$	$\Sigma xy = 181.3$

In the third column of Table 26–13, we tabulate the squares of the abscissas given in column 1, and in the fourth column we list the products of x and y. The sums are given below each column.

Substituting these sums into Eq. 267 and letting $n = 8$, we have

$$\text{slope} = \frac{8(181.3) - 36(36.65)}{8(204) - (36)^2} = 0.3904$$

This is not the only way to fit a line to a data set, but it is widely regarded as the one that gives the "best fit."

and

$$y \text{ intercept} = \frac{204(36.65) - 36(181.3)}{8(204) - (36)^2} = 2.824$$

which agree well with our graphically obtained values. The equation of our best-fitting line is then

$$y = 0.3904x + 2.824$$

Note the close agreement with the values we obtained by eye: slope $= 0.40$ and y intercept $= 2.8$.   ◆◆◆

Exercise 8 ◆ Regression

Correlation Coefficient

Find the correlation coefficient for each set of data.

1.		2.		3.	
-8	-6.238	-20	82.29	-11.0	-65.30
-6.66	-3.709	-18.5	73.15	-9.33	-56.78
-5.33	-0.712	-17.0	68.11	-7.66	-47.26
-4	1.887	-15.6	59.31	-6.00	-37.21
-2.66	4.628	-14.1	53.65	-4.33	-27.90
-1.33	7.416	-12.6	45.90	-2.66	-18.39
0	10.2	-11.2	38.69	-1.00	-9.277
1.33	12.93	-9.73	32.62	0.66	0.081
2.66	15.70	-8.26	24.69	2.33	9.404
4	18.47	-6.8	18.03	4.00	18.93
5.33	21.32	-5.33	11.31	5.66	27.86
6.66	23.94	-3.86	3.981	7.33	37.78
8	26.70	-2.4	-2.968	9.00	46.64
9.33	29.61	-0.93	-9.986	10.6	56.69
10.6	32.35	0.53	-16.92	12.3	64.74
12	35.22	2	-23.86	14.0	75.84

Method of Least Squares

Find the least squares line for each set of data.

4. From problem 1.
5. From problem 2.
6. From problem 3.

Graphics Calculator

7. Many graphics calculators can automatically compute the correlation coefficient for a set of data. Check your manual for the proper instructions, and use your calculator to solve problems 1 through 3 in this exercise set.
8. Some graphics calculators can graph a least squares line for a given set of data, and some can do nonlinear regression. If your calculator has this capability, use it to solve problems 4 through 6 of this exercise set.

Computer

9. Many computer programs, including *Lotus 123, Derive, Maple, MathCad,* and *Minitab,* can do all or some of the following: find a correlation coefficient, make scatter plots, find the least squares line, and do nonlinear regression. If you are able, use your computer to do any of problems 1 through 6 in this exercise set.

••• CHAPTER 26 REVIEW PROBLEMS ••••••••••••••••••••••••••••••••••••••

Label each type of data as continuous, discrete, or categorical.

1. The life of certain radios.
2. The number of houses sold each day.
3. The colors of the cars at a certain dealership.
4. The sales figures for a certain product for the years 1970–80 are as follows:

Year	Number Sold
1970	1344
1971	1739
1972	2354
1973	2958
1974	3153
1975	3857
1976	3245
1977	4736
1978	4158
1979	5545
1980	6493

Make an *x-y* graph of the sales versus the year.

5. Make a bar graph for the data of problem 4.
6. If we toss three coins, what is the probability of getting one head and two tails?
7. If we throw two dice, what chance is there that their sum is 8?

8. A die is rolled twice. What is the probability that both rolls will give a two?

9. At a certain factory, 72% of the workers have brown hair, 6% are left-handed, and 3% both have brown hair and are left-handed. What is the probability that a worker chosen at random will either have brown hair or be left-handed, or both?

10. What is the probability of tossing a five or a nine when two dice are tossed?

11. Find the area under the normal curve between the mean and 0.5 standard deviation to one side of the mean.

12. Find the area under the normal curve between 1.1 standard deviations on both sides of the mean.

13. Find the area in the tail of the normal curve to the left of $z = 0.4$

14. Find the area in the tail of the normal curve to the right of $z = 0.3$.

Use the following table of data for problems 15 through 29.

146	153	183	148	116	127	162	153
168	161	117	153	116	125	173	131
117	183	193	137	188	159	154	112
174	182	144	144	133	167	192	145
162	138	137	154	141	129	137	152

15. Determine the range of the data.

16. Make a frequency distribution using class widths of 5. Show both absolute and relative frequency.

17. Draw a frequency histogram showing both absolute and relative frequency.

18. Draw a frequency polygon showing both absolute and relative frequency.

19. Make a cumulative frequency distribution.

20. Draw a cumulative frequency polygon.

21. Find the mean.

22. Find the median.

23. Find the mode.

24. Find the variance.

25. Find the standard deviation.

26. Predict the population mean with a 68% confidence interval.

27. Predict the population mean with a 95% confidence interval.

28. Predict the population standard deviation with a 68% confidence interval.

29. Predict the population standard deviation with a 95% confidence interval.

30. On a test given to 300 students, the mean grade was 79.3 with a standard deviation of 11.6. Assuming a normal distribution, estimate the number of A grades (a score of 90 or over).

31. For the data of problem 30 estimate the number of failing grades (a score of 60 or less).

32. Determine the quartiles and give the quartile range of the following data:

    118   133   135   143   164   173   179   199   212   216   256

33. Determine the quartiles and give the quartile range of the following data:

    167   245   327   486   524   639   797   853   974   1136   1162   1183

34. The proportion defective for samples of 700 keyboards per day for 20 days is as follows:

Day	Number Defective	Proportion Defective
1	123	0.176
2	122	0.175
3	125	0.178
4	93	0.132
5	142	0.203
6	110	0.157
7	98	0.140
8	120	0.172
9	139	0.198
10	142	0.203
11	128	0.183
12	132	0.188
13	92	0.132
14	114	0.163
15	92	0.131
16	105	0.149
17	146	0.208
18	113	0.162
19	122	0.175
20	92	0.131
Total	2351	

Find the values of the central line, determine the upper and lower control limits, and make a control chart for the proportion defective.

35. Five castings per day are taken from a production line each day for 21 days and are weighed. Their weights, in grams, are as follows:

Day	Measurements of 5 Samples				
1	1134	1168	995	992	1203
2	718	1160	809	432	650
3	971	638	1195	796	690
4	598	619	942	1009	833
5	374	395	382	318	329
6	737	537	692	562	960
7	763	540	738	969	786
0	777	1021	626	786	472
9	1015	1036	1037	1063	1161
10	797	1028	1102	796	536
11	589	987	765	793	481
12	414	751	1020	524	1100
13	900	613	1187	458	661
14	835	957	680	845	1023
15	832	557	915	934	734
16	353	829	808	626	868
17	472	798	381	723	679
18	916	763	599	338	1026
19	331	372	318	304	354
20	482	649	1133	1022	320
21	752	671	419	715	413

Find the values of the central line, determine the upper and lower control limits, and make a control chart for the mean of the weights.

36. Find the values of the central line, determine the upper and lower control limits, and make a control chart for the range of the weights.

37. Find the correlation coefficient and the least squares fit for the following data:

x	y
5	6.882
11.2	−7.623
17.4	−22.45
23.6	−36.09
29.8	−51.13
36.0	−64.24
42.2	−79.44
48.4	−94.04
54.6	−107.8
60.8	−122.8
67.0	−138.6
73.2	−151.0
79.4	−165.3
85.6	−177.6
91.8	−193.9
98	−208.9

38. For the relative frequency histogram for the population of students' grades (Fig. 26–3), find the probability that a student chosen at random will have received a grade between 74 and 86.

39. Find the 68% confidence interval for drawing a king from a deck of cards, for 200 draws from the deck, replacing the card each time before the next draw.

40. In a survey of 120 shoppers, 71 said that they preferred Brand A potato chips over Brand B. Estimate the proportion of shoppers in the entire population that would prefer Brand A, given the 68% confidence interval.

Writing

41. Give an example of one statistical claim (such as an advertisement, a commercial, or a political message) that you have heard or read lately that has made you skeptical. State your reasons for being suspicious, and point out how the claim could otherwise have been presented to make it more plausible.

Team Project

42. A certain diode has the following characteristics:

Voltage across Diode (V)	Current through Diode (mA)
0	0
5	2.063
10	5.612
15	10.17
20	15.39
25	21.30
30	27.71
35	34.51
40	42.10
45	49.95
50	58.11

Linearize the data using the methods of Sec. 20–7. Then apply the method of least squares to find the slope and the y intercept of the straight line obtained. Using those values, write and graph an equation for the current as a function of the voltage. How does your graphed equation compare with the plot of the original points?

Summary of Facts and Formulas

Many mathematics courses cover only a fraction of the topics in this text. Further, some of these formulas are included for reference even though they may not appear elsewhere in the text. We hope that this Summary will provide a handy reference, not only for your current course but for others, and for your technical work after graduation.

	No.				Page
ALGEBRAIC LAWS	1	Commutative Law	Addition	$a + b = b + a$	9
	2		Multiplication	$ab = ba$	12,67
	3	Associative Law	Addition	$a + (b + c) = (a + b) + c = (a + c) + b$	10
	4		Multiplication	$a(bc) = (ab)c = (ac)b = abc$	13
	5	Distributive Law		$a(b + c) = ab + ac$	13,68,202
RULES OF SIGNS	6	Addition and Subtraction		$a + (-b) = a - (+b) = a - b$	8
	7			$a + (+b) = a - (-b) = a + b$	9
	8	Multiplication		$(+a)(+b) = (-a)(-b) = +ab$	14,67
	9			$(+a)(-b) = (-a)(+b) = -(+a)(+b) = -ab$	14,67
	10	Division		$\dfrac{+a}{+b} = \dfrac{-a}{-b} = -\dfrac{-a}{+b} = -\dfrac{+a}{-b} = \dfrac{a}{b}$	18,76,227
	11			$\dfrac{+a}{-b} = \dfrac{-a}{+b} = -\dfrac{-a}{-b} = -\dfrac{a}{b}$	18,76,227
PERCENTAGE	12			Amount = base × rate $\quad A = BP$	42
	13			Percent change $= \dfrac{\text{new value} - \text{original value}}{\text{original value}} \times 100$	44
	14			Percent error $= \dfrac{\text{measured value} - \text{known value}}{\text{known value}} \times 100$	45
	15			Percent concentration of ingredient $A = \dfrac{\text{amount of } A}{\text{amount of mixture}} \times 100$	45
	16			Percent efficiency $= \dfrac{\text{output}}{\text{input}} \times 100$	44,100

	No.				Page
BINARY NUMBERS	17	Largest n-Bit Binary Number		$2^n - 1$	692
	18	Addition		$0 + 0 = 0$ $0 + 1 = 1$ $1 + 0 = 1$ $1 + 1 = 0$ with carry to next column	
	19	Subtraction		$0 - 0 = 0$ $0 - 1 = 1$ with borrow from next column $1 - 0 = 1$ $1 - 1 = 0$ $10 - 1 = 1$	
	20			Subtracting B from A is equivalent to adding the two's complement of B to A	
	21	Multiplication		$0 \times 0 = 0$ $0 \times 1 = 0$ $1 \times 0 = 0$ $1 \times 1 = 1$	
	22	Division		$0 \div 0$ is not defined $1 \div 0$ is not defined $0 \div 1 = 0$ $1 \div 1 = 1$	
	23	Complements	n-Bit One's Complement of x	$(2^n - 1) - x$	
	24			The one's complement of a number can be written by changing the 1s to 0s and the 0s to 1s.	
	25		n-Bit Two's Complement of x	$2^n - x$	
	26			To find the two's complement of a number first write the one's complement, and add 1.	
	27	Negative Binary Numbers		If M is a positive binary number, then $-M$ is the two's complement of M.	

	No.				Page
EXPONENTS	28	Definition		$a^n = \underbrace{a \cdot a \cdot a \cdot \ldots \cdot a}_{n \text{ factors}}$	60,64
	29	Law of Exponents	Products	$x^a \cdot x^b = x^{a+b}$	346,60, 64,68
	30		Quotients	$\dfrac{x^a}{x^b} = x^{a-b} \quad (x \neq 0)$	347,61, 65,76
	31		Powers	$(x^a)^b = x^{ab} = (x^b)^a$	346,62, 65,206
	32		Product Raised to a Power	$(xy)^n = x^n \cdot y^n$	346,62,65
	33		Quotient Raised to a Power	$\left(\dfrac{x}{y}\right)^n = \dfrac{x^n}{y^n} \quad (y \neq 0)$	347,63,65
	34		Zero Exponent	$x^0 = 1 \quad (x \neq 0)$	345,63,65
	35		Negative Exponent	$x^{-a} = \dfrac{1}{x^a} \quad (x \neq 0)$	345,64,65
	36	Fractional Exponents		$a^{1/n} = \sqrt[n]{a}$	351
	37			$a^{m/n} = \sqrt[n]{a^m} = (\sqrt[n]{a})^m$	351
RADICALS	38	Rules of Radicals	Root of a Product	$\sqrt[n]{ab} = \sqrt[n]{a}\,\sqrt[n]{b}$	351
	39		Root of a Quotient	$\sqrt[n]{\dfrac{a}{b}} = \dfrac{\sqrt[n]{a}}{\sqrt[n]{b}}$	352
	40		Root of a Power	$\sqrt[n]{a^m} = (\sqrt[n]{a})^m$	
SPECIAL PRODUCTS AND FACTORING	41	Binomials	Difference of Two Squares	$a^2 - b^2 = (a-b)(a+b)$	71,205
	42		Sum of Two Cubes	$a^3 + b^3 = (a+b)(a^2 - ab + b^2)$	218
	43		Difference of Two Cubes	$a^3 - b^3 = (a-b)(a^2 + ab + b^2)$	218
	44	Trinomials	Test for Factorability	$ax^2 + bx + c$ is factorable if $b^2 - 4ac$ is a perfect square	208
	45		Leading Coefficient = 1	$x^2 + (a+b)x + ab = (x+a)(x+b)$	71,209
	46		General Quadratic Trinomial	$acx^2 + (ad+bc)x + bd = (ax+b)(cx+d)$	71,213
	47		Perfect Square Trinomials	$a^2 + 2ab + b^2 = (a+b)^2$	72,217
	48			$a^2 - 2ab + b^2 = (a-b)^2$	217
	49	Factoring by Grouping		$ac + ad + bc + bd = (a+b)(c+d)$	

	No.					Page
FRACTIONS	50	Simplifying	$\dfrac{ad}{bd} = \dfrac{a}{b}$			226
	51	Multiplication	$\dfrac{a}{b} \cdot \dfrac{c}{d} = \dfrac{ac}{bd}$			229
	52	Division	$\dfrac{a}{b} \div \dfrac{c}{d} = \dfrac{a}{b} \cdot \dfrac{d}{c} = \dfrac{ad}{bc}$			230
	53	Addition and Subtraction	Same Denominators	$\dfrac{a}{b} \pm \dfrac{c}{b} = \dfrac{a \pm c}{b}$		234
	54		Different Denominators	$\dfrac{a}{b} \pm \dfrac{c}{d} = \dfrac{ad}{bd} \pm \dfrac{bc}{bd} = \dfrac{ad \pm bc}{bd}$		236
PROPORTION	55	In the Proportion $a{:}b = c{:}d$	The product of the means equals the product of the extremes.		$ad = bc$	
	56		The extremes may be interchanged.		$d{:}b = c{:}a$	
	57		The means may be interchanged.		$a{:}c = b{:}d$	
	58		The means may be interchanged with the extremes.		$b{:}a = d{:}c$	
	59	Mean Proportional	In the Proportion	$a{:}b = b{:}c$		530
			Geometric Mean	$b = \pm\sqrt{ac}$		530,739
VARIATION	60	$k =$ Constant of Proportionality	Direct	$y \propto x$ or $y = kx$		531
	61		Inverse	$y \propto \dfrac{1}{x}$ or $y = \dfrac{k}{x}$		543
	62		Joint	$y \propto xw$ or $y = kxw$		548

No.   Page

SYSTEMS OF LINEAR EQUATIONS

Algebraic solution

No.		Content	Page
63		$a_1x + b_1y = c_1$ $a_2x + b_2y = c_2$ $\qquad$ $x = \dfrac{b_2c_1 - b_1c_2}{a_1b_2 - a_2b_1}$, and $y = \dfrac{a_1c_2 - a_2c_1}{a_1b_2 - a_2b_1}$ where $a_1b_2 - a_2b_1 \neq 0$	340,273
64		$a_1x + b_1y + c_1z = k_1$ $a_2x + b_2y + c_2z = k_2$ $a_3x + b_3y + c_3z = k_3$ $x = \dfrac{b_2c_3k_1 + b_1c_2k_3 + b_3c_1k_2 - b_2c_1k_3 - b_3c_2k_1 - b_1c_3k_2}{a_1b_2c_3 + a_3b_1c_2 + a_2b_3c_1 - a_3b_2c_1 - a_1b_3c_2 - a_2b_1c_3}$ $y = \dfrac{a_1c_3k_2 + a_3c_2k_1 + a_2c_1k_3 - a_3c_1k_2 - a_1c_2k_3 - a_2c_3k_1}{a_1b_2c_3 + a_3b_1c_2 + a_2b_3c_1 - a_3b_2c_1 - a_1b_3c_2 - a_2b_1c_3}$ $z = \dfrac{a_1b_2k_3 + a_3b_1k_2 + a_2b_3k_1 - a_3b_2k_1 - a_1b_3k_2 - a_2b_1k_3}{a_1b_2c_3 + a_3b_1c_2 + a_2b_3c_1 - a_3b_2c_1 - a_1b_3c_2 - a_2b_1c_3}$	303

Determinants

Value of a Determinant

No.		Content	Page
65	Second Order	$\begin{vmatrix} a_1 & b_1 \\ a_2 & b_2 \end{vmatrix} = a_1b_2 - a_2b_1$	295
66	Third Order	$\begin{vmatrix} a_1 & b_1 & c_1 \\ a_2 & b_2 & c_2 \\ a_3 & b_3 & c_3 \end{vmatrix} = a_1b_2c_3 + a_3b_1c_2 + a_2b_3c_1 - (a_3b_2c_1 + a_1b_3c_2 + a_2b_1c_3)$	300,302
67	Minors	The signed minor of element b in the determinant $\begin{vmatrix} a & b & c \\ d & e & f \\ g & h & i \end{vmatrix}$ is $-\begin{vmatrix} d & f \\ g & i \end{vmatrix}$	301
68		To find the value of a determinant: 1. Choose any row or any column to develop by minors. 2. Write the product of every element in that row or column and its signed minor. 3. Add these products to get the value of the determinant.	301

Cramer's Rule

No.		Content	Page
69		The solution for any variable is a fraction whose denominator is the determinant of the coefficients, and whose numerator is also the determinant of the coefficients, except that the column of coefficients of the variable being solved for is replaced by the column of constants.	303
70	Two Equations	$x = \dfrac{\begin{vmatrix} c_1 & b_1 \\ c_2 & b_2 \end{vmatrix}}{\begin{vmatrix} a_1 & b_1 \\ a_2 & b_2 \end{vmatrix}}$ and $y = \dfrac{\begin{vmatrix} a_1 & c_1 \\ a_2 & c_2 \end{vmatrix}}{\begin{vmatrix} a_1 & b_1 \\ a_2 & b_2 \end{vmatrix}}$	296
71	Three Equations	$x = \dfrac{\begin{vmatrix} k_1 & b_1 & c_1 \\ k_2 & b_2 & c_2 \\ k_3 & b_3 & c_3 \end{vmatrix}}{\Delta}$, $y = \dfrac{\begin{vmatrix} a_1 & k_1 & c_1 \\ a_2 & k_2 & c_2 \\ a_3 & k_3 & c_3 \end{vmatrix}}{\Delta}$, $z = \dfrac{\begin{vmatrix} a_1 & b_1 & k_1 \\ a_2 & b_2 & k_2 \\ a_3 & b_3 & k_3 \end{vmatrix}}{\Delta}$ Where $\Delta = \begin{vmatrix} a_1 & b_1 & c_1 \\ a_2 & b_2 & c_2 \\ a_3 & b_3 & c_3 \end{vmatrix} \neq 0$	303

	No.				Page
SYSTEMS OF LINEAR EQUATIONS (*Continued*)	72	Properties of Determinants	Zero Row or Column	If all elements in a row (or column) are zero, the value of the determinant is zero.	307
	73		Identical Rows or Columns	The value of a determinant is zero if two rows (or columns) are identical.	308
	74		Zeros below the Principal Diagonal	If all elements below the principal diagonal are zeros, then the value of the determinant is the product of the elements along the principal diagonal.	308
	75		Interchanging Rows with Columns	The value of a determinant is unchanged if we change the rows to columns and the columns to rows.	308
	76		Interchange of Rows or Columns	A determinant will change sign when we interchange two rows (or columns).	309
	77		Multiplying by a Constant	If each element in a row (or column) is multiplied by some constant, the value of the determinant is multiplied by that constant.	309
	78		Multiples of One Row or Column Added to Another	The value of a determinant is unchanged when the elements of a row (or column) are multiplied by some factor, and then added to the corresponding elements of another row or column.	310
MATRICES	79	Addition	Commutative Law	$$\mathbf{A} + \mathbf{B} = \mathbf{B} + \mathbf{A}$$	323
	80		Associative Law	$$\mathbf{A} + (\mathbf{B} + \mathbf{C}) = (\mathbf{A} + \mathbf{B}) + \mathbf{C}$$ $$= (\mathbf{A} + \mathbf{C}) + \mathbf{B}$$	323
	81		Addition and Subtraction	$$\begin{pmatrix} a & b \\ c & d \end{pmatrix} + \begin{pmatrix} w & x \\ y & z \end{pmatrix} = \begin{pmatrix} a+w & b+x \\ c+y & d+z \end{pmatrix}$$	
	82	Multiplication	Conformable Matrices	The product $\mathbf{AB}$ of two matrices $\mathbf{A}$ and $\mathbf{B}$ is defined only when the number of columns in $\mathbf{A}$ equals the number of rows in $\mathbf{B}$.	323
	83		Commutative Law	$$\mathbf{AB} \neq \mathbf{BA}$$ Matrix multiplication is *not* commutative.	328
	84		Associative Law	$$\mathbf{A(BC)} = \mathbf{(AB)C} = \mathbf{ABC}$$	324
	85		Distributive Law	$$\mathbf{A(B + C)} = \mathbf{AB} + \mathbf{AC}$$	324
	86		Dimensions of the Product	$$(m \times p)(p \times n) = (m \times n)$$	324
	87		Product of a Scalar and a Matrix	$$k\begin{pmatrix} a & b \\ c & d \end{pmatrix} = \begin{pmatrix} ka & kb \\ kc & kd \end{pmatrix}$$	323
	88		Scalar Product of a Row Vector and a Column Vector	$$(a \quad b \quad \cdots)\begin{pmatrix} x \\ y \\ \vdots \end{pmatrix} = (ax + by + \cdots)$$	325
	89		Tensor Product of a Column Vector and a Row Vector	$(2 \times 1)(1 \times 2) \quad (2 \times 2)$ $$\begin{pmatrix} a \\ b \end{pmatrix} (x \quad y) = \begin{pmatrix} ax & ay \\ bx & by \end{pmatrix}$$	
	90		Product of a Row Vector and a Matrix	$(1 \times 2) \quad (2 \times 3) \qquad\qquad (1 \times 3)$ $$(a \quad b)\begin{pmatrix} u & v & w \\ x & y & z \end{pmatrix} = (au + bx \quad av + by \quad aw + bz)$$	85

No.				Page
			MATRICES (Continued) — **Matrix Multiplication (Continued)**	
91		Product of a Matrix and a Column Vector	$(2 \times 2)\,(2 \times 1) \qquad (2 \times 1)$ $\begin{pmatrix} a & b \\ c & d \end{pmatrix} \begin{pmatrix} x \\ y \end{pmatrix} = \begin{pmatrix} ax + by \\ cx + dy \end{pmatrix}$	328
92		Product of Two Matrices	$(2 \times 3) \qquad (3 \times 2) \qquad\qquad (2 \times 2)$ $\begin{pmatrix} a & b & c \\ d & e & f \end{pmatrix} \begin{pmatrix} u & x \\ v & y \\ w & z \end{pmatrix} = \begin{pmatrix} au + bv + cw & ax + by + cz \\ du + ev + fw & dx + ey + fz \end{pmatrix}$	327
93		Product of a Matrix and Its Inverse	$\mathbf{AA^{-1} = A^{-1}A = I}$	335
94		Multiplying by the Unit Matrix	$\mathbf{AI = IA = A}$	339, 333
95	Solving Systems of Equations	Matrix Form for a System of Equations	$\mathbf{AX = B}$	339
96		Elementary Transformations of a Matrix	1. Interchange any rows. 2. Multiply a row by a nonzero constant. 3. Add a constant multiple of one row to another row.	335
97		Unit Matrix Method	When we transform the coefficient matrix **A** into the unit matrix **I**, the column vector **B** gets transformed into the solution set.	
98		Solving a Set of Equations Using the Inverse	$\mathbf{X = A^{-1}B}$	339
QUADRATICS				
99		General Form	$ax^2 + bx + c = 0$	370, 372
100		Quadratic Formula	$x = \dfrac{-b \pm \sqrt{b^2 - 4ac}}{2a}$	377
101		The Discriminant	If a, b, and c are real, and — if $b^2 - 4ac > 0$ the roots are real and unequal; if $b^2 - 4ac = 0$ the roots are real and equal; if $b^2 - 4ac < 0$ the roots are not real	379
102		Polynomial of Degree n	$a_0 x^n + a_1 x^{n-1} + \cdots + a_{n-1}x + a_n$	535
103		Factor Theorem	If a polynomial equation $f(x) = 0$ has a root r, then $(x - r)$ is a factor of the polynomial $f(x)$; conversely, if $(x - r)$ is a factor of a polynomial $f(x)$, then r is a root of $f(x) = 0$.	391

	No.				Page
INTERSECTING LINES	104		Opposite angles of two intersecting straight lines are equal.		153
	105		If two parallel straight lines are cut by a transversal, corresponding angles are equal and alternate interior angles are equal.		154
	106		If two lines are cut by a number of parallels, the corresponding segments are proportional.		154
QUADRILATERALS	107		Square	Area $= a^2$	163
	108		Rectangle	Area $= ab$	163
	109		Parallelogram: Diagonals bisect each other	Area $= bh$	163
	110		Rhombus: Diagonals intersect at right angles	Area $= ah$	163
	111		Trapezoid	Area $= \dfrac{(a+b)h}{2}$	163
POLYGON	112	n sides	Sum of Angles $= (n-2)\,180°$		156
CIRCLES	113		Circumference $= 2\pi r = \pi d$		164
	114		Area $= \pi r^2 = \dfrac{\pi d^2}{4}$		39,164
	115		Central angle θ (radians) $= \dfrac{s}{r}$		437
	116		Area of sector $= \dfrac{rs}{2} = \dfrac{r^2\theta}{2}$		434
	117		1 revolution $= 2\pi$ radians $= 360°$ $1° = 60$ minute 1 minute $= 60$ seconds		431,174
	118		Any angle inscribed in a semicircle is a right angle.		166
	119		Tangents to a Circle	Tangent AP is perpendicular to radius OA.	165
	120			Tangent $AP =$ tangent BP OP bisects angle APB	166
	121		Intersecting Chords	$ab = cd$	165

	No.				Page
SOLIDS	122		Cube	Volume $= a^3$	168
	123			Surface area $= 6a^2$	168
	124		Rectangular Parallelepiped	Volume $= lwh$	168
	125			Surface area $= 2(lw + hw + lh)$	168
	126		Any Cylinder or Prism	Volume $=$ (area of base)(altitude)	168
	127		Right Cylinder or Prism	Lateral area (not incl. bases) $=$ (perimeter of base)(altitude)	168
	128		Sphere	Volume $= \frac{4}{3}\pi r^3$	168
	129			Surface area $= 4\pi r^2$	168
	130		Any Cone or Pyramid	Volume $= \frac{1}{3}$ (area of base)(altitude)	168
	131		Right Circular Cone or Regular Pyramid	Lateral area $= \frac{1}{2}$ (perimeter of base) $\times$ (slant height)	168
	132		Any Cone or Pyramid	Volume $= \frac{h}{3}(A_1 + A_2 + \sqrt{A_1 A_2})$	168
	133		Right Circular Cone or Regular Pyramid	Lateral area $= \frac{s}{2}$ (sum of base perimeters) $= \frac{s}{2}(P_1 + P_2)$	168
SIMILAR FIGURES	134		Corresponding dimensions of plane or solid similar figures are in proportion.		539
	135		Areas of similar plane or solid figures are proportional to the squares of any two corresponding dimensions.		540
	136		Volumes of similar solid figures are proportional to the cubes of any two corresponding dimensions.		540

	No.			Page
ANY TRIANGLE	137	Areas	Area $= \frac{1}{2}bh$	157
	138		Hero's Formula: Area $= \sqrt{s(s-a)(s-b)(s-c)}$ where $s = \frac{1}{2}(a+b+c)$	157
	139		Sum of the Angles	$A + B + C = 180°$	158,183
	140		Law of Sines	$\dfrac{a}{\sin A} = \dfrac{b}{\sin B} = \dfrac{c}{\sin C}$	411
	141		Law of Cosines	$\begin{aligned} a^2 &= b^2 + c^2 - 2bc\cos A \\ b^2 &= a^2 + c^2 - 2ac\cos B \\ c^2 &= a^2 + b^2 - 2ab\cos C \end{aligned}$	416
	142		Exterior Angle	$\theta = A + B$	158
SIMILAR TRIANGLES	143	If two angles of a triangle equal two angles of another triangle, the triangles are similar.		158
	144	Corresponding sides of similar triangles are in proportion.		158
RIGHT TRIANGLES	145		Pythagorean Theorem	$a^2 + b^2 = c^2$	159,183
	146	Trigonometric Ratios	Sine	$\sin\theta = \dfrac{y}{r} = \dfrac{\text{opposite side}}{\text{hypotenuse}}$	177,183
	147		Cosine	$\cos\theta = \dfrac{x}{r} = \dfrac{\text{adjacent side}}{\text{hypotenuse}}$	177,183
	148		Tangent	$\tan\theta = \dfrac{y}{x} = \dfrac{\text{opposite side}}{\text{adjacent side}}$	177,183
	149		Cotangent	$\cot\theta = \dfrac{x}{y} = \dfrac{\text{adjacent side}}{\text{opposite side}}$	177
	150		Secant	$\sec\theta = \dfrac{r}{x} = \dfrac{\text{hypotenuse}}{\text{adjacent side}}$	177
	151		Cosecant	$\csc\theta = \dfrac{r}{y} = \dfrac{\text{hypotenuse}}{\text{opposite side}}$	177
	152	Reciprocal Relations	(a) $\sin\theta = \dfrac{1}{\csc\theta}$ (b) $\cos\theta = \dfrac{1}{\sec\theta}$ (c) $\tan\theta = \dfrac{1}{\cot\theta}$	494,178
	153		In a right triangle, the altitude to the hypotenuse forms two right triangles which are similar to each other and to the original triangle.	
	154	A and B are Complementary Angles	Cofunctions	(a) $\sin A = \cos B$ (d) $\cot A = \tan B$ (b) $\cos A = \sin B$ (e) $\sec A = \csc B$ (c) $\tan A = \cot B$ (f) $\csc A = \sec B$	186
CONGRUENT TRIANGLES	155		Two angles and a side of one are equal to two angles and the corresponding side of the other (ASA), (AAS).	158
	156	Two Triangles Are Congruent If	Two sides and the included angle of one are equal, respectively to two sides and the included angle of the other (SAS).	158
	157		Three sides of one are equal to the three sides of the other (SSS).	158
COORDINATE SYSTEMS	158		Rectangular	$x = r\cos\theta$	484
	159			$y = r\sin\theta$	484
	160		Polar	$r = \sqrt{x^2 + y^2}$	484
	161			$\theta = \arctan\dfrac{y}{x}$	484

No.   Page

	No.			Page
TRIGONOMETRIC IDENTITIES	162	Quotient Relations	$$\tan \theta = \frac{\sin \theta}{\cos \theta}$$	494
	163		$$\cot \theta = \frac{\cos \theta}{\sin \theta}$$	494
	164	Pythagorean Relations	$\sin^2 \theta + \cos^2 \theta = 1$	495
	165		$1 + \tan^2 \theta = \sec^2 \theta$	495
	166		$1 + \cot^2 \theta = \csc^2 \theta$	495
	167	Sum or Difference of Two Angles	$\sin (\alpha \pm \beta) = \sin \alpha \cos \beta \pm \cos \alpha \sin \beta$	503
	168		$\cos (\alpha \pm \beta) = \cos \alpha \cos \beta \mp \sin \alpha \sin \beta$	503
	169		$$\tan (\alpha \pm \beta) = \frac{\tan \alpha \pm \tan \beta}{1 \mp \tan \alpha \tan \beta}$$	504
	170	Double-Angle Relations	$\sin 2\alpha = 2 \sin \alpha \cos \alpha$	507
	171		(a) $\cos 2\alpha = \cos^2 \alpha - \sin^2 \alpha$ (b) $\cos 2\alpha = 1 - 2 \sin^2 \alpha$ (c) $\cos 2\alpha = 2 \cos^2 \alpha - 1$	508
	172		$$\tan 2\alpha = \frac{2 \tan \alpha}{1 - \tan^2 \alpha}$$	509
	173	Half-Angle Relations	$$\sin \frac{\alpha}{2} = \pm \sqrt{\frac{1 - \cos \alpha}{2}}$$	510
	174		$$\cos \frac{\alpha}{2} = \mp \sqrt{\frac{1 + \cos \alpha}{2}}$$	512
	175		(a) $\tan \dfrac{\alpha}{2} = \dfrac{1 - \cos \alpha}{\sin \alpha}$ (b) $\tan \dfrac{\alpha}{2} = \dfrac{\sin \alpha}{1 + \cos \alpha}$ (c) $\tan \dfrac{\alpha}{2} = \pm \sqrt{\dfrac{1 - \cos \alpha}{1 + \cos \alpha}}$	512, 513
	176	Sum or Difference of Two Functions	$\sin \alpha + \sin \beta = 2 \sin \frac{1}{2} (\alpha + \beta) \cos \frac{1}{2} (\alpha - \beta)$	
	177		$\sin \alpha - \sin \beta = 2 \cos \frac{1}{2} (\alpha + \beta) \sin \frac{1}{2} (\alpha - \beta)$	
	178		$\cos \alpha + \cos \beta = 2 \cos \frac{1}{2} (\alpha + \beta) \cos \frac{1}{2} (\alpha - \beta)$	
	179		$\cos \alpha - \cos \beta = 2 \sin \frac{1}{2} (\alpha + \beta) \sin \frac{1}{2} (\alpha - \beta)$	
	180	Product of Two Functions	$\sin \alpha \sin \beta = \frac{1}{2} \cos(\alpha - \beta) - \frac{1}{2} \cos(\alpha + \beta)$	
	181		$\cos \alpha \cos \beta = \frac{1}{2} \cos(\alpha - \beta) + \frac{1}{2} \cos(\alpha + \beta)$	
	182		$\sin \alpha \cos \beta = \frac{1}{2} \sin(\alpha + \beta) + \frac{1}{2}\sin(\alpha - \beta)$	
	183	Inverse Trigonometric Functions	$$\theta = \text{arcsine } C = \arctan \frac{C}{\sqrt{1 - C^2}}$$	
	184		$$\theta = \arccos D = \arctan \frac{\sqrt{1 - D^2}}{D}$$	

	No.				Page
LOGARITHMS	**185**	Definition of e		$\displaystyle\lim_{k \to \infty} \left(1 + \frac{1}{k}\right)^k = e$	562
	186	Exponential to Logarithmic Form		If $b^x = y$ then $x = \log_b y$ $(y > 0,\ b > 0,\ b \neq 1)$	571
	187	Laws of Logarithms	Products	$\log_b MN = \log_b M + \log_b N$	576
	188		Quotients	$\log_b \dfrac{M}{N} = \log_b M - \log_b N$	576
	189		Powers	$\log_b M^p = p \log_b M$	577
	190		Roots	$\log_b \sqrt[q]{M} = \dfrac{1}{q} \log_b M$	579
	191		Log of 1	$\log_b 1 = 0$	580
	192		Log of the Base	$\log_b b = 1$	580
	193	Log of the Base Raised to a Power		$\log_b b^n = n$	580
	194	Base Raised to a Logarithm of the Same Base		$b^{\log_b x} = x$	581
	195	Change of Base		$\log N = \dfrac{\ln N}{\ln 10} \cong \dfrac{\ln N}{2.3026}$	581

No.   Page

SOME USEFUL FUNCTIONS	**196**	Power Function		$y = ax^n$	535,596
	197	Exponential Function		$y = a(b)^{nx}$	598
	198	Series Approximation	$b^x = 1 + x \ln b + \dfrac{(x \ln b)^2}{2!} + \dfrac{(x \ln b)^3}{3!} + \cdots \quad (b > 0)$		575
	199	Exponential Growth		$y = ae^{nt}$	562,586, 740
	200		Doubling Time or Half-Life	$t = \dfrac{\ln 2}{n}$	586
	201	Exponential Decay		$y = ae^{-nt}$	563,586, 740
	202	Exponential Growth to an Upper Limit		$y = a(1 - e^{-nt})$	563
	203	Time Constant	$T = \dfrac{1}{\text{growth rate } n}$		564
	204	Recursion Relation for Exponential Growth	$y_t = By_{t-1}$		566
	205	Nonlinear Growth Equation	$y_t = By_{t-1}(1 - y_{t-1})$		566
	206	Series Approximations	$e = 2 + \dfrac{1}{2!} + \dfrac{1}{3!} + \dfrac{1}{4!} + \cdots$		569
	207		$e^x = 1 + x + \dfrac{x^2}{2!} + \dfrac{x^3}{3!} + \dfrac{x^4}{4!} + \cdots$		570

No.   Page

208	Logarithmic Function		$y = \log_b x$ $(x > 0,\ b > 0,\ b \neq 1)$	572
209	Series Approximation		$\ln x = 2a + \dfrac{2a^3}{3} + \dfrac{2a^5}{5} + \dfrac{2a^7}{7} + \cdots$ where $a = \dfrac{x-1}{x+1}$	575
210	Sine Wave of Amplitude a		$y = a \sin(bx + c)$	456
211			$\text{Period} = \dfrac{360}{b} \text{ deg/cycle} = \dfrac{2\pi}{b} \text{ rad/cycle}$	454
212			$\text{Frequency} = \dfrac{b}{360} \text{ cycle/deg} = \dfrac{b}{2\pi} \text{ cycle/rad}$	454
213			$\text{Phase shift} = -\dfrac{c}{b}$	457
214	Addition of a Sine Wave and a Cosine Wave		$A \sin \omega t + B \cos \omega t = R \sin(\omega t + \phi)$ where $R = \sqrt{A^2 + B^2} \qquad \text{and} \qquad \phi = \arctan \dfrac{B}{A}$	476,506
215	Series Approximation		$\sin x = x - \dfrac{x^3}{3!} + \dfrac{x^5}{5!} - \dfrac{x^7}{7!} + \cdots$	
216			$\cos x = 1 - \dfrac{x^2}{2!} + \dfrac{x^4}{4!} - \dfrac{x^6}{6!} + \cdots$	

SOME USEFUL FUNCTIONS *(Continued)*

	No.					Page
COMPLEX NUMBERS	217		Powers of j		$j = \sqrt{-1}, \quad j^2 = -1, \quad j^3 = -j, \quad j^4 = 1, \quad j^5 = j,$ etc.	607
	218		Rectangular Form		$a + jb$	606
	219			Sums	$(a + jb) + (c + jd) = (a + c) + j(b + d)$	606
	220			Differences	$(a + jb) - (c + jd) = (a - c) + j(b - d)$	606
	221			Products	$(a + jb)(c + jd) = (ac - bd) + j(ad + bc)$	607
	222			Quotients	$\dfrac{a + jb}{c + jd} = \dfrac{ac + bd}{c^2 + d^2} + j\,\dfrac{bc - ad}{c^2 + d^2}$	609
	223		Trigonometric Form		$a + jb = r(\cos\theta + j\sin\theta)$	613
	224			where	$a = r\cos\theta$	613
	225				$b = r\sin\theta$	613
	226				$r = \sqrt{a^2 + b^2}$	613
	227				$\theta = \arctan\dfrac{b}{a}$	613
	228		Polar Form		$r \,\underline{|\theta} = a + jb$	612
	229			Products	$r \,\underline{|\theta} \cdot r' \,\underline{|\theta'} = rr' \,\underline{|\theta + \theta'}$	614
	230			Quotients	$r \,\underline{|\theta} \div r' \,\underline{|\theta'} = \dfrac{r}{r'} \,\underline{|\theta - \theta'}$	615
	231			Roots and Powers	DeMoivre's Theorem: $(r \,\underline{|\theta})^n = r^n \,\underline{|n\theta}$	615
	232		Exponential Form	Euler's Formula	$re^{j\theta} = r(\cos\theta + j\sin\theta)$	619
	233			Products	$r_1 e^{j\theta_1} \cdot r_2 e^{j\theta_2} = r_1 r_2 e^{j(\theta_1 + \theta_2)}$	620
	234			Quotients	$\dfrac{r_1 e^{j\theta_1}}{r_2 e^{j\theta_2}} = \dfrac{r_1}{r_2}\, e^{j(\theta_1 - \theta_2)}$	620
	235			Powers and Roots	$(re^{j\theta})^n = r^n e^{jn\theta}$	620

No.				Page
PROGRESSIONS				
236	**Arithmetic Progression** Common Difference $= d$	Recursion Formula	$a_n = a_{n-1} + d$	731
237		General Term	$a_n = a + (n-1)d$	732
238		Sum of n Terms	$s_n = \dfrac{n(a + a_n)}{2}$	732
239			$s_n = \dfrac{n}{2}[2a + (n-1)d]$	733
240	**Geometric Progression** Common Ratio $= r$	Recursion Formula	$a_n = ra_{n-1}$	737
241		General Term	$a_n = ar^{n-1}$	737
242		Sum of n Terms	$s_n = \dfrac{a(1 - r^n)}{1 - r}$	738
243			$s_n = \dfrac{a - ra_n}{1 - r}$	738
244		Sum to Infinity	$S = \dfrac{a}{1 - r} \quad$ where $\quad |r| < 1$	744
BINOMIAL THEOREM				
245		Binomial Expansion	$(a+b)^n = a^n + na^{n-1}b + \dfrac{n(n-1)}{2!}a^{n-2}b^2 + \dfrac{n(n-1)(n-2)}{3!}a^{n-3}b^3 + \cdots + b^n$	747
246		General Term	$r\text{th term} = \dfrac{n!}{(r-1)!\,(n-r+1)!}a^{n-r+1}b^{r-1}$	749
247	Binomial Series		$(a+b)^n = a^n + na^{n-1}b + \dfrac{n(n-1)}{2!}a^{n-2}b^2 + \dfrac{n(n-1)(n-2)}{3!}a^{n-3}b^3 + \cdots$ where $|a| > |b|$	752
248			$(1+x)^n = 1 + nx + \dfrac{n(n-1)}{2!}x^2 + \dfrac{n(n-1)(n-2)}{3!}x^3 + \cdots$ where $|x| < 1$	752
STATISTICS AND PROBABILITY				
249	**Measures of Central Tendency**	Arithmetic Mean	$\bar{x} = \dfrac{\sum x}{n}$	767
250		Median	The median of a set of numbers arranged in order of magnitude is the middle value, or the mean of the two middle values.	769
251		Mode	The mode of a set of numbers is the measurement(s) that occurs most often in the set.	769
252	**Measures of Dispersion**	Range	The range of a set of numbers is the difference between the largest and the smallest number in the set.	770
253		Population Variance σ^2	$\sigma^2 = \dfrac{\sum(x - \bar{x})^2}{n}$	772
254		Sample Variance s^2	$s^2 = \dfrac{\sum(x - \bar{x})^2}{n-1}$	772
255		Standard Deviation s	The standard deviation of a set of numbers is the positive square root of the variance.	772
256	**Probability**	Of a Single Event	$P(A) = \dfrac{\text{number of ways in which } A \text{ can happen}}{\text{number of equally likely ways}}$	777
257		Of Two Events Both Occurring	$P(A, B) = P(A)P(B)$	778
258		Of Several Events All Occurring	$P(A, B, C, \ldots) = P(A)P(B)P(C)\cdots$	778
259		Of Either of Two Events Occurring	$P(A + B) = P(A) + P(B) - P(A, B)$	778
260		Of Two Mutually Exclusive Events Occurring	$P(A + B) = P(A) + P(B)$	779
261	Gaussian Distribution		$y = \dfrac{1}{\sigma\sqrt{2\pi}}e^{-(x-\mu)^2/2\sigma^2}$ where $\mu = $ population mean and $\sigma = $ population standard deviation	787

No.							Page

STATISTICS AND PROBABILITY (Continued)

No.				Page
262	Binomial Probability Formula		$P(x) = \dfrac{n!}{(n-x)!\,x!}\,p^x q^{n-x}$	782
263	Standard Error	Of the Mean	$SE_{\bar{x}} = \dfrac{\sigma}{\sqrt{n}} = \dfrac{s}{\sqrt{n}}$	795
264		Of the Standard Deviation	$SE_s = \dfrac{\sigma}{\sqrt{2n}}$	796
265		Of a Proportion	$SE_p = \sqrt{\dfrac{p(1-p)}{n}}$	796
266	Correlation Coefficient		$r = \dfrac{n\,\Sigma xy - \Sigma x\,\Sigma y}{\sqrt{n\,\Sigma x^2 - (\Sigma x)^2}\,\sqrt{n\,\Sigma y^2 - (\Sigma y)^2}}$	807
267	Least Squares Line		Slope $m = \dfrac{n\Sigma xy - \Sigma x\,\Sigma y}{n\,\Sigma x^2 - (\Sigma x)^2}$	809
			y intercept $b = \dfrac{\Sigma x^2\,\Sigma y - \Sigma x\,\Sigma xy}{n\,\Sigma x^2 - (\Sigma x)^2}$	809

BOOLEAN ALGEBRA AND SETS

Boolean Operators

No.		Truth Table	Venn Diagram	Switch Diagram	Logic Gate
268	AND	$A\ B\ A\cdot B$ $0\ 0\ \ 0$ $0\ 1\ \ 0$ $1\ 0\ \ 0$ $1\ 1\ \ 1$	Intersection $A \cap B$ $= \{x : x \in A, x \in B\}$	In Series	AB AND gate
269	OR	$A\ B\ A+B$ $0\ 0\ \ 0$ $0\ 1\ \ 1$ $1\ 0\ \ 1$ $1\ 1\ \ 1$	Union $A \cup B$ $= \{x : x \in A, \text{ or } x \in B\}$	In Parallel	$A+B$ OR Gate
270	NOT	$A\ \ \bar{A}$ $0\ \ 1$ $1\ \ 0$	Complement A^c $= \{x : x \in U, x \in A\}$		$\bar{A}$ Inverter
271	Exclusive OR	$A\ B\ A+B$ $0\ 0\ \ 0$ $0\ 1\ \ 1$ $1\ 0\ \ 1$ $1\ 1\ \ 0$	$A \oplus B = \{x : x \in A$ or $x \in B,$ $x \in A \cap B\}$		$A \oplus B$ Exclusive OR Gate

No. Page

			AND	OR	
	272	Commutative Laws	(a) $AB \equiv BA$	(b) $A + B \equiv B + A$	
	273	Boundedness Laws	(a) $A \cdot 0 \equiv 0$	(b) $A + 1 \equiv 1$	
	274	Identity Laws	(a) $A \cdot 1 \equiv A$	(b) $A + 0 \equiv A$	
	275	Idempotent Laws	(a) $AA \equiv A$	(b) $A + A \equiv A$	
	276	Complement Laws	(a) $A \cdot \overline{A} \equiv 0$	(b) $A + \overline{A} \equiv 1$	
	277	Associative Laws	(a) $A(BC) \equiv (AB)C$	(b) $A + (B + C) \equiv (A + B) + C$	
	278	Distributive Laws	(a) $A(B + C) \equiv AB + AC$	(b) $A + BC \equiv (A + B)(A + C)$	
	279		(a) $A(\overline{A} + B) \equiv AB$	(b) $A + \overline{A} \cdot B \equiv A + B$	
	280	Absorption Laws	(a) $A(A + B) \equiv A$	(b) $A + (AB) \equiv A$	
	281	DeMorgan's Laws	(a) $\overline{A \cdot B} \equiv \overline{A} + \overline{B}$	(b) $\overline{A + B} \equiv \overline{A} \cdot \overline{B}$	
	282	Involution Law	$\overline{\overline{A}} \equiv A$		

BOOLEAN ALGEBRA AND SETS (*Continued*)

Laws of Boolean Algebra

No.					Page
283			Distance Formula	$d = \sqrt{(\Delta x)^2 + (\Delta y)^2} = \sqrt{(x_2 - x_1)^2 + (y_2 - y_1)^2}$	636
284			Slope m	$m = \dfrac{\text{rise}}{\text{run}} = \dfrac{\Delta y}{\Delta x} = \dfrac{y_2 - y_1}{x_2 - x_1}$	636,138
285				$m = \tan(\text{angle of inclination}) = \tan\theta$ $0 \le \theta < 180°$	637
286			General Form	$Ax + By + C = 0$	641
287			Parallel to x axis	$y = b$	643
288			Parallel to y axis	$x = a$	643
289		Equation of Straight Line	Slope-Intercept Form	$y = mx + b$	641,139
290			Two-Point Form	$\dfrac{y - y_1}{x - x_1} = \dfrac{y_2 - y_1}{x_2 - x_1}$	642
291			Point-Slope Form	$m = \dfrac{y - y_1}{x - x_1}$	641
292			Intercept Form	$\dfrac{x}{a} + \dfrac{y}{b} = 1$	
293			Polar Form	$r\cos(\theta - \beta) = p$	
294		If L_1 and L_2 are parallel, then		$m_1 = m_2$	638
295		If L_1 and L_2 are perpendicular, then		$m_1 = -\dfrac{1}{m_2}$	638
296		Angle of Intersection		$\tan\phi = \dfrac{m_2 - m_1}{1 + m_1 m_2}$	639
297		General Second-Degree Equation		$Ax^2 + Bxy + Cy^2 + Dx + Ey + F = 0$	650
298	Any Conic	Translation of Axes		To translate or shift the axes of a curve to the left by a distance h and downward by a distance k, replace x by $(x - h)$ and y by $(y - k)$ in the equation of the curve.	650
299		Eccentricity		$e = \dfrac{\cos\beta}{\cos\alpha}$	
300				$e = 0$ for the Circle $0 < e < 1$ for the Ellipse $e = 1$ for the Parabola $e > 1$ for the Hyperbola	
301		Definition of a Conic		$PF = e \cdot PD$	
302		Polar Equation for the Conics		$r = \dfrac{ke}{1 - e\cos\theta}$	

THE STRAIGHT LINE

CONIC SECTIONS

No.					Page
303			The set of points in a plane equidistant from a fixed point		648
304	Circle of Radius r		Standard Form	$x^2 + y^2 = r^2$	649
305				$(x - h)^2 + (y - k)^2 = r^2$	649
306			General Form	$x^2 + y^2 + Dx + Ey + F = 0$	650
307			The set of points in a plane such that the distance PF from each point P to a fixed point F (the focus) is equal to the distance PD to a fixed line (the directrix) $PF = PD$		655
308	Parabola		Standard Form	$y^2 = 4px$	656
309				$x^2 = 4py$	657
310				$(y - k)^2 = 4p(x - h)$	658
311				$(x - h)^2 = 4p(y - k)$	658
312			General Form	$Cy^2 + Dx + Ey + F = 0$ or $Ax^2 + Dx + Ey + F = 0$	659
313		Focal Width or length of latus rectum		$L = |4p|$	658
314		Area		$Area = \dfrac{2}{3}ab$	

CONIC SECTIONS (Continued)

No.				Page
315			The set of points in a plane such that the sum of the distances PF and PF' from each point P to two fixed points F & F' (the foci) is constant, and equal to the length $2a$ of the major axis $$PF + PF' = 2a$$	664
316			$$\frac{x^2}{a^2} + \frac{y^2}{b^2} = 1$$ $a > b$	667
317			$$\frac{y^2}{a^2} + \frac{x^2}{b^2} = 1$$ $a > b$	668
318			$$\frac{(x-h)^2}{a^2} + \frac{(y-k)^2}{b^2} = 1$$ $a > b$	669
319			$$\frac{(y-k)^2}{a^2} + \frac{(x-h)^2}{b^2} = 1$$ $a > b$	669
320		General Form	$$Ax^2 + Cy^2 + Dx + Ey + F = 0$$ $A \neq C$, but have same signs	670
321		Distance from Center to Focus	$$c = \sqrt{a^2 - b^2}$$	
322		Focal Width or length of latus rectum	$$L = \frac{2b^2}{a}$$	666,672
323			Eccentricity $e = \dfrac{a}{d} = \dfrac{c}{a}$	
324			Area $= \pi ab$	

CONIC SECTIONS (Continued)

Ellipse

Standard Form

No.					Page
325			The set of points in a plane such that the difference of the distances PF and PF' from each point P to two fixed points F & F' (the foci) is constant, and equal to the distance $2a$ between the vertices $$PF' - PF = 2a$$		674
326				$$\frac{x^2}{a^2} - \frac{y^2}{b^2} = 1$$	676
327				$$\frac{y^2}{a^2} - \frac{x^2}{b^2} = 1$$	676
328				$$\frac{(x-h)^2}{a^2} - \frac{(y-k)^2}{b^2} = 1$$	680
329				$$\frac{(y-k)^2}{a^2} - \frac{(x-h)^2}{b^2} = 1$$	680
330			General Form	$Ax^2 + Cy^2 + Dx + Ey + F = 0$ $A \neq C$, and have opposite signs	681
331			Distance from Center to Focus	$c = \sqrt{a^2 + b^2}$	677
332			Slope of Asymptotes	Transverse Axis Horizontal — Slope $= \pm \dfrac{b}{a}$	677
333				Transverse Axis Vertical — Slope $= \pm \dfrac{a}{b}$	677
334			Length of Latus Rectum	$L = \dfrac{2b^2}{a}$	678
335				Axes Rotated 45° $xy = k$	682

CONIC SECTIONS (Continued)

Hyperbola

Standard Form

	No.				Page
	336		Limit Notation	$\lim\limits_{x \to a} f(x) = L$	743
	337		Increments	$\Delta x = x_2 - x_1, \quad \Delta y = y_2 - y_1$	635
	338		Definition of the Derivative	$\dfrac{dy}{dx} = \lim\limits_{\Delta x \to 0} \dfrac{\Delta y}{\Delta x} = \lim\limits_{\Delta x \to 0} \dfrac{f(x + \Delta x) - f(x)}{\Delta x}$ $= \lim\limits_{\Delta x \to 0} \dfrac{(y + \Delta y) - y}{\Delta x}$	
	339		The Chain Rule	$\dfrac{dy}{dx} = \dfrac{dy}{du} \cdot \dfrac{du}{dx}$	
DIFFERENTIAL CALCULUS	**340**	Rules for Derivatives	Of a Constant	$\dfrac{d(c)}{dx} = 0$	
	341		Of a Power Function	$\dfrac{d}{dx} x^n = nx^{n-1}$	
	342		Of a Constant Times a Function	$\dfrac{d(cu)}{dx} = c\dfrac{du}{dx}$	
	343		Of a Constant Times a Power of x	$\dfrac{d}{dx} cx^n = cnx^{n-1}$	
	344		Of a Sum	$\dfrac{d}{dx}(u + v + w) = \dfrac{du}{dx} + \dfrac{dv}{dx} + \dfrac{dw}{dx}$	
	345		Of a Function Raised to a Power	$\dfrac{d(cu^n)}{dx} = cnu^{n-1}\dfrac{du}{dx}$	
	346		Of a Product	$\dfrac{d(uv)}{dx} = u\dfrac{dv}{dx} + v\dfrac{du}{dx}$	
	347		Of a Product of Three Factors	$\dfrac{d(uvw)}{dx} = uv\dfrac{dw}{dx} + uw\dfrac{dv}{dx} + vw\dfrac{du}{dx}$	
	348		Of a Product of n Factors	The derivative is an expression of n terms, each term being the product of $n - 1$ of the factors and the derivative of the other factor.	
	349		Of a Quotient	$\dfrac{d}{dx}\left(\dfrac{u}{v}\right) = \dfrac{v\dfrac{du}{dx} - u\dfrac{dv}{dx}}{v^2}$	

No.				Page

DIFFERENTIAL CALCULUS *(Continued)*

Rules for Derivatives *(cont.)*

No.	Category	Formula
350	Of the Trigonometric Functions	$\dfrac{d(\sin u)}{dx} = \cos u\,\dfrac{du}{dx}$
351		$\dfrac{d(\cos u)}{dx} = -\sin u\,\dfrac{du}{dx}$
352		$\dfrac{d(\tan u)}{dx} = \sec^2 u\,\dfrac{du}{dx}$
353		$\dfrac{d(\cot u)}{dx} = -\csc^2 u\,\dfrac{du}{dx}$
354		$\dfrac{d(\sec u)}{dx} = \sec u \tan u\,\dfrac{du}{dx}$
355		$\dfrac{d(\csc u)}{dx} = -\csc u \cot u\,\dfrac{du}{dx}$
356	Of the Inverse Trigonometric Functions	$\dfrac{d(\operatorname{Sin}^{-1} u)}{dx} = \dfrac{1}{\sqrt{1-u^2}}\dfrac{du}{dx}\qquad -1 < u < 1$
357		$\dfrac{d(\operatorname{Cos}^{-1} u)}{dx} = \dfrac{-1}{\sqrt{1-u^2}}\dfrac{du}{dx}\qquad -1 < u < 1$
358		$\dfrac{d(\operatorname{Tan}^{-1} u)}{dx} = \dfrac{1}{1+u^2}\dfrac{du}{dx}$
359		$\dfrac{d(\operatorname{Cot}^{-1} u)}{dx} = \dfrac{-1}{1+u^2}\dfrac{du}{dx}$
360		$\dfrac{d(\operatorname{Sec}^{-1} u)}{dx} = \dfrac{1}{u\sqrt{u^2-1}}\dfrac{du}{dx}\qquad |u| > 1$
361		$\dfrac{d(\operatorname{Csc}^{-1} u)}{dx} = \dfrac{-1}{u\sqrt{u^2-1}}\dfrac{du}{dx}\qquad |u| > 1$
362	Of Logarithmic and Exponential Functions	(a) $\dfrac{d}{dx}(\log_b u) = \dfrac{1}{u}\log_b e\,\dfrac{du}{dx}$ (b) $\dfrac{d}{dx}(\log_b u) = \dfrac{1}{u \ln b}\dfrac{du}{dx}$
363		$\dfrac{d}{dx}(\ln u) = \dfrac{1}{u}\dfrac{du}{dx}$
364		$\dfrac{d}{dx}b^u = b^u\,\dfrac{du}{dx}\ln b$
365		$\dfrac{d}{dx}e^u = e^u\,\dfrac{du}{dx}$

Graphical Applications

No.	Category	Description
366	Maximum and Minimum Points	To find maximum and minimum points (and other stationary points) set the first derivative equal to zero and solve for x.
367	First-Derivative Test	The first derivative is negative to the left of, and positive to the right of, a minimum point. The reverse is true for a maximum point.
368	Second-Derivative Test	If the first derivative at some point is zero, then, if the second derivative is 1. Positive, the point is a minimum. 2. Negative, the point is a maximum. 3. Zero, the test fails.
369	Ordinate Test	Find y a small distance to either side of the point to be tested. If y is greater there, we have a minimum; if less, we have a maximum.
370	Inflection Points	To find points of inflection, set the second derivative to zero and solve for x. Test by seeing if the second derivative changes sign a small distance to either side of the point.
371	Newton's Method	$x_{n+1} = x_n - \dfrac{f(x_n)}{f'(x_n)}$

Differentials

No.	Category	Formula
372	Differential of y	$dy = f'(x)\,dx$
373	Approximations Using Differentials	$\Delta y \cong \dfrac{dy}{dx}\Delta x$

DIFFERENTIAL CALCULUS (Continued)

No.				Formula	Page
374			The Indefinite Integral	$$\int F'(x)\,dx = F(x) + C$$	
375				$$\int f(x)\,dx = F(x) + C \qquad \text{where} \qquad f(x) = F'(x)$$	
376	The Definite Integral		The Fundamental Theorem	$$\int_a^b f(x)\,dx = F(b) - F(a)$$	
377			Properties of the Definite Integral	$$\int_a^b c\,f(x)\,dx = c\int_a^b f(x)\,dx$$	
378				$$\int_a^b [f(x) + g(x)]\,dx = \int_a^b f(x)\,dx + \int_a^b g(x)\,dx$$	
379				$$\int_a^b f(x)\,dx = -\int_b^a f(x)\,dx$$	
380				$$\int_a^b f(x)\,dx = \int_a^c f(x)\,dx + \int_c^b f(x)\,dx$$	
381	Approximate Area under a Curve		Midpoint Method	$$A \cong \sum_{i=1}^n f(x_i^*)\,\Delta x$$ where $f(x_i^*)$ is the height of the ith panel at its midpoint	
382			Average Ordinate Method	$$A \cong y_{\text{avg}}\,(b - a)$$	
383			Trapezoid Rule	With unequal spacing $$A \cong \tfrac{1}{2}[(x_1 - x_0)(y_1 + y_0) + (x_2 - x_1)(y_2 - y_1) + \cdots + (x_n - x_{n-1})(y_n + y_{n-1})]$$	
384				With equal spacing, $h = x_1 - x_0$: $$A \cong h[\tfrac{1}{2}(y_0 + y_n) + y_1 + y_2 + \cdots + y_{n-1}]$$	
385			Prismoidal Formula	$$A = \frac{h}{3}\,(y_0 + 4y_1 + y_2)$$	
386			Simpson's Rule	$$A \cong \frac{h}{3}\,(y_0 + 4y_1 + 2y_2 + 4y_3 + \cdots + 4y_{n-1} + y_n)$$	
387	Exact Area under a Curve		Defined by Riemann Sums	$$A = \lim_{\Delta x \to 0} \sum_{i=1}^n f(x_i^*)\,\Delta x = \int_a^b f(x)\,dx$$	
388			By Integration	$$A = \int_a^b f(x)\,dx = F(b) - F(a)$$	
389			Areas between Two Curves	$$A = \int_a^b [f(x) - g(x)]\,dx$$	

No.						Page
			Disk Method		Volume $= dV = \pi r^2\, dh$	
390	APPLICATIONS OF THE DEFINITE INTEGRAL	Volumes of Solids of Revolution				
391			Disk Method		$V = \pi \displaystyle\int_a^b r^2\, dh$	
392			Ring Method		$dV = \pi(r_o^2 - r_i^2)\, dh$	
393			Ring Method		$V = \pi \displaystyle\int_a^b (r_o^2 - r_i^2)\, dh$	
394			Shell Method		$dV = 2\pi rh\, dr$	
395			Shell Method		$V = 2\pi \displaystyle\int_a^b rh\, dr$	

No.				Page
396	Length of Arc		$s = \int_a^b \sqrt{1 + \left(\dfrac{dy}{dx}\right)^2}\, dx$	
397			$s = \int_c^d \sqrt{1 + \left(\dfrac{dx}{dy}\right)^2}\, dy$	
398	Surface Area		About x Axis: $S = 2\pi \int_a^b y \sqrt{1 + \left(\dfrac{dy}{dx}\right)^2}\, dx$	
399			About y Axis: $S = 2\pi \int_a^b x \sqrt{1 + \left(\dfrac{dy}{dx}\right)^2}\, dx$	
400	Centroids	Of Plane Area: 	$\bar{x} = \dfrac{1}{A} \int_a^b x(y_2 - y_1)\, dx$	
401			$\bar{y} = \dfrac{1}{2A} \int_a^b (y_1 + y_2)(y_2 - y_1)\, dx$	
402		Of Solid of Revolution of Volume V: 	About x Axis: $\bar{x} = \dfrac{\pi}{V} \int_a^b xy^2\, dx$	
403			About y Axis: $\bar{y} = \dfrac{\pi}{V} \int_c^d yx^2\, dy$	

APPLICATIONS OF THE DEFINITE INTEGRAL (*Continued*)

No.   Page

No.				Formula	
			Thin Strip		
404			$p \bullet \quad A$ (r)	$I_p = Ar^2$	
405		Of Areas	**Extended Area** $dA = y\,dx$	$I_x = \dfrac{1}{3}\int y^3\,dx$	
406				$I_y = \int x^2\,y\,dx$	
407				Polar $I_0 = I_x + I_y$	
408				Radius of Gyration: $r = \sqrt{\dfrac{I}{A}}$	
409			(s) $\bullet A$ / $B \quad A$	Parallel Axis Theorem: $I_B = I_A + As^2$	
410	Moment of Inertia		(r) $p \bullet \quad \bullet M$	$I_p = Mr^2$	
411			Disk: (r, dh)	$dI = \dfrac{m\pi}{2}\,r^4\,dh$	
412			Ring: (r_o, r_i, dh)	$dI = \dfrac{m\pi}{2}\,(r_0^4 - r_i^4)\,dh$	About Axis of Revolution (Polar Moment of Inertia)
413		Of Masses	Shell: (dr, r, h)	$dI = 2\pi m r^3 h\,dr$	
414			**Solid of Revolution:** (r, 0, a, b, h)	Disk Method: $I = \dfrac{m\pi}{2}\displaystyle\int_a^b r^4\,dh$	
415				Shell Method: $I = 2\pi m \displaystyle\int r^3 h\,dr$	
416			(s) $\bullet$ / $B \quad A$	Parallel Axis Theorem: $I_B = I_A + Ms^2$	
417	Average and rms Values		$y = f(x)$ (0, a, b, x)	Average Ordinate: $y_{avg} = \dfrac{1}{b-a}\displaystyle\int_a^b f(x)\,dx$	
418				Root-Mean-Square Value: $rms = \sqrt{\dfrac{1}{b-a}\displaystyle\int_a^b [f(x)]^2\,dx}$	

APPLICATIONS OF THE DEFINITE INTEGRAL (*Continued*)

No.						Page
DIFFERENTIAL EQUATIONS	**First-Order**		419	Variables Separable	$f(y)\,dy = g(x)\,dx$	
			420	Integrable Combinations	$x\,dy + y\,dx = d(xy)$	
			421		$\dfrac{x\,dy - y\,dx}{x^2} = d\left(\dfrac{y}{x}\right)$	
			422		$\dfrac{y\,dx - x\,dy}{y^2} = d\left(\dfrac{x}{y}\right)$	
			423		$\dfrac{x\,dy - y\,dx}{x^2 + y^2} = d\left(\tan^{-1}\dfrac{y}{x}\right)$	
			424	Homogeneous	$M\,dx + N\,dy = 0$ (Substitute $y = vx$)	
			425	First-Order Linear	Form	$y' + Py = Q$	
			426		Integrating Factor	$R = e^{\int P\,dx}$	
			427		Solution	$ye^{\int P\,dx} = \int Q e^{\int P\,dx}\,dx$	
	Second-Order	**Right Side Zero**	428	Form	$ay'' + by' + cy = 0$	
			429	Auxiliary Equation	$am^2 + bm + c = 0$	
				Form of Solution	Roots of Auxiliary Equation	Solution	
			430		Real and Unequal	$y = c_1 e^{m_1 x} + c_2 e^{m_2 x}$	
			431		Real and Equal	$y = c_1 e^{mx} + c_2 x e^{mx}$	
			432		Nonreal	(a) $y = e^{ax}(C_1 \cos bx + C_2 \sin bx)$ or (b) $y = C e^{ax} \sin(bx + \phi)$	
		Right Side Not Zero	433	Form	$ay'' + by' + cy = f(x)$	
			434	Complete Solution	$y = \quad y_c \quad + \quad y_p$ $\qquad\quad \uparrow \qquad\qquad \uparrow$ complementary function $\qquad$ particular integral	
			435	Bernoulli's Equation	$\dfrac{dy}{dx} + Py = Qy^n$ (Substitute $z = y^{1-n}$)	

No.						Page

436		Laplace Transform	Definition		$\mathscr{L}[f(t)] = \int_0^\infty f(t)e^{-st}\, dt$	
437			Inverse Transform		$\mathscr{L}^{-1}[F(s)] = f(t)$	
438			Euler's Method		$x_q = x_p + \Delta x$ $y_q = y_p + m_p\, \Delta x$	
439			Modified Euler's Method		$x_q = x_p + \Delta x$ $y_q = y_p + \left(\dfrac{m_p + m_q}{2}\right)\Delta x$	
440	Numerical Solution		Runge–Kutta Method		$x_q = x_p + \Delta x$ $y_q = y_p + m_{avg}\, \Delta x$ where $\quad m_{avg} = \left(\tfrac{1}{6}\right)(m_p + 2m_r + 2m_s + m_q)$ and $\quad m_p = f'(x_p, y_p)$ $\qquad m_r = f'\!\left(x_p + \dfrac{\Delta x}{2},\, y_p + m_p\dfrac{\Delta x}{2}\right)$ $\qquad m_s = f'\!\left(x_p + \dfrac{\Delta x}{2},\, y_p + m_r\dfrac{\Delta x}{2}\right)$ $\qquad m_q = f'(x_p + \Delta x,\, y_p + m_s\, \Delta x)$	

DIFFERENTIAL EQUATIONS (*Continued*)

No.					Page
441	INFINITE SERIES — Power Series	Notation		$u_1 + u_2 + u_3 + \cdots + u_n + \cdots$	726
442		Tests for Convergence	Limit Test	$\lim_{n \to \infty} u_n = 0$	
443			Partial Sum Test	$\lim_{n \to \infty} S_n = S$	
444			Ratio Test	If $\lim_{n \to \infty} \left| \dfrac{u_n + 1}{u_n} \right|$ (a) is less than 1, the series converges. (b) is greater than 1, the series diverges. (c) is equal to 1, the test fails.	
445		Maclaurin's Series		$f(x) = f(0) + f'(0)x + \dfrac{f''(0)}{2!} x^2 + \cdots + \dfrac{f^{(n)}(0)}{n!} x^n + \cdots$	
446		Taylor's Series		$f(x) = f(a) + f'(a)\,(x-a) + \dfrac{f''(a)}{2!}\,(x-a)^2 + \cdots + \dfrac{f^{(n)}(a)}{n!}\,(x-a)^n + \cdots$	
447			Remainder after n Terms	$R_n = \dfrac{(x-a)^n}{n!}\, f^{(n)}(c)$ where c lies between a and x.	
448	Fourier Series	Period of 2π		$f(x) = a_0/2 + a_1 \cos x + a_2 \cos 2x + a_3 \cos 3x + \cdots + a_n \cos nx + \cdots$ $+ b_1 \sin x + b_2 \sin 2x + b_3 \sin 3x + \cdots + b_n \sin nx + \cdots$	
449			where	$a_0 = \dfrac{1}{\pi} \displaystyle\int_{-\pi}^{\pi} f(x)\, dx$	
450				$a_n = \dfrac{1}{\pi} \displaystyle\int_{-\pi}^{\pi} f(x) \cos nx\, dx$	
451				$b_n = \dfrac{1}{\pi} \displaystyle\int_{-\pi}^{\pi} f(x) \sin nx\, dx$	
452		Period of $2L$		$f(x) = \dfrac{a_0}{2} + a_1 \cos \dfrac{\pi x}{L} + a_2 \cos \dfrac{2\pi x}{L} + a_3 \cos \dfrac{3\pi x}{L} + \cdots$ $+ b_1 \sin \dfrac{\pi x}{L} + b_2 \sin \dfrac{2\pi x}{L} + b_3 \sin \dfrac{3\pi x}{L} + \cdots$	
453			where	$a_0 = \dfrac{1}{L} \displaystyle\int_{-L}^{L} f(x)\, dx$	
454				$a_n = \dfrac{1}{L} \displaystyle\int_{-L}^{L} f(x) \cos \dfrac{n\pi x}{L}\, dx$	
455				$b_n = \dfrac{1}{L} \displaystyle\int_{-L}^{L} f(x) \sin \dfrac{n\pi x}{L}\, dx$	
456		Waveform Symmetries	Odd and Even Functions	(a) Odd functions have Fourier Series with only sine terms (and no constant term). (b) Even functions have Fourier Series with only cosine terms (and may have a constant term).	
457			Half-Wave Symmetry	A waveform that has half-wave symmetry has only odd harmonics in its Fourier Series.	

	No.			Applications		Page
MIXTURES	A1	Mixture Containing Ingredients A, B, C, . . .		Total amount of mixture = amount of A + amount of B + $\cdots$		100
	A2			Final amount of each ingredient = initial amount + amount added − amount removed		100
	A3	Combination of Two Mixtures		Final amount of A = amount of A from mixture 1 + amount of A from mixture 2		101
	A4	Fluid Flow		Amount of flow = flow rate × elapsed time $A = QT$		
WORK	A5			Amount done = rate of work × time worked		248
	A6		Constant Force	Work = force × distance = Fd		
	A7		Variable Force	$\text{Work} = \int_a^b F(x)\, dx$		
FINANCIAL	A8	Unit Cost		$\text{Unit cost} = \dfrac{\text{total cost}}{\text{number of units}}$		
	A9	Interest: Principal a Invested at Rate n for t years Accumulates to Amount y	Simple	$y = a(1 + nt)$		
	A10		Compounded Annually	(a) $y = a(1 + n)^t$		559
				(b) Recursion Relation $y_t = y_{t-1}(1 + n)$		559
	A11		Compounded m times/yr	$y = a\left(1 + \dfrac{n}{m}\right)^{mt}$		560
STATICS	A12		Moment about Point a	$M_a = Fd$		104
	A13	Equations of Equilibrium (Newton's First Law)		The sum of all horizontal forces = 0		104
	A14			The sum of all vertical forces = 0		104
	A15			The sum of all moments about any point = 0		104
	A16		Coefficient of Friction	$\mu = \dfrac{f}{N}$		

No.					Page
	A17		Uniform Motion (Constant Speed)	Distance = rate × time $D = Rt$	248
	A18		Uniformly Accelerated (Constant Acceleration a, Initial Velocity v_0) For free fall, $a = g = 9.807$ m/s^2 $= 32.2$ ft/s^2	Displacement at Time t	$s = v_0 t + \dfrac{at^2}{2}$	
	A19			Velocity at Time t	$v = v_0 + at$	
	A20			Newton's Second Law	$F = ma$	252
	A21	Linear Motion		Average Speed	$\text{Average speed} = \dfrac{\text{total distance traveled}}{\text{total time elapsed}}$	
	A22		Nonuniform Motion	Displacement	$s = \displaystyle\int v\, dt$	
	A23			Instantaneous Velocity	$v = \dfrac{ds}{dt}$	
	A24				$v = \displaystyle\int a\, dt$	
	A25			Instantaneous Acceleration	$a = \dfrac{dv}{dt} = \dfrac{d^2s}{dt^2}$	
	A26		Uniform Motion	Angular Displacement	$\theta = \omega t$	441
	A27			Linear Speed of Point at Radius r	$v = \omega r$	442
	A28	Rotation	Nonuniform Motion	Angular Displacement	$\theta = \displaystyle\int \omega\, dt$	
	A29			Angular Velocity	$\omega = \dfrac{d\theta}{dt}$	
	A30				$\omega = \displaystyle\int \alpha\, dt$	
	A31			Angular Acceleration	$\alpha = \dfrac{d\omega}{dt} = \dfrac{d^2\theta}{dt^2}$	
	A32	Curvilinear Motion	x and y Components	Displacement	(a) $x = \displaystyle\int v_x\, dt$ (b) $y = \displaystyle\int v_y\, dt$	
	A33			Velocity	(a) $v_x = \dfrac{dx}{dt}$ (b) $v_y = \dfrac{dy}{dt}$	
	A34				(a) $v_x = \displaystyle\int a_x\, dt$ (b) $v_y = \displaystyle\int a_y\, dt$	
	A35			Acceleration	(a) $a_x = \dfrac{dv_x}{dt}$ (b) $a_y = \dfrac{dv_y}{dt}$ $= \dfrac{d^2x}{dt^2}$ $= \dfrac{d^2y}{dt^2}$	

MOTION

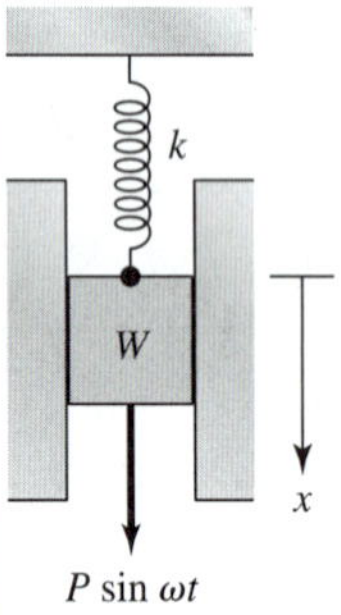

	No.					Page
MECHANICAL VIBRATIONS	A36		**Free Vibrations** $(P = 0)$	Simple Harmonic Motion (No Damping)	$x = x_0 \cos \omega_n t$	
	A37				Undamped Angular Velocity $\quad \omega_n = \sqrt{\dfrac{kg}{W}}$	
	A38				Natural Frequency $\quad f_n = \dfrac{\omega_n}{2\pi}$	
	A39			Under-damped	$x = x_0 e^{-at} \cos \omega_d t$	
	A40				Damped Angular Velocity $\quad \omega_d = \sqrt{\omega_n^2 - \dfrac{c^2 g^2}{W^2}}$	
	A41			Overdamped	$x = C_1 e^{m_1 t} + C_2 e^{m_2 t}$	
	A42		Forced Vibrations	Maximum Deflection	$x_0 = \dfrac{Pg}{W \sqrt{4a^2 \omega^2 + (\omega_n^2 - \omega^2)^2}}$	

In the figure: k, W, x, $P \sin \omega t$, Coefficient of friction $= c$

	No.				Page
MATERIAL PROPERTIES	**A43**	Density		Density = $\dfrac{\text{weight}}{\text{volume}}$ or $\dfrac{\text{mass}}{\text{volume}}$	
	A44	Mass		Mass = $\dfrac{\text{weight}}{\text{acceleration due to gravity}}$	
	A45	Specific Gravity		SG = $\dfrac{\text{density of substance}}{\text{density of water}}$	554
	A46	Pressure	Total Force on a Surface	Force = pressure × area	
	A47		Force on a Submerged Surface	$F = \delta \displaystyle\int y\, dA$	
	A48			$F = \delta \bar{y} A$	
	A49	pH		pH = −10 log concentration	594
TEMPERATURE	**A50**	Conversions between Degrees Celsius (C) and Degrees Fahrenheit (F)		$C = \tfrac{5}{9}(F - 32)$	
	A51			$F = \tfrac{9}{5}C + 32$	
STRENGTH OF MATERIALS	**A52**	Tension or Compression	Normal Stress	$\sigma = \dfrac{P}{a}$	
	A53		Strain	$\epsilon = \dfrac{e}{L}$	
	A54		Modulus of Elasticity and Hooke's Law	$E = \dfrac{PL}{ae}$	
	A55			$E = \dfrac{\sigma}{\epsilon}$	
	A56	Thermal Expansion	Elongation	$e = \alpha L\, \Delta t$	
	A57	Cold	New Length	$L = L_0 (1 + \alpha \Delta t)$	
	A58	Hot	Strain	$\epsilon = \dfrac{e}{L} = \alpha \Delta t$	
	A59		Stress, if Restrained	$\sigma = E\epsilon = E\alpha\, \Delta t$	
	A60	Temperature change = Δt Coefficient of thermal expansion = α	Force, if Restrained	$P = a\sigma = aE\alpha\, \Delta t$	
	A61		Force needed to Deform a Spring	$F =$ spring constant × distance = kx	

No.   Page

	No.			Formula	Page
ELECTRICAL TECHNOLOGY	**A62**	Ohm's Law		$\text{Current} = \dfrac{\text{voltage}}{\text{resistance}}$ $\qquad I = \dfrac{V}{R}$	
	A63	Combinations of Resistors	In Series	$R = R_1 + R_2 + R_3 + \cdots$	
	A64		In Parallel	$\dfrac{1}{R} = \dfrac{1}{R_1} + \dfrac{1}{R_2} + \dfrac{1}{R_3} + \cdots$	
	A65	Power Dissipated in a Resistor		$\text{Power} = P = VI$	
	A66			$P = \dfrac{V^2}{R}$	
	A67			$P = I^2 R$	
	A68	Kirchhoff's Laws	Loops	The sum of the voltage rises and drops around any closed loop is zero	
	A69		Nodes	The sum of the currents entering and leaving any node is zero	
	A70	Resistance Change with Temperature		$R = R_1\,[1 + \alpha\,(t - t_1)]$	
	A71	Resistance of a wire		$R = \dfrac{\rho L}{A}$	
	A72	Combinations of Capacitors	In Series	$\dfrac{1}{C} = \dfrac{1}{C_1} + \dfrac{1}{C_2} + \dfrac{1}{C_3} + \cdots$	
	A73		In Parallel	$C = C_1 + C_2 + C_3 + \cdots$	
	A74	Charge on a Capacitor at Voltage V		$Q = CV$	

No.				Page

ELECTRICAL TECHNOLOGY (Continued)

No.			Sinusoidal Form	Complex Form	Page
A75		Alternating Voltage	$v = V_m \sin(\omega t + \phi_1)$	$\mathbf{V} = V_m\,\phi_1\,\underline{|\phi_1}$	626
A76		Alternating Current	$i = I_m \sin(\omega t + \phi_2)$	$\mathbf{I} = I_m\,\underline{|\phi_2}$	626
A77		Period	$P = \dfrac{2\pi}{\omega}$ seconds		473
A78		Frequency	$f = \dfrac{1}{P} = \dfrac{\omega}{2\pi}$ hertz		473
A79		Current	$i = \dfrac{dq}{dt}$		
A80		Charge	$q = \displaystyle\int i\,dt$ coulombs		
A81	Capacitor	Instantaneous Current	$i = C\dfrac{dv}{dt}$		
A82		Instantaneous Voltage	$v = \dfrac{1}{C}\displaystyle\int i\,dt$ volts		
A83		Current when Charging or Discharging	Series *RC* Circuit	$i = \dfrac{E}{R}e^{-t/RC}$	569
A84		Voltage when Charging		$v = E(1 - e^{-t/RC})$	
A85		Voltage when Discharging		$v = Ee^{-t/RC}$	
A86	Inductor	Instantaneous Current	$i = \dfrac{1}{L}\displaystyle\int v\,dt$ amperes		
A87		Instantaneous Voltage	$v = L\dfrac{di}{dt}$		
A88		Current when Charging	Series *RL* Circuit	$i = \dfrac{E}{R}(1 - e^{-Rt/L})$	568
A89		Current when Discharging		$i = \dfrac{E}{R}e^{-Rt/L}$	
A90		Voltage when Charging or Discharging		$v = Ee^{-Rt/L}$	

	No.					Page
ELECTRICAL TECHNOLOGY (*Continued*)	**A91**	Series *RLC* Circuit	**DC Source**	Resonant Frequency	$\omega_n = \sqrt{\dfrac{1}{LC}}$	
	A92			No Resistance: The Series *L C* Circuit:	$i = \dfrac{E}{\omega_n L}\sin \omega_n t$	
	A93			Underdamped	$i = \dfrac{E}{\omega_d L}e^{-at}\sin \omega_d t$	
	A94				where $\omega_d = \sqrt{\omega_n{}^2 - \dfrac{R^2}{4L^2}}$	
	A95			Overdamped	$i = \dfrac{E}{2j\omega_d L}\left[e^{(-a+j\omega_d)t} - e^{(-a-j\omega_d)t}\right]$	
	A96		**AC Source**	Inductive Reactance	$X_L = \omega L$	
	A97			Capacitive Reactance	$X_C = \dfrac{1}{\omega C}$	
	A98			Total Reactance	$X = X_L - X_C$	196
	A99			Magnitude of Impedance	$|Z| = \sqrt{R^2 + X^2} = \sqrt{R^2 + \left(\omega L - \dfrac{1}{\omega C}\right)^2}$	197
	A100			Phase Angle	$\phi = \arctan \dfrac{X}{R}$	197
	A101			Complex Impedance	$Z = R + jX = Z\,\underline{|\phi} = Ze^{j\phi}$	627
	A102			Steady-State Current	$i_{ss} = \dfrac{E}{Z}\sin(\omega t - \phi)$	
	A103	Ohm's Law for AC			$\mathbf{V = ZI}$	627
	A104	Decibels Gained or Lost			$G = 10\log_{10}\dfrac{P_2}{P_1}\quad$ dB	594

Conversion Factors

UNIT	EQUALS
LENGTH	
1 angstrom	1×10^{-10} meter
	1×10^{-4} micrometer (micron)
1 centimeter	10^{-2} meter
	0.3937 inch
1 foot	12 inches
	0.3048 meter
1 inch	25.4 millimeters
	2.54 centimeters
1 kilometer	3281 feet
	0.5400 nautical mile
	0.6214 statute mile
	1094 yards
1 light-year	9.461×10^{12} kilometers
	5.879×10^{12} statute miles
1 meter	10^{10} angstroms
	3.281 feet
	39.37 inches
	1.094 yards
1 micron	10^4 angstroms
	10^{-4} centimeter
	10^{-6} meter
1 nautical mile (International)	8.439 cables
	6076 feet
	1852 meters
	1.151 statute miles
1 statute mile	5280 feet
	8 furlongs
	1.609 kilometers
	0.8690 nautical mile
1 yard	3 feet
	0.9144 meter

ANGLES	
1 degree	60 minutes
	0.01745 radian
	3600 seconds
	2.778×10^{-3} revolution
1 minute of arc	0.01667 degree
	2.909×10^{-4} radian
	60 seconds
1 radian	0.1592 revolution
	57.296 degrees
	3438 minutes
1 second of arc	2.778×10^{-4} degree
	0.01667 minute

UNIT AREA	EQUALS
1 acre	4047 square meters
	43 560 square feet
1 are	0.024 71 acre
	1 square dekameter
	100 square meters
1 hectare	2.471 acres
	100 ares
	10 000 square meters
1 square foot	144 square inches
	0.092 90 square meter
1 square inch	6.452 square centimeters
1 square kilometer	247.1 acres
1 square meter	10.76 square feet
1 square mile	640 acres
	2.788×10^7 square feet
	2.590 square kilometers

UNIT VOLUME	EQUALS
1 board-foot	144 cubic inches
1 bushel (U.S.)	1.244 cubic feet
	35.24 liters
1 cord	128 cubic feet
	3.625 cubic meters
1 cubic foot	7.481 gallons (U.S. liquid)
	28.32 liters
1 cubic inch	0.01639 liter
	16.39 milliliters
1 cubic meter	35.31 cubic feet
	10^6 cubic centimeter
1 cubic millimeter	6.102×10^{-5} cubic inch
1 cubic yard	27 cubic feet
	0.7646 cubic meter
1 gallon (imperial)	277.4 cubic inches
	4.546 liters
1 gallon (U.S. liquid)	231 cubic inches
	3.785 liters
1 kiloliter	35.31 cubic feet
	1.308 cubic yards
	220 imperial gallons
1 liter	10^3 cubic centimeters
	10^6 cubic millimeters
	10^{-3} cubic meter
	61.02 cubic inches

UNIT	EQUALS
MASS	
1 gram	10^{-3} kilogram
	6.854×10^{-5} slug
1 kilogram	1000 grams
	0.06854 slug
1 slug	14.59 kilograms
	14,590 grams
1 metric ton	1000 kilograms
FORCE	
1 dyne	10^{-5} newton
1 newton	10^{5} dynes
	0.2248 pound
	3.597 ounces
1 pound	4.448 newtons
	16 ounces
1 ton	2000 pounds
AT SEA LEVEL	
1 kilogram	2.205 pounds
VELOCITY	
1 foot/minute	0.3048 meter/minute
	0.011 364 mile/hour
1 foot/second	1097 kilometers/hour
	18.29 meters/minute
	0.6818 mile/hour
1 kilometer/hour	3281 feet/hour
	54.68 feet/minute
	0.6214 mile/hour
1 kilometer/minute	3281 feet/minute
	37.28 miles/hour
1 knot	6076 feet/hour
	101.3 feet/minute
	1.852 kilometers/hour
	30.87 meters/minute
	1.151 miles/hour
1 meter/hour	3.281 feet/hour
1 mile/hour	1.467 feet/second
	1.609 kilometers/hour
POWER	
1 British thermal unit/hour	0.2929 watt
1 Btu/pound	2.324 joules/gram
1 Btu-second	1.414 horsepower
	1.054 kilowatts
	1054 watts

UNIT	EQUALS
POWER (Continued)	
1 horsepower	42.44 Btu/minute
	550 footpounds/second
	746 watts
1 kilowatt	3414 Btu/hour
	737.6 footpounds/second
	1.341 horsepower
	10^3 joules/second
	999.8 international watt
1 watt	44.25 footpounds/minute
	1 joule/second
PRESSURE	
1 atmosphere	1.013 bars
	14.70 pounds/square inch
	760 torrs
	101 kilopascals
1 bar	10^6 baryes
	14.50 pounds-force/square inch
1 barye	10^{-6} bar
1 inch of mercury	0.033 86 bar
	70.73 pounds/square foot
1 pascal	1 newton/square meter
1 pound/square inch	0.068 03 atmosphere
ENERGY	
1 British thermal unit	1054 joules
	1054 wattseconds
1 foot-pound	1.356 joules
	1.356 newtonmeters
1 joule	0.7376 foot-pound
	1 wattsecond
	0.2391 calories
1 kilowatthour	3410 British thermal units
	1.341 horsepowerhours
1 newtonmeter	0.7376 footpounds
1 watthour	3.414 British thermal units
	2655 footpounds
	3600 joules

Source: Adapted from P. Calter, *Schaum's Outline of Technical Mathematics,* McGraw-Hill Book Company, New York, 1979.

Answers to Selected Problems

Note: The graphs in this appendix are so tiny that it is difficult to convey highly accurate information with them. Please look at these graphs for general trends only. For larger graphs, please see the SSM and ISM.

••• CHAPTER 1 ••

Exercise 1

1. $7 < 10$ **3.** $-3 < 4$ **5.** $\frac{3}{4} = 0.75$ **7.** -18 **9.** 13 **11.** 4 **13.** 2 **15.** 5 **17.** 5
19. 2.00 **21.** 55.86 **23.** 2.96 **25.** 278.38 **27.** 745.6 **29.** 0.5 **31.** 34.9 **33.** 0.8 **35.** 7600
37. 274,800 **39.** 2860 **41.** 484,000 **43.** 29.6 **45.** 8370

Exercise 2

1. -1090 **3.** -1116 **5.** 105,233 **7.** 1789 **9.** -1129 **11.** -850 **13.** 1827 **15.** 4931
17. 593.44 **19.** -0.00031 **21.** 78,388 mi^2 **23.** 35.0 cm **25.** 41.1 Ω

Exercise 3

1. $7\overline{3}00$ **3.** 0.525 **5.** $-17,800$ **7.** 22.9 **9.** \$3320 **11.** \$1441 **13.** \$1,570,000 **15.** 17,180 rev
17. 980.03 cm

Exercise 4

1. 163 **3.** -0.347 **5.** 0.7062 **7.** 70,840 **9.** 17 **11.** 371.708 m **13.** 10 **15.** 0.00144
17. -0.00000253 **19.** -175 **21.** 0.2003 **23.** 0.0901 **25.** 0.9930 **27.** 313 Ω **29.** 0.279

Exercise 5

1. 8 **3.** -8 **5.** 1 **7.** 1 **9.** 1 **11.** -1 **13.** 100 **15.** 1 **17.** 1000 **19.** 0.01
21. 10,000 **23.** 0.00001 **25.** 1.035 **27.** -112 **29.** 0.0146 **31.** 0.0279 **33.** 59.8
35. -0.0000193 **37.** 125 W **39.** 878,000 cm^3 **41.** 5 **43.** 7 **45.** -2 **47.** 7.01 **49.** 4.45
51. 62.25 **53.** -7.28 **55.** -1.405 **57.** 4480 Ω

Exercise 6

1. 3340 **3.** -5940 **5.** 5 **7.** 3 **9.** 121 **11.** 27 **13.** 27 **15.** 24 **17.** 12 **19.** 2
21. 30 **23.** 46.2 **25.** 978 **27.** 2.28 **29.** 0.160 **31.** 59.8 **33.** 55.8 **35.** 3.51 **37.** 7.17
39. 3.23 **41.** 0.871 **43.** 7.93

Exercise 7

1. 10^2 **3.** 10^{-4} **5.** 10^8 **7.** 0.01 **9.** 0.1 **11.** 1.86×10^5, 186×10^3
13. 2.5742×10^4, 25.742×10^3 **15.** 9.83×10^4, 98.3×10^3 **17.** 2850 **19.** 90,000 **21.** 0.003667
23. 10 **25.** 10^3 **27.** 10^3 **29.** 10^7 **31.** 10^2 **33.** 400 **35.** 2.1×10^9 **37.** 2×10^6
39. 2×10^{-9} **41.** 7.0×10^5 **43.** 1.070×10^4 **45.** 3.1×10^{-2} **47.** 1.55×10^6 **49.** 2.11×10^4
51. 1.7×10^2 **53.** 9.79×10^6 Ω **55.** 2.72×10^{-6} W **57.** 6.77×10^{-5} F **59.** 120 h

Exercise 8

1. 12.7 ft   **3.** 9144 in.   **5.** 58,000 lb   **7.** 44.8 ton   **9.** 364 km   **11.** 735.9 kg
13. 6.2×10^3 megohms   **15.** 9.348×10^{-3} μF   **17.** 1194 ft   **19.** 32.7 N   **21.** 17.6 liters   **23.** 3.60 m
25. 0.587 acre   **27.** 2.30 m^2 **29.** 243 acre   **31.** 12,720 in.^{3} **33.** 56.4 m^3 **35.** 1.63×10^{-2} in.^{3}
37. 3.31 mi/h   **39.** 107 km/h   **41.** 18.3 births/week   **43.** 19.8¢/m^2 **45.** 236 cents/lb   **47.** 117 gal
49. 14.0 lb   **51.** 8.00 cm, 2.72 cm, 3.15 cm, 3.62 cm, 13.3 cm   **53.** 5.699 m^2 **55.** 2.450 gal

Exercise 9

1. 17   **3.** −37   **5.** 14   **7.** 14.1   **9.** 233   **11.** 8.00   **13.** $3975   **15.** 53.3°C   **17.** $13,266

Exercise 10

1. 372%   **3.** 0.55%   **5.** 40.0%   **7.** 70.0%   **9.** 0.23   **11.** 2.875   **13.** $\frac{3}{8}$ **15.** $1\frac{1}{2}$ **17.** 105 tons
19. 220 kg   **21.** 120 liters   **23.** 1090 Ω   **25.** $1562   **27.** 518   **29.** 65.6   **31.** 100
33. 108 km   **35.** $640   **37.** 45.9%   **39.** 18.5%   **41.** 33.4%   **43.** 11.6%   **45.** 1.5%   **47.** 96.6%
49. 31.3%   **51.** 11%   **53.** 5.99%   **55.** 67.0%   **57.** 73%   **59.** 2.3%   **61.** 112.5 and 312.5 V
63. 37.5%   **65.** 25 liters

Review Problems

1. 83.35   **3.** 88.1   **5.** 0.346   **7.** 94.7   **9.** 5.46   **11.** 6.8   **13.** 30.6   **15.** 17.4
17(a) 179 **(b)** 1.08 **(c)** 4.86 **(d)** 45,700   **19.** 70.2%   **21.** 3.63×10^6 **23.** 12,000 kW   **25.** 3.4%
27. 7×10^{11} bbl   **29.** 10,200   **31.** 4.16×10^{-10} **33.** 2.42   **35.** $-\frac{2}{3} < -0.660$ **37.** 2370
39. −1.72   **41.** 64.5%   **43.** 219 N   **45.** 525 ft   **47.** 22.0%   **49.** 7216   **51.** 4.939   **53.** 109
55. 0.207   **57.** 14.7   **59.** 6.20   **61.** 2   **63.** 83.4   **65.** 93.52 cm   **67.** 9.07   **69.** 75.2%
71. 0.0737   **73.** 7.239   **75.** 121   **77.** 1.21   **79.** 0.30

◆◆◆ CHAPTER 2 ◆◆◆

Exercise 1

1. 2   **3.** 1   **5.** 5   **7.** $b/4$ **9.** $3/2a$

Exercise 2

1. $12x$ **3.** $-10ab$ **5.** $2m - c$ **7.** $4a - 2b - 3c - d - 17$ **9.** $11x + 10ax$ **11.** $-26x^3 + 2x^2 + x - 31$
13. 0   **15.** $20 - 2ay^3$ **17.** $18b^2 - 3ac - 2d$ **19.** $7m^2 - 5m^3 + 15ab - 4q - z$ **21.** $5.83(a + b)$
23. $3a^2 - 3a + x - b$ **25.** $-1.1 - 4.4x$ **27.** $4a^2 - 5a + y + 1$ **29.** $2a + 2b - 2m$ **31.** $6m - 3z$
33. $5w + 2z + 6x$ **35.** $a + 12$ **37.** $22w^2$

Exercise 3

1. 10^5 **3.** x^3 **5.** y^7 **7.** x^7 **9.** w^6 **11.** x^{n+3} **13.** y^3 **15.** x **17.** 100   **19.** x **21.** x^{12}
23. a^{xy} **25.** x^{2a+2} **27.** $8x^3$ **29.** $27a^3b^3c^3$ **31.** $-1/27$ **33.** $8a^3/27b^6$ **35.** $1/a^2$ **37.** $y^3/27$
39. $9b^6/4a^2$ **41.** $2/x^2 + 3/y^3$ **43.** x^{-1} **45.** x^2y^{-2} **47.** $a^{-3}b^{-2}$ **49.** 1   **51.** $\frac{1}{9}$ **53.** 1   **55.** $6w^3$
57. $i^2R/9$

Exercise 4

1. x^5 **3.** $10a^3b^4$ **5.** $36a^6b^5$ **7.** $-8p^4q^4$ **9.** $2a^2 - 10a$ **11.** $x^2y - xy^2 + x^2y^2$
13. $-12p^3q - 8p^2q^2 + 12pq^3$ **15.** $6a^3b^3 - 12a^3b^2 + 9a^2b^3 + 3a^2b^2$ **17.** $a^2 + ac + ab + bc$
19. $x^3 - xy + 3x^2y - 3y^2$ **21.** $9m^2 - 9n^2$ **23.** $49c^2d^4 - 16y^6z^2$ **25.** $x^4 - y^4$ **27.** $a^2 - ac - c - 1$
29. $m^3 - 5m^2 - m + 14$ **31.** $x^8 - x^2$ **33.** $-z^3 + 2z^2 - 3z + 2$ **35.** $a^3 + y^3$ **37.** $a^3 - y^3$
39. $x^3 - px^2 - mx^2 + mpx + nx^2 - npx - mnx + mnp$ **41.** $m^5 - 1$ **43.** $9a^2x - 6abx + b^2x$
45. $x^3 - 3x - 2$ **47.** $x^3 - cx^2 + bx^2 - bcx + ax^2 - acx + abx - abc$ **49.** $-c^5 - c^4 + c + 1$

51. $4a^6 - 10a^4c^2 + 14a^4 - 25a^2c^2 + 10a^2$ **53.** $a^3 + a^2c - ab^2 + b^2c - b^3 + 2abc - ac^2 + bc^2 - c^3 + a^2b$
55. $n^5 - 34n^3 + 57n^2 - 20$ **57.** $a^2 + 2ac + c^2$ **59.** $m^2 - 2mn + n^2$ **61.** $A^2 + 2AB + B^2$
63. $14.4a^2 - 16.7ax + 4.84x^2$ **65.** $4c^2 - 12cd + 9d^2$ **67.** $a^2 - 2a + 1$ **69.** $y^4 - 40y^2 + 400$
71. $a^2 + 2ab + b^2 + 2ac + 2bc + c^2$ **73.** $x^2 + 2x - 2xy - 2y + y^2 + 1$ **75.** $x^4 + 2x^3y + 3x^2y^2 + 2xy^3 + y^4$
77. $a^3 + 3a^2b + 3ab^2 + b^3$ **79.** $p^3 - 9p^2q + 27pq^2 - 27q^3$ **81.** $a^3 - 3a^2b + 3ab^2 - b^3$ **83.** $3a + 3b + 3c$
85. $p - q$ **87.** $31y - 5z$ **89.** $2x - 4$ **91.** $7 - 5x$ **93.** $LW - 3.0L + 4.0W - 12$
95. $x^2 + 7.00x + 12.2$ **97.** $4.19r^3 - 45.2r^2 + 163r - 195$

Exercise 5

1. z^2 **3.** $-5xyz$ **5.** $2f$ **7.** $-2xz$ **9.** $-5xy^2$ **11.** $3c$ **13.** $-5y$ **15.** $-3w$ **17.** $7n$ **19.** $4b$
21. b^2 **23.** r^{m-n} **25.** $4s$ **27.** -24 **29.** $-5z/xy$ **31.** $-5n^2/m^2x$ **33.** $3ab$ **35.** $3x^5$ **37.** $42/y$
39. $-4x$ **41.** $2a^2 - a$ **43.** $7x^2 - 1$ **45.** $3x^4 - 5x^0$ **47.** $2x^2 - 3x$ **49.** $a + 2c$ **51.** $ax - 2y$
53. $2x + y$ **55.** $ab - c$ **57.** $xy - x^2 - y^2$ **59.** $1 - a - b$ **61.** $a - 3b + c^2$ **63.** $x - 2y + y^2/x$
65. $m + 2 - 3m/n$ **67.** $x + 8$ **69.** $a + 5$ **71.** $9a^2 + 6ab + 4b^2$ **73.** $x - 6 + 15/(x + 2)$
75. $5x + 13 - 35/(3 - x)$ **77.** $a - b + c$ **79.** $x + y - z$ **81.** $x - 2y - z$

Review Problems

1. $b^6 - b^4x^2 + b^4x^3 - b^2x^5 + b^2x^4 - x^6$ **3.** 3.86×10^{14} **5.** $9x^4 - 6mx^3 - 6m^3x + 10m^2x^2 + m^4$
7. $8x^3 + 12x^2 + 6x + 1$ **9.** $6a^2x^5$ **11.** $a - b - c$ **13.** $16a^2 - 24ab + 9b^2$ **15.** $x^2y^2 - 6xy + 8$
17. $4a^2 - 12ab + 9b^2$ **19.** $ab - b^4 - a^2b$ **21.** $16m^4 - c^4$ **23.** $2a^2$ **25.** $24 - 8y$
27. $6x^3 + 9x^2y - 3xy^2 - 6y^3$ **29.** $13w - 6$ **31.** $a^5 + 32c^5$ **33.** $(a - c)^{m-2}$ **35.** $x + y$
37. $b^3 - 9b^2 + 27b - 27$ **39.** $x^2/2y^3$ **41.** $x^2 + 2x - 8$ **43.** $ab - 2 - 3b^2$ **45.** $5a - 10x - 2$
47. 8.01×10^6 **49.** $10x^3y^4$ **51.** $x^3 + 3x^2 - 4x$ **53.** 1.77×10^8 **55.** $6a^3b^3$
57. $x^4 + x + 1 + R(x^2 + x + 1)$ **59.** $0.13x + 540$ **61.** $4.02r^2$ ft

••• CHAPTER 3 •••

Exercise 1

1. 33   **3.** 4/5   **5.** 1/2   **7.** 1/2   **9.** 4   **11.** 3   **13.** -1.38 **15.** 1   **17.** $-1/2$ **19.** 2
21. 5   **23.** 28   **25.** 3   **27.** 3   **29.** 35   **31.** $-5/23$ **33.** 5/6   **35.** 19/5   **37.** 1   **39.** 3/25

41. 0   **43.** -0.0549 **45.** $3/a$ **47.** $5/c + 1$ **49.** $\dfrac{3b + c - a}{b}$

Exercise 2

1. $5x + 8$ **3.** x and $83 - x$ **5.** x and $6 - x$ **7.** A, $2A$, and $180 - 3A$ **9.** $82x$ km   **11.** 5
13. 57, 58, 59   **15.** 14   **17.** 6   **19.** 4   **21.** 20   **23.** 14   **25.** 81

Exercise 3

1. 8 technicians   **3.** \$130,137   **5.** 57,000 gal   **7.** \$64.29 to brother, \$128.57 to uncle, \$257.14 to bank
9. \$94,000   **11.** \$60,000

Exercise 4

1. 333 gal   **3.** 485 kg of 18% alloy, 215 kg of 31% alloy   **5.** 1380 kg   **7.** 1.29 liters   **9.** 60.0 lb
11. 59.1 lb

Exercise 5

1. 4.97 ft   **3.** $R_1 = 753$ lb downwards; $R_2 = 31\overline{0}0$ lb upwards   **5.** 339 cm   **7.** 6445 lb at right, 8425 lb at left

Review Problems

1. 12   **3.** 5   **5.** 0   **7.** $10\frac{1}{2}$ **9.** 2   **11.** 4   **13.** 4   **15.** 6   **17.** −5/7   **19.** −6   **21.** 9.27
23. 7   **25.** 6   **27.** 47 cm, 53 cm   **29.** 12 kg   **31.** $22,500   **33.** 15,000 ft   **35.** −13/5   **37.** 13

39. −5/7   **41.** −5/2   **43.** −7/3   **45.** −6/b   **47.** $\dfrac{a + 10}{2}$ **49.** $\dfrac{5a - c + 8}{a}$ **51.** $4608

53. 6 masons   **55.** 766 kg   **57.** 16.8 gal   **59.** 49.7 gal   **61.** $398 for computer, $597 for printer
63. $58,093 at 8.56%, $165,728 at 5.94%   **65.** $107,248 at 10.25%, $18,567 at 4.25%   **67.** 5.40 gal

••• CHAPTER 4 ••

Exercise 1

1. Is a function   **3.** Not a function   **5.** Not a function   **7.** Yes   **9.** $y = x^3$ **11.** $y = x + 2x^2$
13. $y = (2/3)(x - 4)$ **15.** The amount by which five exceeds x **17.** Twice the cube root of x **19.** $A = bh/2$
21. $V = 4\pi r^3/3$ **23.** $d = 55t$ mi   **25.** $H = 125t - gt^2/2$ **27.** Domain $= -10, -7, 0, 5, 10$; Range $= 3, 7, 10, 20$
29. $x \geq 7, y \geq 0$ **31.** $x \neq 0$, all y **33.** $x \neq 9, y \neq 1$ **35.** $x < 1, y > 0$ **37.** $x \geq 1, y \geq 0$
39. $-5 \leq y \leq 70$

Exercise 2

1. Explicit   **3.** Implicit   **5.** x is independent, y is dependent   **7.** x and y are independent, w is dependent

9. x and y are independent, z is dependent   **11.** $y = 2/x + 3$ **13.** $y = (2x^2 + x)/3$ **15.** $q = \dfrac{p(p + 5)}{2}$

17. $e = PL/aE$ **19.** 6   **21.** −21   **23.** 12.5   **25.** −5   **27.** 2.69   **29.** −15   **31.** 5/4
33. $2a^2 + 4$ **35.** $5a + 5b + 1$ **37.** $9\frac{1}{2}$ **39.** 142   **41.** 20   **43.** −52   **45.** 18   **47.** 8.66
49. $1.06 \times 10^4\ \Omega, 1.08 \times 10^4\ \Omega, 1.11 \times 10^4\ \Omega$ **51.** 797 in., 1040 in., 1410 in.

Exercise 3

1. $2x^2 + 3$ **3.** $1 - 6x$ **5.** $4 - 3x^3$ **7.** −77   **9.** $y = (8 - x)/3$
11. $y = (x - 6)/5$ **13.** $y = -x - 3$ **15.** $y = (x - 10)/2$

Review Problems

1(a) Is a function **(b)** Not a function **(c)** Not a function   **3.** $S = 4\pi r^2$
5.(a) Explicit, y independent, w dependent **(b)** Implicit   **7.** $w = (3 - x^2 - y^2)/2$ **9.** $7x^2$ **11.** 6   **13.** 28

15. 13/90   **17.** y is 7 less than five times the cube of x **19.** $x = \pm\sqrt{(6 - y)/3}$ **21.** $y = 8 - x$
23. $q = p(p + 7)/5$ **25.** $28\frac{2}{3}$ **27.** $y = (35 - x)/21$ **29.** $y = (x - 5)/10$ **31.** 104
33. $63x^2 - 210x + 175$

••• CHAPTER 5 ••

Exercise 1

1. Fourth   **3.** Second   **5.** Fourth   **7.** First and fourth   **9.** $x = 7$
11. $E(-1.8, -0.7), F(-1.4, -1.4), G(1.4, -0.6), H(2.5, -1.8)$
13. $E(-0.3, -1.3), F(-1.1, -0.8), G(-1.5, 1), H(-0.9, 1.1)$
15.   **17.**   **19.** $(4, -5)$

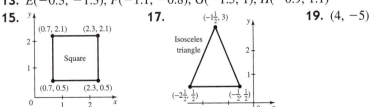

Exercise 2

1.

3.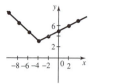

5.

7.

9.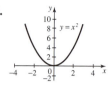

11.

13.

15.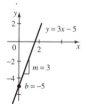

Exercise 3

1. 2   **3.** −1   **5.** 1   **7.** −11/7   **9.**   **11.**   **13.**

15.   **17.**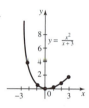

Exercise 4

9. 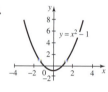   **11.**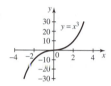

Exercise 5

1.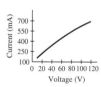

3.

5.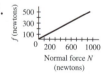

7.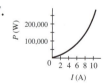

9.   **11.**

Exercise 6

1.
3.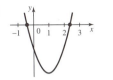
5.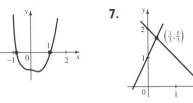
7.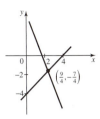
9.

Review Problems

1.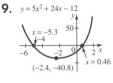
3.
5.
7.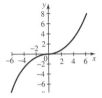

9. $y = 5x^2 + 24x - 12$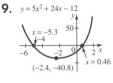
11. $y = 6x^3 + 3x^2 - 14x - 21$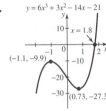
13.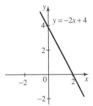
15.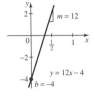

••• CHAPTER 6 ••

Exercise 1

1(a) 62.8° **(b)** 64.6° **(c)** 37.7°   **3.** 5.05   **5.** $A = C = 46.3°; B = D = 134°$

Exercise 2

1. $4405   **3.** 605 cm   **5.** 75.0 ft, 117 ft, 205 ft   **7.** 35.4 m   **9.** 35.4 m   **11.** 15.5 ft   **13.** 35.4 ft
15. 0.577 in.   **17.** 39.0 in.   **19.** 43.5 acres   **21.** 2.62 m   **23.** 5.2 mm

Exercise 3

1. 423 square units   **3.** 248 square units   **5.** $3042   **7.** $348   **9.** $3220   **11.** $861

Exercise 4

1. 7.16 m   **3.** 15.8 in.   **5.** 104 m   **7.** 318 acres   **9.** 0.738 unit   **11.** 44.7 units   **13.** 218 cm
15. 1.200 in.   **17.** 247 cm

Exercise 5

1. 6.34×10^7 in.^{3} **3(a)** 4.2 m **(b)** 150 m^3 **5.** 176 in.^{3} **7.** 0.583 in.   **9.** 0.99 in.   **11.** $76\overline{0}$ in.^{2}
13. 3800 lb   **15.** 38.1 ft^3 **17.** 1.72×10^9, 6.96×10^6 **19.** 15.8, 1.12   **21.** 230 lb

Review Problems

1. 1440 mi/h   **3.** 2.88 m   **5.** 13 m   **7.** 175,000 square units   **9.** 43.1 in.   **11.** 17.5 m   **13.** 161°
15. 213 cm^3 **17.** 1030 in.^{2} **19.** 1150 m^2 **21.** 87,800 cm^3 **23.** 11.0 ft^3 **25.** 114 cm^3

••• CHAPTER 7 ••

Exercise 1

1. 0.485 rad   **3.** 0.6152 rad   **5.** 3.5 rad   **7.** 0.7595 rev   **9.** 0.191 rev   **11.** 0.2147 rev   **13.** 162°
15. 171°   **17.** 29.45°   **19.** 244°57′45″   **21.** 161°54′36″   **23.** 185°58′19″

Exercise 2

1. 0.50, 0.87, 0.58, 1.73, 1.15, 2.00   **3.** 0.24, 0.97, 0.25, 4.01, 1.03, 4.13   **5.** 53°   **7.** 63°
9. 0.528, 0.849, 0.622, 1.61, 1.18, 1.89   **11.** 0.447, 0.894, 0.500   **13.** 0.832, 0.555, 1.50
15. 0.491, 0.871, 0.564   **17.** 0.9793   **19.** 3.0415   **21.** 0.9319   **23.** 1.0235   **25.** 0.955, 0.296, 3.230
27. 0.546, 0.838, 0.652   **29.** 0.677, 0.736, 0.920   **31.** 2.05   **33.** 2.89   **35.** −10.9

Exercise 3

1. 30.0°   **3.** 31.9°   **5.** 28.9°   **7.** 3/5, 4/5, 3/4   **9.** 5/13, 12/13, 5/12   **11.** 75.0°   **13.** 40.5°
15. 13.6°

Exercise 4

1. $B = 47.1°, b = 167, c = 228$ **3.** $b = 1.08, A = 58.1°, c = 2.05$ **5.** $B = 0.441$ rad, $b = 134, c = 314$
7. $a = 6.44, c = 11.3, A = 34.8°$ **9.** $a = 50.3, c = 96.5, B = 58.6°$ **11.** $c = 470, A = 54.3°, B = 35.7°$
13. $a = 2.80, A = 35.2°, B = 54.8°$ **15.** $b = 25.6, A = 46.9°, B = 43.1°$ **17.** $b = 48.5, A = 40.4°, B = 49.6°$
19. $a = 414, A = 61.2°, B = 28.8°$ **21.** cos 52°   **23.** cot 71°   **25.** tan 26.8°   **27.** cot 54°46′
29. cos 1.10 rad

Exercise 5

1. 402 m   **3.** 285 ft   **5.** 64.9 m   **7.** 39.9 yd   **9.** 128 m   **11.** 21.4 km, 14.7 km
13. 432 mi, S58°31′W   **15.** 156 mi North, 162 mi East   **17.** 30.2°, 20.5 ft   **19.** 37.6°   **21.** 139, 112
23. 63.0°, 117°   **25.** 125 cm   **27.** 2.55 in.   **29.** 9.3979 cm   **31.** 5.53 in.   **33.** 0.866 cm, 0.433 cm

Exercise 6

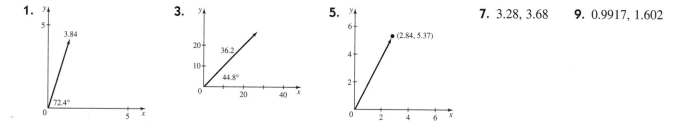

1. 3.84, 72.4°   **3.** 36.2, 44.8°   **5.** (2.84, 5.37)   **7.** 3.28, 3.68   **9.** 0.9917, 1.602

11. 616, 51.7°   **13.** 8811, 56.70°   **15.** 2.42, 31.1°   **17.** 616, 38.3°   **19.** 69.2, 32.5°   **21.** 9.14, 54.8°
23. 1090, 34.0°   **25.** 6.97, 57.1°

Exercise 7

1. 12.3 N   **3.** 1520 N   **5.** 25.7°   **7.** 4.94 tons   **9.** 119 km/h   **11.** 5.70 m/min, 1.76 min
13. 115 mi/h   **15.** $X = 354\ \Omega, Z = 372\ \Omega$ **17.** 7.13 Ω, 53.7°

Review Problems

1. 38°12′, 0.667 rad, 0.106 rev   **3.** 157.3 deg, 157°18′, 0.4369 rev
5. 0.919, 0.394, 2.33, 0.429, 2.54, 1.09, 66.8°   **7.** 0.600, 0.800, 0.750, 1.33, 1.25, 1.67, 36.9°
9. 0.9558, 0.2940, 3.2506, 0.3076, 3.4009, 1.0463   **11.** 0.8674, 0.4976, 1.7433, 0.5736, 2.0098, 1.1528   **13.** 34.5°
15. 70.7°   **17.** 65.88°   **19.** $B = 61.5°, a = 2.02, c = 4.23$ **21.** 356, 810   **23.** 473, 35.5°   **25.** 7.27 ft
27. 0.5120   **29.** 1.3175   **31.** 2.1742   **33.** 60.1°   **35.** 46.5°   **37.** 46.5°   **39.** 16.6   **41.** 3.79

43.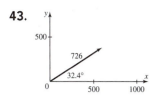

45.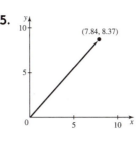

47. 32.2, 57.4°   **49.** $AC = 74.98$ ft, N28°18′E

••• CHAPTER 8 •••

Exercise 1

1. $y^2(3 + y)$ **3.** $x^3(x^2 - 2x + 3)$ **5.** $a(3 + a - 3a^2)$ **7.** $5(x + y)[1 + 3(x + y)]$ **9.** $(1/x)(3 + 2/x - 5/x^2)$
11. $(5m/2n)(1 + 3m/2n - 5m^2/4)$ **13.** $a^2(5b + 6c)$ **15.** $xy(4x + cy + 3y^2)$ **17.** $3ay(a^2 - 2ay + 3y^2)$
19. $cd(5a - 2cd + b)$ **21.** $4x^2(2y^2 + 3z^2)$ **23.** $ab(3a + c - d)$ **25.** $L_0(1 + \alpha t)$ **27.** $R_1[1 + \alpha(t - t_1)]$
29. $t(v_0 + at/2)$ **31.** $(4\pi D/3)(r_2^3 - r_1^3)$

Exercise 2

1. $(2 - x)(2 + x)$ **3.** $(3a - x)(3a + x)$ **5.** $4(x - y)(x + y)$ **7.** $(x - 3y)(x + 3y)$
9. $(3c - 4d)(3c + 4d)$ **11.** $(3y - 1)(3y + 1)$ **13.** $(m - n)(m + n)(m^2 + n^2)$ **15.** $(m^n - n^m)(m^n + n^m)$
17. $(a^2 - b)(a^2 + b)(a^4 + b^2)(a^8 + b^4)$ **19.** $(5x^2 - 4y^3)(5x^2 + 4y^3)$ **21.** $(4a^2 - 11)(4a^2 + 11)$
23. $(5a^2b^2 - 3)(5a^2b^2 + 3)$ **25.** $\pi(r_2 - r_1)(r_2 + r_1)$ **27.** $4\pi(r_1 - r_2)(r_1 + r_2)$ **29.** $m(v_1 - v_2)(v_1 + v_2)/2$
31. $\pi h(R - r)(R + r)$

Exercise 3

1. Not factorable   **3.** Factorable   **5.** Factorable   **7.** $(x - 7)(x - 3)$ **9.** $(x - 9)(x - 1)$
11. $(x + 10)(x - 3)$ **13.** $(x + 4)(x + 3)$ **15.** $(x - 7)(x + 3)$ **17.** $(x + 4)(x + 2)$ **19.** $(b - 5)(b - 3)$
21. $(b - 4)(b + 3)$ **23.** $2(y - 10)(y - 3)$ **25.** $3(w + 4)(w + 8)$ **27.** $(x + 7y)(x + 12y)$
29. $(a + 3b)(a - 2b)$ **31.** $(a + 9x)(a - 7x)$ **33.** $(a - 24bc)(a + 4bc)$ **35.** $(a + 48bc)(a + bc)$
37. $(a + b + 1)(a + b - 8)$ **39.** $(x^2 - 5y^2)(x - 2y)(x + 2y)$ **41.** $(t - 2)(t - 12)$ **43.** $(x - 10)(x - 1)$

Exercise 4

1. $(a^2 + 4)(a + 3)$ **3.** $(x - 1)(x^2 + 1)$ **5.** $(x - b)(x + 3)$ **7.** $(x - 2)(3 + y)$ **9.** $(x + y - 2)(x + y + 2)$
11. $(m - n + 2)(m + n - 2)$ **13.** $(x + 3)(a + b)$ **15.** $(a + b)(a - c)$ **17.** $(2 + x^2)(a + b)$
19. $(2a - c)(3a + b)$ **21.** $(a + b)(b - c)$ **23.** $(x - a - b)(x + a + b)$

Exercise 5

1. $(4x - 1)(x - 3)$ **3.** $(5x + 1)(x + 2)$ **5.** $(3b + 2)(4b - 3)$ **7.** $(2a - 3)(a + 2)$ **9.** $(x - 7)(5x - 3)$
11. $3(x + 1)(x + 1)$ **13.** $(3x + 2)(x - 1)$ **15.** $2(2x - 3)(x - 1)$ **17.** $(2a - 1)(2a + 3)$
19. $(3a - 7)(3a + 2)$ **21.** $(7x^3 - 3y)(7x^3 + 5y)$ **23.** $(4x^3 + 1)(x^3 + 3)$ **25.** $(5x^{2n} + 1)(x^{2n} + 2)$
27. $3[(a + x)^n + 1][(a + x)^n - 2]$ **29.** $(2t - 9)(8t - 5)$

Exercise 6

1. $(x + 2)^2$ **3.** $(y - 1)^2$ **5.** $2(y - 3)^2$ **7.** $(3 + x)^2$ **9.** $(3x + 1)^2$ **11.** $9(y - 1)^2$ **13.** $4(2 + a)^2$
15. $(7a - 2)^2$ **17.** $(x + y)^2$ **19.** $(aw + b)^2$ **21.** $(x + 5a)^2$ **23.** $(x + 4y)^2$ **25.** $(z^3 + 8)^2$
27. $(7 - x^3)^2$ **29.** $(a - b)^2(a + b)^2$ **31.** $(2a^n + 3b^n)^2$

Exercise 7

1. $(4 + x)(16 - 4x + x^2)$ **3.** $2(a - 2)(a^2 + 2a + 4)$ **5.** $(x - 1)(x^2 + x + 1)$ **7.** $(x + 1)(x^2 - x + 1)$
9. $(a + 4)(a^2 - 4a + 16)$ **11.** $(x + 5)(x^2 - 5x + 25)$ **13.** $8(3 - a)(9 + 3a + a^2)$

15. $(7 + 4x)(49 - 28x + 16x^2)$ **17.** $(y^3 + 4x)(y^6 - 4xy^3 + 16x^2)$ **19.** $(3x^5 + 2a^2)(9x^{10} - 6x^5a^2 + 4a^4)$
21. $(2a^{2x} - 5b^x)(4a^{4x} + 10a^{2x}b^x + 25b^{2x})$ **23.** $(4x^n - y^{3n})(16x^{2n} + 4x^ny^{3n} + y^{6n})$ **25.** $(4\pi/3)(r_2 - r_1)(r_2^2 + r_2r_1 + r_1^2)$

Review Problems

1. $(x - 5)(x + 3)$ **3.** $(x^3 - y^2)(x^3 + y^2)$ **5.** $(2x - 1)(x + 2)$
7. $(a - b + c + d)[(a - b)^2 - (a - b)(c + d) + (c + d)^2]$ **9.** $(x/y)(x - 1)$ **11.** $(y + 5)(x - 2)$
13. $(x - y)(1 - b)$ **15.** $(y + 2 - z)(y + 2 + z)$ **17.** $(x - 3)(x - 4)$
19. $(4x^{2n} + 9y^{4n})(2x^n + 3y^{2n})(2x^n - 3y^{2n})$ **21.** $(3a + 2z^2)^2$ **23.** $(x^m + 1)^2$ **25.** $(x - y - z)(x - y + z)$
27. $(1 + 4x)(1 - 4x)$ **29.** $x^2(3x + 1)(3x - 1)$ **31.** $(p - q - 3)[(p - q)^2 + 3(p - q) + 9]$
33. $(3a - 2w)(9a^2 + 6aw + 4w^2)$ **35.** $4(2x - y)^2$ **37.** $(3b + y)(2a + x)$ **39.** $2(y^2 - 3)(y^2 + 3)$
41. $(1/R)(V_2 - V_1)(V_2 + V_1)$

◆◆◆ CHAPTER 9 ◆◆◆

Exercise 1

1. $x \neq 0$ **3.** $x \neq 5$ **5.** $x \neq 2, x \neq 1$ **7.** 0.5833   **9.** 0.9375   **11.** 3.6667   **13.** 7/16   **15.** 11/16

17. 7/9   **19.** -1 **21.** $d - c$ **23.** $\dfrac{2}{3}$ **25.** $\dfrac{15}{7}$ **27.** $a/3$ **29.** $3m/4p^2$ **31.** $(2a - 3b)/2a$

33. $x/(x - 1)$ **35.** $\dfrac{x + 2}{x^2 + 2x + 4}$ **37.** $\dfrac{n(m - 4)}{3(m - 2)}$ **39.** $\dfrac{2(a + 1)}{a - 1}$ **41.** $\dfrac{x + z}{x^2 + xz + z^2}$ **43.** $\dfrac{a - 2}{a - 3}$

45. $(x - 1)/2y$ **47.** $\dfrac{2}{3(x^4y^4 - 1)}$ **49.** $\dfrac{x + y}{x - y}$ **51.** $(b - 3 + a)/5$

Exercise 2

1. 2/15   **3.** 6/7   **5.** $2\frac{2}{15}$ **7.** 75/8   **9.** $1\frac{1}{2}$ **11.** $a^4b^4/2y^{2n}$ **13.** $7p^2/4xz$ **15.** $x - a$

17. $ab/(x^2 - y^2)$ **19.** $cd/(x^2 - y^2)^2$ **21.** $\dfrac{(x - 1)(x + 4)}{(x + 3)(x - 5)}$ **23.** $\dfrac{(x + 1)(x + 2)}{(x - 4)(x - 3)}$ **25.** $\dfrac{(2x - 3)(x + 1)}{(x - 1)(x - 6)}$

27. $1\frac{15}{16}$ **29.** 9/128   **31.** 5/18   **33.** $5\frac{1}{4}$ **35.** b^2 **37.** $8a^3/3dxy$ **39.** $(3an + cm)/(x^4 - y^4)$

41. $1/(c - d)$ **43.** $1/(x + 2)$ **45.** $\dfrac{6(x - 1)}{x + 2}$ **47.** $\dfrac{(p + 2)^2}{(p - 1)^2}$ **49.** $\dfrac{(z - 1)(z + 4)}{(z + 3)(z - 5)}$ **51.** $1\frac{2}{3}$

53. $2\frac{5}{12}$ **55.** $9\frac{2}{5}$ **57.** $x + 1/x$ **59.** $\dfrac{2}{x} - \dfrac{1}{2}$ **61.** $3m/4\pi r^3$ **63.** $2P/(a + b)h$ **65.** $4\pi d^3/3$

Exercise 3

1. 1   **3.** 1/7   **5.** 5/3   **7.** 7/6   **9.** 19/16   **11.** 7/18   **13.** 13/5   **15.** $-92/15$
17. 67/35   **19.** 8/3   **21.** 13/4   **23.** 91/16   **25.** 6/a   **27.** $(a + 3)/y$ **29.** x **31.** $x/(a - b)$

33. $19/(a + 1)$ **35.** $19a/10x$ **37.** $\dfrac{2ax - 3x + 12}{6x}$ **39.** $(5b - a)/6$ **41.** $\dfrac{9 - x}{x^2 - 1}$ **43.** $\dfrac{2b(2a - b)}{a^2 - b^2}$

45. $\dfrac{2x^2 - 1}{x^4 - x^2}$ **47.** $\dfrac{2(3x^2 + 2x - 3)}{x^3 - 7x - 6}$ **49.** $\dfrac{x^2 - 2x}{(x - 3)(x + 3)}$ **51.** $\dfrac{5x^2 + 7x + 16}{x^3 - 8}$ **53.** $\dfrac{-d}{(x + 1)(x + d + 1)}$

55. $\dfrac{d(x^2 + dx - 1)}{x(x + d)}$ **57.** $(x^2 + 1)/x$ **59.** $2/(x^2 - 1)$ **61.** $(3a - a^2 - 2)/a$ **63.** $(abx + a + b)/ax$

65. $4\frac{7}{12}$ mi   **67.** 2/25 min   **69.** 11/8 machines   **71.** $\dfrac{2h(a + b) - \pi d^2}{4}$ **73.** $\dfrac{V_1V_2d + VV_2d_1 + VV_1d_2}{VV_1V_2}$

Exercise 4

1. 85/12 **3.** 65/132 **5.** 69/95 **7.** $\dfrac{3(4x + y)}{4(3x - y)}$ **9.** $y/(y - x)$ **11.** $\dfrac{5(3a^2 + x)}{3(20 + x)}$ **13.** $\dfrac{2(3acx + 2d)}{3(2acx + 3d)}$

15. $\dfrac{2x^2 - y^2}{x - 3y}$ **17.** $\dfrac{(x + 2)(x - 1)}{(x + 1)(x - 2)}$ **19.** 1 **21.** $\dfrac{4}{3(a + 1)}$ **23.** 1 **25.** $\dfrac{x + y + z}{x - y + z}$ **27.** $\dfrac{\rho L_1 L_2}{A_2 L_1 + A_1 L_2}$

Exercise 5

1. 12 **3.** 20 **5.** 14 **7.** 72 **9.** 24 **11.** 7 **13.** 24 **15.** 24 **17.** $\frac{5}{2}$ **19.** 12 **21.** 17
23. $\frac{3}{13}$ **25.** $-\frac{2}{3}$ **27.** $-\frac{11}{4}$ **29.** $\frac{1}{2}$ **31.** No solution **33.** -2 **35.** 4 **37.** 8 **39.** 2

Exercise 6

1. 150 **3.** 5 and 6 **5.** 15 and 35 **7.** 26 h **9.** 240 mi **11.** 2 days **13.** $5\frac{5}{6}$ days **15.** 2.4 h
17. 4.2 h **19.** 12.4 h **21.** 7.8 weeks **23.** 5.6 winters **25.** 6.1 h **27.** $11,360 **29.** $1500, $2000

Exercise 7

1. $bc/2a$ **3.** $\dfrac{bz - ay}{a - b}$ **5.** $\dfrac{a^2 d + 3d^2}{4ac + d^2}$ **7.** $\dfrac{b + cd}{a^2 + a - d}$ **9.** $2z/a + 3w$ **11.** $(b - m)/3$

13. $(c - m)/(a - b - d)$ **15.** b **17.** $\dfrac{2c + bc + 3b}{3 - 2b + c}$ **19.** $\dfrac{w}{w(w + y) - 1}$ **21.** $(p - q)/3p$

23. $(5b - 2a)/3$ **25.** $\dfrac{a^2(c + 1)(c - a)}{c^2}$ **27.** $\dfrac{ab}{a + b}$ **29.** $\dfrac{ab}{b - a}$ **31.** $md/(m + n)$ and $nd/(m + n)$

33. $\dfrac{anm}{nm - m - n}$ **35.** $\dfrac{L - L_0}{L_0 \alpha}$ **37.** PL/Ee **39.** $(kAt_1 - qL)/kA$ **41.** $\dfrac{RR_1}{R_1 - R}$ **43.** $\dfrac{R_1(1 + \alpha t) - R}{R_1 \alpha}$

45. $\dfrac{M - Fx}{R_1 - F}$ **47.** $\dfrac{m_2(25 - x)}{x - 10}$ **49.** $E/(gy + \frac{1}{2}v^2)$

Review Problems

1. $2ab/(a - b)$ **3.** $1/(3m - 1)$ **5.** $a^2 + 1 + 1/a^2$ **7.** a/c **9.** $4/(4 - x^2)$ **11.** 0

13. $\dfrac{a^2 + ab + b^2}{a + b}$ **15.** $(2a - 3c)/2a$ **17.** $3a/(a + 2)$ **19.** $\dfrac{x - y - z}{x + y - z}$ **21.** $\dfrac{a + b - c - d}{a - b + c - d}$

23. $(b + 5)/(c - a)$ **25.** 1 **27.** 11 **29.** $(r - pq)/(p - q)$ **31.** $q(p + 1)/(p^2 + r)$
33. 7 **35.** 22 **37.** -5 **39.** 3 **41.** 6.35 days **43.** $19\frac{29}{45}$ **45.** $2ax^2 y/7w$ **47.** $14\frac{15}{28}$

◆◆◆ **CHAPTER 10** ◆◆◆

Exercise 1

1. $(2, -1)$ **3.** $(1, 2)$ **5.** $(-0.24, 0.90)$ **7.** $(3, 5)$ **9.** $(-\frac{3}{4}, 3)$ **11.** $(-3, 3)$ **13.** $(1, 2)$
15. $(3, 2)$ **17.** $(15, 6)$ **19.** $(3, 4)$ **21.** $(2, 3)$ **23.** $(1.63, 0.0971)$ **25.** $m = 2, n = 3$
27. $w = 6, z = 1$ **29.** $(9.36, 4.69)$ **31.** $v = 1.05, w = 2.58$

Exercise 2

1. $(60, 36)$ **3.** $(87/7, 108/7)$ **5.** $(15, 12)$ **7.** $m = 4, n = 3$ **9.** $r = 3.77, s = 1.23$ **11.** $(1/2, 1/3)$
13. $(1/3, 1/2)$ **15.** $(1/10, 1/12)$ **17.** $w = 1/36, z = 1/60$

21.   **23.**   **25.** $\left(\dfrac{bs - cr}{a(b - c)}, \dfrac{r - s}{b - c}\right)$ **27.** $(-3q, -2p)$

Exercise 3

1. 8 and 16   **3.** 4/21   **5.** 82   **7.** 7 and 41   **9.** 9 × 16   **11.** $x = 2.10$ mi, $y = 0.477$ mi
13. 7.00 ft/s and 5.25 ft/s   **15.** 5.3 mi/h and 1.3 mi/h   **17.** $v_0 = 0.381$ cm/s, $a = 3.62$ cm/s^2 **19.** \$2500 at 4%
21. \$2684 at 6.2% and \$1716 at 9.7%   **23.** \$6000 for 2 years   **25.** 2740 lb mixture, 271 lb sand
27. 5.57 lb peat, 11.6 lb vermiculite   **29.** $T_1 = 490$ lb, $T_2 = 185$ lb   **31.** Carpenter: 25.0 days, helper: 37.5 days
33. 18,000 gal/h and 12,000 gal/h   **35.** 92,700 people and 162,000 people   **37.** $I_1 = 13.1$ mA, $I_2 = 22.4$ mA
39. $R_1 = 27.6\ \Omega$, $\alpha = 0.00511$

Exercise 4

1. (15, 20, 25)   **3.** (1, 2, 3)   **5.** (5, 6, 7)   **7.** (1, 2, 3)   **9.** (3, 4, 5)   **11.** (6.18, 13.7, 18.1)
13. (2, 3, 1)   **15.** (1/2, 1/3, 1/4)   **17.** $(1, -1, 1/2)$ **19.** $(a/11, 5a/11, 7a/11)$
21. $[(a + b - c), (a - b + c), (-a + b + c)]$ **23.** 876   **25.** 361   **27.** 6.38 mA, 0.275 mA, -2.53 mA
29. $F_1 = 426$ N, $F_2 = 1080$ N, $F_3 = 362$ N

Review Problems

1. (3, 5)   **3.** $(5, -2)$ **5.** $(2, -1, 1)$ **7.** $\left(\dfrac{-9}{5}, \dfrac{54}{5}, \dfrac{-21}{5}\right)$ **9.** (8, 10)   **11.** (2, 3)

13. $[(a + b - c)/2, (a - b + c)/2, (b - a + c)/2]$ **15.** (7, 5)   **17.** (6, 8, 10)   **19.** (2, 3, 1)
21. (13, 17)   **23.** (2, 9)   **25.** 8 days

◆◆◆ CHAPTER 11 ◆◆

Exercise 1

1. -14 **3.** 15   **5.** -27 **7.** 17.3   **9.** $-2/5$ **11.** $ad - bc$ **13.** $\sin^2\theta - 3\tan\theta$
15. $3i_1\cos\theta - i_2\sin\theta$ **17.** (3, 5)   **19.** $(-3, 3)$ **21.** (1, 2)   **23.** (3, 2)   **25.** (15, 6)   **27.** (3, 4)
29. (2, 3)   **31.** (1.63, 0.0971)   **33.** $m = 2, n = 3$ **35.** $w = 6, z = 1$ **37.** $(-6/13, 30/13)$

39. $\left(\dfrac{dp - bq}{ad - bc}, \dfrac{aq - cp}{ad - bc}\right)$ **41.** (3, 4)   **43.** \$31,550, \$22,986   **45.** $x = 90.1$ kg, $y = 8.98$ kg

Exercise 2

1. 11   **3.** 45   **5.** 48   **7.** -28 **9.** (5, 6, 7)   **11.** (15, 20, 25)   **13.** (1, 2, 3)   **15.** (3, 4, 5)
17. (6, 8, 10)   **19.** $(-2.30, 4.80, 3.09)$ **21.** (3, 6, 9)   **23.** $(\tfrac{1}{2}, \tfrac{1}{3}, \tfrac{1}{4})$ **25.** 1.54, -0.377, -1.15

Exercise 3

1. 2   **3.** 18   **5.** -66 **7.** $x = 2, y = 3, z = 4, w = 5$ **9.** $x = -4, y = -3, z = 2, w = 5$
11. $x = a - c, y = b + c, z = 0, w = a - b$ **13.** $x = 4, y = 5, z = 6, w = 7, u = 8$
15. $v = 3, w = 2, x = 4, y = 5, z = 6$ **21.** $-4.31, 7.57, 6.27, 3.71$ **23.** 102 kg, 98.0 kg, 218 kg, 182 kg

Review Problems

1. 20   **3.** 0   **5.** -3 **7.** 0   **9.** 18   **11.** 15   **13.** 0   **15.** -29 **17.** 133
19. $x = -4, y = -3, z = 2, w = 1$ **21.** (3, 5)   **23.** (4, 3)   **25.** (5, 1)   **27.** (2, 5)   **29.** (2, 1)
31. (3, 2)   **33.** (9/7, 37/7)   **35.** (3.19, 1.55)   **37.** $(-15.1, 3.52, 11.4)$
39. $0.558, -1.34, -0.240, -0.667$ **41.** $x = -0.972, y = 2.28, z = 1.38, u = -0.385, v = 2.48$

◆◆◆ CHAPTER 12

Exercise 1

1. A, B, D, E, F, I   **3.** C, H   **5.** I   **7.** B, I   **9.** F   **11.** 6   **13.** 4×3 **15.** 2×4

17. $\begin{pmatrix} 0 & 0 \\ 0 & 0 \\ 0 & 0 \\ 0 & 0 \end{pmatrix}$ **19.** If $y = 6$, $x = 2$, $w = 7$, and $z = 8$

Exercise 2

1. (12 13 5 8)   **3.** $\begin{pmatrix} 14 \\ -10 \\ 6 \\ 14 \end{pmatrix}$ **5.** $\begin{pmatrix} 7 & 1 & -4 & -7 \\ -9 & -4 & 9 & 2 \\ 3 & 7 & -3 & -9 \\ -1 & -4 & -5 & 6 \end{pmatrix}$ **7.** $(-15 \ -27 \ 6 \ -21 \ -9)$

9. $\begin{pmatrix} -12 & -6 & 9 & 0 \\ 3 & -18 & -9 & 12 \end{pmatrix}$ **11.** $7\begin{pmatrix} 3 & 1 & -2 \\ 7 & 9 & 4 \\ 2 & -3 & 8 \\ -6 & 10 & 12 \end{pmatrix}$ **13.** $\begin{pmatrix} 6 & 16 & 3 \\ -16 & 8 & 19 \\ 31 & -14 & 15 \end{pmatrix}$ **17.** 35   **19.** -61

21. $(10 \ 27 \ -14 \ 15 \ -7)$ **23.** $(-60 \ 18 \ 10)$ **25.** $(15 \ 2 \ 40 \ 2 \ 34)$ **27.** $\begin{pmatrix} 19 & -13 \\ 17 & -43 \end{pmatrix}$

29. $\begin{pmatrix} -20 & 10 \\ 8 & -2 \\ -4 & 16 \end{pmatrix}$ **31.** $\begin{pmatrix} 6 & -10 \\ 3 & -5 \end{pmatrix}$ **33.** $\begin{pmatrix} 0 & 2 \\ 2 & 0 \\ -10 & -6 \\ -5 & 0 \end{pmatrix}$ **35.** $\begin{pmatrix} 2 & 0 & -2 & 1 \\ 8 & 8 & -1 & 2 \\ 2 & 2 & 2 & 6 \end{pmatrix}$

37. $AB = \begin{pmatrix} 1 & 2 & -1 \\ -2 & 9 & 2 \\ -1 & -2 & 1 \end{pmatrix}$ $BA = \begin{pmatrix} 5 & -2 & 2 \\ 4 & -2 & 0 \\ -2 & 3 & 8 \end{pmatrix}$ **39.** $\begin{pmatrix} 778 & 1050 \\ 1225 & 1653 \end{pmatrix}$ **41.** $\begin{pmatrix} 5 \\ 1 \end{pmatrix}$ **43.** $\begin{pmatrix} 13 \\ -1 \\ 10 \end{pmatrix}$

45. \$1,305   **47.** 29, 70.3, 22.5, 19.9   **49.** (a) $\begin{pmatrix} 67 & 44 & 41 \\ 74 & 59 & 75 \end{pmatrix}$ (b) $\begin{pmatrix} 3 & -8 & 3 \\ -8 & -7 & -11 \end{pmatrix}$ (c) MP $= \begin{pmatrix} 802.26 \\ 1233.58 \end{pmatrix}$

TP $= \begin{pmatrix} 782.05 \\ 957.79 \end{pmatrix}$

Exercise 3

1. $\begin{pmatrix} 0 & -0.200 \\ 0.125 & 0.100 \end{pmatrix}$ **3.** $\begin{pmatrix} 0.118 & 0.059 \\ -0.020 & 0.157 \end{pmatrix}$ **5.** $\begin{pmatrix} 0.074 & 0.168 & 0.057 \\ 0.148 & -0.164 & 0.115 \\ 0.090 & 0.094 & -0.041 \end{pmatrix}$

7. $\begin{pmatrix} 0.500 & 0 & -0.500 & 0 \\ 10.273 & -0.545 & -13.091 & 1.364 \\ -4.727 & 0.455 & 5.909 & -0.636 \\ -4.591 & 0.182 & 5.864 & -0.455 \end{pmatrix}$

Review Problems

1. $(2 \ 12 \ 10 \ -2)$ **3.** 86   **5.** $\begin{pmatrix} 0.063 & 0.250 \\ -0.188 & 0.250 \end{pmatrix}$ **7.** $\begin{pmatrix} 7 & 2 & 6 \\ 9 & 0 & 3 \end{pmatrix}$ **9.** $x = -4$, $y = -3$, $z = 2$, $w = 1$

11. $(2, -1, 1)$ **13.** $(3, 5)$ **15.** -0.557 mA, 0.188 mA, -0.973 mA, -0.271 mA   **17.** \$101,395, \$124,281

◆◆◆ CHAPTER 13

Exercise 1

1. $3/x$ **3.** 2   **5.** $a/4b^2$ **7.** a^3/b^2 **9.** p^3/q **11.** 1   **13.** $1/(16a^6b^4c^{12})$ **15.** $1/(x + y)$

17. $m^4/(1 - 6m^2n)^2$ **19.** $2/x + 1/y^2$ **21.** $1/(3m)^3 - 2/n^2$ **23.** $x^{2n} + 2x^n y^m + y^{2m}$ **25.** $x/2y^7$ **27.** p
29. x^{m^2} **31.** $b/3$ **33.** $9y^2/4x^2$ **35.** $27q^3y^3/8p^3x^3$ **37.** $9a^8b^6/25x^4y^2$ **39.** $4^n a^{6n} x^{6n}/9^n b^{4n} y^{2n}$

41. $3p^2/2q^2x^4z^3$ **43.** $5^p w^{2p}/2^p z^p$ **45.** $25n^4x^8/9m^6y^6$ **47.** $\dfrac{(1 + a)^2(b - 1)}{(1 - a + a^2)(b + 1)^2}$ **49.** $x^7(x - y)^2$

51. $9a^{2n} + 6a^n b^n + 4b^{2n}$ **53.** $\dfrac{(x + 1)^2(y - 1)}{(y + 1)^2(x^2 - x + 1)}$ **55.** $(7/15)a^{n-m+2}x^{2-n}y^{n-3}$ **57.** $a^3(5a^2 - 16b^2)$

59. $(a^{4-m}b^{n-3})/5$ **61.** $x^3(x^2 + y^2)$ **63.** $(a + b)^4/(a^2 + b^2)^2$ **65.** $R = PI^{-2}$ **67.** $2I^2R$

Exercise 2

1. $\sqrt[4]{a}$ **3.** $\sqrt[3]{z^3}$ **5.** $\sqrt{m - n}$ **7.** $\sqrt[3]{y/x}$ **9.** $b^{1/2}$ **11.** y **13.** $(a + h)^{1/n}$ **15.** xy **17.** $3\sqrt{2}$
19. $3\sqrt{7}$ **21.** $-2\sqrt[3]{7}$ **23.** $a\sqrt{a}$ **25.** $6x\sqrt{y}$ **27.** $2x^2\sqrt[3]{2y}$ **29.** $6y^2\sqrt[5]{xy}$ **31.** $a\sqrt{a - b}$
33. $3\sqrt{m^3 + 2n}$ **35.** $x\sqrt[3]{x - a^2}$ **37.** $(a - b)(a + b)\sqrt{a}$ **39.** $\sqrt{21}/7$ **41.** $\sqrt[3]{2}/2$ **43.** $\sqrt[3]{6}/3$
45. $\sqrt{2x}/2x$ **47.** $a\sqrt{15ab}/5b$ **49.** $\sqrt[3]{x}/x$ **51.** $x\sqrt[3]{18}/3$ **53.** $(x - y)\sqrt{x^2 + xy}$

55. $\dfrac{(m + n)\sqrt{m^2 - mn}}{m - n}$ **57.** $\dfrac{(2a - 6)\sqrt{6a}}{3a}$ **59.** $Z = (R^2 + X^2)^{1/2}$ **61.** $c = a\sqrt{10}$

63. $S = 4t\sqrt{16t^2 - 120t + 325}$

Exercise 3

1. $\sqrt{6}$ **3.** $\sqrt{6}$ **5.** $25\sqrt{2}$ **7.** $-7\sqrt[3]{2}$ **9.** $-5\sqrt[3]{5}$ **11.** $-\sqrt[4]{3}$ **13.** $10x\sqrt{2y}$ **15.** $11\sqrt{6}/10$
17. $(a + b)\sqrt{x}$ **19.** $(x - 2a)\sqrt{y}$ **21.** $-4a\sqrt{3x}$ **23.** $(5a + 3c)\sqrt[3]{10b}$ **25.** $(a^2b^2c^2 - 2a + bc)\sqrt[5]{a^3bc^2}$

27. $6\sqrt{2x}$ **29.** $\dfrac{25}{18y}\sqrt{3x}$ **31.** 45 **33.** $16\sqrt{5}$ **35.** $24\sqrt[3]{5}$ **37.** $6\sqrt[6]{72}$ **39.** $10\sqrt[6]{72}$

41. $6x\sqrt{3ay}$ **43.** $2\sqrt[5]{x^3y^3}$ **45.** $\sqrt[4]{a^2b}$ **47.** $2xy\sqrt[6]{x^4yz^3}$ **49.** $2\sqrt{15} - 6$ **51.** $x^2\sqrt{1 - xy}$
53. $a^2 - b$ **55.** $16x - 12\sqrt{xy} - 10y$ **57.** $9y$ **59.** $9x^3\sqrt[3]{4x}$ **61.** $9 - 30\sqrt{a} + 25a$ **63.** $250x\sqrt{2x}$
65. $250ax$ **67.** $3/4$ **69.** $\sqrt[6]{18}/4$ **71.** $3\sqrt[3]{2a}/a$ **73.** $4\sqrt[3]{18}/3 + \sqrt[3]{4} + 3$ **75.** $3/2$

77. $\dfrac{2\sqrt[6]{a^5b^2c^3}}{ac}$ **79.** $\dfrac{1}{2c}\sqrt[12]{a^2bc^{10}}$ **81.** $\dfrac{8 + 5\sqrt{2}}{2}$ **83.** $4\sqrt{ax}/a$ **85.** $6\sqrt[3]{2x}/x$ **87.** $-\dfrac{30 + 17\sqrt{30}}{28}$

89. $\dfrac{a^2 - a\sqrt{b}}{a^2 - b}$ **91.** $\dfrac{x - 2\sqrt{xy} + y}{x - y}$ **93.** $\dfrac{5\sqrt{3x} - 5\sqrt{xy} + \sqrt{6xy} - \sqrt{2y}}{6x - 2y}$

95. $\dfrac{3\sqrt{6a} - 4\sqrt{3ab} + 6\sqrt{10ab} - 8\sqrt{5b}}{9a - 8b}$

Exercise 4

1. 36 **3.** 4 **5.** 13.8 **7.** 8 **9.** 8 **11.** 7 **13.** 4.79 **15.** 3 **17.** 8 **19.** 0.347
21. No solution   **23.** 7 **25.** 14.1 cm, 29.0 m   **27.** 3.91 m, 4.77 m   **29.** $C = 1/(\omega^2 L \pm \omega\sqrt{Z^2 - R^2})$

Review Problems

1. $2\sqrt{13}$ **3.** $3\sqrt[3]{6}$ **5.** $\sqrt[3]{2}$ **7.** $9x\sqrt[4]{x}$ **9.** $(a - b)x\sqrt[3]{(a - b)^2x}$ **11.** $\dfrac{\sqrt{a^2 - 4}}{a + 2}$ **13.** $x^2\sqrt{1 - xy}$

15. $-\dfrac{1}{46}(6 + 15\sqrt{2} - 4\sqrt{3} - 10\sqrt{6})$ **17.** $\dfrac{2x^2 + 2x\sqrt{x^2 - y^2} - y^2}{y^2}$ **19.** $26\sqrt[3]{x^2}$ **21.** $\sqrt[6]{a^3b^2}$

23. $9 + 12\sqrt{x} + 4x$ **25.** $7\sqrt{2}$ **27.** $25\sqrt{2}/12$ **29.** $\sqrt{2b}/2b$ **31.** $72\sqrt{2}$ **33.** 10 **35.** 14
37. 15.2 **39.** $x^{2n-1} + (xy)^{n-1} + x^n y^{n-2} + y^{2n-3}$ **41.** $27x^3y^6/8$ **43.** $8x^9y^6$ **45.** $3/w^2$ **47.** $1/3x$
49. $1/x - 2/y^2$ **51.** $1/9x^2 + y^8/4x^4$ **53.** $p^{2a-1} + (pq)^{a-1} + p^a q^{a-2} + q^{2a-3}$ **55.** p^2/q **57.** r^2/s^3
59. 1 **61.** $V = 36\pi r^3$ **63.** $B = \sqrt{A}\sqrt{C}$

••• CHAPTER 14 ••

Exercise 1

1. 0, 2/5   **3.** 0, 9/2   **5.** ±3   **7.** ±1   **9.** 0, 0.355   **11.** 3, −5   **13.** 5, −4   **15.** 2, −1
17. 5/2, −1   **19.** 3/2, −4   **21.** 1/5, −3   **23.** 1/2, −2/3   **25.** 9, −4   **27.** 1, −1/6   **29.** −a, −3a
31. −2b, − 4b/3   **33.** $x^2 − 11x + 28 = 0$ **35.** $15x^2 − 19x + 6 = 0$ **37.** 6, 15/4
39. 4 (−21 doesn't check)

Exercise 2

1. $4 ± \sqrt{14}$ **3.** $(−7 ± \sqrt{61})/2$ **5.** $(3 ± \sqrt{89})/8$ **7.** $(−3 ± \sqrt{21})/3$ **9.** $(−7 ± \sqrt{129})/8$
11. 0.485, −6.74

Exercise 3

1. 3.17, 8.83   **3.** 3.89, −4.87   **5.** 1.96, −5.96   **7.** 0.401, −0.485   **9.** 0.170, −0.599   **11.** 2.87, 0.465
13. 0.907, −2.57   **15.** 5.87, −1.87   **17.** 12.0, −14.0   **19.** 5.08, 10.9   **21.** 2.81, −1.61   **23.** 8.37, −2.10
25. 0.132, −15.1   **27.** 2.31, 0.623   **29.** Real, unequal   **31.** Not real   **33.** Not real

Exercise 4

1. 2/3 or 3/2   **3.** 4 and 11   **5.** 5 and 15   **7.** 4 × 6 m   **9.** 15 × 30 cm   **11.** 162 × 20$\overline{0}$ m
13. 2.26 in.   **15.** 6.47 in.   **17.** 146 mi/h   **19.** 8$\overline{0}$ km/h   **21.** 9 h and 12 h   **23.** 7 m/day
25. At each end   **27.** 0.630 s and 8.38 s   **29.** 125 Ω and 655 Ω   **31.** 0.381 A and 0.769 A
33. −0.5 A and 0.1 A   **35.** −1.1 V

Exercise 5

1.   **3.**   **7.**   **9.** 20 cm   **11.** 6.33 ft

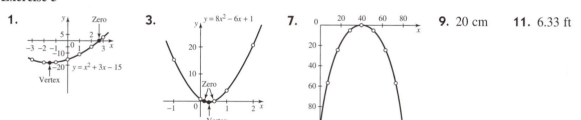

Exercise 6

1. $\sqrt[3]{2}$, $\sqrt[3]{4}$ **3.** 1   **5.** 1 (0 doesn't check)   **7.** −1, −1/27   **9.** 1, −64   **11.** $\sqrt[3]{2}$/8 (8.69 doesn't check)
13. 625 (256 doesn't check)   **15.** 496 (243 doesn't check)

Exercise 7

1. 3, −5   **3.** 2, −1   **5.** 3/2, −4   **7.** 4/3, −8/3   **13.** −0.930, 0.157, 0.773   **15.** −0.961, 0.293, 0.668
17. 0.629, 3.32

Exercise 8

1. (6, 2)   **3.** (3, 2), (−5, 6)   **5.** (2, 3), (−46, 15)   **7.** (2.61, −1.20), (−2.61, −1.20)
9. (4.86, 0.845), (4.86, −0.845)   **11.** (4, −7), (7, −4)   **13.** (2.04, 2.01), (−0.838, −2.57)
15. (1.26, 1.07), (0.491, −0.381)

Review Problems

1. 6, −1   **3.** ±2, ±3   **5.** 0, 5   **7.** 0, −2   **9.** 5, −2/3   **11.** 1, 1/2   **13.** ±3   **15.** 1.79, −2.79
17. 4, 9   **19.** 0.692, −0.803   **21.** ±$\sqrt{10}$ **23.** 1/2, −2   **25.** 0, 9   **27.** 1, 512   **29.** ±5
31. 3.54, −2.64   **33.** 0.777, −2.49   **35.** 0.716, −2.32   **37.** 1.49, −1.26   **39.** (7.41, −0.41), (−0.41, 7.41)
41. 2, 4, −3   **43.** (1.68, 1.16), (−2.68, 3.34)   **45.** 202 bags   **47.** 15 × 3$\overline{0}$ ft   **49.** 3 mi/h   **51.** 1$\frac{1}{2}$ s
53. 5.0 × 16 in. or 8.0 × 1$\overline{0}$ in.   **55.** 3.56 km/h   **57.** 12 ft × 12 ft

••• CHAPTER 15 •••

Exercise 1

1. II **3.** IV **5.** I **7.** II **9.** II or III **11.** IV **13.** III **15.** pos **17.** pos

19. pos **21.** neg

	sin	cos	tan		r	sin	cos	tan	cot	sec	csc
23.	+	−	−	**29.**	12.65	0.949	−0.316	−3.00	−0.333	−3.16	1.05
25.	−	+	−	**31.**	17.0	−0.471	−0.882	0.533	1.88	−1.13	−2.13
27.	+	−	−	**33.**	5.00	0	1.00	0	—	1.00	—

	sin	cos	tan		sin	cos	tan		sin	cos	tan
35.	−3/5	−4/5	3/4	**41.**	0.9816	−0.1908	−5.145	**47.**	0.8090	−0.5878	−1.376
37.	$-2/\sqrt{5}$	$-1/\sqrt{5}$	2	**43.**	−0.4848	0.8746	−0.5543	**49.**	0.9108	−0.4128	−2.206
39.	2/3	$-\sqrt{5}/3$	$-2/\sqrt{5}$	**45.**	−0.8898	0.4563	−1.950	**51.**	−0.2264	−0.9740	0.2324

	sin	cos	tan
53.	−0.1959	0.9806	−0.1998
55.	−0.8480	−0.5299	1.600

57. 0.8192 **59.** 1.906 **61.** 1.711 **63.** 1.111 **65.** 135°, 315°

67. 33.2°, 326.8° **69.** 81.1°, 261.1° **71.** 90°, 270° **73.** 219.5°, 320.5° **75.** 60.1°, 240.1°

77. 227.4°, 312.6° **79.** 34.2°, 325.8° **81.** 102.6°, 282.6° **83.** 1/2 **85.** 0 **87.** 1 **89.** 0 **91.** 0

93. −1 **95.** 0 **97.** 1/2 **99.** 1 **101.** 0

Exercise 2

	ANGLES			SIDES		
	A	**B**	**C**	**a**	**b**	**c**
1.	32.8°		101°			413
3.	108°	29.2°		29.3		
5.		71.1°			8.20	6.34
7.	104°			21.7		15.7
9.	23.0°			42.8	89.7	
11.	106.4°			90.74		29.55
13.			150°		34.1	82.0

Exercise 3

	ANGLES			SIDES		
	A	**B**	**C**	**a**	**b**	**c**
1.	44.2°	29.8°				21.7
3.		48.7°	63.0°	22.6		
5.	56.9°	95.8°				70.1
7.		25.5°	19.5°	452		
9.	45.6°	80.4°	54.0°			
11.		30.8°	34.2°	82.8		
13.	69.0°	44.0°				8.95
15.		10.3°	11.7°	3.69		
17.	66.21°	72.02°				1052
19.	69.7°	51.7°	58.6°			

Exercise 4

1. 30.8 m and 85.6 m **3.** 32.3°, 60.3°, 87.4° **5.** N48.8° W **7.** S59.4°E **9.** 598 km **11.** 28.3 m

13. 77.3 m, 131 m **15.** 107 ft **17.** 337 mm **19.** 33.7 cm **21.** 73.4 in. **23.** 53.8 mm and 78.2 mm

25. 419 **27.** $A = 45.0°, B = 60.0°, C = 75.0°, AC = 1220, AB = 1360$ **29.** 21.9 ft

Exercise 5

1. 521 at 10.0°   **3.** 87.1 at 31.9°   **5.** 6708 at 41.16°   **7.** 37.3 N at 20.5°
9. 121 N at N 59.8° W   **11.** 1720 N at 29.8° from larger force   **13.** 44.2° and 20.7°
15. Wind: S 37.4° E, plane: S 84.8° W   **17.** 413 km/h at N 41.2° E   **19.** 632 km/h, 3.09°   **21.** 26.9 A, 32.3°
23. 39.8 at 26.2°

Review Problems

1. $A = 24.1°$, $B = 20.9°$, $c = 77.6$ **3.** $A = 61.6°$, $C = 80.0°$, $b = 1.30$ **5.** $B = 20.2°$, $C = 27.8°$, $a = 82.4$
7. IV   **9.** II   **11.** neg   **13.** neg   **15.** neg   **17.** -0.800 -0.600 1.33   **19.** $12\overline{0}0$ at 36.3°
21. 22.1 at 121°   **23.** 1.11 km

	sin	cos	tan
25.	0.0872	-0.9962	-0.0875
27.	0.7948	-0.6069	-1.3095

29. 0.4698   **31.** 1.469   **33.** S 3.5° E   **35.** 130.8°, 310.8°   **37.** 47.5°, 132.5°   **39.** 80.0°, 280.0°
41. 695 lb, 17.0°   **43.** -224 cm/s, 472 cm/s   **45.** 22.42

◆◆◆ CHAPTER 16 ◆◆◆

Exercise 1

1. 0.834 rad   **3.** 0.6149 rad   **5.** 9.74 rad   **7.** 0.279 rev   **9.** 0.497 rev   **11.** 0.178 rev   **13.** 162°
15. 21.3°   **17.** 65.3°   **19.** $\pi/3$ **21.** $11\pi/30$ **23.** $7\pi/10$ **25.** $13\pi/30$ **27.** $20\pi/9$ **29.** $9\pi/20$
31. $22\frac{1}{2}°$ **33.** 147°   **35.** 20°   **37.** 158°   **39.** 24°   **41.** 15°   **43.** 0.8660   **45.** 0.4863
47. -0.3090 **49.** 1.067   **51.** -2.747 **53.** -0.8090 **55.** -0.2582 **57.** 0.5854   **59.** 0.8129
61. 1.337   **63.** 0.2681   **65.** 1.309   **67.** 1.116   **69.** 0.5000   **71.** 0.1585   **73.** 19.1 in.^{2}
75. 485 cm^2 **77.** 3.86 in.   **79.** 2280 cm^2

Exercise 2

	r	θ	s
1.			6.07 in.
3.			$23\overline{0}$ ft
5.			43.5 in.
7.		2.21 rad	
9.		1.24 rad	
11.	0.824 cm		
13.	125 ft		

15. 3790 mi   **17.** 589 m   **19.** 24.5 cm   **21.** 2.0×10^8 mi   **23.** 0.251 m
25. r = 355 mm, R = $71\overline{0}$ mm, $\theta = 60.8°$ **27.** 6210 mi   **29.** 14.1 cm   **31.** 87.6064 ft

Exercise 3

1. 194 rad/s; 11,100 deg/s   **3.** 12.9 rev/min; 1.35 rad/s   **5.** 8.02 rev/min; 0.840 rad/s   **7.** 4350°
9. 0.00749 s   **11.** 353 rev/min   **13.** 1.62 m/s   **15.** 1040 mi/h   **17.** 57.6 rad/s

Review Problems

1. 77.1°   **3.** 20°   **5.** 165°   **7.** 2.42 rad/s   **9.** $5\pi/3$ **11.** $23\pi/18$ **13.** 0.3420   **15.** 1.000
17. -0.01827 **19.** 155 rev/min   **21.** 336 mi/h   **23.** 1.1089   **25.** -0.8812 **27.** -1.0291
29. 0.9777   **31.** 0.9097   **33.** 222 mm   **35.** 2,830 cm^3

••• CHAPTER 17 •••

Exercise 1

1. $P = 2$ s, $f = \frac{1}{2}$ cycles/s, amplitude $= 7$

3.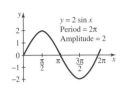

5.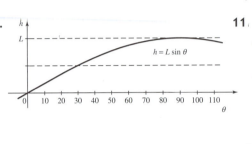

7.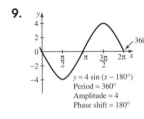

9.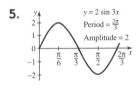

11.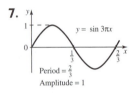

Exercise 2

1.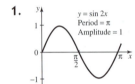

3.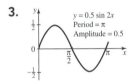

5.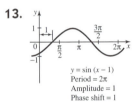

7.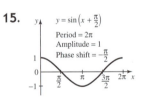

9.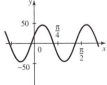

11.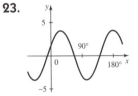

13.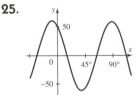

15.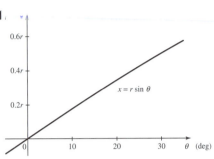

17a. $y = \sin 2x$ **17b.** $y = \sin(x - \pi/2)$ **17c.** $y = 2 \sin(x - \pi)$ **19.** $y = -2 \sin(x/3 + \pi/12)$

21.

23.

25.

27.

29.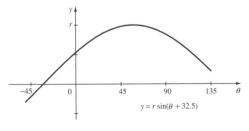
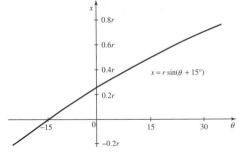
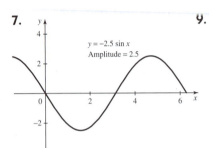
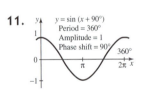

Exercise 4

1.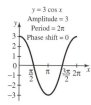
$y = 3 \cos x$
Amplitude = 3
Period = 2π
Phase shift = 0

3.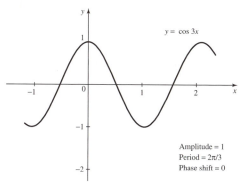
$y = \cos 3x$
Amplitude = 1
Period = $2\pi/3$
Phase shift = 0

5.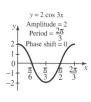
$y = 2 \cos 3x$
Amplitude = 2
Period = $\frac{2\pi}{3}$
Phase shift = 0

7.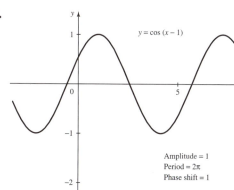
$y = \cos (x - 1)$
Amplitude = 1
Period = 2π
Phase shift = 1

9.
$y = 3 \cos \left(x - \frac{\pi}{4}\right)$
Amplitude = 3
Period = 2π
Phase shift = $\frac{\pi}{4}$

11.
$y = 2 \tan x$

13.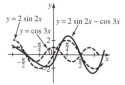
$y = 3 \tan 2x$

15.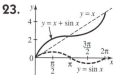
$y = 2 \tan (3x - 2)$

17.

19.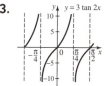

21.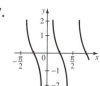

23.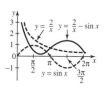
$y = x$
$y = x + \sin x$
$y = \sin x$

25.
$y = \frac{2}{x}$ $y = \frac{2}{x} - \sin x$
$y = \sin x$

27.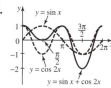
$y = \sin x$
$y = \cos 2x$
$y = \sin x + \cos 2x$

29.
$y = 2 \sin 2x$
$y = \cos 3x$
$y = 2 \sin 2x - \cos 3x$

31.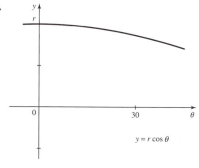
$y = r \cos \theta$

33.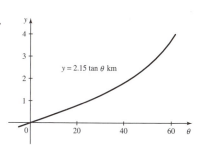
$y = 2.15 \tan \theta$ km

35.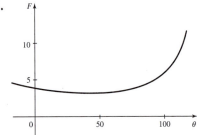

37.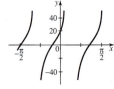

39.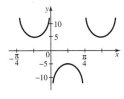

Exercise 5

1. $P = 0.0147$ s, $\omega = 427$ rad/s   **3.** $P = 0.0002$ s, $\omega = 31,400$ rad/s   **5.** $f = 8$ Hz, $\omega = 50.3$ rad/s   **7.** 3.33 s

9. $P = 0.0138$ s, $f = 72.4$ Hz   **11.** $P = 0.0126$ s, $f = 79.6$ Hz   **13.** $P = 400$ ms, amplitude $= 10$, $\phi = 1.1$ rad

15. $y = 5 \sin(750t + 15°)$ **17.** 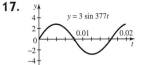   **19.**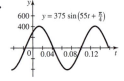

21. $y = 35.7 \sin(\omega t + 45.4°)$ **23.** $y = 843 \sin(\omega t - 28.2°)$ **25.** $y = 9.40 \sin(\omega t + 51.7°)$

27. $y = 660 \sin(\omega t + 56.5°)$ **29.**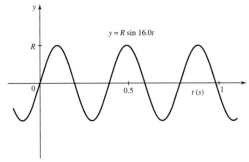

31. $v_{max} = 4.27$ V, $P = 13.6$ ms, $f = 73.7$ Hz, $\phi = 27° = 0.471$ rad, $v(0.12) = -2.11$ V

33. $i = 49.2 \sin(220t + 63.2°)$ mA

Exercise 6

1.
3.
5.
7.
9.
11.

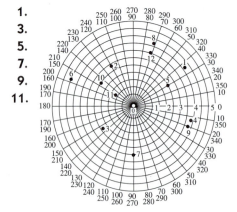

27.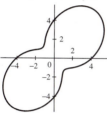

29.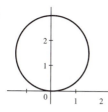

31.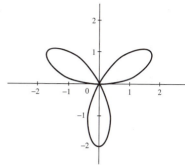

33.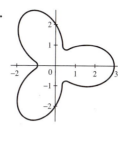

35. $(6.71, 63.4°)$ **37.** $(5.00, 36.9°)$ **39.** $(7.61, 231°)$ **41.** $(597, 238°)$ **43.** $(3.41, 3.66)$
45. $(298, -331)$ **47.** $(2.83, -2.83)$ **49.** $(12.3, -8.60)$ **51.** $(-7.93, 5.76)$ **53.** $x^2 + y^2 = 2y$
55. $x^2 + y^2 = 1 - y/x$ **57.** $x^2 + y^2 = 4 - x$ **59.** $r \sin \theta + 3 = 0$ **61.** $r = 1$ **63.** $\sin \theta = r \cos^2 \theta$
65.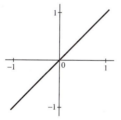

67.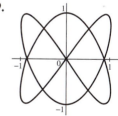
$r = 2.35 \sin^2 \theta + 2.94 \cos^2 \theta$

Exercise 7

7.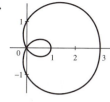

9.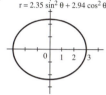

11.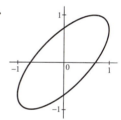

13.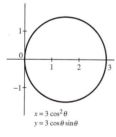
$x = 3 \cos^2 \theta$
$y = 3 \cos \theta \sin \theta$

15.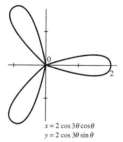
$x = 2 \cos 3\theta \cos \theta$
$y = 2 \cos 3\theta \sin \theta$

17.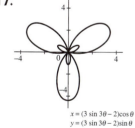
$x = (3 \sin 3\theta - 2)\cos \theta$
$y = (3 \sin 3\theta - 2)\sin \theta$

19.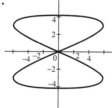

21.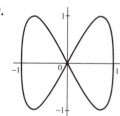

Review Problems

1.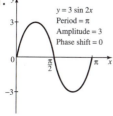
$y = 3 \sin 2x$
Period $= \pi$
Amplitude $= 3$
Phase shift $= 0$

3.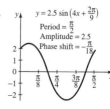
$y = 1.5 \sin \left(3x + \frac{\pi}{2}\right)$
Period $= \frac{2\pi}{3}$
Amplitude $= 1.5$
Phase shift $= -\frac{\pi}{6}$

5.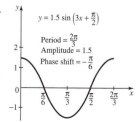
$y = 2.5 \sin \left(4x + \frac{2\pi}{9}\right)$
Period $= \frac{\pi}{2}$
Amplitude $= 2.5$
Phase shift $= -\frac{\pi}{18}$

7. $y = 5 \sin(2x/3 + \pi/9)$ **9.**

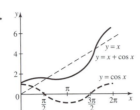

11, 13.

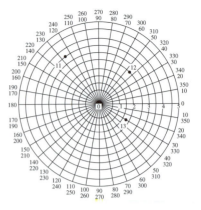

15.

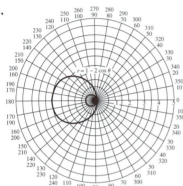

17. $(6.52, 144°)$ **19.** $(2.54, 2.82)$ **21.** $(2.73, -2.64)$ **23.** $x^2 + y^2 = 2x$

25. $r = 1/(5 \cos \theta + 2 \sin \theta)$ **27.** $P = 0.00833$ s, $\omega = 754$ rad/s   **29.** 3.33 s   **31.**

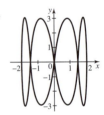

33. (a) $y = 101 \sin(\omega t - 15.2°)$ **33. (b)** $y = 3.16 \sin(\omega t + 56.8°)$ **35.**

37.

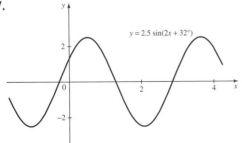

39.

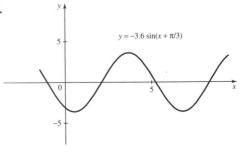

••• CHAPTER 18 •••

Exercise 1

1. $(\sin x - 1)/\cos x$ **3.** $1/\cos \theta$ **5.** $1/\cos \theta$ **7.** $-\tan^2 x$ **9.** $\sin \theta$ **11.** $\sec \theta$ **13.** $\tan x$
15. $\cos \theta$ **17.** $\sin \theta$ **19.** 1 **21.** 1 **23.** $\sin x$ **25.** $\sec x$ **27.** $\tan^2 x$ **29.** $-\tan^2 \theta$

Exercise 2

1. $\frac{1}{2}(\sqrt{3} \sin \theta + \cos \theta)$ **3.** $\frac{1}{2}(\sin x + \sqrt{3} \cos x)$ **5.** $-\sin x$ **7.** $\sin \theta \cos 2\phi + \cos \theta \sin 2\phi$
9. $(\tan 2\theta - \tan 3\alpha)/(1 + \tan 2\theta \tan 3\alpha)$ **11.** $-\cos \theta - \sin \theta$ **13.** $-\sin x$ **33.** $y = 11{,}200 \sin (\omega t + 41.0°)$
35. $y = 112 \sin (\omega t + 41.4°)$

Exercise 3

1. 1 **3.** $\sin 2x$ **5.** 2

Exercise 5

1. 30°, 150° **3.** 45°, 225° **5.** 60°, 120°, 240°, 300° **7.** 90° **9.** 30°, 150°, 210°, 330°
11. 45°, 135°, 225°, 315° **13.** 0°, 45°, 180°, 225° **15.** 60°, 120°, 240°, 300° **17.** 60°, 180°, 300°
19. 120°, 240° **21.** 0°, 60°, 120°, 180°, 240°, 300° **23.** 135°, 315° **25.** 0°, 60°, 180°, 300°

Exercise 6

1. 22.0° **3.** 281.5° **5.** 81.8° **7.** 75.4° **9.** 33.2°

Review Problems

13. 1 **15.** $2 \csc^2\theta$ **17.** 1 **19.** 1 **21.** 1 **23.** $-\tan \theta$ **25.** 120°, 240° **27.** 45°, 225°
29. 15°, 45°, 75°, 135°, 195°, 225°, 255°, 315° **31.** 90°, 306.9° **33.** 189.3°, 350.7° **35.** 60°, 300°
37. 55.6° **39.** 74.3° **41.** $78.2 \sin(\omega t + 52.5°)$ **43.** 4.2 ft

••• CHAPTER 19 •••

Exercise 1

1. 9/2 **3.** 8/3 **5.** 8 **7.** 8 **9.** $3x$ **11.** $-5a$ **13.** $(x + 2)/x$ **15.** ± 12 **17.** ± 15
19. 30, 42 **21.** 24, 32 **23.** elder = \$40,000, younger = \$35,000 **25.** 9, 11 **27.** 300 turns

Exercise 2

1. 197 **3.** 71.5 **5.** $x = 15$; $y = 55$ **7.** $x = 138$; $y = 140, 152$ **9.** 438 km **11.** 137 **13.** 67 hp
15. 911 m **17.** 1800 cm³

Exercise 3

1. 2050 **3.** 79.3 **5.** 2799 **7.** 66.7 **9.** $x = 27.5$; $y = 8840$ **11.** $x = 122$; $y = 203$
13.
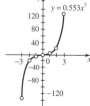
15. 396 m **17.** 4.03 s **19.** 765 W **21.** 1.73 **23.** f8 **25.** f13 **27.** 8.30 ft³

29. 236 metric tons **31.** 640 ft² **33.** 3.67 in. **35.** 28,900 m²

Exercise 4

1. 1420   **3.** y is halved   **5.** $x = 91.5$; $y = 61.5$ **7.** $x = 9236$; $y = 1443$ **9.** 41.4% increase

11.   **13.** 103 lb/in.^{2} **15.** 4.49×10^{-6} dyne   **17.** 79 lb   **19.** 200 lux   **21.** 5.31 m

23. 0.909   **25.** $33\frac{1}{3}$% increase

Exercise 5

1. 352   **3.** 4.2% increase   **5.** $w = 116$ $x = 43.5$ $y = 121$ **7.** 13.9   **9.** $y = 4.08x^{3/2}/w$

11. $13\frac{3}{4}$% decrease   **13.** 1/3   **15.** 9/4   **17.** 169 W   **19.** 3.58 m^3 **21.** 3.0 weeks   **23.** 12,600 lb

25. 394 times/s   **27.** 2.21

Review Problems

1. 1040   **3.** 10.4% increase   **5.** 121 liters/min   **7.** 1.73   **9.** 29 yr   **11.** shortened by 0.3125 in.
13. 8.5   **15.** 22 h 52 min   **17.** 215,000 mi   **19.** 42 kg   **21.** 2.8   **23.** 45% increase   **25.** 469 gal
27. 102 m^2 **29.** 1.1 in.   **31.** 17.6 m^2 **33.** 12.7 tons   **35.** 144 oz   **37.** $608   **39.** 1.5
41. 761 cm^2

••• CHAPTER 20 •••

Exercise 1

1. 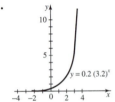   **3.**   **5.**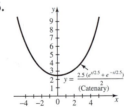

7. $4852   **9. (a)** $6.73, **(b)** $7.33, **(c)** $7.39   **11.** $1823

Exercise 2

1. 284 units   **3.** 12.4 million   **5.** 18.2 million barrels   **7.** 506°C   **9.** 0.101 A   **11.** 1050°F
13. 9.604 in. Hg   **15.** 72 mA   **17.** $30\overline{0}$%

Exercise 3

1. $\log_3 81 = 4$ **3.** $\log_4 4096 = 6$ **5.** $\log_x 995 = 5$ **7.** $10^2 = 100$ **9.** $5^3 = 125$ **11.** $3^{57} = x$
13. $5^y = x$ **15.** 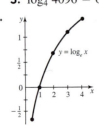   **17.** 1.4409   **19.** 0.7731   **21.** 1.6839   **23.** 2.9222   **25.** 1.4378

27. -2.1325 **29.** 4.5437   **31.** 38.4   **33.** 187,000   **35.** 1.39×10^{-3} **37.** 5.93×10^{-3} **39.** 9530

41. 1.07×10^{-4}　**43.** 3.8774　**45.** 7.7685　**47.** -0.1684　**49.** -4.7843　**51.** -9.9825　**53.** 10.5558
55. 17.22　**57.** 2.408　**59.** 0.6942　**61.** 2.550　**63.** 7.572×10^{6}　**65.** 1.063×10^{-8}

Exercise 4

1. $\log 2 - \log 3$　**3.** $\log a + \log b$　**5.** $\log x + \log y + \log z$　**7.** $\log 3 + \log x - \log 4$
9. $-\log 2 - \log x$　**11.** $\log a + \log b + \log c - \log d$　**13.** $\log 12$　**15.** $\log(3/2)$　**17.** $\log 27$
19. $\log(a^{3}c^{4}/b^{2})$　**21.** $\log(xy^{2}z^{3}/ab^{2}c^{3})$　**23.** $\log[\sqrt{x + 2}(x - 2)]$　**25.** $2^{x} = xy^{2}$　**27.** $a^{2} + 2b^{2} = 0$
29. $p - q = 100$　**31.** 1　**33.** 2　**35.** x　**37.** $3y$　**39.** 549.0　**41.** 322　**43.** 2.735　**45.** 12.58
47. 3.634　**49.** 3.63　**51.** 1.639　**53.** 2.28　**55.** 195　**57.** -8.80　**59.** 5.46

Exercise 5

1. 2.81　**3.** 16.1　**5.** 2.46　**7.** 1.17　**9.** 5.10　**11.** 1.49　**13.** 0.239　**15.** -3.44　**17.** 1.39
19. 0　**21.** $3e^{4.159x}$　**23.** 0.0147 s　**25.** 1.69 s　**27.** 1.5 s　**29.** 28,700 ft　**31.** 4.62 rad　**33.** 2.4 yr
35. 7.95 yr　**37.** 9.9 yr　**39.** 3.4%

Exercise 6

1. 2　**3.** 1/3　**5.** 2/3　**7.** 2/3　**9.** 2　**11.** 3　**13.** 25　**15.** 6　**17.** 3/2　**19.** 47.5
21. -11.0, 9.05　**23.** 10/3　**25.** 22　**27.** 0.928　**29.** 0.916　**31.** 12　**33.** 101　**35.** 50,000
37. $7.\overline{0}$ yr　**39.** $3.\overline{5}$　**41.** 27.36 in. Hg　**43.** $13\overline{0}0$ kW　**45.** -3.01 dB　**47.** 17.5 yr　**49.** 10^{-7}
51. $4\overline{0}$

Exercise 7

1.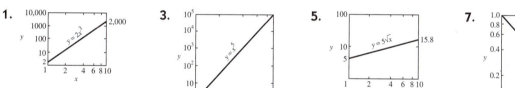

3.

5.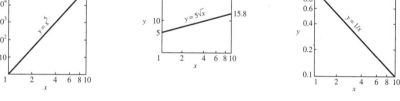

7.

9.

11.

13.

15.

17.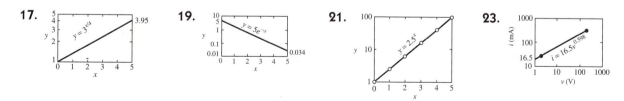

19.

21.

23.

25.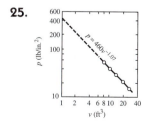

Review Problems

1. $\log_x 352 = 5.2$ **3.** $\log_{24} x = 1.4$ **5.** $x^{124} = 5.2$ **7.** 3/4   **9.** 1/128   **11.** $\log 3 + \log x - \log z$
13. $\log 10$ **15.** $\log(\sqrt{p}/\sqrt[4]{q})$ **17.** 2.5611   **19.** 701.5   **21.** 0.337   **23.** 2.943   **25.** 4.4394
27. -4.7410 **29.** 4.362   **31.** 2.101   **33.** 13.44   **35.** 3.17   **37.** 1.00×10^{-3} **39.** \$2071
41. 828 rev/min   **43.** 23 yr   **45.**

••• CHAPTER 21 ••

Exercise 1

1. $j3$ **3.** $j/2$ **5.** $4 + j2$ **7.** $-5 + j3/2$ **9.** $j5$ **11.** $-1 + j$ **13.** $2a + j2$ **15.** $3/4 + j/6$
17. $4.03 + j1.20$ **19.** j **21.** j **23.** $j14$ **25.** -15 **27.** $-j96$ **29.** -25 **31.** $6 - j8$
33. $8 + j20$ **35.** $36 + j8$ **37.** $42 - j39$ **39.** $21 - j20$ **41.** $2 + j3$ **43.** $p - jq$ **45.** $n + jm$
47. $j2$ **49.** 2   **51.** $2 + j$ **53.** $-5/2 + j/2$ **55.** $11/34 - j41/34$ **57.** $0.833 \pm j1.28$
59. $1.00 \pm j2.24$ **61.** $(x + j3)(x - j3)$ **63.** $(2y + jz)(2y - jz)$

Exercise 2

1, 3, 5, 7, 9

Exercise 3

1. $6.40\underline{/38.7°}$, $6.40(\cos 38.7° + j \sin 38.7°)$ **3.** $5\underline{/323°}$, $5(\cos 323° + j \sin 323°)$
5. $5.39\underline{/202°}$, $5.39(\cos 202° + j \sin 202°)$ **7.** $10.3\underline{/209°}$, $10.3(\cos 209° + j \sin 209°)$
9. $8.06\underline{/240°}$, $8.06(\cos 240° + j \sin 240°)$ **11.** $1.63 + j2.52$, $3\underline{/57°}$ **13.** $-7.79 + j4.50$, $9\underline{/150°}$
15. $3.70 + j4.01$, $5.46\underline{/47.3°}$ **17.** $4.64 + j7.71$, $9(\cos 59° + j \sin 59°)$
19. $4.21 - j5.59$, $7(\cos 307° + j \sin 307°)$ **21.** $15(\cos 40° + j \sin 40°)$ **23.** $16.1(\cos 54.9° + j \sin 54.9°)$
25. $56\underline{/60°}$ **27.** $2(\cos 25° + j \sin 25°)$ **29.** $4.70(\cos 50.2° + j \sin 50.2°)$ **31.** $10\underline{/60°}$
33. $8(\cos 45° + j \sin 45°)$ **35.** $49\underline{/40°}$ **37.** $7.55\underline{/26°}$, $7.55\underline{/206°}$ **39.** $3.36\underline{/24.3°}$, $3.36\underline{/144.3°}$, $3.36\underline{/264.3°}$
41. $13.8 + j7.41$, $-13.8 - j7.41$

Exercise 4

1. $3.61e^{j0.983}$ **3.** $3e^{j0.873}$ **5.** $2.5e^{j\pi/6}$ **7.** $5.4e^{j\pi/12}$ **9.** $-4.95 + j0.706$, $5\underline{/172°}$, $5(\cos 172° + j \sin 172°)$
11. $0.156 + j2.19$, $2.2\underline{/85.9°}$, $2.2(\cos 85.9° + j \sin 85.9°)$ **13.** $18e^{j6}$ **15.** $21e^{j4}$ **17.** $3.6e^{j7}$ **19.** $3e^{j3}$
21. $11e^{j3}$ **23.** $3e^{j}$ **25.** $9e^{j10}$ **27.** $8e^{j3}$

Exercise 5

1. $4.59 + j5.28$, $7.00\underline{/49.0°}$ **3.** $54.8 + j185$, $193\underline{/73.5°}$ **5.** $-31.2 + j23.3$, $39.0\underline{/143°}$ **7.** $8 - j2$
9. $38.3 + j60.2$ **11.** $8.67 + j6.30$ **13.** $29 - j29$ **15.** $12(\cos 32° - j \sin 32°)$ **17.** $3.08\underline{/32°}$

Exercise 6

1. $177\underline{/25°}$　**3.** $40\underline{/-90°}$　**5.** $102\underline{/0°}$　**7.** $212 \sin \omega t$　**9.** $424 \sin(\omega t - 90°)$　**11.** $11 \sin \omega t$
13. $155 + j0$, $155\underline{/0°}$　**15.** $0 - j18$, $18\underline{/270°}$　**17.** $72 - j42$, $83.4\underline{/-30.3°}$　**19.** $552 - j572$, $795\underline{/-46.0°}$
21. (a) $v = 603 \sin(\omega t + 85.3°)$　**(b)** $v = 293 \sin(\omega t - 75.5°)$

Review Problems

1. $3 + j2$, $3.61\underline{/33.7°}$, $3.61(\cos 33.7° + j \sin 33.7°)$, $3.61e^{j0.588}$
3. $-2 + j7$, $7.28\underline{/106°}$, $7.28(\cos 106° + j \sin 106°)$, $7.28e^{j1.85}$　**5.** j　**7.** $9 + j2$　**9.** $60.8 + j45.7$
11. $11 + j7$　**13.** $12(\cos 38° + j \sin 38°)$　**15.** $33/17 - j21/17$　**17.** $2(\cos 45° + j \sin 45°)$
19, 21

23. $7 - j24$　**25.** $125(\cos 30° + j \sin 30°)$　**27.** $32\underline{/32°}$　**29.** $0.884\underline{/-45°}$

❖❖❖ CHAPTER 22

Exercise 1

1. 3　**3.** 9　**5.** 6.60　**7.** 3.34　**9.** 6.22　**11.** -7.90　**13.** 26　**15.** 9.06　**17.** 4.66　**19.** 1.615
21. -1.080　**23.** $61.5°$　**25.** $110°$　**27.** $93.8°$　**29.** $-1/2$　**31.** 0.541　**33.** 0.1862　**35.** 0.276
37. $166°$　**39.** $164°$　**41.** $35.5°$　**43.** $125°$　**45.**

47.

49. $2.25x - y - 1.48 = 0$　**51.** $5.52x + 0.390y - 22.9 = 0$　**53.** $2x + y + 7 = 0$　**55.** $3x - y + 6 = 0$
57. $2x - 3y - 6.9 = 0$　**59.** $3x + 4y - 8 = 0$　**61.** $4x - y - 17 = 0$　**63.** $3x + y - 19 = 0$
65. $0.472x + y + 3.52 = 0$　**67.** $x - 0.400y + 4.30 = 0$　**69.** $x = -3$　**71.** $2x - y - 2 = 0$
73. $x + 2y - 2 = 0$　**75.** 78.6 mm　**77.** 2008 ft　**79.** 1260 m　**81.** $0.911°$　**83.** $33.7°$
85. $F = kL - kL_0$　**87.** 10.5 m/s, 64.3 m/s　**89.** 150.1 Ω　**93.** $P = 20.6 + 0.432x$; 21.8 ft
95. $t = -0.789x + 25.0$; $x = 31.7$ cm; $m = -0.789$　**97.** $y = P + t(S - P)/L$; $5555

Exercise 2

1. $x^2 + y^2 = 49$　**3.** $(x - 2)^2 + (y - 3)^2 = 25$　**5.** $(x - 5)^2 + (y + 3)^2 = 16$　**7.** $C(0, 0)$, $r = 7$
9. $C(2, -4)$, $r = 4$　**11.** $C(3, -5)$, $r = 6$　**13.** $C(4, 0)$, $r = 4$　**15.** $C(5, -6)$, $r = 6$　**17.** $C(-3, 1)$, $r = 5$
19. $x^2 + y^2 - 2x - 2y - 23 = 0$　**21.** $4x + 3y = 25$　**23.** $4x - 3y + 10 = 0$　**25.** $(3, 0), (2, 0), (0, 1), (0, 6)$
27. $(2, 3)$ and $(-3/2, -1/2)$　**31.** 8.02 ft　**33.** $(x - 6)^2 + y^2 = 100$; 7.08 ft

Exercise 3

1. $F(2, 0)$, $L = 8$　**3.** $F(0, -3/7)$, $L = 12/7$　**5.** $x^2 + 8y = 0$　**7.** $3y^2 = 4x$
9. $V(3, 5)$, $F(6, 5)$, $L = 12$, $y = 5$　**11.** $V(3, -1)$, $F(3, 5)$, $L = 24$, $x = 3$
13. $V(2, -1)$, $F(13/8, -1)$, $L = 3/2$, $y = -1$　**15.** $V(3/2, 5/4)$, $F(3/2, 1)$, $L = 1$, $x = 3/2$
17. $y^2 - 8x - 4y + 12 = 0$　**19.** $y^2 + 4x - 2y + 9 = 0$　**21.** $2y^2 + 4y - x - 4 = 0$　**23.** $y^2 = 3(10^8)x$
25. 12 ft　**27.** $x^2 = 109y$; 22.9 m　**29.** 0.563 m　**31.** $x^2 = -1200(y - 75)$　**33.** 0.0716 ft

Exercise 4

1. $V(\pm 5, 0)$, $F(\pm 3, 0)$　**3.** $V(\pm 2, 0)$, $F(\pm 1, 0)$　**5.** $V(0, \pm 4)$, $F(0, \pm 2)$　**7.** $\dfrac{x^2}{25} + \dfrac{y^2}{9} = 1$

9. $\dfrac{x^2}{36} + \dfrac{y^2}{4} = 1$ **11.** $\dfrac{x^2}{169} + \dfrac{y^2}{25} = 1$ **13.** $\dfrac{3x^2}{115} + \dfrac{7y^2}{115} = 1$

15. $C(2, -2)$; $V(6, -2)$, $(-2, -2)$; $F(4.65, -2)$, $(-0.65, -2)$ **17.** $C(-2, 3)$; $V(1, 3)$, $(-5, 3)$; $F(0,3)$, $(-4, 3)$

19. $C(1, -1)$; $V(5, -1)$, $(-3, -1)$; $F(4, -1)$, $(-2, -1)$ **21.** $C(-3, 2)$; $V(-3, 7)$, $(-3, -3)$; $F(-3, 6)$, $(-3, -2)$

23. $\dfrac{x^2}{9} + \dfrac{(y - 3)^2}{36} = 1$ **25.** $\dfrac{(x + 2)^2}{12} + \dfrac{(y + 3)^2}{16} = 1$ **27.** $(1, 2)$, $(1, -2)$ **29.** 60.8 cm   **31.** $\overline{20}$ cm

33. 6.9 ft   **35.**

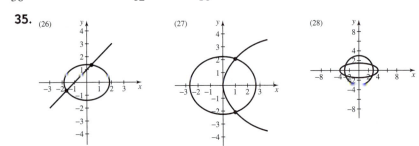

Exercise 5

1. $V(\pm 4, 0)$, $F(\pm\sqrt{41}, 0)$, $a = 4$, $b = 5$, slope $= \pm 5/4$ **3.** $V(\pm 3, 0)$, $F(\pm 5, 0)$, $a = 3$, $b = 4$, slope $= \pm 4/3$

5. $V(\pm 4, 0)$, $F(\pm 2\sqrt{5}, 0)$, $a = 4$, $b = 2$, slope $= \pm 1/2$ **7.** $\dfrac{x^2}{25} - \dfrac{y^2}{144} = 1$ **9.** $\dfrac{x^2}{9} - \dfrac{y^2}{7} = 1$

11. $\dfrac{y^2}{16} - \dfrac{x^2}{16} = 1$ **13.** $\dfrac{y^2}{25} - \dfrac{3x^2}{64} = 1$

15. $C(2, -1)$; $a = 5$, $b = 4$; $F(8.4, -1)$, $F'(-4.4, -1)$; $V(7, -1)$, $V'(-3, -1)$; slope $= \pm 4/5$

17. $C(2, -3)$; $a = 3$, $b = 4$; $F(7, -3)$, $F'(-3, -3)$; $V(5, -3)$, $V'(-1, -3)$; slope $= \pm 4/3$

19. $C(-1, -1)$; $a = 2$, $b = \sqrt{5}$; $F(-1, 2)$, $F'(-1, -4)$; $V(-1, 1)$, $V'(-1, -3)$; slope $= \pm 2/\sqrt{5}$

21. $\dfrac{(y - 2)^2}{16} - \dfrac{(x - 3)^2}{4} = 1$ **23.** $\dfrac{(x - 1)^2}{4} - \dfrac{(y + 2)^2}{12} = 1$ **25.** $xy = 36$ **27.** $\dfrac{x^2}{324} - \dfrac{y^2}{352} = 1$

29. $pv = 25,000$

Review Problems

1. 4   **3.** -1.44 **5.** $147°$ **7.** $-2b/a$ **9.** $m = 3/2$, $b = -7/2$ **11.** $7x - 3y + 32 = 0$

13. $5x - y + 27 = 0$ **15.** $7x - 3y + 21 = 0$ **17.** $y = 5$ **19.** -3 **21.** $x + 3 = 0$ **23.** 23.0

35. Parabola, $V(-3, 3)$, $F(-3, 2)$ **37.** Ellipse, $C(4, 5)$; $a = 10$, $b = 6$; $F(4, 13)$, $(4, -3)$; $V(4, 15)$, $(4, -5)$

39. Circle, $C(0, 0)$, $r = 3$ **41.** $x^2 + y^2 - 8x + 6y = 0$ **43.** $(x - 1)^2 - 9(y - 1)^2 = 16$

45. $625x^2 - 84y^2 = 10,000$ **47.** $(y + 7)^2 = 4(x - 3)$ **49.** $(-1, -3)$ and $(-3, -1)$ **51.** $\dfrac{x^2}{6.25} - \dfrac{y^2}{36} = 1$

53. 12.3 ft and 83.7 ft

◆◆◆ CHAPTER 23 ◆◆◆

Exercise 1

1. 2   **3.** 3   **5.** 6   **7.** 13   **9.** 12   **11.** 5   **13.** 103   **15.** 119   **17.** 101   **19.** 10

21. 0100 1000   **23.** 0101 1101   **25.** 1 0001 0010   **27.** 0111 0110   **29.** 10 0000 1011 0111

31. 1 0100 0011 0011 0100 **33.** 0.1 **35.** 0.11 **37.** 0.0100 1100 **39.** 0.1000 1100
41. 0.111 **43.** 0.0110 0011 **45.** 0.5 **47.** 0.25 **49.** 0.5625 **51.** 101.1 **53.** 100.011
55. 11 1011 0100.0111 1000 **57.** 1.5 **59.** 2.25 **61.** 25.40625

Exercise 2

1. D **3.** 9 **5.** 93 **7.** D8 **9.** 92A6 **11.** 9.3 **13.** 1.38 **15.** 0110 1111 **17.** 0100 1010
19. 0010 1111 0011 0101 **21.** 0100 0111 1010 0010 **23.** 0101.1111 **25.** 1001.1010 1010
27. 242 **29.** 51 **31.** 14,244 **33.** 62,068 **35.** 3.9375 **37.** 2748.8671875 **39.** 27 **41.** 399
43. AB5 **45.** 6C8

Exercise 3

1. 6 **3.** 7 **5.** 63 **7.** 155 **9.** 111 **11.** 01 0110 **13.** 1 1001 0011 **15.** 1010 1010 0011
17. 0110 0110 0110 1000

Exercise 4

1. 0110 0010 **3.** 0010 0111 0100 **5.** 0100 0010.1001 0001 **7.** 9 **9.** 61 **11.** 36.8

Review Problems

1. 6.125 **3.** 0100 0101 **5.** 1264 **7.** 1 1010.111 **9.** 348 **11.** BE **13.** 20AA **15.** 134
17. 0100 1110 0101 1101 **19.** 0100 1010

••• CHAPTER 24

Exercise 1

1. unconditional, nonlinear **3.** conditional, nonlinear **5.** conditional, linear **7.** $x > 5$ and $x < 9$
9. $x > -11$ and $x \leq 1$ **11.** $5 < x < 8$ **13.** $-1 \leq x < 24$ **15.**

17.   **19.**   **21.**   **23.**

25.   **27.**   **29.**   **31.**

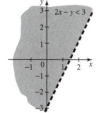

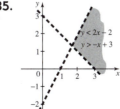

33.   **35.**   **37.**   **39.**

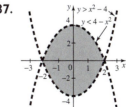

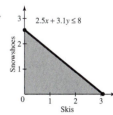

Exercise 2

1. $x > 9$ **3.** $-3 \leq x \leq 3$ **5.** $x < 1$ or $x > 3/2$ **7.** $x \geq 40/7$ **9.** $-2 < x < 5$ **11.** $-3 \leq x \leq 1$
13. $x > 3$ or $x < -3$ **15.** $x > 8/5$ or $x < -8/5$ **17.** $x \geq 10/3$ or $x \leq -2$ **19.** $x < 5$
21. $-1/2 \leq x \leq 7/8$ **23.** $x \leq -5$ or $x \geq 1$ **25.** $x \leq -2$ or $x \geq 3$ **27.** $25x + 155 \leq 350$, 7.8 h
29. 106 lenses per week **31.** $-1.52 < x < 1.52$ **33.** $-1.84 < x < 1.35$

Exercise 3

1. $x = 1.5$, $y = 3.5$ **3.** $x = 2.78$, $y = 3.89$ **5.** 69 pulleys, 68 sprockets   **7.** 8 disk drives, 4 monitors

Review Problems

1. $x > -11$ and $x \le 1$ **3.** [number line: 3 4 5 6 7 8 9, x; $(-\infty, 8)$]   **5.** [number line: $x > 8$, 8, x]   **7.** $x > 8$

9. $x < -3$ or $x > 11$ **11.** $x = 2.5$, $y = 0$ **13.** $x \le -5$ or $x \ge -3$ **15.** $x \le -7$ or $x \ge 2$
17. $-1.73 < x \le 8.82$ **19.** $0.253 < x < 1.18$

◆◆◆ CHAPTER 25 ◆◆

Exercise 1

1. $3 + 6 + 9 + 12 + 15 + \cdots + 3n + \cdots$ **3.** $2 + 3/4 + 4/9 + 5/16 + 6/25 + \cdots + (n + 1)/n^2 + \cdots$
5. $u_n = 2n$; 8, 10   **7.** $u_n = 2^n/(n + 3)$; 32/8, 64/9   **9.** $u_n = u_{n-1} + 4$; 13, 17
11. $u_n = (u_{n-1})^2$; 6561, 43,046,721   **13.** Diverges   **15.** Converges

Exercise 2

1. 46   **3.** 43   **5.** 49   **7.** $x + 24y$ **9.** 3, $7\frac{1}{3}$, $11\frac{2}{3}$, 16, $20\frac{1}{3}$, . . .   **11.** 5, 11, 17, 23, 29, . . .
13. -7 **15.** 234   **17.** 225   **19.** 15   **21.** 10, 15   **23.** $-6\frac{3}{5}$, $-7\frac{1}{5}$, $-7\frac{4}{5}$, $-8\frac{2}{5}$ **25.** 3/14
27. 7/11, 7/13   **29.** \$4400   **31.** \$94,000   **33.** 1610 ft

Exercise 3

1. 80   **3.** -9375 **5.** 5115   **7.** $-292,968$ **9.** ± 15 **11.** ± 30 **13.** 24, 72   **15.** ± 20, 80, ± 320
17. $e^{1/2}$ or 1.649   **19.** 3.2°F   **21.** 8.3 yr   **23.** 28.8 ft   **25.** 62   **27.** 2.0   **29.** 1.97   **31.** \$184,202
33. \$7776

Exercise 4

1. $5 - c$ **3.** $3/(c + 4)$ **5.** 288   **7.** 12.5   **9.** 19/33   **11.** $7\frac{15}{22}$ **13.** 45.5 in.

Exercise 5

1. 720   **3.** 42   **5.** 35   **25.** $512,512x^{10}$ **27.** $330x^4a^7$ **29.** $6435x^{14}$

Review Problems

1. 35   **3.** 6, 9, 12, 15   **5.** $2\frac{2}{5}$, 3, 4, 6   **7.** 85   **9.** $a^5 - 10a^4 + 40a^3 - 80a^2 + 80a - 32$ **11.** 48
13. $9 - x$ **15.** 8   **17.** 5/7   **19.** 49/90   **21.** 1215   **23.** -1024 **25.** 20, 50   **27.** 1440
29. 252   **31.** $128x^7y^{14} - 448x^{13/2}/y + 672x^6y^8 - 560x^{11/2}/y^5 + \ldots$ **33.** $1 - a/2 + 3a^2/8 - 5a^3/16 + \cdots$
35. $-26,730a^4b^7$

◆◆◆ CHAPTER 26 ◆◆

Exercise 1

1. Discrete   **3.** Categorical   **5.** Categorical   **7.** [line graph: Population (thousands) vs Year, 1920–1930]   **9.**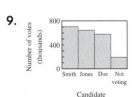

Exercise 2

1.

1(a) Range = 172 − 111 = 61

Class Midpt.	Class	Limits	Abs. Freq.	Rel. Freq. (%)
			1(b)	1(c)
113	110.5	115.5	3	7.5
118	115.5	120.5	4	10.0
123	120.5	125.5	1	2.5
128	125.5	130.5	3	7.5
133	130.5	135.5	0	0.0
138	135.5	140.5	2	5.0
143	140.5	145.5	2	5.0
148	145.5	150.5	6	15.0
153	150.5	155.5	7	17.5
158	155.5	160.6	2	5.0
163	160.5	165.5	5	12.5
168	165.5	170.5	3	7.5
173	170.5	175.5	2	5.0

3.

3(a) Range = 972 − 584 = 388

Class Midpt.	Class	Limits	Abs. Freq.	Rel. Freq. (%)
			3(b)	3(c)
525	500.1	550	0	0.0
575	550.1	600	1	3.3
625	600.1	650	2	6.7
675	650.1	700	4	13.3
725	700.1	750	3	10.0
775	750.1	800	7	23.3
825	800.1	850	1	3.3
875	850.1	900	7	23.3
925	900.1	950	4	13.3
975	950.1	1000	1	3.3

5.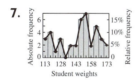

7.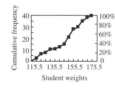

9.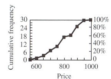

11.

Class Midpt.	Class	Limits	Cumulative Freq. Abs.	Rel. (%)
			11(a)	11(b)
47.5	45.05	50.05	5	16.7
52.5	50.05	55.05	7	23.3
57.5	55.05	60.05	10	33.3
62.5	60.05	65.05	11	36.7
67.5	65.05	70.05	14	46.7
72.5	70.05	75.05	23	76.7
77.5	75.05	80.05	23	76.7
82.5	80.05	85.05	25	83.3
87.5	85.05	90.05	28	93.3
92.5	90.05	95.05	29	96.7
97.5	95.05	100.05	30	100.0

13.

15.

17.

3								
4	8.5	8.9	9.6	8.4	9.4			
5	4.8	9.3	9.3	9.3	0.3			
6	1.4	9.3	6.3	9.3				
7	2.5	1.2	1.4	4.5	3.6	1.4	4.9	2.7	2.8
8	8.2	4.6	9.4	5.7	3.8			
9	9.2	2.4						
10								

Exercise 3

1. 77   **3.** 145 lb   **5.** $796   **7.** 81.6   **9.** 157.5   **11.** none   **13.** 59.3 min   **15.** 76   **17.** 149 lb
19. $792   **21.** 116, 142, 158, 164, 199   **23.** 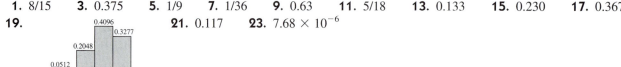   **25.** 83   **27.** 4.28, 7.40 and 10.02; 5.74
29. 697.4, 26.4   **31.** 201, 14.2

Exercise 4

1. 8/15   **3.** 0.375   **5.** 1/9   **7.** 1/36   **9.** 0.63   **11.** 5/18   **13.** 0.133   **15.** 0.230   **17.** 0.367
19.   **21.** 0.117   **23.** 7.68×10^{-6}

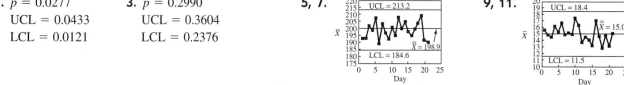

Exercise 5

1. 0.4332   **3.** 0.2119   **5.** 620 students   **7.** 36 students   **9.** 1

Exercise 6

1. 69.47 ± 0.34 inches   **3.** 2.35 ± 0.24 inches   **5.** 164.0 ± 2.88 **7.** 16.31 ± 2.04 **9.** 0.250 ± 0.031

Exercise 7

1. $\bar{p} = 0.0277$
     UCL = 0.0433
     LCL = 0.0121

3. $\bar{p} = 0.2990$
     UCL = 0.3604
     LCL = 0.2376

5, 7.

9, 11.

Exercise 8

1. 1.00   **3.** 1.00   **5.** $y = -4.79x - 14.5$

Review Problems

1. Continuous   **3.** Categorical   **5.**   **7.** 5/36   **9.** 0.75   **11.** 0.1915   **13.** 0.3446
15. 81

17.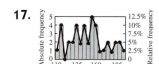

19.

Data	Cumulative Absolute Frequency	Cumulative Relative Frequency
Under 112.5	1	1/40 (2.5%)
Under 117.5	5	5/40 (12.5%)
Under 122.5	5	5/40 (12.5%)
Under 127.5	7	7/40 (17.5%)
Under 132.5	9	9/40 (22.5%)
Under 137.5	13	13/40 (32.5%)
Under 142.5	15	15/40 (37.5%)
Under 147.5	19	19/40 (47.5%)
Under 152.5	21	21/40 (52.5%)
Under 157.5	26	26/40 (65%)
Under 162.5	30	30/40 (75%)
Under 167.5	31	31/40 (77.5%)
Under 172.5	32	32/40 (80%)
Under 177.5	34	34/40 (85%)
Under 182.5	35	35/40 (87.5%)
Under 187.5	37	37/40 (92.5%)
Under 192.5	39	39/40 (97.5%)
Under 197.5	40	40/40 (100%)

21. $\bar{x} = 150$ **23.** 137 and 153   **25.** $s = 22.1$ **27.** 150 ± 7.0

29. 22.1 ± 4.94 **31.** 13   **33.** $Q_1 = 407$, $Q_2 = 718$, $Q_3 = 1055$; Quartile range = 648

35. 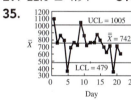   **37.** -1.00; $y = -2.31x + 18.1$ **39.** 0.0769 ± 0.0188

INDEX TO APPLICATIONS

INDEX TO GRAPHICS CALCULATOR INSTRUCTIONS AND EXERCISES

INDEX TO CAS INSTRUCTIONS
AND EXERCISES

INDEX TO WRITING QUESTIONS

INDEX TO TEAM PROJECTS

GENERAL INDEX